Advances in Neutron Capture Therapy

Edited by

Albert H. Soloway, Rolf F. Barth, and
David E. Carpenter
The Ohio State University
Columbus, Ohio

Plenum Press • New York and London

Library of Congress Cataloging in Publication Data

Advances in neutron capture therapy / edited by Albert H. Soloway, Rolf F. Barth, and David E. Carpenter.
p. cm.
"Proceedings of the Fifth International Symposium on Neutron Capture Therapy, held September 14–17, 1992, in Columbus, Ohio"—T.p. verso.
"International Society for Neutron Capture Therapy; UICC; The Arthur G. James Cancer Hospital and Research Institute"—T.p. verso.
Includes bibliographical references and index.
ISBN 0-306-44567-0
1. Boron-neutron capture therapy—Congresses. 2. Cancer—Radiotherapy—Congresses. I. Soloway, Albert H. II. Barth, Rolf F. III. Carpenter, David E. IV. International Society for Neutron Capture Therapy. V. International Union against Cancer. VI. Arthur G. James Cancer Hospital and Research Institute. VII. International Symposium on Neutron Capture Therapy (5th: 1992: Columbus, Ohio)
[DNLM: 1. Neoplasms—radiotherapy—congresses. 2. Neutrons—therapeutic use—congresses. 3. Boron Compounds—therapeutic use—congresses. QZ 269 A2438 1992]
RC271.N48A36 1993
616.99'40642—dc20
DNLM/DLC 93-37103
for Library of Congress CIP

THE OHIO STATE UNIVERSITY

THE ARTHUR G. JAMES CANCER HOSPITAL AND RESEARCH INSTITUTE

International Society for Neutron Capture Therapy

UICC

Proceedings of the Fifth International Symposium on Neutron Capture Therapy, held September 14–17, 1992, in Columbus, Ohio

ISBN 0-306-44567-0

A Division of Plenum Publishing Corporation
233 Spring Street, New York, N.Y. 10013

Printed in the United States of America

Advances in Neutron Capture Therapy

DEDICATION

It is with great pleasure that we dedicate the proceedings of this symposium to Professor Hiroshi Hatanaka on the occasion of his sixty-first birthday. Dr. Hatanaka received his medical degree and surgical and neurosurgical training at Tokyo University, following which he was Fulbright Scholar and Surgical Research Fellow in Neurosurgery at the Massachusetts General Hospital. Inspired by Dr. William H. Sweet, he developed a strong interest in Boron Neutron Capture Therapy, and upon his return to Japan in 1967, he initiated a clinical trial that has continued to the time of this meeting. As Professor of Neurosurgery at Teikyo University, he has pioneered in the clinical development of BNCT and its role in the treatment of patients with brain tumors, including the most malignant of them all, glioblastomas. He has done this despite the criticism by some neurosurgeons and radiation therapists, who viewed his efforts with skepticism because of the clinical failures in the United States in the early 1960s. Dr. Hatanaka realized that it was not the concept of BNCT that was flawed, but rather the lack of more selective tumor-targeting compounds than those used in the previous trials. In spite of many difficulties, he has continued to use the best available technology in BNCT to treat brain tumor patients who otherwise would face almost certain death. He has done this without any consideration of his own financial gain, but rather for these patients who believed that BNCT was their last hope. It is due to his persistence, at times transcending what appeared to be insurmountable odds, that has stimulated clinicians, engineers and scientists to further develop BNCT. We salute him for his encouragement of our own efforts and those of his many colleagues throughout the world, for his enduring friendship, and especially for his indomitable spirit, which has overcome many adversities.

Albert H. Soloway
Rolf F. Barth

PREFACE

Binary systems for the treatment of cancer potentially are among the most attractive of the new therapeutic modalities that currently are under investigation. The basic concept is to selectively destroy malignant cells while concomitantly sparing normal tissue. Neutron capture therapy (NCT) is the binary system that has been the subject of the Fifth International Symposium on Neutron Capture Therapy, which was held September 13-17, 1992, in Columbus, Ohio, under the auspices of the International Society for Neutron Capture Therapy. Its objective was to bring together researchers from throughout the world and to provide a forum at which they could present the latest advances in the development of Neutron capture therapy.

Neutron capture therapy has largely, but not exclusively, focused on the use of boron-10 as the target nuclide. Boron neutron capture therapy (BNCT) is based on the nuclear reaction that occurs when the stable isotope, boron-10, absorbs low-energy non-ionizing thermal neutrons to yield alpha particles and recoiling lithium-7 nuclei. The size and energy of these high linear energy transfer (LET) particles result in their being confined largely to the cells in which the capture reaction occurs. For BNCT to be successful, a sufficient number of ^{10}B atoms must be localized within neoplastic cells, and enough thermal neutrons must be delivered and absorbed by the ^{10}B to produce a lethal $^{10}B(n,\alpha)^{7}Li$ reaction. Two major problems must be surmounted. First, how can a sufficient number of ^{10}B atoms be delivered with differential selectively to cancer cells? Second, how can an adequate number or "fluence" of neutrons ($n \cdot cm^{-2}$) be delivered to the tumor volume to initiate the neutron capture reaction? The papers presented at the Fifth International Symposium have attempted to provide answers to these questions. As will become evident to the reader of these proceedings, advances have been made in each of the major areas that are essential for the ultimate success of BNCT - neutron sources, boron delivery systems, *in vitro* and *in vivo* studies to evaluate efficacy, and finally, clinical studies and future treatment plans.

Currently, nuclear reactors are the primary source of neutrons for BNCT. The radiation produced from the nuclear fission reaction within the reactor core is a mixture of gamma rays, fast or high energy neutrons, and lower energy epithermal and thermal neutrons. The poor tissue penetrating properties of thermal neutrons have led to the development of epithermal neutron beams, which have a greater depth of penetration and should be clinically more effective than thermal neutrons in destroying more deep seated tumors. The properties of these beams have been described by a number of participants at this symposium, and tangible progress has been made at several nuclear reactor facilities towards the development of neutron beams that would be suitable for clinical trials. Consideration also has been given to alternative neutron sources such as accelerators and radionuclides (californium-252), and these ultimately may significantly broaden the possibilities of using BNCT in hospitals, rather than only at special reactor facilities.

Turning to boron delivery systems, a major effort has been directed towards the synthesis of boron-containing compounds with greater selectivity for neoplastic cells compared to normal cells. Success has been achieved in incorporating boron into carrier molecules such as MoAbs and porphyrins, both of which are capable of targeting tumors, or alternatively into chemical compounds that are biochemical precursors of cellular constituents. A number of these compounds currently are being evaluated in both *in vitro* and *in vivo* systems, but still the major focus is on sodium borocaptate (BSH) and p-boronophenylalanine (BPA). The former has been used clinically in Japan in over 120 patients with brain tumors, the majority of which were glioblastomas, and the clinical results probably have been the single most important stimulus to the current interest in BNCT. Boronophenylalanine also has been used in Japan to treat patients with cutaneous melanomas by means of BNCT, and this experience has stimulated a number of investigators to explore the use of BPA as a possible capture agent for metastatic intracerebral and ocular melanomas, as well as primary brain tumors using a variety of animal models. In addition, interest has developed in the possible use of gadolinium-157 as a capture agent, and its neutron reaction ($^{157}Gd(n,\gamma)^{158}Gd$), which yields gamma rays, internal conversion and Auger electrons. Although ^{157}Gd may not be as well suited for NCT as ^{10}B, it still may be effective as a tumoricidal agent.

Numerous *in vitro* and *in vivo* studies are reported in the proceedings of this symposium that further establish the potential efficacy of BNCT as a therapeutic modality. Among these are the studies that have been carried out in canines, and the work of several groups using rodent models. Finally, clinical updates have been presented from Japan on the treatment of patients with primary brain tumors and cutaneous melanomas. Our European colleagues have made significant progress towards the initiation of a clinical trial of BNCT for glioblastomas with BSH as the capture agent, using the reactor at the Joint Research Center in Petten, The Netherlands, as a source of epithermal neutrons. In the United States, the group at the New England Medical Center/Massachusetts Institute of Technology have laid the groundwork for a Phase I Study for cutaneous melanoma using BPA as the capture agent.

We would like to make some mention of the organization of the meeting and the published proceedings. Each major subject has been introduced by a plenary speaker who has attempted to provide an integrative overview of a particular area. These have been followed by papers presented at either Oral or Poster-Discussion sessions, which have been arranged thematically in this volume and begin with neutron sources and end with clinical studies. The volume ends with a roundtable discussion held by a group of distinguished radiation oncologists and four summary papers that attempt to provide an integrative summary of the four major research areas that were the focus of this meeting.

Progress has been made in almost all areas of BNCT research since the Fourth International Symposium, which was held in December 1990 in Sydney, Australia, but it is evident that more must be done in order to unequivocally establish the therapeutic efficacy of BNCT. The requirements for more widespread clinical use now have been defined. These include a better definition of normal brain tolerance for epithermal neutrons and other radiation components of the reactor beam, and the development of boron compounds that have greater selectivity for tumor compared to normal tissues. As can be seen from these proceedings, appropriately titled "Advances in Neutron Capture Therapy," rapid progress is being made worldwide. Now, the question no longer is whether clinical trials will be initiated in the United States and in Europe, but when? These should only occur when all of the complex questions alluded to in the preceding discussion have been reasonably addressed. A premature trial resulting in failure would adversely effect future BNCT research and clinical trials, and for this reason, caution and restraint must be exercised. If both safety and efficacy can be demonstrated, then Neutron Capture Therapy will become a powerful tool to treat certain types of human cancer that currently are incurable.

Albert H. Soloway
Rolf F. Barth

Columbus, Ohio
April 15, 1993

ACKNOWLEDGEMENTS

The Fifth International Symposium on Neutron Capture Therapy was dedicated to the memory of our dear friend and colleague, Dr. Ralph G. Fairchild, who was one of the principal organizers of the first meeting of the International Society for Neutron Capture Therapy, which was held in Cambridge, Massachusetts, October, 1983. Spanning almost 30 years, his scientific career and unbounded enthusiasm for research have been a major stimulus to the continuing development of Boron Neutron Capture Therapy. His death in December, 1990, has profoundly affected all of us who have an interest in BNCT.

We would like to thank The Ohio State University Vice President for Research; the Deans of the Colleges of Medicine, Pharmacy, Veterinary Medicine and Engineering; the Arthur G. James Cancer Hospital and Research Institute; and the Comprehensive Cancer Center for their financial support in helping to make this meeting a success. We also thank our corporate sponsors: U.S. Borax Research, Boron Biologicals, Inc., and Vestar, Inc.; and acknowledge a donation from Neutron Technology.

The symposium and publication of this monograph have been generously supported by grant DE-FG02-92ER61501 from the United States Department of Energy, and P13 CA57334 from the National Cancer Institute, Department of Health and Human Services and by the International Union Against Cancer, under whose auspices this meeting was held.

Finally, we would like to thank both the Scientific Program Committee and the Local Organizing Committee for their assistance in planning this meeting, as well as recognizing the help of postdoctoral fellows, students, and staff members from many departments at The Ohio State University in conducting it. Lastly, we would like to express our gratitude to David Carpenter, who has tirelessly given of himself through all stages of this endeavor from the very beginning to its final conclusion, and to Ruth Melville and Nancy Jones who have assisted him.

CONTENTS

NUCLEAR PHYSICS AND ENGINEERING

Accelerator and Isotopic Neutron Sources

DOSIMETRY

Treatment Planning and Beam Evaluation

ANALYTICAL

Quantitative Measurements

Radiographic and Miscellaneous Analytical Studies

BIOLOGICAL STUDIES

Small Animal Studies: Pharmacokinetics

CLINICAL STUDIES

Treatment of Brain Tumors and Arteriovenous Malformations

Treatment of Melanoma

Pharmacokinetic and Biodistribution Studies

Planning for Clinical Trials

IN SUMMARY

REVIEW OF REACTOR-BASED NEUTRON BEAM DEVELOPMENT FOR BNCT APPLICATIONS

R.L. Moss

High Flux Reactor Division
Joint Research Centre
Petten, 1755 ZG, The Netherlands

INTRODUCTION

The revival over the last decade of BNCT as a potentially successful form of radiotherapy, especially for the treatment of otherwise untreatable, deep-seated tumours such as glioma, has led to a variety of neutron beam designs becoming available and being proposed at a number of nuclear research centres throughout the world. Present designs focus on producing beams of predominately epithermal neutrons which can penetrate healthy tissue to give the tumour at depth lethal doses whilst leaving the healthy tissue relatively unscathed. As well as the established design route using nuclear research reactors[1,2,3,4], recent work is also looking at accelerators (see Yanch, Shefer, these proceedings) and radioisotope sources (see Kim, these proceedings). Despite this proliferation of designs for epithermal neutron beams, BNCT is performed currently, albeit in Japan only, using thermal neutron beams[5,6]. Epithermal neutron beams have therefore, as yet, still not been used for the treatment of patients.

Nevertheless, it is expected that within the next 1-2 years, clinical trials using epithermal neutrons will begin in both Europe (see Fankhauser, Gavin, these proceedings) and the USA (see Zamenhof, these proceedings). In the USA, the two well-known reactor centres at Brookhaven and MIT, which were essentially the pioneers of neutron beam development for BNCT applications, upgraded their facilities a number of years ago, and are currently undergoing review to formulate procedures prior to clinical trials. In Europe, within the last 5 years, an epithermal neutron beam has been designed and installed at the High Flux Reactor in Petten and is the only such beam available outside the USA.

This paper reviews those issues that confront the beam designer in developing a neutron beam, including : the merits of epithermal neutrons over thermal neutrons; design methods utilising the nuclear characteristics of moderating and filter materials to optimise the beam parameters; and the engineering aspects of installing and operating such beams. With a beam available, it is essential that the beam is characterised. The dosimetry techniques used in performing this task are presented accordingly and shown how the measured beam parameters both in the free beam and phantoms, are used to qualify or rank a beam.

A brief history of beam development for BNCT applications is also given, with further

Advances in Neutron Capture Therapy, Edited by
A.H Soloway *et al.*, Plenum Press, New York, 1993

comments and concluding remarks on the current status and future prognoses for beam development at other centres throughout the world.

NEUTRON CAPTURE REACTIONS IN BORON AND TISSUE

The isotope ^{10}B has a high cross-section or probability to capture thermal neutrons. Therefore, when thermal neutrons, ie. neutrons with energies < 1eV, interact with boron nuclei, they tend to be absorbed or "captured", resulting in the production of short range, high Linear Energy Transfer (LET) particles. The nuclear reaction is :

$$^{10}B + n_{th} \longrightarrow [^{11}B] \begin{cases} \longrightarrow {}^{7}Li(1.01\ MeV) + {}^{4}He(1.78\ MeV) \quad (6.3\%) \\ \longrightarrow {}^{7}Li(0.84\ MeV) + {}^{4}He(1.47\ MeV) \quad (93.7\%) \\ \qquad \hookrightarrow \gamma\ (0.48\ MeV) \end{cases}$$

The ^{7}Li atom and the stripped ^{4}He atom, or α-particle, have a maximum range in tissue of approximately 5 and 9 μm respectively. Thus, in 93.7% of the disintegrations, a total energy of 2.31 MeV is deposited within the range of one cell diameter. Other nuclides that are present in tissue (eg. hydrogen, nitrogen, carbon, oxygen) possess such low cross sections for thermal neutron capture, that already modest amounts of boron (in the order of several tens of microgram boron per gram tissue) suffice to deliver a substantially increased dose to the boron-containing tissue.

Higher energy neutrons tend not to be captured but to lose their energy through interaction with other nuclei. The predominate interaction is elastic scatter with the hydrogen nuclei. Hence, to achieve low enough energies to be captured by the ^{10}B nuclei, a neutron must undergo a sufficient number of scattering events to reduce its energy. Neutrons of high energy utilise the overlying tissue, ie. the healthy tissue above the tumour, to lose energy and become thermalised. If one therefore has a beam of neutrons with a broad energy range, the quantity of thermalised neutrons at depth varies with position. From nuclear calculations, it can be shown that the preferred energy range of the incident neutrons is between 1 eV and 20 keV, ie. the so-called epithermal range. The thermal neutron flux density as a function of depth in tissue for an epithermal neutron beam and a thermal neutron beam is shown in Figure 1. It can be seen most significantly that an epithermal neutron beam has a thermal neutron flux peak at 2-3 cm depth in the tissue, delivers a much lower thermal neutron dose to the tissue at the surface, ie. the skin, and can penetrate the tissue to give greater doses at depth.

If one now considers the merits of epithermal neutron beams over thermal neutron beams in medical terms, it is even more enlightening to plot the corresponding curves for therapeutic gain, which is defined as the total effective dose in the tumour at depth, ie. including all the doses due to fast neutrons, nitrogen and hydrogen capture, boron capture of course, and the incident gamma rays, to the maximum effective dose in the healthy tissue, see Figure 2. The therapeutic gain, which is deemed significant if greater than one, is satisfied at depth for the epithermal neutron beam, but only at the surface for the thermal neutron beam. On the same plot, the therapeutic gain for the "ideal" beam is indicated and is defined as a pure, non-contaminated beam of 2keV neutrons, showing even more impressive gains at depth. The neutron beam designer must strive therefore towards reducing the contaminants in the beam but at the same time maintain a significant beam intensity.

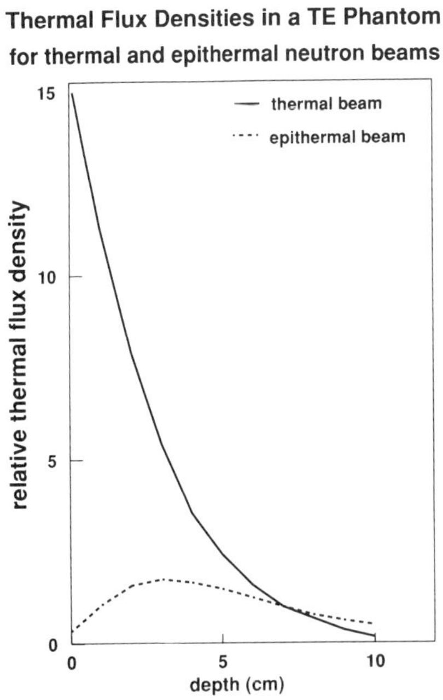

Figure 1. Thermal neutron flux density as a function of depth inside a TE phantom (curves normalised to 7cm depth)

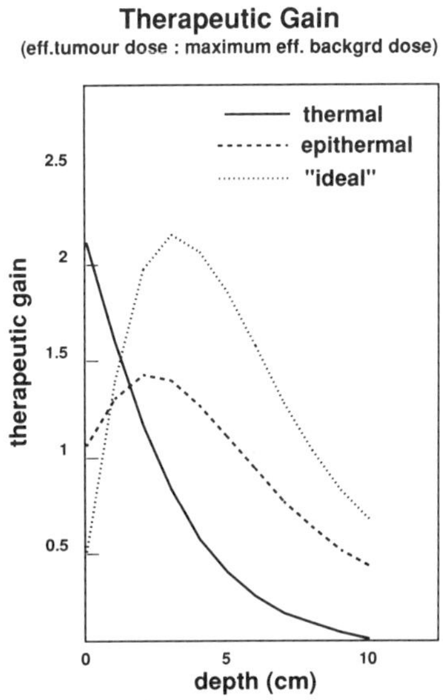

Figure 2. Therapeutic gain as a function of depth for a thermal, epithermal and "ideal" beam

NEUTRON BEAM DESIGN

Assuming that the boron compound has transported a plentiful supply of ^{10}B nuclei to each tumour cell, it is the task of the beam designer to ensure that a sufficient number of neutrons arrive at all the tumour cells, and are subsequently captured by a sufficient number of ^{10}B nuclei to cause death of the cell.

For a reactor-based facility, the neutrons travelling down the beam emanate from the fission process in the reactor core following fission of ^{235}U in the reactor fuel. The initial energy of these neutrons is of the order of 2 MeV. The energy of the neutrons must therefore be reduced by 6-7 orders of magnitude before the neutron interacts with the ^{10}B nuclei in the tumour cell. As an additional consequence of using a reactor-based beam, gamma rays as well as neutrons of all other energies are present in the beam, which themselves can give significant doses to the healthy tissue. These unwanted contaminants must therefore be removed from the beam. This is achieved by taking advantage of the nuclear characteristics of certain materials, referred to as moderators, filters or attenuators. Through interactions and/or collisions in the materials, a neutron either loses some of its energy, is deflected completely out of the beam, is captured by the nuclei of the material, or, at preferred energies, is allowed to traverse the material through characteristic "windows". The neutron beam designer must choose his materials carefully in order to optimise the beam properties. Despite there being a wealth of materials available, it is virtually impossible to produce a pure or "ideal" beam of epithermal neutrons. Examples of materials used at various facilities are shown in Table 1. A compromise is usually reached whereby there is a trade off between the desired epithermal neutron flux intensity and the beam contaminants.

Table 1. Examples of materials used in beam designs at various reactors

Reactors	Moderators	Filters	Attenuators
BMRR	Al_2O_3, Al	Al, Cd	Bi
MITR-II	Al	Al, S, Cd, Li	Bi
HFR	Al	Al, Ti, S, Cd	Ar, poly-B
MURR	Al_2O_3, Al, C	Al	Bi, Pb
GTRR	C, D_2O	Al, Cd, S	Pb, poly-B
Mu.ITR	Al, C	Al	LiF_3, Bi
TRIGA	C	Bi	LiF_3, Bi, Pb

As an example of the type of combination of filter materials that can be used, the beam designers at Petten[7], in order to produce their epithermal neutron beam from the High Flux Reactor, proposed: 15cm Al; 5cm S; 1cm Ti; 0.1cm Cd; and as a gamma attenuator, 150cm liquid Ar.

In addition to the nuclear requirements of a beam, the beam designer must also consider and accurately model the neutron source (reactor core), the beam tube configuration (geometry, materials), the beam port, the irradiation room and the shielding requirements.

The calculational methods that are used by the beam designer are usually based on deterministic or stochastic theories. The most common types of computer codes used are respectively discrete-ordinate methods, eg. DOT1[8], and Monte Carlo methods, eg. MCNP[9]. The latter tends to be the preferred code, as this, amongst other things, contains a precise, continuous cross section library, a full 3-D capability for modelling complex geometries and performs detailed neutron and photon transport through target media. Numerous references of examples of such design work can be found in the literature but is probably most concisely given in the Proceedings of the MIT Workshop held in 1989[10].

NEUTRON BEAM INSTALLATION

It is just as important when designing a beam tube with the correct nuclear characteristics, to consider and resolve many engineering aspects. For example, in the construction of the epithermal neutron beam configuration at Petten[7], the following points had to be taken into account before the beam components could be installed : removal of existing radioactive material; design of filter components to "fit" inside existing geometric constraints of the beam tube; use of liquid components, eg. liquid argon or D_2O, to be contained in pressurised, leak proof tubing or casing; installation of beam shutters with both automated and manual (back-up) operating systems; all designs and drawings to comply with international standards, etc.

Shielding and safety aspects include: designing components to overlap, thus avoiding for example, neutron streaming; full thermo-mechanical strength calculations of all components; inclusion of safety instrumentation for monitoring temperatures and pressures, and interlock systems; and introduction of back-up systems and redundancies with respect to potential component failures.

NEUTRON BEAM CHARACTERISATION

Characterisation of the beam through measurement is the ultimate confirmation of the calculations and serves simultaneously, through measurements at regular intervals, to confirm that the beam is stable, ie. no changes or shifts in the beam intensity and spectrum. The beam itself should be measured at the intended therapy position and without influence from the surroundings, ie. a free-beam or an in-air measurement.

The techniques available to measure the neutron and gamma intensities and the neutron spectrum include: ^{3}He detectors; proton recoil spectrometers; Bonner spheres; activation foils; BF_3 chambers; and for the gammas : thermoluminescence detectors (TLDs); and GM-tubes.

From the variety of beam parameters that can be measured, the most pertinent ones should include: separately, the total, epithermal, thermal and fast neutron flux densities (neutrons/cm^2 sec); separately, the fast neutron dose and gamma dose per epithermal neutron (Gy/n-cm^2); and for further detail, can also include; the fast neutron dose per thermal neutron (Gy/n-cm^2); and separately, the absorbed dose from fast neutrons and gammas in free air (Gy/hr). In Table 2, the beam parameters, where known, for various epithermal neutron beams are compared and shown. As indicated earlier, depending on the beam configuration, the flux intensity varies as a trade-off with the other contaminants in the beam.

Short of actual patient treatment, the final proof of beam performance is the measurement of the neutron and gamma fields in phantoms. A variety of phantoms have been used at various centres, including patient-like phantoms[11]. Generally, simple shapes are used, which simplify the comparison with calculation. Typical shapes are solid or hollow cylinders and cubes, made of polyethylene and, if appropriate, filled with tissue equivalent liquid. At Petten[12], for example, a beagle-head phantom made of acryllic plastic is also used. The neutron and gamma field in these phantoms are measured by activation foils and TLDs, with in particular, activation wires to give 3D-flux distribution measurements.

A beam's performance can be further qualified or represented[14] when considering its medical merits. Measured parameters can include: advantage factor (ratio of tumour dose to maximum normal tissue dose); therapeutic gain (ratio of effective tumour dose to effective

Table 2. Examples of beam parameters for various epithermal neutron beams

Reactors	epithermal n. flux density (10^9n/cm^{2}s)	fast neutrons per epithermal neutrons (10^{-13} Gy/n-cm^2)	gammas per epithermal neutrons (10^{-13} Gy/n-cm^2)
BMRR	1.2	4.8	1.4
MITR-II	0.2	12.5	-
HFR	0.33	12.6	8.4
MURR	9.5	9.0	0.4
GTRR	4.0	1.5	-
PBF	10.0	2.0	1.0
LVR-15	1.4	-	-
Musashi-TRIGA	0.5	-	-
INEL-TRIGA	7.6	-	-
FiR 1 (TRIGA)	3.5	2.6	1.0

maximum normal tissue dose); advantage depth (position at which the dose to the tumour equals the dose to the normal tissue); advantage ratio (ratio of total therapeutic dose to total dose to the normal tissue at the advantage depth); advantage depth dose ratio (position where total dose rate equals the normal tissue dose rate), and the thermal neutron flux density at peak in tissue (n/cm^2 sec).

REACTOR-BASED NEUTRON BEAMS FOR BNCT

The first clinical trials of BNCT, and in fact, all treatment of patients by BNCT, have until now been performed using beams of thermal neutrons emanating from nuclear research reactors. The initial trials in the States in the 50's and 60's were deemed a failure because the dose delivered to the overlying tissue were much too high, resulting in, amongst other things, unacceptable damage to both the cerebral vasculature and the scalp. Other reasons, such as the use of non-preferentially seeking boron compounds are well documented[13]. In Japan, the above problems are only overcome by removing the overlying tissue when performing brain tumour treatments[6], and performing melanoma treatment where only a viable dose at the surface is required[6]. A summary of the neutron beams at those centres involved in the past, present, imminent and possible future are given in Table 3.

Table 3. Reactor-based neutron beams for BNCT - past, present and future

Early US trials : number of patients treated indicated in [brackets]
1. Brookhaven Graphite Research Reactor (BGRR), 1951-61 [28] 2. Brookhaven Medical Research Reactor (BMRR), 1959-61 [17] 3. Massachusetts Institute of Technology Reactor (MITR), 1959-61, [18]
Japanese reactors used in BNCT (past and present)
1. Hitachi Training Reactor (HiTR), 1968-75, [13] 2. JAERI Research Reactor (JRR-3), 1969, [1] 3. JAERI Research Reactor (JRR-4), 1990-..., [>25] 4. Musashi Institute of Technology Reactor (MuITR), 1977-89, [87] 5. Kyoto Univ. Research Reactor (KURR), 1974, 1981, 1990-...., [>12]
Epithermal Neutron Beams (imminent)
1. High Flux Reactor Petten (HFR), 1993-...., 2. MITR(-II), 1993-...., 3. BMRR, 1994-.....,
Proposed reactor-based epithermal neutron beams (future)
USA 1. Georgia Tech. Research Reactor (GTRR) 2. Missouri Univ. Research Reactor (MURR) 3. Power Burst Facility (PBF) - future uncertain? 4. TRIGA-II, general design available **Europe** 1. LVR-15, Rez, Czechoslovakia 2. FiR 1, Finland 3. TRIGA, Pavia, Italy **Elsewhere** 1. Mu.ITR-II, Japan 2. HIFAR, Lucas Heights, Australia

CONCLUSION

As a plenary talk, this paper has attempted to indicate in general terms the tasks a beam designer is confronted with to produce, in the instance here, an epithermal neutron beam that may be used for the BNCT treatment of glioma patients. With the beam designed and installed, the characterisation of the beam is of paramount importance. Thereafter the programme of biological experiments, such as cell-culture irradiations and healthy tissue tolerance studies may proceed, whereby the preconditions before starting clinical trials are established.

Finally, it should be noted that at a typical research reactor centre, it is not to be expected that an epithermal neutron beam could treat more than 1000 patients per year. In the USA alone, about 16,000 glioma patients per year are diagnosed, inferring that at least 16 treatment centres are required if all patients, if felt appropriate, are to be treated by BNCT. A similar story applies to Western Europe. Therefore there is presently a severe shortfall in the number of facilities that could be used. Nevertheless, the start of clinical trials of BNCT, using epithermal neutron beams emanating from nuclear research reactors is imminent and once proven, other facilities will be required. However, the building or re-opening of other research reactors is extremely unlikely and hence, in the near future, the beam designer must turn his attention to alternative neutron sources. These include most prominently, accelerators and to a lesser extent radioisotope sources.

REFERENCES

1. Fairchild,R.G.,"Development and dosimetry of an epithermal neutron beam for possible use in neutron capture therapy. I. "Epithermal" neutron beam development", Phys.Med.Biol., **10**, 1965
2. Moss, R.L., "Progress towards Boron Neutron Capture Therapy at the High Flux Reactor Petten", Proc. of Neutron Beam Design, Development and Performance for Neutron Capture Therapy, Basic Life Sciences, Vol.54, Plenum Press, 1990
3. Choi,J.R. et al,"Neutron capture therapy beams at the MIT research reactor", idem
4. Brugger,R.M. et al,"Intermediate energy neutron beams from the MURR",idem
5. Hatanaka,H. et al,"Clinical experience of boron-neutron capture therapy for gliomas - a comparison with conventional chemo-immuno-radiotherapy", Boron-Neutron Capture Therapy for Tumors, Ed. H.Hatanaka, Niigata, Japan: Nishimura, 1986
6. Mishima,Y. et al,"Treatment of malignant melanoma by single thermal neutron capture therapy with melanoma-seeking ^{10}B-compound", Lancet, **2**, 1989
7. Moss,R.L. et al,"Design, construction and installation of an epithermal neutron beam for BNCT at the High Flux Reactor at Petten", Proc. on Neutron Capture Therapy for Cancer", Ed. B.J.Allen, Plenum Press, New York, 1992
8. W.A.Rhoades and R.L.Childs, "Updated Version of DOT 4 One- and Two-Dimensional Neutron/Photon Transport Code", Oak Ridge National Laboratory, ORNL-5621, July 1980.
9. Briesmeister, J.F.(Ed), "MCNP - A General Monte Carlo Code for Neutron and Photon Transport", LA-7596-M, Rev.2.
10. Proc. of an Inter. Workshop on "Neutron Beam Design, Development and Performance for Neutron Capture Therapy", MIT Cambridge USA, March 1989, Basic Life Sciences, Vol.54, Plenum Press, 1990
11. Aizawa,O. et al,"Remodelling and dosimetry on the neutron irradiation facility of the Musashi Institute of Technology Reactor for boron neutron capture therapy", J.Nuclear Technology, **48**, 1980
12. R.L.Moss, A.Siefert, P.R.D.Watkins, G.Constantine and K.Philipp, "The Petten BNCT Facility : I - Phantom Dosimetry Techniques used in the Nuclear and Biological Characterisation of the Epithermal Neutron Beam and in the Healthy Tissue Tolerance Studies on the Canine Brain, and II - the Development towards a Treatment Planning System for the Treatment of Glioma Patients by BNCT", Proc. of an Int. Workshop on "Macro and Microdosimetry and Treatment Planning for Neutron Capture Therapy", November 1991, MIT, Cambridge, USA, Plenum Press
13. Slatkin,D.N.,"A history of boron neutron capture therapy of brain tumours", Brain, **114**, 1991

EVALUATION OF NEUTRON IRRADIATION FIELDS FOR BNCT BY USING ABSORBED DOSE IN A PHANTOM

O. Aizawa

Atomic Energy Research Laboratory
Musashi Institute of Technology
Ozenji 971, Asao-ku, Kawasaki 215, Japan

INTRODUCTION

In a previous paper[1], we defined the "irradiation time" as the time of irradiation in which the maximum "total background dose" becomes 2,500 RBE-cGy. In this paper, we have modified the definition a little as the time of irradiation in which the maximum "1μg/g B-10 dose" becomes 3,000 RBE-cGy, because we assumed that normal tissue contained 1μg/g B-10. Moreover, we have modified the dose criteria for BNCT as follows: The "eye dose", "total body dose" and "except-head dose" should be less than 200, 100 and 50 RBE-cGy, respectively. We have added one more criterion for BNCT that the thermal neutron fluence at the tumor position should be over 2.5×10^{12} n/cm^2 at the "irradiation time".

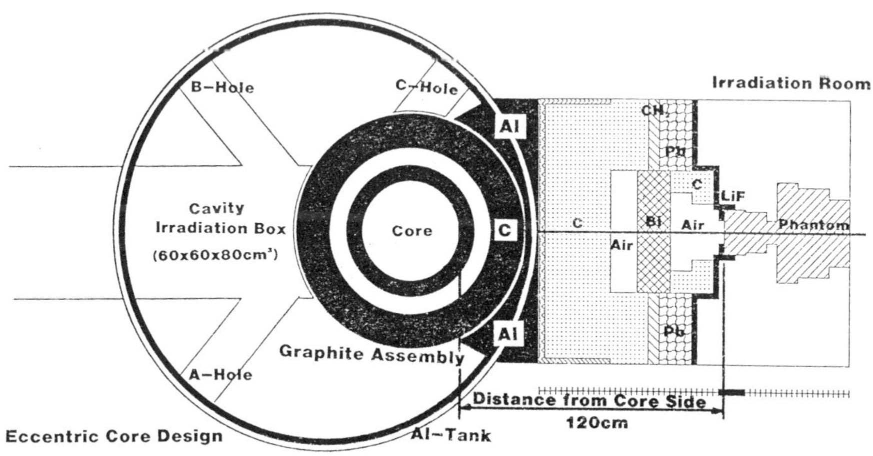

Fig. 1 Conceptional design of eccentric core configuration for TRIGA-II reactor

Advances in Neutron Capture Therapy, Edited by
A.H Soloway *et al.*, Plenum Press, New York, 1993

The distance from the core side to the irradiation port in the "old configuration" of the Musashi reactor (TRIGA-II, 100kW) was 160 cm. We are now planning to design an eccentric core and to move the reactor core nearer to the irradiation port, as shown in Fig. 1. So we have calculated the dose characteristics by changing the distance between the core side and the irradiation port to be 140, 130 and 120cm.

The other assumptions used in this paper are as follows: (1) The B-10 concentrations in tumor are 30 and/or 10μg/g. (2) The depth of the tumor is 5.0 cm to 5.5 cm from the surface. (3) The RBE values used are 1.0 for all gamma rays and 2.3 for $B^{10}(n,\alpha)$ reaction products. (4) The RBE values for neutrons are the following three cases: the first case is using 1.6 for all neutrons; the second one is using 3.2 for non-thermal neutrons and 1.6 for thermal neutrons; the third case is using 4.8 for fast neutrons, 3.2 for faster epithermal and epithermal neutrons, and 1.6 for thermal neutrons.

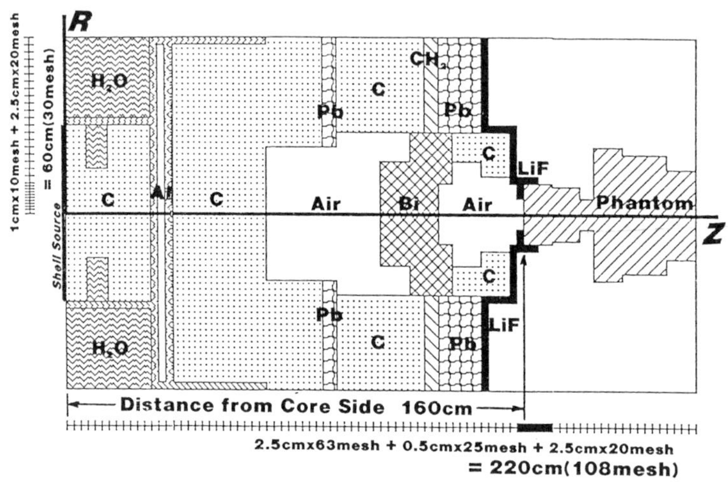

Fig. 2. Old configuration of irradiation field for BNCT at the Musashi reactor

REFERENCE CALCULATIONS

A two-dimensional discrete ordinate transport code DOT3.5 was employed for the following calculations by adopting the S_{12} (166 down) and P_3 approximations. The group constants used are the neutron and gamma-coupled cross sections based on the BUGLE Library[2] (47-neutrons, 20-gammas). The calculation was initiated from the "old configuration" at the Musashi reactor by using a shell source at the core side as shown in Fig. 2. The number of meshes used for the calculations was 30x108 meshes in R-Z geometry as shown in the figure. We can see that the mesh interval is variable, and the smallest z-mesh is 0.5cm in a half of the head phantom. The dose characteristics are shown in Table 1. It can be seen from this table that the "irradiation time" defined here is between 3.66 and 4.23 hours, which depends on the RBE values used for neutrons, and it is clear that this is consistent with the average irradiation time of 4 hours in actual treatments of brain tumors at the Musashi reactor.

Table 1. Dose characteristics for "old configuration" at the Musashi reactor (distance from the core side =160 cm, carbon thickness = 32.5 cm)

Thermal Neutron Flux at the Surface : 1.6×10^9 (n/cm²sec) Cd–ratio at the Surface of the Phantom : 114			
RBE–values★ Used for Neutrons	Fast=1.6 Faster–epi=1.6 Epithermal=1.6 Thermal=1.6	Fast=3.2 Faster–epi=3.2 Epithermal=3.2 Thermal=1.6	Fast=4.8 Faster–epi=3.2 Epithermal=3.2 Thermal=1.6
Irradiation Time Max. Adv. Depth Advantage Ratio	4.23 (hr) 5.25 (cm) 5.21	3.84 (hr) 5.08 (cm) 4.83	3.66 (hr) 5.01 (cm) 4.63
Max./Tumor Thermal Dose	1210 (cGy) 147	1100 (cGy) 134	1050 (cGy) 127
Max./Tumor Non–Thermal Dose	301 (cGy) 94	547 (cGy) 171	663 (cGy) 221
Max./Tumor Gamma Dose	1080 (cGy) 461	978 (cGy) 419	932 (cGy) 400
Max./Tumor 10μg/g Dose	7930 (cGy) 1370	7480 (cGy) 1330	7270 (cGy) 1320
Max./Tumor 30μg/g Dose	18900 (cGy) 2690	17400 (cGy) 2530	16800 (cGy) 2470
Thermal Neutron Fluence at the Tumor(5.0–5.5cm)	nvt 3.4×10^{12} (n/cm²)	nvt 3.1×10^{12} (n/cm²)	nvt 2.9×10^{12} (n/cm²)
Eye Dose Total Body Dose Except–Head Dose	120 (cGy) 71 28	129 (cGy) 76 30	139 (cGy) 80 31

★ RBE Values Used: 2.3 for α–rays and 1.0 for γ–rays.

We can also see that the neutron fluence of about 3×10^{12} n/cm^2 has been attained in the tumor located at the depth of about 5 cm from the surface, and the tumor dose is about 2,500 RBE-cGy in the case of the B-10 concentration of 30μg/g. The "eye dose", "total body dose" and "except-head dose" are about 130, 80 and 30 RBE-cGy, respectively, and they all satisfy the dose criteria for BNCT defined above. We can notice that both the maximum "gamma dose" and "thermal dose" are about 1,000 RBE-cGy, and the "non-thermal dose" depends strongly on the RBE values used for neutrons.

A new index of a cadmium ratio (Cd-ratio) is also shown in this table. As the "Cd-ratio" is well known in reactor physics[3], it is defined as the ratio of the activity of the bare foil to its activity when entirely enclosed in cadmium. Here we are using a gold foil of 100μm in thickness as its foil. We can say that the neutron spectrum is "hard" when the "Cd-ratio" is low, and also the neutron spectrum is "soft" when the "Cd-ratio" is high.

ECCENTRIC CORE DESIGN

If we can move the reactor core to the irradiation port by 40 cm as shown in Fig. 1, we can design a new configuration as shown in Fig. 3, in which the distance from the core side to the irradiation port is shortened by 40 cm from 160 cm to 120 cm. The dose characteristics are shown in Table 2 . We can see from this table that the "Cd-ratio" is a bit lower than with the "old configuration". This means that the faster neutron component is higher than with the "old configuration". We can also see that the "irradiation time" becomes less than 1 hour. The maximum "non-thermal dose" becomes 1,460 RBE-cGy; however, it is acceptable for BNCT treatments.

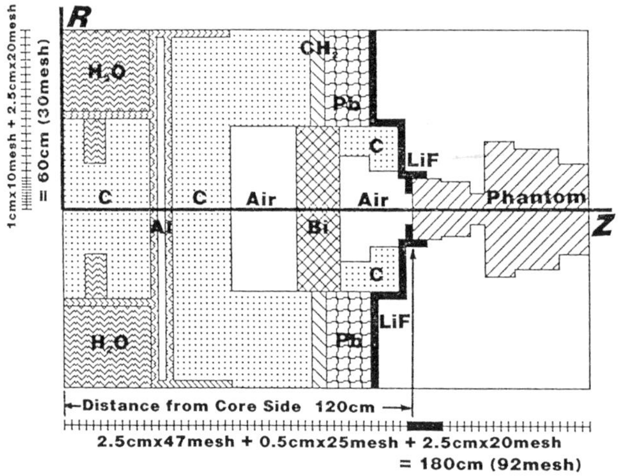

Fig. 3. Eccentric core configuration (distance from the core side = 120 cm, carbon thickness = 20 cm)

CONCLUSIONS

The distance from the core side to the irradiation port is a very important factor in order to design a neutron irradiation field for BNCT. In this paper, we conclude that we can get the acceptable dose for BNCT with only 1-hour irradiation by using a 100kW reactor if we can set the irradiation port at the distance of 120 cm from the core side, for example, by using an eccentric core configuration.

Table 2. Dose characteristics for the eccentric core configuration (distance from the core side = 120 cm, carbon thickness = 20 cm)

Thermal Neutron Flux at the Surface : 5.7×10^9 ($n/cm^2 sec$) Cd-ratio at the Surface of the Phantom : 37.9			
RBE-values★ Used for Neutrons	Fast=1.6 Faster-epi=1.6 Epithermal=1.6 Thermal=1.6	Fast=3.2 Faster-epi=3.2 Epithermal=3.2 Thermal=1.6	Fast=4.8 Faster-epi=3.2 Epithermal=3.2 Thermal=1.6
Irradiation Time Max. Adv. Depth Advantage Ratio	0.92 (hr) 5.28 (cm) 4.40	0.72 (hr) 4.83 (cm) 3.76	0.65 (hr) 4.65 (cm) 3.46
Max./Tumor Thermal Dose	953 (cGy) 143	749 (cGy) 112	673 (cGy) 101
Max./Tumor Non-Thermal Dose	817 (cGy) 242	1280 (cGy) 380	1460 (cGy) 460
Max./Tumor Gamma Dose	913 (cGy) 438	717 (cGy) 344	644 (cGy) 309
Max./Tumor 10μg/g Dose	6890 (cGy) 1470	6060 (cGy) 1350	5740 (cGy) 1330
Max./Tumor 30μg/g Dose	15500 (cGy) 2760	12900 (cGy) 2360	11800 (cGy) 2240
Thermal Neutron Fluence at the Tumor(5.0–5.5cm)	nvt 3.3×10^{12} (n/cm^2)	nvt 2.6×10^{12} (n/cm^2)	nvt 2.3×10^{12} (n/cm^2)
Eye Dose Total Body Dose Except-Head Dose	161 (cGy) 94 39	173 (cGy) 100 41	187 (cGy) 105 43

★ RBE Values Used: 2.3 for α-rays and 1.0 for γ-rays.

REFERENCES

(1) O. Aizawa, "A Theoretical Approach to Whole-body Macrodosimetry for BNCT", International Workshop on Macro and Microdosimetry and Treatment Planning for Neutron Capture Therapy, Held at Massachusetts Institute of Technology, Oct. 31 - Nov. 1 (1991)

(2) R. W. Roussin, "DCL-75/BUGLE-80 Coupled 47-Neutron, 20-Gamma-ray, P3 Cross- Section Library for LWR Shielding Calculations", Oak Ridge National Laboratory (1980)

(3) K. H. Beckurts and K. Wirtz, "Neutron Physics", p. 277, Springer-Verlag (1964)

ANALYSIS OF FILTERED NEUTRON BEAMS AT HIFAR

Baiba V. Harrington[1] and Geoffrey Constantine[2]

[1] Australian Nuclear Science and Technology Organisation
[2] Visiting Research Fellow

INTRODUCTION

The High Flux Australian Reactor (HIFAR) is a D_2O moderated and cooled DIDO class reactor which is operated at 10 MWth. The 25 cylindrical fuel elements are located in rows of 4, 6, 5, 6, 4 elements, on a square pitch of 152.4 mm but with the central row staggered by half a pitch. Each fuel element consists of 4 concentric fuel tubes, an outer aluminium casing and an inner aluminium liner tube. It has been proposed that the 10 inch diameter horizontal (10H) facility in HIFAR could be the basis of an epithermal Neutron Capture Therapy (NCT) facility[1]. Since access to the 10H facility is restricted, experiments were devised[2] which to some extent simulated filtered epithermal beams in 10H. A vertical 50.8 mm diameter voided liner tube in an off-centre fuel element position and an aluminium spectrum shifter were used to extract a neutron beam which was filtered.

A calculational study was undertaken to develop a Monte Carlo model of HIFAR, including the entire core, and to validate it against the experiments. This whole core MCNP[3] model of HIFAR may be applied equally well to predict the performance of filtered beams in the 10H horizontal facility as well as to other HIFAR calculations not necessarily related to NCT.

FILTERED NEUTRON BEAM EXPERIMENTS

During these experiments[2], done at low power, a fuel element adjacent to the central element was replaced by a 'dummy' element in which aluminium was substituted for fuel and the central region was voided by a dry liner. An aluminium spectrum shifter (350 mm height) was lowered down the voided liner tube to the core centre and various combinations of the filter materials Al, Fe, S, Ti, Cd were lowered into the top shielding region of the reactor. In two experiments liquid Ar was placed in a cryostat above the dummy element. The lengths of filter materials given in table 1 include the aluminium containers for the sulphur and liquid argon. The spectrum shifter and filter materials were 50.8 mm in diameter. Reaction rates were measured in a fission chamber sandwiched between various thicknesses of polythene (and positioned 5435 mm above core centre) in an attempt to assess the neutron spectrum of the beam.

Advances in Neutron Capture Therapy, Edited by
A.H Soloway *et al.*, Plenum Press, New York, 1993

Table 1. Filter Combinations Tested

Beam	Reactor Power (kW)	Filter Materials (mm) Al	S	Fe	Cd	Ar	Ti
1	10	265	100	-	1	1066	-
2		355	-	-	1	1066	-
3		160	100	100	1	-	-
4		160	100	100	1	-	10
5		160	100	100	1	-	20
6	100	260	100	100	1	-	-
7		260	100	100	1	-	10
8		260	100	100	1	-	20

CALCULATIONS

Analysis of the experiments involved four distinct sets of calculations which are described in more detail in the subsections below. The first was to generate argon point cross sections and from these the factors for exponential attenuation through the argon. The second set were MCNP calculations to generate the detector response as a function of incident neutron energy for each fission chamber and polythene geometry. Thirdly, the neutron source distribution in HIFAR was calculated using the 3D diffusion code POW3D[4] of the AUS[5] modular neutronics code scheme. Finally, data from the preceding calculations were used in a full core MCNP source calculation of HIFAR with tallies which included detector response functions and attenuation through the filter materials.

(1) Argon Cross Sections

The code NJOY[6] was used to generate argon point cross sections from ENDL–84[7] data in which the resonance region was replaced by resonance parameters from Mughabghab et. al.[8]. The calculated argon cross sections (figure 1) are higher than those of Mughabghab[9], but the considerably better agreement with measurements at 56 keV suggests that the Mughabghab data does not include ^{36}Ar. Even though the relative abundance of ^{36}Ar is only 0.337% it has a high cross section and contributes significantly to natural argon cross sections. The calculated argon data were used to derive exponential attenuation of neutrons as a function of energy for 1066 mm of argon.

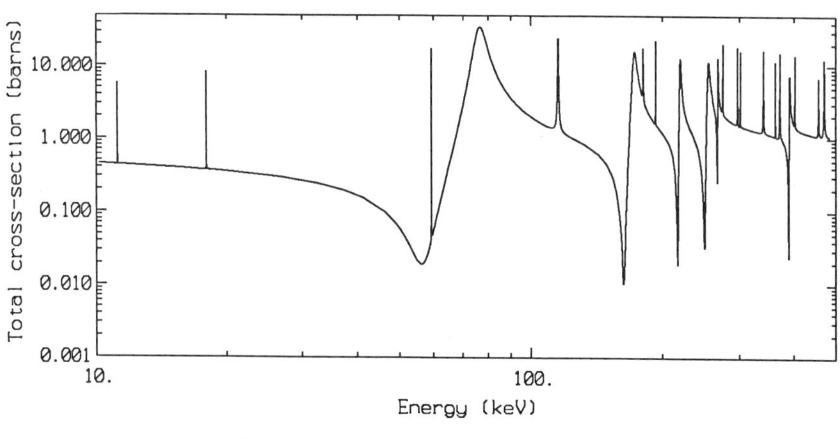

Figure 1. Argon total cross section

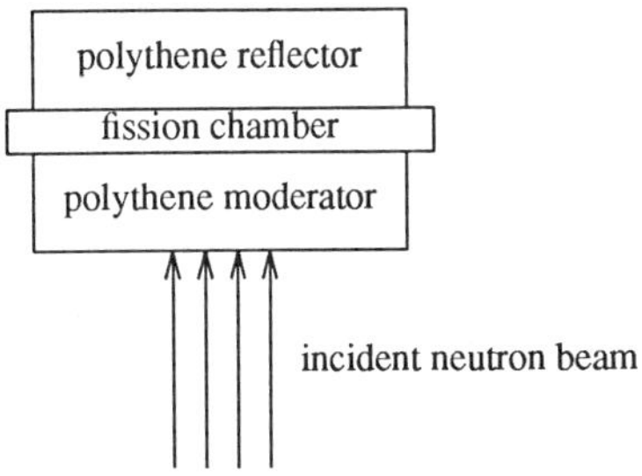

Figure 2. Neutron detector arrangement

Table 2. Configurations of Detectors

Detector	Moderator Thickness (mm)	Reflector Thickness (mm)
1	9.6	9.6
2	16.0	16.0
3	25.6	25.6
4	48.0	48.0
5	67.2	48.0
6	99.2	48.0

(2) Detector Response Functions

The six detector assemblies used for the beam measurements consisted of a ^{235}U fission chamber sandwiched between 150×150 mm stacks of polythene slabs (figure 2) of varying thickness (table 2). For each detector assembly a series of MCNP calculations, was used to generate microscopic ^{235}U fission reaction rates as a function of incident neutron energy. The incident, 6 cm diameter, parallel neutron beams were assumed to be normal to and centred on the lower face of the 'moderator' polythene layer. As for Bonner spheres, the response functions depend on the thickness of polythene between the source and detector. An increase in polythene thickness results in an increase in the relative response to higher energy neutrons (figure 3).

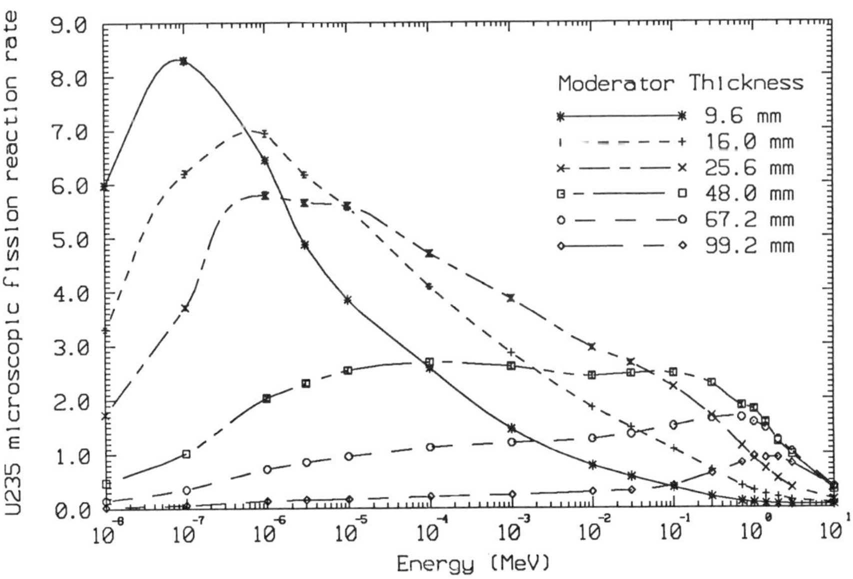

Figure 3. Response functions for Polythene/Fission Chamber Detectors

(3) Neutron Source Distribution in HIFAR

The POW3D[4] three-dimensional diffusion model of HIFAR developed by Robinson[10], including fuel burnup, fission products and the higher actinides, was used to calculate an average axial neutron source distribution as well as a distribution over the fuel elements. The 3D HIFAR model was based on earlier 2D models[11,12] but included the six signal-arm control blades which are lowered between the fuel elements at an angle.

Five group cross sections as a function of burnup were generated for the core by repeated execution, at specified time steps, of the four AUS modules MIRANDA[13], ANAUSN[14], EDITAR[15] and CHAR[16], essentially as described in previous HIFAR studies[11,17]. The AUS module MIRANDA was used to prepare the multigroup cross-section data required for this study from the AUS general purpose 200 group cross section library (generated essentially from ENDF/B-IV data). The 1-D S_N ANAUSN fuel cell calculation of flux was followed by the editing module EDITAR, which averaged the cross sections over space and energy. The burnup module CHAR used the calculated spatial flux to generate material compositions with fission products included, for the next burnup time step. These material compositions were then used in MIRANDA to generate burnup dependent cross sections, and the cycle through the AUS modules was repeated. The 5 group core cross sections were condensed further to 2 groups using both real and adjoint flux weighting from an RZ diffusion calculation of HIFAR. The 2 group cross sections were used in the coarse mesh 3-D HIFAR model[10] which was both accurate and fast.

The calculated axial neutron source distribution was averaged over the fuel elements adjacent to the dummy element. To calculate the power (and hence source) distribution over the fuel elements, the specific composition of each fuel element at the time of the experiment was used and the dummy element and aluminium spectrum shifter were included. Because of the differing degrees of burnup the ^{235}U loadings varied between 72 and 156 gms for the 60% enriched ^{235}U fuel elements[18].

(4) Filtered Epithermal Beams at HIFAR

The full core MCNP model of HIFAR included the dummy fuel element, the spectrum shifter, the D_2O facilities, and extended to the outer surface of the graphite reflector and vertically to the top shielding subsurface, 1250 mm above core centre. The explicit neutron source distribution for the core state at the time of the experiment was used, but a uniform fuel composition in the MCNP model of 110 gm ^{235}U per fuel element was satisfactory. The point detector tally option was used to calculate the detector responses in count rates per second for each beam (table 3). The filters were included as exponential line-of-sight attenuators (using MCNP cross section data for all except argon) and the detector response functions as multipliers. For the two experiments with argon filters the attenuation through argon, from (1) above, was folded into the detector response functions. A multiplication factor was used to account for reactor power, beam diameter, the conversion of the response functions to the total fission reaction rate and the sensitivity of the fission chamber. The MCNP statistical errors for the calculated responses were less than 5% for all beams except for beam 8 (approx. 6%).

RESULTS AND CONCLUSION

The calculated responses of the six detectors for each of the eight experimental filtered beams and the corresponding ratios of measured to calculated response are given in table 3 . The agreement for the argon filtered beams 1 and 2 is excellent for all except

Table 3. Calculated count rates and Measured/Calculated Count Rate Ratios

Detector / Beam	Count rates (cps)						Measured/Calculated ratios					
	1	2	3	4	5	6	1	2	3	4	5	6
1	91.6	148	184	121	58	12.4	1.00	1.06	0.97	0.97	0.97	1.26
2	66.9	110	138	92	44	9.6	0.97	1.00	0.98	1.03	1.07	1.15
3	7.8	19.6	34.5	32.1	18.1	4.7	0.98	0.94	0.90	0.95	0.99	1.10
4	1.0	2.8	5.3	5.7	3.7	1.3	1.44	1.41	1.33	1.26	1.32	1.24
5	0.5	1.5	2.9	3.4	2.3	0.9	-	-	-	-	-	1.36
6	47.5	120	209	191	105	25.7	0.97	0.94	0.94	0.97	1.02	1.10
7	5.1	13.2	24.3	24.6	14.8	4.5	1.25	1.17	1.18	1.22	1.26	1.48
8	2.1	5.7	11.1	12.1	7.7	2.6	1.67	1.40	1.37	1.39	1.49	1.51

detector 6, which is relatively most sensitive to high energy neutrons. Since the difference between calculated and measured values for detector 6, in absolute terms, is only several counts per second and since the accuracy of measurements is uncertain, a further investigation was not warranted. The calculated responses for beams 3 and 6 which include Fe (with 0.42 wt% Mn), agree well with the measured results which are only about 5% higher. The ENDF/B-IV data used for Fe has not been examined in detail.

The calculated response values for beams 4,7,8 which include Al, Fe and Ti as filter materials, are consistently and significantly low. The Al/Fe filters generate spectra in which neutrons close to 24 keV dominate while addition of titanium suppresses the 24 keV component resulting in beams with a harder spectrum. The calculated Ti filtered beams could be in error due to inaccuracies in the high energy (>100 keV) ENDF/B-IV cross section data for Ti, which are significantly different to the ENDF/B-V data (not available in Australia).

In these calculations to achieve the excellent agreement for the argon filtered beams, the detail of the generated argon point cross sections with resonances and the inclusion of ^{36}Ar were essential. Also a whole core HIFAR model was necessary. A model with reflective boundaries and including only the dummy element and the surrounding fuel elements did not suffice for two main reasons. Firstly, the k_∞ spectrum for the reflective boundary condition model, cannot be expected to be the same as the k_{eff} spectrum. Secondly, in a model with reflective boundaries it is impossible to determine the absolute value of the source for normalisation purposes since there is no way of attributing a power distribution to the fuel elements which are included.

In conclusion the excellent agreement achieved particularly for the Ar but also the Fe filtered beams provides confidence in the correctness of the whole core MCNP model of HIFAR. This model may be used to predict the performance of epithermal neutron beams in the 10H horizontal facility and for other HIFAR calculations.

ACKNOWLEDGEMENTS

G.S. Robinson and B.E. Clancy provided advice and assistance in implementing codes and developing models. B.J. Allen organised the Visiting Research Fellowship at Ansto.

REFERENCES

1. Storr, G.J. Allen, B.J. Harrington, B.V. Davis, L.R. Elcombe, M.M. & Meriaty, H. (1992) in: *Progress in Neutron Capture Therapy for Cancer* , eds., Allen, B.J. Moore, D.B. Harrington B.V. (Plenum Press), pp. 79-81.
2. Carolan, M. Constantine, G. Godfrey, R. Meriaty, H. Storr, G. Allen, B. A Filtered Neutron Beam at HIFAR, (These proceedings).
3. Briesmeister, J.F. (1986) MCNP - A General Monte Carlo Code for Neutron and Photon Transport, LA 7396-M Rev. 2.
4. Barry, J.M. Harrington, B.V. Pollard, J.P. AUS module POW3D - A general purpose 0,1,2 and 3D, multigroup neutron diffusion code including feedback-free kinetics, (ANSTO/E report to be published).
5. Robinson, G.S. (1987) A guide to the AUS Modular Neutronics Code System, AAEC/E645.
6. MacFarlane, R.E. (1990) NJOY89, A Code System for Producing Pointwise and Multigroup Neutron and Photon Cross Sections from ENDF/B Evaluated Nuclear Data, ORNL/RSIC/PSR-171.
7. Cullen, D.E. McLaughlin, P.K. (1985) ENDL-84, The Lawrence Livermore National Laboratory Evaluated Nuclear Data Library in the ENDF-V Format, IAEA-NDS-11, Rev. 4
8. Mughabghab, S.F. Divadeenam, M. Holden, N.E. (1981) Neutron Cross Sections (Academic Press), Volume 1, Part A, p. 18-1.
9. McLane, V. Dunford, C.L. Rose, P.F. (1988) Neutron Cross Sections (Academic Press), Vol 2 p. 113.
10. Robinson, G.S. (1991) 3D Diffusion calculations of HIFAR including the coarse control arms and their burnup, ANSTO/E703.
11. Harrington, B.V. (1983) Neutronic Models for the HIFAR Reactor, AAEC/E571.
12. Storr, G.J. (1989) HIFUME - A fuel management Code for HIFAR, ANSTO/NTP/TN 132.
13. Robinson, G.S. (1986) MIRANDA - a module based on multiregion resonance theory for generating cross sections within the AUS neutronics code system, AAEC/E626.
14. Clancy, B.E. (1982) ANAUSN - a one dimensional multigroup SN transport theory module for the AUS reactor neutronics system, AAEC/E539.
15. Robinson, G.S. (1986) EDITAR - a module for reaction rate editing and cross section averaging within the AUS neutronics code system, AAEC/E621.
16. Robinson, G.S. (1986) CHAR and BURNMAC - burnup modules of the AUS neutronics code system, AAEC/E621.
17. Robinson, G.S. Harrington, B.V. (1985) Neutronic calculations for reduced enrichment fuel in HIFAR, IAEA-TECDOC-324, p. 123.
18. Connolly, J.W. & Clancy, B.E. The Science and Engineering of HIFAR Safety (ANSTO/E report to be published).

FILTERED NEUTRON BEAM MEASUREMENTS AT HIFAR

M Carolan[1,3], G Constantine[2], R Godfrey[3], H Meriaty[3], G Storr[3], B J Allen[3]

1 Department of Physics, University of Wollongong
PO Box 1144, Wollongong, NSW 2500, Australia
2 Visiting Ansto Research Fellow
3 Australian Nuclear Science and Technology Organisation
Private Mail Bag 1, Menai, NSW 2234, Australia

INTRODUCTION

A series of filtered epithermal neutron beam experiments have been done at the High Flux Australian Reactor (HIFAR) in support of a proposal to install an epithermal NCT facility. Previous epithermal neutron beam studies for application to BNCT at Ansto have been carried out by calculation only[1,2]. The objectives of these experiments were to verify the reliability of MCNP models of complex filtered neutron beam arrangements[3] and develop experience in epithermal neutron beam characterisation at Ansto. In these experiments sulphur, aluminium, titanium, iron and argon filters were used in various combinations. Access to the 10" horizontal port (10H) which would be used for the proposed patient treatment facility is not possible at present because it is being used for other experiments. For this reason a filtered beam was extracted through the top plate of HIFAR in a configuration similar to that described by Ross for the DIDO reactor at Harwell[4].

EXPERIMENTAL METHOD

The HIFAR reactor is a D_2O moderated and cooled reactor normally operated at a power of 10MW. The cylindrical aluminium reactor vessel contains 25 fuel elements each containing 170 g of 60% enriched ^{235}U in four concentric aluminium tubes with a central 60 mm diameter space for irradiation experiments.

A central fuel element position (C4) was loaded with a dummy element during a reactor shutdown. A dry liner tube was inserted inside the dummy element to exclude D_2O from its central region.The increased reactivity associated with such an in-core void has safety implications, and to maintain this below the required limit of 0.3% dk/k it was necessary to minimise its volume by loading a solid aluminium into the dry liner tube at the core midplane.

This aluminium cylinder was 350 mm long and served as a spectrum shifter. The aluminium cross section has a large scattering resonance at 35 keV and a cross section minimum at 25 keV. Its effect is to scatter neutrons in this energy range out of the core and up the dry liner tube. The spectrum shifter was suspended by a fine nickel wire from a 150 mm long aluminium filter section supported by the shielding step in the dry liner tube. This configuration is shown in Fig.1a.

Advances in Neutron Capture Therapy, Edited by
A.H Soloway *et al.*, Plenum Press, New York, 1993

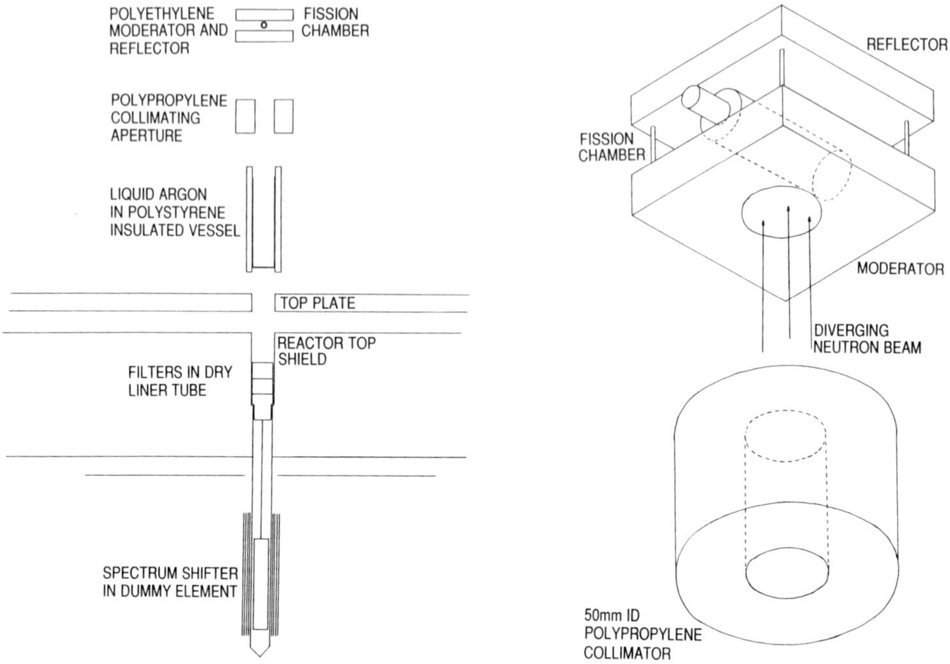

Figure 1 (a) Schematic of the HIFAR top plate filtered beam and (b) the polyethylene moderated fission chamber detector

Further 100 mm long filter components of aluminium steel and sulphur were fabricated to fit within the dry liner tube resting on top of the 150 mm aluminium section. The sulphur was cast into an aluminium canister and a lid crimped on to prevent spillage. The 10 mm titanium filter was placed over the access hole on the top plate. A 1 mm layer of cadmium could also be placed in the beam at this point. The liquid argon component of the filter was contained within an aluminium cryostat attached to an aluminium frame above the C4 access hole in the reactor top plate. The cryostat could be swung out of the beam for the purpose of refilling or aligning the assembly. The argon cryostat consisted of an aluminium vessel sheathed in expanded polystyrene foam for insulation. The bottom of the vessel was left uncovered to avoid the neutron scatter which would have resulted. Unfortunately this led to high boil-off rate, which necessitated frequent refilling to minimise variation in argon column length. The total length of the argon filter was 1066 mm although this could have varied by up to 60 mm during the course of the counting period. The filters used included combinations of 150, 250 and 350 mm of Al with 100 mm of S and Fe, 1 mm of Cd, 1066 mm of Ar and 10 or 20 mm of Ti.

Above the argon cryostat a polypropylene collimator was used to define the beam before it was incident on the undersurface of the detector. This collimator had an internal diameter of 50 mm. All the filter elements were aligned with the aid of a plumbob so that the beam was incident centrally on the base of the detector.

The detector consisted of a fission chamber (Series A type FC165 from 20th Century Electronics) having an active length of 57.2 mm, sandwiched between stacks of 3.2 mm thick polyethylene sheets as shown in Fig.1b. These were 150 mm square. The thickness of the

Table 1. Polyethylene thicknesses for neutron detectors 1-6

Moderator Thickness (mm)	Reflector Thickness (mm)
9.6	9.6
16.0	16.0
25.6	25.6
48.0	48.0
67.2	48.0
99.2	48.0

polyethylene stacks could thus be altered by the addition or removal of sheets. The stacks of polyethylene were bound together with the aid of tape and fastened at the corners using steel screws. The thicknesses of polyethylene used are shown in Table 1.

The principle of operation of the detector is similar to that of the Bonner Sphere. The thin polyethylene configurations are most sensitive to lower energy neutrons, fast neutrons passing through unmoderated and not reacting in the fission chamber. The detectors with thicker sections of polyethylene remove thermal neutrons from the beam and moderate the fast neutron component giving a peak response for fast neutrons. To the thin moderator detector configurations with highest sensitivities to lower energy neutrons reflectors were added to improve detector efficiency. The reflector thickness was matched to the moderator thickness, so that the energy response functions of the detector configurations were as disparate as possible. The energy response functions calculated using MCNP are discussed more fully in a theoretical companion paper[3].

Sensitivity studies of the detector response functions were done using a preliminary MCNP model to determine how sensitive these response functions would be to small changes in geometry and alignment of the detector such as might occur throughout the course of the experiment. A number of parameters were varied. These included the density of the polyethylene, position of the fission chamber between the polyethylene stacks, diameter of the neutron beam and its point of impact on the detector surface. The effects of filling the void surrounding the fission chamber and increasing the size of the polyethylene squares were also investigated.

The density of the polyethylene used for the construction of our detector was measured as 0.919(+-0.0005) gcm^{-3}. Considering that the density of polyethylene usually ranges between ~0.92 and 0.96 gcm^{-3}, the density was increased in the sensitivity study to 0.93 gcm^{-3}. This should give an estimate of the possible error due to inhomogeneities in the polyethylene density. Variations in detector response introduced by this change in density are ~1% for the thin detectors and up to 6% for the thick detectors with fast neutrons.

It is conceivable that the fission chamber could move slightly during the experiment. We would expect this effect to be most significant for the thinner detectors and it is seen to be less than 5% for the 9.6 mm detector. The effect of varying the beam diameter is also most significant for the thin detector. The detector response changes by 20 - 40% for 10 eV and 3 MeV neutrons, respectively, if the beam diameter falls from 6 cm to 4 cm. The thicker moderator detectors show a lesser dependence on the beam diameter due to the larger number of scattering events undergone by a neutron before reaching the fission chamber. Similarly, the effect of changing the position of incidence of the beam on the detector surface is most significant for the thinner detectors. Filling the void around the fission chamber with polyethylene seems to have the effect of preferentially increasing the high energy response of the detectors as well as improving their overall sensitivity. These results are summarised in Table 2. It should be noted when considering these variations in detector response, that although large changes are seen for fast neutrons on thin detectors, as parameters are varied the contribution of fast neutrons is only a small fraction of the final detector count.

Table.2. Variation of Detector energy response with variation in the experimental parameters.

Parameter	Change	Energy	Detector Response*		
			Thin (0.96 cm)	Medium (4.8 cm)	Thick (9.92 cm)
Polyethylene Density	1.1%	10 eV	-1%	<-1%	-6%
		1 keV	<1%	-2.5%	-4%
		3 MeV	<1%	6%	
Position of FC	1 cm off axis	10 eV	-5%	<-1%	-3%
		1 keV	-6%	-3%	-3.5%
		3 MeV	-1%	6%	-1%
Diameter of Beam	4 cm (-33%)	10 eV	25%		12%
		1 keV	20%		9%
		3 MeV	43%		5%
	5 cm (-16%)	10 eV	13%		8%
		1 keV	9.8%		6%
		3 MeV	20%		1.6%
	7 cm (16%)	10 eV	-11.4%		-3.3%
		1 keV	-11.5%		-8.6%
		3 MeV	-21.8%		-8.4%
	8 cm (33%)	10 eV	-22%		1.4%
		1 keV	-20%		-17.7%
		3 MeV	-37%		-9%

*The uncertainties in the MCNP calculations for the detectors are: 0.96 cm detector ~1-2%, 4.8 cm detector ~2.5-3%, 9.92 cm detector ~5-7% (3-4% for 3 MeV)

CONCLUSION

The detector counts for each filter combination were compared with those predicted by MCNP modelling and were found to be in good agreement[3]. The experiment successfully demonstrated that MCNP models are able to predict spectra from filtered epithermal neutron beams. Experience was gained in the fabrication of filtered epithermal beam setups and the experimental methods of characterising them. The polyethylene type detector used was shown to be effective in coarse characterisation of neutron beams. Its lack of symmetry as compared with the Bonner Sphere however leads to more critical alignment requirements and the response function of the detector would have to be recalculated for any new experiment where the neutron beam was not 6 cm in diameter and normally incident. Future characterisation of the 10H HIFAR beam will therefore probably rely on foil activation and proton recoil techniques. These experiments can proceed now that confidence has been established in MCNP calculations and experimental techniques.

REFERENCES

1. B. V. Harrington, Optimisation of an Epithermal Beam in HIFAR for Boron Neutron Capture Therapy. ANSTO/E662. 1987.
2. G. J. Storr, et al, Design Considerations for the Proposed HIFAR Thermal and Epithermal Neutron Capture Therapy Facilities in *Progress In Neutron Capture Therapy for Cancer,* eds., B. J. Allen, D. E. Moore, B. V. Harrington, Plenum Press New York, 1992.
3. B. V. Harrington, G Constantine. Analysis of Filtered Neutron Beams at HIFAR. These proceedings.
4. D. Ross et al., Modelling an Epithermal Neutron Beam for a DIDO Type Reactor using MCNP - A Monte Carlo Code in *Progress In Neutron Capture Therapy for Cancer,* eds., B. J. Allen, D. E. Moore, B. V. Harrington, Plenum Press New York, 1992.

PRECLINICAL TESTS FOR NEUTRON CAPTURE THERAPY ON THE LVR-15 REACTOR

J. Burian[1], M. Marek[1], J Rataj[1], V. Mares[2], P. Stopka[3], O. Strouf[3]

[1]Nuclear Research Institute, Rez, Czechoslovakia
[2]Institute of Physiology and [3]Institute of Inorganic Chemistry, Czechoslovak Academy of Sciences, Prague, Czechoslovakia

A facility designed to produce the beam of epithermal neutrons for neutron capture therapy was realized on the LVR-15 reactor[1] in NRI Rez. The first biological and chemical preclinical research activities were started.

FACILITY

A description of the final configuration can be seen in Fig.1. It is composed of cylindrical layers, diameter of 100 cm: 25 cm of Al_2O_3, 60 cm of Al and 15 cm of S. To improve the neutron - gamma ratio 5 + 10 cm of Pb and 10 cm Bi are included. A thermal neutron filter (6 mm of B_4C) reduces the final thermal flux and therefore the production of gamma rays in Pb is also decreased. The beam diameter is 11.5 cm.

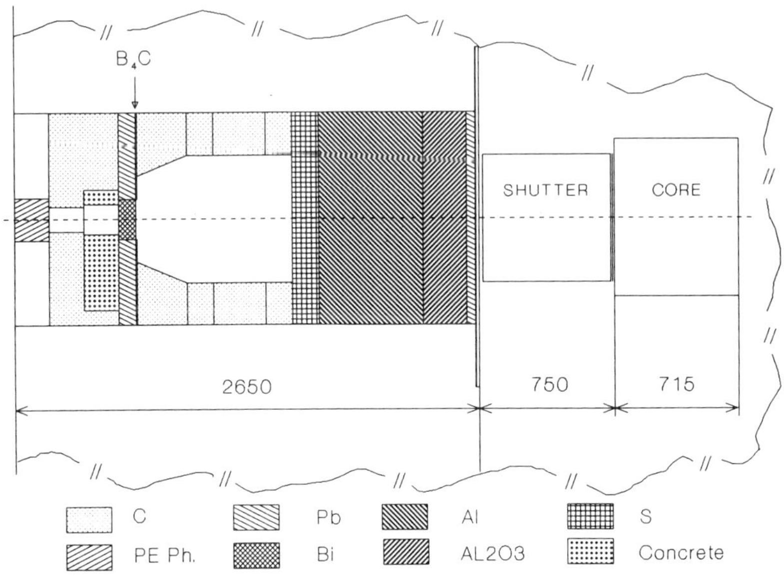

Figure 1. The BNCT facility on the LVR-15 reactor

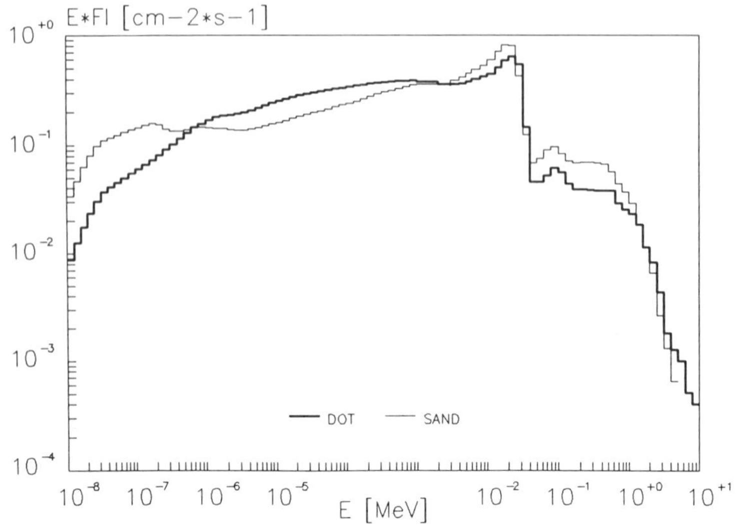

Figure 2. Neutron free beam spectra

A calculational analysis of a two-dimensional model was performed by the DOT 4.2 transport code (discrete ordinates method) in cylindrical (r-z) geometry. The coupled cross section library SAILOR (47 neutron, 20 gamma groups) was used for this purpose.

Experimental data for neutrons and gamma rays were obtained both for the free beam geometry and inside a cylindrical polyethylene phantom and a human phantom. The following experimental methods were used: Bonner spheres, a scintillation spectrometer, a set of activation detectors, thermoluminescent detectors, Si-diodes and a small semiconductor detector Si(Li).

The DOT neutron spectrum and the free-in-air spectrum unfolded by SAND code are shown in Fig.2.

For a reactor power of 5 MW, the facility will give at the treatment position (2.5 cm inside the cylindrical phantom) a neutron beam with an average energy of 6.8 keV, and a thermal neutron flux 1.6 * 10^9 neutrons per cm^2 per s. When the usual relative biological effectiveness factors of 1.6 and 2.3 for the fast neutron and $^{10}B(n,\alpha)$ reaction respectively are assumed and with an intratumoral ^{10}B concentration 30 μg/g, a therapeutic dose of 20 Gy is delivered in 34 minutes. Healthy tissue (3 μg/g of ^{10}B) obtains 32% of the dose under these conditions.

The parameters of our neutron beam are supposed to be improved. Above all thermal neutron absorbing slabs with ^6Li will be completed to decrease gamma-ray production.

CHEMISTRY

In our study of BNCT the question of oxygen radical generation and the role of dissolved oxygen in the course of BNCT arose. We tested the Sodium mercaptoborate (NaMHB), which is generally used in BNCT, Tetramethylammonium mercaptoborate (TMA-MHB), which is often used for BNCT research, and Sodium tetraborate ($Na_2B_4O_7$), for which the organic radicals generation under the conditions of BNCT has been described[3].

The studied solutions of NaMHB, TMA-MHB and $Na_2B_4O_7$ were saturated with gaseous mixtures of O_2/N_2, and then irradiated by neutrons. The identification of generated radicals was made by the ESR method after rapid freezing and by the spin trapping method using Tiron as a specific spin trap for superoxide anion radicals O_2^-.

In the case of NaMHB and TMA-MHB the anisotropic spectrum of O_2^- has been found ($g_{\parallel}$=2.0328, $g_{\perp}$=2.0044) using the method of rapid freezing with ESR spectroscopy. The quantity of generated O_2^- proportionally depends on the concentration of dissolved oxygen, see Fig. 3.

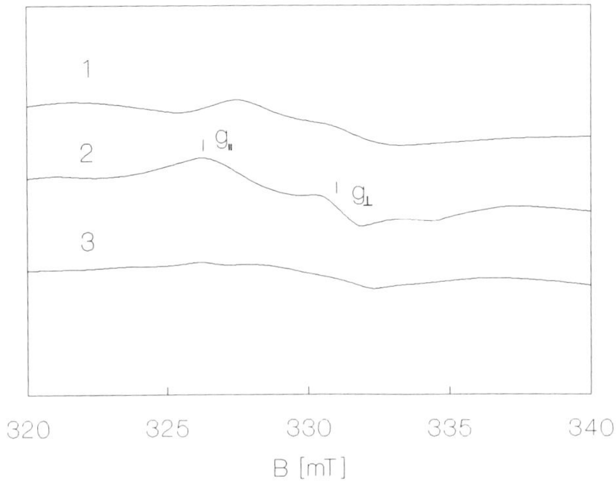

Figure 3. ESR spectra of frozen solution of sodium mercaptoborate irradiated by thermal neutrons for 30 minutes at room temperature. The solutions were saturated: 1 - by air, 2 - by oxygen, 3 - by nitrogen with 1% of oxygen. Microwave power - 10 mW, modulation amplitude 0.2 mT, microwave frequancy - 9.306 GHz, temperature 77 K.

In the case of $Na_2B_4O_7$ only the weak spectrum of O_2^- has been found for oxygen saturated solution ($g_{\parallel}$=2.0332, $g_{\perp}$=2.0044). The solutions with lower concentrations of oxygen were ESR silent.

The generation of O_2^- was proved by the spin trapping method and the positive presence has been found in all cases in which the anisotropic O_2^- spectrum was measured at 77K.

We have found and identified the superoxide anion radicals O_2^- in the course of BNCT and the dependence of quality of generated superoxide anion radicals on the concentration of dissolved oxygen under the conditions of boron neutron capture reactions. The formation of O_2^- can raise the therapeutical effect of BNCT by destroying of tumor cells. There is some similarity with the role of oxygen for oxygen radicals generation and/or singlet oxygen generation in the mechanism of photodynamic therapy of skin cancer.

BIOLOGY

Uptake of mercapto-undecahydro-*closo*-dodecaborate ($Na_2{}^{10}B_{12}H_{11}SH$), NaMHB, synthesized by Plesek [4], was studied in 2- and 7-day-old cultures of the diploid human lung fibroblasts (LEP 19), heteroploid E7 mouse neuroblastoma C1300, C6 rat glioma and HeLa cells. The cells were incubated with 200 μg of NaMHB (cca 124 μg ^{10}B) per ml media for 1 to 7 h and content of 10B determined by $^{10}B(n,\alpha)^7Li$ reaction.

The content of ^{10}B progressively increased within the initial 1-2 h incubation interval in all lines studies. In the heteroploid cell lines at the exponential phase of growth (2-day-old cultures) the increase was often steeper and more prolonged (at least up to 7 h) than in the stationary (7-day-old) cultures. Irrespective of in vitro age, the accumulation of NaMHB was found relatively highest in C6 glioma cells. In 2-day-old cultures the final NaMHB content/mg protein was 5 to 7 times higher than in other cell lines.

Comparison of the total content of ^{10}B in the individual cell line samples at 4 h post treatment interval with that calculated for a volume aliquot of culture medium corresponding to the volume of a measured cell sample (R_1 ratio) was found highest (cca 9.52) in 2-day-old C6 glioma cultures. In other 2 days cell line cultures it ranged from 0.98 to 2.02. In 7 days cultures the R_1 ratio was consistently lower (0.11 to 1.47); however, in C6 glioma cells it was again higher than in the other cell lines (1.47). The data indicate the accumulation of ^{10}B against the concentration gradient in most of the exponentially growing cell lines, namely the C6 glioma cells[5].

The preliminary experiments in which C6 glioma cell cultures were treated with 200 μg NaMHB/ml and irradiated with 10^8 thermal neutrons/cm^{2}s for 20 min showed inhibition of growth of the population. After irradiation with 10^9 thermal neutrons/cm^{2}s for 30 min necrosis of cells appeared in approx. 50% of samples within the subsequent 2 days.

CONCLUSIONS

The designed filter modeling the epithermal neutron beam of 11.5 cm in diameter has been found to be fairly acceptable for BNCT purposes. The generation of the superoxide anion radicals O_2^- and its concentration dependence on the initial oxygen concentration in studied solutions have been found. The accumulation of ^{10}B against the concentration gradient was found in cells growing at the higher rate, namely the C6 glioma cells. These cells in culture appeared a suitable model for fast screening of the neutron capture effect.

REFERENCES

1. Burian J., Marek M., Rataj J.: The parameters of epithermal neutron beam for BNCT at the LVR-15 reactor. EC Project Workshop, Pavia, June 1991.
2. 1990 Recommendations of the International Commission on Radiological Protection. ICRP Publication 60, November 1990.
3. Carmichael A.J., Arroyo C.M., Moore M.L.: *Radiat. Phys. Chem.* 32:227 (1988).
4. Plesek et al.: Czechosl. Author's certificates and patents no. 288254, 1985 and 242064, 1985
5. V.Mares et al.: Accumulation of ^{10}B in cell cultures exposed to ($Na_2{}^{10}B_{12}H_{11}SH$) used for the NCT of brain tumors. *J. Pharmacol. and Exp. Therapy,* 1992, in press.

STUDY ON REMODELING THE HEAVY WATER FACILITY OF THE KYOTO UNIVERSITY REACTOR FOR NEUTRON CAPTURE THERAPY FROM THE CONCEPT OF NEUTRON ENERGY SPECTRUM CONTROL

Tooru Kobayashi, Yoshinori Sakurai and Keiji Kanda

Research Reactor Institute, Kyoto University,
Kumatori-cho, Sennan-gun, Osaka 590-04, Japan

INTRODUCTION

In August 1988, there was heavy water leakage from the thermocouple guide pipe of the heavy water tank adjacent to the core of the Kyoto University Research Reactor (KUR). The need for a fundamental reexamination of the Heavy Water Thermal Neutron Facility (HWTNF) has been recognized since the guide pipe was repaired. Clinical irradiation was restarted in February 1990, and as of September 1, 1992, 24 clinical treatments had been carried out. Some improvements for the clinical treatment have been requested by clinicians and other users, especially the ability to use the facility while the reactor is in continuous operation.

From the viewpoints of reactor engineering and medical physics, there are four goals for remodeling the present HWTNF: (1) to simplify and secure maintenance, overhaul, and repair, (2) to enable the facility to be used during continuous operation of the KUR,[1] (3) to improve the performance of the irradiation field of thermal neutrons for biomedical uses,[2,3] and (4) to control neutron energy spectrum. In this paper, an outline for the redesign of the facility is reported.

OPTIMIZATION OF THE PRESENT HWTNF

Figure 1 shows the present HWTNF of the KUR. At present, the heavy water layer, graphite layer and bismuth-lead layer are 136.2 cm, 48 cm and 15 cm in thickness, respectively. Survey calculations for the heavy water layer, bismuth layer and graphite layer were performed. It was found that the optimized thicknesses of the heavy water layer and bismuth layer of 60 cm diameter would be 100 cm and 15 cm, respectively. It was also found that the graphite layer does not affect the beam quality of thermal neutrons, i.e. contamination of gamma rays and epithermal or fast neutrons.

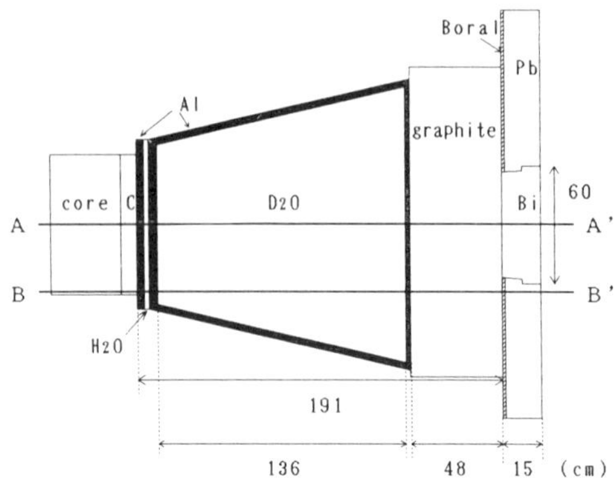

Figure 1 The present Heavy Water Thermal Neutron Facility of the KUR.

From that optimization, the space between the heavy water and lead–bismuth layer, about 82 cm, is to be used for a shutter system, described below. From the calculation results, the available thermal neutron flux could be more than 5 times that of the present facility, without changing the beam quality.

A PROPOSED SHUTTER SYSTEM

A design for a shutter system for the HWTNF during 5 MW continuous operation is discussed. The design requirement is that a total dose rate outside the shutter should be less than 100 μSv/h (10 mR/h) during the 5 MW KUR operation. This value has been decided on the basis of clinical experiences of irradiations at the KUR. This shutter system consists of the following three parts in order from the core. First a shield against fast and epithermal neutrons (water, polyethylene, iron), the next a shield against thermal neutrons (boral, cadmium), and the last a shield against gamma rays (lead or lead + bismuth).

Parameter survey calculations were performed for various thickness of each layer. The calculating procedure was as follows: coupled 100 groups of neutron and 20 groups of gamma-ray cross sections were made by RADHEAT-V3 with the ENDF/B-IV Library; the cross section data of bismuth were based on JENDL-3; the (γ,n) reactions of heavy water were calculated and evaluated according to the method of Aoki et al..[4] In the second step, the cross sections were collapsed by a one-dimensional transport code named ANISN-JR to 21 groups of neutron and 9 groups of gamma-ray cross sections. Finally, neutron and gamma-ray fluxes were calculated by ANISN-JR.

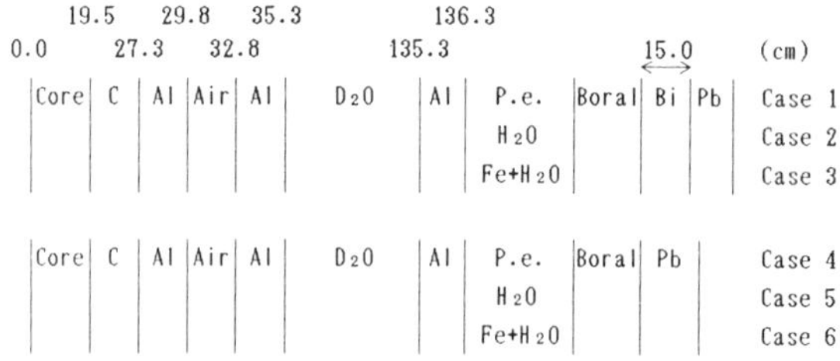

Figure 2 Calculation models for parametric survey of shutter systems.
Case 1 and 4: polyethylene, case 2 and 5: water, and case 3 and 6: iron and water.
The left and right boundary conditions are reflection and vacuum, respectively.

Figure 2 shows geometries for the calculations. The structural limitation of the facility is that the length between the core tank and the lead–bismuth layer is unchanged, i.e. 191 cm (see Fig. 1). Calculations were done both for the central part of the irradiation field includingthe bismuth layer (A–A' in Fig. 1) and for the surrounding part including the lead layer (B–B' in Fig. 1). In Fig. 2, cases 1–3 correspond to the bismuth layer and cases 4–6 correspond to the lead layer. Cases 2 and 5 for the water shutter shown in Fig. 3 are described below.

A parameter survey of the thickness of the water layer (case 2) was performed when the thickness of boral, bismuth, and lead layers were 1 cm, 15 cm, and 20 cm, respectively. It was found that the thickness of the water layer must be more than 80 cm to satisfy the limit of 100 μSv/h. The boral layer is not necessary when the water layer is 80 cm in thickness, because neutrons are sufficiently moderated and absorbed by the 80 cm water layer. It was found that the thickness of the lead layer should be more than 20 cm for case 2.

A parameter survey of the thickness of the lead layer surrounding the bismuth layer (case 5) was carried out with the thickness of water and boral layers as 80 cm and 1 cm, respectively. It was found that the thickness of the lead layer surrounding the bismuth layer should be more than 32 cm.

For the rest of the cases, parameter surveys were performed in the same way. From those results it was found that (1) the minimum thicknesses of the shutter layers to shield against fast and epithermal neutrons were 80 cm for water, 60 cm for polyethylene, and 64 cm for iron plus water (36 cm iron + 28 cm water), respectively; (2) a 1 cm thick of boral layer was enough to shield thermal neutrons for all cases, and (3) the minimum thickness for lead layers was 20 cm for case 1–3 (having 15 cm bismuth layer) and 32 cm for case 4–6. Figure 3 shows a proposed HWTNF of the KUR.

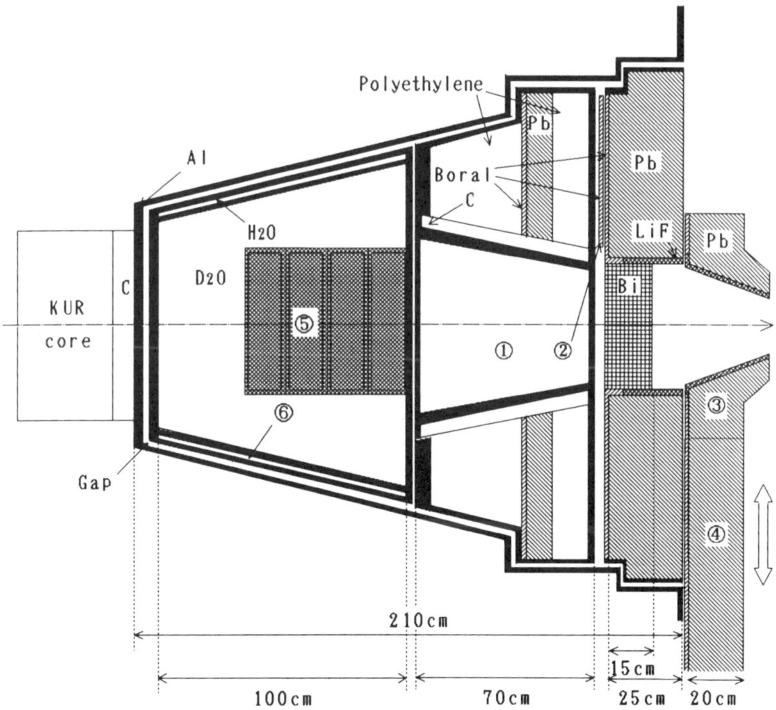

Figure 3 A proposed Heavy Water Thermal Neutron Facility of the KUR.
1: Water shutter, 2: Boral shutter, 3: Clinical collimator,
4: Lead shutter, 5: Spectrum shifter, and 6: Water jacket.

CONCLUSION

From the results of a one-dimensional transport code, ANISN, it was confirmed that by optimizing the present facility, the thermal neutron flux could be increased by more than 5 times, to ~10^{10} n/cm^2/sec. Also, by optimizing the thickness of the heavy water and graphite layers, the present thickness of ~82 cm could be reduced. Using this extra room for a shutter system, the dose rate could be reduced to less than 100 μSv/h in the working area in the irradiation room during the 5 MW KUR operation. On the basis of the structural design and handling convenience, it can be said that only a shutter system with replaceable water, as shown in Fig. 3, is feasible. Using this shutter system, we could come into the irradiation room and handle a patient during the 5 MW KUR operation.

When this remodeling is completed, the following advantages can be expected for boron neutron capture therapy:

(1) shutdown and restart times can be reduced,
(2) clinical irradiation can be carried out at the same time as other reactor experiments,
(3) a system of measuring ^{10}B concentrations with prompt gamma-rays from (n,γ) reactions, which requires operating the reactor, can be used so that ^{10}B concentrations in tissue can be known shortly before and/or after clinical irradiation,
(4) irradiation time after injecting boron compounds can be freely decided, and
(5) freedom of usage increases for all experiments.

There are three tasks for the future to improve the neutron beam quality: (i) to control the neutron energy spectrum which will probably be done by removing some parts of heavy water from the heavy water tank, as shown in Fig. 3, (ii) to control the ratio of the incident gamma rays and thermal neutrons, and (iii) to control the directional components of neutron beam.

ACKNOWLEDGEMENTS

The authors wish to acknowledge Dr. Seiji Shiroya and Mr. Hironobu Unesaki of Kyoto University Research Reactor Institute for their suggestions on the calculation method. The authors thank the Nuclear Data Center of Japan Atomic Energy Research Institute and the Sigma Committee of Japan for preparing the evaluated cross section of bismuth.

The calculation was performed at the Data Processing Center of Kyoto University.

REFERENCES

1. Y.Sakurai, T.Kobayashi and K.Kanda, Analysis for remodeling of the KUR Heavy Water Facility for biomedical uses, *Annu. Rep. Res. Reactor Inst. Kyoto Univ.* 24:90 (1991).
2. S.Fujihara, T.Kobayashi and K.Kanda, Reevaluation of thermal neutron field of the KUR Heavy Water Facility for biomedical uses, *Annu. Rep. Res. Reactor Inst. Kyoto Univ.* 23:13 (1990).
3. T.Kobayashi, T.Kozuka, H.Chatani, K.Kanda, and T.Shibata, Experimental study on increasing of thermal neutron flux in the KUR Heavy Water Facility for effective use of exposure tubes and a U-235 Fission Converter, *Annu. Rep. Res. Reactor Inst. Kyoto Univ.* 18:133 (1985).
4. K.Aoki, S.Shiroya, T.Kobayashi, K.Kanda and T.Shibata, Technique to calculate the effect (γ,n) reaction of heavy water moderator on beam quality of biomedical irradiation facility, *Annu. Rep. Res. Reactor Inst. Kyoto Univ.* 15:27 (1982).

IMPROVEMENT OF THERMAL NEUTRON FLUENCE FOR BNCT TREATMENT OF DEEP-SEATED TUMORS: SIMULATION CALCULATION AND PHANTOM EXPERIMENTS

Yoshinori Sakurai, Tooru Kobayashi, Koichi Ono, Takashi Sato and Keiji Kanda

Research Reactor Institute, Kyoto University,
Kumatori-cho, Sennan-gun, Osaka 590-04, Japan

INTRODUCTION

It is important to get sufficient thermal neutron flux at the tumor for the purpose of increasing the efficiency of NCT. To improve the effect of the treatment for a tumor deeply seated in a human body, the distribution of thermal neutron fluence should be improved. In this paper, we studied on the methods for increasing the thermal neutron doses at the deep location from the viewpoint of the physics and engineering.

The following three methods are taken into consideration: (1) the adjustment of the directional components of the thermal neutron beam,[1] (2) the making of a void in a human body[2] and (3) the partial replacement of body water (i.e. light water) with heavy water in a human body.[3] Phantom experiments and simulation calculations were performed in order to confirm the effects of these methods and to study their characteristics.

PHANTOM EXPERIMENTS

Three kinds of experiments were performed using the KUR Heavy Water Facility. The clinical collimator and a cubic phantom were used in the experiments. The outer dimensions of the phantom case were 20 × 20 × 20 cm. The case consisted of acrylic plates of 5 mm in thickness and it was filled with water. The experimental geometries for method 1, 2 and 3 are shown in Figs. 1–a), 1–b) and 1–c), respectively. In the method 1, the directional component of the neutron beam was changed by removing the inner reflector (thermo-plastic of 1 cm thickness) from the cone shaped collimator (^{6}LiF sheet of 1 cm thickness).

SIMULATION CALCULATIONS

Simulation Calculations for Phantom Experiments were performed in P3S8 using the two-dimensional transport code, DOT 3.5. Using the ENDF/B-IV, JENDL-3 and POPOP-4 libraries, coupled 11 neutron group and 7 gamma-ray group cross sections were generated by RADHEAT-V3 for DOT 3.5.

In order to observe the difference among the thermal neutron flux distributions due to the directional characteristics of the beam, simulation calculations were performed in P3S16 using DOT 3.5. The calculational geometry is shown in Fig. 2. The calculations were performed for eight kinds of thermal neutron directional components. The size of the irradiation field is fixed to 10 cm diameter.

The Monte Carlo calculations using MCNP−V3A are performed to observe the effect of the size of the incident beam. The calculational geometry is almost same as Fig. 2. The calculations were performed for an isotropic and a parallel beam. The size of the irradiation field was varied by changing the inner diameter of the ^{6}LiF collimator (i.e. 2, 4, 6, 8, 10, 15 and 20 cm). The number of random histories calculated was one million for each case.

RESULTS AND DISCUSSION

The Experimental and Calculational Results for the Three Methods

For the method 1, the depth dose distributions of the thermal neutron flux along the central axis are shown in Fig. 3. The calculated values were normalized at the phantom surface, and the calculated and experimental values were normalized at the depth of 2 cm from the phantom surface. Thermal neutrons at that depth are expected to attenuate according to an exponential function. The calculational and experimental values were in good agreement except near the phantom surface.

For each method, the comparison of two or three cases was done at the depth of 5 cm in the phantom, which is the depth limit presently considered for clinical treatment with a thermal neutron beam. For the method 1, the thermal neutron flux without the reflector is about 1.4 times larger than with it. For the method 2, it is found that the thermal neutron

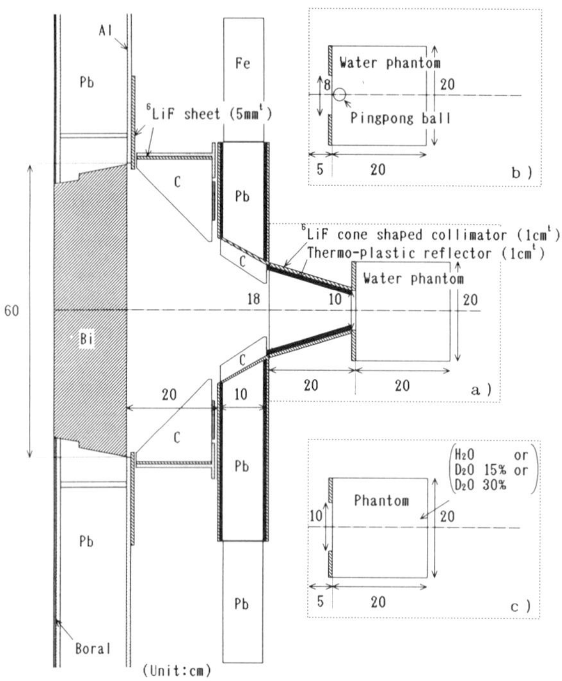

Figure 1 Experimental geometries.

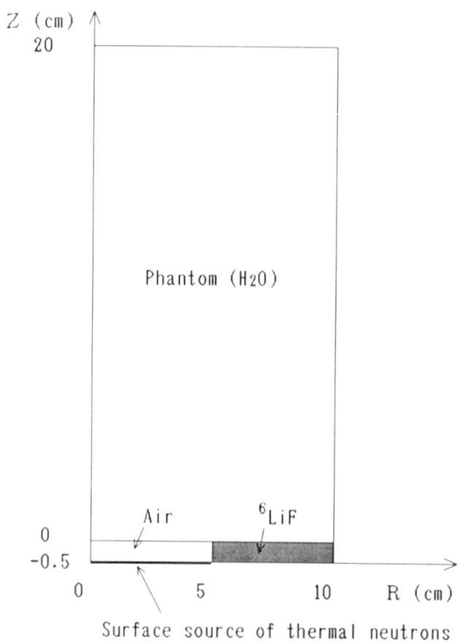

Figure 2 Calculational geometry for the directional components of the incident thermal neutron beam.

flux with a pingpong ball is about 2.7 times larger than without it. For the method 3, the thermal neutron fluxes for the 15 and 30% heavy water are about 7 and 36% larger than that of the natural water. The maximum substitution ratio of heavy water for clinical treatment is expected to be about 30 mol percent.

Directional Characteristics and the Sizing of the Thermal Neutron Beam

Figures 4 shows the DOT calculated results of the thermal neutron flux distributions along the central axis. The sideward components increases as the case–number increases. Case 1 and 8 correspond almost to parallel and isotropic source, respectively. The calculated values were normalized at the phantom surface.

It is found that the thermal neutron flux of the case 1 (parallel source) is about 2 times larger than that of the case 8 (isotropic source) at 5 cm depth. Thermal neutrons penetrate deeper according as the forward component of the neutron beam increases. Thus, it is confirmed that the directional characteristics of the thermal neutron beam considerably affect the depth dose distribution of thermal neutrons.

Figures 5–a) and b) show the MCNP calculated results of the thermal neutron flux distributions along the central axis for isotropic and parallel sources, respectively. The calculated values were normalized at the phantom surface. Thermal neutrons decrease more rapidly in the water phantom as the diameter of the collimator decreases and that a collimator diameter of more than 15 cm hardly affects the depth dose distribution of thermal neutrons.

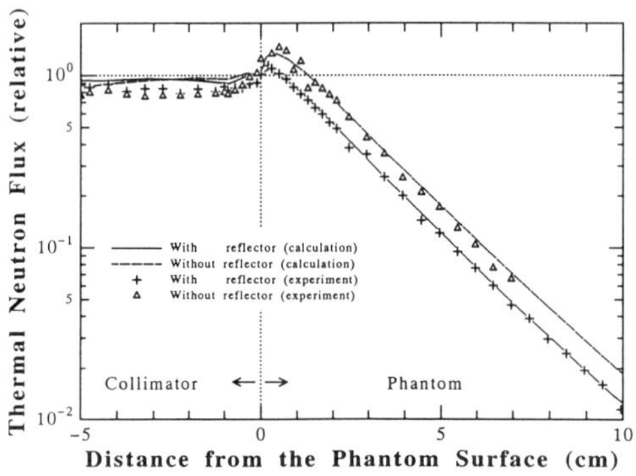

Figure 3 Results for the method 1.

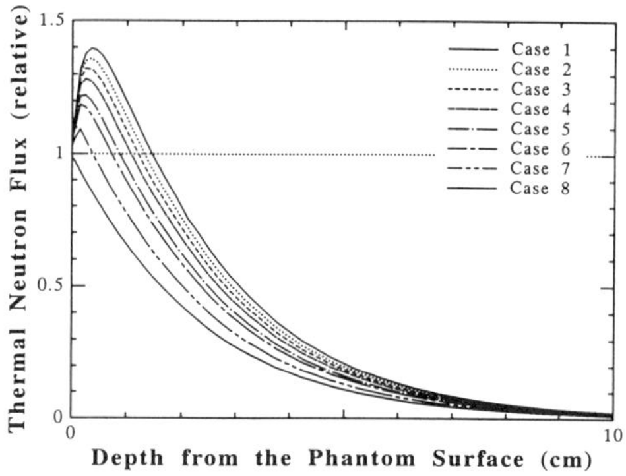

Figure 4 Calculation results for the directional components of the incident beam.

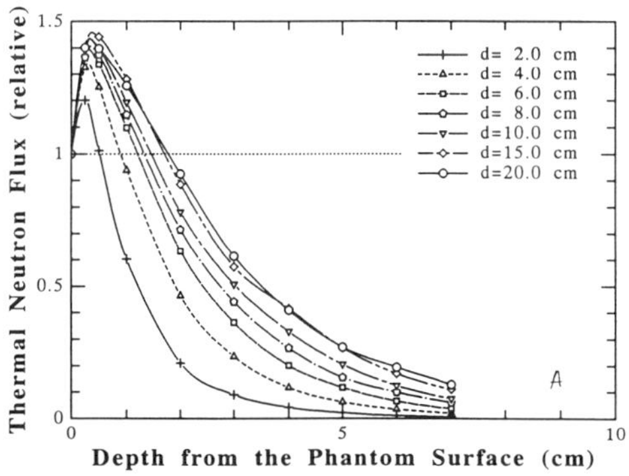

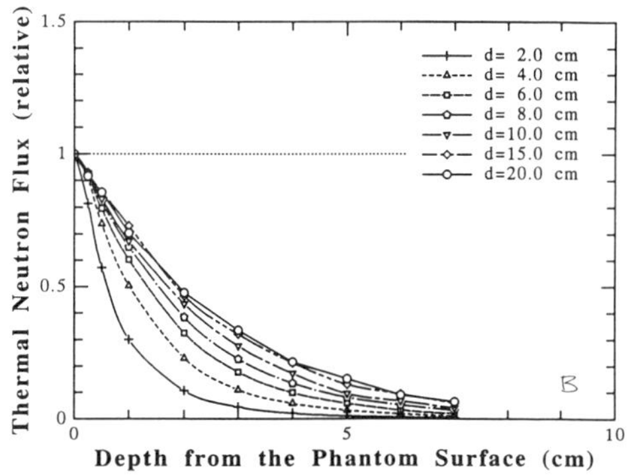

Figure 5 Calculation results for the size of irradiation field: a) parallel source, b) isotropic source.

Comparing the collimator diameter sizes of 2 cm and 10 cm at the depth of 5 cm in the phantom, the thermal neutron flux of the latter is about 10 times larger than that of the former for both parallel and isotropic sources. It follows that the size of the irradiation field also considerably affect the depth dose distribution of thermal neutrons.

CONCLUSION

From the results of the three kinds of phantom experiments and calculations, it was confirmed that each method has the effect of increasing the depth thermal neutron dose. In the actual treatment, it is thought that either one of these three methods will be selected or some methods will be combined according to the patient circumstances.

It was also seen that the thermal neutron flux changes rapidly near the phantom surface. Hence, it is confirmed that further examination needs to be performed for the monitoring position and its system. From the results of the simulation calculations, it is expected that the depth dose distribution can be skillfully controlled by the changes of the directional components and the irradiation field size of the incident beam.

ACKNOWLEDGEMENTS

The authors wish to acknowledge Dr. S. Shiroya and Mr. H. Unesaki of Kyoto University Research Reactor Institute (KURRI) for their suggestions on the calculational method. The authors thank Prof. H. Hatanaka of Teikyo University Hospital, Prof. Y. Mishima of Kobe University School of Medicine for their invaluable discussion of clinical treatment. The authors thank the Nuclear Data Center of Japan Atomic Energy Research Institute and the Sigma Committee of Japan for preparing the evaluated cross section of bismuth. The calculations were performed at the Data Processing Center of Kyoto University.

REFERENCES

1. T.Kobayashi, S.Fujihara, and K.Kanda, Effective thermal neutron collimation for neutron capture therapy using neutron scattering and absorption reactions, *in*:"Progress in Neutron Capture Therapy," B.J.Allen, *et al.*, ed., Plenum, NY (1992).
2. H.Hatanaka, S.Kamano, K.Amano, S.Hojo, K.Sano, S.Egawa, and H.Yasukochi, Clinical experience of boron–neutron capture therapy for gliomas, *in*:"Boron–Neutron Capture Therapy for Tumors," H.Hatanaka, ed., Nishimura, Niigata (1986).
3. T.Kobayashi, M.Ono, and K.Kanda, Measurement and analysis on neutron flux distributions in a heavy water phantom using the KUR neutron guide tube for BNCT, *Strahlentherapie und Onkologie* 165:101–103 (1989).

THE SENSITIVITY OF RESEARCH REACTOR BASED BNCT IRRADIATION FACILITY PERFORMANCE TO CORE CONVERSION FROM HIGH TO LOW ENRICHED URANIUM FUELS

Lisa A. Heimberger,[1] and Tunc Aldemir[2]

[1]Westinghouse Hanford Co.
Richland, WA 99352 U.S.A.
[2]The Ohio State University
Columbus, OH 43210 U.S.A.

INTRODUCTION

A critical requirement for boron neutron capture therapy (BNCT) is a strong source of thermal (i.e. < 1eV) or epithermal (1 eV - 10 keV) neutrons. Nuclear reactors are the currently available most intense sources of such neutrons, however, the neutron field in nuclear reactor cores is heavily contaminated by gamma radiation. Nuclear reactors also produce fast (i.e. > 10 keV) neutrons which can cause damage to the healthy tissue through H(n,n′)H (i.e. proton recoil) reactions and lead to the generation of additional gamma photons through (n,γ) reactions with hydrogen. The effectiveness of reactor based BNCT strongly depends on how well the fast neutron and gamma dose to healthy tissue is reduced through efficient irradiation facility design. Optimal facility design, in turn, is a strong function of the core neutron spectrum.

A large number of nuclear reactors in the world which are suitable as neutron sources for BNCT use highly enriched uranium (HEU) fuels (i.e. 93 atom % ^{235}U). In view of the restrictions on the export of HEU from U.S. since 1979 to prevent nuclear weapons proliferation and also the 1986 U.S. Nuclear Regulatory Commission ruling, these reactors are in the process of being converted to use low enriched uranium (LEU) fuels (i.e. < 20 atom % ^{235}U). The conversion process usually leads to the hardening of the core spectrum (i.e. a shift to higher neutron energies) due to the need to increase the fuel-to-moderator ratio to compensate for the increased parasitic neutron absorption in ^{238}U. This paper presents the results of a 3-dimensional Monte Carlo study which investigates the potential impacts of such a core spectrum hardening on a hypothetical irradiation facility utilizing the reactor as a thermal neutron source for BNCT.

SYSTEM DESCRIPTION AND COMPUTATIONAL MODEL

The facility concept chosen for the study is the one used at the Musashi Institute of Technology Research Reactor[1] (MITRR) in Japan for the clinical implementation of

Advances in Neutron Capture Therapy, Edited by
A.H Soloway *et al.*, Plenum Press, New York, 1993

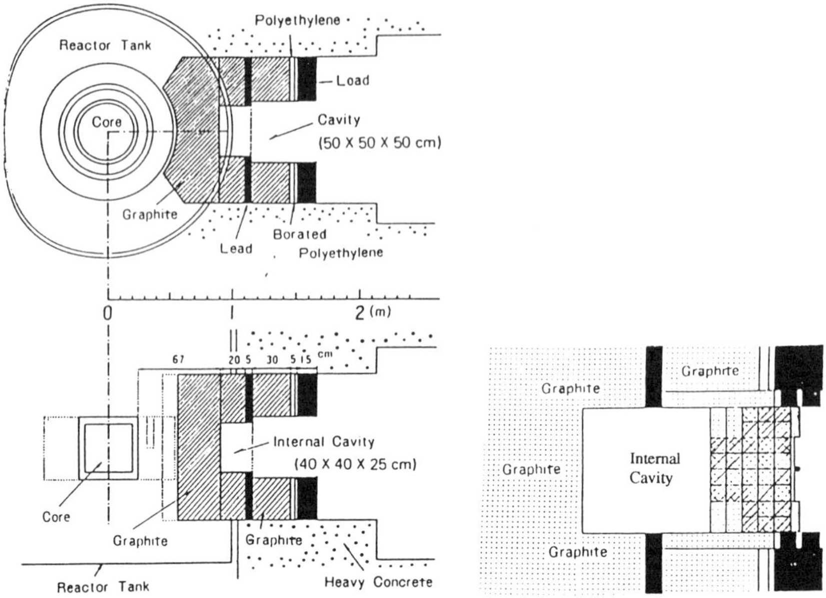

Figure 1. Top and side cutaway views of the MITRR thermal column and side cutaway view of the facility

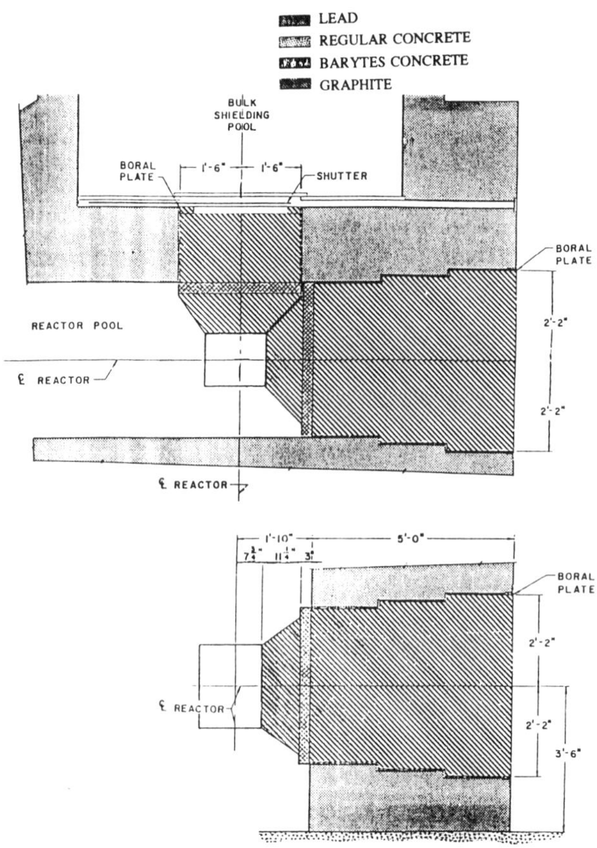

Figure 2. Top and side cutaway views of the OSURR thermal column

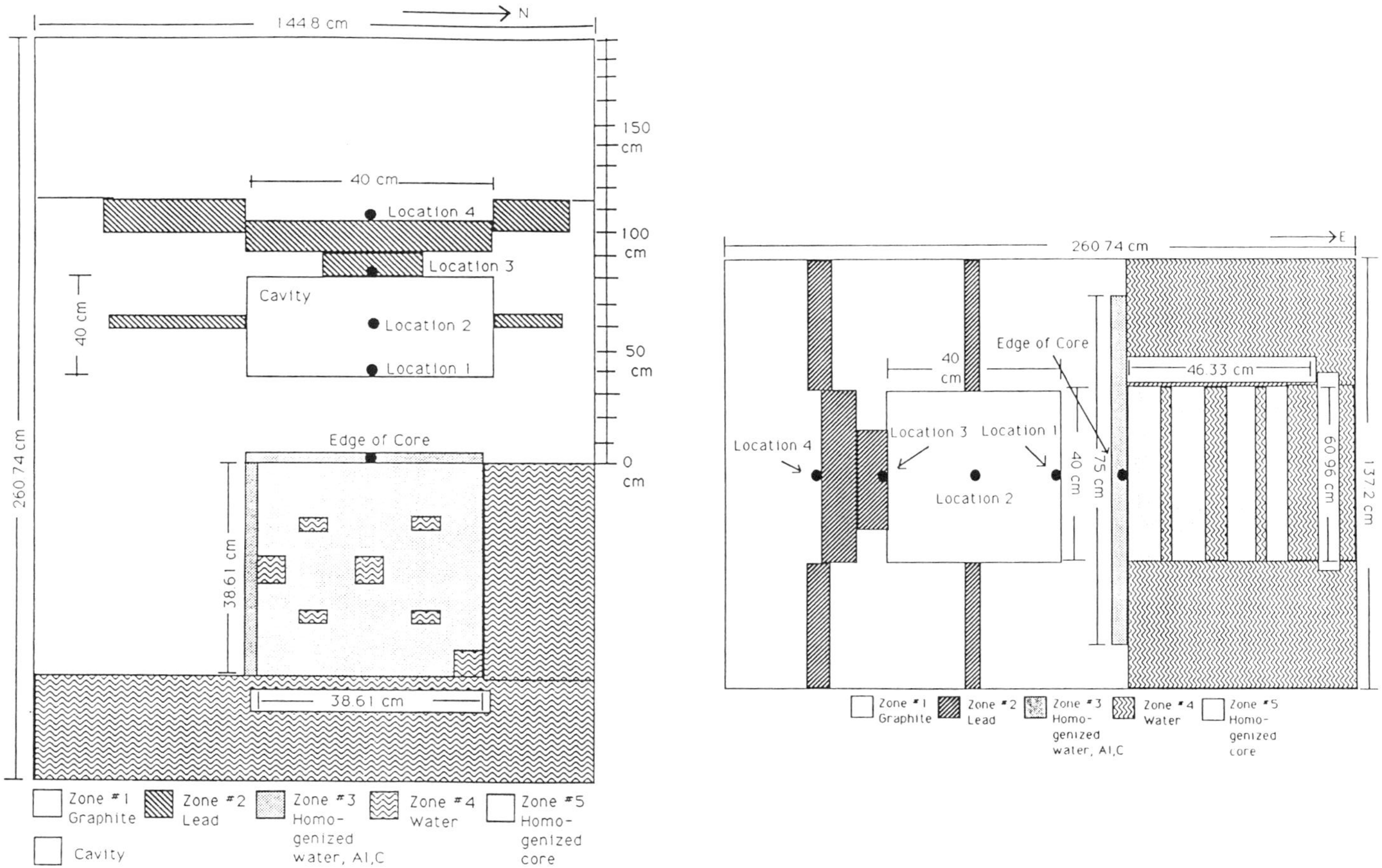

Figure 3. Facility Model For Calculations

BNCT. The MITRR is a 100 kW TRIGA Mark II. Figure 1 shows the top and side cutaway views of the MITRR thermal column where the facility is located and the side cutaway view of the facility which consists of a 40 × 40 × 25 cm internal cavity and a 50 × 50 × 50 cm outer cavity. There is 37 cm thick graphite between the reactor core and the internal cavity. The internal cavity and the outer cavity are separated by 25 cm thick bismuth gamma shield. The patient is positioned behind the bismuth shield. The MITRR facility has been extensively used since 1976.

For the assessment of the impacts of LEU conversion, the facility was assumed to be located in the main thermal column of The Ohio State University Research Reactor[2] (OSURR). The OSURR is the first university reactor to convert from HEU to 19.75 atom % enriched U_3Si_2 fuel,[2] which is expected to be the standard fuel for the future LEU core conversions. Figure 2 shows the top and side cutaway views of the OSURR thermal column and Fig. 3 schematically shows the computational model and the detector locations used in the calculations. Location 4 is the irradiation location. The material zone #2 at Locations 3 and 4 in Fig.1 represent the bismuth shield.

The MORSE code[3] with 47 neutron group BUGLE-80 library was used for the study to estimate the variation in the fast (ϕ_F), epithermal (ϕ_E) and thermal (ϕ_T) neutron fluxes following OSURR conversion from the HEU to the LEU core[2] at the 4 locations shown in Fig.3. The upper energy limits for ϕ_F, ϕ_E and ϕ_T were 17 MeV, 7.11 keV and 0.414 eV, respectively. For the MORSE runs, the neutrons were assumed to be born with prompt fission spectrum along an axial line at the center of each fuel element in the core (24 elements for the HEU core and 20 for the LEU core, including the control elements). The probability of a neutron originating in a given element was assumed to be equal to the power fraction generated by the element as determined from previous diffusion calculations,[4] with a truncated cosine distribution in the axial direction. Boundary crossing detectors with 2.5 cm radius were used for all Locations 1-4. This computational scheme was found to yield good results against the $(\phi_F+\phi_E)/\phi_T$ ratios measured in the OSURR thermal column at 40 cm and 75 cm away from the core (see Table 1). Lead was substituted for the bismuth shield in the study due to the lack of bismuth neutron cross sections in the BUGLE-80 library.

Table 1. Comparison of the measured and computed fluxes for model verification.

Distance From the Core (cm)	Computed $(\phi_F+\phi_E)/\phi_T$	Measured $(\phi_F+\phi_E)/\phi_T$
40	0.499±0.119	0.4±0.1
75	0.055±0.023	0.062±0.093

RESULTS

Table 2 summarizes the results of the study obtained by using 1.5 million neutron histories for each MORSE run. The results show that substantial changes occur in ϕ_F and ϕ_T near the irradiation location following conversion to LEU fuel (i.e. Locations 3 and 4 in Fig.3). Table 2 indicates that ϕ_F increases by 30% at Location 3 and ϕ_T decreases by 35% and 25% at Locations 3 and 4, respectively. The change in ϕ_E near the irradiation location is seen to be relatively small (about 8% decrease at Location 3). Changes in ϕ_F and ϕ_E at Location 4 are statistically inconclusive in view of the large standard deviations of the results.

Table 2. MORSE Results For Fast, Epithermal and Thermal Neutron Fluxes

	Flux Per Source Neutron (± Standard Deviation in Percent)					
	HEU Core			LEU Core		
Location	ϕ_F	ϕ_E	ϕ_T	ϕ_F	ϕ_E	ϕ_T
1	7.50E-07 (27)	4.82E-06 (26)	2.03E-05 (9)	1.42E-06 (30)	3.73E-06 (15)	1.74E-05 (11)
2	1.27E-07 (71)	4.97E-07 (40)	2.37E-06 (17)	1.69E-07 (52)	2.50E-07 (42)	1.56E-06 (20)
3	1.04E-06 (32)	2.08E-06 (21)	1.81E-05 (10)	1.37E-06 (30)	1.92E-06 (23)	1.17E-05 (12)
4	1.05E-07 (100)	9.94E-08 (73)	6.38E-07 (37)	0	2.26E-07 (64)	4.81E-07 (37)

CONCLUSION

The results of this study indicate BNCT facilities designed for using HEU fueled reactors as thermal neutron sources may suffer a penalty in required irradiation times following conversion of the core to LEU fuel and may need to be reoptimized for the new core spectrum. However, in view of the case specific nature of the study, further work is needed to investigate the generality of the results.

REFERENCES

[1] O. Aizawa, K. Kanda, T. Nozaki, T. Maksumo, "Remodeling and Dosimetry on the Neutron Irradiation Facility of the Musashi Institute of Technology Reactor for Boron Neutron Capture Therapy", *Nucl. Technol. 48: 150 (1980).*

[2] T. Aldemir, J. W. Talnagi and D. W. Miller, "Low-Enriched Uranium Conversion/Power Upgrade of The Ohio State University Research Reactor", *Nucl. Technol. 86: 248 (1989).*

[3] M. B. Emmet, "The MORSE Monte Carlo Radiation Transport Code System", ORNL-4972, Oak Ridge National Laboratory (1973).

[4] M. D. Seshadri, "Neutronic Design of of a Low Enriched Uranium U_3Si_2–Al Core for The Ohio University Research Reactor," M.S. Thesis, The Ohio State University (June 1988).

NEUTRON SPECTRUM DETERMINATION FOR THE BORON NEUTRON CAPTURE THERAPY PROGRAM IN THAILAND

T. Vilaithong,[1] W. Hoffmann,[2] G. Hoyes,[1] S. Dangtip,[1] S. Singkarat,[1] P. Suanpoot[1]

[1]Fast Neutron Research Facility, Department of Physics, Chiang Mai University, Chiang Mai 50002, Thailand

[2]Department of Physics, University of Wuppertal, 5600 Wuppertal 1, Germany

INTRODUCTION

In 1988, Chiang Mai University Hospital alone reported more than 500 cases with cervical cancer[1]. Most of these patients were in an advanced stage. The result of conventional irradiation in Thailand, using external beams (X or gamma rays) and implants of ^{226}Ra, ^{137}Cs in advanced cervical cancer, is rather poor with the 3 year survival rate in stage IIIB ranging from 25 to 34 percent [2,3]. Carcinoma of the cervix is now believed to be the number one female cancer in Thailand.

Brachytherapy using ^{252}Cf sources has been used in many centers [4,5]. In particular, Maruyama *et al.* [6] reported a 45 percent 5 year survival with 4 percent complications and 65 percent local control in stage IIIB carcinoma of the cervix.

At Chiang Mai University, a ^{10}B neutron therapy project has been initiated which aims at the establishment and evaluation of new techniques for the treatment of uterine carcinomas using ^{252}Cf. In this context, interest is focussed on the use of thermal neutron capture in ^{10}B as an adjuvant treatment modality, for enhancement of the therapeutic effect of neutrons. In order to prepare such applications, a multitude of challenging problems in basic and applied science have to be addressed. Our work reported here concerns the calculation and measurement of the energy spectra of neutrons in a water phantom irradiated with an internal ^{252}Cf source.

MEASUREMENTS AND CALCULATIONS

The neutron detector used in our measurements is a 1.4 cm diam spherical glass chamber of NE213 liquid scintillator viewed through a quartz light guide by a 10-stage R647-01 Hamamatsu photomultiplier tube (PMT). The photomultiplier base

(PMB) employs a tapered resistor chain to reduce the space charge effect. The PMT and PMB are housed in a 1.9 cm diam by 19.1 cm long mu-metal magnetic shield. The detector was developed and tested by Oyama *et al.* [7]. The neutron detector was placed inside a 1 mm thick plastic tube in a 30 x 30 x 30 cm^3 water phantom. A 20 μg Cf-252 (dated 9 September 1988) was also housed inside a thin plastic tube and suspended vertically in the water phantom. The separating distance between the source and detector could be varied to an accuracy of about 0.1 cm. Electronic circuitry for the pulse height measurement of the recoil proton employs commercial (ORTEC and CANBERRA) electronic modules and includes a pulse shape discriminating (PSD) circuit to discriminate against gamma radiations.

The pulse height from the neutron detector was calibrated periodically with a series of radioactive gamma sources and a standard pulser. The Compton half-height in the gamma ray spectrum was used as a calibrating point. The corresponding electron energies were determined from Monte Carlo calculations [7]. The calibration was performed with the detector inside the water phantom to reduce gain shift due to temperature variations. The operating voltage for the PMB was chosen to be 900 volts. At this voltage the detector system is linear above about 300 keV electron energy throughout the region of interest. The discriminating threshold was set at ^{22}Na first peak which corresponds to about 1.5 MeV proton energy. At this bias the neutron peak is completely separated from the gamma peak in the PSD spectrum. Admitting some gamma radiations into the pulse height spectrum gives rise to a peak at about 3-4 MeV in the resulting unfolded neutron spectrum.

The pulse height spectrum of the recoil protons was unfolded using the FORIST code [8]. The neutron response matrix has been calculated by a Monte Carlo code and normalized to the measured values [9].

For comparison the neutron spectra were also calculated for Cf-252 point source using the MORSE-CG computer code provided by KFA Julich in Germany [10,11]. The code takes into account elastic neutron nucleus collisions including hydrogen collisions, non-elastic neutron nucleus collisions and inelastic collisions as well as capture and fission. The transport of gammas produced is calculated including capture, fission, inelastic and elastic reactions. The multigroup cross - sections library was derived from the ENDF/B-IV data base [12]. The number of energy bins was taken to be 118 between 10^{-5}eV and 14.9 MeV. The Cf-252 source spectrum was taken as Maxwellian with an average energy of 2.13 MeV (T = 1.42) [13]. The code was installed on a VAX-station 3200 and the number of events was chosen so that the neutron flux was calculated with a statistical error of less than 6 percent. The required computing time was 40 hours to get the spectra of various water depths in a hemisphere of 0.8 cm thickness around the point source.

RESULTS

The neutron spectra were determined at 3, 4, 5, 10 and 14 cm depths in water in the center plane orthogonal to the source axis. The measured unfolded spectra at those positions are shown in Fig.1. For comparison, the fission - spectrum of Cf-252 is given as a Maxwellian [$N(E) = \sqrt{E}\exp(-E/1.42)$] with an average energy of 2.13 MeV. The spectra change considerably with increasing distance from the source. At 3.5 MeV for example, thay are 18% more neutrons at 3 cm than at 14 cm, whereas at 7 MeV there are 57% fewer neutrons at 3 cm than at 14 cm. This effect can be described as neutron hardening with increasing distance from the source.

The spectra at various distances were compared with the spectra derived from the MORSE Monte Carlo calculation. Fig.2 gives as an example the spectrum at 14 cm distance from the source. The agreement is good within the statistical error chosen for the calculations. For other distances the agreement is of the same quality.

The neutrons are slowed down mainly by elastic scattering off the hydrogen atoms, therefore the energy will be transferred to neutrons of lower energy and show up in the low energy part of the neutron spectrum. Since the calculations can cover also the energy range below the measurement threshold, Fig.4 gives the calculated spectra The reduction of flux above 2 MeV due to interactions with water was measured from 3 to 19 cm distance from the source. Fig.3 shows the neutron flux multiplied by the square of the distance to compensate for the pure geometrical dependence. The measurements are again in good agreement with the MORSE calculation results.

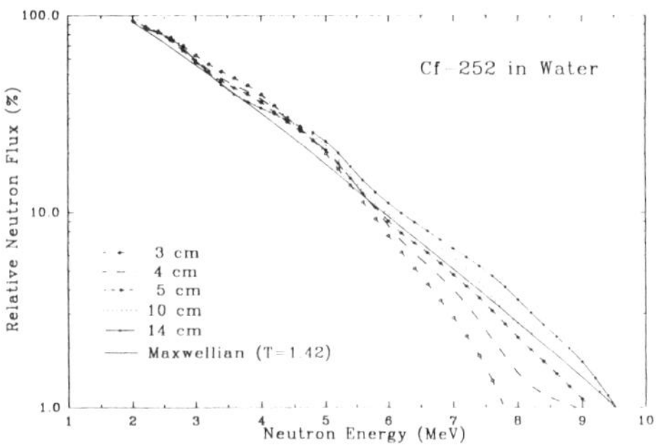

Fig.1 Differential neutron flux distribution in water above 2 MeV as a function of distance from the center of a Cf-252 source (normalized to flux at 2 MeV). Maxwellian distribution with T = 1.42 MeV is given for comparison.

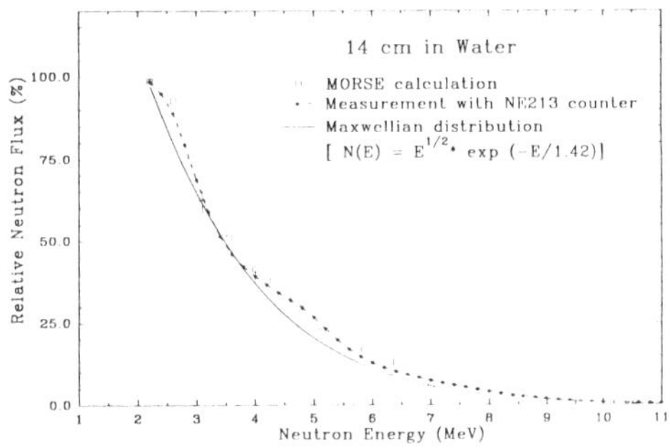

Fig.2 Measured and calculated relative flux distribution at 14 cm distance from the Cf-252 source in water.

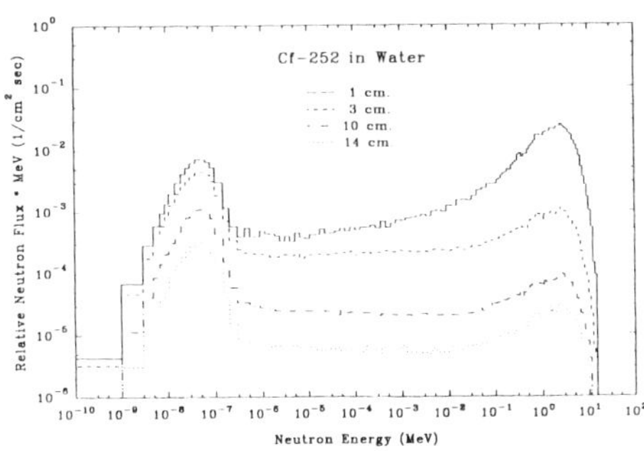

Fig.3 Neutron flux above 2 MeV as a function of water depth. Flux is multiplied by $(\text{distance})^2$ to demonstrate interaction in water.

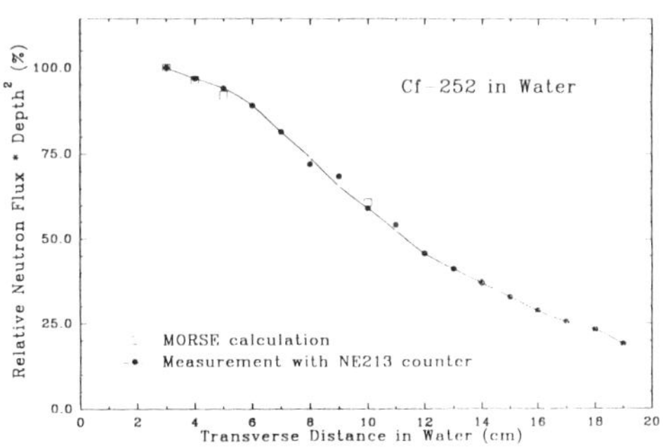

Fig.4 Calculated Cf-252 neutron energy flux (neutron flux * energy) as a function of energy for various distance from the center of the source perpendicular to the source axix.

of the energy flux [N(E)xE] down to thermal energies. As one would expect, the ratio of thermal energy flux to the high energy flux is increasing with increasing water depth.

CONCLUSION

It was found by measurement with a proton recoil spectrometer and by Monte Carlo calculations that if a Cf-252 source is placed in water or tissue, the shape of the neutron spectrum above 2 MeV changes considerably as a function of distance from the source. This is of particular importance to Cf-252 dosimetry, because normally dosemeters sensitive to high enery neutrons show a strong energy dependant response. Since more than one third of the total dose in tissue is contributed from neutrons of an energy above 2 MeV, the use of non-tissue-equivalent dosemeters like TLD's requires consideration of the spectral changes. For treatment planning, the effect of decreasing RBE with increasing distance has to be taken into account.

ACKNOWLEDGMENTS

The authors wish to thank Dr.Y. Oyama of JAERI for providing the neutron detector and helping with the use of the FORIST code. This work has been supported by the Institute for Science and Technology Research and Development, Chiang Mai University, the International Program in Physical Science, Uppsala University, and the German Academic Exchange Service DAAD.

REFERENCES

[1] Statistical Report, Cancer Unit, Faculty of Medicine, Chiang Mai University (1988).

[2] P. Kraipibul and S. Srisupundit,"Carcinoma of the Uterine Cervix: Preliminary Result of Treatment in Stage IB-IIIB," p.52 in Proceedings of the 8th International Working Party and Conference for the Diagnosis and Treatment of the Carcinoma of the Cervix in Developing Areas, Bangkok (1985).

[3] P. Tepmongkol,"Radiation Therapy for Carcinoma of Cervix Uteri: A Clinical Study in Siriraj Hospital," p.56, *ibid.*

[4] "Californium-252 Brachytherapy and Fast Neutron Beam Therapy," Proceedings of the Workshop held in Lexington, Kentucky, Eds. Y. Maruyama, J. Lawrence Beach and J. M. Feola, Harwood Academic Publishers, New York (1986).

[5] "International Neutron Therapy Workshop," Lexington, Kentucky, Ed. Y. Maruyama, Harwood Academic Publishers, New York (1991).

[6] Y. Maruyama *et al.*, Int. J. Rad. Oncol. Biol. Phys.11, 1475 (1985).

[7] Y. Oyama *et al.*, Nucl. Instr. and Meth. A256, 333 (1987).

[8] FORIST Spectra Unfolding Code, Radiation Shielding Information Center, Oak Ridge National Laboratory, PSR-92 (1975).

[9] Y. Oyama, private communication (1992).

[10] M. B. Emmett, "The MORSE Monte Carlo Radiation Transport Code System," ORNL-4972 (1975).

[11] P.Cloth *et al.*, "HERMES, A Monte Carlo Program System for beam Materials Interaction Studies," KFA-Report Jul-2203, May (1988).

[12] R.Kinsey, "ENDF/B-IV Cross Section Measurement Standards," Report BNL-NCS-50496 (1979).

[13] J.W. Boldemann, "Reviews of Measurements of the Prompt Fission Spectrum from the Spontaneous Fission of Cf-252," in Properties of Neutron Sources, IAEA-TECDOC-410 (1987).

A STANDARDIZED METHOD FOR BEAM DESIGN IN NEUTRON CAPTURE THERAPY

Gregory J. Storr and Baiba V. Harrington

Nuclear Technology Program, Australian Nuclear Science & Technology Organisation, PMB 1, Menai 2234, NSW, Australia

INTRODUCTION

A desirable end point for a given beam design for Neutron Capture Therapy (NCT) should be a quantitative description of tumour control probability and normal tissue damage[1]. Achieving this goal will ultimately rely on data from NCT human clinical trials. Traditional descriptions of beam designs have used a variety of assessment methods to quantify proposed or installed beam designs. These methods include measurement and calculation of "free field" parameters, such as neutron and gamma flux intensities and energy spectra, and figures-of-merit in tissue equivalent phantoms. We propose here a standardized method for beam design in NCT. This method would allow all proposed and existing NCT beam facilities to be compared equally.

The traditional approach to determining a quantitative description of tumour control probability and normal tissue damage in NCT research may be described by the following path:

Beam design → dosimetry → macroscopic effects → microscopic effects

Methods exist that allow neutron and gamma fluxes and energy dependence to be calculated and measured to good accuracy[2]. By using this information and intermediate dosimetric quantities such as kerma factors for neutrons and gammas, macroscopic effects (absorbed dose) in geometries of tissue or tissue-equivalent materials can be calculated. After this stage, for NCT the data begins to become more sparse and in some areas ambiguous. Uncertainties in the Relative Biological Effectiveness (RBE) of some NCT dose components[3,4] means that beam designs based on assumptions considered valid a few years ago may have to be reassessed. A standard method is therefore useful for comparing different NCT facilities.

THE FIRST CONDITION - FREE FIELDS

In experiments, good results may be obtained using reactor and radiation dosimetry techniques to measure radiation intensities and spectra. High confidence can be placed on results from accurate reactor physics calculations, which have been shown to compare extremely well with measurement[2,5,6].

The first condition in the standardized method for NCT beam design is to compare the calculated and measured intensities of neutron and gamma fluxes, their spectral dependence and isotropy in air. Free field parameters are a measure of the quality of radiation[8], and are an important gross measure of the quality of a beam design for NCT. One free field method used in NCT beam design is the "average energy" concept[9], where neutron kerma rates in water are used to obtain a single value energy parameter. A similar useful free field parameter is $\langle k_n \rangle$,[15] the kerma in tissue per incident epithermal neutron (E > 1eV). For epithermal beam designs the gamma absorbed dose per epithermal neutron, and fast neutron absorbed dose per epithermal neutron are useful for measuring beam contamination.

Using the standard method for beam design, the following parameters would be assessed in a free field:

- Neutron and gamma spectra and intensities
- Absorbed dose-rates in air, or with water kerma factors
- Gamma and fast neutron dose per epithermal neutron

THE SECOND CONDITION - PHANTOMS

The second condition for the standard method is the measurement and calculation of thermal neutron and gamma fields in tissue-equivalent phantoms. As in a well designed NCT beam the thermal flux is the major component of the absorbed dose in a phantom through ^{10}B and nitrogen capture, this is a good method to assess beam design quality.

Another method used to measure a beam design with phantoms are the "figures-of-merit"; specifically advantage depth (AD), advantage ratio (AR) and advantage depth dose rate (ADDR). The definitions of AD and AR are given elsewhere[7]. Results from our two-dimensional DORT[10] phantom calculations[11,12] lead us to recommend that the use of components of absorbed dose in the form of iso-dose contours, and absorbed dose values at specified locations on standard phantoms, are both necessary and sufficient to evaluate beam designs for NCT. To support this consider the following arguments.

(1) A complete evaluation of a design requires an understanding of absorbed dose behavior, and how the assumed RBE factors for each radiation component, and the ^{10}B concentrations in tumour and tissue affect RBE-dose distributions and hence the figure-of-merit values. Figure 1 shows the RBE-dose-rates for tissue (^{10}B 3μg/g) and tumour (^{10}B 30μg/g) along the centre-line of a cylindrical phantom for a 15-111 keV beam, using RBE values of 2.3 for ^{10}B reactions, 1.0 for gamma reactions and 1.6 for other neutron reactions. Also shown are RBE-dose-rates when the hydrogen recoil component is changed to 4.0[4]. The hydrogen recoil dose at the surface becomes important in this case. To properly compare beam designs, it would be best to compare absorbed dose components prior to comparing beams with RBE-doses or parameters based on RBE-dose.

(2) AD and AR take no account of the radial dose distribution in a phantom as they are derived from centre-line tissue and tumour RBE doses. As an alternative iso-dose contours give a quantitative description of dose distributions radially and axially.

(3) If the phantom assessment volume is too large the value of the figures-of-merit (and absorbed and RBE-dose) will be incorrect. Values of dose are approximated well by using small assessment volumes, or poorly by using large assessment volumes. Figure 2 is a plot of the apparent AD against assessment volume for a wide and narrow ideal 2-7 keV beam (*i.e.* no photon contamination and neutrons in the specified energy range) impinging on a cylindrical phantom. It has been shown[11] that as the assessment volume is increased AD (and also AR) are overestimated for narrow beams and underestimated for wide beams. In phantoms where the assessment volume is too large designs comparing only figures of merit will not indicate an error.

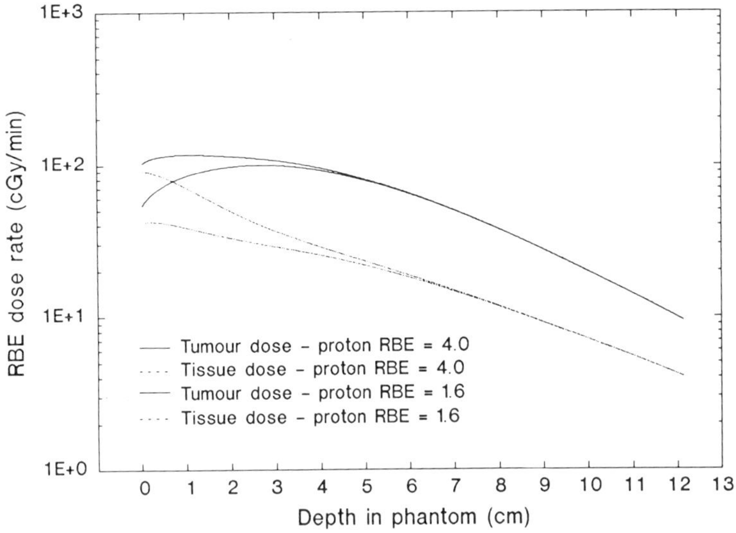

Figure 1. Centre-line RBE-dose rates for a 15-111 keV ideal beam

Comparison of Therapeutic Gain (TG) along the phantom centre-line is sufficient to establish the position at which tumour dose equals the maximum tissue dose. Still more information is made available by considering TG values at other radial and axial locations in the phantom. However, as TG is derived from RBE-dose, and the value of RBE factors remains a controversial issue, then beam design comparisons using TG as a measurement parameter should only be done with consistent RBE factors. A more useful approach is to use absorbed dose TGs with assumed ^{10}B ratios in tumour and healthy tissue when comparing beam designs for the same and different facilities.

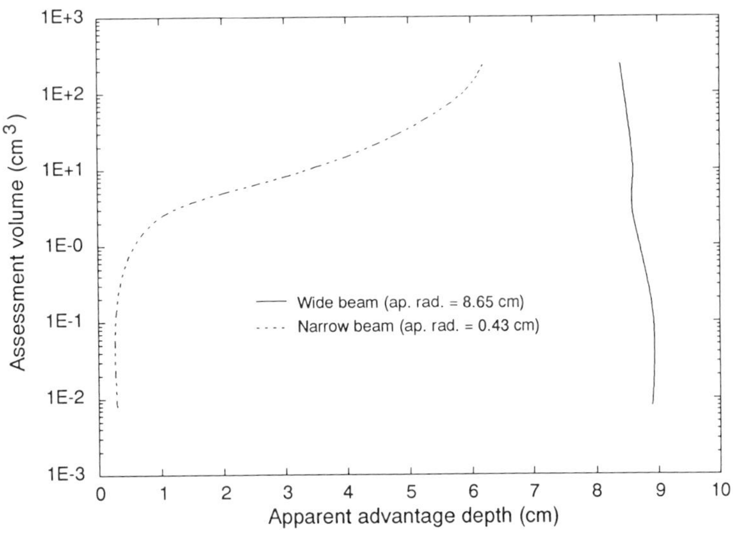

Figure 2. Apparent AD for different phantom assessment volumes

The advantage volume (AV) has been used[11] as a measure of the volume in a phantom where the tumour dose exceeds the maximum healthy tissue dose. AV values for varying beam parameters give a better indication of the variation of radial dose in the phantom, than do the traditional figures-of-merit. However, AV can also be misrepresented if the assessment volume for determining dose within the phantom is too large.

We support the proposal[13] to use standard phantoms, and in the standard method recommend that a cylindrical phantom of an easily modeled and manufactured tissue-equivalent material be used, with dimensions of 16.5 cm diameter and 25.0 cm axial. The rationale for using this approach is :

- the geometry is consistent with many phantom studies
- practical (experimentally reproducible and easily made)
- Monte Carlo and Discrete Ordinates calculations possible
- calculation and measurement comparison possible

The standard phantom could be named the "Fairchild standard phantom"[14].

CONCLUSIONS

The standard method for beam design in Neutron Capture Therapy would use or assess the following :

- In air neutron and gamma flux spectra and intensities
- Absorbed dose-rates
- Contaminant doses per epithermal neutron
- The use of a standard cylindrical phantom
- Iso-dose contours with assumed ^{10}B concentrations and ratios
- Therapeutic gains at standard locations
- Thermal flux calculation and measurement comparison
- If required, figures-of-merit that account for radial distribution

We believe this standardized approach provides a consistent and rigorous method for NCT beam design and beam comparison.

ACKNOWLEDGEMENTS

The authors gratefully acknowledge the assistance of Idaho National Engineering Laboratory staff and resources in preparing this paper.

REFERENCES

1. E.L. Travis, "Primer of Medical Radiobiology",Second Edition,Year Book Medical Publishers,Inc.,Chicago,(1989).
2. F.J. Wheeler, D.K. Parsons, B.L. Rushton, and D.W. Nigg, Epithermal neutron beam design for neutron capture therapy at the Power Burst Facility and the Brookhaven Medical Research Reactor,*Nucl.Tech.*92:106(1990).
3. J.A Coderre, D.D. Joel, M.S. Makar, P.L. Micca, M.M. Nawrocky, and D.N. Slatkin, Paradoxical outcomes of in vitro predictive tests for boron neutron capture therapy efficacy in vivo. *Presented at the International Workshop on Macro and Microdosimetry and Treatment Planning for Neutron Capture Therapy,Cambridge,MA,(1991).*
4. P.R. Gavin, F.J. Wheeler, R. Huiskamp, A. Siefert, S.L. Kraft, and C.E. DeHaan, Large animal model studies of normal tissue tolerance using an epithermal beam and borocaptate sodium. *Presented at the International workshop and plenary meeting: Towards clinical trials of glioma with BNCT,* Petten, The Netherlands,(1991).
5. M. Carolan, G. Constantine, R. Godfrey, H. Meriaty, G. Storr, and B. Allen, A filtered neutron beam at HIFAR, *These Proceedings.*
6. B.V. Harrington and G. Constantine, Analysis of filtered neutron beams at HIFAR, *These Proceedings.*

7. R.M. Brugger, G. Constantine, O.K. Harling, and F.J. Wheeler, Rapporteurs' Report, *in:* "Neutron Beam Design, Development and Performance for Neutron Capture Therapy." O.K. Harling, J.A. Bernard, and R.G. Zamenhof, eds.,Plenum,New York,(1990).
8. ICRU Report 26: "Neutron Dosimetry for Biology and Medicine," ICRU Publications, Washington,(1977).
9. P. Watkins, G. Constantine, F. Stecher-Rasmussen, W Freudenreich, R.L. Moss, and R. Ricchena, MCNP calculations for the design and characterisation of the Petten BNCT epithermal neutron beam, *in:* "Progress in Neutron Capture Therapy for Cancer," B.J. Allen, D.E. Moore, and B.V. Harrington,eds.,Plenum,New York,(1992).
10. W.A. Rhoades and R.L. Childs, DORT - Two-Dimensional Discrete Ordinates Transport Code, Oak Ridge National Laboratory, CCC-484,(1989).
11. G.J. Storr, "Assessment of ideal neutron beams for Neutron Capture Therapy, *Radiat.Res.,*131:235,(1992).
12. B.V. Harrington and G.J. Storr, Calculated NCT dose components in surface tissue. *Presented at the International Workshop on Macro and Microdosimetry and Treatment Planning for Neutron Capture Therapy,Cambridge,MA,(1991).*
13. V. Benary, R.G. Fairchild, J. Kalef-Ezra, D. Greenberg, Y. Kamen, S. Fairman, and L. Wielopolski, Dose measurements and calculations in epithermal neutron beam at the Brookhaven Medical Research Reactor (BMRR), *in:* "Progress in Neutron Capture Therapy for Cancer," B.J. Allen, D.E. Moore, and B.V. Harrington,eds.,Plenum,New York,(1992).
14. R.G. Fairchild, Development and dosimetry of an 'epithermal' neutron beam for possible use in Neutron Capture Therapy. I. 'Epithermal' neutron beam development, *Phys.Med.Biol.*,10:491,(1965).
15. T.E. Blue, T-X.B. Qu, R.N. Christensen, P. Guo, and J.W. Blue, An integrated neutronic and thermal-hydraulic design for an accelerator neutron irradiation facility, *in:* "Progress in Neutron Capture Therapy for Cancer," B.J. Allen, D.E. Moore, and B.V. Harrington,eds.,Plenum,New York,(1992).

APPLICATION OF FRICKE DOSIMETRY FOR BNCT

Bartlomiej Ciesielski, Lucian Wielopolski

Brookhaven National Laboratory
Radiation Therapy Facility
Upton, NY 11973

INTRODUCTION

Ferrous sulfate dosimeters, introduced for dosimetry by Hugo Fricke in 1927, are recognized for their accuracy, reproducibility, and stability among other chemical systems for dose measurements. Their response to different qualities of ionizing radiations has been extensively studied and their physical characteristics together with the analytical methods are well documented in number of publications [1-4]. The sensitivities of these dosimeters to different types of ionizing radiation have been well described in terms of the radiation chemical yields, G, thus making them absolute dosimeters.

A Fricke standard solution is composed of: 1 mmol/l ferrous sulfate or ferrous ammonium sulfate, 0.4 mol/l sulfuric acid, and 1 mmol/l sodium chloride. Fricke dosimetry is based on the oxidation of ferrous ions (Fe^{2+}) to ferric ions (Fe^{3+}) by the by products of water radiolysis: $OH^{\cdot}$, H_2O_2, $HO_2^{\cdot}$, e_{aq}. The resulting concentration of the ferric ions is measured employing spectrophotometry at 224 or 304 nm. Although measurements at 304 nm are less sensitive than at the 224 nm, they are preferred due to their lower susceptibility to impurities. The measured difference in the absorbance ΔA, between irradiated and unirradiated solutions, is converted to an absorbed dose as follows:

$$D = \frac{\Delta A}{\epsilon \, l \, \rho \, G} \qquad (1)$$

where: ϵ is the molar linear absorption coefficient, l is the length of the light path in the photometric cell, and ρ is the density of the Fricke solution. Insensitivity of Fricke solutions to thermal neutrons is overcome by adding boric acid. Although studies with sensitized Fricke solutions have been reported [5,6], no comprehensive study regarding

Advances in Neutron Capture Therapy, Edited by
A.H Soloway *et al.*, Plenum Press, New York, 1993

reproducibility, radiation chemical yield, G, for the ionizing products of $^{10}B(n,\alpha)^7Li$ reaction have been published.

The most relevant advantages of Fricke dosimetry are: 1) relatively simple equipment needed for chemical analysis, 2) independent of dose rate up to 10^6 Gy/s, 3) the dose conversion from the dosimeter solution to water, or tissue, is close to unity for most radiations and much less energy dependent than those for gaseous or solid detectors, 4) due to its similarity to water and tissue, very small perturbations in the radiation transport are introduced.

The aim of this study is to test the applicability of the boronated Fricke dosimetry for the measurements of thermal neutrons contribution to the total dose in a mixed radiation field at the thermal port of the BMRR. This might play a major role in elucidating explicitly the boron contribution to the dose in BNCT.

EXPERIMENTAL METHODS

The total absorbed dose, in a mixed field, is delivered by the three following components: gamma rays, fast neutrons and thermal neutrons. In standard Fricke solutions the measured absorbance ΔA_F is mainly due to gamma and fast neutron components and can be expressed as:

$$\Delta A_F = \epsilon \; l \; \rho \; (D_\gamma G_\gamma + D_F G_F) \tag{2}$$

where: D_γ denotes the gamma dose due to the external field and produced in the sample by (n,γ), mainly in the hydrogen. D_F is the dose due to fast neutrons, G_γ and G_F are the radiation chemical yields for gammas and fast neutrons, respectively. In boronated Fricke solution, the measured absorbance ΔA_{BF} can be written as:

$$\Delta A_{BF} = \epsilon \; l \; \rho \; (D_\gamma G_\gamma + D_F G_F + D_T G_T) \tag{3}$$

where: D_T and G_T are dose and chemical yield for the thermal neutron component, respectively. These equations neglect the contributions from thermal neutrons captured by N, Cl and Na, which were estimated to be less than 0.1% of the total dose. The difference between these two absorbancies yields:

$$\Delta A_{BF} - \Delta A_F = \epsilon \; l \; \rho \; D_T \; G_T \tag{4}$$

which enables to determine the value of D_T. The dose delivered through boron capture is related to the thermal neutrons fluence Φ (n/cm^2) by:

$$D_T = N_A \; E \; \sigma \; c \; \Phi \; / \; \rho \quad \text{(Gy)} \tag{5}$$

where: N_A is Avogadro number, $E = 3.65\times10^{-13}$ J is the energy released in the $^{10}B(n,\alpha)^7Li$ reaction [7], $\sigma = 3837\times10^{-24}$ cm^2 neutron capture cross section of ^{10}B, c is the concentration of ^{10}B in mol/m^3.

In the present calculations following values were used: $\epsilon = 220.0$ m^2/mol, $\rho = 1024$ kg/m^3, $G_T = 4.4\ 10^{-7}$ mol/kgGy [5], $G_\gamma = 1.62\ 10^{-6}$ mol/kgGy [5], and $G_F = 7.0\times10^{-7}$ mol/kgGy [3].

RESULTS

Spectrometric measurements at 304 nm of unirradiated standard and boronated Fricke solutions were not affected by the presence of boric acid, enriched 94.6% in ^{10}B, of concentrations up to 0.5 mol/l. Subsequently, these solutions were exposed to 6 and 25 MVp X-ray beams; the results agreed within +/- 3% with ion chamber dosimetry.

Fricke sensitivity to neutrons as a function of boron concentration is shown in Fig. 1. These results were obtained by irradiating 1.5 cc polyethylene vials with 1 cc solutions in the thermal port of BMRR at 1 MW for 12 min. The measured doses and thermal neutrons fluences together with corresponding data obtained with ion chambers, TLDs, and gold foils [8], under identical irradiation conditions, are summarized in Table I. The measurable range of the neutron fluence and its dependence on boron concentration is shown in Fig. 2.

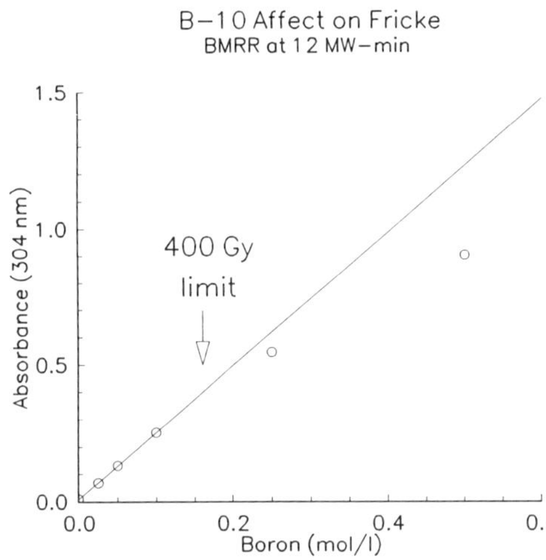

Fig. 1. Variation in the absorbance of the boronated Fricke solutions versus boron concentration at a constant theramal neutron fluence.

DISCUSSION

It was demonstrated that neither absorbance nor the chemical yields for gamma radiation were affected by the presence of boric acid in standard Fricke solutions. In neutron fields the enhancement of the ferric ions production was linear with boron concentration up to about 400 Gy of the total absorbed dose. The two points in Fig. 1 below the straight line correspond to high boron concentrations (0.25 and 0.5 mol/l) which resulted in doses exceeding the 400 Gy linearity range. Comparison of the results with those quoted in the literature, Tab. I, demonstrates the applicability of the method for direct boron dosimetry. No attempt at present was made to separate between doses due to fast neutrons and gammas, thus only a range of the combined doses were reported in Tab. I. The fluence range covered by the Fricke/Fricke-boron dosimetry can be altered by choosing an appropriate concentration of ^{10}B, over a range from 10^{10} up to 10^{15} n/cm^2, thus it might be used as personnal dosimeter for the entire treatment.

Table I. Comparison of beam characterization at BMRR, thermal port, obtained with Fricke/Fricke-boron solutions and standard dosimetry (in cGy per 1 MW-min).

Radiation component	Fricke dosimetry	Standard dosimetry[a]
Gamma rays	(15 - 54)[b] (Gamma rays and fast neutrons combined)	6
Fast neutrons		13
Thermal flux[c]	2.67×10^{11}	2.8×10^{11}
Dose from $^{10}B(n,\alpha)^7Li$ (per 1 ppm of Boron-10)	2.25	2.4

[a] from Ref.8.
[b] The lower limit was calculated with G value for gammas, the higher limit assuming G value for fast neutrons of energy 0.65 MeV (the average energy in the thermal port).
[c] in (n/cm^2)

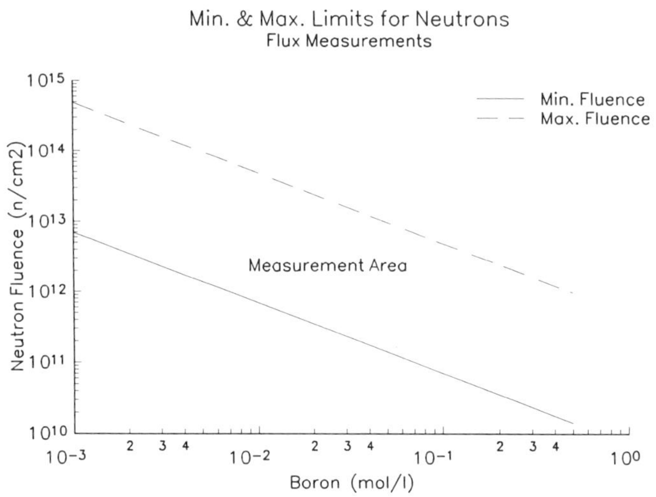

Fig. 2. The range of neutron fluences that can be measured versus ^{10}B concentration. The lower limit is determined by precision of the spectrophotometer, where the upper one is dictated by the 400 Gy, the linear range of the dosimeter.

In summary, it has been demonstrated that the difference in the absorbances between standard and boronated Fricke solutions reflect directly the dose due to neutrons captured in boron, regardless of their energy spectrum, in a physical enviroment similar to that of tissue.

ACKNOWLEDGEMENTS

The work was supported by Fogarty International Fellowship # I FO5 TW04614 - 01 ICP(5)

REFERENCES

1. A. O. Fregene, Calibraton of the ferrous sulfate dosimeter by ionometric and calorimetric methods, Nature 19, 818-819 (1966).
2. R. C. Lawson, D. Porter, The response of ferrous sulfate dosemeter to neutrons, Phys. Med. Biol. 20(3), 420-430 (1975).
3. S. Egusa, K. Ishigure, S. Tagawa, Y. Tabata, K. Oshima, Absorbed dose measurements in a mixed field of fast neutrons and γ rays by the Fricke and ceric sulfate dosimeters, Radiat. Phys. Chem. 11, 129-134 (1978).
4. R. Katz, G.L. Sinclair, M.P.R. Waligorski, The Fricke dosimeter as a 1-hit detector, Nucl. Traks Radiat. Meas. 11(6), 301-307 (1986).
5. R. Broszkiewicz, "Chemiczne Metody Dozymetrii Promieniowania Jonizujacego", Wydawnictwa Naukowo-Techniczne, Warszawa, (1971) (in Polish).
6. Y. U. Teliks, K. D.Tone, V.Y. Kal'kis, Chemical dosimetry at the modernized nuclear reactor IRT-200, in Radiation Physics II. Neutron and Gamma Dosimetry, AEC-tr-6604, 35-43 (1965).
7. D. Gabel, R. Fairchild, B. Larsson, H.G. Borner, The relative effectiveness in V79 chinese hamster cells of the neutron capture reactions in boron and nitrogen, Radiat. Res. 98, 307-316 (1984).
8. R. Fairchild, S.B. Kahl, B.H. Laster, J. Kalef-Ezra, E.A. Popenoe, In vitro determination of uptake, retention, distribution, biological efficacy, and toxicity of boronated compounds for neutron capture therapy: a comparison of porphyrins with sulfhydryl boron hydrides, Cancer Res. 50. 4860-4865 (1990).

NUCLEAR CHARACTERISATION OF THE HFR PETTEN BNCT FACILITY

P. Watkins[1], Y. Harker[2], C. Amaro[2], W. Voorbraak[3], F. Stecher-Rasmussen[3], H. Verhagen[3], C. Perks[4], H. Delafield[4], G. Constantine[1] and R.L. Moss[1]

[1]Institute for Advanced Materials
JRC Petten, The Netherlands
[2]Idaho National Engineering Laboratory
Idaho Falls, USA
[3]The Netherlands Energy Research Foundation
ECN Petten, The Netherlands
[4]AEA Technology, Harwell, UK

INTRODUCTION

The major work at Petten over the last few years has been the design and construction of an epithermal neutron beam facility in the HB11 beam tube. Following installation of the filter assembly in 1990, and the first operation of the beam at full reactor power in June 1991, it was finally possible to start the planned series of irradiation experiments. These include phantom irradiations, cell culture experiments and a healthy tissue tolerance study using beagle dogs. Prior to this work a considerable amount of neutron and photon dosimetry was performed to measure various parameters of the beam. Apart from providing definite values for such data, these measurements were essential to validate the calculational models which were used in the design and later for the treatment planning.

The campaign of measurements at HB11 has continued in parallel with the other experiments noted above. In this paper the various measurements that have been performed are reviewed both for the free beam and also within phantoms. These were undertaken as part of an intensive collaborative effort between INEL, AEA Harwell, ECN and JRC Petten and included: stacked resonance-energy activation foils for measuring the neutron spectrum in the thermal energy range; threshold foils and proton recoil proportional counters for the fast neutron energies; and thermoluminescence detectors (TLD) for the gamma field.

Advances in Neutron Capture Therapy, Edited by
A.H Soloway *et al.*, Plenum Press, New York, 1993

In addition to measurements the foil configurations and phantoms have been modelled using the Monte Carlo code MCNP. Wherever possible the results from these calculations are compared with the measured values and the reasons for any discrepancies are discussed.

STANDARD DOSIMETRY MEASUREMENTS

After some initial, tentative experiments it became clear that it was important to create a set of well defined dosimetry measurements which would be repeated at regular intervals. Such an approach minimises the possible sources of variation and has the additional benefit of providing data which shows the time dependent behaviour of the beam. This latter point was considered to be of very great importance, first because one of the requirements for future clinical trials would be evidence to demonstrate the stability of the beam in time and secondly, because the HFR is a "working" reactor with many experiments within the core which vary between reactor cycles. The variation of the beam parameters with reactor cycle is therefore of considerable importance.

The possible dosimetric measurements that could be included in such a database is immense but a decision was taken to define a minimum set of simple measurements which could be extended as required. This basic set of measurements has already been supplemented by a number of separate dosimetry measurements which were needed to determine specific features of the beam.

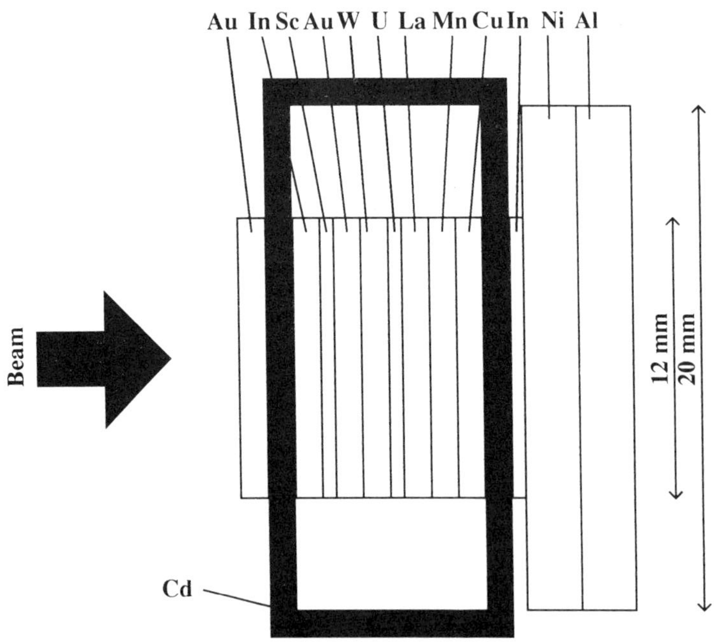

Figure 1. Free beam foil set.

The standard neutron measurement is that of a stack of resonance and threshold foils, Fig 1, which is positioned at the beam centre immediately on exit from the final aperture. The foils used are responsive to the thermal, low epithermal and fast neutron energies. The important epithermal region from 1 keV to about 0.5 MeV is relatively untouched by this measurement. This is recognised as a major deficiency and is being remedied by the development of a technique which uses resonance foils within a moderating medium. However the existing foil set has the advantage of being straightforward to irradiate and is capable of detecting any gross changes in the beam parameters.

The photon component of the beam is measured using 5 TLDs placed at the same position as the foil set and arranged in the form of a cross. The TLDs are of the Harshaw type 700 using enriched ^{7}Li and are encapsulated in perspex boxes for the irradiations.

CALCULATED SPECTRUM AND FOIL RESPONSES

To compare the neutron measurements with calculations the foil sets were modelled using the Monte Carlo code MCNP. The MCNP model is shown in Fig 2. The comparisons with measurement are shown in Table 1 under the column C/M where it can be seen that in most cases the agreement with measurement is quite good except for the fast neutron energy range which is discussed below. The stability of the beam as a function of time is also shown in the table and is seen to be very good.

Table 1. Free beam foil set results over time, normalised to 1.0 for 2/4/92 and ratio of calculated to measured values(C/M).

Foil	Composition	Energy Response	Date of Irradiation 3/7/91	2/4/92	21/5/92	10/7/92	C/M
^{197}Au	1% Au in Al	Thermal	1.04	1.00	0.95	0.95	0.96
^{115}In	0.2% In in Al	Epi	0.96	1.00	0.94	0.94	0.69
^{45}Sc	99.99% Sc	Epi	0.99	1.00	0.95	0.97	0.78
^{197}Au	1% Au in Al	Epi	1.01	1.00	0.95	0.96	0.94
^{186}W	1% W in Al	Epi	0.96	1.00	0.94	0.95	1.36
^{238}U	23% U in Al	Epi	0.98	1.00	0.94	0.95	1.08
^{139}La	5% La in Al	Epi	0.98	1.00	0.95	0.96	1.07
^{55}Mn	1% Mn in Al	Epi	1.00	1.00	0.92	0.96	1.43
^{63}Cu	10% Cu in Al	Epi	1.01	1.00	0.94	0.95	1.16
^{115}In	100% In	Fast	1.22	1.00	0.87	0.92	2.43
^{58}Ni	100% Ni	Fast	2.63	1.00	0.93	0.94	3.24
^{27}Al	100% Al	Fast	----	1.00	0.88	1.03	5.26

Initial calculations using an MCNP model of the complete HFR core and HB11 facility had been used to design the filter components and to generate neutron and photon sources just behind the final aperture. These sources have subsequently been used in the treatment planning models and for the foil and phantom simulations reported here.

When the beam was first opened in 1991 a series of measurements were undertaken which suggested that the very fast component of the beam (> 0.5 MeV) was greater than predicted by the MCNP calculations. In the light of this result the calculated spectrum was adjusted to give a "standard" spectrum that has been used in the work reported here.

In retrospect these early measurements have been shown to be misleading. One of the major components of the filter, the argon cryostat, was very unstable during this initial operational period. Considerable changes have since been made and the argon system now behaves in a far more consistent manner. However as a consequence of these improvements all the measurements now show a very much reduced fast neutron component, indicating a spectrum much closer to that of the original calculations. For this reason the data in Table 1 is normalised to the measured values of April 1992 and the earlier values can be considered as atypical for the existing configuration.

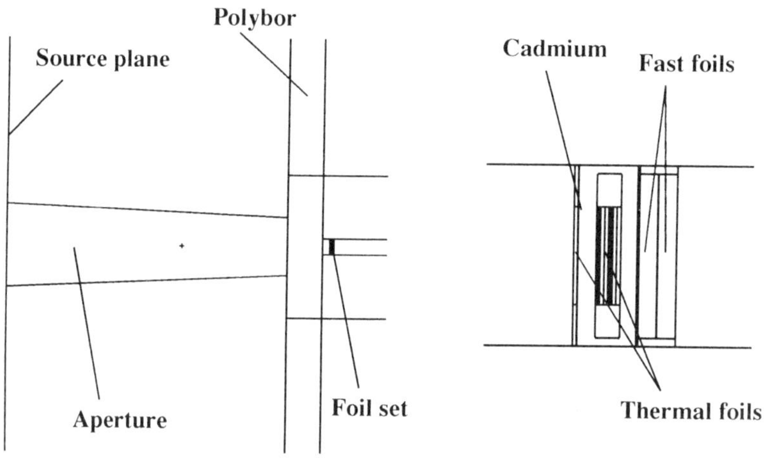

Figure 2. MCNP model of the free beam foil set.

The "standard" spectrum used in the calculations was not changed as a result of these revised measurements and consequently there remains a significant discrepancy with the calculated values for those foils responding to fast neutrons. However the effect of this reduction in flux above 0.5 MeV on the fast neutron dose is minimal since flux levels in this energy range are very low.

The results from the TLD irradiations show the same sort of stability over time although there is a dependency upon the condition of the argon cryostat. For fresh, unirradiated argon the gamma dose in the free beam is measured at about 1 Gy hr^{-1}. However after operation the argon becomes activated and the gamma contamination of the free beam rises by about 25 %.

PHANTOM MEASUREMENTS AND CALCULATIONS

In addition to the free beam measurements several different phantoms have also been used to investigate the beam parameters. These included two lucite phantoms constructed by collaborators from INEL, a full sized beagle head phantom and a cylindrical phantom. Both phantoms have provisions for the insertion of neutron activation wires or TLDs.

Various hollow perspex phantoms have also been used. These can be filled with water or tissue equivalent liquid (TEL) and are designed to accept simple activation foil packs at a number of positions throughout their volume. Many of the cell irradiation studies at Petten have used these same phantoms.

Two further phantoms are available which have also been used in conventional photon and neutron therapy. The first of these is a cuboidal tank of water into which BF_3, ionisation or fission counters may be inserted. A tracking device allows these detectors to be moved remotely thus providing a means of rapidly determining flux contours within the tank. A second phantom is constructed from a solid cube of polythene with holes bored at a number of positions to accept ionisation or fission counters. In both cases these phantoms are being used to validate the treatment planning calculations[2]. The polythene block phantom is also being developed as a means of rapidly checking the epithermal component of the beam prior to every irradiation.

In all cases both measurements and calculations have been performed with these phantoms. The beagle phantom is now described in more detail. This phantom is of a size comparable to that of a beagle dog and contains a number of finely bored holes that can accept Au/Cu wires for neutron activation analysis or TLDs for the determination of the gamma component. The phantom was provided by INEL and formed part of an intensive measurement campaign performed at Petten in 1991[2]. Since that time the beagle phantom has been irradiated on several occasions and forms a central part of the validation programme for the treatment planning model. It has the important benefit of providing a means of measuring the thermal neutron flux at depth in a body that is of similar size and composition to the beagles used in the healthy tissue tolerance study.

As with the free beam measurements the beagle phantom was also modelled with MCNP, Fig 3. From the measured copper and gold activation rates estimates were made of the thermal neutron flux as a function of depth. Fig 4 compares the measured and calculated estimates of thermal neutron flux where the agreement is seen to be very good, especially as no renormalisation of measured nor calculated values has been performed.

PROTON RECOIL MEASUREMENTS

Free beam measurements of the neutron spectrum have also been performed by collaborators from AEA Technology using gas filled proton recoil counters[4]. These measurements have been be undertaken at low reactor power to avoid saturating the counters. Potentially these measurements could provide spectral information from energies of about 10 keV upwards. the results from these measurements agree with the calculated spectrum and also with the foil measurements. However there remain some discrepancies with regard to the absolute intensity which have yet to be resolved.

Unfortunately it has not been possible to perform these measurements very frequently and there remains a paucity of proton recoil data for the HB11 beam, especially for the time period after the improvements to the argon system.

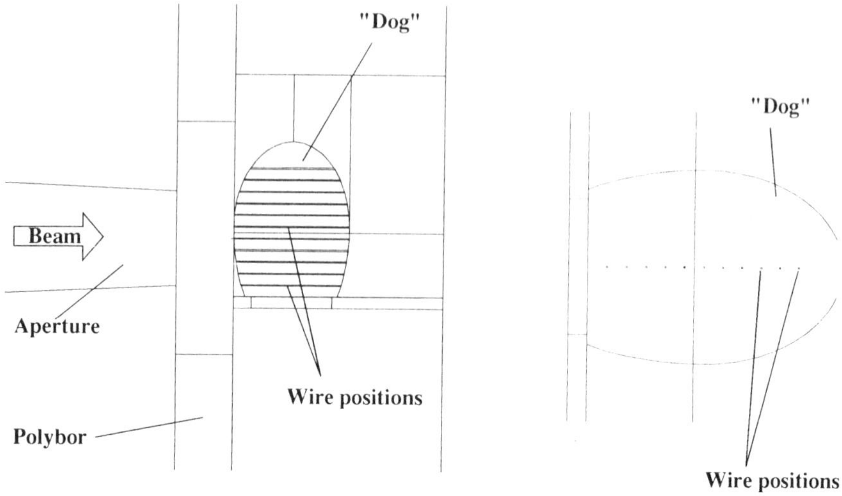

Figure 3. MCNP model of the lucite beagle phantom.

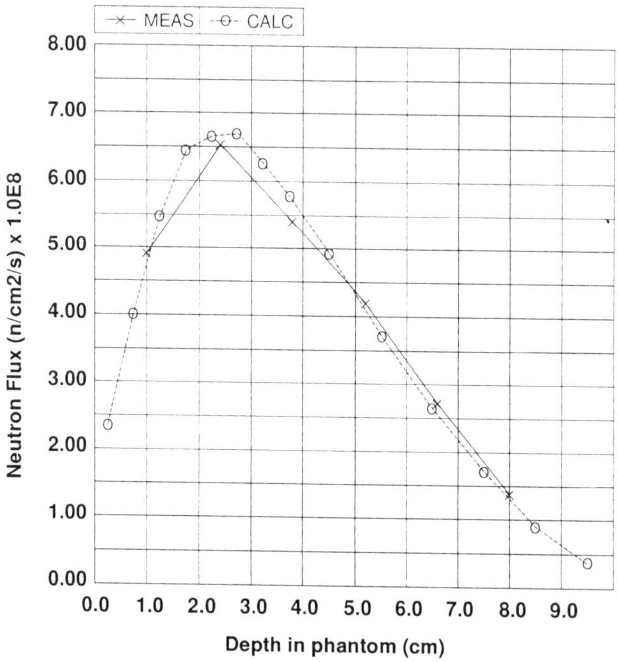

Figure 4. Comparison of thermal neutron flux from measurement and as calculated from MCNP.

CONCLUSIONS

The present series of free beam dosimetry measurements at the HFR Petten show that the BNCT beam remains very stable. No strong cycle-to-cycle variations have been observed. However the existing neutron foil measurements are incapable of responding to neutrons in the important epithermal energy range and further measurements are being developed to investigate this region.

Calculations with MCNP have been shown to reproduce the measured values reasonably well although there are some unacceptable discrepancies for some of the free beam foil materials. Measured and calculated values of the thermal neutron flux within the various phantoms agree well with each other.

It is intended to obtain further measured data at most neutron energies in the near future, particularly using the proton recoil method. These will be used in a spectrum adjustment procedure, currently under development, to correct the calculated spectrum.

REFERENCES

1. Briesmeister, J.F(Ed), "MCNP - A General Purpose Monte Carlo Code for Neutron and Photon Transport", LA-7396-M, Revision 2, September 1986.

2. Konijnenberg, M., et al, "Measurements and Calculations for Boron Neutron Capture Therapy Treatment Planning", This meeting.

3. Harker,Y., et al , "Neutron Spectrum Measurements at the Petten HFR and at Brookhaven BMRR", This meeting.

4. Perks, C., amd Delafield, H.J., "Neutron Spectrometry Measurements of the Petten HFR, HB11 Neutron Beam", International Workshop and Plenary Meeting, 'Towards Clinical Trials of Glioma with BNCT', Petten, The Netherlands, September 1991.

A COMPARISON OF EPITHERMAL NEUTRON BEAMS FOR NCT: EFFECT OF BEAM COLLIMATION ON THERAPY PARAMETERS

J.C. Yanch and O.K. Harling

Nuclear Engineering Department and Nuclear Reactor Laboratory
Massachusetts Institute of Technology, Cambridge, MA

INTRODUCTION

Collimation of neutron beams has a large effect on a number of therapy parameters of interest in BNCT. The effects of collimation include changes in dose rate, thermal neutron penetration and the degree of dose uniformity throughout the patient volume. Whether or not collimation of a neutron beam will improve a given therapy parameter depends on the size of the beam relative to the size of the phantom. This paper describes the results of a computer simulation study undertaken to examine the dosimetric effects of collimating a number of different beam configurations; the result of the study is a series of guidelines to aid in both epithermal and thermal neutron beam design.

METHOD

The study of collimation effects was carried out via Monte Carlo simulation using MCNP, a continuous energy coupled neutron-photon code developed at Los Alamos National Laboratory.[1] The dosimetric effects of monoenergetic neutron beams ranging from 0.025 eV to 800 keV incident on a brain-and-skull phantom were investigated. Beams were modeled as uniform discs with diameters 6.0, 12.0, 16.5 or 33.0 cm. Neutron flux was kept constant for all beam diameters. Maximum angle of emission was varied to examine the effects of neutron collimation. Neutrons emerging normal to the plane of emission simulates the case of well collimated beam (such as the Petten beam or the proposed Power Burst Facility beam). Neutrons allowed to leave the emission plane at all angles up to $\pm 90°$, a purely isotropic beam, simulates the situation in which a large neutron scatterer (e.g. bismuth) is placed very close to the patient position (as is the case with the BNL and MIT Reactor beams as well as with many of the proposed moderated accelerator beams).

Figure 1 illustrates the geometry of the brain-and-skull phantom[2] and provides an indication of the beam sizes modelled relative to the phantom. The phantom was divided into several concentric cylinders and 0.9 cm thick discs resulting in a large number of cells, typically 0.73 cm^3 in volume. Thermal neutron, fast neutron and photon doses were based on kerma factors available in the literature[3,4] and were calculated as a function of depth every 0.9 cm up to 10.8 cm; radially, doses were calculated every 0.9 cm up to 5.4 cm from the central axis in both the y and z directions. In this way, the dosimetric effect of each combination of beam energy, beam size, and maximum angle of emission could be determined as a function of both depth and radial distance from the center of the phantom. Each beam was assessed in terms of penetrability, dose rate, and uniformity of dose within the phantom.

Advances in Neutron Capture Therapy, Edited by
A.H Soloway *et al.*, Plenum Press, New York, 1993

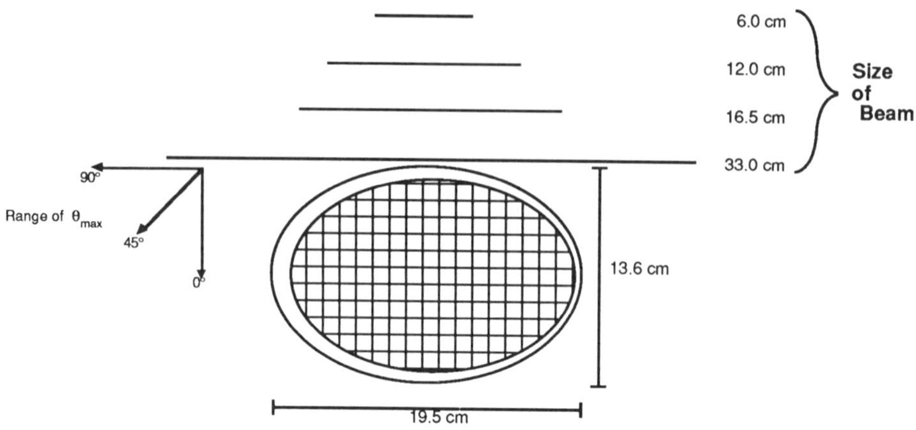

Figure 1. Illustration of the elliptical head phantom used and the four beam sizes investigated.

RESULTS

The effect of increasing beam collimation (reducing the maximum angle of neutron emission) on beam uniformity, penetration and dose rate depends critically upon whether the beam size is small or large compared to the size of the phantom.

When the beam is small (e.g. the 6.0 diameter beam) both thermal and fast neutron dose become less uniform as the beam becomes more collimated. For example, Figure 2 shows thermal neutron dose [$^{14}N(n,p)^{14}C$] versus depth from a 6.0 cm diameter beam of 100 eV neutrons. Data have been normalized to 1.0 at the maximum dose/neutron in the phantom. As the maximum emission angle decreases (i.e. as collimation is improved) the individual dose curves lie farther and farther apart, indicating a wider variation in dose throughout the phantom.

Thermal neutron penetration is also somewhat reduced as the beam becomes more collimated. Using a well-collimated beam (θ_{max} = 0°) thermal neutron dose in the central cylinder falls to 10% of its maximum at a depth of 8 cm in the phantom. For a less collimated beam (Figure 2a), dose falls to 10% at 8.5 cm representing a somewhat enhanced penetration of thermal neutron fluence. Because of improved beam uniformity with the isotropic beams, enhanced penetration is even more pronounced in the outer cylinders.

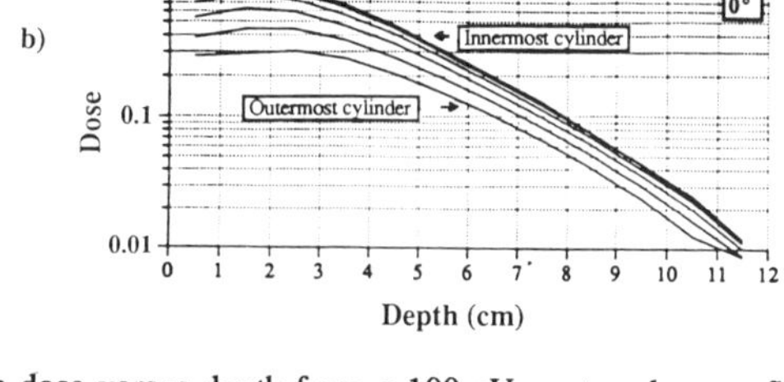

Figure 2. Thermal neutron dose versus depth from a 100 eV neutron beam. Source diameter is 6.0 cm. a) θ_{max} = 45°, b) θ_{max}= 0°. The uppermost curve represents the dose in the innermost cylinder while each of the successively lower curves represents dose averaged over cylindrical annuli at greater and greater distances from the center of the phantom.

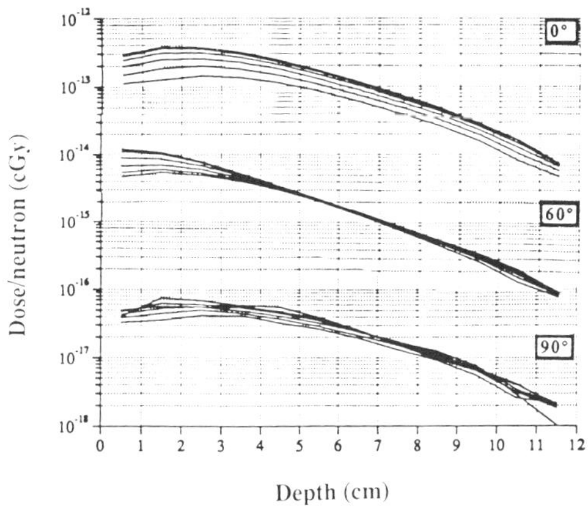

Figure 3. Thermal neutron dose versus depth from a 2 keV neutron beam. Source diameter is 6.0 cm. Data are shown for θ_{max} of 0°, 60° and 90°.

The effect of improving collimation, however, is to also improve dose rate. Figure 3 shows radial-depth-dose curves of thermal neutron dose due to an incident 6.0 cm diameter, 2 keV beam. The data are not normalized and illustrate the large reduction in dose rate as the beam becomes more and more isotropic. A difference of almost four orders of magnitude is seen in going from a perfectly collimated (0°) to a completely isotropic (90°) beam.

Opposite effects (regarding both beam uniformity and penetrability) are observed when the beam is larger rather than smaller than the phantom. Both uniformity of dose and beam penetration improve as the beam becomes more collimated. However, changes in collimation have little effect on dose rate in the phantom as illustrated in Figure 4.

The effects of beam collimation on advantage depth are expected consequences of changes in beam penetrability and again depend on the relative sizes of the beam and phantom. When the beam is small, advantage depth decreases as the beam is more and more collimated. When the beam is large, improved collimation causes advantage depth to increase. These trends are illustrated in Figure 5.

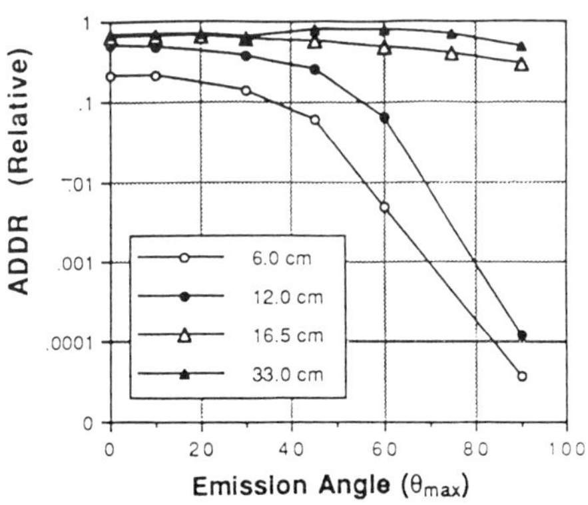

Figure 4. Relative advantage-depth-dose-rate versus maximum angle of neutron emission for four beam diameters. Beam energy equals 2 keV.

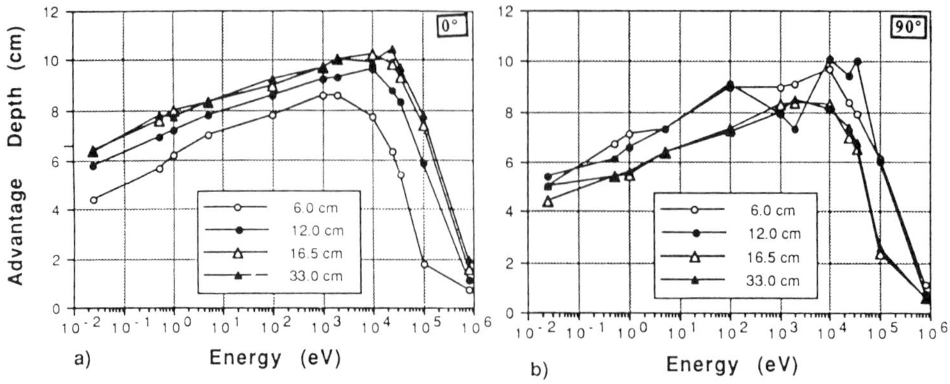

Figure 5. Advantage depth vs. beam energy for four different beam diameters. a) $\theta_{max} = 0°$, b) $\theta_{max} = 90°$.

SUMMARY

In summary, the effect of neutron beam collimation on therapy parameters of interest in BNCT depend on the size of the neutron beam relative to the size of the object being irradiated. When the beam is small, increasing collimation will reduce fast and thermal neutron dose uniformity, reduce neutron penetration (and hence advantage depth), but will greatly increase dose rate. In contrast, when the beam is large compared to the size of the irradiated object, increasing collimation will improve dose uniformity, increase neutron penetration (and advantage depth) but will have no effect on the rate of dose delivery.

ACKNOWLEDGMENT

The work reported here was supported by the U.S. Department of Energy under contracts DE-FG02-87ER-60600 and DE-FG07-80ER-10770.A015.

REFERENCES

1. J.F. Briesmeister, *MCNP—A General Monte Carlo Code for Neutron and Photon Transport.* LA 7396-M Rev. 2, Los Alamos National Laboratory, 1986.

2. J.C. Yanch, X-L. Zhou and G.L. Brownell, A Monte Carlo investigation of the dosimetric properties of monoenergetic neutron beams for neutron capture therapy. *Rad. Res.* **126**: 1-20, 1991.

3. R.S. Caswell, J.J. Coyne and M.L. Randolph, Kerma factors of elements and compounds for neutron energies below 30 MeV. *Int. J. Appl. Radiat. Isot.* **33**: 1227-1262, 1982.

4. R.G. Zamenhof, B.W. Murray, G.L. Brownell, G.R. Wellum and E.I. Tolpin, Boron neutron capture therapy for the treatment of cerebral gliomas. I. Theoretical evaluation of the efficacy of various neutron beams. *Med. Phys.* **2**: 47-60, 1975.

PHYSICS PARAMETERS FOR AN EPITHERMAL-NEUTRON BEAM AT THE GEORGIA INSTITUTE OF TECHNOLOGY

David W. Nigg, Gregory J. Storr[1], and Floyd J. Wheeler

National Center for BNCT Measurement and Development
Idaho National Engineering Laboratory
EG&G Idaho, Inc.
P.O. Box 1625
Idaho Falls, ID 83415

Boron Neutron Capture Therapy (BNCT) research in the United States has focused on the use of epithermal (0.5 eV to 10 keV) neutron beams for treatment of certain highly-malignant primary brain tumors and possibly for treatment of metastatic malignant melanoma. Various types of epithermal neutron sources for BNCT have been proposed over the years. Reactor-based sources, accelerator-based sources, and radioactive neutron sources have all been extensively examined. The first practical, large-scale, epithermal neutron beam for BNCT was installed at the Brookhaven Medical Research Reactor (BMRR)[1]. This beam was designed and constructed in a cooperative effort between Brookhaven National Laboratory (BNL) and the Idaho National Engineering Laboratory (INEL). It has been used extensively for BNCT research activities conducted by INEL, BNL, and others.

Although the BMRR beam has served the BNCT research community well, and will continue to be a valuable research tool for the foreseeable future, there are also reasons to seriously consider the construction of an advanced epithermal neutron beam facility at the Georgia Institute of Technology Research Reactor (GTRR). The neutronic characteristics of the GTRR are such that a properly-designed epithermal neutron filter in the existing large horizontal medical irradiation port at this facility would offer certain significant improvements in performance relative to the BMRR beam.

The GTRR is a heavy-water moderated and cooled plate-fuel-type reactor with a graphite neutron reflector. The proposed GTRR epithermal neutron beam facility developed in this work would be installed in the existing large biomedical irradiation port as shown in Figure 1. In this conceptual design an in-vessel filter/moderator structure, composed of 90% aluminum and 10% D_2O by volume, occupies the space between the core and the reactor vessel on the side of the reactor adjacent to the biomedical beam port. Outside the vessel, the existing large reflector and shield penetration is lined with a laminated bismuth-lead-cadmium filter housing. Within this housing is a region of dry aluminum plates occupying a volume fraction of 90%. This region is approximately 0.7 m in thickness. It is followed by a 0.076 m region of lithiated-aluminum plates and a 0.01 m titanium plate for spectral shaping in the 0-1 eV and the 10-100 keV ranges, respectively. Finally, there is a 0.076 m bismuth-lead gamma shield. The movable concrete shield block is also lined with laminated bismuth, lead and cadmium to help provide collimation of the neutron beam and to suppress contamination from concrete capture gamma radiation. A bismuth and lithiated-polyethylene thermal neutron and gamma shield is positioned on the exit port wall of the currently-existing treatment room. Provision can be made for a variety of beam delimiter inserts to control the size and shape of the beam exit port.

---

[1] On attachment from the Australian Nuclear Sciences and Technology Organisation (ANSTO), PMB 1, Menai, NSW, Australia.

Advances in Neutron Capture Therapy, Edited by
A.H Soloway *et al.*, Plenum Press, New York, 1993

Calculations to estimate the neutronic performance of the proposed epithermal beam at the GTRR were performed by a combination of two- and three-dimensional discrete-ordinates techniques using local adaptations of the well-known DORT[2] and TORT[3] computer codes, with cross sections taken from the BUGLE80 coupled neutron-gamma library[4]. These calculations were supplemented by confirmatory Monte Carlo analyses performed using the MCNP[5] code. This approach is basically the same as was successfully employed for physics design of the BMRR epithermal neutron beam, although extra emphasis is being placed on accurately calculating the fast neutron component of the beam since all methods tend to underestimate this component if one is not careful. As an added check, the exit port spectrum was also calculated using the DORT code and an 80-neutron, 20-gamma group cross section library that was independently-generated for this application using the AUS[6] system. The results were in agreement with the results obtained using the BUGLE-80 library.

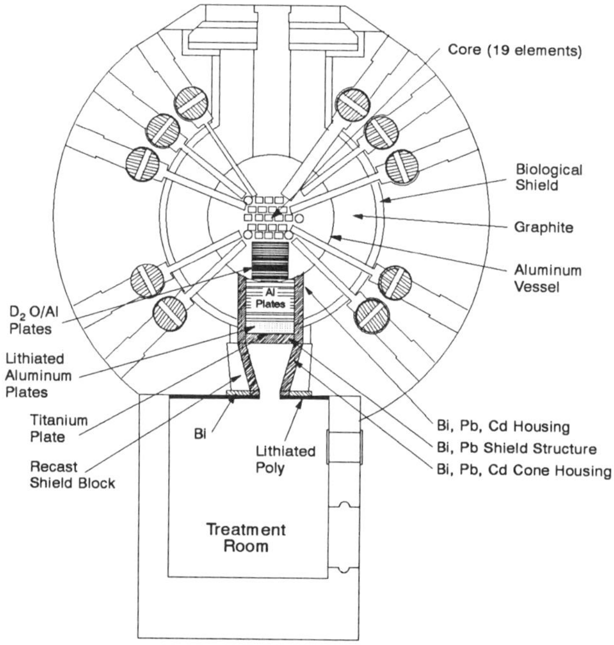

Figure 1. Conceptual Epithermal Neutron Facility for Boron Neutron Capture Therapy at the Georgia Institute of Technology Research Reactor.

The calculated exit port neutron flux spectra for the BMRR and the GTRR neutron beams are compared in Figure 2. The integrated epithermal intensity of the GTRR beam (about 4×10^9 n/cm^2-s at the rated reactor power of 5 MW) is about a factor of 2 greater than that of the BMRR beam at its rated power of 3 MW. The calculated fast neutron component for the GTRR beam (about 1.5×10^{-11} cGy per unit epithermal flux) is, radiobiologically, very significantly smaller than that of the current BMRR beam. The proposed GTRR beam is also well-collimated, with a current to flux ratio of about 0.85 compared to 0.6 for the BMRR beam. A well-collimated beam produces less undesirable healthy-tissue dose near the surface relative to the desirable boron capture dose at depth.

The results of clinically realistic phantom dosimetry calculations based on the methods outlined in References 7 and 8 show that the desirable boron dose at depth that can be generated with the GTRR beam is generally about a factor of three times larger per MW of reactor power than is the case for the BMRR beam. For a given total therapeutic dose, the treatment time for GTRR is therefore about a factor of 5 shorter at the rated reactor power of 5 MW than is the case for BMRR at its rated power of 3 MW. In addition, the fast neutron dose component of the GTRR beam relative to that of the BMRR beam is much reduced, as expected. The GTRR beam also offers a more favorable boron dose-depth profile than the BMRR beam. When the various calculated space-dependent physical dose components that are generated within the irradiation volume are weighted by appropriate Relative Biological Effectiveness (RBE) factors, it can be shown that the so called "advantage depth" for the GTRR beam is about 1 cm greater than that of the BMRR beam for clinically realistic tumor boron concentrations. This is radiobiologically very significant.

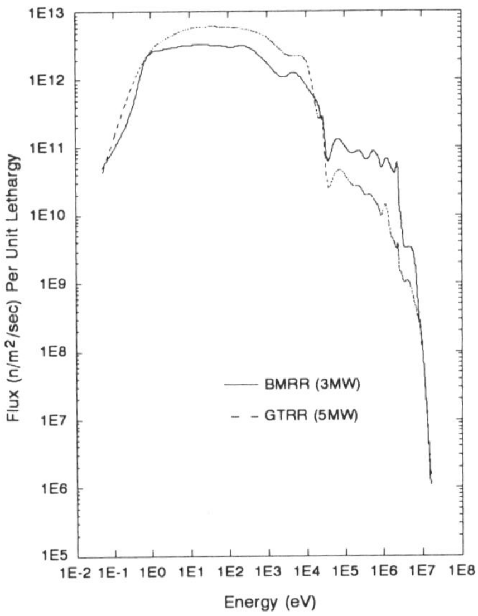

Figure 2. Comparison of calculated BMRR and GTRR neutron beam spectra.

When considered on the combined basis of physics, radiobiology, economics, and geographic location, the proposed GTRR beam offers, overall, the most desirable known source of epithermal neutrons for BNCT that is available, or can be made available, in the U.S., using present reactor or accelerator technology. The best currently operating epithermal neutron source in the U.S. is the BMRR beam, which will continue to see heavy use for BNCT research for the foreseeable future. The BMRR beam, however, has a radiobiologically significant level of fast neutron contamination, which has led to surface tissue damage in some canine experiments[9]. With the GTRR beam, this contamination can be suppressed to a level such that dose near the surface is not treatment-limiting. This would allow some types of advanced animal experimentation to be done without surgically reflecting the scalp. Also, extrapolation of animal BNCT irradiation data for the purpose of deciding whether to proceed to human clinical trials would be facilitated by the availability of the proposed GTRR beam because this beam has spectral qualities that are more typical of what would be required for routine human treatment should BNCT become a widespread modality.

ACKNOWLEDGMENT

This study was performed under the auspices of the U.S. Department of Energy, Office of Energy Research, DOE Field Office, Idaho, Contract Number DE-AC07-76ID01570.

REFERENCES

1. F.J. Wheeler, et.al., Epithermal neutron beam design for neutron capture therapy at the power burst facility and the Brookhaven Medical Research Reactor, Nuclear Technology 92, 106, October, 1990.

2. W.A. Rhoades, and R.L. Childs, An updated version of the DOT-4 one- and two-dimensional neutron/photon transport code, Oak Ridge National Laboratory, ORNL-5851, April 1982.

3. W.A. Rhoades, and R.L. Childs, The TORT three-dimensional discrete-ordinates neutron/photon transport code, Oak Ridge National Laboratory, ORNL-6268, 1987.

4. R.W. Roussin, BUGLE-80 coupled 47-neutron, 20 gamma-ray P3 cross section library, Radiation Shielding Information Center, DLC-75, 1980.

5. MCNP Monte Carlo neutron and photon transport code systems, Radiation Shielding Information Center, CC-200A.

6. G.S. Robinson, A guide to the AUS modular neutronics code system, AAEC/E645, (Australian Atomic Energy Commission), April 1987.

7. D.W. Nigg, et. al., Demonstration of three-dimensional deterministic radiation transport theory dose distribution analysis for Boron Neutron Capture Therapy, *Medical Physics,* 18:1, 43-53, January 1991.

8. F.J. Wheeler and D.W. Nigg, Three-dimensional radiation dose distribution analysis for neutron capture therapy, *Nuclear Science and Engineering,* 110, 16-31, January 1992.

9. P.R. Gavin, et. al., Large animal model - the biologic effects of mixed irradiation fields utilizing BSH and epithermal neutron irradiation, *Proceedings of International Workshop on Macro- and Microdosimetry and Treatment Planning for Neutron Capture Therapy,* MIT, October 1991, (To be published).

ENHANCEMENT OF THE EPITHERMAL NEUTRON BEAM AT THE BROOKHAVEN MEDICAL RESEARCH REACTOR

Hungyuan B. Liu[1], Robert M. Brugger[1], and David C. Rorer[2]

[1] Medical Department, and [2] Reactor Division
Brookhaven National Laboratory, Upton, NY 11973, U.S.A.

INTRODUCTION

The Brookhaven Medical Research Reactor (BMRR)[1] became operational in 1959 to support Boron Neutron Capture Therapy (BNCT) research in the Medical Department of the Brookhaven National Laboratory (BNL). The research was interrupted in 1961 after the failure of initial BNCT trials to show efficacy.

In 1988, moderators to produce an epithermal beam were designed and installed in one of the shutters at the BMRR. The horizontal section of the BMRR extending from the core out to the epithermal port is depicted in Figure 1, including the reactor core, graphite reflector, inner bismuth shield, moderator tanks in C, A and B, beam shutter, outer bismuth shield, and Li_2CO_3 in polyethylene (Li-poly) shield at the irradiation port. The moderator region C consists of two empty spaces. Al and Al_2O_3, which were identified by Less and Brugger[2] as good moderator to produce an epithermal beam, were selected by Wheeler and Fairchild[3] as the primary moderator

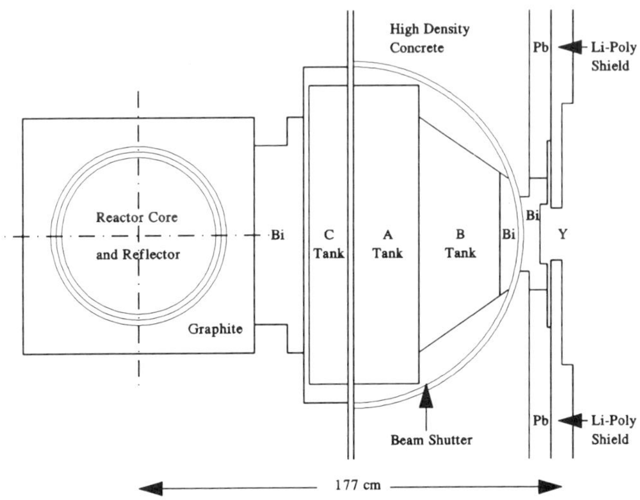

Figure 1. The Present BMRR Epithermal Beam.

Advances in Neutron Capture Therapy, Edited by
A.H Soloway *et al.*, Plenum Press, New York, 1993

Table 1. Comparisons of MCNP Calculations and Dosimetric Measurements After One New Fuel Element Added in the Core in June 1992.

	Comments	Difference
MCNP	1. Epithermal flux	+ 18%
	2. Beam neutron and gamma doses proportionally increased	
	3. No perceptible change in neutron spectrum	
Measure	1. Epithermal flux in air	+ 16%
	2. Thermal flux at different depths in the lucite phantom	+ 18%
	3. Fast neutron dose rate in air	+ 17%
	4. Gamma dose rate in air	+ 14%

moderator. Plates of Al and Al_2O_3 were used to fill in the A and B regions of the existing shutter to produce the present epithermal neutron beam. A 3.8 cm-thick Li-poly shield was added in 1991 around the bismuth port to reduce the neutron flux coming from outside the port.

An irradiation point for future reference is labelled as the "Y" point in Figure 1. There are now 30 partially spent fuel elements in the core. The reactor may be operated at 3 MW power. Since the moderators were assembled in 1988, neutron spectra[4] and dosimetric measurements and calculations in air and in phantoms have been made[5-7] at the epithermal port. The BMRR beam is the best available now to proceed with BNCT. The purpose of this study is to show how this beam can be further improved based on MCNP calculations which are confirmed by dosimetric measurements.

POSSIBLE CHANGES

To improve an epithermal beam, the major target is to increase the epithermal flux and to reduce the relative fast neutron dose in the beam. Possible changes of the BMRR beam are in the areas including the core, moderator assembly, and irradiation port. The graphite reflector and inner bismuth would be very difficult to modify at this time. MCNP calculations of the BMRR beam were made to predict improvements of the neutron beam at the epithermal port. After each change was made, the calculated results at the "Y" point were normalized by the calculated values of the present design. Finally, optimized changes in each area were combined to predict an improved epithermal beam.

Reactor Core One way to increase the epithermal flux at the irradiation port is to shift the fuel elements in the core toward the port while still retaining a critical but controlled condition. In the present core, there are two spaces facing the epithermal port which are not fueled. One is a graphite space block and the other is a tube for sample irradiation. In June 1992, the BMRR was operated with one new fuel element replacing the graphite space block. While the element was in position, flux and dose measurements at the epithermal port were made to compare to the previous measured values and MCNP calculations. The results are listed in Table 1. The MCNP calculation predicts increased fluxes and doses by 18% for such a change; no significant change in neutron spectrum was predicted. Different dosimetric measurements in a lucite phantom and in air verified the calculated results. The thermal flux measured by bare and Cd-covered gold foil activation at different depths in the lucite phantom went up 18% on average while the fast neutron and gamma dose rates measured by paired ionization chambers went up no more than 18%.

Table 2. Comparisons of Beam Parameters at the Irradiation Point "Y".

	Power MW	Φ_{epi} x 10^9 n/cm^2/s	D_n/n_{epi} x 10^{-11}cGy /(n/cm^2)	D_γ/n_{epi} x10^{-11}cGy /(n/cm^2)	J/Φ
BMRR*† May 1992	3	1.2	4.8	1.4	0.67
Stage 1A*† June 1992	3	1.4	4.8	1.4	0.67
Stage 1B* Oct. 1992	3	1.7	4.8	1.4	0.67
Stage 2* FY 1993	3	2.2	3.4	1.7	0.72

* Calculated by MCNP; † Measured

These results indicate that most of the epithermal neutrons reaching the irradiation port come from fission neutrons produced in the front rows of fuel elements on the epithermal side of the core. By changing the core configuration while maintaining criticality and control, the beam intensity at the epithermal port can be effectively increased without reducing the beam quality. An alternative is to place new fuel elements in the first or even second rows facing the epithermal port to shift the power distributions in the core toward the epithermal port. Thus, the beam intensity at the epithermal port will be significantly increased. This part of proposed change is now being made at the BMRR. The increase of epithermal flux is predicted to be more than 40% (Table 2, Stage 1B).

Moderator Tanks C, A & B The two empty spaces in moderator tank C can be filled with aluminum pellets to move the moderator assembly toward the core. The packing density of aluminum pellets was assumed to be 60% in the MCNP calculation. Then, space left in the beam shutter can be designed to accommodate the outer bismuth (or lead plus ^{6}Li) shield and this will allow an air indentation at the irradiation port so that the beam can be partially collimated. MCNP calculations also indicate that the beam can be enhanced by optimizing the moderator configuration and thicknesses of Al and Al_2O_3 to increase the epithermal flux and reduce the fast neutron dose per epithermal neutron.

Bismuth Shields The outer bismuth blocks shield the patient from the neutron induced gamma rays from Al and construction material. An alternative is to use lead plus 0.05% atomic number density of ^{6}Li to replace bismuth. The ^{6}Li captures the thermal neutrons while the lead appears to more effectively moderate fast neutrons than bismuth.

Combined Changes With a combination of feasible changes, the calculated results of new design are compared to the calculated values of the present design and listed in Table 2 as Stage 2. The final design is seen in Figure 2. At stage 2, the major changes include: (1) Four new fuel elements are placed in the first row facing the epithermal port; (2) The C tank is filled with aluminum pellets; (3) A piece of 14.6 cm-thick Al followed by a piece of 41.2 cm-thick Al_2O_3 fill in A and B tanks; (4) Bismuth shield at the irradiation port is replaced by 13.4 cm-thick lead plus 0.05% atomic number density of ^{6}Li; (5) A 15.8 cm air indentation is created at the irradiation port from the face of lead shield to the "Y" point.

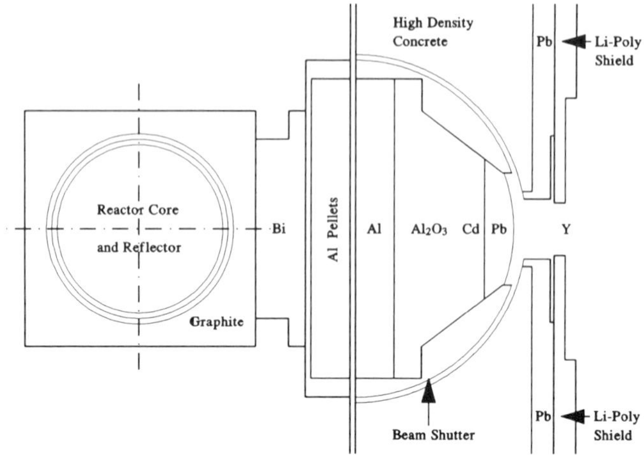

Figure 2. The New Design.

The final result shows, with a relatively insignificant increase of the gamma dose per epithermal neutron, the fast neutron dose per epithermal neutron can be reduced by 30% and the epithermal flux can be increased by 80%. With the new design, the epithermal flux at the "Y" point is 2.2×10^9 n/cm^2/sec at 3 MW power and the neutron dose per epithermal neutron is 3.4×10^{-11} cGy/(n/cm^2). The neutron current to flux ratio can also be increased by 7% from 0.67 to 0.72 to gain more penetration into the head.

CONCLUSIONS

Improvements for the BMRR epithermal neutron beam have been evaluated by MCNP calculations and measurements. Different dosimetric measurements have been made after one new fuel element was in place of the graphite space block stringer in the core. Measurements show an 18% increase of beam intensity without reducing the beam quality. These results are consistent with the predictions of an MCNP calculation.

Major changes to enhance the beam include rearranging the fuel elements in the core, placing aluminum pellets in the moderator tank C, redesigning the moderator assembly, replacing the outer bismuth by lead plus 0.05% atomic number density of ^{6}Li, and modifying the irradiation port to accommodate an air indentation. The MCNP calculated values for the present and new designs were compared to demonstrate the improvements. The fast results show that the epithermal flux can be increased by 80% at the irradiation port. The fast neutron dose per epithermal neutron can be reduced by 30%. The beam directionality can be improved by 7%.

ACKNOWLEDGEMENT

This research was supported by the U.S. Department of Energy contract DE-AC02-76CH00016.

REFERENCES

1. J. B. Godel, "Description of Facilities and Mechanical Components, Medical Research Reactor (MRR)," BNL 600 (T-173), Brookhaven National Laboratory, Feb. 1960.
2. T. J. Less, R. M. Brugger, "Reactor Moderated Intermediate Energy Neutron Beams for Neutron Capture Therapy", Strahlentherapie und Onkologie, Feb./March 1989.
3. F. J. Wheeler, D. K. Parsons, D. W. Nigg, D. E. Wessol, L. G. Miller, and R. G. Fairchild, "Physics Design for the Brookhaven Medical Research Reactor Epithermal Neutron Source", Neutron Beam Design, Development and Performance for Neutron Capture Therapy, O. Harling and R. Zamenhoff, eds. Mar. 1989, Plenum Press, New York, 1990.
4. Y. D. Harker, R. A. Anderl, G. K. Becker, and L. G. Miller, "Spectral Characterization of the Epithermal Neutron Beam at the Brookhaven Medical Research Reactor", Journal of Nuclear Science and Engineering, Vol. 110, 1992.
5. S. K. Saraf, J. Kalef-Ezra, R. G. Fairchild, B. H. Laster, S. Fiarman, E. Ramsey," Epithermal Beam Development at the BMRR: Dosimetric Evaluation," Neutron Beam Design, Development, and Performance for Neutron Capture Therapy, O. Harling and R. Zamenhof, eds., Mar. 1989, Plenum Press, New York, 1990.
6. H. B. Liu, " Isodose Curves and Treatment Planning for Boron Neutron Capture Therapy," Dissertation in Nuclear Engineering Program of University of Missouri-Columbia, Aug. 1992.
7. Y. D. Harker, C. R. Omevo, P. D. Randolph, and F. J. Wheeler, "Neutron Measurements in Canine and Cylindrical Phantoms at the Brookhaven Medical Research Reactor and the High-Flux Reactor (Petten)", Journal of Nuclear Science and Engineering, Vol. 110, 1992.

EPITHERMAL BNCT NEUTRON BEAM DESIGN FOR A TRIGA II REACTOR

Iiro Auterinen and Pekka Hiismäki

Technical Research Centre of Finland (VTT)
Reactor Laboratory
SF-02151 Espoo, Finland

INTRODUCTION

In Finland a collaborative effort by Helsinki University Central Hospital, MAP Medical Technologies Inc. and VTT Reactor Laboratory has started aiming at BNCT of glioma patients. For this the capabilities of the FiR-1 TRIGA II 250 kW research reactor have been evaluated.

The FiR-1 is located in the middle of the Otaniemi campus eight kilometers from the center of Helsinki and four kilometers from the Central Hospital. The power of the reactor was increased in 1967 to 250 kW and the instrumentation modernised in 1981. It is a pool reactor with graphite reflector and a core loading of 3 kg 20w% ^{235}U in the special TRIGA uranium-zirconium hydride fuel (8-12 w% U, 91% Zr, 1% H).

The advantages of using a TRIGA reactor for BNCT have already been pointed out earlier by Whittemore[1] and have been verified in practice by the thermal neutron treatment work done at the Musashi 100 kW reactor[2]. The advantages include a wide core face area and a wide spatial angle covered by the thermal-epithermal column system, large flux-per-Watt feature and inherent safety of the TRIGA fuel.

Because of its wider applicability and less stringent requirements for clinical operation conditions, an epithermal neutron beam has been selected as the design goal. The epithermal flux should be sufficient for glioblastoma patient treatment: 10^9 epithermal neutrons/cm^2/s with low enough fast neutron ($< 3 \cdot 10^{-13}$ Gy/epithermal n/cm^2) and gamma contamination[3].

SELECTING THE BEST SPECTRUM SHIFTER MATERIAL

Our basic idea in the design was to fill the thermal column space starting from the very bottom with a suitable moderating material; additionally we got interested in the neutron converter concept proposed by professor Rief from JRC-Ispra, Italy[4]. The two geometries are shown in Figure 1. Another key feature in this design is the widening of the patient space for increased freedom in patient positioning and closer access to the core.

The one-dimensional discrete ordinates transport code ANISN/PC[5] was used together with the BUGLE-80 cross section data[6] to find an optimal material composition for moderating reactor core neutrons down to epithermal energies. A spherical geometry was chosen to resemble the wide column design and to emphasise the moderation properties of the materials and not only scattering. So the situation is different from narrow beam geometries where the scattering filtering effect is used for tailoring the neutron spectrum. The ANISN model included a spherical, homogenous core surrounded by a graphite reflector shell and a 150 cm or 100 cm thick moderating shell. The P3 scattering approximation and S8 quadrature were used.

Advances in Neutron Capture Therapy, Edited by
A.H Soloway *et al.*, Plenum Press, New York, 1993

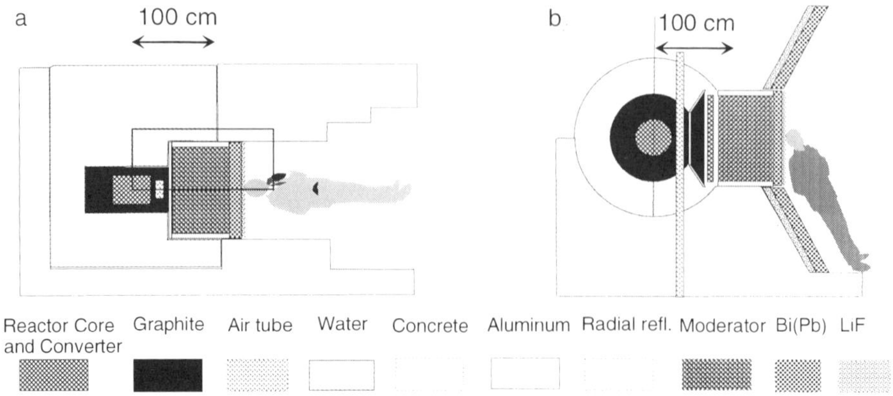

Figure 1. a) Side view of the basic epithermal beam concept. The rectangle shows the volume modelled in TORT. **b)** Horizontal view of the neutron converter concept.

Nearly all elements with reasonably low thermal neutron absorption (16 elements) and many of their mutual compounds were tested. First the attenuation of the fast flux (7.1 keV - 17.3 MeV) relative to the epithermal (0.414 eV - 7.1 keV) was studied. Fluorine and many fluorides showed better performance than oxides, which are more commonly used or proposed as moderators for BNCT. The advantages of the AlF_3 as a moderator have been indicated also in an earlier work by Ross *et al.*[7]

As the next step the fast neutron dose per unit epithermal flux was calculated based on the ICRU-26 dose function[8] that was extended using log-log extrapolation down to thermal energies (ICRU(ext)). A comparison of the 100 cm and the 150 cm moderator results showed that the required thickness of each material can be found by looking at the epithermal flux and fast neutron dose values inside the moderator.

The final criterion for selecting the moderator was the epithermal flux at the design limit of $2.6 \cdot 10^{-13}$ Gy/epithermal n/cm^2 for the fast neutron dose. Also here all the tested fluorides (AlF_3, PbF_2, CaF_2, $CaSiF_6$, MgF_2) are among the best candidates. The optimised mixture of aluminum fluoride and aluminum (60% AlF_3 + 40% Al) clearly outperforms the others, giving an epithermal flux ($1.9 \cdot 10^{10}$ epith.n/cm^2/s) that is 50% higher than what is achieved with the optimised mixture of aluminum oxide and aluminum (30% Al_2O_3 + 70% Al). The advantage of the aluminum fluoride is even more profound if a lower fast neutron dose limit is set; 90% higher epithermal flux ($6.4 \cdot 10^9$ epith.n/cm^2/s) at $1.0 \cdot 10^{-13}$ Gy/epith.n/cm^2. A glass with aluminum, lead and beryllium fluorides[9] and pure lead fluoride give the second highest epithermal flux with the set fast neutron dose limit.

THREE-DIMENSIONAL MODELLING OF THE NEUTRON BEAM DESIGN

A basic three-dimensional model of the proposed neutron beam design was calculated using the TORT code[10] with the BUGLE-80 data. The TORT model (see Figure 1) included the core, the graphite reflector with a tangential beam tube and irradiation ring, part of the water tank and concrete, the column filled with moderator, surrounded by a 5 cm reflector (Ni or Bi), the 15 cm Bi- gamma shield, LiF- thermal neutron absorber, LiF+polyethylen+Pb-radiation shields at the sides and an ellipsoidal brain tissue[11] phantom with dimensions: x(depth): 17cm, y(width):20cm, z(height):14cm.

A 1/4 geometry was used with reflecting conditions for three planes through the core center. Grid spacings were mainly 2 to 3 cm in the core, 5 cm in the moderator column and Bi-shield, 1 (depth) and 2 cm (width and height) in the phantom and 10 cm in concrete. The P3 scattering approximation and S8 quadrature were used.

Moderator columns with aluminum fluoride and aluminum oxide were calculated. The AlF_3 was modelled either as a 70% relative density powder with 1% by volume LiF(nat) or a mixture of this powder and aluminum. The Al_2O_3 was as a full density material with 1%v LiF(nat) or as a mixture with aluminum metal. The mixing ratios used were close to optimum according to the ANISN model. The column length was 69 or 89 cm.

Figure 2 shows that the required fast neutron limit can be reached with an aluminum oxide column not much thicker than the calculated one. With the mixture of aluminum oxide and aluminum a thicker column would be required. Based on the ANISN calculations the shorter aluminum fluoride column was expected to be sufficient, but the thicker column was required.

The wide, uniform but highly isotropic beam produces a deep and wide boron dose distribution which peaks at about 3 cm in the phantom (Figure 3). The epithermal flux at the phantom surface was in the case of Figure 3 $1.1 \cdot 10^9$ n/cm²/s. With the 69 cm AlF_3(70%v) it was $4.9 \cdot 10^9$ n/cm²/s, with 89 cm AlF_3(70%v) $7 \cdot 10^8$ n/cm²/s, with 69 cm Al_2O_3 $8.7 \cdot 10^8$ n/cm²/s and with 69 cm Al_2O_3(40%v) + Al(60%v) $4.7 \cdot 10^9$ n/cm²/s. The three figures of merit proposed by the MIT BNCT group[12] are listed in Table 1 for four of the calculated cases. These results, still preliminary and not optimised, are in the best cases even better than those for the BMRR 3 MW beam in a comparable wide beam geometry[13]. If the porosity of the aluminum fluoride could be overcome that material could produce better results than shown here for the powder.

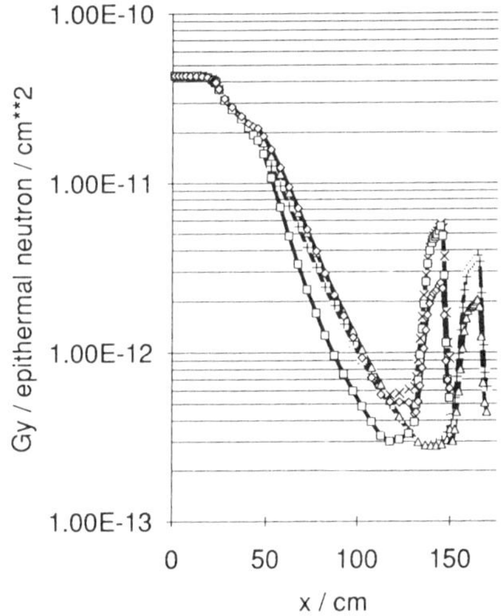

Figure 2. Relative fast neutron dose ICRU(ext) at column centerline for the calculated cases: (×) 69 cm, (+) 89 cm AlF_3(70%v), (Δ) 89 cm AlF_3(50%v) + Al(30%v), (□) 69 cm Al_2O_3, (◊) 69 cm Al_2O_3(40%v) + Al(60%v).

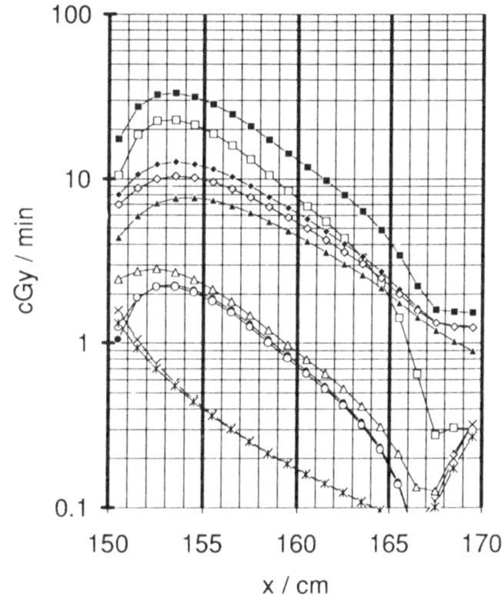

Figure 3. Physical dose at phantom centerline for the 89 cm AlF_3(50%v) + Al(30%v) column case. The curves are from top to bottom: (■) total dose to tumor, (□) B-10 dose for 30 ppm, (◆) total background, (◊) non B background, (▲) gamma dose, (Δ) total neutron dose, (●) B-10 dose for 3 ppm, (O) thermal neutron dose, (×) ICRU(ext) fast neutron dose, (✱) ICRU(>7.1keV) fast neutron dose

NEUTRON PRODUCTION WITH AN URANIUM CONVERTER

Advantages of using a neutron converter, either of highly enriched uranium plates[4] or of low enriched, high concentration, hydrogen-free TRIGA fuel elements, outside the core as well as a partial high power fuel loading in the core for improved neutron production were also studied. Results obtained with KENO and TIMOC 3-dimensional Monte Carlo modelling are reported by Rief *et al.*[4] A comparison of those results with the TORT results reported here indicates that there is no clear advantage or need for using a neutron converter if the reactor power is kept the same in both cases. Also the technical and licensing requirements caused by additional fissile material suggest that, in the first stage of deployment, a converter is not practical.

Table 1. Advantage Depths (in cm), Advantage Ratios and Advantage Depth Dose Rates for some of the calculated cases; doses in cGy/min or RBE*cGy/min. The MIT 1989 RBE's are those suggested for the MIT Workshop[11,12,13]: B-10 2.3, n 1.6, γ 1.0; the Eur. 1992 RBE's[14] are for BSH: B-10 in tumor 2.3, B-10 background 0.33, fast n 4.5, thermal and epithermal n 2.7, γ 1.0.

		Maximum			Minimum		
		AD	AR	ADDR	AD	AR	ADDR
69 cm AlF_3(70%v)	no RBE	9.5	2.8	39.7	8.5	2.4	48.2
	MIT 1989 RBE	11.5	4.4	47.5	10.5	3.3	58.2
	Eur. 1992 RBE	10.5	3.4	63.4	9.5	3.3	77.4
89 cm AlF_3(50%v)	no RBE	10.5	2.9	11.8	9.5	2.5	14.2
+ Al(30%v)	MIT 1989 RBE	12.5	4.7	13.9	10.5	3.5	21.1
	Eur. 1992 RBE	11.5	3.8	18.2	11.5	3.7	18.2
69 cm Al_2O_3	no RBE	9.5	2.9	6.5	8.5	2.5	8.0
	MIT 1989 RBE	11.5	4.7	7.6	9.5	3.5	11.5
	Eur. 1992 RBE	9.5	3.9	12.2	8.5	3.0	15.3
69 cm Al_2O_3(40%v)	no RBE	10.5	2.9	31.3	9.5	2.5	37.9
+ Al(60%v)	MIT 1989 RBE	12.5	4.7	36.5	10.5	3.5	55.7
	Eur. 1992 RBE	10.5	3.8	59.0	10.5	3.6	59.0

CONCLUSIONS

The modelling calculations done so far indicate that using optimal moderating material selections a 250 kW TRIGA reactor can produce an epithermal beam suitable for human patient glioma treatment. No additional fissile material as a neutron booster is necessarily required.

REFERENCES

1. Whittemore, W.L. (1992) in *Progress in Neutron Capture Therapy for Cancer*, eds. Allen B.J. *et al.* (Plenum Press, New York), pp. 57-62.
2. Matsumoto, T., Aizawa, O., Nozaki, T. & Sato, T. (1989) Pigm. Cell Res. **2**:240-245.
3. Wheeler, F.J., Parsons, D.K., Rushton, B.L. & Nigg, D.W. (1990) *Nucl. Tech.* **92**, 106-114.
4. Rief, H., Van Heusden, R. & Perlini, G. (1992) Generating Epithermal Neutron Beams for BNCT at TRIGA Reactors, (This conference).
5. Parsons, D.K. (1987) ANISN/PC Manual, EGG-2500 (EG&G Idaho, Inc., Idaho Falls, Idaho).
6. Roussin, R,W. (1980) BUGLE-80 Coupled 47-Neutron, 20 Gamma Ray P3 Cross Section Library, DLC-75 (Radiation Shielding Information Center, Oak Ridge National Laboratory).
7. Ross, D., Constantine, G. & Weaver, D.R. (1992) in *Progress in Neutron Capture Therapy for Cancer*, eds. Allen B.J. *et al.* (Plenum Press, New York), pp. 67-69.
8. Neutron dosimetry for biology and medicine (1977) ICRU-report 26.
9. Sun, K.-H. & Callear, T.E. (1949) U.S. Pat. 2,466,506.
10. Rhodes, W.A. & Childs, R.L. (1987) The TORT Three-Dimensional Discrete Ordinates Neutron/Photon Transport Code, ORNL-6268 (Oak Ridge National Laboratory).
11. Zamenhof, R.G., Clement, S.D., Harling, O.K., Brenner, J.F., Wazer, D.E., Madoc-Jones, H. & Yanch, J.C. (1990) in *Neutron Beam Design, Development, and Performance for Neutron Capture Therapy*, eds. Harling O.K. *et al.* (Plenum Press, New York), pp. 283-305.
12. Clement, S.D., Chai, J.R., Zamenhof, R.G., Yanch, J.C. & Harling, O.K. (1990) in *Neutron Beam Design, Development, and Performance for Neutron Capture Therapy*, eds. Harling O.K. *et al.* (Plenum Press, New York), Basic Life Sciences Vol. 54 pp. 51-69.
13. Fairchild, R.G., Kalef-Ezra, J., Saraf, S.K., Fiarman, S., Ramsay, E., Wielopolski, L., Laster, B.H. & Wheeler, F.J. (1990) in *Neutron Beam Design, Development, and Performance for Neutron Capture Therapy*, eds. Harling O.K. *et al.* (Plenum Press, New York), pp. 185-199.
14. Healthy Tissue Tolerance Studies for BNCT (1992) Eur. Coll. BNCT of Tumors Newsletter 12, 8-11.

GENERATING EPITHERMAL NEUTRON BEAMS FOR NEUTRON CAPTURE THERAPY IN SMALL REACTORS

H. Rief, R. Van Heusden and G. Perlini

Joint Research Centre of the CEC
Ispra Establishment
I-2102 Ispra (Va), Italy

INTRODUCTION

In an approach different from other proposals dealing with BNCT therapy facilities, the possibility of using a ^{235}U converter plate in the thermal column of 250 kW TRIGA reactors has been investigated. The study is based on many years of experience with the use of the JRC neutron irradiation facility EURACOS[1,2,3] which was built around a highly enriched converter made up of a circular uranium-aluminum alloy plate (80 cm diameter, 1.83 cm thickness) containing a total of 4.2 kg of ^{235}U.

In the case of a converter facility thermal neutrons from the graphite surrounding the core impinge on the surface-area (~5000 cm^2) of the converter which in turn emits fission neutrons. In Euracos they were used to measure neutron propagation in thick layers of power reactor shielding materials.

Instead of the shielding materials we propose to place a composition of weakly moderating and filtering materials in front of the converter plate. This composition has the function of 'shifting' a relatively large fraction of fission neutrons into the epithermal energy range between ~10 keV and 1 eV and of preventing fast and thermal neutrons from reaching the therapy position.

It will be shown that converter facilities installed at the thermal column of small reactors, e.g. TRIGA's, are able to produce high quality epithermal neutron beams having all the characteristics required in neutron capture therapy if their thermal fission power exceeds 5 kW.

DESIGN PROPOSALS FOR EPITHERMAL NEUTRON BEAM GENERATORS

The use of fission neutrons as a source for an epithermal shifter arrangement is quite similar to an early INEL project[4] which employs the Idaho Power Burst Facility emitting neutrons with a very hard spectrum from the core surface. For this reason we initiated our parametric studies with a neutron shifter configuration almost identical to this project. A perspective design of the proposal is given in Figure 1. It shows the core, the graphite reflector, the thermal column (partially voided), the fast neutron converter and the neutron spectrum shifter. It consists of layers of aluminium (89%), heavy water (10%) and lithium (1%) filling a volume of 90x90x78 cm^3. It is followed by a cadmium sheet, a bismuth gamma shield of 15 cm thickness and a collimator made of nickel. In order to minimize lateral neutron losses the spectral shifter configuration was surrounded by 5 cm of nickel.

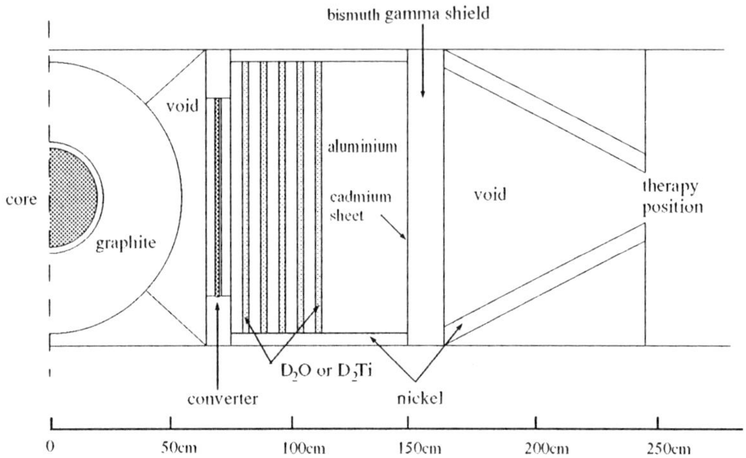

Figure 1. Schematic horizontal cross section of a spectral shifter - filter arrangement installed at a TRIGA reactor.

The calculations dealing with neutron propagation from the converter plate towards the therapy position were carried out by the Monte Carlo code TIMOC [5] assuming a measured (cosinusoidal) radial source distribution in the ^{235}U disk.

Table 1. Epithermal neutron flux in the therapy position using a 4.6 kW converter plate.

Material Composition in Spectral Shifter	Epithe. flux $n/cm^2/s$	fast/epithe. flux-ratio
63cm Al, 5x2 cm D_2O	$1.06 \cdot 10^9$	.027
63cm Al, 5x2 cm D_2Ti	$.77 \cdot 10^9$	.017
63cm Al, 5x2 cm AlF_3	$1.86 \cdot 10^9$	.116
63cm Al, 5x2 cm TiF_4	$1.86 \cdot 10^9$	.088
58cm AlF_3	$.92 \cdot 10^9$	.022
63cm AlF_3, 3x2 cm Al, 2x2 cm Ti	$.49 \cdot 10^9$	.0025

The reference composition is made-up of 63 cm aluminium and five layers, 2 cm each, of D_2O as shown in Figure 1. In this case one source neutron originating in the converter leads to an average epithermal flux of $2.75 \cdot 10^{-6}$ n cm^{-2} in the range of 1eV to 10keV at the therapy position. This means that $1.45 \cdot 10^{14}$ fissions per second are needed to reach an epithermal flux of 10^9 $n \cdot cm^{-2}s^{-1}$ as required in medical applications. It corresponds to a thermal power release of 4.6 kW in the converter.

As is well known the epithermal neutron beam has to be carefully tailored to compromise between enhanced penetration depth and minimal damage to healthy tissue. The reduction of high energy neutrons is of particular importance. For this reason we performed a series of spectrum calculations using different material compositions and cross section libraries. In an attempt to improve of this layout further we studied the effect of replacing the heavy water by titanium-deuteride (TiD_2). This rather unconventional material was considered for two reasons: titanium replaces, so to speak, the oxygen of D_2O (which does not serve the purpose of tailoring the neutron spectrum), and it adds to the filtering effect since it has, like aluminium, a low neutron cross section in the epithermal energy range with the additional advantage of "closing" the cross section windows of aluminium above 10 keV. Figure 2 shows spectra calculations with two different sets of multigroup cross sections (28 and 218 energy groups) taken from the SCALE-3[6] library. For comparison the spectrum determined for the INEL[4] project is also included. It can be seen that there is a reasonable agreement between the three curves. In particular it should be noticed that the "contamination" caused by neutrons with energies above 10 keV is generally lower for the TRIGA estimates.

Finally aluminiumfluoride as a moderating material was investigated in different geometrical compositions. The results of these (still preliminary) studies are listed in Table 1. Because of the low power of the TRIGA reactor as well as the converter plate, contamination by photons is less than in other epithermal beams currently being studied.

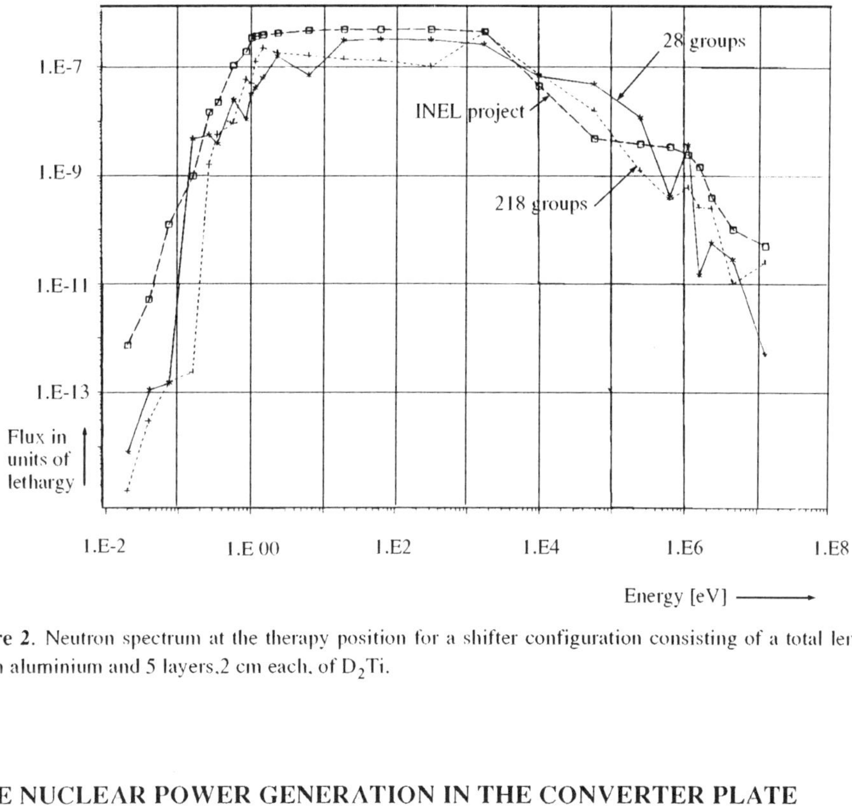

Figure 2. Neutron spectrum at the therapy position for a shifter configuration consisting of a total length of 63 cm aluminium and 5 layers,2 cm each, of D_2Ti.

THE NUCLEAR POWER GENERATION IN THE CONVERTER PLATE

The neutron source intensity (nuclear power generation) was calculated for two different converter configurations by the Monte Carlo code KENO-4[7] using a precise 3D geometry description of the proposed lay-outs. One configuration is the highly enriched disk-shaped converter used in the Euracos facility. The other is a design proposal of a single array of D_2O cooled cylindrical 20% enriched fuel elements in an aluminium matrix. Our calculations (partially financed by the Finnish VTT) confirmed that a converter power of

~4.5 kW, as required in the preceding paragraph, can be reached for both types of converters by a 250 kW TRIGA reactor. It however requires the thermal column to be voided and the converter to be placed at a position 70 cm from the core centre (leaving the cylindrical graphite reflector around the core untouched).

In the case of the disk-shaped converter the cooling can be achieved by forced air circulation. For a system of 1000 m^3/h air throughput a detailed thermodynamic estimate was carried out. It led to the following main results: Maximum surface temperature of the converter ~194°C; Temperature increase of the cooling air: 22° K. Both these values are rather conservative since heat loss by radiation to the aluminium-D_2O spectral shifter and the heavy shielding structures was neglected. But even with the quoted figures perfectly safe operation of the converter would be possible, considering the fact that the Al-U alloy is stable up to 640°C and its aluminium canning up to 500°C.

CONCLUSIONS

From this (still preliminary) feasibility study a number of relevant conclusions can be drawn: - A full scale epithermal BNCT facility satisfying all medical requirements can be installed at inexpensive low power reactors such as TRIGAs. - Because of the small fuel consumption and the long lifetime (several decades) of the converter plate the operation costs of the proposed irradiation plant are comparatively low. - There are quite a few TRIGA reactors in university environments, and therefore at places where new therapy facilities, which require multi-disciplinary cooperation, are most likely to be used efficiently.

REFERENCES

1. R. Nicks, G. Perlini, and H. Rief, Project and performance of the EURACOS II irradiation facility. Technical Note No. 1.05.00.85.45. JRC Ispra (1985).
2. R. Nicks, G. Perlini, and H. Rief, Fission neutron penetration in iron and sodium-I. Activation measurements, *Ann. nucl. Energy* 15:437(1988).
3. G. Perlini, and H. Rief, Fission neutron penetration in iron and sodium-II. Neutron spectrometry, *Ann. nucl. Energy* 16:189(1989).
4. D. K. Parsons, F.J. Wheeler, B.L. Rushton, and D.W. Nigg, Neutronics design of the INEL facility for BNCT clinical trials, Proceedings of "1988 Int. Reactor Physics Conf." Jackson Hole. (1988).
5. H.Rief, TIMOC - a Monte Carlo sensitivity code. (prepared for publication).
6. SCALE-3: A modular code system for performing standardized computer analysis for licensing evaluation. NUREG/CR-0200 (ORNL/NUREG/CSD-2),(1984).
7. C.M. Petrie, and N.F. Cross, KENO-IV-an improvedMonte Carlo Criticality program. ORNL-4938.(1975).

EPITHERMAL AND FAST NEUTRON SPECTRUM MEASUREMENT AT THE BROOKHAVEN MEDICAL RESEARCH REACTOR (BMRR)

Gad Shani

Department of Nuclear Engineering
Ben-Gurion University
POB 653, Beer-Sheva, Israel 84105

INTRODUCTION

The main purpose of the Brookhaven Medical Research Reactor (BMRR) is for boron neutron capture therapy (BNCT). In particular, an epithermal neutron beam was developed to provide a neutron beam where most of the neutrons are in the epithermal range. It was concluded in earlier studies[1] that the ideal beam for BNCT is in the energy range of 1-20 keV. Several Monte Carlo and other calculations of the neutron spectrum were made in the BMRR epithermal neutron beam.[2] There are considerable differences between the various calculated spectra. One indirect spectrum measurement was published using foil activation.[3] The measured spectrum did not agree well with any of the calculated spectra. All calculated and measured spectra show that in spite of the careful design of neutron and gamma filtering (made of bismuth, Al_2O_2, and D_2O),[4] the high energy neutrons still make up a considerable part of the neutron spectrum. These high energy neutrons increase the dose to the irradiated sample. According to the ICRP 6O,[5] the quality factor of neutrons of energy between 200 keV and 2 MeV is 20 times higher than that of gamma radiation and 4 times higher than thermal neutrons. The importance of the fast neutron contribution to the RBE was demonstrated by Laster[6] using a biological dosimeter. The conclusion of this introduction is that a direct reliable measurement of the epithermal and fast neutron spectra is needed.

EPITHERMAL NEUTRON SPECTRUM MEASUREMENT

The neutron spectrum was measured in two parts: first the keV range flux and second the MeV range flux. For the epithermal (keV) range a neutron spectrometer was

Advances in Neutron Capture Therapy, Edited by
A.H Soloway *et al.*, Plenum Press, New York, 1993

designed and built. The spectrometer was made of He^3 gas and two surface barrier detectors, all contained in a sealed box. Neutron interaction with He^3 nucleons creates a proton, a triton, and 764 keV of energy ($He^3 + n = p + H^3 + 764$ keV). The neutron's kinetic energy is added to the 764 keV reaction energy and the total energy is divided between the two outcoming particles according to their masses, 1/4 to the H^3 and 3/4 to the proton. The electronic system sums up the two pulses from the two detectors to give a resulting pulse proportional to En + 764 keV. A good resolution system is needed to be able to see a detailed spectrum in the low keV range. The resolution in the present system was better that 3%. It depends strongly on the gas pressure and the temperature of the solid state detectors. The detector efficiency is also strongly dependent on the gas pressure, so that a compromise has to be found between low pressure for good resolution and high pressure for higher efficiency. One hundred millimeters of Hg was found to be a reasonable pressure, where 3% efficiency and 3% resolution were obtained at room temperature. Details of the detector development and investigation have been described elsewhere.[7] The detector is shown schematically in Figure 1. The electronic system is shown in Figure 2. It is made of two branches for each detector, energy and timing. The timing branch is used to verify that the two pulses summed originated from the same $^3He(n,p)^3H$ reaction. The measured neutron spectrum in the keV range up to 300 keV is shown in Figure 3. The measured spectrum is compared with the Monte Carlo calculated spectrum. The fit seems to be good.

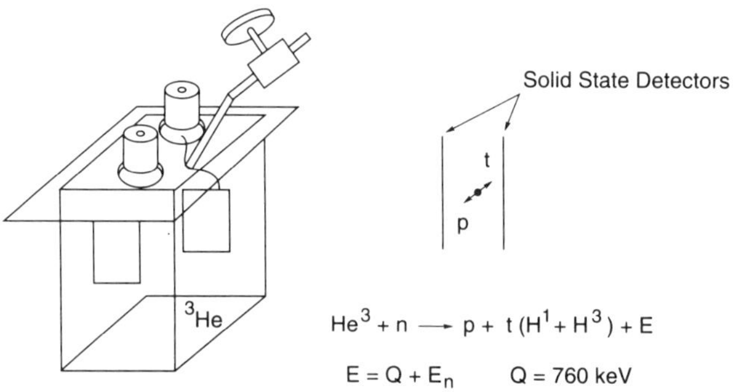

Figure 1. The neutron spectrometer

FAST NEUTRON SPECTRUM MEASUREMENT

The neutron spectrum in the MeV range was measured with a 2" x 2" liquid scintillator, NE 213. Neutron detection is done by proton recoil in the scintillator so that the recoiled proton energy is related to the neutron energy according to the scattering angle

$$E_p = E_n \cos^2\Theta,$$

where E_p is the recoiled proton energy, E_n is the incoming neutron energy, and Θ is the angle between the incoming neutron direction and the recoiled proton direction. The reaction is isotropic, so that a monoenergetic neutron beam produces a flat distribution of recoiled protons up to $E_p = E_n$. The neutron spectrum is obtained from the recoiled proton distribution by unfolding. A computer code, FORIST,[9] containing the detector response

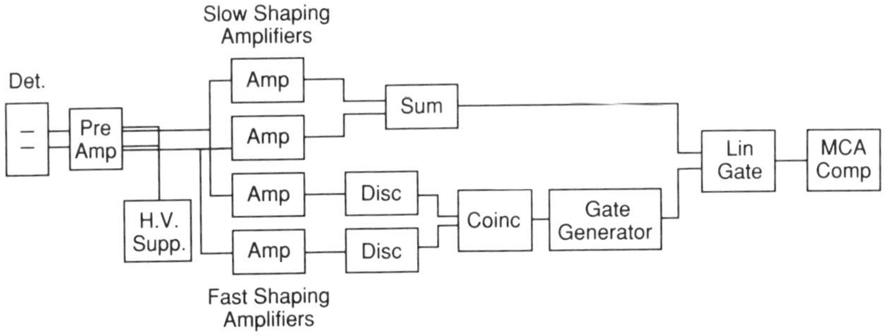

Figure 2. The electronic system

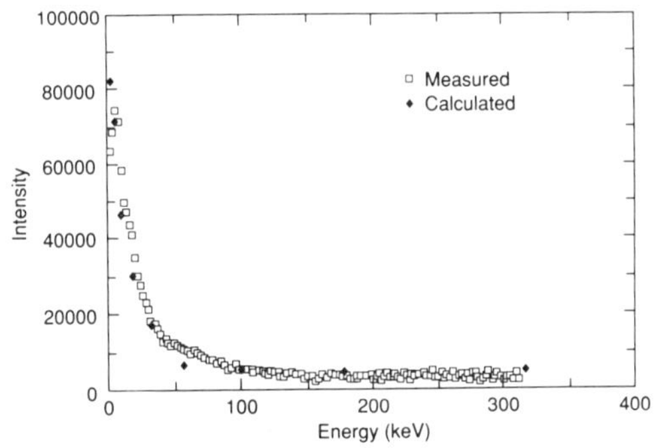

Figure 3a. Epithermal neutron spectrum

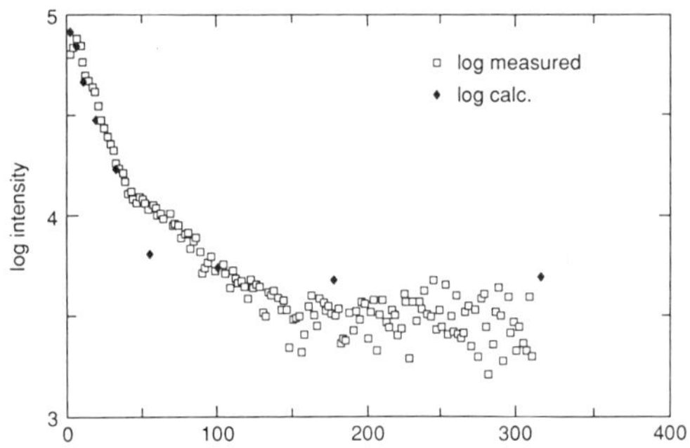

Figure 3b. Epithermal neutron spectrum (log scale)

matrix to 80 groups of monoenergetic neutron beams and a least square fitting program, was used to analyze the measured spectrum.

Another important characteristic of the NE 213 scintillator is that it provides a different pulse rise time for a neutron and a gamma ray, thus enabling separation of neutron counts from gamma counts. A pulse shape discrimination system provides the pure neutron spectrum. The electronic system used for the neutron spectrum measurement is shown in Figure 4 and has been described in detail elsewhere.[8]

Several neutron spectrum measurements were made at low reactor power (10-100W). The detector was placed 30 cm from the port window, opposite the window

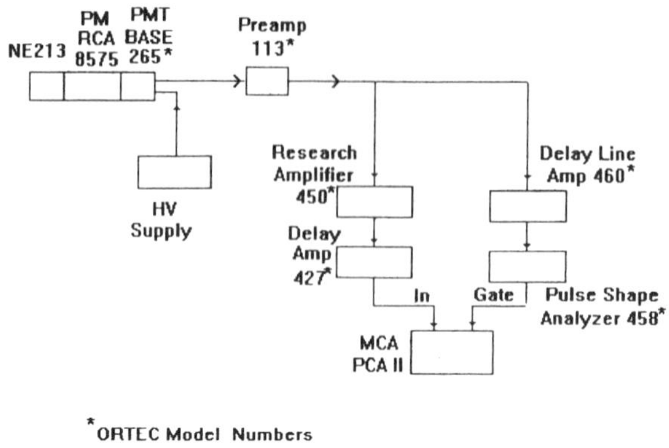

Figure 4. Fast neutron spectrometer electronic system

center. A typical measured neutron spectrum in the energy range 0.8-12 MeV is shown in Figure 5. Some undulations are seen in the spectrum. The intensity falls off at about 2 MeV and at about 11 MeV. In the range 4-6 MeV some peaks are seen. Smaller undulations are seen in the range above it. All these can be explained by neutron absorption in the structural materials. The basic shape is that of the upper part of the fission spectrum.

These measurements provide a relative flux. It is compared to the INEL measured and calculated flux. The present measurement is normalized to the INEL measurement at 8 MeV. The two measured spectra fit relatively well, although the present one has a more detailed structure. The calculated INEL spectrum is slightly lower.

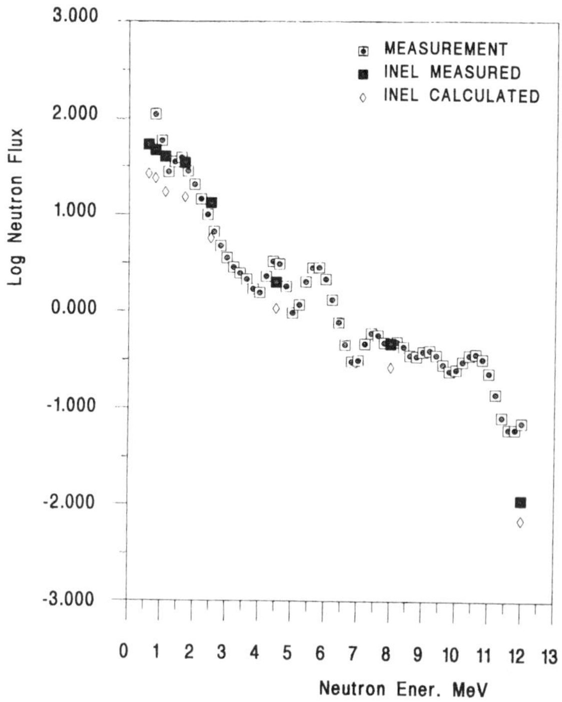

Figure 5. Fast neutron spectrum at the BMRR epithermal port

REFERENCES

1. R.M. Brugger et al., *in:* "Neutron Beam Design, Development and Performance for Neutron Capture Therapy," O.K. Harling et al., ed., Plenum, New York (1990).
2. H.B. Liu, private communication.
3. G.K. Becker et al., *in:* "Neutron Beam Design," Harling et al., ed.
4. R.G. Fairchild et al., ibid.
5. ICRP 60, "1990 Recommendation of the International Commission on Radiation Protection," Pergamon, New York (1991).
6. B.H. Laster et al., A comparison of the dose · RBE and the biological dosimetry approaches for treatment planning in BNCT, *in:* these proceedings.
7. G. Shani, Development of He^3 solid state detectors epithermal neutron spectrometer, *Nucl. Ins. Meth.* A314:553 (1992).
8. A. Tsechanski and G. Shani, System preparation and fast neutron spectra measurement in a graphite stack, *Nucl. Tech.* 62:227 (1983).
9. R.H. Johnson, "FORIST, Neutron Spectrum Unfolding Code," PSR-92, Oak Ridge National Laboratory (1978).

ACCELERATOR-BASED EPITHERMAL NEUTRON BEAMS FOR NEUTRON CAPTURE THERAPY

J.C. Yanch[1], R.E. Shefer[2], B.J. Hughey[2] and R.E. Klinkowstein[2]

[1]Department of Nuclear Engineering, MIT, Cambridge, MA
[2]Science Research Laboratory, Inc., Somerville, MA

INTRODUCTION

Collaborative work on the development of an accelerator-based epithermal neutron beam is underway between the Massachusetts Institute of Technology, Cambridge, MA and Science Research Laboratory, Inc., Somerville, MA. Design of the major hardware components of the accelerator, a 4 mA tandem electrostatic device, commenced one year ago in July 1991. Completion of this machine is expected to require approximately one more year. A low-current prototype accelerator (maximum current 1 mA) has been constructed by SRL and is being used to carry out experimental testing of a moderator/reflector epithermal beam-line assembly. This paper provides an overview of the progress made to date (September 1992) toward the goal of a clinically-useful accelerator-based epithermal neutron beam.

THE TANDEM CASCADE ACCELERATOR

Construction of a 4 mA, 2.5 MeV Tandem Cascade Accelerator (TCA) for the production of neutrons via the $^{7}Li(p,n)^{7}Be$ nuclear reaction is currently nearing completion at SRL. The TCA is a tandem electrostatic accelerator which utilizes a high current negative ion source in conjunction with a high current solid state cascade multiplier power supply to provide a compact, low cost, proton accelerator capable of supplying multimilliampere currents at several million electron volts. The accelerator design and operating parameters have been described elsewhere.[1] The inherent simplicity and flexibility of this accelerator provide several features which are desirable for both laboratory and clinical BNCT neutron sources. First, the proton current and beam energy can be varied continuously over a large range. In addition, the beam aperture of the accelerating tube is large so that critical focusing and alignment of the beam (which is usually energy dependent) is not required. Second, both the ion beam source and target are at ground potential during operation. This configuration allows different moderator geometries to be incorporated easily into the target design. Finally, a high degree of compactness is achieved by a patented SRL design which allows the power supply to be mounted directly onto the accelerator column thereby eliminating the need for external power supply chassis. The physical characterstics of the 2.5 MeV TCA are compatible with installation in a research laboratory or clinical facility. The accelerator weighs less than 2,000 lbs., has an overall length of approximately 2.6 m, and requires approximately 25 kW of electrical power. A 3.7 MeV, 1 mA TCA is currently operating at Science Research Laboratory and is pictured in Figure 1.

Advances in Neutron Capture Therapy, Edited by
A.H Soloway *et al.*, Plenum Press, New York, 1993

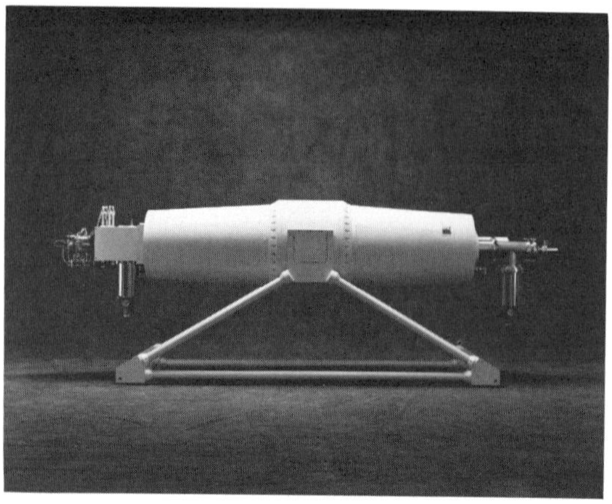

Figure 1. The fully operational 3.7 MeV tandem cascade accelerator at Science Research Laboratory (maximum proton current: 1 mA).

LITHIUM TARGET YIELD MEASUREMENTS

The 3.7 MeV TCA has been operated at a reduced beam energy with a water cooled pure Li metal target to verify the calculated neutron yield and angular distribution of the $^7Li(p,n)^7Be$ reaction. The lithium target used in these experiments was produced by pressing a short length of lithium wire onto a thin aluminum substrate in an argon atmosphere. The flat face of a small cylindrical piston was used to apply uniform pressure over a 2.5 cm diameter region of the substrate surface thereby producing a uniform lithium layer of the same diameter. Good adhesion of the lithium to the substrate surface was achieved by cleaning the surfaces of the wire and substrate with hexane prior to application. The lengh of Li wire was chosen so as to produce a layer approximately 90 μm in thickness. This thickness is sufficient to reduce the energy of a 2.5 MeV proton to the $^7Li(p,n)^7Be$ reaction threshold (1.88 MeV). The use of a thin lithium layer is advantageous in that it facilitates efficient heat transfer to the substrate material. While this is not a critical consideration for the low current experiments reported here, it is extremely important in the design of a target capable of withstanding bombardment by a multimilliampere proton beam.

Neutron flux measurements were performed with a calibrated neutron dose meter (NCR Industries' AN/PDR-70 SNOOPY NP-2 Neutron Survey Meter). The absolute response of the dose meter to the theoretical $^7Li(p,n)$ neutron angular distribution and energy spectra was calculated using the manufacturer's energy calibration and including finite solid angle effects at a distance of 1 meter from the target. The expected dose meter response for a 10 μA, 2.5 MeV proton beam on target is shown as a solid curve in Figure 2.

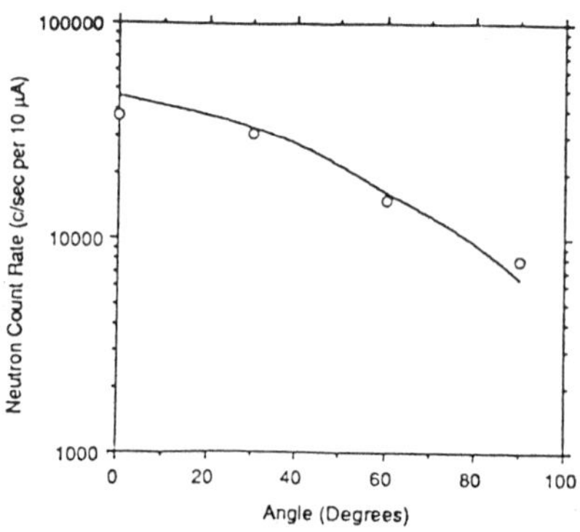

Figure 2. Comparison of calculated (solid curve) and measured (open circles) neutron count rate as a function of angle.

Experimental measurements of the neutron count rate at 0°, 30°, 60° and 90° relative to the incident proton direction are also plotted in Figure 2. These measurements were performed at a proton beam current of 0.2 μA and scaled to 10 μA. Overall, very good agreement is observed between the measured and predicted neutron yield and angular distribution; the measured yield at 0° is approximately 18% lower than the predicted yield with a corresponding higher than predicted yield at 90°. This "flattening" of the angular distribution at large angles can be attributed to neutrons scattering in the target housing.

MODERATOR DESIGN

The Monte Carlo code MCNP (Monte Carlo for Neutron and Photon transport)[2] was used to design a configuration of materials that would be effective in reducing the energy of $^{7}Li(p,n)$ neutrons down to therapeutically useful levels. The design procedure and results have been described elsewhere.[3] Briefly, the final design consisted of a cylindrical D_2O moderator, surrounded by a 0.02 cm layer of ^{6}Li and an 18 cm thick lead reflector. The dimensions of the moderator can be altered somewhat to produce a harder or softer beam. A combination of D_2O-^{6}Li may be placed around the reflector to moderate and absorb neutrons leaving the assembly. The dosimetric effect of the neutron beam produced by this assembly (consisting of a 24 cm diameter, 19 - 27 cm length D_2O moderator with a 0.02 cm ^{6}Li absorber at the irradiation port) was assessed within a 2 cm diameter cylinder along the central axis of a 16.0 cm diameter x 16.0 cm length cylindrical head phantom composed of brain-equivalent material.[3] Figure 3 shows dose versus depth along the central axis of the phantom from each of the individual dose components as well as the total dose to tumour and healthy tissue for a 19 cm D_2O moderator length and 30 μg/g ^{10}B in the tumour. Dose rate was determined by multiplying the dose/neutron by 2.15×10^{14}, the number of neutrons generated per minute from a ^{7}Li target by a 2.5 MeV, 4 mA proton beam. RBE values of 4.1, 4.0, and 1.0 have been applied to the boron dose, neutron dose and photon dose components, respectively.[4] Dose rate at a depth of 7.0 cm in the phantom is 35.8 RBE cGy/minute; the maximum advantage depth (i.e. assuming no ^{10}B in healthy tissue) is 8.0 cm.

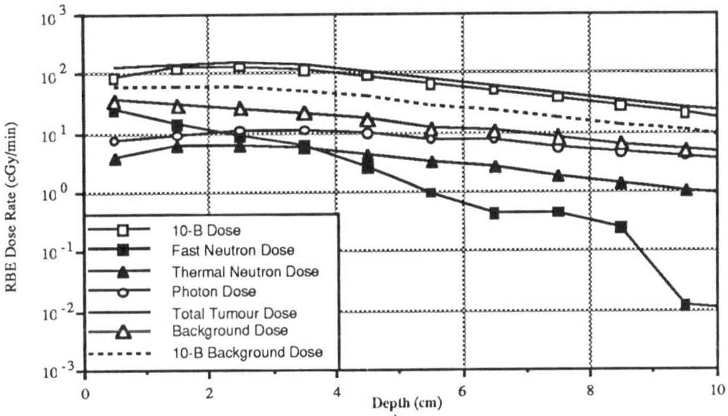

Figure 3. Dose versus depth in phantom assuming ^{10}B concentrations of 30 ppm in the tumour.

The moderator/reflector configuration designed by the MCNP study has been constructed and installed on the existing TCA at SRL for experimental testing of the MCNP predictions. A photograph of the moderator/reflector assembly on the lithium target beamline of the TCA is shown in Figure 4. The moderator consists of a D_2O-filled cavity within the cylindrical lead reflector which is sealed by a thin lucite window. During actual operation, the lucite window will be covered by a thin layer of ^{6}Li-carbonate to eliminate thermal neutrons from the beam. Due to the expense and difficulty of obtaining ^{6}Li, a commercially available borated silicon polymer absorber with similar thermal neutron attenuation characteristics was used in the experiments reported here. The moderator cavity can accommodate a maximum axial thickness of 27 cm of D_2O. This thickness can be reduced in 1 cm steps to 17 cm by inserting lead rings into the cavity. When additional rings are inserted, the moderator is repositioned so that the lithium target sits just within the D_2O volume.

Figure 4. The moderator and reflector installed on the low current TCA at Science Research Laboratory.

The total neutron dose equivalence rate was measured at the moderator irradiation port for D_2O thicknesses of 19 and 27 cm with the calibrated neutron dose meter described previously. Measurements were performed with a 2.5 MeV proton beam at very low current (0.04 μA) in order to produce neutron dose equivalence rates within the measurement range of the instrument and then scaled to the design current of 4 mA. In order to provide a direct comparison with the MCNP calculations, the measured dose-equivalence rates were converted to cGy/min using the calibration curve supplied by the manufacturer and NCRP quality factors.[5] Table 1 shows both the predicted and measured neutron dose rates at the irradiation port of the moderator/reflector for 19 cm and 27 cm D_2O thicknesses. In both cases, the measured dose rates are in very good agreement with the MCNP predictions. More detailed experiments to measure the dose components as a function of depth in a brain-equivalent phantom are planned in order to verify the predicted dose rates to the entire phantom volume.

Table 1. Comparison of measured neutron dose rate to predicted dose rate from MCNP simulations. Measured dose rates have been scaled to a proton current of 4 mA.

D_2O Thickness (cm)	Measured Dose Rate (cGy/min)	Predicted Dose Rate (cGy/min)
19	20.3 ± 2.5	18.3
27	5.6 ± 0.56	6.4

REFERENCES

1. R.E. Shefer, R.E. Klinkowstein, J.C. Yanch and G.L. Brownell, "An epithermal neutron source for BNCT using a tandem cascade accelerator." In *Progress in Neutron Capture Therapy for Cancer* (B.J. Allen, D.E. Moore and B.A. Harrington, Eds.). New York: Plenum Press, pp. 119 - 122, 1992.
2. J.F. Briesmeister, *MCNP—A General Monte Carlo Code for Neutron and Photon Transport,* LA7396-M Rev. 2, Los Alamos National Laboratory (1986).
3. J.C. Yanch, X.L. Zhou, R.E. Shefer and R.E. Klinkowstein, "Accelerator-based epithermal neutron beam design for neutron capture therapy," *Medical Physics*, **19**(3): 709-721, 1992.
4. O.K. Harling, R. Rogus, J.R. Choi, D.J. Moulin, R.G. Zamenhof and J.C. Yanch, "Dosimetry and dose control for clinical trials of neutron capture therapy at the MITR-II reactor," to be published in Proceedings of International Workshop on Macro and Microdosimetry and Treatment Planning for Neutron Capture Therapy, MIT, Cambridge, MA, November 1991.
5. NCRP Report No. 38, p. 16.

EFFECT OF VARIATIONS IN THE RBE ON THE DESIGN OF AN AENIF FOR BNCT

T.E. Blue,[1] T-X B. Qu,[1] N. Gupta,[2] and R.A. Gahbauer[2]

[1]Nuclear Engineering Program
[2]Radiation Oncology Department
The Ohio State University
Columbus, OH 43210

1. INTRODUCTION

The purpose of this paper is to examine the impact of various expressions for the Relative Biological Effectiveness (RBE) of low energy neutrons, as a function of neutron energy, on the choice of moderator and reflector materials for an Accelerator Epithermal Neutron Irradiation Facility (AENIF) for BNCT. Specifically, we will present the RUFTED, where we calculate $<RBE_n>$ in the denominator of the RUFTED consistently with the neutron energy spectra for the moderator assembly of the specified moderator thickness and moderator-reflector material composition. We will place special emphasis on the comparison between the D_2O-MgO and the BeO-MgO moderator-reflector combinations, since the less neutronically suitable D_2O-MgO combination is advantageous, because D_2O is more available than BeO, and because D_2O is toxic only in so far as it is slightly contaminated with tritium.

2. BACKGROUND

Previously, we have optimized a moderator assembly for an AENIF for BNCT on the basis of two figures of merit (FOMs)[1]. One of these FOMs is the useful neutron fluence (Φ_u), which is defined as the integral, over neutron energies greater than 1 eV, of the energy dependent neutron fluence per source neutron, where by source neutron we mean a neutron created by the ^{7}Li(p,n)^{7}Be reaction in the Li-7 target of our AENIF. The second of these FOMs is the RUFTED, which is the ratio of Φ_u to the dose equivalent per source neutron for a small volume of tissue at the irradiation point in air (a point on the center-line of the cylindrically symmetric moderator assembly 2 cm from the moderator assembly exit port). It is defined by the expression

$$\text{RUFTED} = \left\{ \frac{K_g}{\Phi_u} + \frac{\Phi}{\Phi_u} < RBE_n >< k_n > \right\}^{-1} \quad , \qquad (1)$$

where in the definition of the RUFTED, $<RBE_n>$ is the average neutron RBE, $<k_n>$ is the average neutron kerma factor and K_g is the gamma ray absorbed dose, for a differential volume of tissue at the irradiation point in air.

In following sections of this paper, we calculate the RUFTED using Eqn. 1. We calculate <k_n> and <RBE_n> in the denominator of the RUFTED consistently with the neutron energy spectra for D_2O-MgO and BeO-MgO moderator assemblies. For both the BeO-MgO and D_2O-MgO moderator assemblies, the geometry is given by Fig. 1. The moderator is assumed to be 20 cm thick for the BeO-MgO moderator-reflector material combination, and 25 cm thick for the D_2O-MgO moderator assembly.

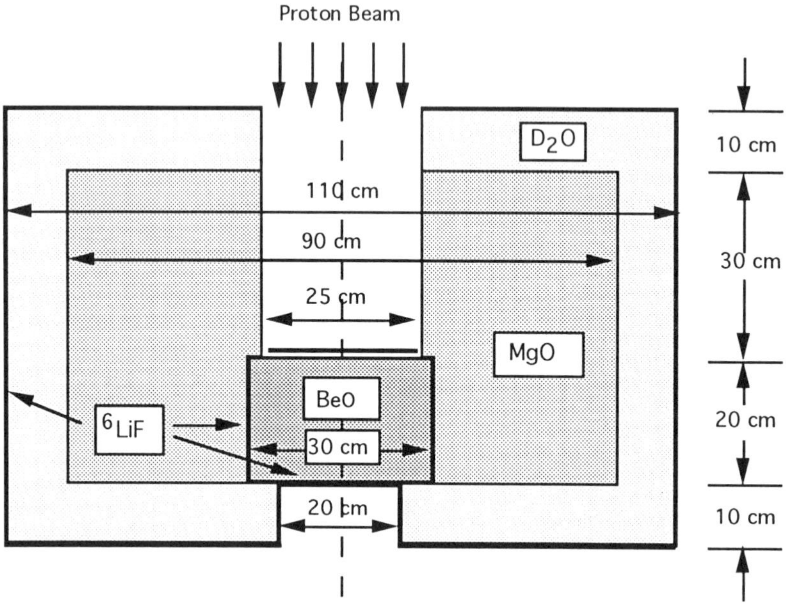

Figure 1. The BeO-MgO moderator assembly

The term in the RUFTED representing the biological effects of neutrons is <RBE_n> <k_n>Φ where

$$< RBE_n >< k_n > \Phi = \int_0^\infty RBE(E_n)k_n(E_n)\Phi(E_n)dE_n . \qquad (2)$$

The meaning of <RBE_n> in Eqn. 2 is calculated from Eqn. 2 to be

$$< RBE_n > \equiv \frac{\int_0^\infty RBE(E_n)k_n(E_n)\Phi(E_n)dE_n}{\int_0^\infty k_n(E_n)\Phi(E_n)dE_n} \qquad (3)$$

Similarly, the meaning of <k_n> in Eqn. 2 is calculated from Eqns. 2 and 3 to be

$$< k_n > = \frac{\int_0^\infty k_n(E_n)\Phi(E_n)dE_n}{\int_0^\infty \Phi(E_n)dE_n} \qquad (4)$$

and the meaning of the total neutron fluence (Φ) in Eqn. 2 is calculated from Eqns. 2, 3 and 4 to be

$$\Phi = \int_0^\infty \Phi(E_n)dE_n \qquad (5)$$

where $\Phi(E_n)$ dE_n is the fluence of neutrons with energies in dE_n about E_n.

3. ANALYSIS

Fairchild suggested [2] assuming that the RBE of recoil protons is a constant (RBE_p) from their birth energies to a cutoff energy (E_c), below which the protons do not have any biological effect. Assuming further

$$p(E_n \rightarrow E_p) = 1 / E_n \qquad (6)$$

where $p(E_n\text{->}E_p)dE_p$ is the probability that a neutron of energy E_n will, in a single collision, create a recoil proton with energy in dE_p about E_p,

$$\begin{aligned} RBE(E_n) &= RBE_p(E_n - E_c)^2 / E_n^2 && E_c < E_n \qquad (7)\\ &= 0 && 35eV < E_n < E_c \\ &= RBE_p\,(605/615) && E_n < 35eV \end{aligned}$$

Figure 2 compares the neutron fluence spectra ($\Phi(u) = E\,\Phi(E)$, in units of neutrons per cm^2 per second per unit lethargy) for AENIF moderator assemblies with D_2O-MgO and BeO-MgO moderator -reflector material combinations. The D_2O and BeO moderator thicknesses are 25 and 20 cm, respectively. In Fig. 2, the neutron kerma factor (in units of cGy-cm^2) is also plotted versus the neutron energy; since, according to Eqn. 4, $<k_n>$ should be calculated by weighting $k_n(E)$ by $\Phi(E_n)$.

Figure 3 compares the neutron kerma spectra ($K_n(u)$, in units of cGy per source neutron per unit lethargy) for AENIFs with D_2O-MgO and BeO-MgO moderator assemblies with moderator thicknesses of 25 and 20 cm, respectively. The areas beneath the curves for different intervals of neutron energy, are proportional to the neutron kerma rate integrated over the energy intervals. In Fig. 3, the neutron RBE as given by Eqn. 7 is also plotted versus the neutron energy; since, according to Eqn. (3), the average neutron RBE should be calculated by weighting $RBE(E_n)$ by $K_n(E_n)$. In plotting $RBE(E_n)$, we have assumed that in Eqn. 7 RBE_p = 3.5 and E_c = 10 keV, as indicated by Alpen[3]. Recent experiments by Gavin et al [4] with an endpoint of late necrosis of the skin of dogs indicate the neutron RBE for the Brookhaven Medical Research Reactor (BMRR) epithermal neutron field is 4.5. Averaging $RBE(E_n)$ over the BMRR neutron kerma

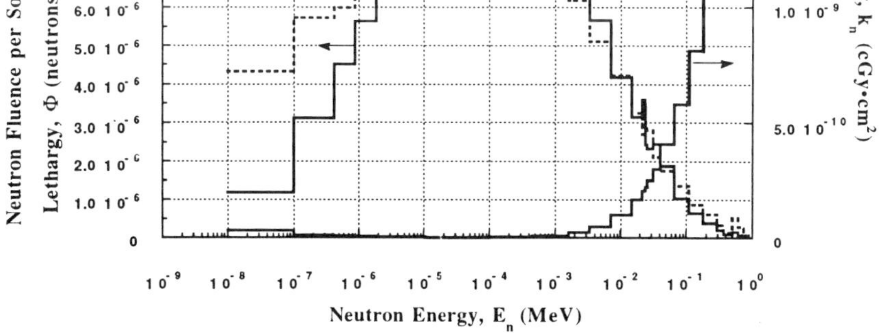

Figure 2. Neutron Fluence, $\Phi(u)$, and Neutron Kerma Factor, k_n, vs. Neutron Energy, E_n

spectra, in the manner shown in Eqn. 3, we conclude that RBE_p in Eqn. 7 must equal 6.6 , in order for $\langle RBE_n \rangle$, as calculated by Eqn. 3 for the BMRR, to equal the RBE of 4.5 measured by Gavin. We average $RBE(E_n)$, as given by Eqn. 7 with $RBE_p = 6.6$, over the neutron kerma spectra for the D_2O-MgO and the BeO-MgO moderator assemblies with moderator thicknesses of 25 and 20 cm, respectively.

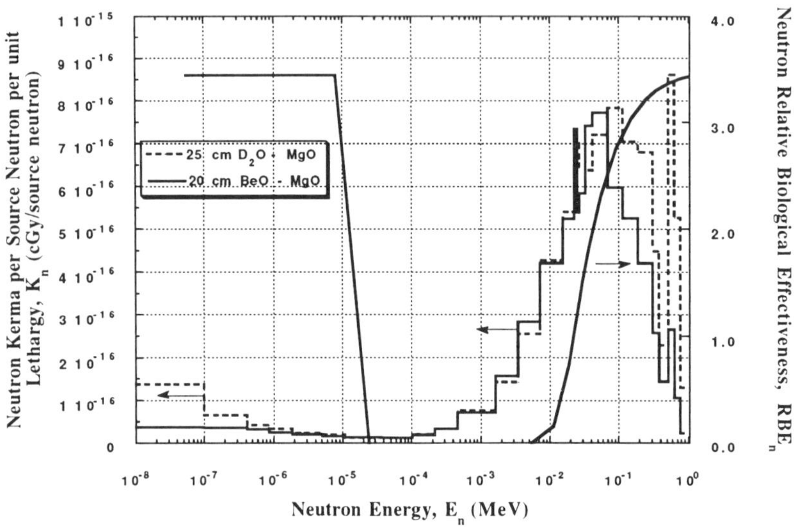

Figure 3. Neutron Kerma, $K_n(u)$, and Neutron Relative Biological Effectiveness, RBE_n vs. Neutron Energy, E_n

4. RESULTS

For an AENIF with a D_2O-MgO moderator assembly with a 25 cm thick D_2O moderator, $\langle k_n \rangle = 3.5 \pm 0.1 \times 10^{-11}$cGy-cm^2 . For an AENIF with a BeO-MgO moderator assembly with a 20 cm thick BeO moderator, $\langle k_n \rangle = 3.2 \pm 0.1 \times 10^{-11}$cGy-cm^2. Calculating $\langle RBE_n \rangle$ in the denominator of the RUFTED consistently with the neutron energy spectra for the moderator assemblies, assuming Fairchild's model for $RBE(E_n)$, with $RBE_p = 6.6$; $\langle RBE_n \rangle = 3.9 \pm 0.2$ for the D_2O-MgO moderator assembly and $\langle RBE_n \rangle = 3.4 \pm 0.2$ for the BeO-MgO moderator assembly. For an AENIF with the D_2O-MgO moderator assembly, RUFTED = $4.3 \pm 0.2 \times 10^9$ neutrons/cm^2-RBE cGy, assuming $\langle RBE_n \rangle = 3.9 \pm 0.2$. For an AENIF with the BeO-MgO moderator assembly, RUFTED = $5.5 \pm 0.3 \times 10^9$ neutrons/cm^2-RBE cGy, assuming $\langle RBE_n \rangle = 3.4 \pm 0.2$. In conclusion, the RUFTED is 30% larger for the BeO-MgO moderator assembly as compared to the D_2O-MgO moderator assembly.

ACKNOWLEDGMENTS

This work was supported by the United States Department of Energy under contract DE-AC02-76CH00016. Also, the authors acknowledge the careful typing and editing of J. Eric Denison.

REFERENCES

1. C.K. Wang, T.E. Blue and R.A. Gahbauer, "A Neutronic Study of an Accelerator-Based Neutron Irradiation Facility for BNCT," *Nucl. Tech. 84*, 93 (1989).
2. R.G. Fairchild and V.P. Bond, "Current Status of ^{10}B NCT: Enhancement of Tumor Dose via Beam Filtration and Dose Rate and the Effects of These Parameters on Minimum Boron Content," *International Journal of Radiation Oncology, Biology and Physics* 11:831 (1985).
3. E.L. Alpen, "Biological Effectiveness of Recoil Protons from Neutrons of 10 keV to 100 keV Energy," *Proceedings of 5th International Symposium on NCT*, Columbus, OH (1992, in press).
4. P.R. Gavin, *et al*, "Large Animal Model-The Biologic Effects Mixed Irradiation Fields Utilizing BSH and Epithermal Neutron Irradiation," *Proceedings of International Workshop on Macro- and Microdosimetry and Treatment Planning for NCT*, MIT, 31 Oct.1 - Nov. 1991 (to be published).

LOW ENERGY ACCELERATOR-BASED NEUTRON SOURCES FOR NEUTRON CAPTURE THERAPY

Terry H. Wu,[1], Robert M. Brugger, and Jay F. Kunze

Nuclear Engineering Program
University of Missouri
Columbia, MO 65211
1) Currently at St. Louis University Hospital
Rad.Therapy Dept., Vista @ Grand Blvd.
St. Louis, MO 63110

INTRODUCTION

Neutron beams for neutron capture therapy (NCT) have been studied by a number of investigators,[1,2,3] using high flux research reactors as the source of neutrons. There are very few research reactors in the world, probably less than ten, that can meet the conditions of both flux intensity and neutron energy spectrum. Hence the availability of NCT for patients with malignant brain tumors would be quite limited, even if all of these reactors would be able to accommodate the necessary beam design and hospital room atmosphere into the operational program.

For a number of years it has been suggested that serious design work should be undertaken to develop accelerator-based sources for the neutrons, using compact accelerator designs. The relatively new RFQ (radiofrequency-quadrupole) linear accelerators have the promise of achieving this objective.

This study investigates the optimization of the moderator and reflector system that would be used between the accelerator and the patient. The study is similar to that of Blue et al at Ohio State University[4], and Dolan[5] at INEL, but goes further in attempting to optimize the useful neutron flux for boron NCT with respect to the unwanted dose from neutrons and gamma rays. This study also investigates both the Li-7(p,n) and H-3(p,n) proton accelerator target reactions. Though these two targets are studied separately, and each would involve an entirely different target design, a combination target of tritiated lithium hydride deserves to receive attention in future designs.

The yields from the targets were calculated from the differential cross sections and stopping powers in the literature, assuming thick targets for the incident protons. The yields were divided into five angular groups, which were used as input to the MCNP Monte Carlo computer code. Neutron fluxes, neutron doses, and gamma doses were calculated at the patient position (skull) and at a tumor at various depths.

Advances in Neutron Capture Therapy, Edited by
A.H Soloway *et al.*, Plenum Press, New York, 1993

BASIC SYSTEM DESIGN AND COMPUTATIONAL DESIGN TOOLS

The basic system design that was eventually adopted, after a number of design iterations, is shown in Figure 1. The moderator adopted was 20 cm of BeO, for the Li reaction, and 34 cm of Al_2O_3 for the ^3H reaction. The moderator and the 3 cm diameter accelerator vacuum (drift) tube were surrounded by the reflector of MgO for the Li-7 reaction, Al_2O_3 for the tritium reaction. These selections were made on the basis both of maximizing desired neutron flux and of minimizing other neutron and gamma doses (see following section). Total diameter of the outside of the reflector was 85 cm, and it extended 33 cm back along the accelerator drift tube.

The goal of this design was to produce neutrons, at the patient, predominantly in the energy range of 1 eV to 10 keV. Our studies, and those of other investigators, had shown that this range would optimize the treatment for tumors at a 3 to 7 cm depth from the surface of the skull. The reference tumor used for the calculations was 6 cm. It was assumed that the tumor contained 30 μg of boron-10 per gram of tissue, and that there was a 10:1 ratio between tumor and healthy tissue, i.e. the healthy tissue had 3 μg/g concentration of B-10.

The neutron source reactions studied were from the two reactions shown in Table 1. The proton energy on target was limited to a maximum value that would not produce neutrons above nominally 1 MeV, since higher energy neutrons are much more difficult to moderate down to the desired intermediate energy (epithermal) neutron energy range desired at the patient position. Since the threshold proton energy for the two reactions is not the same, the needed maximum proton energy differs between the two reactions, being only 1.77 MeV for the tritium target, but 2.5 MeV for the Li-7 target. Furthermore, the neutron yield for the hypothetical pure tritium target is nearly twice that of a pure Li-7 target. However, the target design to obtain the full utilization of the proton beam is not as obvious for the tritium target as it is for the lithium target.

Table 1. - The neutron-producing proton target reactions employed in the design, and the limiting conditions.

	$H^3(p,n)He^3$	$Li^7(p,n)Be^7$
Q-Value (MeV)	-0.764	-1.645
Threshold Energy(MeV) in lab	1.019	1.881
Neutron Energy(MeV) @ threshold	0.064	0.029
Incident Proton Energy (MeV)	1.77	2.50
Maximum Neutron Energy (MeV)	1.00	0.787
Total neutron yield, n/p	2.54×10^{-4}	1.48×10^{-4}

DOSE RESULTS

All dose calculations were done with reference to a thin tissue-equivalent cell at the patient irradiation position, used to determine the neutron and gamma fluxes that would be present at the surface of the skull. The intent was to optimize the desired intermediate neutron flux, but also to minimize the doses to the patient from the other neutrons and gamma rays.

The selection was done on a somewhat arbitrary basis, using plots similar to that of Figure 2, which shows the selection of reflector material for the lithium target reaction. The dotted lines indicate the minimum acceptable intermediate neutron flux, and the minimum ratio of that flux to total neutron dose to healthy tissue.

The gamma dose was an item of major concern, and this was minimized by adding a small amount of Li-6 neutron absorber to both moderator and reflector. The Li-6 absorbed the thermal neutrons, reducing their direct dose, and also reduced the radiative capture in the moderator and reflector materials, thus reducing the gamma dose. Figure 3 is a plot which resulted in the selection of 0,05 atom percent as the "optimum" amount of Li-6 poison to recommend.

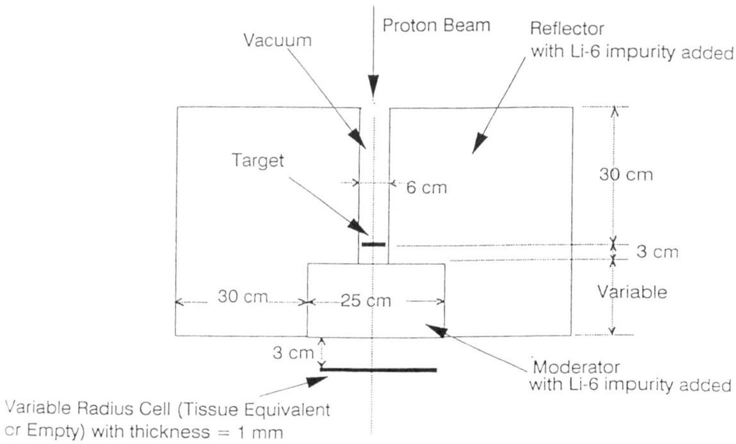

Figure 1. Moderator-Reflector Assembly Design. Moderator is BeO. Reflector is MgO for Li-7 target and Al_2O_3 for the tritium target.

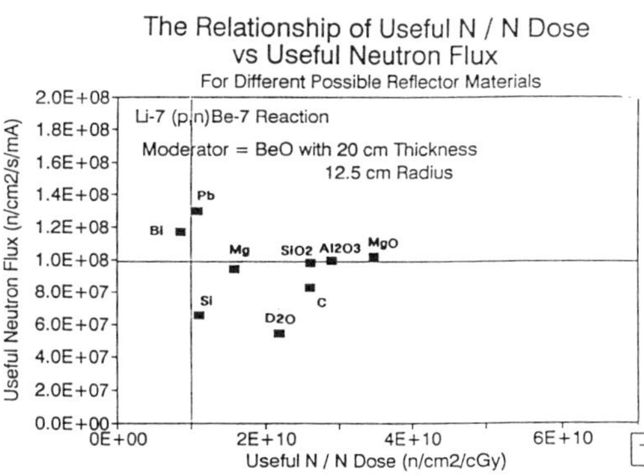

Figure 2. Selection of reflector material for Li-6 target.

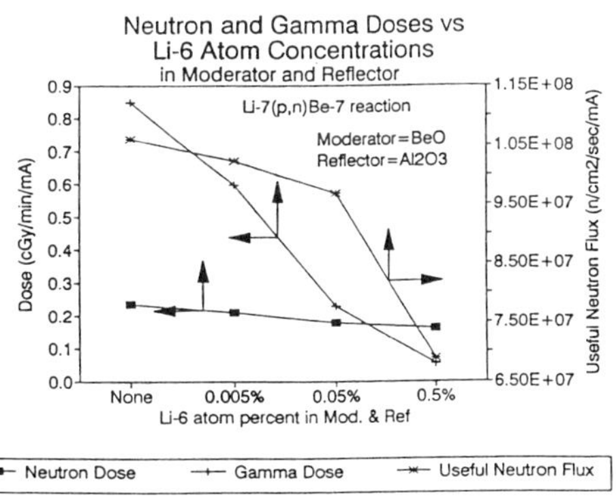

Figure 3. Selecting the amount of Li-6 poison to use.

OVERALL RESULTS and DISCUSSION

Table 2 lists the overall results from the design, for the desired (useful) intermediate energy neutron flux, and the undesirable doses to healthy tissue. Results are quoted with respect to one milliampere of proton beam on the target. Present RFQ machine technology has achieved proton beam currents of a few milliamps, with up to 20 milliamps believed to be achievable with current state-of-the art operation.[6]

These beam currents would give the desired treatment, 2000 cGy (≈5000 cSv or Rem) to a tumor having 30 μg of B-10/g of tissue, in approximately 45 minutes for the H-3 target (55 minutes for Li-6 target). This is a factor of 4 to 5 times longer than for beams designed for some of the higher powered research reactors, such as that at the University of Missouri. However, the 45 minute treatment period appears to be within the tolerable range for a patient treatment time.

TABLE 2 - Calculated Results, per mA of Proton Beam

	H-3 target	Li-7 target
AT PATIENT IRRADIATION POSITION IN AIR		
Total Neutron flux ø ($n/cm^2 sec$)	1.29E+8	1.31E+8
Epithermal Neutron flux ø(1eV-10keV)	1.10E+8	0.94E+8
Neutron dose in cGy/epithermal n/cm^2	4.1E-11	4.3E-11
Gamma dose in cGy/epithermal n/cm^2	2.8E-11	2.7E-11
Thermal neutron ø ($n/cm^2 sec$)	1.14E+8	1.00E+8
Total dose to tumor in cGy/minute	2.23	1.78
AT 6 cm DEPTH TUMOR WITH 30 μg B-10/g of TISSUE		
Fraction of dose from B-10(n,α)	65%	67%
Fraction of dose from Fast neutrons	4%	5%
Fraction of dose from N-14(n,p)	1%	1%
Fraction of dose from gammas	30%	27%
"Advantage Depth"*	9 cm	8 cm

*Depth at which B-10-caused dose is equal to the neutron and gamma doses to the tumor.

REFERENCES

1. F.J.Wheeler, D.K.Parson, B.L.Rushton, and D.W.Nigg, Epithermal Neutron Beam Design for Neutron Capture Therapy at the Power Burst Facility and the Brookhaven Medical Research Reactor Epithermal Neutron Source", Nuclear Technology, 92, 106 (1990)
2. R.G. Fairchild, J.Kalef-Ezra, S.K.Saraf, S. Fiarman, E.Ramsey, L.Wielopolski, B.H.Laster, F.J.Wheeler, "Installation & Testing of an Optimized Epithermal Neutron Beam at the Brookhaven Medical Research Reactor," in Neutron Beam Design, Dev., and Perf.for Neutron Capture Therapy, O.K.Harling, ed., Plenum Press, N.Y.(1990)
3. R.M. Brugger & W.H. Herleth, "Intermediate Energy Neutron Beams from the MURR,", op.cit, page 153
4. C.K.C.Wang, T.E.Blue, R.Gahbauer, "A Neutronic Study of an Accelerator-based Neutron Irradiation Facility for Boron Neutron Capture Therapy," Nuclear Technology, 84, 93 (1989)
5. T.J.Dolan et al, Idaho National Engineering Lab, EG&G, Inc., internal report, private communication, (1989)
6. T.P.Wangler, J.E.Stovall, T.S.Bhatia, C.K.Wang, T.E.Blue, R.A.Gahbauer, "Conceptual Design of an RFQ Accelerator-based Source for NCT", Proceedings of the 1989 IEEE PArticle Accelerator Conf., Accelerator Science and Technology, Chicago, IL, CH2669-0, 678 (1989)

AN EPITHERMAL NEUTRON BEAM FROM THE MURR AND FROM AN ACCELERATOR SOURCE COMPARED TO THE BEAM AT THE BMRR

Hungyuan B. Liu and Robert M. Brugger

Medical Department, Brookhaven National Laboratory
Upton, NY 11973

INTRODUCTION

An ideal neutron beam for BNCT is a beam of epithermal neutrons, forward directed, and free of gamma rays and thermal and fast neutrons. There are now two reactor-based epithermal beams[1,2] in service in the USA and one in The Netherlands[3]. Several designs for reactor- and accelerator- based epithermal beams have been proposed. Of these neutron beams, three neutron beams have been selected, evaluated, and compared: (1) the operating Brookhaven Medical Research Reactor (BMRR) epithermal beam[1], (2) the designed Missouri University Research Reactor (MURR) epithermal beam[4], and (3) the accelerator-based epithermal neutron beam designed by Wu[5]. The horizontal sections of these three neutron beams are depicted in Figure 1 to Figure 3. Irradiation points for future reference are labelled as "Y" in each figure. These neutron beams are compared with respect to the neutron spectra, neutron and gamma fluxes and doses, and beam directionality.

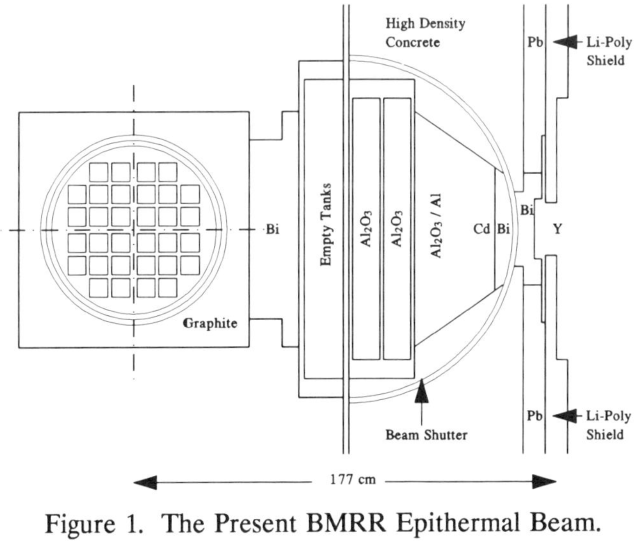

Figure 1. The Present BMRR Epithermal Beam.

Advances in Neutron Capture Therapy, Edited by
A.H Soloway *et al.*, Plenum Press, New York, 1993

Table 1. Comparisons of Beam Parameters at the Irradiation Point "Y".

	Power MW	Φ_{epi} x 10^9 n/cm^2/s	D_n/n_{epi} x 10^{-11}cGy /(n/cm^2)	D_γ/n_{epi} x10^{-11}cGy /(n/cm^2)	J/Φ
BMRR*†	3	1.2	4.8	1.4	0.67
MURR*	10	9.5	2.9	0.4	0.82
Accelerator*	10 mA	0.9	4.3	2.7	0.66

* Calculated by MCNP † Measured

NEUTRON SPECTRA

In BNCT, for good beam penetration and low skin neutron dose, the beam will have neutrons with energies between 0.4 eV and 10 keV. But, there are always thermal and fast neutrons associated the epithermal neutrons. The spectra at the irradiation port of these three neutron beams are shown in Figure 4. The spectra of

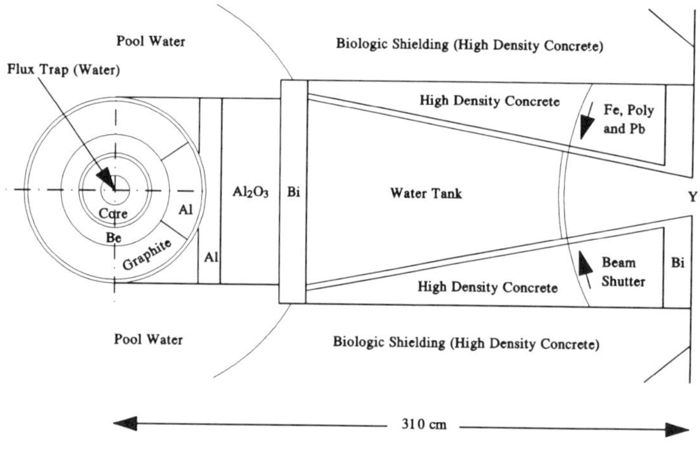

Figure 2. The MURR Epithermal Beam.

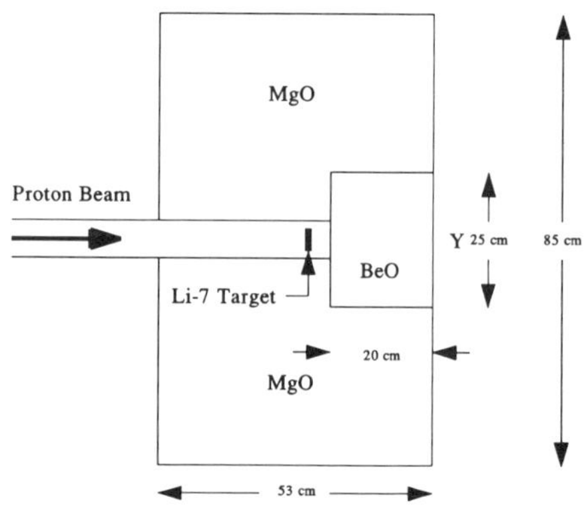

Figure 3. The Accelerator-Based Epithermal Beam.

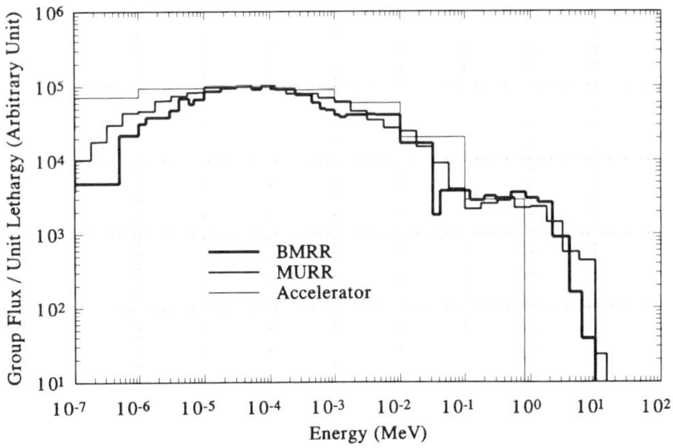

Figure 4. Comparisons of the Neutron Spectra Normalized at the Peaks.

the BMRR and MURR are similar in shape. The spectrum of the accelerator-based neutron beam has fewer high energy fast neutrons because the maximum bombarding proton energy is limited to 2.5 MeV to minimize the gamma rays produced by the higher energy protons bombarding the ^7Li target[5]. The neutrons coming from the $^7Li(p,n)^7Be$ reaction have a maximum energy of 0.79 MeV. However, because the fast neutrons between 10 keV and 100 keV are greater than the MURR and BMRR beams, the fast neutron dose coming from these fast neutrons compensates for the missing high energy neutrons. Besides, the spectrum of the accelerator-based neutron beam shows higher thermal and low energy epithermal neutrons in the beam which could be very important if the ^{10}B compound concentrates in the scalp as well as the tumor.

Comparisons of the beam parameters for these neutron beams are listed in Table 1. The BMRR has less Al in the moderator and more Al_2O_3. The Al and Al_2O_3 plates were assembled in the A and B regions of the BMRR to produce the epithermal beam. The design for the MURR, as seen in Figure 2, replaces two graphite wedges with Al wedges. Thus, the fission neutrons are directly moderated by the Al moderator. But for the BMRR beam, the fission neutrons from the core have to pass through 14 cm of graphite reflector and 19 cm of bismuth before they reach the moderator tanks, which greatly reduces the beam intensity. For 1 MW power, the BMRR has an epithermal flux 4.0×10^8 n/cm^2/s at "Y" (177.0 cm from the core), while in contrast the MURR beam has a more intense epithermal flux of 9.5×10^8 n/cm^2/s at "Y" (310.0 cm from the core).

FLUXES AND DOSES

The BMRR is operating at 3 MW power while the MURR is operating at 10 MW power. The accelerator-based neutron beam might be operated at 10 mA current. Under the condition of 10 ppm ^{10}B uniformly distributed through a head model, it takes about 54 minutes to deliver 2.5×10^{12} n/cm^2 thermal neutron fluence to the center of the head model[6], about 7 cm deep. It takes only 5 minutes for the MURR beam to reach this amount of thermal neutron fluence. For an accelerator-based neutron beam, it becomes 84 minutes to reach the goal. The irradiation time can be important but not critical, but a higher flux beam is preferred to reduce the irradiation time.

Another important comparison of the neutron beams is the fast neutron and gamma doses associated the beam. As seen in Table 1, the MURR beam has the lowest fast neutron and beam gamma doses per epithermal neutron. The fast neutron dose is the major dose delivered to the skin, especially due to high RBE of fast neutrons. The BMRR beam has the highest fast neutron dose per epithermal neutron of the three neutron beams, but still is a satisfactory beam for patient trials at this time. It appears that the dose to the normal brain reach the tolerance dose limit first using the BMRR beam for irradiation, so the skin dose becomes tolerable, or less than the dose limit[6]. In other words, the skin dose can be controlled under the tolerable limit.

Compared to the fast neutron dose, the beam gamma dose is insignificant. Taking the BMRR beam as an example, the ratio of the gamma dose per epithermal neutron to the fast neutron dose per epithermal neutron is 1 to 3.4. In addition, the ratio of the RBE is 1 to 3.8, so the RBE gamma dose per epithermal neutron is about 1/13 of the RBE fast neutron dose per epithermal neutron.

BEAM DIRECTIONALITY

The design for the MURR beam includes a long air indentation so the beam is directed. A neutron current to flux ratio of 0.82 was calculated for the MURR beam as compared to 0.67 calculated for the BMRR beam at "Y". To evaluate the effects of beam directionality when the head model was irradiated by the BMRR beam from the temple side, MCNP calculations were made by using different neutron current to flux (J/Φ) ratios with the same spectrum.

The forward beam ($J/\Phi=1$) delivers a peak thermal flux 30% higher than a beam with J/Φ equal to 0.6. In the center of the brain, the difference is about 20%. Also, the forward beam gives a deeper penetration. Another way of comparison is the "neutron flux gain", which is defined as the ratio of the peak thermal flux in the head to the incident epithermal flux in air. This neutron flux gain is directly related to the beam directionality. The forward beam is expected to have a gain of about 3. The MURR beam has a gain of 2.4 while the BMRR beam has a gain of 1.8 and the accelerator-based neutron beam has a gain of 1.7. In addition, the forward beam delivers 34% less fast neutron dose in the skin compared to a beam of J/Φ equal to 0.6. Because of high RBE associated fast neutrons, this reduction of fast neutron dose becomes another significant benefit of a forward beam.

CONCLUSIONS

Three epithermal neutron beams based on the BMRR, MURR, and an accelerator were inter-compared for different beam parameters in air at the irradiation point. The BMRR beam has the highest fast neutron plus gamma doses per epithermal neutron among these neutron beams but is satisfactory for patient trials by BNCT at the present time. The RBE dose delivered to the normal brain reaches the tolerance dose limit before the skin RBE dose reaches its limit, so the skin dose can be controlled under the limit. Generally speaking, a treatment can be completed in 54 minutes using the BMRR beam for irradiation at a full-power operation of the reactor.

The MURR beam has better beam parameters, including lower fast neutron and gamma doses per epithermal neutron, higher in intensity, and also directed. The irradiation time could be 5 minutes to complete a treatment. The accelerator-based neutron beam which has shown promising beam parameters similar to the BMRR beam could be a choice in hospitals. However, a complete system at the required power has not yet been demonstrated.

ACKNOWLEDGEMENTS

This research was supported by U.S. Department of Energy contract DE-AC02-76CH00016.

REFERENCES

1. R. G. Fairchild, J. Kalef-Ezra, S. K. Saraf, S. Fiarman, E. Ramsey, L. Wielopolski, B. Laster, F. Wheeler, "Installation and Testing of an Optimized Epithermal Neutron Beam at the Brookhaven Medical Research Reactor," Neutron Beam Design, Development, and Performance for Neutron Capture Therapy, O. Harling and R. Zamenhof, eds., Mar. 1989, Plenum Press, New York, 1990.
2. J. R. Choi, R. G. Zamenhof, J. C. Yanch, O. K. Harling, "Performance of the Currently Available Epithermal Neutron Beam at the Massachusetts Institute of Technology Reactor (MITR-II)," Progress in Neutron Capture Therapy, B. J. Allen, D. E. Moore, B. V. Harrington, eds., Plenum Press, New York, 1992.
3. G. Constantine, P. Watkins, F. Stacher-Rasmussen, W. Freudenreich, R. L. Moss, R. Ricchence, "MCNP Calculations for the Design and Characterization of the Petten BNCT Epithermal Neutron Beam," Progress in Neutron Capture Therapy, B. J. Allen, D. E. Moore, B. V. Harrington, eds., Plenum Press, New York, 1992.
4. R. M. Brugger, J. A. Shih, H. B. Liu, "An Epithermal Neutron Beam for Neutron Capture Therapy at the Missouri University Research Reactor," Nuclear Technology, 98, pp. 322-332 (1992).
5. H. T. Wu, "A Low Energy Accelerator-Based Neutron Sources for Boron Neutron Capture Therapy," Dissertation in Nuclear Engineering Program of University of Missouri-Columbia, Aug. 1992.
6. H. B. Liu, "Isodose Curves and Treatment Planning for Boron Neutron Capture Therapy," Dissertation in Nuclear Engineering Program of University of Missouri-Columbia, Aug. 1992.

MONTE CARLO SHIELDING ANALYSIS FOR AN ACCELERATOR EPITHERMAL NEUTRON IRRADIATION FACILITY (AENIF) IN A CONVENTIONAL X-RAY IRRADIATION ROOM

Tanxia Qu,[1] Thomas E. Blue,[1] Kevin Herminghuysen,[1]
Christos Kanellitsas,[2] Reinhard Gahbauer[2]

[1] Nuclear Engineering Program

[2] Radiation Oncology Department
The Ohio State University

INTRODUCTION

An Accelerator Epithermal Neutron Irradiation Facility (AENIF) was proposed for BNCT[1]. The proposed AENIF consists of a low-energy high-current proton accelerator, a high energy beam transport system, a target assembly, and a moderator assembly. The accelerator generates a 30 mA–2.5 MeV proton beam. The protons strike the ^7Li target and create neutrons via the $^7Li(p,n)^7Be$ reaction. The source neutrons, having an average energy of about 400 keV, are too energetic for BNCT. Therefore a moderator assembly is used to degrade the energy of the neutrons to epithermal energies (1 eV- 1 keV). It is the purpose of this study to investigate the dose equivalent to radiation therapy personnel, if the moderator assembly is placed inside a conventional x-ray irradiation room.

BACKGROUND

The optimized moderator assembly of the proposed AENIF is shown in Figure 1 for a 25 cm diameter target[2]. The target consists of a thin layer of ^7Li on a 25 cm diameter copper disk. The moderator assembly consists of a cylindrical moderator made of BeO, which is 25 cm in diameter and 20 cm in height, surrounded by a 30-cm-thick magnesium oxide reflector. A loading of 0.05 g/cm^2 of ^6Li is placed at the interface between the moderator and reflector regions. A loading of 0.01 g/cm^2 of ^6Li is placed at the front window of the irradiation port to reduce the thermal neutron contamination of the epithermal neutron field. In addition, a 10-cm-thick layer of D_2O surrounds the reflector, except at the irradiation port, and the D_2O is in turn surrounded by an outer skin of ^6LiF. These two layers function as a neutron shield to reduce the patient's whole body dose. The total (4π) neutron source strength for the $^7Li(p,n)^7Be$ reaction is 2.76 x 10^{13} neutrons per second per 30 mA of proton current[1]. This source strength was used in our analysis.

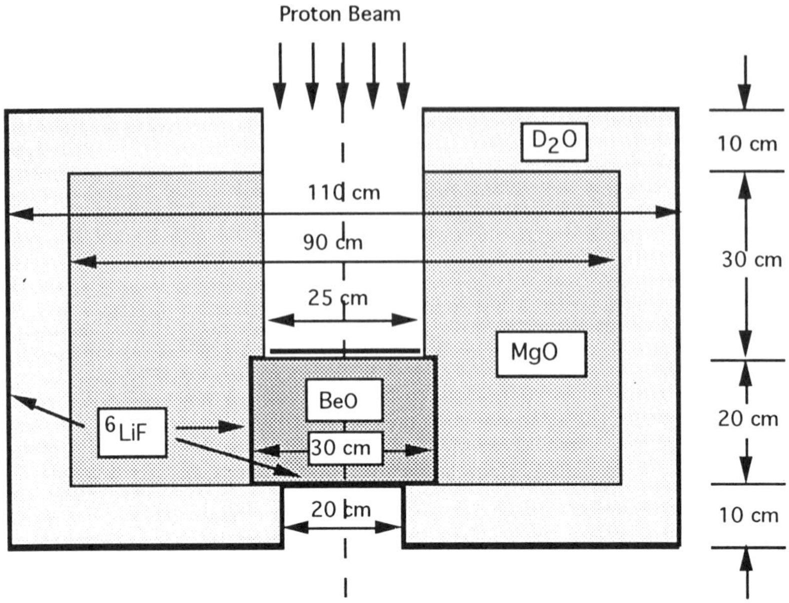

Figure 1. The Configuration of the Moderator Assembly for the AENIF

METHODOLOGY

The Monte Carlo code MCNP [3] was used with its associated pointwise continuous – energy cross section data. The cross section data used are from the ENDF/B-V based RMCCS and MCPLIB libraries for neutrons and photons, respectively. The calculational setup is shown in Figure 2. The head phantom is a cylinder, 16.7 cm in diameter and 24 cm high. The body phantom is a cylinder, 28.3 cm in diameter and 150 cm high[4]. The head phantom was located on the moderator assembly central axis, 2 cm from the irradiation port. The compositions and densities of the patient's body are from reference 5. The room configuration is that of a 6 MV linear accelerator x-ray irradiation room in the Arthur James Cancer Hospital at OSU. It is assumed that the ceiling, floor and walls of the room are made of ANSI/ANS ordinary concrete with a density of 2.35 g/cm^3 [6].

The dose equivalent rates were calculated at the point where the radiation therapy technician works (henceforth called the radiation therapy technician site) using MCNP F5 (next-event-to-a-point) tallies. The ANSI phantom-related response functions were applied to the calculated neutron and photon fluxes to calculate the dose equivalent rate.[6] NCRP No. 38 [7] "Protection against Neutron Radiation" was used to evaluate the adequacy of the shielding. NCRP No. 38 states "the maximum permissible prospective dose equivalent (MPDE) for whole body irradiation from all occupational sources shall be 5 rem in any one year." This corresponds to 100 mrem/week or 2.5 mrem/hr.

RESULT AND DISCUSSION

Two calculations were performed to calculate the dose equivalent rate due to neutrons and photons at the radiation therapy technician site. The results, normalized to a proton beam current of 30 mA, are listed in Table 1. One calculation was performed as an analog simulation[3]. It sampled 452,000 histories on a SUNSparc IPC workstation for a CPU time of 2,700 min. The Fractional Standard Deviations (FSD)[3] for the neutron and photon tallies are 12% and 39%, respectively. The total dose equivalent rate was calculated as 260 ± 40 mrem/hr. A second calculation was performed using the weight window biasing method.[3] It sampled 110,000 histories on a CRAY-YMP/8 supercomputer for a CPU time of 375 min. The FSDs for the neutron and photon tallies are 12% and 4%, respectively. The total dose equivalent rate was calculated as 190 ± 20 mrem/hr. Since the analog and the weight window biasing results are not inconsistent, it is reasonable to average them. The weighted means of the neutron and photon dose equivalent rates are 160 ± 15 mrem/hr and 52 ± 2 mrem/hr, respectively. The weighted mean of the total dose equivalent rate is 210 ± 20 mrem/hr, which is about 85 times larger than the 2.5 mrem/hr MPDE suggested by NCRP No. 38.

The analog MCNP calculation required four times as many histories to achieve the same neutron tally FSD as the weight window biasing calculation. Even with four times as many histories, the FSD of the photon tally for the analog calculation is worse than the FSD using weight window biasing. The concrete wall is about 100 cm thick and hence the calculation is a deep-penetration shielding problem. For the photon dose equivalent rate, the FSD of 4% is good. For the neutron and the total dose equivalent rates, the FSDs of

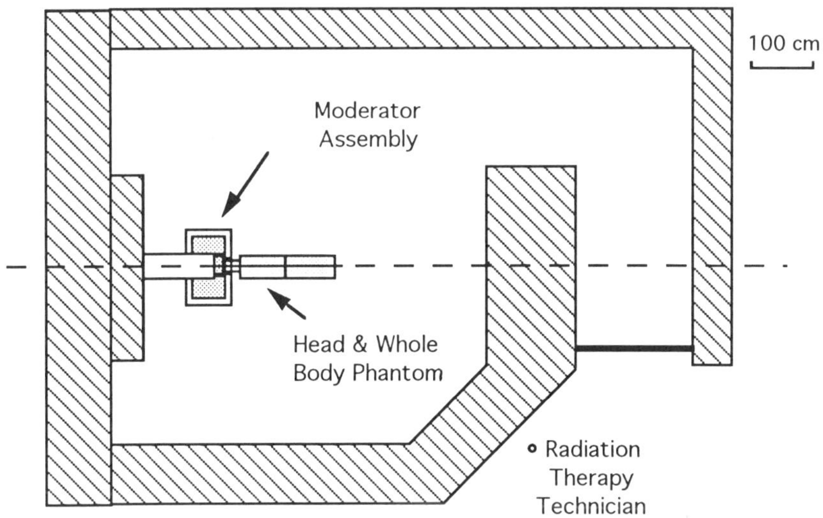

Figure 2. Schematic diagram of an x-ray irradiation room in The Arthur James Cancer Hospital. The diagram is drawn to scale except for the detector point, which is ~1 cm away from the wall which is 100 cm thick.

Table 1. Dose equivalent rates (mrem/hr) at the radiation therapy technician site, normalized to a proton current of 30 mA

	Analog	Weight Window	Weighted Mean
Neutron	190 ± 12%	140 ± 12%	160 ± 9%
Photon	70 ± 39%	52 ± 4%	52 ± 4%
Total	260 ± 15%	190 ± 9%	210 ± 8%

9% and 8%, respectively, may indicate a lack of precision. These results may be incorrect by as much as a few times their true values [3]. However, even admitting that the calculation may be in error by this much, one can conclude that the shielding provided by this conventional x-ray irradiation room is inadequate for BNCT.

CONCLUSION

A moderately shielded conventional 6 MV linear accelerator x-ray irradiation room built with ordinary concrete may not provide adequate shielding for a radiation therapy technician working outside of the irradiation room, with the proposed AENIF operating at a proton beam current of 30 mA. As future work, we intend to determine if the shielding of a conventional 18 MV linear accelerator X-ray irradiation room is adequate, and if it is not, how the shielding of the room could be modified to make it so.

ACKNOWLEDGMENT

This work was supported by the U.S. Department of Energy under contract DE-AC02-76CH00016

REFERENCES

1. C.K. Chris Wang, T. E. Blue, and R. A. Gahbauer, "A Neutronic Study of an Accelerator-Based Neutron Irradiation Facility for Boron Neutron Capture Therapy," *Nucl. Tech.* 84, 93 (1989).
2. T. Qu, "An Advanced Design Study of an Accelerator Neutron Irradiation Facility for BNCT," *Ph.D. Dissertation*, The Ohio State University (1993).
3. J. F. Briesmeister, ed., "MCNP - A General Monte Carlo Code for Neutron and Photon Transport," *LA-7396-M, Los Alamos National Laboratory* (1989).
4. J. E. Turner, "Atoms, Radiation, and Radiation Protection, " Pergamon Press (1986).
5. J.J. Ritts, M.Solomito, and P.N. Steven, "The Calculation of Neutron-induced Physical Dose in Human Tissues, " *ORNL-TM-2991* (1970).
6. A. B. Chilton, "Principles of Radiation Shielding," Prentice-Hall, Inc. (1984).
7. NCRP Report 38, "Protection against Neutron Radiation" (1971).

ACCELERATOR NEUTRON IRRADIATION FACILITY FOR HOSPITAL-BASED NEUTRON CAPTURE THERAPY

C-K C. Wang,[1] P. E. Eggers[2] and H. L. Crawford[2]

[1] Mechanical Engineering, Georgia Institute of Technology
Atlanta, Georgia 30332
[2] Eggers & Associates, 5366 Reserve Drive, Dublin, Ohio 43017

INTRODUCTION

An accelerator-based source for intense epithermal neutrons has been under development by the authors over the past five years[1-9]. This research has led to the design of an irradiation facility capable of producing a neutron flux having the energy spectrum, intensity and levels of unwanted radiation suitable for hospital-based neutron capture therapy (NCT). This neutron irradiation facility design utilizes a radiofrequency quadrupole accelerator to produce proton beam levels of 10 to 50 mA. Although both lithium and beryllium targets were investigated, the lithium target was found to offer significant benefits in terms of the quality of the neutron beam energy spectrum, viz, reduced levels of higher energy neutrons (E > 10 keV) and secondary gamma radiation. A key feature of this irradiation facility design provides is a rotating disk cooling system which allows a 10 cm X 10 cm beam of 2.5 Mev protons to impinge on the lithium target at heating levels of up to 125,000 watts while maintaining the lithium target at or below its melting point. A proprietary target module design hermetically sealed containment for the lithium which (1) isolates the lithium from the air during facility installation or maintenance and (2) prevents the possible loss of lithium during operation near its melting point while exposed to centrifugal forces associated with target rotation. Another feature of this irradiation facility design is (1) the absence of any hydrogenous materials in the primary pathway of the neutron beam, (2) the use of beryllia oxide moderating means, (3) magnesium oxide reflecting means and (4) filtering means to maximize the flux rate of neutrons in the energy range of 1 eV to 10 keV. In addition, collimation and shielding means is provided to maintain unwanted radiation and associated whole-body dose to acceptable levels.

METHODOLOGY

The development of the high flux accelerator-based neutron irradiation facility (HFANIF) was directed toward meeting four principle goals:

1. Patient safety--minimize levels of nontherapeutic neutron and gamma radiation.
2. Efficiency--maximize the beam fluence rate of therapeutic neutrons (1 eV to 10 keV) per milliamp of proton current.
3. Reliability--operate cooling means well below critical heat flux; contain and hermetically seal lithium to isolate from atmosphere and rotational effects.

Advances in Neutron Capture Therapy, Edited by
A.H Soloway *et al.*, Plenum Press, New York, 1993

4. Expandability--design target, cooling system and shielding/collimation means to permit operation over the range of proton current levels from 10 mA to 50 mA.

The analysis and design of the HFANIF included (1) three-dimension, multiphase heat transfer analysis (TRUMP[10]), (2) neutron transport analysis (MCNP-4[11-12]), (3) shielding analysis (MCNP-4 and QAD), and (4) structural analysis[13] of rotating target and evacuated containment system. This design of the moderator subsystem was based, in part, on lithium target experiments performed using a Van de Graaff accelerator [2,3,5,6].

HIGH FLUX ACCELERATOR-BASED NEUTRON IRRADIATION FACILITY DESIGN

The product of over five years of research is the irradiation facility design (patent pending) illustrated in Figure 1. The HFANIF system including shielding but excluding the RFQ accelerator measures 4.75 meters X 4.75 meters X 4.0 meters in height. The RFQ accelerator produces 10 to 50 mA of 2.5 MeV protons (depending on therapeutic fluence rate requirements) and features octupole correction magnets for beam uniformity and duodecapole correction magnets for beam confinement. The lithium target is hermetically sealed within 10 cm X 30 cm modules which will contain lithium even in the event the melting point is exceeded. The target cooling system incorporates a dual chamber rotating disk with D_2O primary coolants in an outer closed annulus (in primary neutron (seam path) and H_2O secondary coolant within inner annulus of hollow disk. The H_2O secondary coolant circulated through an external two-phase heat exchanger. The beryllium oxide moderator surrounded by a magnesium oxide reflector are optimized to produce a neutron fluence in the energy range from 1 eV to 10 keV. Shielding the patient and staff from unwanted radiation is accomplished through a combination of borated polyethylene, bismuth, boron carbide and structural concrete floors and walls. A control system maintains preselected levels of target rotation speed, secondary coolant flow, target temperature, pressure (vacuum) level and proton beam current.

COMPARISON WITH OTHER ACCELERATOR-BASED SYSTEMS

The research group at MIT and Science Research Laboratories (SRL) proposed using a 4 mA, 2.5 MeV proton beam produced by a tandem cascade accelerator (TCA) to bombard a Li target, and claimed that the system will produce a epithermal neutron field with spectrum and flux intensity suitable for NCT[14]. treatment[14]. Based on our research, this claim appears optimistic for the following reasons. First, MIT/SRL's design does not include the cooling system for the Li target. One should allow at least 5 to 10 cm of space between the Li target and the moderator. Second, MIT/SRL's design uses 19 cm thick D_2O (instead of 20 cm thick BeO proposed in our design) as the neutron moderator. Based on our studies[4] of the moderating property of D_2O we found that although a D_2O moderator results in a higher neutron flux (per unit proton current) at the beam port, the BeO moderator results in a preferred neutron spectrum. For instance, our study indicated that for a 20 cm thick D_2O moderator the epithermal neutron flux at the beam port is 60% higher than from a 20 cm thick BeO moderator. However, the fast neutron flux for a 20 cm D_2O moderator is 2.5 times of that for a 20 cm BeO moderator. Most of the fast neutrons have energies between 10 keV and 200 keV. According to a recent report, the RBE value may be as high as 5 or more for fast neutrons in this energy range[15]. Therefore, the neutron spectrum resulting from a 19 cm thick D_2O moderator most likely will be too hard for neutron capture therapy. Third, MIT/SRL's design does not include any neutron beam collimator and the head phantom is effectively in direct contact with the external surface of the moderator[14]. Our research indicates that 5 to 15 cm of collimation should be provided between the moderator surface and the patient to allow enough room for the dose monitoring device (e.g., an ion chamber). In addition, a neutron collimator of 20 cm thick borated polyethylene or equivalent neutron shielding is essential to reduce the patient's whole-body dose to an acceptably safe level. The current MIT/SRL's design[14] will deliver at least several hundred cSv of whole-body dose equivalent to a patient which is well beyond acceptable limits. In view of the above considerations, the neutron flux intensity of the MIT/SRL will be reduced by a factor of 6 or more. In contrast to the MIT/SRL design, our proposed design of using a 10 to 50 mA RFQ accelerator, beryllia

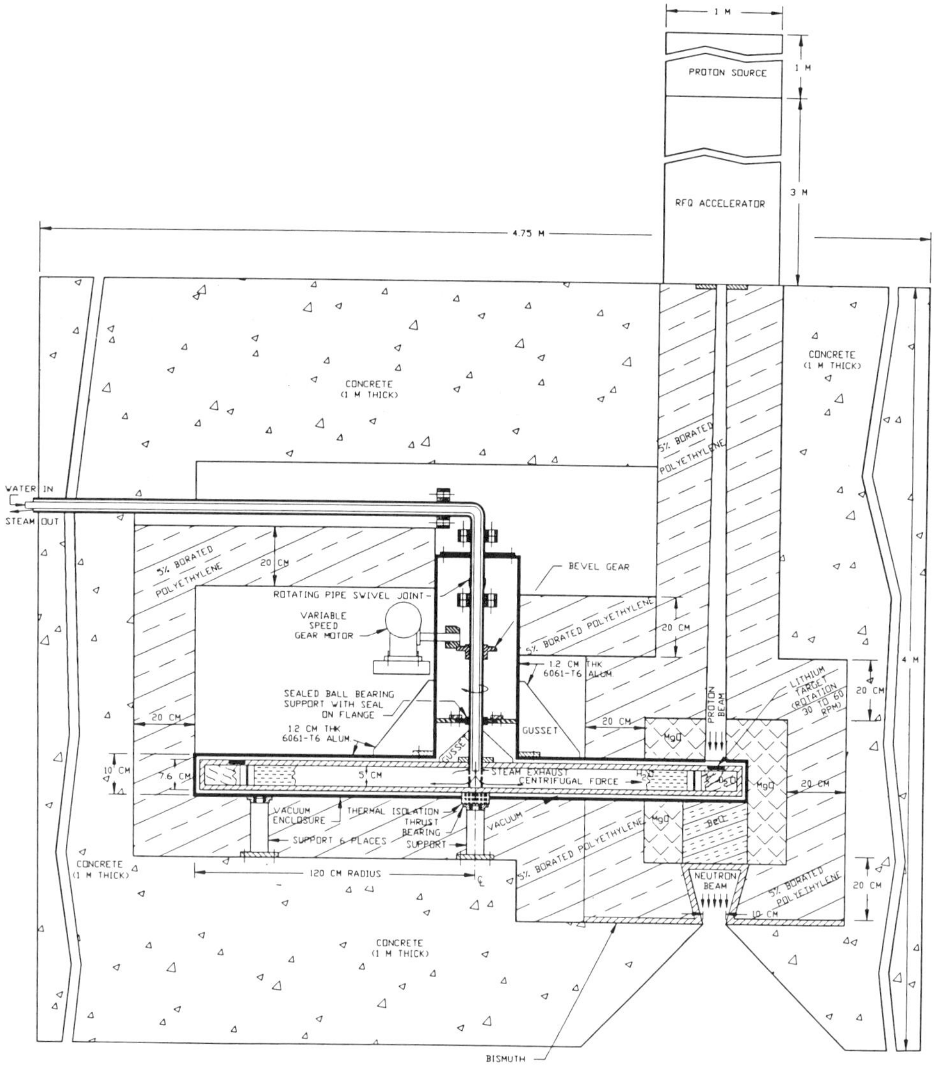

FIGURE 1. HIGH-FLUX ACCELERATOR-BASED NEUTRON IRRADIATION FACILITY

moderator and collimation will provide the desired epithermal neutron beam of 1 eV to 10 keV while providing neutron fluence rates both safe and sufficiently intense for practical NCT. The characteristics of the proposed HFANIF system are summarized in Table 1 as compared with the characteristics of the Brookhaven Medical Research Reactor neutron source.

TABLE 1. COMPARISON OF PROPOSED ACCELERATOR IRRADIATION FACILITY WITH REACTOR-BASED FACILITY

PARAMETER	PROPOSED ACCELERATOR NEUTRON IRRADIATION FACILITY		BROOKHAVEN MEDICAL RESEARCH REACTOR
	10 mA beam	50 mA beam	
1. Epithermal-neutron fluence rate for energies of 1eV to 10keV (x10E+8 neutrons/sq.cm.-sec.)	1.6	7.8	18.0
2. Absorbed dose rate from fast neutrons >10 keV, free in air (cGy/min.)	0.96	4.8	5.3
3. Absorbed dose rate from gamma rays, free in air (cGy/min.)	0.22	1.10	1.20
4. Absorbed fast neutron dose per epithermal neutron fluence (x10E-11 cGy-sq.cm./neutron)	6.6	6.6	4.9
5. Absorbed gamma ray dose per epithermal neutron fluence (x10E-11 cGy-sq.cm./neutron)	1.4	1.4	1.1
6. Thermal neutron fluence rate at 2-cm depth in phantom (x10E+8 neutrons/sq.cm.-sec.)	1.8	8.8	19.0
7. Whole body dose rate to patient (cGy/min.)	0.06	0.32	Data Not Available
8. RUFTED--Ratio of Useful Neutron Fluence To Entrance Dose equivalent	6.8	6.8	9.2
9. Neutron current/fluence at beam port surface	0.94	0.94	0.58
10.Treatment time based on therapeutic fluence of 1.0E+12 thermal neutrons/sq.cm. (minutes)	94	19	9
11.Treatment time based on therapeutic fluence of 5.0E+12 thermal neutrons/sq.cm. (minutes)	470	94	44

CONCLUSIONS

The following conclusions are based on the results of five years of research in the field of accelerator-based neutron irradiation systems:

1. Higher proton beam current levels (10 to 50 milliamps) may be essential to produce epithermal neutron fluence rates adequate for neutron capture therapy while safe in terms of whole body radiation dose.
2. Lithium target provides significant advantages over beryllium target based on neutron beam quality and level of whole-body dose of non-therapeutic radiation.
3. Dual chamber, rotating target design maintains lithium target below melting point for beam currents of up to 50 milliamp; rotating target mass is less than 350 kg and rotation speed is less than 60 revolutions per minute.
4. Neutron beam collimator is essential to reduce patient's whole body dose to safe level.
5. Whole body dose (including secondary gamma dose resulting from neutron/H_2O interaction) to the patient can be maintained below 4 cGy/hour at 10 milliamp beam current and below 20 cGy/hour at 50 milliamp beam current.
6. Neutron source can be adapted to produce neutron energy spectrum required through interchange of target, moderator, reflector, collimator and filter components.
7. Neutron source can provide safe therapeutic epithermal beam (1 eV To 10 keV) at fluence rates of 1.6 X 10^8 neutrons/cm^2-sec (10 mA) to 7.8 X 10^8 neutrons/cm^2-sec (50 mA).
8. Cost of target, cooling system, moderator and shielding (excluding hospital site costs for walls, floor, hvac, utilities) estimated to be under $200,000.
9. Cost of RFQ accelerator estimated to be in range of $3 to $6 million for 10 to 50 milliamp proton current levels.

10. Commercialization of NCT using proton beam currents below 10 mA through dose fractionation may be limited due to the cumulative effects of NCT on pathway of targeting agent to tumor, cost of targeting drugs and cost of neutron irradiation therapy session.

ACKNOWLEDGEMENTS

This research was funded, in part, by a grant from the National Cancer Institute (Grant No. 1R43CA54646-01). The authors also wish to acknowledge participation in our research by AccSys Technology, Inc. and in particular Dr. Robert W. Hamm.

REFERENCES

1. WANG, C.K. and B.R. Moore (1992) "A Calculational Study of an Intense Epithermal Neutron Source Based on $^{9}Be(p,n)^{9}B$ Reaction for Neutron Capture Therapy," presented at the Fifth International Symposium on Neutron Capture Therapy.
2. WANG, C.K., T.E. Blue, and J.W. Blue (1990) "An Experimental Study of the Moderator Assembly for a Low-Energy Proton Accelerator Neutron Irradiation Facility for BNCT" in Neutron Beam Design, Development and Performance for Neutron Capture Therapy, pp 271-280 (O.K. Harling, Ed.) New York; Plenum Press.
3. WANG, C.K. and T.E. Blue (1990) "A Neutron Spectrometer for Neutron Energies Between 1 eV and 10 keV," Nuclear Instruments and Methods, A290, pp 237-241.
4. WANG, C.K., Blue, T.E. and Gahbauer, R. (1989a) "A Neutronic Study of an Accelerator-Based Neutron Irradiation Facility for Boron Neutron Capture Therapy", Nuclear Technol., 84(1) p 93-107.
5. WANG, C.K., Blue, T.E. and Blue, J.W. (1989b) "An Experimental Study of the Moderator Assembly for a Low-Energy Proton Accelerator Neutron Irradiation Facility for BNCT," in, Neutron Beam Design, Development, and Performance for Neutron Capture Therapy, p 271-280 (O.K. Harling, Ed.), New York: Plenum Press.
6. WANG, C.K. (1989c) "A Pilot Study of an Epithermal Neutron Source Based on a Low-Energy Proton Accelerator for Neutron Capture Therapy," Ph.D. dissertation, The Ohio State University.
7. WANGLER, T.P. and Wang, C.K., (1989) et al. "Conceptual Design of an RFQ Accelerator-Based Neutron Source for Boron Neutron Capture Therapy," Proceedings of the 1989 Particle Accelerator Conference in Chicago.
8. QU, T.B., T.E. Blue and C.K. Wang (1990) "A Neutronic Study of an Accelerator-Based Neutron Irradiation Facility for the Treatment of Superficial Melanoma by Boron Neutron Capture Therapy," Nuclear Technology, Vol. 91, No. 3, pp 404-412.
9. Blue, T.E. and C.K. Wang (1989) "An Equivalent Thermal Neutron Fluence Detector for Boron Neutron Capture Therapy," Physics in Medicine and Biology, Vol. 34, No. 9, pp 1301-1307.
10. EDWARDS, A.L. (1972) "TRUMP: A Computer Program for Transient and Steady State Temperature Distributions in Multidimensional Systems," UCRL-14754, Rev. 3, Lawrence Livermore Laboratory, University of California, Livermore, California 94550.
11. BREISMEISTER, J.F. (1990) ed., "MCNP-A General Monte Carlo Code for Neutron and Photon Transport, Version 4.0., "Los Alamos National Laboratory, LA-7396-M, Rev. 3.
12. LASL GROUP X-6 (1979) "MCNP - A General Monte Carlo Code for Neutron and Proton Transport," LA-7396-M, distributed as CCC-2000 by the Radiation Shielding Information Center of Oak Ridge National Laboratory.
13. ASME Boiler and Pressure Vessel Code (1989), Section VIII-1, American Society of Mechanical Engineers (ASME), New York.
14. YANCH, J.C., X-L Zhou, R.E. Klinkowstein (May/June 1992), "Accelerator-Based Epithermal Neutron Beam Design for Neutron Capture Therapy," Medical Physics, Vol. 19, No. 3, pp 709-721.
15. ALPEN, E.L. (1992), "Biological Effectiveness of Recoil Protons from Neutrons of 10 keV to 100 keV," presented at the 5th International Symposium on Neutron Capture Therapy in Columbus, OH.

ON THE STUDY OF ENERGY SPECTRA AND ANGULAR DISTRIBUTIONS OF THE NEUTRONS EMITTED FROM A BERYLLIUM TARGET BOMBARDED WITH 4-MeV PROTONS FOR NEUTRON CAPTURE THERAPY

C-K Chris Wang and Brian R. Moore

Health Physics Program
School of Mechanical Engineering
Georgia Institute of Technology
Atlanta, Georgia 30332-0225

INTRODUCTION

Due to the recent advancement of the technologies in high-current proton accelerators, neutron sources based on 2.5 MeV proton bombarding a lithium (Li) target have been vigorously studied in the past few years[1,2] and was thought to be feasible for neutron capture therapy (NCT). Neutrons are generated based on the $^7Li(p,n)^7Be$ reactions and then moderated with beryllia (BeO) or heavy water (D_2O). For a treatment time of 1 hour it was determined that a proton current of 30 mA might be necessary. This intense proton beam is achievable with the newly available radio-frequency quadrupole (RFQ) accelerator[3]. However, because Li has a low melting point (179°C) and a low thermal conductivity (45 W/m°C), the removal of the immense heat (75 kW) from the Li target imposes a big challenge on the target design. In addition, Li is very reactive chemically, and it thus further complicates the target design in order to provide proper access for system maintenance[4].

The easiest way to solve the design problems associated with a Li target is simply not to use lithium, and to use beryllium (Be) instead. Be has two obvious advantages over Li as a target material: (1) the melting point and thermal conductivity of Be (1278°C and 130 W/m°C) is much higher than those of Li; and (2) Be is chemically more inert than Li is (especially at high temperatures). However, a disadvantage is that the neutron yield for a Be target bombarded with 2.5 MeV protons is only about 1/6 of that of a Li target. Neutron yield of a Be target may be increased by bombarding with more energetic protons but the resulting neutron spectrum may also become harder. For instance, the neutron yield of a Be target bombarded with 4-MeV protons is comparable to that of a Li target bombarded with 2.5 MeV protons. 4 MeV corresponds to the maximum achievable proton energy for a RFQ. With proton energy at 4 MeV, the neutrons emitted from a Be target may have energies as high as 2.2 MeV. Whereas the maximum energy of neutrons from a Li target bombarded with 2.5 MeV protons is only 800 keV. To find out if Be is an alternative target material, the characteristics of the neutrons emitted from a Be target must be known. This study determines the neutron yield, neutron energy spectra and angular distributions for a thick Be target bombarded with 4 MeV protons.

INTERACTIONS AND CROSS SECTIONS

The experimentally measured neutron production cross section data for the proton-Be reactions is less available and less accurate than that for the proton-Li reactions. Unlike the case of proton-Li reactions, for which the neutron production is solely dominated by the $^{7}Li(p,n)^{7}Be$ reaction, the neutrons produced in the case of proton-Be reactions are due to two competing reactions:

$$^{9}Be + {}^{1}H \rightarrow \quad {}^{9}B + {}^{1}n - 1.852 \text{ MeV} \qquad (1)$$

$$\qquad\qquad {}^{8}Be + {}^{1}n + {}^{1}H - 1.73 \text{ MeV} \qquad (2)$$

Reaction (1) is referred as the (p,n) reaction. The threshold energy for the proton for the ground state of reaction (1) is 2.059 MeV. Reaction (2) is a three-body process, and is referred as the (p,p'n) and/or (p,p')(n) reactions, and the corresponding threshold energy for the proton is 1.85 MeV. Although the threshold energy of the (p,p'n) reaction is below the threshold energy of the (p,n) reaction, the neutron yield for the (p,p'n) reaction below the (p,n) threshold is very low[5], indicating an extremely small cross section for the (p,p'n) reaction in this energy region. Some of the measured neutron spectra[5] suggest that the fractional neutron yields for (p,n) and (p,p'n) reactions are 90% and 10% respectively for Be bombarded with 2.5 MeV protons. However, when the bombarding proton energy increases to 4 MeV, the fractional neutron yield of (p,p'n) reaction equals that of the (p,n) reaction. In general, the neutron spectrum due to the (p,p'n) reactions is more continuous and softer than that due to the ground-state (p,n) reactions. This is because the neutron produced in the ground-state (p,n) reaction receives 90% of the total excess energy of the reaction in the center-of-mass (COM) system, whereas the neutron produced in the (p,p'n) reaction equally shares with a proton the 80% of the total excess energy of the reaction in the COM system. Therefore, although the maximum neutron energy increases with the bombarding proton energy, the overall neutron spectrum for a Be target may actually become softer as the bombarding proton energy increases. The "softness" of the neutron spectrum may be enhanced even more as the proton energy exceeds a threshold of 4.64 MeV, at which a new neutron energy group based on the first-excited-state $^{9}Be(p,n)^{9}B^{*}$ reactions starts to contribute. The spectrum of the new neutron group is significantly softer than that of the ground-state neutron group because $^{9}B^{*}$ is left in a 2.3-MeV excited state. The fractional neutron yield for the $^{9}Be(p,n)^{9}B^{*}$ reaction, however, is not available from literature. This study focuses on the neutron spectrum and angular distributions for a thick Be target bombarded with 4 MeV protons. Therefore, one only needs to consider the ground-state (p,n) reaction and the (p,p'n) reaction.

METHOD

The total neutron production cross section used in this study was obtained from Gibbons[6]. To calculate the energy spectrum and angular distribution of neutrons emitted from the ground-state (p,n) reactions, two assumptions were made: (1) the neutron angular distribution in the COM system is isotropic, and (2) the neutron receives 90% of the total excess energy in the COM system as the result of momentum balance. To calculate the energy spectrum and angular distribution for neutrons emitted from the (p,p'n) reactions, the angular distribution of neutrons in the COM system was also assumed to be isotropic, but the neutron energy in the COM system was assumed to be randomly distributed between 0 and 80% of the total excess energy in the COM system. The energy spectra and angular distributions were then converted from COM system to LAB system. A computer program based on the Monte Carlo method was written to implement the calculational procedures. The calculations of neutron spectrum and angular distribution were done separately for (p,n) and (p,p'n) reactions for a thin Be target bombarded with various proton energies ranging from 2 to 4 MeV, and the results for thin targets were then combined based on the predetermined fractional neutron yields to obtain the neutron energy spectrum and angular distribution for a thick Be target. The fractional neutron yields for various proton energies were obtained by interpolating or extrapolating the two given data points: 0.9 for the (p,n) reaction and 0.1 for the (p,p'n) reaction for protons at 2.5 MeV, and 0.5 for both (p,n) and (p,p'n) reactions for protons at 4 MeV. The fractional neutron yields for various proton energies were then weighted based on the corresponding differential path lengths, and used to obtain the neutron yield, energy spectrum, and angular distribution for the thick Be target. The differential path lengths for a 4-MeV proton slowing down in Be were calculated by a computer program using the stopping powers provided by Anderson[7].

RESULTS AND DISCUSSIONS

The total neutron yield for a thick Be target bombarded with 4 MeV protons was calculated to be 1.35 x 10^{12} neutrons/mA, which is 50% more than the neutron yield for a thick Li target bombarded with 2.5 MeV protons. The two calculated neutron spectra for the ^{9}Be(p,n) and the ^{9}Be(p,p'n) reactions are shown in Figure 1. It is quite obvious that the neutron spectrum for the ^{9}Be(p,p'n) reaction is significantly softer than that of the ^{9}Be(p,n) reaction. Figure 2 shows the neutron angular neutron distributions for the ^{9}Be(p,n) and the ^{9}Be(p,p'n) reactions. As it is shown, the neutrons due to the ^{9}Be(p,p'n) reactions is slightly more forwardly-directed than that due to the ^{9}Be(p,n) reactions. The fraction of neutrons emitted in the forward direction (i.e. 0° - 9°) of either reaction is about 60% which is significantly lower than that of the ^{7}Li(p,n) reaction. 80% of the neutrons produced in the ^{7}Li(p,n) reactions are in the forward direction. For comparison, the corresponding distributions for the ^{7}Li(p,n) reaction are also shown in Figure 1 and 2. Figure 3 shows the complete neutron spectra for three emitting angular intervals (0° - 45°, 45° - 90°, 90° - 180°) for 4-MeV protons bombarding a thick Be target. As it is shown, the neutron energies are widely distributed between 0 and 2 MeV with 1.2 MeV being the peak of the distribution. Also shown in figure 3 is that the spectrum for the most forward interval (0° - 45°) is also the hardest among the three spectra, and the spectrum for the backward interval (90° - 180°) is the softest.

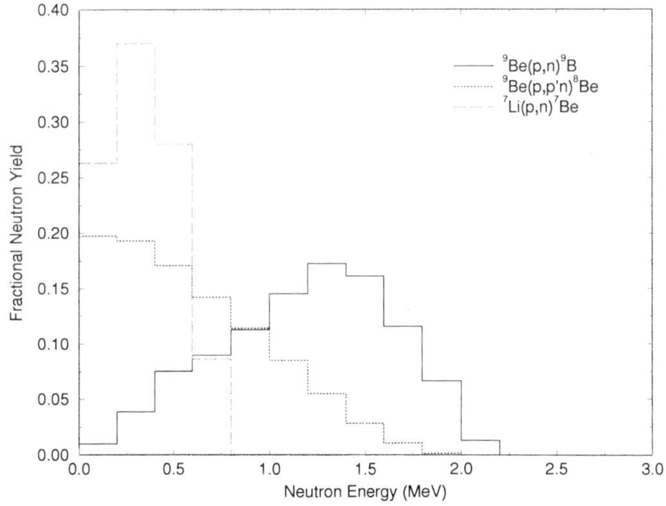

Figure 1. The distributions of fractional neutron yield vs. neutron energy for neutrons emitted from the ^{9}Be(n,p)^{9}B reactions and from the ^{9}Be(p,p'n)^{8}Be reactions for a thick Be target bombarded with 4-MeV protons, and for neutrons emitted from the ^{7}Li(p,n)^{7}Be reactions for a thick Li target bombarded with 2.5 MeV protons.

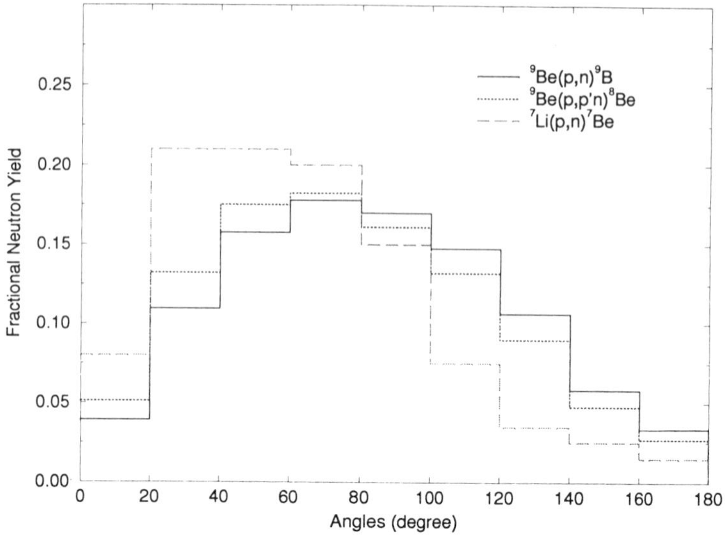

Figure 2. The distributions of fractional neutron yield vs. angular intervals for neutrons emitted from the ^{9}Be(n,p)^{9}B reactions and from the ^{9}Be(n,p'n)^{8}Be reactions for a thick Be target bombarded with 4-MeV protons, and for neutrons emitted from ^{7}Li(p,n)^{7}Be reactions for a thick Li target bombarded with 2.5 MeV protons.

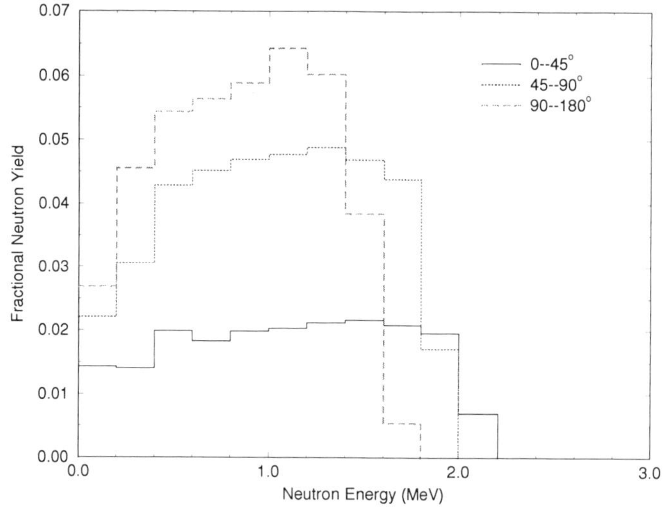

Figure 3. The distributions of fractional neutron yield vs. energy of neutrons emitted from a thick Be target bombarded with 4-MeV protons. Three angular intervals are presented separately in this figure: 0° - 45°, 45° - 90°, and 90° - 180°.

CONCLUSIONS

The calculational results of the total neutron yield, neutron spectra, and neutron angular distributions for a thick Be target bombarded with 4-MeV protons have been presented. The results show that the total neutron yield for a Be target bombarded with 4 MeV protons is 50% more than that for a Li target bombarded with 2.5 MeV protons. But the neutron spectrum for the Be target is significantly harder than that for a Li target. In addition, the neutrons emitted from a Be target is less forwardly-directed than those emitted from a Li target. Therefore, from the standpoint of neutron spectrum and angular distribution, this study reassures that Be is less favorable for NCT than Li as a target material. However, our preliminary neutron transport calculations indicate that a useful epithermal neutron beam for NCT can be developed using the neutrons from a Be target. In addition, because of the good thermal conductivity and high melting temperature of Be, it should be straightforward to design, build, and operate an accelerator neutron source based on a Be target. Thus, Be may serve as a backup target material depending upon how successful a Li target can be built and operated. Finally, a thorough experimental study should be performed to confirm the calculational results presented in this paper.

REFERENCES

1. C.K. Wang, T.E.Blue, and R. Gahbauer, A neutronic study of an accelerator-based neutron irradiation facility for boron neutron capture therapy, Nuclear Technology, 84:93 (1989).
2. J.C. Yanch, X-L. Zhou, R.E. Shefer, and R.E. Klinkowstein, Accelerator-based epithermal neutron beam design for neutron capture therapy, Medical Physics, 19:709 (1992).
3. Private communication with R.W. Hamm, AccSys Technology, Inc., 1177A Quarry Lane, Pleasanton, CA 94566.
4. P.E. Eggers, C.K. Wang, and H. Crawford, A design study of a low-energy proton accelerator neutron irradiation facility consisting a rotating target cooling system for neutron capture therapy, this Proceedings.
5. J.B. Marion and J.S. Levin, Investigation of the $^{9}Be(p,n)^{9}B$ and $^{9}Be(p,\alpha\gamma)^{6}Li$ reactions, Physics Review, 115:144 (1959).
6. J.H. Gibbons and R.L. Macklin, Total neutron yields from light elements under proton and alpha bombardment, Physical Review, 114:571 (1959).
7. H.H. Anderson and J.F. Ziegler. "Hydrogen: Stopping Powers and Ranges of Ions in Matter," Pergamon Press, Inc., Maxwell House, NY (1977).

CALIFORNIUM-BASED EPITHERMAL NEUTRON BEAMS FOR NEUTRON CAPTURE THERAPY

J.K. Kim*, J.C. Yanch and M.J. Wilson

Department of Nuclear Engineering, MIT, Cambridge, MA
*Current Address: Department of Nuclear Engineering, Hanyang University, Seoul, South Korea

INTRODUCTION

The requirement of a safe and effective epithermal neutron beam for boron neutron capture therapy (BNCT) has led to the investigation of a variety of different types of neutron sources. A small number of existing nuclear reactors have already developed epithermal beams that are predicted to provide safe and rapid neutron capture therapy[1,2]. However, in the event that proposed clinical trials of neutron capture therapy in the U.S. prove successful, it is anticipated that hospital-based neutron sources will be required and to this end development of accelerator-based neutron beams has begun. Alternatively, it may be possible for hospital based sources of epithermal neutrons to be provided by an appropriately moderated or filtered beam from a neutron-producing radionuclide. One potential isotope is Californium-252. This paper describes the results of a design study carried out to determine the feasibility of using ^{252}Cf as a source of epithermal neutrons for BNCT. Results are discussed in terms of both clinical requirements and ^{252}Cf availability.

CALIFORNIUM-252

Californium-252 is an artificial isotope with a half-life of 2.6 years. It decays primarily by alpha emission but undergoes spontaneous fission with a frequency of 3.1%. The total neutron emission rate from one gram of ^{252}Cf is roughly 2.3 x 10^{12} neutrons/second with a mean fission neutron energy of 2.14 MeV. A number of gamma photons are also given off both with the initial decay of the isotope itself (average energy 0.8 MeV), and from the fission products. Currently the only facility in the Western world set up to produce ^{252}Cf is the High Flux Isotope Reactor at Oak Ridge National Laboratory, with a production rate of roughly 0.4 grams/year[3].

^{252}CF NEUTRON BEAM DESIGN

Method

The Monte Carlo computer code MCNP[4] was used in the design of the ^{252}Cf-based epithermal neutron beam-line assembly. MCNP is a general purpose Monte Carlo radiation transport code for three-dimensional, continuous energy, time-dependent neutron, photon and electron transport. The MCNP version used in this study, Version 4.1, was installed and run on SUN SPARC workstations in the Whitaker College Biomedical Imaging and Computation Laboratory at MIT. The neutron energy spectrum for the ^{252}Cf source was modeled as a Watt fission spectrum using coefficients provided with the MCNP code[4]. A separate run was made for source gamma emission, the energy spectrum for which was obtained from Mannhart[5].

MCNP was used to examine the suitability of a variety of different materials as possible moderator, filter or reflector components for the ^{252}Cf beam-line assembly. Ability to maximize epithermal (4 eV - 40 keV) neutron flux or current at the patient position was used as the initial criterion for choosing a material (and its size and shape). As the design procedure progressed, MCNP was used to assess the dosimetric effect of the modified ^{252}Cf neutron spectrum in an elliptical phantom consisting of skull- and brain-equivalent materials. Further details regarding the design procedure, phantom dimensions and dosimetry are available elsewhere[6].

The phantom was divided into a number of small "cells" in order that dose could be measured as functions of both depth and radial distance from the beam axis. Fast neutron, thermal neutron and photon fluxes were assessed in each cell; appropriate kerma values were applied to estimate the contribution of each component to total dose. The ^{10}B contribution was estimated by modification of the thermal neutron fluence with fluence-to-kerma conversion factors listed by Zamenhof et al[7].

A tumour ^{10}B concentration of 40 μg/g[8] and a tumour-to-healthy tissue ratio of 4:1[9] were assumed. Neutron beams were evaluated in terms of Advantage Depth (beam penetrability), advantage ratio and dose rate at a depth of 7.0 cm in brain equivalent tissue. Three different sets of RBE values were used in the evaluation of the neutron beam; these three sets represent a range of reasonable candidates for the true RBE values[10] and are as follows for ^{10}B reaction products, neutrons and photons: 2.3, 1.6, 0.5 (Set 1); 4.0, 2.3, 0.5 (Set 2); and 4.0, 4.0, and 1.0 (Set 3).

Results

The Monte Carlo design study indicated that a cylindrical heavy water moderator, followed by an extensive aluminum filter was the optimal configuration for producing an epithermal neutron beam[6]. A neutron reflector surrounding the moderator/filter configuration is needed to maximize neutron flux at the patient position. A schematic illustration of the final design is shown in Figure 1. The californium source is placed along the central axis of a 26 cm long, 11 cm radius D_2O cylindrical moderator, at a position 6 cm from the end. At the opposite end of the moderator is placed an aluminum filter, 64 cm in length. Surrounding the D_2O/Al assembly on three sides is a 35 cm wide lead reflector. Between the moderator and filter 0.1 cm ^{6}Li is placed to limit the number of (n,γ) reactions in the Al. Layers of ^{6}Li and extra moderating material (such as D_2O or plastic) can also be placed around the lead reflector to reduce radiation leakage from the assembly.

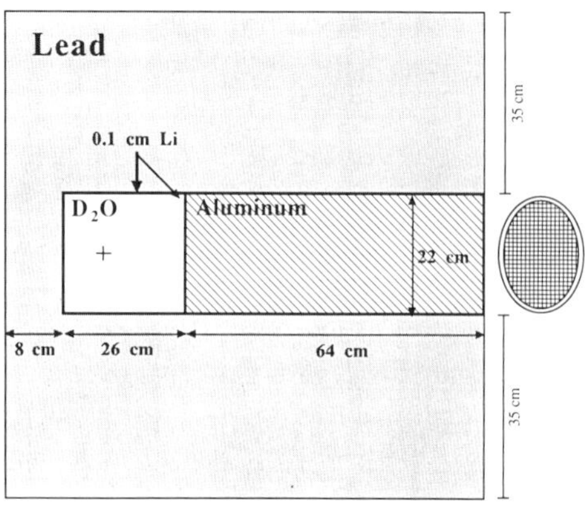

Figure 1. Illustration of the moderator/filter design; dimensions of components given in the text above

Figure 2 illustrates the dosimetric effect of this beam on the brain-equivalent phantom. A source strength of 1.0 g ^{252}Cf is assumed. Shown are the individual dose components as well as the total dose to tumour and healthy tissue for ^{10}B ratios of 40:10 using the RBE values designated earlier as Set 3. Advantage depths, advantage ratios and dose rate at a depth of 7.0 cm for the three sets of RBEs are provided in Table 1.

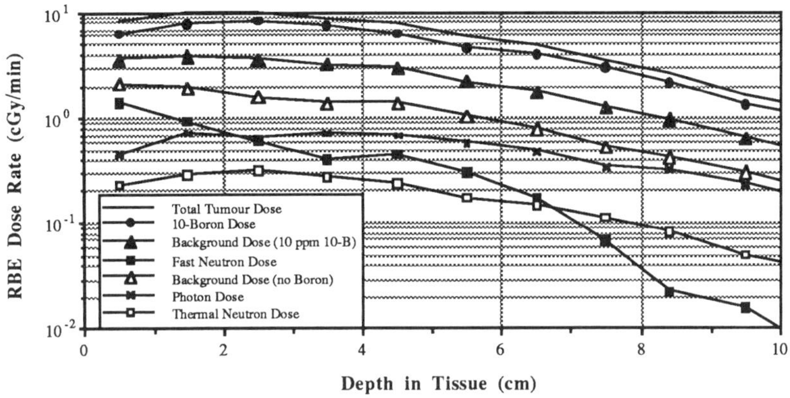

Figure 2. Dose versus depth in the head phantom assuming a tumour ^{10}B concentration of 40 μg/g and a 4:1 ^{10}B ratio in tumour to healthy tissue. RBE values of 4.0, 4.0 and 1.0 are used for ^{10}B reaction products, neutrons and photons, respectively.

Table 1. Advantage parameters for the ^{252}Cf epithermal beam assuming 40 ppm ^{10}B in the tumour and a 4:1 ratio of ^{10}B in tumour and healthy tissue. RBE values for the various scenarios are the same as given earlier in the text.

	40 : 10 ppm ^{10}B		
	Set 1	Set 2	Set 3
AD_{max} (cm)	9.4	10.2	9.0
AD_{min} (cm)	7.3	7.8	7.0
Dose at 7 cm (cGy/min)	2.2	3.8	4.1
AR	6.8	7.8	5.5

DISCUSSION

The ^{252}Cf-based epithermal beam-line assembly described above appears to be capable of providing a therapeutic dose rate of 2.0 - 4.0 cGy per minute per gram of ^{252}Cf at a depth of 7.0 cm in the brain. This dose rate is factor of roughly three to five times lower than the dose rate provided by existing reactor beams[1,2]. Further, with a 2.6 year half-life a ^{252}Cf-based neutron source would need to be periodically supplemented.

Currently, the only facility producing ^{252}Cf in the western world is at Oak Ridge National Laboratory. However, the Savannah River Site once performed a high-flux demonstration irradiation that produced a calculated 3 grams of ^{252}Cf[11]. At the HFIR an average of 0.4 g/year is produced but an estimated yield of 0.8 - 1.0 g/year could be reached with appropriate financial and feed material resources[11]. In addition, the

Department of Energy is planning a new research reactor (the Advanced Neutron Source) which will have a design capability of 1.5 grams of ^{252}Cf per year.

Unfortunately, the DOE has reduced activities at nuclear materials production sites and the future of the Advanced Neutron Source is still under debate. The unavoidable conclusion is that epithermal neutron beams based on a one-gram source of Californium are not feasible. On the other hand, a number of medical centers maintaining sources 0.1 gram in size is a far more realistic possibility. A 0.1 gram source of ^{252}Cf used in the beam-line assembly described above would result in a dose rate of only 0.2 to 0.4 cGy/minute. This is admittedly far too low to consider for patient irradiations. However, this dose rate can be greatly enhanced by the use of a subcritical multiplying assembly as described below[12].

Incorporation of a subcritical multiplying assembly (SMA) into the design described above involves placing fissionable material (^{235}U) into the moderator to provide a secondary source of neutrons. Neutrons emitted by the ^{252}Cf produce fission in ^{235}U resulting in neutron population multiplication via the chain reaction. For subcriticality, the assembly is designed with an "effective multiplication factor" (K_{eff}) less than 1.0 (which avoids most of the safety, environmental, siting and operational problems inherent in fission reactors). For an SMA, the neutron population is multiplied by a factor of $1/(1 - K_{eff})$. Thus, for values of K_{eff} between 0.5 and 0.99 provide neutron intensity increases by factors of 2 to 100, respectively. Neutron source multiplications of up to 90 have been shown to satisfy the non-reactor criterion.

With this approach, use of 0.1 g of ^{252}Cf now becomes feasible as a source of epithermal neutrons for BNCT[12]. The predicted dose rate of 0.2 - 0.4 cGy/min can be dramatically improved using the SMA approach. Design calculations are currently underway to optimize the configuration of source, moderator and ^{235}U for an epithermal neutron beam for BNCT.

REFERENCES

1. J.R. Choi, S.D. Clement, O.K. Harling and R.G. Zamenhof, Neutron capture therapy beams at the MIT research reactor, in *Neutron Beam Design, Development and Performance for Neutron Capture Therapy* (O K Harling, J A Bernard and R G Zamenhof, Eds.), New York: Plenum Press (1990).
2. R.G. Fairchild, J. Kalef-Ezra, S.K. Saraf, S. Fiarman, E. Ramsey, L. Wielpolski, B.H. Laster and F.J. Wheeler, Installation and testing of an optimized epithermal neutron beam at the Brookhaven Medical Research Reactor (BMRR), in *Neutron Beam Design, Development and Performance for Neutron Capture Therapy* (O K Harling, J A Bernard and R G Zamenhof, Eds.), New York: Plenum Press (1990).
3. J.B. Knauer, C.W. Alexander and J.E. Bigelow, ^{252}Cf: Properties, production, source fabrication and procurement, *Nuclear Science Applications* **4** 3-17 (1991).
4. J.F. Briesmeister, *MCNP—A General Monte Carlo Code for Neutron and Photon Transport*, LA7396-M Rev. 2, Los Alamos National Laboratory (1986).
5. Mannhart, IAEA-TEDOC-410 (1987).
6. J.C. Yanch, J.K. Kim and M.J. Wilson, "Design of a Californium-based Epithermal Neutron Beam for Neutron Capture Therapy," submitted to *Physics in Medicine and Biology*, March 1992.
7. R.G. Zamenhof, B.W. Murray, G.L. Brownell, G.R. Wellum and E.I. Tolpin, Boron neutron capture therapy for the treatment of cerebral gliomas. I: Theoretical evaluation of the efficacy of various neutron beams *Med. Phys.* **2** 47–60 (1975).
8. J.A. Coderre, D.N. Slatkin, P.L. Micca and J.R. Ciallella, Boron neutron capture therapy of a murine melanoma with p-Boronophenylalanine: dose response analysis using a morbidity index *Rad. Res.* **128** 177-185 (1991).
9. G. Solares , R. Zamenhof, S. Saris, D. Wazer, S. Kerley, M. Joyce, H. Madoc-Jones, L. Adelman and O. Harling, Biodistribution and pharmacokinetics of p-Boronophenylalanine in C57 Bl/6 mice with GL262 intracerebral tumors, and survival following neutron capture therapy, in *Progress in Neutron Capture Therapy for Cancer* (B Allen, D E Moore and B V Harrington, Eds.) New York: Plenum Press (1992).
10. O.K. Harling , R. Rogus, J.R. Choi, D.J. Moulin, R.G. Zamenhof and J.C. Yanch, Dosimetry and dose control for clinical trials of neutron capture therapy at the MITR-II reactor, to be published in Proceedings of the International Workshop on Macro and Microdosimetry and Treatment Planning for Neutron Capture Therapy, (O K Harling and R G Zamenhof, Eds.) New York: Plenum Press (1992).
11. Bigelow, J., ORNL, personal communication.
12. Hopkins, G.R., J.C. Yanch, J.M. Moran and E.T. Cheng, "Epithermal neutron flux enhancement for NCT using subcritical multiplying assemblies." abstract submitted to 5th International Symposium on Neutron Capture Therapy for Cancer, Columbus, OH, September 1992.

Cf-252 NEUTRONS FOR THE TREATMENT OF SUPERFICIAL CARCINOMAS

Yosh Maruyama[1,2] Jacek Wierzbicki[1], Jose M. Feola[1]

[1]Radiation Therapy Oncology Center
University of Kentucky Medical Center
Lexington, KY 40536-0084 USA

[2]Currently at
Radiation Oncology
Harper Hospital
Gershenson Cancer Center
Wayne State University
Detroit, MI 48201, USA

INTRODUCTION

The use of broad beams of thermal/epithermal neutrons from reactors or accelerators used in conjunction with chemical agents containing Boron-10 (^{10}B) or Gadolinium-157 has been postulated to allow selective therapy in tumor and the target volume. We have proposed that neutron emitting radioactive sources such as Cf-252 implanted in tumor can be used to deliver selective and localized neutron therapy to the tumor and that ^{10}B or Gd-157 enhancement can be used to augment the high LET radiation dose(1,2,3). Neutron capture therapy using chemical agents can be used to enhance local and regional anti-tumor therapy. For Cf-252, the thermal neutrons result from the tumor/tissue interactions of the fast neutrons. The treatment of superficial carcinomas represents a situation where considerable radioresistance can be encountered with recurrent tumors, e.g. melanoma, breast, head and neck cancers and sarcomas, especially after prior surgery and radiation. Both Cf-252 neutron therapy and BNCT have potential for improving the local control and palliation of these tumors(4). In this paper we describe plaque therapy methods using Cf-252 for the treatment of superficial carcinomas.

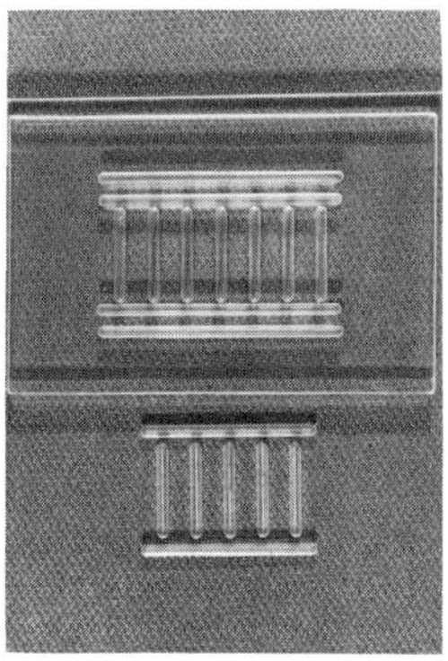

Figure 1. Plaques used for body surface ^{252}Cf therapy (two shown in figure 1). The slots are loaded with sealed tubes of ^{252}Cf and applied to the body surface.

Advances in Neutron Capture Therapy, Edited by
A.H Soloway *et al.*, Plenum Press, New York, 1993

IRRADIATION PLAQUE

A plaque for Cf-252 neutron therapy is shown in Figure 1. This plaque is loaded with multiple sealed sources of Cf-252 measuring 25 x 2.8 mm in size. The dose can be determined using along and away tables and computer treatment planning. The distribution of sources follows Paterson and Parker rules, and different size plaques can be constructed to match the dimensions of the lesion. The plaque is positioned against the body surface lesion, thereby exposing it to neutrons for a period adequate to deliver a therapeutic dose.

The depth dose pattern for the plaque is shown in Figure 2. The biologically effective dose is a mixture of fast neutrons and photons which exhibit different absorption characteristics. While the fast neutron dose decreases rapidly with depth, the photon components decrease more slowly. A third component, i.e. the thermal neutrons not shown on the figure, begins to appear as the fast neutrons lose energy tumour tissue interactions[1,2]. The thermal neutron component has been determined to be ~ 15-20%, and the increase with distance from the sources at about 5 cm. Thus Cf-252 has potential for treating the peripheral regions of the tumor using the BNCT reaction, with adjuvant therapy using ^{10}B[5] or Gd-157 loaded agents.

BORON NEUTRON CAPTURE THERAPY (BNCT)

Thermal neutrons offer the possibility of enhancing the selective local dose delivered to the target brachytherapy volume if the volume can be loaded with ^{10}B. The thermal neutrons deliver additional local therapy by means of the emitted alpha and ^7Li particles. This potential was studied in collaboration with Matalka et al[5] using the MRA-27 human melanoma implanted in nude rats after loading the rats with ^{10}B-BPA. Tumor bearing rats were irradiated with Cf-252 neutrons using plaques as shown (Fig. 1) with or without ^{10}B-BPA therapy. Experimental studies revealed a greatly enhanced efficacy (x2) of Cf-252 neutrons when BPA was given before irradiation.

For different tumors whose sensitivities could be greater to photons and electrons, Gd-157 BNCT could be the agent selected for adjuvant therapy.

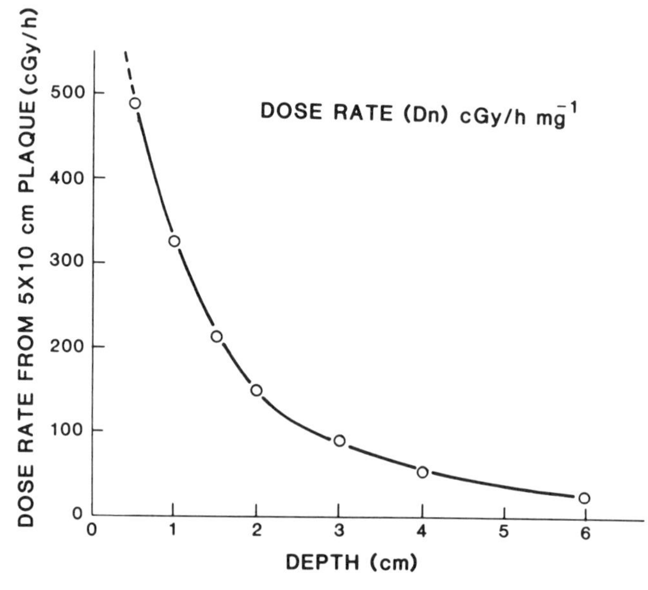

Figure 2. Depth dose curve plotted as dose rate (cGy/h) below a 5X10 cm plaque in cm.

DISCUSSION

Body surface tumors are common and usually dealt with using surgical methods. Certain tumors recur frequently on the body surface, such as melanoma, breast, and head and neck cancers. With prior surgery and/or radiotherapy these can be radioresistant and it is difficult to get local control or palliation. Using Cf-252 neutron plaques, treatment should be simple and more effective than further photon irradiation, chemotherapy, or hyperthermia. Neutron plaque therapy deserves further careful evaluation and study. By interposing (or not interposing) a moderator, different mixtures of fast and thermal neutrons could be applied. With thermal neutrons there are minimal biological effects unless ^{10}B or Gd-157 loaded compunds are present in the tissue/tumor. The isodose surface treated by BNCT or GdNCT would augment local as well as deep and peripheral dose. The dose appearance would resemble a "halo" or "shell" of hopefully targeted high LET dose surrounding the treated volume, which would treat regional tissues at risk for local spread.

The selectivity, versatility, and accuracy of neutron dose deliverable with Cf-252 brachytherapy allows many possible ways in which present therapy for cancers which are resistant to conventional therapies can be improved. Further study of ^{10}B-BNCT enhanced Cf-252 neutron brachytherapy needs to be a target for the future.

ACKNOWLEDGEMENTS

^{252}Cf sources were provided on a research contract from the U.S. Department of Energy. We thank Ms. Caroline Jones for expert clerical assistance.

REFERENCES

1. J. Wierzbicki, Y. Maruyama, C. W. Alexander. Cf-252 for teletherapy and thermalized Cf-252 neutrons for brachytherapy. Nucl. Sci. Appl., A4:361 (1991).

2. J. Wierzbicki, Y. Maruyama, C. Alexander. Boron neutron capture enhancement in Cf-252 brachytherapy, in: "Progress in Neutron Capture Therapy", B. J. Allen, D. E. Moore, B. V. Harrington (eds). Plenum, N.Y.:187 (1992).

3. Y. Maruyama, J. Wierzbicki, M. Ashtari, R. J. Yaes, J. L. Beach, J. Yanch, R. Zamenhof, C. B. Schroy. Cf-252 neutron capture therapy and teletherapy. Int. J. Radiat. Oncol. Biol. Phys., 23:255 (1992).

4. Y. Maruyama, (ed) "International Neutron Therapy Workshop." Harwood, Chur. Nucl. Sci. Appl. A4 (1991).

5. K. Z. Matalka, R. F. Barth, M. Q. Bailey, Y. Maruyama, J. M. Feola, J. Wierzbicki. Californium-252 brachytherapy of intracerebral melanoma with and without administration of boronophenylalanine (BPA) utilizing a nude rat model. (Elsewhere in these proceedings).

CF-252 NEUTRON THERAPY AND NEUTRON CAPTURE THERAPY

J. Wierzbicki,[1,3] C. Alexander,[2] Y. Maruyama[1,3]

[1]University of Kentucky Medical Center
Lexington, Kentucky 40536-0084
[2]Oak Ridge National Laboratory
Oak Ridge, Tennessee 37831-6015
[3]Present Address: Wayne State University
Department of Radiation Oncology
Harper Grace Hospital
Detroit, MI 48201

INTRODUCTION

The potential of fast fission neutrons in the treatment of malignant tumors has been studied by Koester et al[1]. These studies have been followed by clinical trials of superficial tumor treatment at the Munich research reactor[2] and by further studies of possible use of neutron capture reaction (NCT). For clinical trials of fast fission neutron therapy or NCT, nuclear reactors are the best source of neutrons. However, for routine therapy nuclear reactors are too expensive and hazardous to be based in a hospital or in an urban area.

Another source of fission neutrons is Cf-252. It has been used with encouraging results in the intracavitary and interstitial treatment of different cancers[3]. The clinical trials of Cf-252 brachytherapy have been successful and led to cures of cervical and other cancers with minimal complications. With the availability of Cf-252 sources of higher activity, the opportunity has arisen to use Cf-252 as a source for external neutron beam therapy and for combined photon-neutron-neutron capture therapy. In this paper we analyze the potential of Cf-252 sources for treatment of superficial, as well as deep seated carcinomas.

EXPERIMENTS

Experiments have been done at the Oak Ridge National Laboratory. We used a 6234 μg Cf-252 source. The experimental set-up is shown in Figure 1. The total dose in a cylindrical phantom consists of three components: fast neutron, gamma, and neutron capture dose if a capture agent (B-10 or Gd-152) is added locally. For fast neutron and gamma dose measurements we used paired chambers. Neutron capture doses have been calculated for a concentration of B-10 equal to 50 μg per cc of water. The basis for these calculations were determined using thermal flux densities measured by the gold foil method (gold foils were irradiated twice, with and without 0.5 mm thick cadmium envelopes).

DISCUSSION

Depth doses for fast neutrons gamma radiation and boron-neutron capture reactions are shown in Figure 2. Fast neutron depth dose,affected by inverse square law and neutron attenuation falls rapidly with depth. The gamma depth dose is a combination of californium photon dose and dose created by photons from the (n,γ) reaction. Therefore, it changes very slightly with depth. Boron-neutron capture dose also changes very slightly with depth.

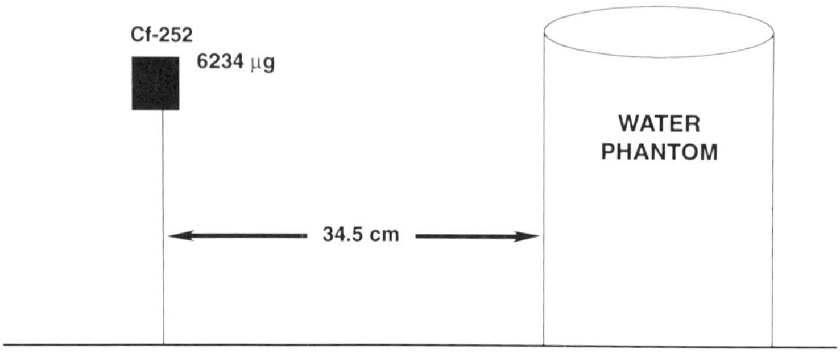

Figure 1. The experimental set-up

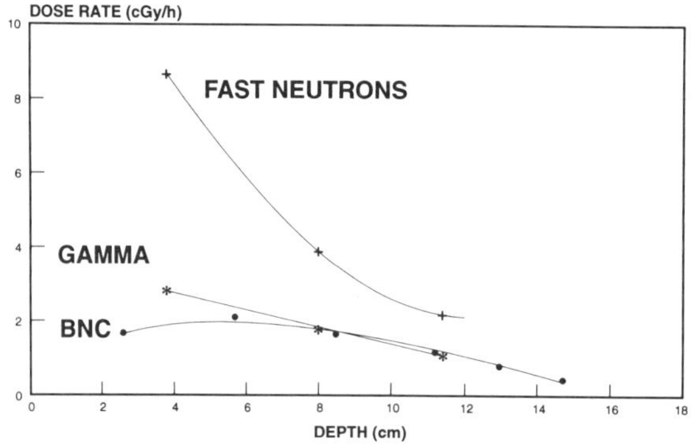

Figure 2. Depth doses for: fast neutrons, gamma radiation and boron-neutron capture reactions. B-10 concentration of 50 mg/cc.

As a result of these depth dose characteristics, the geometry of our experiment (Cf-252 sources, non-collimated and non-moderated beam) was suitable for treatment of superficial carcinomas only. Fast neutron dose was the dominant component of treatment dose. BNC dose was less than 20% of total dose. Relative Biological Effectiveness (RBE) for fast fission neutrons and products of the BNC reaction is between 2 and 3. Therefore, the photon dose, where RBE equals one, has no therapeutical meaning.

At depths of more than 10 cm, fast neutron dose and BNC dose have comparable values. Therefore, even using a moderator of thickness of about 10 cm, there may be therapeutical efficacy for even deep seated tumors. The open Cf-252 beam, as used in our experiment, is not suitable for treatment of deep seated tumors. Doses to the skin and soft tissue would be about two times higher than dose to the tumor. Treatment of deep seated tumors by Cf-252 neutrons requires significant reduction of the fast neutron component. This can be achieved by proper moderation (tailoring) of the beam. Study of tailored Cf-252 beam for NCT of deep seated tumors was done by Zamenhof et al[4]. The major drawback of Cf-252 as a source for epithermal NCT is the large amount of isotope required to achieve a reasonable dose rate for therapy. The solution may be the use of subcritical multiplying assembles[5,6].

We have determined that the absolute dose rate for superficial tumors given by 6234 μg of Cf-252 was about 12 cGy/h. The clinical use of 100 mg source does create special safety or technical problems but the dose rate would rise to about 200 cGy/h. In addition, by changing source-phantom-distance it is possible to get significantly increased dose rates (up to 800 cGy/h). The RBE dose rate for treatment of superficial tumors would then rise to about 2500 cGy/h and therefore be effective for clinical use.

Smaller source-phantom-distance would then also affect the shape of depth dose curve, making them much steeper and effective for superficial cancer therapy.

CONCLUSIONS

Cf-252 offers a simple, relatively inexpensive and readily usable method for hospital based neutron therapy of superficial tumors. Therapeutical effects of fast neutrons can be augmented by the neutron capture reactions of thermal neutrons which result from the moderation of fast neutrons in tissue. A hospital based Cf-252 teletherapy unit would greatly facilitate study of thermal neutron capture therapy in a hospital setting. It would increase the variety of tumors treated, allow the use of many different schedules, number of treatments, and allow the rapid testing of many new pharmacological and biological agents for NCT. It would permit a team of many medical specialists to cooperate in the delivery of NCT for many different cancers.

REFERENCES

1. L. Koester, A. Breit, G. Burger. "The Munich Therapy Project RENT, Treatment Planning for External Beam Therapy with Neutrons," Editor, G. Burger, Urban and Schwarzenberg (1981).

2. F.M. Wagner, L. Koester, T. Auberger, W. Reusch, M. Mayr, P. Kneschourek, A. Breit, H. Schraube. "Fast Reactor Neutrons for the Treatment of Superficial Carcinomas," Nucl. Sc. and Eng., 110:32-17 (1992).

3. Y. Maruyama. "Cf-252 Neutron Brachytherapy: An Advance for Bulky Localized Cancer Therapy," Nucl. Sc. Appl. 1:677-748 (1989).

4. R.G. Zamenhof, B.N. Murray, G.L. Brownell, G.R. Wellum, E.I. Tolpin. "Boron Neutron Capture Therapy For the Treatment of Cerebral Gliomas," Med. Phys. 2:47-60 (1975).

5. J.K. Kim J.C. Yanch, M.J. Wilson. "Californium Based Epithermal Neutron Beams for Neutron Capture Therapy," (These proceedings).

6. G.R. Hopkins, J.C. Yanch, J.C. Mourn, J.M., Cheng. "Epithermal Neutron Flux Enhancement for NCT Using Subcritical Multiplying Assembles," (These proceedings).

ANALYTICAL DOSIMETRY AND TREATMENT PLANNING FOR BNCT

David W. Nigg

National Center for BNCT Measurement and Development
Idaho National Engineering Laboratory
EG&G, Idaho, Inc.
P.O. Box 1625
Idaho Falls, Idaho 83415

Calculation of radiation dose distributions for Boron Neutron Capture Therapy (BNCT) is an essential supporting function for *in vivo* BNCT radiobiological research. Such calculations will also ultimately be routinely required for human treatment planning. Radiation transport and dose distribution analysis for BNCT is much more complex than is the case for standard photon therapy. There are several different physical radiation dose components associated with BNCT, each of which has it's own characteristic spatial distribution and Relative Biological Effectiveness (RBE). Most of these radiation dose components result from neutron interactions that take place after the incident neutrons from the treatment beam have undergone several scattering interactions. Accordingly, many of the simplifying assumptions that work well for radiation transport calculations associated with treatment planning for standard photon therapy are not appropriate for BNCT. Full three-dimensional calculations with an explicit treatment of particle scattering and tissue heterogeneities are required.

In standard photon therapy, the absorbed radiation dose is produced by energetic electrons that result from various well-known interactions between photons and tissue components. In BNCT, on the other hand, most of the absorbed dose is deposited by High Linear Energy Transfer (High LET) secondary charged particles produced by various neutron interactions in tissue and in the incorporated boron. The most important of these is, of course, the therapeutically-desirable boron (n,alpha) interaction. All of the other neutron-induced dose components arise from neutron interactions that take place with the normally-present tissue constituents. These components are non-selective and are therefore not desirable. The most significant of these is the proton recoil dose produced by elastic scatter of fast (>10 keV) neutrons on hydrogen. This component is believed to have a relatively high RBE for acute normal tissue damage. The proton recoil dose component can be suppressed by the use of a properly-designed epithermal neutron beam but, as a practical matter, it cannot be totally eliminated. Other significant unavoidable neutron-induced background dose components include the dose that is deposited as a result of the nitrogen (n,p) interaction and several other smaller components. There is also a photon dose component in BNCT that is produced largely as a result of the radiative capture of neutrons by hydrogen as well as from the small incident photon component of the incident treatment beam itself.

Figure 1 shows the essential elements that must be present in any calculational system used for BNCT analytic dosimetry and treatment planning. At the heart of the system is a module for calculating the solution of the three-dimensional Boltzman transport equation for neutral particles, given certain descriptive information as input. This descriptive information consists of a geometric model of the irradiation volume (ideally constructed directly from medical image data), a mathematical description of the treatment beam, and a complete set of coupled neutron-photon cross

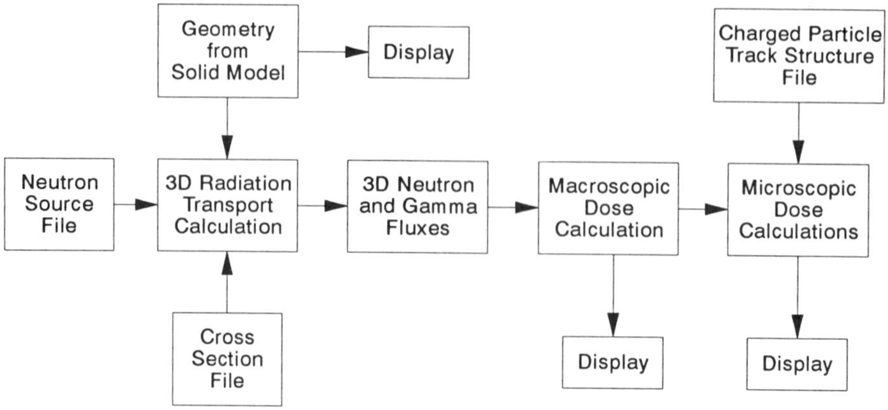

Figure 1. Flow diagram for BNCT neutral particle transport and microdosimetry analysis.

section data for all elements that are present. Geometric descriptions may be constructed by any of three different methods. Each of these methods will be described in detail below. The beam description consists of the spectrum, angular distribution, and the spatial intensity distribution across the plane of incidence, of both the neutron and the photon flux components of the treatment beam. This information is usually constructed from a combination of calculational and experimental data pertinent to the beam of interest. Neutron and photon cross sections are typically taken from standard data collections such as the Evaluated Nuclear Data File[1] and preprocessed into an appropriate multigroup or continuous-energy format.

Given the calculated neutron and photon fluxes throughout the treatment volume, it is a relatively simple matter to construct the resulting macroscopic absorbed dose distribution by folding the calculated flux data with flux-to-dose conversion factor data for each radiation dose component, integrating over the neutron or photon energy range as appropriate, and summing all components. This method is acceptably accurate for all neutron-induced dose components in BNCT since the charged particles that result from these interactions have very short ranges in tissue (on the order of a few micrometers). Thus it is appropriate to assume that, on a macroscopic scale, the spatial distribution for each of these components follows the weighted spatial distribution of the original neutron flux. For the photon-induced dose components, this approximation is, for well-known reasons, less-accurate, especially near voids and material interfaces, but it is currently commonly used in BNCT since the photon dose is typically small compared to the total dose. Standard void and interface correction methods could be used for this dose component in situations where greater accuracy is desired.

As mentioned above, the ranges of the charged particles that produce most of the absorbed dose in BNCT are comparable to typical tissue cell dimensions. As a result it is also necessary to take microdosimetric considerations into account when correlating radiation dose with observed radiobiological response and when attempting to estimate the RBE of the various BNCT dose components under different conditions. For example, the RBE of the boron (n,alpha) dose component can vary widely depending on the intracellular distribution of the particular boron-carrying pharmaceutical that is employed. It is not enough just to know the macroscopic absorbed dose. One must look at what is occurring on a microscopic scale in order to realistically estimate cell survival fractions under various conditions of irradiation. This situation is further complicated by the fact that there are relatively few boron interactions needed per cell to produce a therapeutic dose in BNCT. Thus the statistical nature of neutron interactions with matter must also be explicitly considered. Some cells will experience radiobiologically-significant deviations from the average absorbed dose simply because of this phenomenon. Accordingly, the capability to perform charged particle transport calculations on a microscopic scale and to combine the results with various postulated radiobiological response functions in a statistically correct manner must also be included in calculational systems used for BNCT dosimetry analysis.

Two fundamentally different methods are available for computing the basic neutron and photon flux distributions needed for BNCT. The Monte Carlo stochastic simulation method is the current method of choice, since the complex geometries that are characteristic of biological systems can be very accurately represented using this technique and one can obtain integral information about the radiation dose distribution (e.g. integrated dose over a particular tissue compartment) in quite reasonable (few-minute) computing times. Deterministic solution methods are also feasible. These methods, by their nature, automatically provide very detailed spatial dose distributions as well as integral data, but this is (currently) accomplished at the expense of somewhat longer computing times and some loss of geometric precision. The two different methods therefore complement each other and each has its own appropriate uses.

Three basic methods have been developed for constructing the geometric description of the irradiation volume for use in BNCT dosimetric modeling. The first, and computationally the simplest (and therefore the fastest-running), method is to approximate the irradiation volume using combinations of various geometric primitives (spheres, cubes, cylinders, etc.) described by simple surface equations. This method is incorporated into the well-known MCNP code[2] and has been widely used for all types of radiation transport calculations over the years. In BNCT applications, however, the geometric primitive method has largely been superseded by two independent methods for constructing the necessary geometric descriptions directly from medical image data.

The first of the direct geometric reconstruction methods employs a "voxel reconstruction" technique whereby each plane of medical image data is partitioned into square (typically 1-centimeter) areas, each of which is assigned material properties based on CT density or some other appropriate criterion[3,4]. One then mathematically stacks the square regions generated in each medical image plane to construct a three-dimensional array of parallelpipeds that can be input to a standard Monte Carlo code such as MCNP. The various tissue compartments are represented as combinations of adjacent parallelpipeds. This method permits reconstructions to be generated relatively quickly and it is very amenable to automation. In addition the same method can be used in exactly the same form to construct geometry for a deterministic radiation transport calculation if this is desired. The primary disadvantage of the method stems from the fact that complex curved surfaces are represented by a series of orthogonal voxel facets. This can adversely affect the accuracy of radiation transport calculations based on this technique, unless one employs a relatively large number of small voxels and accepts the longer computer execution times that result.

The second method for direct construction of geometry from medical images is based on the calculation of non-uniform rational B-spline fits to the various tissue compartment (or any desired subcompartment) surfaces. With this method, one first electronically outlines the regions of interest (e.g. skin, skull, brain, various tumor regions, etc.) on each medical image plane. These outlines are then mathematically combined to produce detailed B-Spline equations describing the three-dimensional surfaces that enclose each volume of interest. These surface equations completely describe the geometry and can subsequently be used in a Monte Carlo radiation transport calculation. The spline surfaces can also be combined with geometric primitive surfaces to further specify the calculational geometry if desired. A spline surface reconstruction capability has been incorporated into the special-purpose bnct_edit/rtt_MC code sequence for medical dosimetry[5,6]. (The rtt_MC code will also accept parallelpiped arrays constructed using the voxel technique if desired). The primary advantage of the surface equation reconstruction technique is that it is very efficient and accurate. The complex curved surfaces of the various anatomical structures of interest in BNCT dosimetry are very faithfully represented and particle tracking can be accomplished with the minimum number of boundary crossings that is consistent with the number of tissue compartments modeled. The primary disadvantage is that, unlike the situation with voxel-based reconstructions, one does not automatically obtain spatially-detailed dose distributions. This problem has been overcome without significant loss of calculational efficiency, however, with the development of the so-called "Subelement Monte-Carlo" technique that is available in the rtt_MC code.[5] A secondary disadvantage of the surface reconstruction method is that it may not be easily adaptable to deterministic methods for calculating radiation transport. This remains to be seen.

In conclusion, it may be stated that the technology for performing radiation dose distribution analysis and treatment planning calculations for BNCT is well into the advanced stage. The necessary three-dimensional radiation transport calculations can be accurately performed using any of several different combinations of geometric models and mathematical techniques for obtaining the solution to the governing equations. Currently-ongoing efforts have focused on providing detailed, understandable output displays (dose contours superimposed on medical image data, dose-volume histograms, 3-D visualizations, etc.) , medical image manipulations, tumor location capabilities and other such "user-friendly" features, similar to those that are available with advanced photon treatment planning systems.

ACKNOWLEDGMENTS

The author wishes to express his gratitude to Dr. R. Barth for the opportunity to present this Plenary Lecture at the Fifth International Symposium on Neutron Capture Therapy and to Dr. R. Zamenhof, Dr. L. Reinstein, and Dr. H. Liu for providing some of the visual aides that were used in the oral presentation.

This study was performed under the auspices of the U.S. Department of Energy, Office of Energy research, DOE Field Office, Idaho, under Contract Number DE-AC07-76ID01570.

REFERENCES

1. R. Kinsey, Data formats and procedures for the evaluated nuclear data file ENDF, Brookhaven National Laboratory, BNL-NCS-50496, 1979.

2. MCNP monte carlo neutron and photon transport code systems, Radiation Shielding Information Center, Oak Ridge National Laboratory, U.S.A., CCC-200A.

3. R. Zamenhof, et. al., Treatment planning for neutron capture therapy of glioblastoma multiforme using epithermal neutron beam from the MITR-II research reactor and monte carlo simulation, in: "Progress In Neutron Capture Therapy for Cancer," B.J. Allen, et. al. (ed), Plenum Press, New York, NY (1992).

4. L.E. Reinstein, et. al., SBNCT-PLAN: A 3-dimensional treatment planning system for boron neutron capture therapy, *Proceedings of the Fifth International Symposium on Neutron Capture Therapy*, OSU, September 1992, (To be published).

5. F.J. Wheeler and D.W. Nigg, Three-dimensional radiation dose distribution analysis for boron neutron capture therapy, *Nuclear Science and Engineering*, 110, 16-31, January 1992.

6. D.E. Wessol and F.J. Wheeler, Methods for creating and using free-form geometries in monte carlo particle transport, *Nuclear Science and Engineering*, (To be published).

POTENTIAL MCNP ENHANCEMENTS FOR NCT

Guy P. Estes and William M. Taylor

Radiation Transport Group
Los Alamos National Laboratory
Los Alamos, NM 87545

INTRODUCTION

MCNP,[1] a Los Alamos National Laboratory Monte Carlo radiation transport code, is currently widely used in the medical community for a variety of purposes including treatment planning, diagnostics, beam design, tomographic studies, and radiation protection. This is particularly true in the Neutron Capture Therapy (NCT) community as evidenced by the numerous uses of MCNP in various publications.[2] The current widespread medical use of MCNP after its general public distribution in about 1980 attests to the code's general versatility and usefulness, particularly since its development to date has not been influenced by medical applications. This paper discusses enhancements to MCNP that could be implemented at Los Alamos for the benefit of the NCT community. These enhancements generally fall into two categories, namely 1) those that have already been developed to some extent but are not yet publically available, and 2) those that seem both needed based on our current understanding of NCT goals, and achievable based on our working knowledge of the MCNP code.

MCNP is a general, coupled neutron/photon/electron Monte Carlo code developed and maintained by the Radiation Transport Group at Los Alamos. It has been used extensively for radiation shielding studies, reactor analysis, detector design, physics experiment interpretation, oil and gas well logging, radiation protection studies, accelerator design, etc. over the years. It is estimated to have about 300 person-years of development effort to its credit, and traces it roots back to the World War II Manhattan Project and scientists such as Fermi, von Neumann, Ulam and others.

MCNP is a three-dimensional (3-D) geometry, continuous energy physics code capable of modeling complex geometries, specifying material regions such as organs by the intersections of analytical surfaces. An example of the geometric complexity possible is shown in Fig. 1, where a partial MCNP model of the MIRD human phantom[3] is presented. This figure is a SABRINA code[4] 3-D picture of the MCNP model, where the flesh, ribs, and some organs have been omitted from the plot for clarity. This model will be discussed later in the context of potential medical imaging studies.

NEAR-TERM ENHANCEMENTS

There are a number of patches, techniques, etc. that have been developed in support of Los Alamos programs or for one-of-a-kind studies that have never been

Advances in Neutron Capture Therapy, Edited by
A.H Soloway *et al.*, Plenum Press, New York, 1993

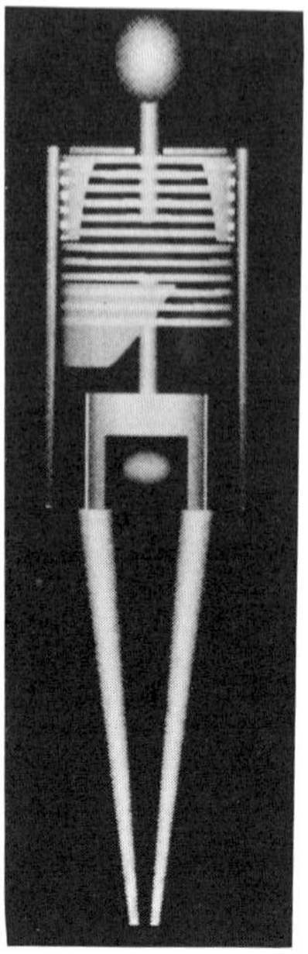

Figure 1. SABRINA-generated three-dimensional plot of the MCNP geometry model for the MIRD human phantom (some bones and organs not shown).

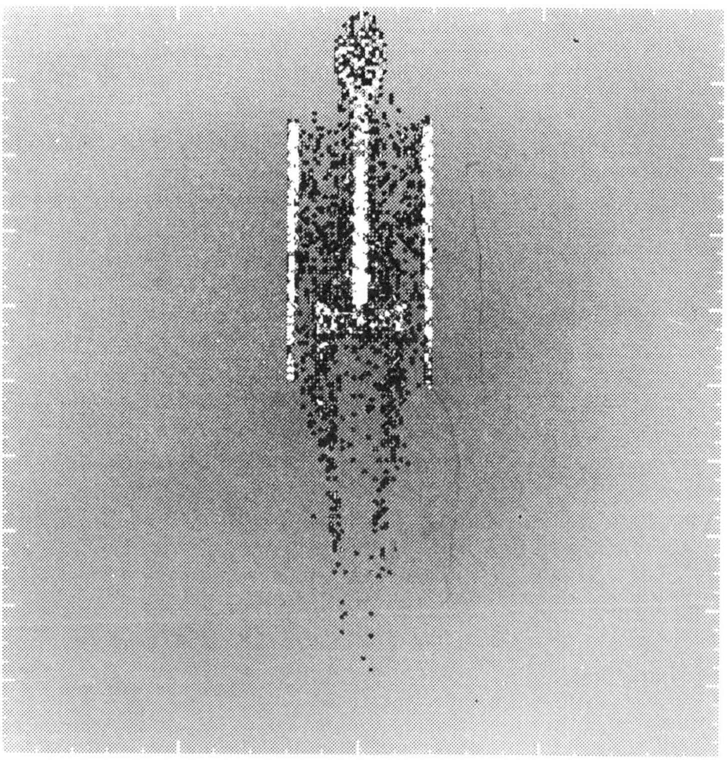

Figure 2. MCNP-computed image (simulating a nuclear-medicine image) of the MIRD phantom model with ^{99m}Tc in the major bones (arms, legs, pelvis, spine, and skull).

incorporated into MCNP because of lack of programmatic incentive. Most of this work exists in the form of patches which may or may not be consistent with the current version of MCNP, or as options in other codes. Examples of this category discussed below are the pinhole "camera" imaging capability, the lattice geometry capability, and the MCNP/SABRINA particle track display capability. While some of these capabilities already exist informally, getting them into the public code would require more than a trivial amount of work because they would have to be carefully integrated with all existing options in the code consistent with our high quality assurance standards.

The first example given in this category is the pinhole "camera" imaging capability that could in principle be used to generate the equivalent of medical images, such as those produced by Anger cameras, uniformly redundant arrays, or coded apertures. Such images could be used for the purpose of helping interpret radioisotope distributions and concentrations in regions of the body from actual medical images. This capability is illustrated using the MIRD human phantom model of Fig. 1. Photon-emitting radioisotopes were assumed to be present in the major bones (skull, spine, arms, pelvis, and legs) of the model, and were imaged through an ideal pinhole in an opaque medium as shown in Fig. 2. The bones with sources stand out clearly in the figure, and the effect of scattered radiation in flesh and other bones can be seen in the interstitial regions. Since scattered and unscattered radiation can be segregated quite easily in MCNP, this technique could be used to computationally quantify scattered components in actual measured medical images, subtract them out, and thereby produce more unambiguous images for medical diagnoses. This will be discussed further in a later section in connection with creating patient-specific MCNP models from CT and MRI images.

The second potential near-term capability discussed is the use of the lattice capability, developed for reactor studies, to model the many cells needed for accurate dose calculations in the head (or body). It is maintained by some radiation oncologists that more accuracy is needed in treatment planning to insure complete tumor kill, especially near material interfaces (e.g., brain/sinuses) where more approximate radiation transport calculations break down. This lattice geometry capability has been tested for a brain geometry with several million cells, but work is needed to provide for automatically varying the material composition of each cell. An example of the geometry created is shown in Fig. 3 where the 0.15 cm mesh is typical of MRI scan resolutions.

The third near-term potential category is that of using the SABRINA particle track display capability for qualitative scoping studies of neutron beam suitability for specific tumor treatment. The MIRD phantom model head is shown in Fig. 4 with particle tracks shown. In this particular case, the energy of the incoming neutrons is too low, and most of them are thermalized before reaching the simulated tumor. Since this example was created with only 100 neutrons in the beam, such qualitative studies could be easily done in a quick scoping manner to determine the beam characteristics needed for a particular treatment scenario.

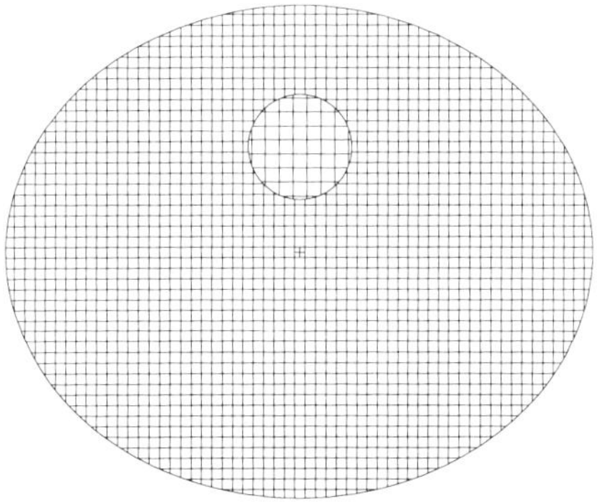

Figure 3. Plot of an MCNP geometry with the same voxel size as a typical MRI scan. The geometry description required only 23 lines of input using the MCNP lattice option, including the simulated tumor.

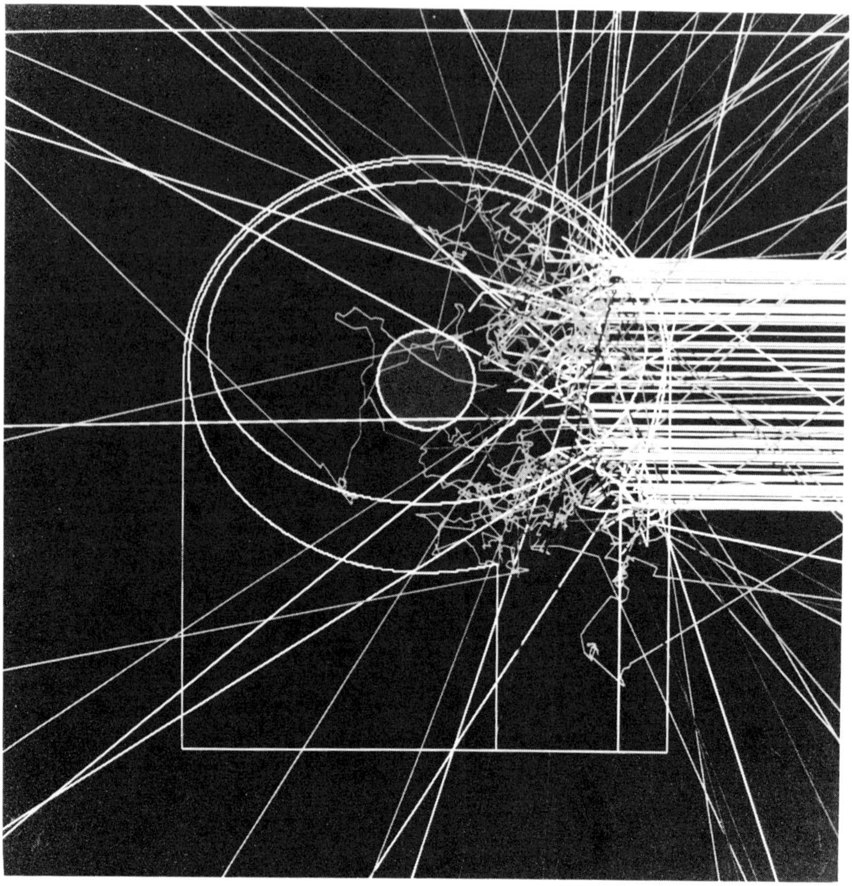

Figure 4. SABRINA-generated plot of neutron paths and scatterings inside the MCNP model of the MIRD phantom head.

POSSIBLE FUTURE ENHANCEMENTS

With regard to longer range enhancements, it is felt that a number of new MCNP features could be developed that would be useful to the NCT community in the future. This assertion is based on the fact that the code was not developed with medical use in mind, and that improvements in efficiency, physics and ease of use for this purpose are reasonably attainable at Los Alamos based on our working knowledge of MCNP. The most attractive of these seems to be the potential for performing patient-specific MCNP computations for diagnostic test interpretation and treatment scenarios, and for improving computational efficiency in order to perform near-real time treatment planning with these models.

By 1) generating patient-specific body geometry from CAT and/or MRI scans and 2) simulating diagnostic images (e.g., CAT, Anger camera, SPECT) using these patient models, 3) high-accuracy dose distributions, radioisotope concentration assay, and scatter corrections would be possible for these patients. Such patient-specific calculations would undoubtedly improve the quality of care for patients, as well as lead to a more quantitative understanding of the effects of dose on tumors and healthy tissue function.

Ultimately, in order for MCNP to be used for routine clinical treatment planning, it will be necessary to speed up the calculations by several orders of magnitude. This can be accomplished by several means. First, the regular rectangular mesh that results from creating patient-specific geometries from CT and MRI scans would be much simpler to calculate than the general geometry that is now built into MCNP. It has already been demonstrated at Los Alamos that providing special coding for regular mesh decreased computational time by a factor of 10 for similar problems. Secondly, computation time is becoming increasingly affordable, in particular with regard to desktop workstations. A Sun Sparc 2 now runs MCNP at about 1/4 the speed of one Cray YMP processor. The Sparc 10 that should be available in 1993 will be equivalent to about 4 Cray processors, a factor of 16 increase in computing power. Further advances are inevitable. Finally, Los Alamos and others are working on the "multi-tasking" of MCNP on workstations on a given network in order to fully utilize available CPU's when the workstation owner is not doing so. At Los Alamos there is a cluster of 16 IBM RISC 6000's, which, when multi-tasked with MCNP, will equal about 10 Cray processors.

SUMMARY AND CONCLUSIONS

In conclusion, it appears that patient-specific 3-D treatment planning and diagnosis is possible if a number of relatively straightforward improvements in MCNP can be implemented. Use of this capability for near real-time therapy planning should be practical in the near future when factoring in the likely advances in workstation computing power and networking. With a coordinated effort from the NCT community, it should be possible to obtain support from DOE Defense Programs dual-use technology transfer initiatives for Los Alamos to provide the described MCNP enhancements described for the NCT community.

REFERENCES

1. Briesmeister, J. F., ed. (1986), *MCNP - A General Monte Carlo Code for Neutron and Photon Transport - Version 3A*, (Los Alamos National Laboratory report LA-7396-M).
2. Harling, O. K., Bernard, J. A., and Zamenhof, R. G., eds. (1989), *Neutron Beam Design, Development, and Performance for Neutron Capture Therapy*, (Plenum Press, New York).
3. Synder, W. S., Ford, M. R., Warner, G. G., and Watson, S. B., (1974), *A Tabulation of Dose Equivalent per Microcurie-Day for Source and Target Organs of an Adult for Various Radionuclides*, (Oak Ridge National Laboratory report ORNL-5000).
4. West, J. T., III (1986), "*SABRINA: An Interactive Three-Dimensional Geometry-Modeling Program for MCNP*," (Los Alamos National Laboratory report LA-10688-M).

POST TREATMENT DOSE DISTRIBUTION EVALUATION FOR A RECENT NCT PATIENT

Gregory J. Storr,[1] and Floyd J. Wheeler

Idaho National Engineering Laboratory, EG&G Idaho, PO Box 1625,
Idaho Falls, ID 83415-1575, USA
[1]On attachment from Nuclear Technology Program, Ansto, PMB 1,
Menai 2234, NSW, Australia

INTRODUCTION

As a prelude to establishing, (1) criteria for selection of Boron Neutron Capture Therapy (BNCT) epithermal beam clinical trial patients and (2) treatment planning strategies, post-treatment evaluations of dose distributions in several BNCT patients are being done using the Idaho National Engineering Laboratory (INEL) BNCT Program's patient treatment planning software (*bnct_edit* and *rtt_MC*)[1,2].

The objectives of this work are :

- calculate the dose delivered to the BNCT treated patients
- calculate the possible dose distribution for the patients using the proposed Georgia Institute of Technology's reactor epithermal beam[3]
- correlate calculated dose with effect

PATIENT DATA

We have data for seven recent Australian and US BNCT patients treated in Japan and at this stage we report in detail on one patient study. As well as individual patient medical images, accurate calculations of dose also relies on knowledge of boron concentrations in the tumor, tissue surrounding the tumor, normal tissue, and blood, heavy water concentrations in the irradiation volume, and dose modification factors (e.g. radiation component RBE's). The patient (# 1) we report on was diagnosed with Glioblastoma Multiforme in the left occipital lobe in February 1991, was treated with BNCT in May 1991 at the JAERI Research Reactor No. 2 (JRR-2), and at September 1992 was surviving.

Table 1 lists the information on patient 1's treatment at JRR-2. The boron tumor concentration compares well with three Australian patients blood boron levels, (measured between 12μg/g and 35μg/g during irradiation) as does the D_2O concentration of 9 wt% in four Australian patients[4].

Table 1. Patient 1 BNCT treatment data

Irradiation time (min)	270.0
Reactor Power (MW)	10.0
Max. th. flux - dura (n cm^{-2} s^{-1})	7.39×10^8
^{10}B concentration - tumor (μg/g)	18.0
^{10}B concentration - blood (μg/g)	0.3
D_2O concentration (wt %)	9.0
Tumor dose - brain surface (cGy)	4600.0
Tumor dose - 5cm deep (cGy)	746.0

MODELING STRATEGY

Figure 1 shows the modeling strategy used to obtain the dose distributions. The method adopted in this work was to prepare burnup dependent smeared fuel element cross sections and all other material cross sections with the AUS neutronics system[5]. As all the Japanese reactors used for BNCT have beams of mostly thermal energy neutrons, the cross section energy group structure selected comprised 26 neutron groups (10 groups < 0.414 eV) and 4 photon groups. These cross sections were then used in discrete ordinates models, using DORT[6], of the reactor core (XY) and filter (RZ). The calculated neutron and gamma angular fluxes at the patient irradiation position were then used as a source input to the patient treatment Monte Carlo model.

PATIENT MODELING

Modeling the irradiation volume of a patient is achieved by using the two modules of the INEL patient treatment planning software; (1) *bnct_edit*[1], a geometry reconstruction

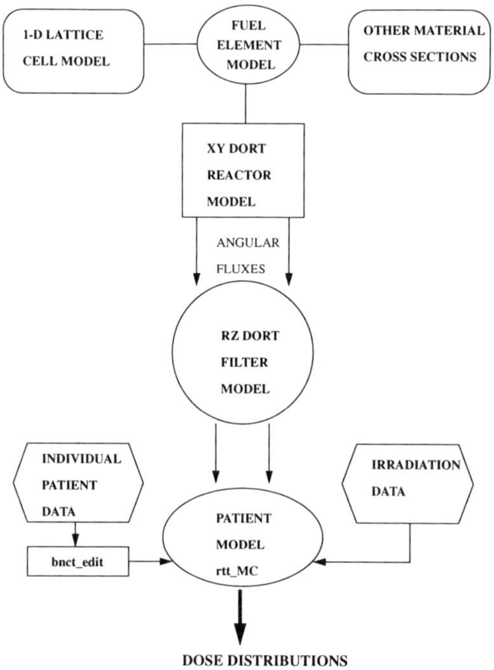

Figure 1. Modeling strategy for dose distribution analysis

package for modeling individual patient geometries as a series of surface equations, and (2) *rtt_MC*^{2}, a neutron and photon Monte Carlo code developed for BNCT dose distribution calculations. The modules are linked through the surface geometry description of the treatment volume, in this case the patient's head. The angular flux data from the filter model edit plane (the beam hole plane) and geometry descriptions for the beam and the patient are combined to form the overall problem. Material assignments are made from standard libraries and for patient 1 a special brain material that includes 9wt% D_2O replacement was used.

Four bodies were constructed in the three-dimensional geometry description of patient 1, corresponding to the head, brain, area of edema and swelling (designated as the target) and the tumor. The body representing the head was modified to include the surgical craniotomy for the patient. The surgical debulking is not represented in this model however, the approximation that the tumor still exists at the time of therapy is valid provided the material composition and boron concentration in this region are represented accurately.

RESULTS

JRR-2 irradiation

The calculated thermal flux was normalized to the measured value at the position adjacent to the treatment position - i.e. 1.0×10^9 n cm^{-2} s^{-1}. When the calculated total thermal flux was normalized to the reactor power, the thermal flux was approximately twice the measurement. Comparison of in-core thermal fluxes showed intensities consistent with measurement in this reactor. There are uncertainties in the exact dimension and composition of the graphite and lead shielding layers adjacent to the reactor aluminium tank and

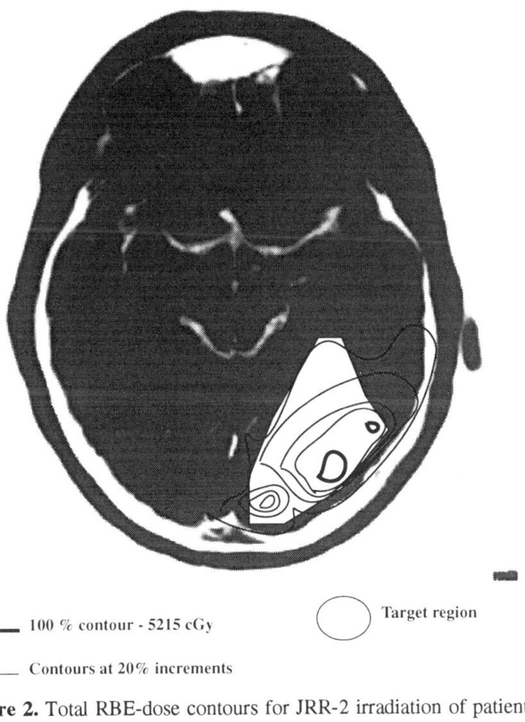

Figure 2. Total RBE-dose contours for JRR-2 irradiation of patient 1.

between the patient treatment position. We believe this to be the cause of the difference between the measurement and calculation. Scoping calculations varying the thickness of the graphite and lead layers outside the reactor tank, showed that while neutron intensity is affected neutron spectral changes are slight for various lead and graphite combinations. The large heavy water moderator layer principally determines the neutron spectrum however, the activation gamma's from the lead and graphite layers change significantly dependent on the layer thicknesses. The bismuth shield helps to reduce this effect significantly.

Figure 2 shows an axial MR image of patient 1 close to the center-line of the target and tumor volumes, with calculated RBE-dose contours overlaying the image. The maximum RBE-dose of 8500 cGy lies within the tumor volume. The target volume which would have contained edema, normal brain and presumably tumor, appears on the MR images as a single structure. Variability in boron concentrations within the tumor and target volumes is well recognized,[7] and this presented us with a range of choices for assigning boron concentrations in the target volume. In Figure 2 the RBE-doses correspond to a target boron level half that measured in the tumor. Note that this calculation gives a secondary peak in the target and outside the tumor volume of approximately 5300 RBE cGy. Because of the presumed mixture of normal tissue, edema and tumor in the target this value is considered in this calculation to be the peak dose to healthy tissue. At the maximum depth of the target volume the RBE-dose is calculated to be approximately 1000 cGy.

Table 2 gives a comparison of physical dose from the calculation reported here and data from Japan. The total physical dose comparison is excellent, however there is a significant difference when comparing the gamma and boron dose components. The possibility exists that the Japan gamma dose does not include the induced gamma component.

GTRR calculation

Patient 1's geometry was modified from the thermal beam case to fill in the "craniotomy", the input source was changed to the proposed Georgia Tech Research Reactor (GTRR) epithermal neutron beam[3] and the D_2O was removed from the tissue materials. Table 3 shows a comparison of percentage of dose for the major dose components at increasingly deep locations in patient 1's reconstructed geometry. In Table 3 100% dose at a specified location equals the total RBE-dose calculated with the JRR-2 beam. The epithermal beam RBE-dose peak is deeper than for the thermal beam however, for this patient whose tumor was located close to the brain surface, a favorable dose distribution is achieved from the thermal beam. The deepest portions of the target volume, which would have been likely to contain tumor cells have a larger dose for the epithermal beam than the thermal beam. Normal brain at depth receives less radiation with the thermal beam. A larger advantage for the epithermal beam in patients with deep tumors is expected.

A correlation of dose with effect will only be possible when more patients are studied. However, 14 months post-BNCT patient 1 underwent a full neurological exam, MR and positron emission tomography (PET) studies and the evidence from the MR study states, " .. no evidence of recurrent tumor."[8] As this patient was diagnosed with Glioblastoma Multiforme, and the early prognosis was 6 months of life, it seems obvious that therapeutic benefit was gained from this patient's decision to undergo Neutron Capture Therapy.

Table 2. Tumor Physical Dose Comparison - Patient 1

	B-10	Gamma	N-14	Fast	Total
rtt_MC	2949	901	278	364	4492
Japan	3700	162	238	463	4563

Table 3. Percentage of JRR-2 RBE dose components with location, for modeled thermal and epithermal beams

	JRR-2				GTRR			
	B-10	Gamma	N-14	Fast	B-10	Gamma	N-14	Fast
Tumor	80	6	6	8	63	22	5	1
Target (1)	65	13	10	12	55	35	8	2
Target (2)	59	19	9	13	73	47	11	1
Brain	4	46	17	33	6	125	25	1

CONCLUSIONS

This study has modeled the fuel, reactor and filter of the JRR-2 reactor in Japan, and applied the calculated neutron and gamma fluxes to a model of a BNCT patient treated at that reactor. The physical dose distribution compares well with data from Japan. The calculated doses for this patient show a favorable distribution using a thermal energy neutron beam. A comparison with the proposed GTRR epithermal neutron beam, showed that dose at depth was larger for the higher energy beam for the same peak healthy tissue dose, even where the patient model did not include a craniotomy or tissue loaded with D_2O.

ACKNOWLEDGEMENTS

Ron Dorn MD, from the Mountain States Tumor Institute in Boise, Idaho, provided the patient dose information. Dave Madden MD, from the Eastern Idaho Regional Medical Center assisted with patient tumor and target location. Robert Weaver of EG&G Idaho provided software and computer assistance crucial to this work. This study was performed under the auspices of the U.S. Department of Energy, Office of Energy Research, DOE Field Office, Idaho, under Contract Number DE-AC07-76IDO1570.

REFERENCES

1. D. Wessol, J. Cobb, F. Wheeler, E. Cohen, B. Cobb, and D. Barber, Interactive Generation of 3-D B-Spline Objects from Planar Image Data for Multidimensional Analysis, *presented at Society of Exploration Geophysicists 60th Ann. Int. Mtg. and Exposition, San Francisco, CA,(1990).*
2. F.J. Wheeler and D.W. Nigg, Three-Dimensional Radiation Dose Distribution Analysis for Boron Neutron Capture Therapy, *Nucl.Sci.Eng.*110:16(1992).
3. D.W. Nigg, G.J. Storr, and F.J. Wheeler, Physics Parameters for an Epithermal- Neutron Facility at the Georgia Institute of Technology Research Reactor, *These proceedings.*
4. N. Blagojevic, G. Storr, B. Allen, H. Hatanaka, and H. Nakagawa, Role of Heavy Water in Boron Neutron Capture Therapy, *presented at International Workshop on Macro and Microdosimetry and treatment planning for Neutron Capture Therapy, Cambridge, MA,(1991).*
5. G.S. Robinson, "A Guide to the AUS Modular Neutronics Code System," AAEC/E645 (1987)
6. W.A. Rhoades and R.L. Childs, DORT - Two-Dimensional Discrete Ordinates Transport Code, Oak Ridge National Laboratory, CCC-484,(1989).
7. D. Haritz, Results of Continued Clinical Investigations of BSH in Patients with Malignant Glioma, *These proceedings.*
8. W. Sweet, Massachusetts General Hospital, [private communication].

MEASUREMENTS AND CALCULATIONS FOR BORON NEUTRON CAPTURE THERAPY TREATMENT PLANNING

M.W. Konijnenberg[1], L.G.H. Dewit[1], B.J. Mijnheer[1], C.P.J. Raaijmakers[1], F. Stecher-Rasmussen[2], and P.R.D. Watkins[3]

[1]The Netherlands Cancer Institute, Amsterdam, the Netherlands
[2]Netherlands Energy Research Foundation ECN, Petten, the Netherlands
[3]Institute for Advanced Materials JRC, Petten, the Netherlands

INTRODUCTION

Treatment planning aims at optimizing the radiation dose delivered over the target volume, while keeping the dose delivered to the healthy tissue at a minimum. Most efforts in treatment planning research for BNCT tend to concentrate on the second part of this aim [1, 2], as inhomogeneities in the sub- and inter-cellular boron distribution may cause considerable dose inhomogeneity. Homogeneity of the thermal and epithermal neutron fluence to the target volume should be an important goal of BNCT research.

The boron dose is directly related to the thermal neutron fluence, while the skin dose is mainly determined by the fast neutrons. The measurements and calculations therefore are dedicated to determine the distributions of the thermal (with an energy cut-off at 0.414 eV) neutron fluence and total neutron dose within a phantom. The measurements were performed at two different beams.

Therefore we have performed measurements of the thermal and intermediate energy neutron fluence and neutron and gamma-ray dose distributions in various phantom types in the therapy beam HB11 at the High Flux Reactor (HFR) and in an epithermal neutron beam of the Low Flux Reactor (LFR) in Petten. Both experimental set-ups have been modeled with the Monte Carlo neutron and photon transport programme MCNP [3]. Calculations assessed the thermal and fast neutron fluence spatial distributions to be compared to the experimental values. Multiple field irradiations in an ellipsoid head phantom with a large boron loaded target area have been determined.

MATERIALS AND METHODS

In the LFR an epithermal neutron beam has been installed with a maximum thermal neutron fluence rate of 2.7 10^7 $cm^{-2}s^{-1}$ (at 30 kW reactor power). The removal of the thermal neutrons from the beam was achieved with two 1 mm thick Cd-sheets, separated at a distance of 1.30 m. and a 5 mm thick B_4C plate at the beam exit. The beam was collimated to a

Advances in Neutron Capture Therapy, Edited by
A.H Soloway *et al.*, Plenum Press, New York, 1993

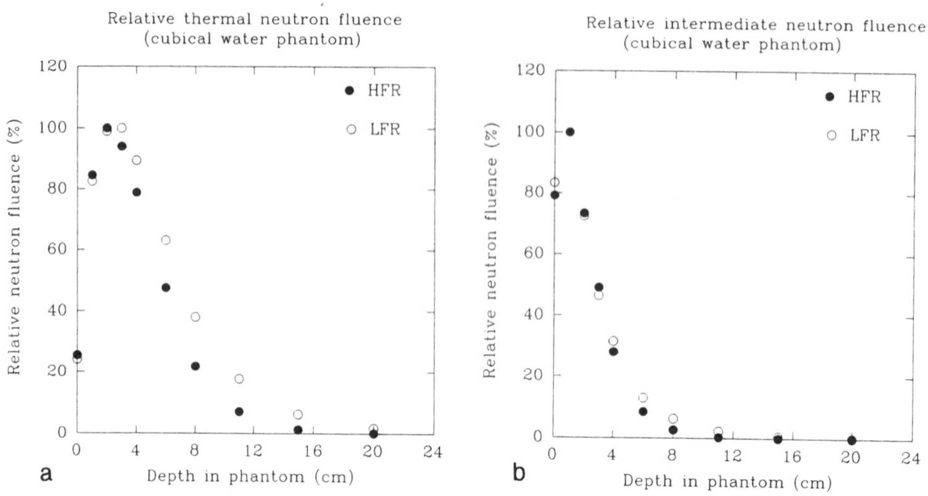

Figure 1. Comparison of the relative neutron fluences as measured at various depths in the rectangular water phantom at the HFR (•) and at the LFR (∘). The thermal neutron fluence (a) was determined from the activity of Mn-foils, whereas the intermediate neutron fluence (b) was assessed from Au-foils placed at these depths.

circular opening of 8 cm diameter, like the collimator used in HB11. The experimental setup as well as the MCNP model describing its exact geometry have been reported before [5, 6, 7].

All experiments were repeated at the HB11 therapy beam of the HFR. The thermal neutron fluences were measured with Au/Mn foil sets at fixed positions within the phantoms. The dose rate for neutrons and photons was determined with two ionization chambers (Exradin). The γ-ray and neutron dose was measured with a tissue-equivalent (TE) plastic chamber flushed with TE gas of the composition: 3.2 % N_2, 32.9 % CO_2 and 63.9 % CH_4. The photon dose was assessed from the readings of a Mg chamber flushed with Ar gas, which is relatively insensitive to neutrons. The photon doses were also determined with TLDs at the same fixed points of the foil measurements.

All calculations were performed on a SUN Sparc2 workstation with MCNP version 3b. The neutron dose and the fluence distributions were calculated over a 1 cm^3 element mesh superimposed over the phantom volume. Besides the models for a cylindrical and a rectangular phantom [7] an anthropomorphic head phantom consisting of two nonconcentric ellipsoids for the skull and brain [4] was modeled. A large target area was defined in the left hemisphere of the brain with its contours described mathematically by the equation:

$$\left(\frac{x+2.5}{2.5}\right)^2 + \left(\frac{y-1}{4}\right)^2 + \left(\frac{z}{2.5}\right)^2 = 1, \tag{1}$$

with a uniform boron concentration of 50 ppm ^{10}B. In the surrounding healthy brain tissue the boron load was modeled to be 5 ppm ^{10}B, whereas the skull had a higher concentration of 10 ppm ^{10}B, due to the larger blood amount in the skin and the bone marrow. All tissue was assumed to be uniformly distributed matter with compositions according to the ICRP recommendations [8] regardless of any anatomical detail. It took approximately 15 hours of CPU-time to calculate the fluences and doses in all mesh elements for each irradiation field.

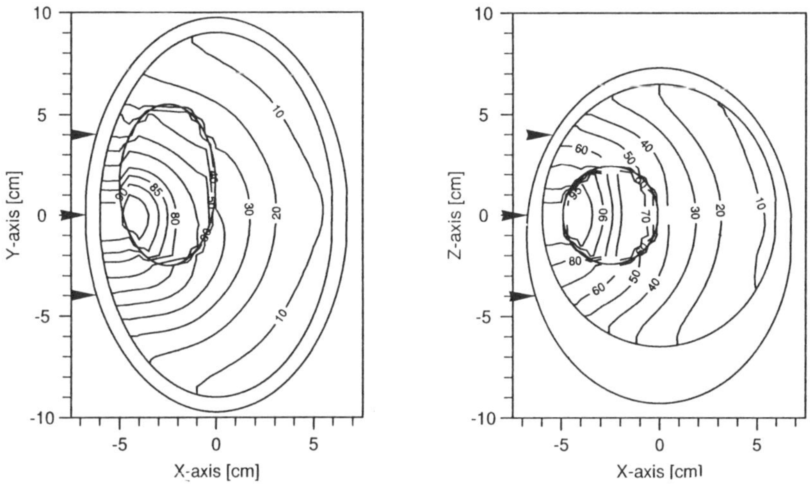

Figure 2. Relative thermal neutron fluence (100% $\equiv$ 4.8 10^6 n/cm^2s) distribution in an ellipsoid phantom for the HB11 beam with a diameter of 8 cm coming in along the arrows indicated. The ^{10}B concentration in the skull was 10 ppm, in the brains 5 ppm and in the target volume 50 ppm.
a. Horizontal plane b. Coronal plane

RESULTS AND DISCUSSION

The rectangular water phantom (with dimensions $50 \times 60 \times 40$ cm^3 and PMMA walls 8 mm thick) was irradiated by both neutron beams. The measured relative neutron fluence depth curves measured with foils, shown in figure 1, demonstrate the similarity of the two neutron beams. The thermal neutron fluence depth curve of figure 1(a) shows the LFR-beam to produce a relatively larger contribution at depth. This effect is caused by thermalization of the larger fast neutron component in the LFR-beam compared to the HFR.

When a large target volume with a high boron load is placed within the human head-like phantom the calculated thermal fluence distribution shows a 20% decrease in the thermal fluence rates behind the tumor (figure 2). Within the target volume the thermal neutron fluence distribution decreases from its maximum down to 40%. The corresponding fast neutron dose distribution shows no significant influence of the ^{10}B loaded target volume.

In an attempt to improve the homogeneity of the thermal neutron distribution over the target area a 4-field irradiation set-up was used (figure 3). In this situation the thermal neutron fluence within the target volume drops from a maximum to 70%. Adding even a fifth field, coming in at $-45°$ with the X-axis further improvement can be achieved, its clinical relevance however can be questioned. The fast neutron dose distribution benefits also from this 4-field technique, (figure 4). The neutron dose is now distributed over four skin areas with a dose reduction by 40% compared to a single-field irradiation, relative to an equal thermal neutron flux in the target volume.

CONCLUSIONS

A homogeneously distributed thermal neutron fluence over a realistic target volume in a human head-like phantom has been achieved using a multiple field technique, as otherwise

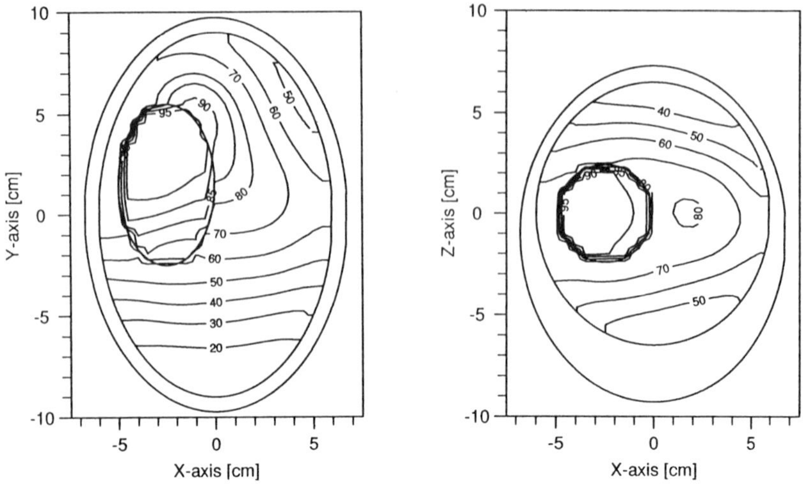

Figure 3. Relative thermal neutron fluence (100% $\equiv$ 1.02 10^7 n/cm^2s) distribution in the head-like phantom for four irradiation fields.

a. Horizontal plane b. Coronal plane

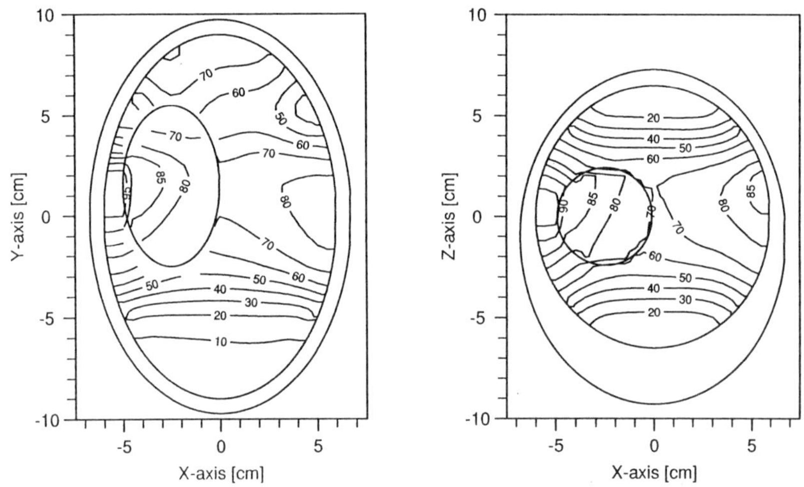

Figure 4. Relative neutron dose distribution (100% $\equiv$ 0.31 Gy/h) for the four-field technique.

a. Horizontal plane b. Coronal plane

its variation would be an unacceptable 60%. According to the article by Yanch et al.[9] a larger field size for a monoenergetic neutron beam will produce a more homogeneous thermal fluence distribution in a single-field irradiation. Further calculations will be performed to investigate its implication for the HFR beam. The influence of different boron concentrations in tumour and healthy tissue remains to be investigated as well.

ACKNOWLEDGEMENTS

This work was financially supported by the Netherlands Cancer Society (NKB grant NKI 90-03).

REFERENCES

[1] D.W. Nigg, P.D. Randolph and F.J. Wheeler,Demonstration of three-dimensional deterministic radiation transport theory dose distribution analysis for neutron capture therapy,*Med.Phys.18(1991)43–53*

[2] H.B. Liu and R.M. Brugger,Dose calculations and treatment planning for the Brookhaven Medical Research Reactor (BMMR) NCT facility, contribution to this symposium.

[3] J.F. Briesmeister (ed.),MCNP, A general Monte Carlo code for neutron and photon transport.*Report LA-7396-M, revision 2 (1987)*

[4] O.L. Deutch and B.W. Murray,Monte Carlo dosimetry calculation for boron neutron capture therapy in the treatment of brain tumors,*Nuclear Tech. 25(1975)320–339*

[5] C.P.J. Raaijmakers, M.W. Konijnenberg, B.J. Mijnheer, F. Stecher-Rasmussen and H. Verhagen,Determination of the gamma-ray dose in an epithermal neutron beam,contribution to the workshop "Dose components and biological effects in conventional radiotherapy and neutron capture therapy.", Essen (1992),*Strahlentherapie und Onkologie*, (in press).

[6] M.W. Konijnenberg, C.P.J. Raaijmakers, L. Dewit, B.J. Mijnheer, R.L. Moss, F. Stecher-Rasmussen and P.R.D. Watkins,A Monte Carlo technique for treatment planning of Boron Neutron Capture Therapy, contribution to the 7th symposium on neutron dosimetry 1991 Berlin,*Rad. Prot. Dos. 44(1992)443–446*

[7] M.W. Konijnenberg, B.J. Mijnheer, C.P.J. Raaijmakers, F. Stecher-Rasmussen and P.R.D. Watkins,An investigation of the possibilities of BNCT treatment planning with the Monte Carlo method, contribution to the Essen workshop, Essen, Germany (1992), *Strahlentherapie und Onkologie*, (in press).

[8] ICRP "Report of the task group on reference man."*Report 23(1975)*, Oxford, Pergamon.

[9] J.C. Yanch, X-L. Zhou and G.L. Brownell:A Monte Carlo investigation of the dosimetric properties of monoenergetic neutron beams for neutron capture therapy.*Radiat. Res. 126(1991)1–20*

DOSE CALCULATION AND TREATMENT PLANNING FOR THE BROOKHAVEN NCT FACILITY

Hungyuan B. Liu and Robert M. Brugger

Medical Department, Brookhaven National Laboratory, Upton, NY 11973

INTRODUCTION

Consistency of the calculated to measured fluxes and doses in phantoms is important for confidence in treatment planning for Boron Neutron Capture Therapy (BNCT) at the Brookhaven Medical Research Reactor (BMRR)[1]. Two phantoms have been used to measure the thermal and epithermal flux and gamma dose distributions for irradiations at the BMRR and these are compared to MCNP calculations. Since MCNP calculations in phantoms or models would be lengthy if the calculations started each time with fission neutrons from the reactor core, a neutron source plane, which was verified by spectrum and flux measurements at the irradiation port, was designed. Measured doses in phantoms are especially important to verify the simulated neutron source plane. Good agreement between the calculated and measured values has been achieved and this neutron source plane is now used to predict flux and dose information for oncologists to form treatment plans as well as designing collimator and room shielding. In addition, a program using MCNP calculated results as input has been developed to predict reliable flux and dose distributions in the central coronal section of a head model for irradiation by the BMRR beam. Dosimetric comparisons and treatment examples are presented.

PHANTOM CALCULATIONS

The horizontal section of the BMRR epithermal beam is depicted in Figure 1. To speed up the MCNP calculations, a neutron source plane was designed to start neutrons at the Li-poly face. The source plane combines information from both measurements and MCNP calculations, including the neutron spectrum, energy-dependent angular distributions, beam intensity profiles along the Li-poly face, and flux and dose measurements at the irradiation port. MCNP calculations with this source plane include enough backup material of bismuth and Li-poly shield behind the plane to account for scattered neutrons in the backward direction. Calculations were made to determine the flux and dose distributions in phantoms for comparisons with the measurements. Several important features in MCNP, including the thermal neutron scattering of material, hydrogen content, streaming effect, and neutron flux vs. activation relation, were carefully considered in the calculations.

Advances in Neutron Capture Therapy, Edited by
A.H Soloway *et al.*, Plenum Press, New York, 1993

PHANTOM MEASUREMENTS

One of the phantoms is a lucite ellipsoid which is represented by the equation: $(X/9.8)^2 + (Y/7.5)^2 + (Z/7.5)^2 = 1$. The dimensions of X, Y, and Z are in centimeters. Lucite rods with slits and holes to place dosimeters, like bare and Cd-covered gold foils and TLDs, can be inserted into the phantom for flux and dose measurements. The phantom can be used to evaluate the fluxes and doses at different depths along the central line and 4 cm off center for irradiation from two orthogonal directions. The other phantom is a cylindrical lucite container which can be filled with water. This phantom is 16.6 cm in diameter and 23 cm in length. Lucite rods with slits and holes to place dosimeters can be inserted along the central axis of the phantom for measurements. Both phantoms have been used to measure the thermal and epithermal fluxes by bare and Cd-covered gold foils (0.00127 cm thick and an average diameter of 0.7 cm) and gamma doses by TLDs (0.1 cm x 0.1 cm x 0.6 cm).

COMPARISONS OF PHANTOM DOSIMETRY

Measurements at the epithermal port taken over a 6 month time spread have shown differences of less than 3%. The comparisons of calculated to measured results are shown in Figure 2. Up to 8 cm into both phantoms, the differences between calculated and measured thermal and epithermal fluxes at different depths are within 10%. With the simulated neutron source plane, good agreement has been obtained between calculations and measurements for two shapes of phantoms with different compositions. Thus, the neutron source plane used in the MCNP calculations was verified to be appropriate for flux and dose prediction in models.

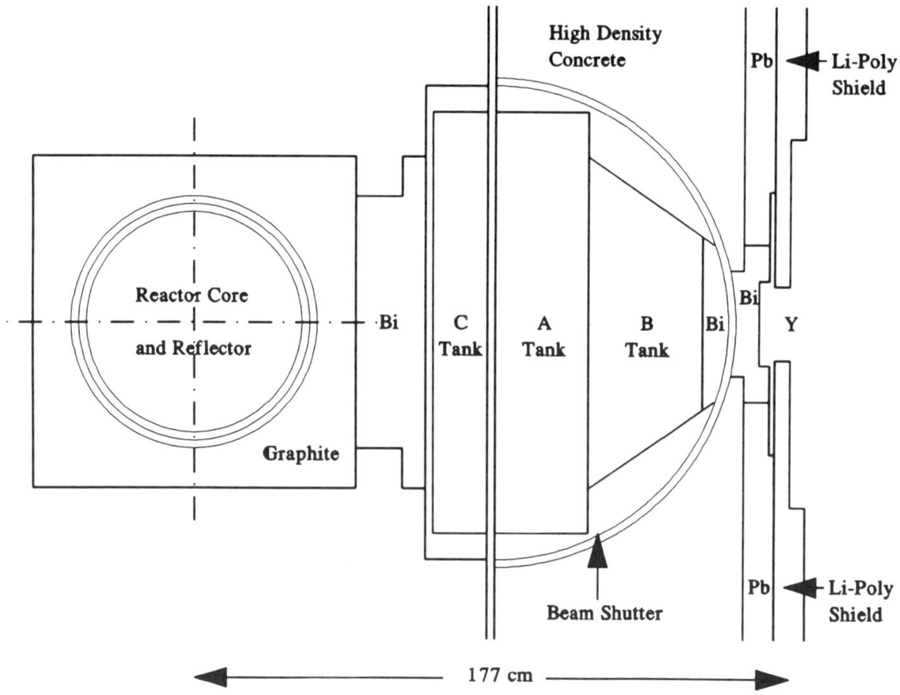

Figure 1. The Present BMRR Epithermal Beam.

TREATMENT PLANNING

Based on the consistency of phantom dosimetry, the neutron source plane was used with a head model to calculate the flux and dose distributions. This model has been used to develop treatment planning for BNCT by Zamenhof et al.[2] Besides the two ellipsoids separating the brain and skull areas, a layer of tissue equivalent material with higher hydrogen density is attached to the skull to evaluate the skin dose, especially the fast neutron dose.

The central coronal section of the head model was divided into many cells in the MCNP calculations. In Figure 3 is shown the horizontal section of the head model calculation setup in MCNP simulating a patient who is lying on the treatment table and facing up. The fluxes and doses in each cell were calculated by MCNP. When the BMRR beam comes from the left side of the head, the direction is designated

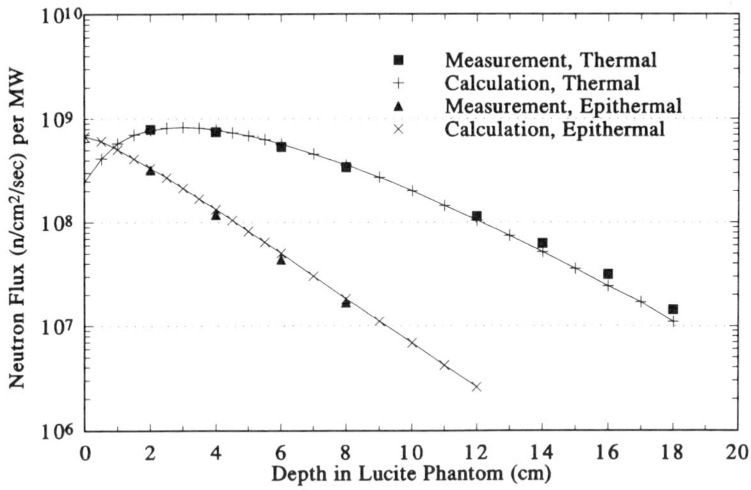

Figure 2. Measured and Calculated Fluxes in the Central Rod of Lucite Phantom.

as 0 degree. When the beam comes from the top of the head, it is a 90 degree irradiation. Rotation of the head to the right side for irradiation, makes the beam incident at 180 degrees. Several MCNP runs were completed with statistical error of thermal flux calculation in each cell reduced to less than 3%. Then, a program was written to use polynomial equations fitting the flux and dose information in each cell as a function of angle. The program is based on these polynomial equations to calculate the flux and dose distributions in each cell in the central coronal section of the head model. Different combinations of head orientations and irradiation weighting factors can be input to optimize an irradiation setup. The RBE's factors of different dose components can be input to calculate the total RBE dose in each cell, so the irradiation time can be calculated based on tolerance dose limit in critical areas.

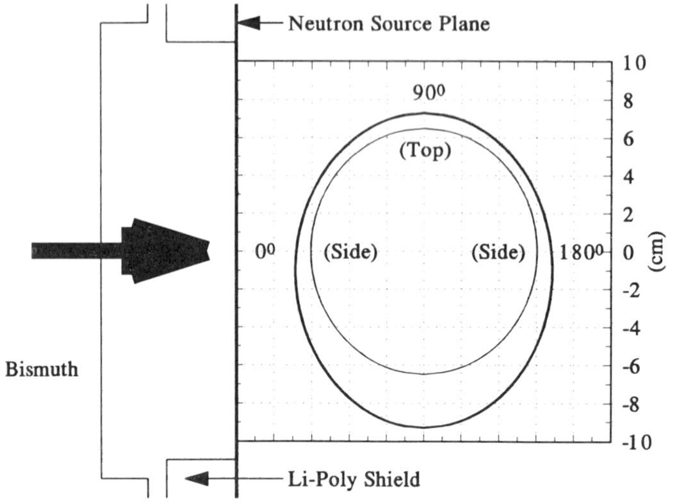

Figure 3. The Calculation Arrangements for the Head Model in the MCNP Calculation.

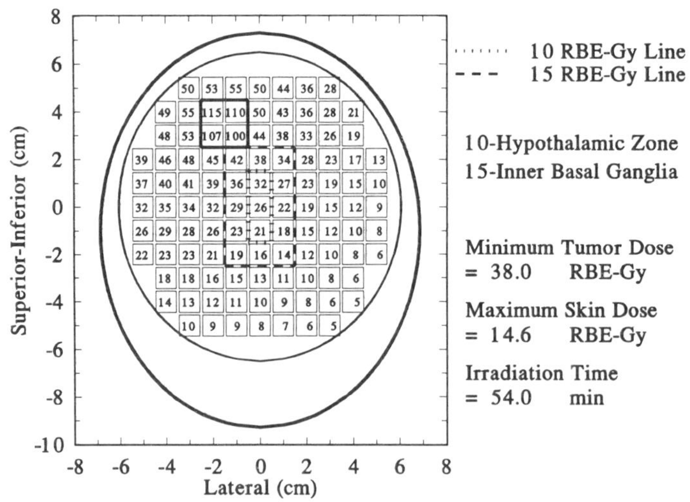

Figure 4. The Dose Calculations for the Assumed Tumor in the Head Model.

Figure 4 shows an example of dose calculations for treatment planning. After the tumor area and boron concentration and distribution are determined or approximated, the total RBE dose in each cell is calculated for irradiation by a 60 degree beam. Then, the total RBE dose in each cell is normalized to the Minimum Tumor Dose (MTD) which is identified as 100%. By applying the tolerance dose limit, the MTD can be determined. The Maximum Skin Dose (MSD) can be determined separately and also the irradiation time. As seen in Figure 4, a 2 x 2 x 2 cm^3 tumor with a uniformly distributed ^{10}B concentration of 30 ppm is assumed and the rest of the head has a ^{10}B level of 10 ppm. The tumor is irradiated by the BMRR beam coming from 60 degrees. Based on 10 RBE-Gy tolerance dose limit in the hypothalamic zone[3], the MTD is calculated as 38 RBE-Gy with a 54-min irradiation at 3 MW power of the BMRR. The MSD is 14.6 RBE-Gy. Here the RBEs were assumed 2.5 for $^{10}B(n,\alpha)^{7}Li$ reactions, 4.5 for proton recoil dose of scattering reactions of fast neutrons and $^{14}N(n,p)^{14}C$ reactions, and 1 for gamma ray reactions.

CONCLUSIONS

Two phantoms, an ellipsoidal lucite phantom and a cylindrical water phantom, have been used for dosimetric measurements to verify the MCNP calculated results using a simulated neutron source plane. Repeated measurements have been made with reliable reproducibility. The agreements between the measured and calculated values are good. Thus, based on the consistency of comparisons, the simulated neutron source plane was used with the head model to predict flux and dose distributions in the head model. A program was created to calculate the flux and doses of each component in the central coronal section of the head model for irradiation by the BMRR beam. Different combinations of head orientations and irradiation weighting factors can be combined to better predict the dose distribution. Once the tumor area and ^{10}B concentration and distribution were determined, a simple but reliable treatment planning can be made to predict a preliminary irradiation setup.

ACKNOWLEDGEMENT

This research was supported by the US Department of Energy contract DE-AC02-76CH00016.

REFERENCES

1. R. G. Fairchild, J. Kalef-Ezra, S. K. Saraf, S. Fiarman, E. Ramsey, L. Wielopolski, B. Laster, F. Wheeler, "Installation and Testing of an Optimized Epithermal Neutron Beam at the Brookhaven Medical Research Reactor," Neutron Beam Design, Development, and Performance for Neutron Capture Therapy, O. Harling and R. Zamenhof, eds., Mar. 1989, Plenum Press, New York, 1990.
2. R. Zamenhof, J. Brenner, J. Yanch, D. Wazer, H. Madoc-Jones, S. Saris, O. Harling, "Treatment Planning Techniques for Neutron Capture Therapy of Glioblastoma Multiforme Using MCNP & NCTPLAN and an Epithermal Neutron Beam from the MITR-II Research Reactor," Progress in Neutron Capture Therapy, B. Allen, D. Moore, B. Harrington, eds., Plenum Press, New York, 1992.
3. D. N. Slatkin, "A History of Boron Neutron Capture Therapy of Brain Tumours-Postulation of a Brain Radiation Dose Tolerance Limit," Brain 114:1609-1629, 1991.

SBNCT PLAN: A 3-DIMENSIONAL TREATMENT PLANNING SYSTEM FOR BORON NEUTRON CAPTURE THERAPY

Lawrence E. Reinstein, Eric B. Ramsay, Janusz Gajewski, Sundaresan Ramamoorthy and Allen G. Meek

Department of Radiation Oncology
University Hospital
SUNY at Stony Brook
Stony Brook, NY 11794-7028

INTRODUCTION

The need for accurate and comprehensive 3-dimensional treatment planning for boron neutron capture therapy (BNCT) has been debated for the past several years. Although many argue against the need for elaborate and expensive treatment planning programs which mimic conventional radiotherapy planning systems, it is clear that in order to realize significant gains over conventional fractionated radiation therapy, patients must be treated to the edge of normal tissue tolerance. Just how close to this edge is dictated by the uncertainties in dosimetry. Hence the focus of BNCT planning is the determination of dose distribution throughout normal tissue volumes.

Although precise geometric manipulation of the epithermal neutron beam is not achievable, the following variables play an important role in BNCT optimization: patient orientation, dose fractionation, number of fields, megawatt-minutes per fraction, use of surface bolus, and use of collimation. Other variables are not as easily adjustable and would not, therefore, be part of treatment planning optimization, include external patient contour, internal patient heterogeneities, boron compound distribution, and RBE's. The boron neutron capture therapy planning system developed at SUNY Stony Brook (SBNCT-Plan) was designed as an interactive graphic tool to assist the radiation onocologist in generating the optimum plan for a neutron capture treatment.

METHODS AND MATERIALS

The SBNCT planning system is composed of the following 3 modules: 1) patient data acquisition, 2) flux, dose, RBE dose calculations, and 3) graphic display of dose distribution and anatomy.

Patient Data Acquisition Module

CT Data. A dense CT scan data set of from 50 to 100 slices is obtained on the GE9800 scanner and transferred via magnetic media to an IBM RISC 6000 workstation.

Advances in Neutron Capture Therapy, Edited by
A.H Soloway *et al.*, Plenum Press, New York, 1993

Patient Orientation. Optimization of the treatment is an iterative process. A guess of the best patient orientation with respect to the beam port is made for the initial calculation which can then be evaluated using the 3-D dose distribution display. Further improvements can be made by rotating and translating the representation of patient anatomy with respect to the beam port. This process is facilitated by the display seen in Figs. 1a, b which consists of a 3-D perspective surface rendering of the patient's head in the vicinity of the beam port. Although obviously not represented by this figure, the head can be interactively manipulated in order to determine the beam entrance approach that best covers the target volume.

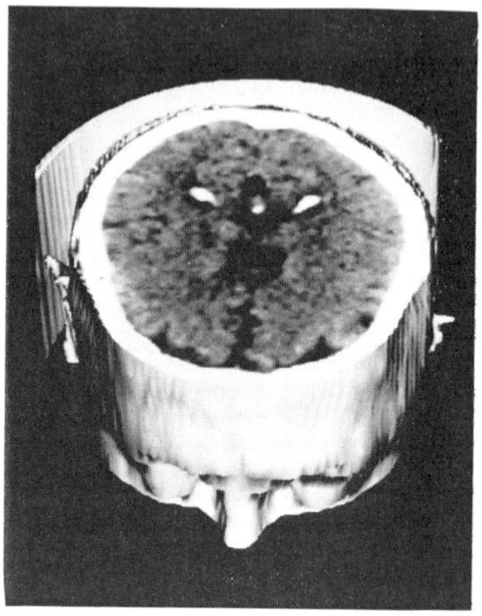

Fig. 1A. 3-Dimensional surface reconstruction of a patient's head using multi-level CT scan input. The single 2-D cutting plane can be interactively scanned throughout the volume. Graphic output is generated by a combination of in-house software and a commercially available graphics package.

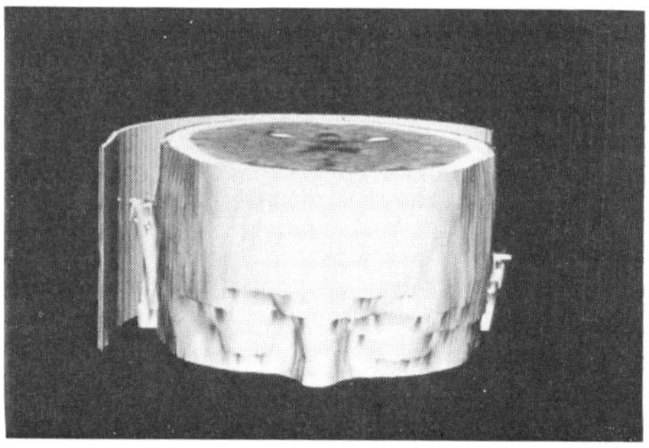

Fig. 1B. Another view of the same patient.

Creation of MCNP Cells. The 3-dimensional calculation window and cell size must be selected by the user. Faster and somewhat less precise "pre-planning" calculations are performed by choosing larger cell dimensions. Once MCNP cell boundaries have been identified by the program, its contents are separated into 3 tissue components: air, soft tissue and bone. The computer assigns a tissue type to each pixel in the MCNP cell on the basis of Hounsfield number. A weighted average of the elemental mass for each tissue component was computed for: H, O, N, C, Ca, P, Na, Cl and K. Values used for the percentage composition by weight of each element are taken from Table 1 of Zamenhof et al.[1] On the basis of the CT data particular to the individual patient and several table-lookup routines the software is able to automatically generate the complete input file properly formatted for the MCNP computation. An illustration of a typical MCNP cell distribution is seen in Fig. 2.

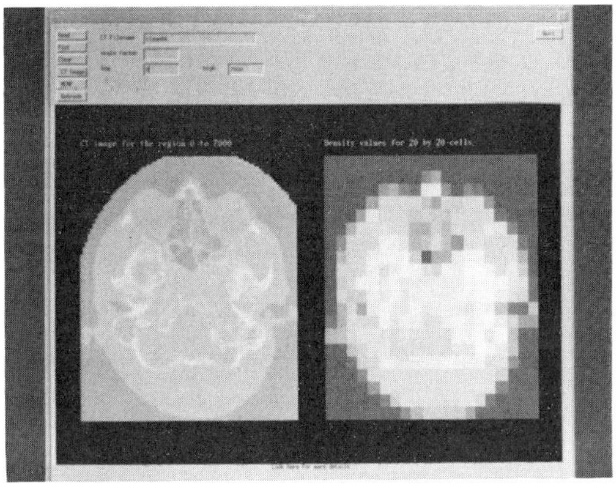

Fig. 2. A single level CT scan illustrating the MCNP calculation cells used as the basis for MCNP calculation.

Calculation Module

The dose calculation "engine" which drives the treatment planning system consists of the MCNP Monte Carlo program running on the BM RISC 6000 UNIX workstation. The reactor geometry was carefully specified using 90 basic cells with 19 different representing 26 nuclides. MCNP performance was further optimized using weight windows variance reduction techniques. Another method of reducing MCNP calculation time is the creation of a source plane at the port face. Using MCNP the source plane consists of approximately 700,000 stored particle tracks which were created from 1.5 million histories originating from the cores. Use of the source plane reduces the run time by about a factor of 10 for equivalent precision. Currently, a single beam run takes approximately two-hours to calculate all components of flux and dose with a 5-10% precision across the dimensions of the patient head for thermal neutrons and photons using a 1 cm^3 cell size. Further details and in-depth description of the dose calculations as well as measurements to verify these calculations have been described in the literature.[2,3]

RESULTS

Visualization of 3-Dimensional Dose Distribution

Several approaches are being investigated for the interactive display of calculated results which provide the radiation oncologist with a tool for comparing the relative merit of alternative treatment plans. Since patient anatomy as well as the calculated dose distribution is represented by a complete 3-dimensional data set, a 2-D "cutting plane" can be used to facilitate interpretation. By interactively moving the plane (at an arbitrary angle) throughout the 3-D data set, the radiation oncologist can make use of the 2-D representation to gain an appreciation of the 3-dimensional situation.

In a given plane the user sees a gray scale rendering of the CT anatomy overlaid by color representations of the calculated dose distribution. At present 3 different color techniques are used for this representation: semi-transparent color wash, color banding, and isodose map. For the purpose of this paper the dose distribution is displayed at only one level. Only thermal dose components (including Boron-10 reactions) are displayed, no RBE factors are included, and the semitransparent color wash technique is used.

The case of a 4cm x 2cm tumor situated at a left lateral depth of approximately 5cm is considered. It is assumed that the normal tissue Boron-10 uptake is 3 ppm while the tumor uptake is 30 ppm. The resulting thermal dose distribution from a single left lateral epithermal irradiation is seen in Fig. 3, with the normalization value

(100%) assigned to the maximum dose in the tumor. (Color wash is, unfortunately, rendered in gray scale for this publication.)

This is better illustrated in Fig. 4 which is a cell by cell graph of the thermal dose along the central axis of the epithermal beam. Since the patient can only be treated to normal brain tolerance, it is helpful to represent the dose distribution such that the maximum dose received by the normal tissue is 100%. Under these treatment conditions, and because of this steep dose gradient the ratio of the minimum tumor-dose to normal-tissue-dose is only 1.8 and not the almost 3.5 seen at the shallow surface of the tumor.

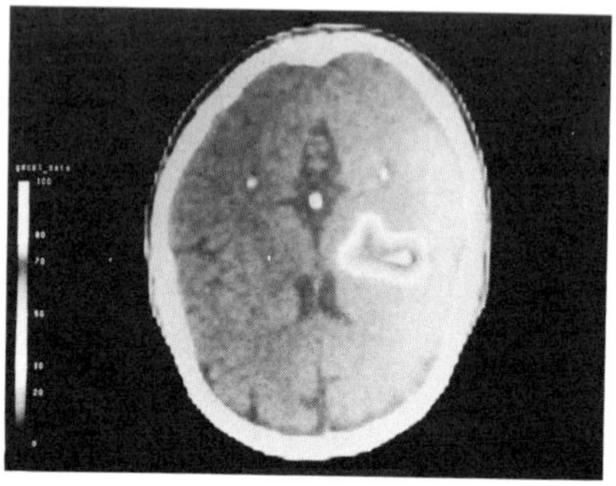

Fig. 3 Thermal and B-10 dose distribution from a left lateral epithermal beam shown in a single cutting plane of the CT data set.

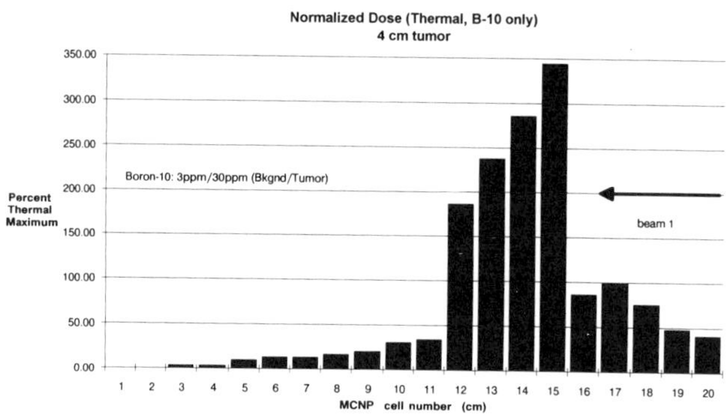

Fig. 4 Bar graph showing central axis dose distribution from a single left lateral epithermal beam. The target seen in this illustration consists of four 1 cm^3 cells situated at 5 cm depth. Note the normalization for this graph is 100% at the normal tissue thermal dose maximum.

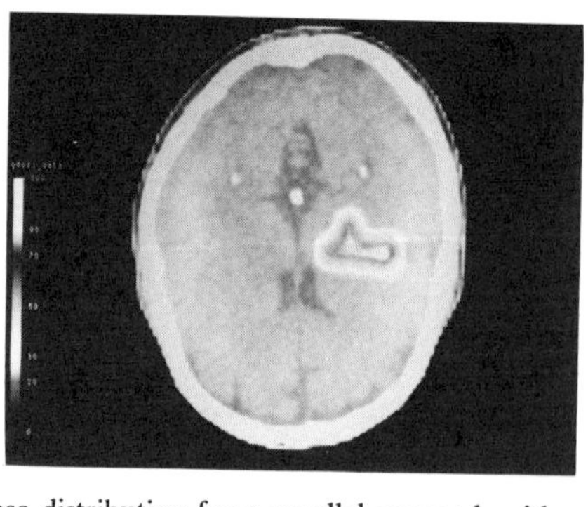

Fig. 5 Thermal and B-10 dose distribution for a parallel opposed epithermal beams. The color wash display is shown overlaying a single cutting plane of the CT data set and is normalized to the maximum dose in the tumor volume.

As in conventional radiation therapy some improvement can be achieved by adding a second parallel opposed beam as illustrated in Fig. 5. Although this is not an "optimized" plan it is readily seen that the high tumor dose is more uniformly distributed throughout the target than in the single beam case.

The central axis depth dose distribution (Fig. 6) clearly demonstrates improved tumor dose homogeneity. With the 100% dose the normalization point at the thermal maximum in normal tissue it is seen that the tumor dose varies from approximately 290% to 380%. In other words the minimum tumor-dose/normal-tissue-dose is now increased to almost 3:1. This is a significant improvement (about 70%) over the single beam situation. Optimized beam weightings might lead to further improvement.

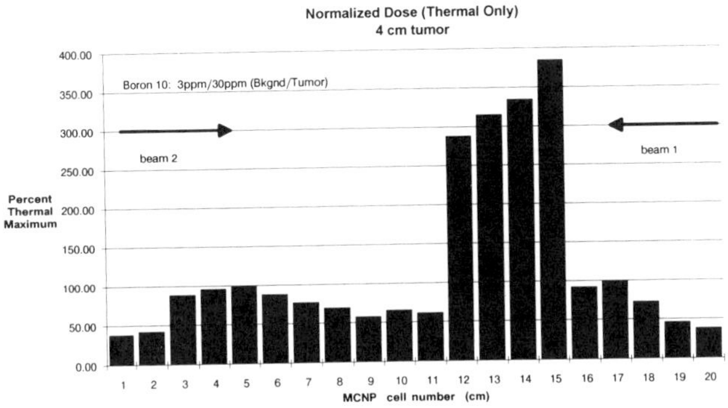

Fig. 6 Bar graph showing central axis depth dose distribution or parallel opposed epithermal beams. The tumor along the axis consists of four 1 cm^3 cells containing 30 ppm boron and the normalization is 100% to the thermal maximum in normal tissue (3 ppm) boron.

CONCLUSION AND DISCUSSION

Although the example above contains several simplifications, the conclusion remains unchanged: It is the normal tissue tolerance that "drives" the treatment plan and necessitates the use of computerized 3-dimensional treatment planning software. The SBNCT planning system provides a tool for optimizing the physical dose distribution by achieving a more uniformed dose across a pre-defined tumor volume and thereby maximizing the therapeutic ratio.

REFERENCES

1. R. Zamenhof, S.D. Clement, O.K. Harling, J.F. Brenner, D.E. Wazer, H. Madoc-Jones and J.C. Yanch, Monte Carlo based dosimetry and treatment planning for neutron capture therapy of brain tumor, in: "Neutron Beam Design, Development, and Performance for Neutron Capture Therapy," O.K. Harling, J.A. Brennan and R.G. Zamenhof, ed., Plenum Press, New York City (1990).

2. R. Zamenhof, J. Brenner, J. Yanch, D. Wazer, H. Madoc-Jones, S. Saris, O. Harling, Treatment planning for neutron capture therapy of Glioblastoma multiforme using an epithermal neutron beam from the MITR-II research reactor and Monte Carlo simulation, in: "Progress in Neutron Capture Therapy for Cancer," B.J. Allen, D.E. Moore, B.V. Harrington, ed., Plenum Press, New York City (1992).

3. J. Gajewski, E.B. Ramsay, L.E. Reinstein, Refinements of Monte Carlo calculations for BNCT at the Brookhaven Medical Research Reactor, presented at the 5th Int'l. Symposium on Neutron Capture Therapy, 1992.

REMOVAL-DIFFUSION THEORY FOR CALCULATION OF NEUTRON DISTRIBUTIONS IN BNCT

John Niemkiewicz,[1] Thomas E. Blue[2]

[1]Department of Radiation Oncology
Riverside Methodist Hospitals
3535 Olentangy River Road
Columbus, OH 43214
[2]Nuclear Engineering Program
The Ohio State University
206 W. 18th Avenue
Columbus, OH 43210-1107

INTRODUCTION

The routine use of Boron Neutron Capture Therapy (BNCT) for the radiation treatment of humans will require the planning of an acceptable RBE dose distribution for each patient. The first step in the calculation of a dose distribution is the determination of the neutron flux distribution. Neutron flux distributions for BNCT have been calculated by the Monte Carlo[1,2] simulation and the Discrete Ordinates[3] techniques. Use of removal-diffusion theory to calculate neutron distributions for BNCT offers some advantages over these techniques including, most notably, the relatively short amount of computer time necessary to calculate a neutron flux distribution. This work is an examination of the potential use of removal-diffusion theory to adequately describe the neutron flux distribution in a human head by first considering the calculation of the neutron flux distribution for the simpler geometry of a rectangular parallelopiped.

Diffusion theory for the calculation of neutron flux distributions is usually developed as an approximation to transport theory.[4] The assumptions that are made for steady-state conditions include: the angular neutron flux is at most linearly dependent on the scattering angle cosine; all the neutrons in an energy group are the same; and, neutron sources are isotropic. In removal-diffusion theory, source neutrons are transported in straight lines until they suffer their first collision. The resulting spatial and energy distribution of the once-collided neutrons is then used as the neutron source for a subsequent diffusion calculation.[5]

Advances in Neutron Capture Therapy, Edited by
A.H Soloway *et al.*, Plenum Press, New York, 1993

METHODS

The first step in the implementation of removal-diffusion theory for the calculation of neutron flux distributions in phantoms was the generation of the flux distribution along the central axis for four energy groups assuming a semi-infinite water material geometry. The neutron source was a parallel beam perpendicularly incident onto the water medium. The beam neutrons had a 1/energy intensity spectrum for the highest three energy groups, and there were no incident neutrons of thermal energy. MCNP,[6] a Monte Carlo computer code, was set-up for this geometry and run on a Sun Sparkstation. A one-dimensional, four group analytical solution to the removal-diffusion equation was then fit to the MCNP calculated neutron distribution by adjusting the diffusion theory parameters in the analytical solution using the Levenberg-Marquardt fitting method.[7]

The removal-diffusion equation that was solved analytically is;

$$-D_g \frac{d^2\phi_g}{dx^2} + \Sigma_{R,g}\phi_g = \sum_{g'=1}^{g-1} \Sigma_{s,g'\to g}\phi_{g'} + \sum_{g'=1}^{g} I_{g'}\Sigma_{s,g'\to g} \exp(-\Sigma_{t,g'}x) \qquad (1)$$

where: D_g = the diffusion coefficient for group g (cm),

ϕ_g = the neutron flux for group g (n/cm^2-sec),

$\Sigma_{R,g}$ = the macroscopic removal cross section for group g (cm^{-1}),

$\Sigma_{s,g'\to g}$ = the macroscopic scattering cross section from group g' to g (cm^{-1}),

I_g = the incident neutron source intensity for group g (n/cm^2-sec),

$\Sigma_{t,g'}$ = the macroscopic total interaction cross section for group g' (cm^{-1}).

The diffusion theory parameters which were adjusted are the diffusion coefficients and the macroscopic cross sections for the four energy groups.

Finally, again using the Sun Sparkstation, neutron flux distributions were calculated for a 3-D, finite geometry using both MCNP and removal-diffusion theory implemented with the code DIF3D,[8] and compared. The same energy group boundaries and external neutron beam source described above were used for these calculations. The adjusted diffusion theory parameters were used for the 3-D, finite geometry removal-diffusion calculations. The neutron flux distributions were determined for both a 17x17x15 cm^3 and a 14x14x15 cm^3 water phantom.

RESULTS

The neutron flux distributions calculated for the two phantom sizes using removal-diffusion theory agreed well with the corresponding distributions calculated using the Monte Carlo technique. Somewhat better agreement was obtained for the fourth energy group flux distribution when the diffusion theory parameters were obtained by fitting to a finite phantom flux distribution on the central axis using equation (1) with a correction for neutron leakage. The MCNP calculations required thousands of hours of cpu time, while the DIF3D calculations were completed in under ten minutes. Due to space limitations only a few representative comparisons of the MCNP results with the DIF3D results can be given. Figure 1 shows the neutron flux distribution along the central axis of the 17x17x15 cm^3 water phantom for the first energy group. Error bars on the MCNP data points indicate plus and minus one standard deviation.

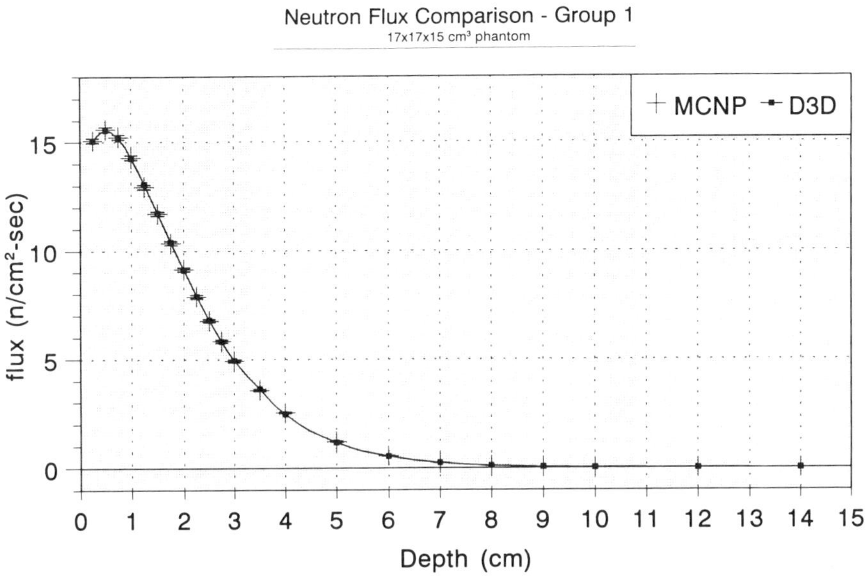

Figure 1. Neutron flux comparison along the central axis in the 17x17x15 cm^3 water phantom for neutrons in energy group 1.

Figures 2 and 3 show the neutron flux calculated along the central axis and along profiles lateral to the central axis, respectively, for the fourth (thermal) energy group.

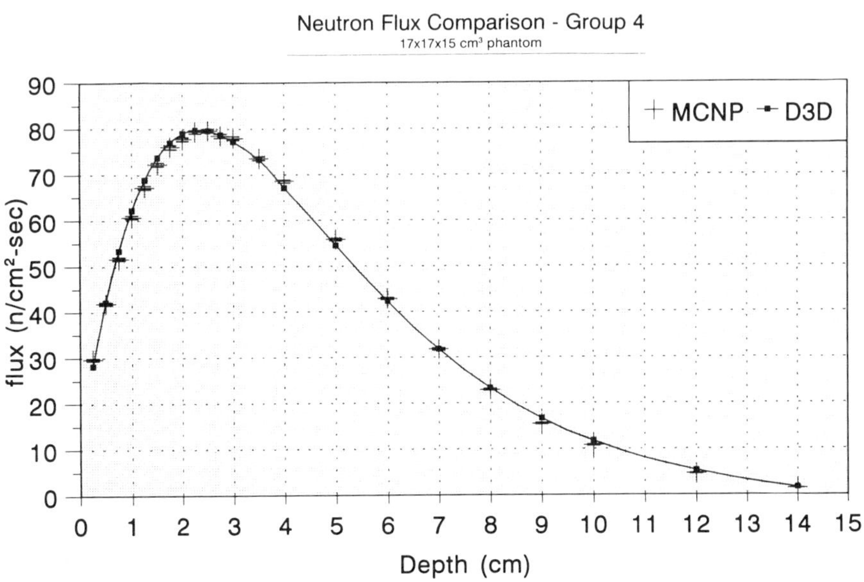

Figure 2. Neutron flux comparison along the central axis in the 17x17x15 cm^3 water phantom for neutrons in energy group 4.

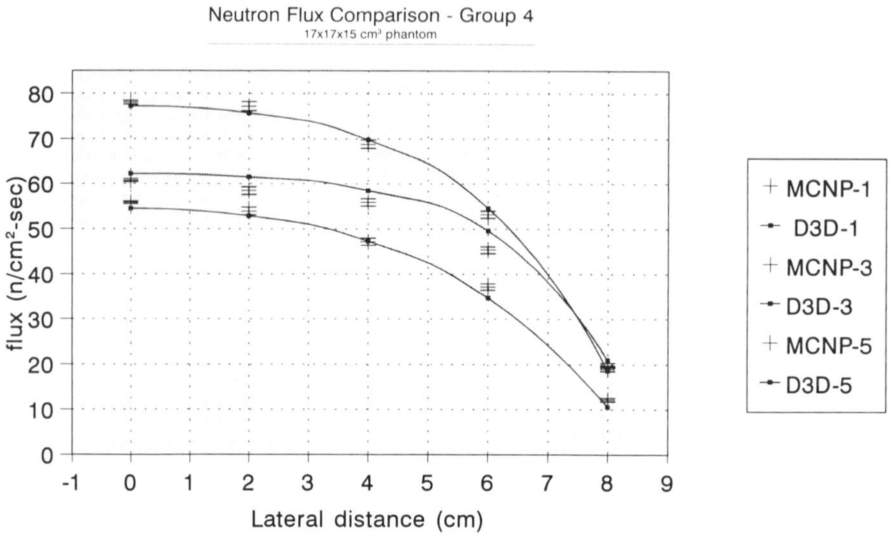

Figure 3. Neutron flux comparison along lateral profiles at 1, 3, and 5 cm. depth in the 17x17x15 cm^3 water phantom for neutrons in energy group 4.

The comparison for the second and third energy groups is similar to the group 1 results. Also, the results for the 14x14x15 cm^3 phantom are similar to the 17x17x15 cm^3 phantom results.

REFERENCES

1. F.J. Wheeler, M.L. Griebenow, D.E. Wessol, and D.W. Nigg, Analytical modeling for neutron capture therapy, *Strahlentherapie und Onkologie* 165, 186-188 (1989).

2. R.G. Zamenhoff, S.D. Clement, O.K. Harling, J.F. Brenner, D.E. Wazer, H. Madoc-Jones, and J.C. Yanch, Monte carlo based dosimetry and treatment planning for neutron capture therapy of brain tumors, *in*: "Neutron Beam Design, Development, and Performance for Neutron Capture Therapy," O.K. Harling, J. Bernard, and R.G. Zamenhoff, eds., Plenum Press, New York (1990).

3. D.W. Nigg, P.D. Randolph, and F.J. Wheeler, Demonstration of three-dimensional deterministic radiation transport theory dose distribution analysis for boron neutron capture therapy, *Medical Physics* 18(1):43-53 (1991).

4. J.J. Duderstadt and L.J. Hamilton, Neutron transport, *in*: "Nuclear Reactor Analysis," John Wiley & Sons, Inc., New York (1976).

5. J.R. Lamarsh, Radiation shielding, sec 10.9, *in*: "Introduction to Nuclear Engineering," 2nd edition, Addison-Wesley, Menlo Park (1983).

6. Los Alamos Monte Carlo Group, "MCNP-A General Monte Carlo code for neutron and photon transport," J. Briesmeister ed., Los Alamos Natl. Lab., LA-7396-M, Rev 2, New Mexico (1986).

7. W.H. Press, B.P. Flannery, S.A. Teukolsky, and W.T. Vetterling, Modeling of Data, pp. 574-580, *in*: "Numerical Recipes in Pascal," Cambridge University Press, Cambridge (1989).

8. K.L. Derstine, "DIF3D: A code to solve one-, two-, and three-dimensional finite difference diffusion theory problems," Argonne Natl. Lab., ANL-82-64, Argonne (1984).

POSSIBILITY OF APPLICATION OF ACOUSTIC PULSE GENERATION TO DOSE DISTRIBUTION MONITORING DURING NEUTRON CAPTURE THERAPY

Y. Hayakawa,[1] J. Tada,[1] T. Inada,[1] and H. Hatanaka.[2]

[1]Inst.Bas.Med.Sci, Univ.Tsukuba,Tsukuba,Ibaraki-ken 305, Japan
[2]Teikyo University School of Medicine, Itabashi-ku,Tokyo 173, Japan

INTRODUCTION

We have previously reported on simultaneous monitoring of neutron fluence rate during boron-neutron capture therapy[1-5]. In these methods, only neutron fluence rate in the locations of the detectors can be monitored.Although the in vitro and in vivo measurements of boron-10 content in blood and tissue by prompt gamma rays give additional information[6-7], the estimated dose is obtained rather indirectly. Direct dose distribution measurement is more desirable. Acoustic pulse generation phenomenon seems to be one of the candidates for the purpose.

ACOUSTIC PULSE GENERATION BY PULSED PROTON BEAM

Acoustic pulse generation due to pulsed beam irradiation seems to be applicable on dose distribution monitoring during irradiation of patients[8-11]. Basic experimental results using pulsed proton beam irradiation were achieved, with the result showing that dose monitoring of patients with the dose rate of 0.2 cGy/Pulse seems to be possible[11].

ACOUSTIC PULSE GENERATION BY PULSED NEUTRON BEAM

Figure 1 shows the schematic view of application of pulsed neutron beam for dose distribution monitoring during neutron capture therapy.

Pulsed neutron beam less than 10μs should be used to ensure the spatial accuracy of 1.5 cm due to the fact that sound velocity is 1500m/s in tissue. Pulsed neutron beam can be obtained with pulsed reactor. The pulse duration, however, is on the order of 1ms, resulting in unacceptably large spatial resolution of more than 1.5m. Accelerator based neutron source can yield shorter pulse. There still remain problems due to behavior of neutrons in the tissue.

The mean life time of thermal neutrons in water is approximately 200μs[12], resulting spatial resolution larger than 30cm, which seems to be unacceptable. The solution to obtaining shorter pulsed signal is to utilize pulsed epithermal neutron source, as the mean slowing down time of these neutrons in water to thermal energy is less than 10μs[12].
As the energy dependence of 10-B(n,α)7-Li crosssection is inversely proportional to the velocity[13], the rate of the reaction during unit time is independent of neutron energy.

Therefore, the acoustic pulse generation by 10-B(n,α) 7-Li reaction during slowing down of neutrons cannot be expected for dosimetric purposes. One can use the acoustic pulse generation during slowing down of epithermal neutrons by administering indium and/or rhodium in the boron compound, as these elements resonantly capture epithermal neutrons[13].

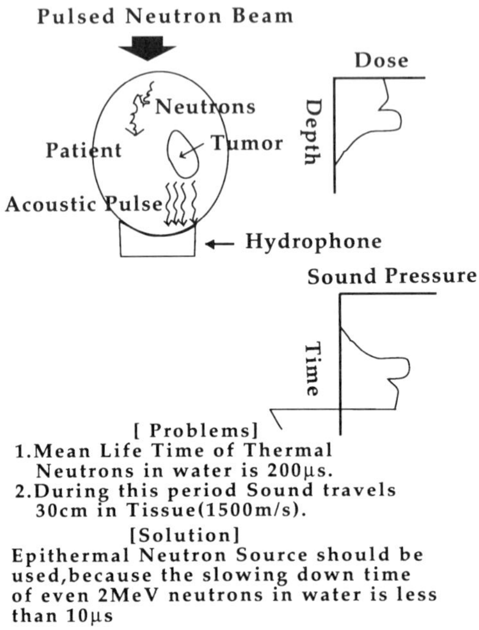

Figure 1. Schematic view of acoustic pulse generation by pulsed epithermal neutron irradiation

REFERENCES

1.Y.Hayakawa, S.Harasawa, A.Nakamoto, K.Amano, H.Hatanaka, and J.Egawa, "Simultaneous monitoring system of thermal neutron flux for boron neutron capture therapy," Radiation Research 75 243-251 (1978).

2. S.Harasawa, A.Nakamoto, Y.Hayakawa, J.Egawa, O.Aizawa, T.Nozaki, T.Minobe, and H.Hatanaka, "Simultaneous monitoring system of thermal neutron flux for boron neutron capture therapy," Radiation Research 75 243-251 (1978).

3. S.Harasawa, Y.Hayakawa, T.Nozaki, O.Aizawa, and H.Hatanaka, "Dosimetry in boron neutron capture therapy," In Biomedical Dosimetry: Physical Aspects, Instrumentation, Calibration, pp.173-186, IAEA-SM-249/20, International Atomic Energy Agency, Vienna, 1981.

4.Y.Hayakawa, T.Inada, S.Harasawa, and H.Hatanaka, "A system for simultaneous monitoring of thermal and epithermal neutron flux during boron neutron capture therapy," Proc. First Internat. Symp. Neutr. Capt. Therapy 77-87(1983).

5.Y.Hayakawa, T.Inada, S.Harasawa, and H.Hatanaka, "A system for simultaneous monitoring of thermal and epithermal neutron flux during boron neutron capture therapy(Chapter 21)," In Boron-neutron Capture Therapy for Tumors. pp.283-292, Nishimura Co. Ltd., Niigata, 1986.

6.K.Kanda, T.Kobayashi, and K.Aoki, "In vivo measurement of time dependent boron-10 concentration in tumor," Proc. First Internat. Symp. Neutr. Capt. Therapy 99-105(1983).

7.R.G.Fairchild, D.Gabel, B.H.Laster, D.Greenberg, W.Kiszenick, and P.L.Micca, "Boron-10 analysis by prompt-gamma and track-etching techniques," Proc. First Internat. Symp. Neutr. Capt. Therapy 114-119(1983).
8.Y.Hayakawa, J.Tada T.Inada et al., "Acoustic pulse generation in excised muscle by pulsed proton beam irradiation and the possibility of clinical application to radiation therapy." J.Acoust.Soc Jpn.(E) 9, 255(1988).
9.Y.Hayakawa, J.Tada, T.Inada et al.,"Acoustic pulse generation in water by pulsed proton beam irradiation and its possible application to radiation therapy." Jpn.J.Appl.Phys.28,Suppl.28-1,217(1989).
10.J.Tada, Y.Hayakawa, T.Inada et al. "Time resolved properties of acoustic pulse generated in water and in soft tissue by pulsed proton beam irradiation - A possibility of dose distribution monitoring in proton radiation therapy," Med.Phys.18,1100-1104(1991).
11.Y.Hayakawa, J.Tada,T.Inada K.Hosono, T.Wagai, H.Tsuji, and H.Tsujii. "Acoustic pulse generated in a living rabbit by pulsed proton beam irradiation" Submitted to Med.Phys.
12.S.Glasstone and M.Edlund, "The elements of nuclear reactor theory," D.Van Nostrand Co. Inc., New York, 1952.
13.D.I.Garber and R.R.Kinsey, "Neutron cross sections, Volume 2:Curves (BNL 325)" ,Brookhaven National Laboratory, New York, 1976.

REFINEMENTS OF MONTE CARLO CALCULATIONS FOR BNCT AT THE BROOKHAVEN MEDICAL REACTOR

J. Gajewski, E. B. Ramsay, L. E. Reinstein

State University of New York at Stony Brook, New York 11794, U.S.A.

INTRODUCTION

The Brookhaven Medical Research Reactor (BMRR) and its epithermal neutron filter have been modeled in detail for the Monte-Carlo code MCNP[1] as a part of the Stony Brook neutron capture therapy planning system (SBNCT-Plan).

While the MCNP calculated neutron spectrum at the reactor port face is reasonably consistent with the measured spectrum, differences of up 40% and 25% at points in the fast and epithermal flux respectively have motivated further refinement of the reactor model (basic model of BMRR).

These include:

- more detailed description of the core geometry including exact positions of fuel elements and moderators.
- use of the latest available cross-section data.
- inclusion of fission fragments produced in the reactor core.
- fine tuning of the bias energy of the source together with adjustment of energy weight windows (a variance reduction technique to improve calculation precision).
- creation of source-surfaces to increase precision and reduce calculation times.

The impact of these refinements on the calculated spectrum are presented and a discussion of possible sources of the remaining discrepancies.

METHOD

When one compares the calculated spectrum with the spectrum measured at the port face in air (figure 2 of ref. 2), one finds that the ratio of epithermal to fast neutrons is measured to be 18 but calculated to be 39.6. Here, we are defining the epithermal range to be from 0.4 eV to 10 keV with fast neutrons being greater than 10 keV. Also, when one compares the thermal flux variation with depth in a tissue equivalent phantom (figure 3 of ref. 2), we find good agreement when a layer of ^{6}Li is placed between the beam and the phantom but poor agreement when the ^{6}Li is removed. In the latter case, the thermal flux build up is too high, followed by a fall off that is too rapid as compared to measurement. In investigating the possible causes of these discrepancies, we will separate our efforts between trying to improve the precision of the monte-carlo calculations and trying to improve the accuracy. The distance from the reactor core to the port face is long and involves several thin layers of aluminum and alumina and as a result, the problem from a monte-carlo point of view is difficult. It can be classified as a deep-penetration problem with a relatively fast change of density of matter indicating that one of the most useful variance reduction technique is that using weight windows. Also, our energy spectrum of neutrons in the port has a special structure; the fast and thermal components are an order of magnitude lower than the epithermal component, and in order to separate them with good precision, we need to use the energy weight windows technique. An additional difficulty comes from the shape of the reactor casing and shutter which fit together as two steel half cylinders with 1/4" air gap between them. This air gap creates a streaming effect for fast neutrons which one would like to treat using angle-energy weight windows. Unfortunately, the MCNP code is not able to do this at the present time. Starting from a guess of importances, we calculated average weights for tracks in each cell and we next used these weights as the input weights for the weight windows generator. The energy-weight window generator was used in those energy regions where the precisions were significantly different. Finally, we biased the energy spectrum of the reactor core. All of the above allowed us to achieve precisions on the order of a few percent for the thermal, epithermal and fast neutrons components.

Table 1. Summary of BMRR core investigation for 1MW of reactor power.

Neutron flux (n/cm^2s)	**Fast**	**Epithermal**	**Thermal**	**Ratio Epith./Fast**
INEL measurement	**3.44e7***	**6.20e8**	**-**	**18.0**
standard power homog.density no ref.plane r=25.83cm	**2.04e7**	**8.08e8**	**1.71e7**	**39.6**
reflected st.power by refl.plane r=25.83cm	**2.70e7**	**1.05e9**	**1.89e7**	**38.9**
standard power homog.density refl.plane r=29.81cm	**1.65e7**	**7.70e8**	**1.60e7**	**46.6**
homog.power homog.density r=25.83cm or r=29.81cm/st.pow.	**1.43e7**	**5.88e8**	**1.02e7**	**41.1**
empty core stand.power.distr.	**9.80e7**	**3.88e9**	**6.40e7**	**39.6**
standard power separated fuel elem./refl.plane/	**2.55e7**	**9.12e8**	**1.78e7**	**35.7**

*** xey = $x*10^y$**

On the accuracy level, the epithermal filter and the reactor facility were thoroughly checked against all the available documentation for BMRR.
Despite this, some ambiguities in terms of materials and distances remained. For example, whether or not boral plates were still present in the reactor and what the exact distance is from the core center to the port. By assuming different core to port face distances within the range of ambiguity, we found that this parameter did not affect in any significant way the discrepancies outlined previously. Similarly, the presence or lack thereof of the Boral plates did not affect the discrepancies with which we are concerned.

Next, the latest available cross-section data from the full set of ENDF-B/V nuclear data library were used (MCNPDAT,package DLC-105C). Because the fuel elements are old and have a high percentage of consumed uranium, we took into account fission products of uranium ^{235}U for which the cross sections are available in the MCNPDAT package and its amount can be estimated from the consumed and unconsumed fraction of uranium. This correction did not change the neutron spectrum at the port significantly. The way the core was treated was also modified by changing its density, correcting the radius and also separating the fuel elements which were previously not described, the core being taken as homogeneous with a power distribution. Options with and without a reflecting plane which passes through the core center were also tried. The results of all these calculations are shown in Table 1. The lowest epithermal to fast flux component ratio that was achieved is equal to 35.7 which is for the case where the fuel elements were separated. The table shows also the case of a reflected power distribution created by the reflecting plane as well as a run with an empty core. We conclude that the power distribution and the density of matter inside the reactor core, although very important for criticality of the system, are not important for the shape of the neutron energy spectrum in the port face because it is far away (170.9cm) from the center of the core. That shape, from the point of view of the core, depends only on the neutron energy spectrum, which is a Watt fission spectrum.

As an additional test of our model, measurements of neutron fluxes at the port face were performed using gold foil activation. The MCNP calculations of the gold foil activity in the center of the port face gives a value 1.35 times higher than measurement. The same method was used to measure gold foil activity inside a water tank. The results are shown in Figure 1 where we can see that agreement is good after scaling the calculations by factor 1.35.

For BNCT applications it is convenient to have the neutron spectrum already at the port face rather than starting calculations from the reactor core every time. Therefore, the source-surface technique, which is available in the MCNP code, was developed for this complex geometry.

For the source surface, which can be arbitrarily chosen, all tracks in some direction are stored. The tracks are represented here by sets of numbers: weight, energy, cosine directions and coordinates. In the source-surface run, the random number generator works the first time for the next-event estimator subroutine which calculates the first collision point. Care must be taken to include enough geometry in the source-surface run because of albedo effects. The source surface should be chosen as a boundary of the system in such a way that the tracks crossing this surface do not make any more collisions. This choice excludes second order albedo effects in the source-surface run. Also, because the history of collisions inside the source surface area is lost, we should choose a source surface as far as possible from the points of interest.

In order to check how this technique works for the BMRR model, we chose a source surface at the last plane of the port face and we compared neutron fluxes in front of the port in the void for the original and the source-surface run; the results were exactly the same for tallies and errors with acceptable statistical uncertainty of the angular flux computed for the source-surface (1.8% for epithermal and 7% for thermal and fast energy regions). Then, the source surface was chosen at the port face, and the phantom was placed out of the bismuth indentation in the reactor port face.
Although we expected some bias of the results by the next-event estimator, no such effect was observed and the results are statistically the same because the source surface is the boundary of two spaces and the first of them has a short free path length (bismuth).
Similar results were achieved when we chose a source surface inside the reactor filter or just before it. In all these cases second order of albedo effect was eliminated.

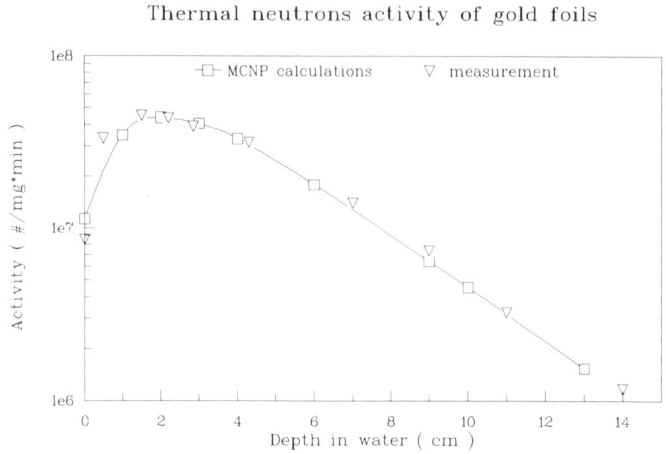

Figure 1. Thermal activity of gold inside water tank.

CONCLUSIONS

- From the reactor core investigation we can observe that the separation of fuel elements inside the core decreased the epithermal to fast component ratio at the port face to 35.7. This ratio does not depend significantly on the density of the core and its power distributions. The existing discrepancy compared with measurement (ratio=18) can possibly come from the cross section for aluminum which has a very complicated shape above 10keV and uncertainties for these resonance functions are relatively large.
- The fission products of uranium ^{235}U does not change results significantly because of the high atomic density ratio between hydrogen and uranium in the reactor core and the long distance from the core to the port.
- The bias energy of the fission source used together with the energy weight windows technique improved the precision of results significantly, 1.5% for epithermal (0.4eV-10keV), 8% and 6% for thermal (below 0.4eV) and fast (above 10keV) components during 16 hours (1.4mln histories) of IBM RISC 6000/320 CPU time (before 2%,11% and 23% respectively).
- The scaling factor 1.35 for total energy neutron spectrum can reproduce the shape of the thermal flux inside the water tank with acceptable accuracy. This means that the thermal spectrum in water is not very sensitive to the incident neutron spectrum.
- The source-surface technique truncated the time of calculations from 16 hours to 50 minutes keeping the same results and precision.

This allows us to put most of the computation time into calculations of a few thousands cells inside a human brain with desired precision.

REFERENCES

1) J.F. Briemeister, ed,"MCNP 4 - Monte Carlo Neutron and Photon Transport Code System", Version 4.2, Los Alamos National Laboratory, 1991.
2) J.Gajewski, E.B.Ramsay, L.E.Reinstein, S.Saraf, R.Fairchild, "Monte Carlo Calculation Results for The Epithermal Neutron Beam at The Brookhaven Medical Research Reactor", ed. B.J.Allen et al., Plenum Press, NY 1992.

3) G.K Becker, Y.D. Harker, L.G. Miller, R.A. Anderl, F.J. Wheeler, "Neutron Spectrum Measurements in the Aluminum Oxide Filtered Beam Facility at the Brookhaven Medical Research Reactor" in Neutron Beam Design, Development, and Performance for Neutron Capture Therapy, O.K. Harling, J.A. Bernard and R.G. Zamenhof, eds, Plenum Press, N.Y.,235 (1990).

THE EFFECT OF INCIDENT NEUTRON ENERGY AND TUMOR DEPTH IN BNCT STUDY

M.K. Marashi[1] and A. Pazirandeh[2]

[1] Nuclear Research Center
Atomic Energy Organization of Iran
P.O.Box 11365/8486
Tehran, Iran

[2] Physics Department
Tehran University
P.O.Box 19395-1943
Tehran, Iran

INTRODUCTION

Since its birth in the 1950s Boron Neutron Capture Therapy has steadily progressed to be a competitive and reliable curing technique. Since the 1970s in virtue of sophisticated dosimetry and neutron flux filtration and more suitable chemical compound as antibody agents, many more laboratories have been investigating the proper neutron beam energy band to maximize the absorbed dose to tumors with minimal effect on surrounding tissues[1,2]. This technique is based on the $^{10}B(n,\alpha)^{7}Li$ reaction induced by thermal neutrons. For deep seated tumors the incident thermal neutron flux attenuates considerably in reaching the tumor, and the skin absorbed dose increases. In order to overcome this problem, an epithermal neutron beam has been studied. Most of the epithermal neutrons slow down along the travelling distance to the tumor.

In this paper, the results of our calculations leading to an optimized neutron spectrum and the effect of tumor depth on the effective dose rate will be presented.

CALCULATIONAL METHOD OF DOSE RATE IN A HEAD PHANTOM

To calculate the absorbed dose rate in a tumor and normal cells in a phantom head, a slab of 20 cm width loaded with a tumor 5 cm thick was considered. The effect of depth on the total absorbed dose was estimated by varying the tumor depth from 1 cm to 5.5 cm (Fig. 1).

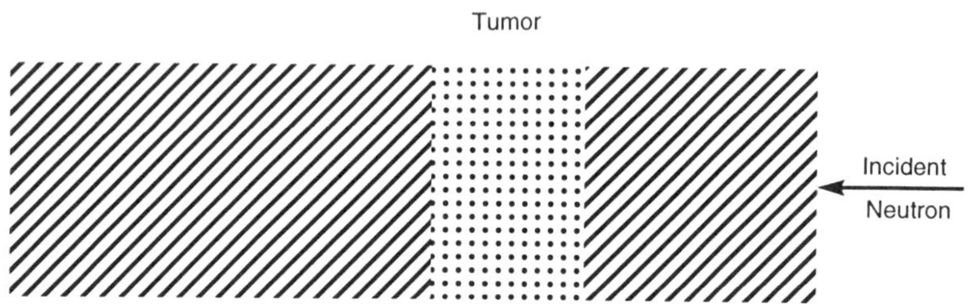

Figure 1. Simulation of phantom head

Table 1. Concentration of elements in phantom head

element	w/o in normal	w/o in cell tumor
H	10.7	10.7
O	71.4	71.4
C	12.1	12.1
Na	4.5	4.5
N	1.3	1.3
B-10	10^{-3}	3.10^{-3}

The interior of the phantom was composed of elements H,O C, N, Na, and B-10, the concentrations of which in tumor and surrounding tissues are given in Table 1.

The absorbed dose rate calculations was performed in tumor and surrounding tissues using thirteen energy group structure, as given in Table 2.

The incident neutron beam to the phantom head, one of the seven energy groups was selected. The incident monoenergetic neutrons are slowed down in the phantom head by neutron scattering and as a result of interactions with the medium nuclei, a mixed field of neutron and gamma spectra is produced.

The absorbed doses in tumor and in surrounding tissues in the phantom head arise from the major reactions i.e., (n,α), (n,p), (n,γ), and (n,n') by neutrons and (γ,γ) by gamma rays.

In order to estimate the spectra of the scattered neutrons and gamma rays, ANISN computer code[3] was used. This code has been designed to solve the one-dimensional linear Boltzmann transport equation for neutrons and gamma rays using discrete ordinate Sn method. The calculated neutron and gamma spectra in discrete ordinate mode were obtained in six energy groups for gamma rays and seven energy groups for neutrons. The calculations were carried out with the P3 and S8 approximation using a 13-group coupled neutron-gamma cross sections library, IRAN4.LIB[4].

Table 2. Energy group structure

neutron		gamma	
group	upper energy(ev)	group	upper energy(ev)
1	1.7333E+7	8	1.4E+7
2	5.2205E+6	9	1.0E+7
3	1.0026E+6	10	5.0E+6
4	4.9787E+5	11	2.0E+6
5	9.8037E+4	12	1.0E+6
6	9.1188E+3	13	5.0E+5
7	5.3156E-01		

The calculated neutron and gamma spectra from ANISN output were used in estimating differential and total absorbed doses in a phantom head using a DOSE/PC computer program[5].

In order to have a measure of the absorbed dose effectiveness in tumor with minimal effect in surrounding tissues the ratio of absorbed dose rate due to $^{10}B(n,\alpha)^{7}Li$ reaction in tumor to the total neutron and gamma absorbed dose in the phantom head was determined. This dose ratio was determined for each group of the incident neutrons and for five different tumor depth from 1 cm to 5.5 cm. The results are shown in Fig 2.

CONCLUSIONS

The results of the calculations shown in Figure 2 indicate that for tumors located at about 1 cm below the scalp the absorbed dose ratio is highest for incident neutrons with energies less than 10kev. But for greater depths the maximum absorbed dose ratio appears for the incident neutrons of energy group six, i.e., 0.532ev to 9.12kev, no matter how deep the tumor is located, in the range studied. This means that the maximum absorbed dose effectiveness is somewhat independent of the tumor depth location. We argue that this is because, 10kev neutrons become thermalized at the front edge of the tumor, while neutronswith higher energies, such as 100kev neutrons, will have their thermalization peak inside the tumor. As the energy of the neutron increases thermalization peak moves to the deeper part. As a result, the total of $^{10}B(n,\alpha)^{7}Li$ reactions decreases. One can see this point from the slope of the line connecting group 5 to group 6, since for deeper tumors, group five with higher energy due to longer travelling distance has more chance to thermalize; contributes more in reactions. It is seen from Fig.2 that for a tumor located at 5.5 cm depth or deeper, the maximum dose ratio moves slightly towards the higher energies.

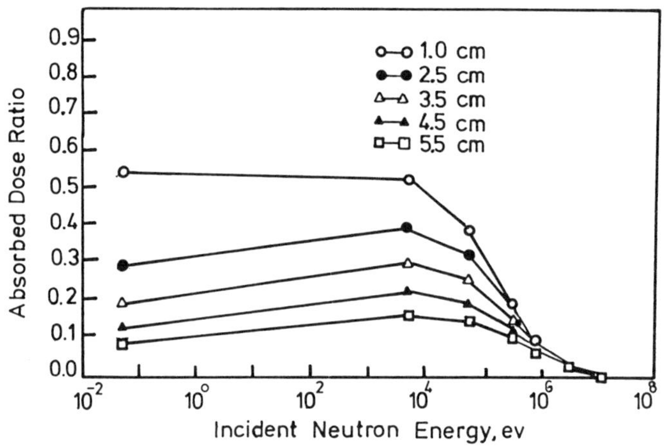

Figure 2. The ratio of absorbed dose rate due to $^{10}B(n,\alpha)^{7}Li$ in tumor to the total absorbed dose rate in phantom head as a function of incident neutron energy at various tumor depths.

REFERENCES

1. Barth R.F., A.H. Soloway and R.G. Fairchild, "Boron neutron capture therapy for cancer" Sci. Am., 100 (Oct. 1990).
2. Hatanaka H., Ed., Proc. 2nd Int. Symp. Neutron Capture Therapy, Tokyo,October 18-21 (1985).
3. Pearson K., ANISN/PC Manual EGG-2500 (April 1987) EG&G Idaho,Inc.,Idaho Falls,Idaho,USA.
4. Marashi M.K., Maiorino J.R., Mendonca A.G. and Santos A., IRAN.LIB(Improved Range of ANISN/PC Library):A P3 coupled neutron-gamma cross-section library in ISOTXS format to be used by ANISN/PC(CCC-0514/02), Ann. Nucl. Energy, Vol. 18, No. 10, pp. 597-602(1991).
5. Marashi M.K., A DOSE/PC program, internal report ,Nuclear Physics Section, Atomic Energy Organization of Iran (1992).

MODIFIED ADVANTAGE DEPTH AND MAXIMUM THERAPEUTIC DEPTH IN THERMAL NEUTRON CAPTURE THERAPY

B J Allen
Biomedicine and Health
Ansto PMB 1 Menai NSW 2234 Australia

INTRODUCTION

A quantitative framework for the calculation of the effective therapeutic depth is derived and applied to thermal neutron capture therapy with sodium borocaptate (BSH) and boronophenylalanine (BPA). These compounds are currently in clinical use in Japan for the intra-operative treatment of brain tumours (BSH) and subcutaneous melanoma (BPA). Boron concentration data are evaluated for biodistribution studies in human patients for both compounds and corresponding maximum therapeutic depths are determined for typical boron tumour to blood ratios.

The equivalent dose (EqGy) for each radiation dose component (Gy) can be expressed in terms of dose modification factors[1]. Of these, the capillary dose factor (CDF) is of major importance, as it implies sparing of capillary endothelial cells. This factor is influenced by the boron concentration in the surrounding tissues, and is thus dependent on the pharmacokinetics of the specific boron compound.

In BNCT, there are concomitant low and high linear energy transfer (LET) radiations and the limiting normal tissue dose needs to be carefully defined. The total dose (DT) arising from reactor and neutron induced gamma rays, and nitrogen and boron neutron capture reactions, is therefore calculated for capillaries (ca), normal tissue (nt) and tumour (tu). The total radiation dose is the sum of all radiation components.

BIODISTRIBUTIONS OF BSH AND BPA

BSH: Boron concentrations in blood and various grades of brain tumours after iv or ic administration have been reported in human patients by a number of authors[2]. Only in a small number of cases was the tumour to blood boron ratio found to be in excess of 2. For glioblastoma multiforme and anaplastic astrocytoma, the average value and SD is estimated to be 1.4 (0.4).

The boron-10 concentrations [B] in the capillaries are assumed to be 1.0 or 0.5 and in normal tissue cells 0.1 relative to that in the tumour (RC). The relative nitrogen concentration in the capillaries is 1 and in normal tissue cells 0.5 relative to the tumour.

The capillary dose factor for the tumour is unity because of the dose equilibrium resulting from comparable blood and tumour cell boron concentrations, whereas it is 0.3 for boron and 1 for nitrogen in the capillaries[3].

BPA: There are three studies of BPA in human patients[2]. In most cases melanoma to blood boron ratios were about 4, although in a few cases the results were much higher. Eliminating the extreme upper and lower values, the mean and SD are 4.3 (1.8). Comparable blood and normal tissue concentrations were found. Note that use of the active L-isomer should lead to improved boron tumour to blood ratios.

The relative boron-10 concentrations in the capillaries and normal tissue cells are therefore assumed to be 0.25 or 0.125. The relative nitrogen concentrations in the capillaries and

Advances in Neutron Capture Therapy, Edited by
A.H Soloway *et al.*, Plenum Press, New York, 1993

normal tissue cells are 1 and 0.5 resp. The capillary dose factor is 1 because the blood boron concentration is equal to that of normal tissue. These assumptions are summarised for BSH and BPA in Table 1.

RESULTS

Matsumoto et al.[4] measured the absorbed dose components as a function of depth in a phantom for the MuITR. The maximum equivalent dose to the tumour, for RBE = 1, and the therapeutic ratio (TR) are calculated on the basis of the capillary tolerance absorbed dose of 15 EqGy[1].

Table 1. Dose modification factors for BSH and BPA

Factor		BSH cap	BSH tissue	BPA cap	BPA tissue
RC	boron	1	0.1	0.25	0.25
	boron	0.5	0.1	0.125	0.125
	nitrogen	1	0.5	1	0.5
CDF	boron	0.3	1	1	1
	nitrogen	1	1	1	1

Table 2. Maximum equivalent doses, therapeutic ratios and irradiation times

	RC_{ca}	DT_{ca}	DT_{nt}	DT_{tu}	TR	time (h)
RBE = 1.0						
BSH	1.0	15	7.7	33.7	2.25	1.88
BPA	0.25	15	14.9	42.6	2.84	2.38
RBE = 2.3						
BSH	1.0	15	8.4	45.2	3.00	1.16
BPA	0.25	15	12.7	40.5	2.70	1.04

The effective biological dose can be used to set the normal tissue dose limit and to determine the TR value. Results are shown in table 2, using RBE = 2.3 for the boron and nitrogen neutron capture reactions, and RBE = 1 for gamma and adopting the tolerance dose of 15 EqGy to the capillaries. The time given in table 2 is the irradiation time at MuITR for the 15 Gy tolerance dose to capillaries[1] to be reached at the surface of the brain. This time increases if the blood boron concentration decreases relative to that of the tumour.

For BSH, inclusion of the RBE increases the equivalent dose to the tumour which is delivered in a shorter time, satisfying the tolerance dose to capillaries. The therapeutic ratio also increases. While for BPA the irradiation time is also reduced, the tumour dose and TR show a small decrease.

MAXIMUM THERAPEUTIC DEPTH

Consistent with Hatanaka[5], the maximum therapeutic depth is defined as the depth at which the tumour dose equals 30 EqGy without exceeding the normal tissue tolerance dose of 15 EqGy. The modified advantage depth (MAD) is defined to be the depth at which the tumour dose equals the tolerance dose to normal tissue when corrections such as the capillary dose factor are included. At this depth the therapeutic ratio is unity, and tumour control should not be expected. These treatment depths are given in Table 3 for a 15 cm beam aperture at the MuITR NCT facility, 20 ppm boron in the tumour, and 15 EqGy tolerance limits to capillary endothelium. The dependence of the therapeutic ratio on depth in the phantom is shown in figure 1 for typical tumour to blood boron concentration ratios for BSH and BPA.

Table 3. Therapeutic depths (cm) for BSH and BPA

BSH RC_{ca}	CDF	MAD	MTD	BPA RC_{ca}	CDF	MAD	MTD
1.0	1.0	0.2		0.25	1.0	3.9	1.5
	0.3	3.0	0.6				
0.5	1.0	1.2		0.125	1.0	5.4	3.0
	0.3	5.0	2.6				

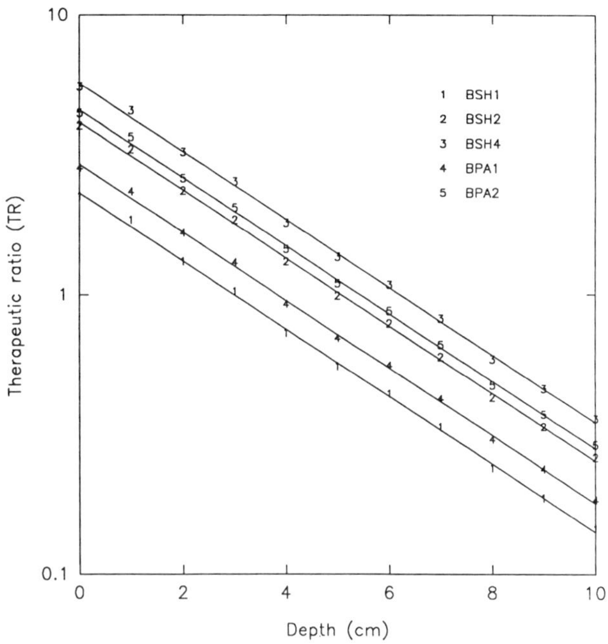

Figure 1. Therapeutic ratios for BSH (curves 1,2,3 for boron tumour to blood ratios of 1,2 4 respectively) and BPA (curves 4,5 for ratios of 4,8 respectively) are shown as a function of phantom depth.

Thus the capillary dose factor has an important effect on the maximum therapeutic depth in thermal BNCT, as shown in table 3. While the MAD and MTD are insensitive to the tolerance dose limit for capillary endothelium, a higher tolerance dose of 19 Gy [5] would increase all doses.

DISCUSSION

The different biological distributions and radiation processes involved with BSH and BPA have a marked effect on the dose calculations. However, as reported previously[6], the final results for practical blood to tumour boron ratios are quite similar. For average tumour to blood boron ratios, the expected MAD and MTD for BSH and BPA are about 4 cm and 1.5 cm respectively. To achieve a lethal dose to the tumour of about 30 EqGy at 4 cm, tumour to blood boron ratios would need to exceed 2 for BSH and 8 for BPA. This is an unlikely condition in view of the existing biodistribution data.

In order to increase the penetration of thermal neutrons and the MTD, Hatanaka[5] administered heavy water to patients for up to 5 days before surgery. Typical heavy water concentrations of 15% were achieved[7], which are calculated to increase the thermal neutron advantage depth by

0.5 cm. However, the increased penetration gives patient dose-depth distributions comparable to the MuITR phantom results for a 15 cm aperture[2]. Thus our results are representative of patient treatments with heavy water and beam apertures smaller than 15 cm. As such, they can be regarded as upper limits if heavy water concentrations lower than 15% are achieved.

CONCLUSIONS

The Maximum Therapeutic Depth is an empirical value based on human and animal irradiations. In this paper the effect of the capillary dose factor and other dose modification factors on the maximum therapeutic depth is quantified for both BSH and BPA. Our results suggest that for thermal BNCT with heavy water administration, the modified advantage depth is 3-5 cm for BSH and 4-5 cm for BPA; the maximum therapeutic depth being 0.6-2.6 cm for BSH and 1.5-3cm for D,L BPA. These results apply for typical boron and heavy water concentrations. In individual patient treatments, these quantities must be determined before dose planning can be completed.

REFERENCES

1 B J Allen, Dose modification factors in boron neutron capture therapy, Strahlenther Onkol, 169 1 29-33 (1993)

2 B J Allen, Maximum therapeutic depth in thermal neutron capture therapy, Strahlenther Onkol, 169 1 34-41 (1993)

3 B J Allen, M Bilek, D E Charlton, Monte Carlo Calculation of Microdosimetry for Boron Neutron Capture Therapy, these proceedings

4 T Matsumoto, O Aizawa, T Nozaki, T Sato, Musashi Institute of Technology Reactor as a Medical Facility with reference to Dosimetry and in situ Boron Concentration Measurement. In: Hatanaka H: Boron Neutron Capture Therapy for Tumours, Nishimura Niigata (1986) 235-242

5 H Hatanaka, Boron Neutron Capture Therapy for Tumours, Karin and Laws: Glioma, Springer-Verlag Berlin 1991 234-249

6 P R Gavin, F J Wheeler, R Huiskamp, C E deHaan, K H I Phillipp, Large animal normal tissue tolerance with boron neutron capture: tolerance of empirical results with a microdosimetric model. In: Proceeding International Workshop on Micro and Macrodosimetry and Treatment Planning in Neutron Capture Therapy, MIT, Oct 1991, in press

7 N Blagojevic, G Storr, B J Allen, H Hatanaka, H Nakagawa, Role of heavy water in boron neutron capture therapy, Proceedings International Workshop on Micro and Macrodosimetry and Treatment Planning in Neutron Capture Therapy, MIT, Oct 1991, in press

BIOLOGICAL EFFECTIVENESS OF RECOIL PROTONS FROM NEUTRONS OF ENERGY 5 keV TO 5 MeV

Edward L. Alpen[1]
Kenneth A. Frankel[2]

[1]Neutron Technology Corp., Berkeley, CA 94708
[2]Lawrence Berkeley Laboratory, University of Calif., Berkeley CA 94720

INTRODUCTION

Several authors, namely Auxier (1968) and McGregor and Allen (1986), have convincingly suggested, on physical grounds, that for proton energies below 15 keV the average energy required to produce an ion pair in aqueous media approaches infinity. Physical theory and some experimental data indicate that the charge state of the proton is so close to zero that nearly all energy transfer is in the form of molecular excitations (Janni, 1982). Ziegler (1985) has also provided data on range energy relationships and stopping power for low energy protons. These latter estimates use codes in which the effective charge on the proton, Z*, is assumed to be 1.0, so no additional light is shed by the Ziegler table on the physical basis for deducing the biological effectiveness of protons below 15 keV. It is our objective to review both the physical background and to compare theory with experiment to estimate the relative biological effectiveness of low energy protons by examining the literature on biological effects of very low energy protons and to make comparisons with the physical data. At the same time, since we were convinced that a careful analysis of late effects RBEs for recoil protons in the energy range 10 keV to 100 keV had not been previously done, this task would also be undertaken.

The objectives of the study were to examine the LET and charge state of low energy recoil protons and to estimate the appropriate RBE for late effects in human normal tissue systems for recoil protons from 10 to 100 keV. We will also extend the study to include RBE recommendations for all recoil proton energies appropriate for BNCT applications, that is, energies up to 5 MeV.

We will develop an RBE matrix from literature sources and physical data for the range of proton energies of importance in Boron Neutron Capture Therapy for the dose delivered in a single fraction, four fractions, and thirty fractions. It is not a realistic scenario that real treatments will be delivered in thirty fractions, but these values allow comparison with conventional supervoltage therapy. All of the clinical experience with patients to date has been with single fraction treatments, but for many radiation oncologists the use of four treatment fractions holds some attraction for safety reasons.

EXPERIMENTAL APPROACH

We will determine the gamma ray tolerance dose for gamma rays from patient and animal data, correcting from the 20-40 fraction regime common in radiotherapy using the nomogram of Liebel and Sheline (1991) for brain (and also vascular endothelium) and for spinal cord and skin from the data of Hopewell (1990). This data appears in Table 1.

For neutrons of E > 1 MeV the simple ratio of fast neutron tolerance doses taken from Larimore and Austin-Seymour (1992) are listed in Table 2 for tissues of relevance

Advances in Neutron Capture Therapy, Edited by
A.H Soloway *et al.*, Plenum Press, New York, 1993

to this report. The ratio of these values for tissue tolerance doses for fast neutrons to the relevant values in Table 1 provides the RBEs for critical tissues of interest in BNCT. As suggested in the following text, adjustment of the tolerance doses downward is required for fraction numbers of four or less.

A review of the extensive literature comparing the RBE values for critical tissues as a function of neutron energy (see, for example, Hornsey, 1991) clearly indicates that there is only moderate variation in tolerance doses for normal tissues as a function of

Table 1. Threshold doses for gamma ray late effects in key tissues involved in boron neutron capture therapy.

Tissue at Risk	One Fraction Dose (Gy)	Four Fractions Dose (Gy)	Thirty Fractions Dose (Gy)
Brain	12-15	35	40-45
Spinal Cord	10-15	35	50
Vascular endothelium	10-15	35	50
Skin	22	35	60

Table 2. The average tolerance doses for critical tissues for fast neutrons as selected from Laramore and Seymour (1992).

CRITICAL TISSUE	TOLERANCE DOSE (Gy)
Skin	18
Brain	10
Spinal Cord	8

Table 3. Estimates of late damage thresholds and RBE values for single and fractionated exposures to 1 to 5 MeV neutrons.

	SKIN (dose, Gy)	BRAIN (dose, Gy)	SP CORD (dose, Gy)
ONE FRACTION			
Neutron dose (Gy)	12 Gy	8 Gy	6 Gy
RBE*	1.8	1.5	1.8
FOUR FRACTIONS			
Neutron dose (Gy)	18	10 Gy	8 Gy
RBE*	2.0	3.5	3.8
THIRTY FRACTIONS			
Neutron dose (Gy)	18	10 Gy	8Gy
RBE*	3.3	4.0	6.0

*Using the reference radiation values of Table 1 for the appropriate fraction number.

neutron energy for energies in the range 1 to 20 MeV. It is possible then to construct a table of RBE values for the three critical tissues in Table 3. Hornsey's data also indicate a modest increase in effectiveness of fast neutrons for fraction numbers of less than four. Some downward adjustment has therefore been made for the single fraction neutron doses to correct for the fact that Laramore and Seymour are reporting on data from multi-fraction human exposures.

For E from 50 keV to 1 MeV no data on human late tolerance exists. A synthesis of tolerances doses was made from mostly animal and some cell data, using the data from Broerse et al (1967), Hall et al (1975), Belli et al (1989), Blakely et al (1985) and Rodriguez et al (1992). The latter being for animal experiments and most of the former for cells. The synthesized table of RBE estimates is shown in Table 4.

Table 4. Estimates of RBE for acute and late effects for protons in the energy range, 50 TO 500 keV.

PROTON ENERGY (keV)	LET* (keV/μm)	ACUTE RBE	LATE EFFECTS RBE
50	97	2.2–3.5	3.5 +/- 0.5
100	95	2.2–3.5	3.5 +/- 0.5
200	65	1.4–3.0	2.5 +/- 0.5
500	41	1.1–1.4	1.5 + 0.5, – 0

*from Janni (1982)

For neutrons of energy less than 50 keV there is even less experimental information than for the foregoing energy region. It is often assumed that because of charge neutralization, there is little associated biological effect (See, for example Auxier, 1968). As we have mentioned in the Introduction there is no clear consensus among physicists as to the charge state of protons with energy of a few tens of keV. Janni (1982) insists that both experimental and theoretical evidence support a very low value for the effective charge, Z*, on these low energy protons. He reports that at 20 keV Z* for the proton in water is 0.173, and at 5 keV the Z* in the same medium is 0.095. The widely used TRIM code of Ziegler (unpublished) provides stopping power and range estimates also, but this code assumes a Z* of 1.0 at all energies within the scope of the code. Summarizing the physical understanding of the nature of energy transfer for low energy fast neutrons and their derived recoil protons, there is significant contention among physical scientists as to the effective charge and stopping power of the recoil protons at energies of a few keV to a few tens of keV. Brandt (in Ziegler, 1985) suggests that it is not presently possible, by either theory or experiment, to determine the answer as to the effective charge of low energy protons, while Janni (1982) contends strongly that energy transfer in this region must be by molecular excitation mechanisms, not ionization.

There clearly must be some energy of the recoil proton that indeed satisfies Janni's contention about effective charge, but no conclusive statements can be drawn at this time. Finally, we can quote from Ziegler (1985) as follows, "As discussed below, this (referring to the effective charge on a low energy proton) is a messy subject." The parenthetical is inserted by us. To quote further from the same reference, "...........that polarization of the medium indicates that for velocities below about twice the Bohr velocity a hydrogen projectile would have an effective charge less than one." At least there is physical consensus that at energies of about 5 keV and less, the effectiveness of a proton to produce ionization must be trivial, and we make the assumption that there is little ionization damage to biological systems from recoil protons of energy less than 5 keV.

None of the studies of the physics of low energy protons have explicitly considered the possibility of elastic collisions with neutrons in which the orbital electron remains with the proton after collision. The general theory of such an interaction is well known, and predicts that the velocity of the recoiling nucleus must exceed the velocity of the orbital electron for stripping to occur.

For recoil protons in the energy range 5 to 50 keV, we must depend upon some elegant experimental data that has been developed by several authors using the neutron iron antiresonance beam at 24.5 keV developed by Constantine et al (1985, 1986) One of a number of reports on cellular effects is that of Morgan et al (1986). The various reports provide cellular acute effects RBE values of 3.5 to 6.5. An approximation, based on our knowledge of the sometimes unreliable extrapolation of RBE values from acute survival studies to late damage, brings us to recommend an RBE value of 3.5 for comparison with single dose gamma rays. Appropriate adjustment for comparison with fractionated gamma rays gives rise to the data in the last cell of the following table. We assume no biological effectiveness through ionization for energies below 5 keV.

Table 5 is a summary of RBE values for all the energy ranges considered in this report, including the just outlined values for the lowest energies.

Table 5. Estimated RBE values as a function of energy for neutrons of the designated energy range.

1MeV < E < 5 MeV	SKIN	BRAIN	SP CORD
RBE – SINGLE FRACTION	1.8	1.5	1.8
RBE – FOUR FRACTIONS	2.0	3.5	3.8
RBE –THIRTY FRACTIONS	3.3	3.5	3.5
500 keV< E < 1 MeV			
RBE – SINGLE FRACTION	1.8	1.8	1.8
RBE – FOUR FRACTIONS	2.4	2.6	2.6
RBE –THIRTY FRACTIONS	3.0	3.2	3.2
200 keV< E < 500 keV			
RBE – SINGLE FRACTION	2.5	2.5	2.5
RBE – FOUR FRACTIONS	3.3	3.5	3.5
RBE –THIRTY FRACTIONS	4.2	4.6	4.6
100 keV< E < 200 keV			
RBE – SINGLE FRACTION	3.5	3.5	3.5
RBE – FOUR FRACTIONS	4.2	4.2	4.2
RBE –THIRTY FRACTIONS	6.0	6.0	6.0
50 keV< E < 100 keV			
RBE – SINGLE FRACTION	3.5	3.5	3.5
RBE – FOUR FRACTIONS	4.2	4.2	4.2
RBE –THIRTY FRACTIONS	6.0	6.0	6.0
5 keV< E < 50 keV			
RBE – SINGLE FRACTION	3.5	3.5	3.5
RBE – FOUR FRACTIONS	4.5	4.5	4.5
RBE –THIRTY FRACTIONS	6.2	6.2	6.2

CONCLUSIONS

1. Reliable patient data is available to assign RBE values for late tissue tolerance for fast neutron energies down to about 500 keV.

2. It has been possible to construct RBE estimates for the energy range from 100 keV to 1 MeV based mostly on acute cell survival and animal tissue responses. Adjustments have been estimated to arrive at late tissue tolerance RBE values.

3. Physics modeling and theoretical calculations have suggested that effective charge on recoil protons below a few tens of keV might approach zero, and most energy transfer should be by chemical excitation.

4. In spite of the conclusions of the physics models, experiments with 24 keV neutrons strongly support an RBE for the energy range from 5 to 50 keV of 5.0 or more.

5. A comprehensive table of RBE values is provided with adjustments for fractionation at one fraction, four fractions and thirty fractions.

6. It is clear, because of the dose-rate effect associated with fractionated gamma ray exposure, that RBE values are quite large for thirty fraction exposures. In the end, only the real value of tolerance dose rather than RBE will drive treatment planning for BNCT as it does for conventional radiotherapy.

REFERENCES

J. A. Auxier, W. S. Snyder and T. D. Jones, (1968). Neutron interactions and penetration in tissue. In, "Radiation Dosimetry, Volume I, Fundamentals," Attix, Roesch and Tochilin Eds,. Academic Press, New York, pp 275–317.

M. Belli, R. Cherubini, S. Finotto, G. Moschini, O. Sapora, G. Simone andM. A. Tabocchini (1989). RBE-LET relationship for inactivation of V79 Chinese hamster cells irradiated by proton beams. *Int. J. Radiat. Biol.* **55**, 93–104.

E. A. Blakely, F. Q. H. Ngo, S.B. Curtis and C. A. Tobias (1984). Heavy ion radiobiology: Cellular studies. *Adv. in Radiat. Biol.* **11**, 295–389.

J. J. Broerse, G. W. Barendsen and G. R. van Kersen (1967). Survival of cultured human cells after irradiation with fast neutrons of different energies in hypoxic and oxygenated conditions. *Int. J. Radiat. Biol.* **13**, 559–572.

G. Constantine, L. J. Baker , LP. G. F. Moore and N. P. Taylor (1985). The depth enhanced neutron source, In, "Neutron Capture Therapy: Proceedings of the Second International Symposium on Neutron Capture Therapy" (H. Hatanka, Ed.), pp. 209–222. MTP Press, Norwell, Maine.

G. Constantine, L. J. Baker and N. P. Taylor (1986). Improved methods for the generation of 24.5 keV neutron beams with possible application to boron neutron capture therapy. *Nuclear Instr. & Methods in Phys. Res.* **A250**, 565–572.

E. J. Hall, J. K. Novak, A. M. Kellerer, H. H. Rossi, S. Marino and L. J. Goodman (1975). RBE as a function of neutron energy. I. Experimental observations. *Radiat. Res.* **64,** 245–255.

J. W. Hopewell, D. W. H. Barnes, M. E. C. Robbins, M. Corp, J. M. Sanson, C. M. A. Young and G. Wiernik (1990). The relative biological effectiveness of fractionated doses of fast neutrons (42 $MeV_{d\longrightarrow Be}$) for normal tissues in the pig. II. Late effects on cutaneous and subcutaneous tissues. *Brit. J. Radiol.* **63**, 760–770.

S. Hornsey (1991) Experimental central nervous system injury from fast neutrons. In, "Radiation Injury to the Nervous System" P. H. Gutin, S. A. Leibel and G. E. Sheline, Eds. pp. 137-148. Raven Press, New York.

ICRU (International Commission on Radiological Units, 1992). "Photon, Electron, and Neutron Interaction Data for Body Tissues", Report 46 Bethesda, MD.

J. F. Janni (1982). Proton range-energy tables, 1 keV to 10 GeV. *Atomic Data and Nuclear Data Tables,* **27**, 147-339.

G. E. Laramore and M. Austin-Seymour (1992). Fast neutron radiotherapy in relation to the radiation sensitivity of human organ systems. In. "Relative Radiosensitivities of Human Organs Systems" vol III.. K. I. Altman and J. Lett., Eds. Academic Press, New York (in press).

S. A. Leibel and G. E. Sheline (1991). Tolerance of the brain and spinal cord to conventional irradiation. In, "Radiation Injury to the Nervous System" P. H. Gutin, S. A. Leibel and G. E. Sheline, Eds. pp. 239–256. Raven Press, New York.

B. J. McGregor and B. J. Allen (1986). Filtered-beam dose distribution for boron neutron capture therapy of brain tumours. In, "International Symposium on Neutron Capture Therapy, Proceedings" Fairchild and Brownell, Eds, pp. 14–25. Brookhaven National Laboratory, Upton, New York.

G. R. Morgan, C. J. Roberts, and P. D. Holt (1989). The radiobiology of a nearly pure, high intensity beam of 24 keV neutrons. *Strahlenther. Onkol.* **165**, 196–198.

G. R. Morgan, A. J. Mill, C. J. Roberts, S. Newman and P.D. Holt (1988). The radiobiology of 24 keV neutrons. *Brit. J. Radiol.* **61**, 1127–1135.

A. Rodriguez, E. L. Alpen and P. Powers-Risius (1992). The RBE-LET relationship for rodent intestinal crypt cell survival, testes weight loss and multicellular spheroid cell survival after heavy ion irradiation. *Radiat. Res.* **132**, 184-192.

J. F. Ziegler, J. P. Biersack and U. Littmark (1985) "The Stopping Power and Range of Ions in Solution", J. F. Ziegler, Ed. Vol. 1, Pergamon Press, New York.

ON THE DEVELOPMENT OF A SCINTILLATION SPECTROMETER FOR COMPARISON OF NEUTRON ENERGY SPECTRA AT HIGH AND LOW FLUX

J. F. Crawford[1] and F. Stecher-Rasmussen[2]

[1]P.S.I., CH-5232 Villigen, Switzerland
[2]ECN, Petten, The Netherlands

INTRODUCTION

The energy spectrum of the HB11 neutron beam at the High Flux Reactor (HFR), Petten, has been measured by proton and α recoil in hydrogen and helium gas proportional counters at power levels of 3-8 kW[1,2,3]. This low power is required because gas counters are limited to kHz rates; it is four orders of magnitude less than the 45 MW required for therapeutic fluxes. Because of the significantly different conditions of the reactor at these two very different power levels, some doubt exists whether the spectrum remains the same at both. In order to test this point, a prototype scintillation detector, small enough to handle the high rates expected at full power, has been constructed at the Paul Scherrer Institute, Villigen, Switzerland (PSI)[4]. It is similar to a device developed at the University of Birmingham, England[5]. The initial objective is to compare amplitude spectra at high and low powers in HB11, to check that they are in fact the same shape.

NE102A plastic scintillator material (made by Nuclear Enterprises, Sighthill, Edinburgh, Scotland) contains some 5.28×10^{22} protons per cm^3. The reaction detected is the same as in a hydrogen gas recoil counter: a neutron scatters elastically on a stationary proton, which recoils with on average half the neutron energy. The recoiling charged proton ionizes and excites atoms near its path; the ions recombine, and the excited atoms de-excite, with the emission of UV or visible light. Wavelength shifters in the scintillator absorb the UV and emit visible light.

The visible photons are brought by plexiglass light pipes to the photo-cathodes of two photo-multiplier tubes (PMT's), where they are converted to photo-electrons; the efficiency of each of these steps is of the order of 0.1[6]. The multiplier section of the PMT then amplifies the photo-cathode signal by a factor of the order of 10^7, to the point that a single photo-electron produces a pulse large enough to be used in subsequent electronic circuits. The limiting stage in the chain of events is just after the photo-cathodes, at which point the amplitude information is at its most meagre. In practice, several tens of photo-electrons will be emitted from the cathode per MeV deposited in the scintillator[7].

The recoil protons, of a range of 0.02 mm at 1 MeV, will usually stop in scintillators of the size considered here. The intrinsic response time of plastic scintillator is of the order of nanoseconds, a good match to the speed of fast PMT's. The combination is four or more orders of magnitude faster than gas proportional counters.

Advances in Neutron Capture Therapy, Edited by
A.H Soloway *et al.*, Plenum Press, New York, 1993

RATES AND EFFICIENCIES

If the neutron flux is ϕ and the mean free path of the neutrons in scintillator is λ, then the neutron event rate in a small scintillator of volume V will be $V\phi/\lambda$, assuming that λ is large compared to the dimensions of the scintillator. For scattering on protons of the keV to MeV neutrons of interest in BNCT, λ varies from 10 mm at 10 keV to 47 mm at 1 MeV. Other interactions in the scintillator deposit very little energy, and can be neglected. It follows that even at the very high fluxes required for BNCT, a scintillator of mm dimensions will experience a neutron event rate of under 10^6 sec^{-1} mm^{-3}. Such rates can be handled by state-of-the-art fast electronics, even neglecting the fact that low-energy scatters will not be detected by the PMT's.

Gammas will also interact in the scintillator, but at a rate set by the much longer Compton mean free path of about 150 mm at 1 MeV. The resulting electrons, whose ranges are stated by the manufacturer to be of the order of some mm, are unlikely to stop in the scintillator, and will emit little light. In consequence of these two effects, the scintillator can be expected to be an order of magnitude less sensitive to γ's than to neutrons, despite the saturation effects of slow protons[7].

RESULTS

In accordance with these ideas, scintillator detectors of mm size have been built and tested at the Low Flux Reactor (LFR) and in the HB11 beam at HFR. As calculated by MCNP, the LFR beam is harder than the HB11 beam. The detectors consisted in essence of a plastic scintillator of mm dimensions, connected by plexiglass light-pipes to two XP2230B PMT's (Philips, Boschbyk, Eindhoven, Nederland). These, which are state-of-the-art conventional tubes, show a clear single photo-electron line, i.e. events arising from single photo-electrons from the photo-cathode, due either to very low light levels or to tube noise. As discussed below, this allows a calibration of the tube outputs in terms of the average amount of deposited energy in the scintillator required to liberate one photo-electron from a tube cathode. Two tubes are used to reduce some backgrounds by demanding coincidences, and to increase the total photo-electron yield.

The fast signals from the PMT's were integrated and amplified in EGG 572 units (EG&G, Munich, Germany) and then fed to ND 581 digitizers (Nuclear Data, Schaumberg, IL) run in coincidence. Dead-time was about 6.5μsec per event, and the coincidence resolving time was about 0.1μsec. For the test, the LFR was run at powers of only a few hundred watts.

For test purposes, several scintillators were tried. Typical spectra are shown in Figure 1, which shows 2-dimensional histograms of the pulse heights observed in the two PMT's. The horizontal axes display the pulse height information, while the vertical axis displays logarithmically the number of counts corresponding to a pair of pulse heights. Units are arbitrary. The ridge between the horizontal axes, corresponding to events which deposit similar amounts of light in the two PMT's, is attributed to signals from proton recoils after neutron scattering events in the scintillator. The attribution was checked by inserting a 50 mm lead block in the beam. Analysis of the data confirms the impression from Figure 1 that the event rate decreased by a factor of about 3-4, consistent with the mean free path of 30-60 mm of neutrons in lead. In addition, spectra (not shown) were taken without a scintillator; they showed no sign of the events attributed to neutron scatters in the scintillator. Residual events in these spectra, which were located close to the horizontal axes, deposit very different amounts of light in the two PMT's; they are attributed to γ's from neutron capture on protons in the plexiglass light pipes.

One-dimensional amplitude spectra were also taken, at higher gain settings to locate the single photo-electron peaks in the two PMT's. By comparison with the 2-dimensional data, and with MCNP calculations of the neutron spectrum expected in the LFR beam, it could be

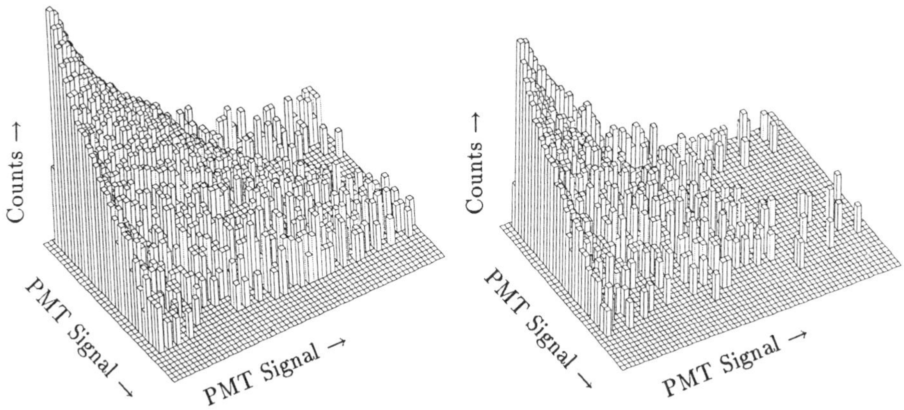

Figure 1. Two-dimensional amplitude spectra from LFR; the vertical axis is logarithmic. The right-hand spectrum was taken with a 50 mm lead block in the beam; while the overall shape is similar, the total rate is lower with the lead in place.

estimated that one photo-electron corresponded, in both tubes, to an energy of about 25 keV deposited in the scintillator. By standard probability arguments, it then follows that, even when coincidences are demanded, the device will detect events arising from 20 keV neutrons with an efficiency of about 12%, and is still sensitive at the 4% level to events from neutrons of only 10 keV. If background rates could be reduced to the point that single events could be used, these efficiencies would become 31% and 17% respectively.

For a test run in the HB11 beam at HFR, the main change in the system was to use faster electronic modules. PMT anode signals were brought, *via* LeCroy 612A fixed gain fast amplifiers, to 2249A ADC channels, and also to a 2228A TDC channel (all from LeCroy, Meyrin, Geneva). This meant that in effect a 1024 x 1024 2-dimensional ADC was available (as compared with 64 x 64 in the LFR test); and that nanosecond timing could be done between the two PMT signals (which was not available in LFR). A coincidence between the two anode signals, timed on one PMT, was used as an event trigger; it started the TDC, which was then stopped by the other PMT signal. The trigger was also used to gate the two ADC channels. Events, each consisting of the TDC and two ADC signals, were read out *via* PSI IO 502 and KineticSystems 3989 units, by a Tulip Vision Line dt 486 dx/e PC into a disk file for later analysis.

While singles rates in the two PMT's were several MHz, presenting significant problems, some useable timing and amplitude spectra (shown in Figures 2 and 3) were obtained from two layouts which differed only in the amount of scintillator material present. Because of the much larger number of ADC channels, the horizontal axes in Figure 3 are also plotted logarithmically. While neutrons do not show up as clearly as in Figure 1, there are significant differences between the layouts, suggesting that neutron scatters in the scintillator are being detected at full HFR power.

FUTURE DEVELOPMENTS

At present the MHz singles rates are attributed to γ's from neutron capture on protons in the plexiglass light pipes. If this attribution is correct, a minor re-design will reduce the singles rates considerably without affecting the real neutron events, perhaps to the point where single counts could be used directly without coincidences being demanded. This would allow the detection of lower energy neutrons.

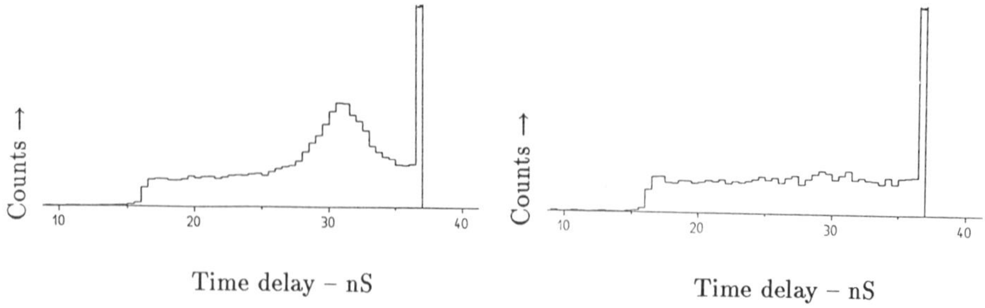

Figure 2. Timing spectra, taken in HB11, for two different sizes of scintillator: 128 mm^3 and 4 mm^3. The broad peak, present in one spectrum but not in the other, is evidence that neutrons are being detected. The peak at 37 nS is an artefact.

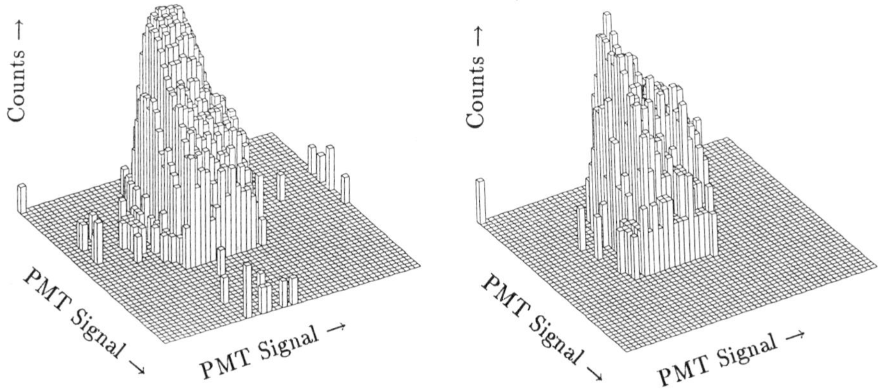

Figure 3. Amplitude spectra corresponding to the timing spectra of Figure 2. Events plotted are those in the region of the prominent peak in Figure 2. The difference between the two plots is again evidence that neutrons are being detected.

It is of course possible to unfold the spectra from a scintillator detector to derive the neutron energy spectrum, as has been done in the gas counter measurements[2,3]. However, this will require a careful study of the considerable non-linearities in the scintillation processes involved in stopping slow protons. In the short term, it is not necessary, since the initial objective of this work is to compare amplitude spectra at high and low powers in HB11, to check that they are in fact the same shape.

Devices based on the ideas presented in this paper could be used for routine monitoring of BNCT beams, in both flux and energy.

The authors would like to thank the crews of the LFR and HFR, Petten, and G. Constantine, for their competence and help during the tests.

REFERENCES

[1] Watkins, P. R. D., 'A review of the calculations performed for the design and shielding of the Petten BNCT facility', given at the International Workshop 'Towards Clinical Trials of Glioma with BNCT', 18-20 September 1991, Petten, The Netherlands.

[2] Perks, C. A., & H. J. Delafield, 'Neutron energy spectrometry measurements in the Petten HFR HB11 neutron beam', ibid.

[3] Perks, C. A., & J. A. B. Gibson, 'Neutron spectrometry and dosimetry for BNCT', Seventh Symposium on Neutron Dosimetry, Berlin, 14-18 October 1991.

[4] Crawford, J. F., *et al*, Workshop on NCT, Essen, FRG, 28-29 February 1992; Strahlentherapie & Onkologie, in press.

[5] Beynon, T. D., University of Birmingham, School of Physics and Space Research, Edgebaston, Birmingham B15 2TT, UK.

[6] Particle Data Group, Phys. Lett. B 239 (1990), III.22.

[7] Wishart, L. P., et al., NIM 57 (1967), 237-244.

MICRODOSIMETRY AND COMPOUND FACTORS FOR NEUTRON CAPTURE THERAPY

Guido R. Solares,[1] Robert G. Zamenhof,[1] and Guillermo Cano[2]

[1]Department of Radiation Oncology, New England Medical Center and Tufts University School of Medicine, Boston, MA, 02111 USA
[2]Department of Neurosurgery, New England Medical Center and Tufts University School of Medicine, Boston, MA, 02111 USA

INTRODUCTION

It has been suggested[1] that in order to inactivate a cell, the nucleus must receive one or more direct hits from a heavy charged particle. This principle could be extended to the α particles and the ^{7}Li recoil ions produced in the $^{10}B(n,\alpha)^{7}Li$ reaction, which is central to the principle of boron neutron capture therapy (BNCT). As a result, the microdistribution of a boron compound in tissue can drastically change the dose needed to inactivate malignant cells. Therefore, being able to determine microscopically the effect of different boron compounds through their microdistributions is clearly of major importance to BNCT. Previous microdosimetric analyses of the $^{10}B(n,\alpha)^{7}Li$ reaction have been accomplished by assuming mathematical models of cells and calculating the absorbed doses due to the α particle and ^{7}Li recoil ion based on assumed boron microdistribution.[2-7]

The approach used here consists of using actual histological and microdistribution data to assess the microdosimetric effect of various boron distributions in tissue, without any assumptions relating to either geometrical modeling of the tissues or to boron distributions. Using a combination of Monte Carlo analysis, analytical calculations, and image processing techniques, several parameters of importance to the microdosimetry of BNCT are considered, including relative energy deposition, number of nuclear hits, average linear energy transfer (LET) in nucleus, etc. These are used to estimate "Compound Factors" which can be used to assist with the evaluation of various potential compounds for NCT.

MATERIALS AND METHODS

Histology and Boron Distribution

A unique form of high resolution quantitative alpha track autoradiography (HR-QAR) has been developed in our laboratory.[8] This technique, which has a spatial resolution of 1-2 microns, allows simultaneous observation of tissue histology and α and ^{7}Li tracks. Contrary to other common autoradiographic methods, the Lexan detector used in this technique does not register proton tracks produced in the $^{14}N(n,p)^{14}C$ reaction.

Advances in Neutron Capture Therapy, Edited by
A.H Soloway *et al.*, Plenum Press, New York, 1993

Figure 1a shows an autoradiograph of a tumor-bearing mouse infused with *p*-boronophenylalanine (BPA). Alpha particles and ^{7}Li recoil ions appear as small black dots. Brain tissue is seen in the background.

Image Processing

The image of an autoradiograph is digitized with a CCD camera and then analyzed using image processing techniques.[8] Applying Fourier transforms to the image, the tracks are extracted and a detailed record of their size and location is obtained (Figure 1b). Use of edge detection algorithms, found in the program Image Analyst™ (Automatix Corp., Billerica, MA), allow segmenting of an image into several areas of interest; i.e., nuclei, cytoplasm, etc. Figure 1c shows the original image (Fig 1a) after image analysis. Nuclei are shown gray, cytoplasm white, and artifacts black.

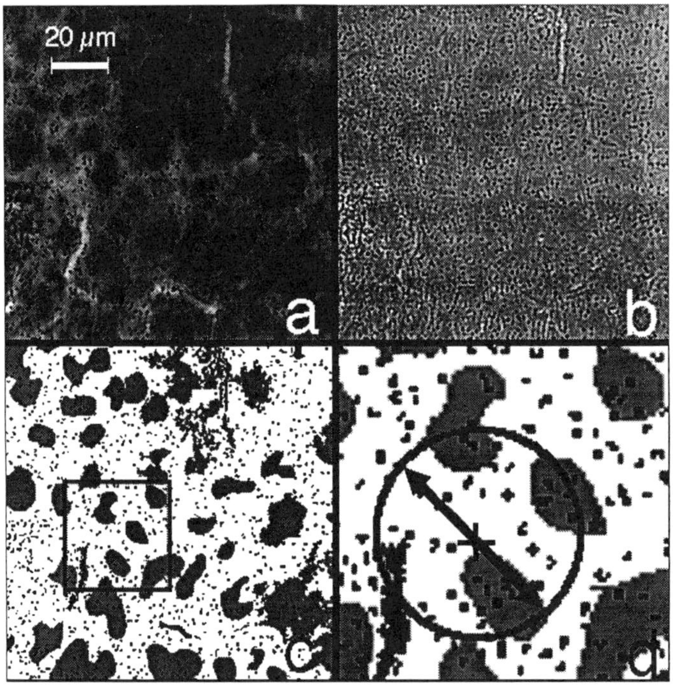

Figure 1. (a) 1 micron-thick autoradiograph of a murine brain tumor containing BPA; α particles and ^{7}Li ions appear as small black dots. (b) same image after computer processing in which tracks have been extracted. (c) histologically segmented image; nuclei shown gray, cytoplasm white and artifacts black. (d) magnified image of the square shown in 1c. The circle represents the range of an alpha particle produced in the $^{10}B(n,\alpha)^{7}Li$ reaction.

Microdosimetry

The microdosimetric analysis is achieved using Monte Carlo methods and analytical calculations. The energy deposition of each particle along its path length and in each region of interest (i.e., nucleus, cytoplasm, etc.) is calculated using its linear energy transfer (LET) function generated by the computer program TRIM.[9] Since the particles are emitted isotropically, each particle in turn is followed over 600 randomly-selected angles (Figure

1d). The energy deposition from the measured boron distribution is then compared to the energy deposition from a computer generated homogeneous boron distribution with the same ^{10}B concentration. In addition, using the concept of "starters," "stoppers," "crossers," "insiders," and "outsiders,"[10] the number of nuclear hits is scored for each α particle and ^{7}Li recoil ion with a minimum LET of 5 KeV/μm . The average LET and an averaged dose-weighted LET (defined below) are calculated for each particle along each path length. Based on these calculated parameters, five proposed "compound factors" are defined as follows:

1) Stochastic Processes Factor (H): the ratio of all nuclear hits that produce a cell kill to the total number of nuclear hits. The number of nuclear hits that produce a cell kill is calculated using a step hit size effectiveness function (HSEF)[7] with a minimum threshold size of 5 KeV/μm. Factor H, is a measurement of how effective a given compound is at producing nuclear hits that actually kill a cell:

$$H = \left[\Sigma \text{ Hit}_{(N)} \times \text{HSEF}\right] / \Sigma \text{ Hit}_{(N)}$$

2) Spatial Distribution Factor (S): is the ratio of factor H, calculated for an inhomogeneous or actual distribution, to the factor H calculated using a homogeneous distribution with the same ^{10}B concentration. Factor S provides a measure of how a measured compound distribution compares to that of a homogeneous distribution:

$$S = H_{\text{inhomogeneous}} / H_{\text{homogeneous}}$$

3) Localization Factor (L): is defined as the ratio of hits to the nucleus to the total number of hits occurring in the cell. Factor L weights the ability of a compound to deliver hits to cell nuclei:

$$L = \Sigma \text{ Hit}_{(N)} / \Sigma \text{ Hit}_{(\text{cell})}$$

4) Dose-weighted LET Factor (W): is the ratio of dose-weighted LET in each segment along each particle track to the average value of the LET in each segment. By weighting those segments that have a higher LET, this factor relates to the biological effectiveness of a compound in depositing dose in the nuclei:

$$W = \Sigma \text{ LET}^2 / \Sigma \text{ LET}$$

5) Relative Energy Deposition Factor (E): is the ratio of the total energy deposited in the nuclei to the total energy released in the $^{10}B(n,\alpha)^{7}Li$ reaction. Factor E, is an indicator of how effective a compound is at using the available energy for cell killing:

$$E = \Sigma \text{ Energy deposited in nucleus} / \Sigma \text{ Energy released}$$

RESULTS AND DISCUSSION

Table 1 shows, for illustration and not for intercomparison purposes, the results for measured distributions of three different boron compounds in three different tumor models: BPA in an intracranial murine glioma, $Na_2B_{12}H_{11}SH$ (BSH) in a intracranial rat glioma, and Boronated Porphyrin (BOPP) in a rat mammary carcinoma. If two or more compounds are used in the same tumor model, the compound factors calculated as proposed here could provide a very useful comparison of the microdosimetric effectiveness of each compound, providing that enough radiographs are analyzed. We hope to obtain such results in a future study.

Table 1. Compound factors for three different boron compounds.[1]

COMPOUND FACTOR[2]	BPA	BSH	BOPP
Spatial Distribution Factor (S)	0.78	1.11	1.61
Localization Factor (L)	0.39	0.58	0.51
Dose-weighted LET Factor (W)	0.81	0.86	0.82
Relative Energy Deposition Factor (E)	0.38	0.52	0.52

[1]BPA in mouse glioma; BSH in rat glioma; BOPP in rat mammary carcinoma.
[2]Estimated uncertainty for these results is ±10%

The ultimate effectiveness of a potential boron compound for BNCT depends on several other factors not considered in the present analysis. Including the ability of a compound to selectively concentrate in tumor and the concomitant concentration in normal tissues and blood. If dose to the capillaries is to be considered as the limiting factor, their geometry is also very important, in assessing the performance of different potential compounds for BNCT .

ACKNOWLEDGMENTS

We thank Dr. J. Coderre from Brookhaven National Laboratory for supplying the BSH and BOPP samples, Mr. Joe Palano for the help provided, and the support of Automatix Corporation, Billerica, MA. The work reported here was supported by the U.S. Department of Energy under grants #DE-FG02-87ER-60600 and #DE-FG07-80ER-10770.A015.

REFERENCES

1. E. Lloyd, M.A. Gemmell, C.B. Henning, D.S. Gemmell, and B.J. Zabransky, Cell survival following multiple-track alpha particles irradiation, *Int. J. Radiat. Biol.* 35:23 (1979).
2. M.A. Davis, J.B. Little, K.M. Ayyangar, and A.R. Reddy, Relative biological effectiveness of the $^{10}B(n,\alpha)^{7}Li$ reaction in HeLa cell. *Radiat. Res.* 43:534 (1970).
3. K. Kitao, A method for calculating the absorbed dose near the interface from the $^{10}B(n,\alpha)^{7}Li$ reaction. *Radiat. Res.* 61:304 (1975).
4. R.A. Rydin, O.L. Deutsh, and B.W. Murray, The effect of geometry on capillary wall dose for boron neutron capture therapy. Phys. Med. Biol. 21:134 (1976).
5. T. Kobayashi, and K.Kanda, Analytical calculation of boron-10 dosage in cell nucleus for boron neutron capture therapy, *Radiat. Res.* 91:77 (1982).
6. F.J. Wheeler, M.L. Griebenow, D.E. Wessol, D.W. Nigg, and R.A. Anderl, A stochastic model for high-LET response for boron neutron capture therapy (BNCT), Basic Life Sci. 50:165 (1989).
7. D. Gabel, S. Foster, and R.G. Fairchild, The Monte Carlo simulation of the biological effect of the $^{10}B(n,\alpha)^{7}Li$ reaction in cells and tissues and its implication for boron neutron capture therapy. *Radiat. Res.* 111:14 (1987).
8. G. Solares, "High resolution alpha track autoradiography and biological studies of boron neutron capture therapy," Ph.D. Thesis, Department of Nuclear Engineering, Massachusetts Institute of Technology, Cambridge, MA (1991).
9. J. Ziegler, TRIM91, "The Transport of Ions In Matter," IBM Research, Yorktown, NY (1991).
10. R. Caswell, Deposition of energy by neutrons in spherical cavities, *Radiat. Res.* 27:92 (1966).

ANALYTICAL DOSE CALCULATION IN CELL NUCLEUS FROM SPHERICAL ^{10}B CONTAINING REGION AND A CONCEPT OF RBE FOR NEUTRON CAPTURE THERAPY

Tooru Kobayashi and Keiji Kanda

Research Reactor Institute, Kyoto University,
Kumatori-cho, Sennan-gun, Osaka 590-04, Japan

INTRODUCTION

For ordinary radiation therapy, such as X rays, gamma rays, or electron beams, the average absorbed dose over the order of millimeters is sufficient to estimate a therapeutic effect. On the other hand, boron neutron capture therapy (BNCT) utilizes high-LET particles of short range such as α particles and ^{7}Li nuclei produced by neutron reactions with ^{10}B. The range of the local destruction covers cells containing ^{10}B and their adjacent cells, since the ranges of α particles and ^{7}Li nuclei are almost the same as a typical cell diameter of 10 μm. The advantage of this therapy is the selective destruction by neutron reactions of only the tumor into which ^{10}B compound has been previously injected. The therapeutic result is sensitive to variations in the absorbed dose distribution in a cell. In the other word, BNCT has the possibility of selective treatment at the cell level.

The success of clinical irradiation for brain tumors by Hatanaka et al. contributed to the development of the dose estimation technique for BNCT. Calculations of absorbed dose caused by the ^{10}B(n,α)^{7}Li reactions in tissue have been made. We also have reported an analytical method of calculating absorbed energy in a cell nucleus for BNCT using a spherical cell model[1]. The technique has been improved to allow dose calculation regardless of the size or geometrical relationship of a spherical ^{10}B-containing region and a target region. The spherical model is meaningful for the primary modeling of cell systems and also has merit for subjects with many parameter survey calculations. The Monte Carlo calculation method has been widely adopted because its advantages in calculating complicated geometry. The more complicated a cell model becomes, the more realistic. But any calculation technique confronts the problem of modeling cell tissue and ^{10}B concentration. These points have been not clarified yet.

^{10}B concentration distribution in a cell is heterogeneous depending on the boron compound and cell characteristics. So it is essential that boron compounds for BNCT be selectively taken up by tumors and be distributed effectively within the tumor cell in amounts sufficient to provide a therapeutic dose. In the USA, the European Community, and Japan, new boron compounds for BNCT have been developed with these considerations in mind[2]. Knowledge of RBE and a useful scale related to the absorbed dose evaluation for NCT are discussed in relation to the absorbed energy in the cell from heavy charged particle produced by the capture reaction in tissue. The selective absorbed dose of cell nuclei near the spherical ^{10}B containing region such as small organelles in the cell, can be estimated quantitatively using a Boron Selective Dose Ratio (BSDR)[3]. And the following new physical scales are defined using the concept of the BSDR for evaluation of new boron compounds for clinical treatment.

CALCULATION METHOD

The present calculation method has been focused on the influence from the spherical ^{10}B containing region such as small organelles in a cell. Both the spherical target and the spherical ^{10}B containing region can take any size and geometric configuration. The following three physical data are obtained as average expected values:

(1) the particle absorbed energy in a cell nucleus,
(2) the particle number hitting and/or passing through a cell nucleus, and
(3) the particle integrated track length in a cell nucleus.

The above three values can be used as measures of the biomedical effect as well as neutron fluence and absorbed dose. The absorbed energy in the cell nucleus was used as an example in this paper. The number of particles hitting and/or passing through the target cell nucleus, and the integrated particle track length in it are also useful to estimate the affects on the DNA and the membranes of the cell or cell nucleus. The possible conditions of calculating geometries are shown in Fig. 1.

THE SENSE OF RBE FOR NCT

RBE is defined for the ratio of tissue–averaged physical absorbed dose for the X rays to that of the radiation assuming the same biological effect. The RBE for the present NCT is usually handled in the same way as conventional RBE for the cell[4]. It is obvious that the energy release process of the neutron reaction localizes at the cell level. The RBE for NCT needs to be investigated with the physical absorbed doses which were estimated from the condition of energy deposition within a cell.

We want to explain our understanding of the relations among "Neutron RBE", "Reaction RBE", "Particle RBE" and "Clinical RBE" used for the present NCT. Neutrons cannot act

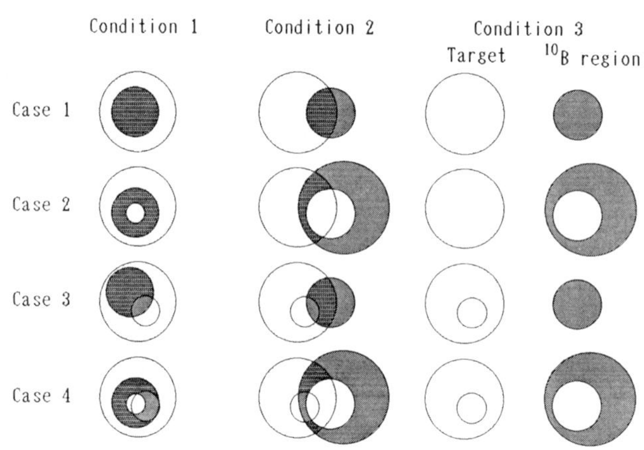

Figure 1 The calculating geometries between ^{10}B containing region and a target.
Condition 1, 2 and 3 are "insider", "overlap" and "outsider", respectively.
Case 1, 2, 3 and 4 show the relation between the target and the ^{10}B region with and/or without a cavity.

to the living tissue before making reactions with the elements of which that tissue is composed. So "Neutron RBE" would be estimated to combine with main reactions yielded in living tissue i.e. $H(n,\gamma)D$, $^{14}N(n,p)^{14}C$ and $^{17}O(n,\alpha)^{14}C$ reactions for thermal neutrons and (n,n') reactions for H, C, O and N for epithermal and fast neutrons. "Reaction RBE" such as $^{10}B(n,\alpha)^{7}Li$, $^{14}N(n,p)^{14}C$, H(n,n')H etc. should be evaluated to combine with particles such as p, α, ^{7}Li, ^{14}C, etc. yielded in the living body. "Particle RBE" has been usually decided for energy dependency of the particles produced by an accelerator. It seems that, compared to "Neutron RBE" and "Reaction RBE", "Particle RBE" is somewhat more useful for NCT; however, we cannot accept that it is entirely suitable for NCT. Generally, RBE of the particle depends on the LET. So "Particle RBE" cannot be strictly dependent upon particle characteristics, because the range–LET relationship of the particle will vary according to its initial energy and the particle will loose the energy in the tissue.

The effect of NCT treatment is now evaluated from both neutron fluence combined with ^{10}B concentration and gamma ray dose. So, for "Clinical RBE" which is the RBE to the NCT treatment, both "Neutron RBE" and "Reaction RBE", which should be evaluated with actual patients data, are still useful. However they should be handled separately from the same RBEs evaluated from the cell experiments until the relations between them are established.

PHYSICAL SCALES RELATED TO DOSE ESTIMATION FOR BNCT

From the understanding for the RBE, we have been considering the necessity of new physical scales instead of averaged absorbed doses in tissue. Three physical scales which are analytically calculated for the particles (1) absorbed energy in cell nucleus (AECN), (2) the number of hitting and/or passing through the cell nucleus and (3) the integrated track length in the cell nucleus, have been studied. These quantities can express the local damages in a cell nucleus. We have studied following scales relating to the BNCT using AECN.

(BSDR : Boron Selective Dose Ratio)

Selective Dose Ratio (SDR) in cell level is defined using AECN for heavy charged particles for NCT. BSDR, which is the SDR for boron, mainly depends on the concentrations of ^{10}B and ^{14}N and their distributions in the cell. So, BSDR could become one of the scales for judging the advantages of boron compounds for BNCT in cell level. The bigger the BSDR, the more effective the neutron irradiation. For example, for a cell nucleus 5 μm in diameter, BSDR for thermal neutrons becomes 2 by about 5 ppm when ^{10}B is present only in cytoplasm and by about 1 ppm when ^{10}B is present only in cell nucleus[3]. Here ^{10}B concentration is tissue averaged value.

$$\mathrm{BSDR} = \frac{\text{AECN by the particles from tissue with } ^{10}\mathrm{B}}{\text{AECN by the particles from tissue without } ^{10}\mathrm{B}}$$

(BTCF : Boron Tumor Curative Factor)

Now, we would like to propose Tumor Curative Factor (TCF), which is defined as the ratio of AECN for the particles for the tumor cell to that for critical normal cell in the clinical irradiation. BTCF, which is the TCF for boron and is rewritten using BSDR as follow, is a scale to judge the advantage of the distribution of the boron compound in the body. BTCF can evaluate the advantages of the boron compounds which have the characteristics more accumulated in the tumor cell and less in the normal cell in critical organ. Excellent compound needs to have large BSDR and BTCF.

$$\mathrm{BTCF} = \frac{\text{BSDR for the tumor cell}}{\text{BSDR for the critical normal cell in the irradiation}}$$

(BTER : Boron Tissue Enhanced Ratio)

We also would like to propose Tissue Enhanced Ratio (TER), which is defined as the ratio of tissue averaged RBE dose for the tissue with neutron enhanced element to that without it. RBE used here is the "Reaction RBE". TER can evaluate the characteristics of the mixed field of neutrons and gamma rays for NCT. BTER, which is the TER for boron, mainly depends on averaged ^{10}B concentrations in the tissue and the contamination ratio of gamma rays to the neutrons at the location. BTER can evaluate the characteristics of depth dose distribution of both gamma rays and neutrons for BNCT. Namely, BTER can evaluate the advantage of the neutron irradiation field for BNCT.

$$\text{BTER} = \frac{\text{Tissue averaged RBE dose for tissue with } ^{10}\text{B}}{\text{Tissue averaged RBE dose for tissue without } ^{10}\text{B}}$$

CONCLUSIONS

The advantages of boron compounds could be evaluated using three scales such as BSDR, BTCF, and BTER. These three coefficients do not point out directly the effect of clinical treatment. We want to stress that BSDR is a useful scale and a concept for boron compound development. Neutron irradiation techniques and neutron enhanced compounds are strongly connected like right and left wheels for a car. Excellent compounds will save thermal neutron fluence for Neutron Capture Therapy. According to a simple assumption, a new boron compound which is ten times more concentrated than present one, will make neutron irradiation fluence one tenth. If we were able to use a new compound for which the BSDR and/or BTCF is 100, many neutron sources such as accelerators, ^{252}Cf and low power research reactors such as those under 10 kW, which are attached to the suitable neutron irradiation facility, could be used for NCT. These relationship are quite interesting in relation to the design for hospital connected neutron sources.

ACKNOWLEDGEMENTS

The authors thank Drs. A. Soloway and R. Barth of the Ohio State University for their invaluable discussion of boron compound development. The authors thank Dr. H. Hatanaka of Teikyo University Hospital, Drs. Y. Mishima and M. Ichihashi of Kobe University School of Medicine and Dr. H. Fukuda of Tohoku University Hospital for their invaluable discussion of clinical treatment.

REFERENCES

1. T.Kobayashi and K.Kanda, "Analytical Calculation of Boron–10 Dosage in Cell Nucleus for Neutron Capture Therapy", *Radiat. Res.*, 91 (1982) 77–94.
2. A.Soloway, A.K.M.Anisuzzaman, L.Liu, R.F.Barth, F.Alam, W.Tjarks, "Carboranyl Precursors of Nucleic Acids -- Potential DNA Probes for BNCT", Progress in Neutron Capture Therapy for Cancer (Ed. B.Allen *et al.*) Plenum NY (1992) 207–210.
3. T.Kobayashi and K.Kanda, "Boron–10 Dosage in Cell Nucleus for Neutron Capture Therapy -- Boron Selective Dose Ratio", Boron–Neutron Capture Therapy for Tumors, (Edit. H.Hatanaka) Nishimura Japan (1986) 293–300.
4. H.Fukuda, T.Kobayashi, T.Matsuzawa, K.Kanda, M.Ichihashi and Y.Mishima, "RBEs of a Thermal Neutron Beam and the ^{10}B(n,α)^{7}Li Reaction on Cultured B–16 Melanoma Cells", *Int. J. Radiat. Biol.*, 51 (1987) 167–175.

MONTE CARLO CALCULATION OF MICRODOSIMETRY IN NITROGEN AND BORON NEUTRON CAPTURE REACTIONS

B J Allen[1], M Bilek[1], D E Charlton[2]

[1]Biomedicine and Health, Ansto PMB 1 Menai NSW 2234
Australia,
[2]Concordia University, Montreal Quebec
Canada H3G 1M8

INTRODUCTION

In Boron Neutron Capture Therapy (BNCT), boron-10 is present in capillaries and normal cells and is selectively taken up by cancer cells which are rapidly dividing. Normal tissue tolerance depends on the dose to the capillary walls resulting from blood borne boron. The single cell capillary wall has a thickness less than the range of the alpha and ^{7}Li reaction products from neutron capture in boron-10 and dose equilibrium depends on extravascular boron concentrations. The intracellular distribution of boron in cells also has a marked effect on the nuclear dose. These effects have been investigated by Gavin et al[1], Gabel et al[2] and others.

We further investigate the effect of inhomogeneous boron distributions and cellular configurations by Monte Carlo calculations. The contribution of contiguous cells to the dose to the nucleus is determined, and we evaluate the capillary dose factor (CDF) for borocaptate (BSH) and boronophenylalanine (BPA), two compounds in current experimental and clinical use.

MODELS

The use of Monte Carlo methods to calculate microdosimetric spectra in cell nuclei from high LET ions is well established. In this paper the dose delivered to the ellipsoidal nuclei is calculated, based on earlier work by Charlton[3] for in vitro irradiation of single cells either in suspension or monolayers, but modified to take into account dose contributions from neighbouring cells. A new geometry for endothelial capillary cells is introduced.

The neutron cross sections for the interactions in nitrogen and boron are well known as are the ion products and their energies. Stopping powers and ranges in liquid water were taken from TRIM[4] and sizes of the nuclei and cells were taken from Stewart[5].

Ions were generated from four zones, viz the cell nucleus, cytoplasm, on the cell surface and external to the cell, for isolated cells in vitro; and the cell nucleus, cytoplasm, lumen and extra-capillary space for capillary wall cells. The concentration of nitrogen and boron in these volumes can be independently varied.

NEAREST NEIGHBOUR SIMULATIONS

The model allows calculations of the effect of nearest neighbour cells, using the program TNC6[3]. These cells can have boron and nitrogen concentrations and a shape and position different to that of the central cell. The energy deposited in the nucleus from the first cell due to interactions in the second cell is calculated for elliptic cells with spherical nuclei. The nucleus was positioned at the centre of the cell.

Advances in Neutron Capture Therapy, Edited by
A.H Soloway *et al.*, Plenum Press, New York, 1993

Table 1. Contiguous cell correction[a]

			dose to nucleus in target cell (Gy)					
n_c	[N]%	[B]ppm	target cell		contiguous cell			totalCCC
			nuc	cyto	nuc	cyto		
1	3.5	0	0.27	0.26	0.038	0.037	0.61	1.14
6					0.23	0.23	0.99	1.87
18					0.68	0.68	1.89	3.57
1	0	20	1.76	1.16	0.018	0.062	3.00	1.03
6					0.108	0.372	3.40	1.16
18					0.324	1.116	4.36	1.49
1	3.5	20	2.03	1.42	0.056	0.099	3.61	1.05
6					0.34	0.59	4.38	1.27
18					1.01	1.78	6.24	1.81

a fluence 2.0×10^{12} n cm^{-2}
cell radii(um)4.7,6.2,9.4; nucleus radius(um)3.8

Results are given in table 1 and the contiguous cell correction (CCC) is the dose enhancement factor to the nucleus resulting from contiguous cells.

Six or more contiguous cells would be present in cell clusters in vivo, indicating that the contiguous cell contributions from nitrogen and boron would be at least 1.9 and 1.2 times the respective central cell contributions. Up to 18 contiguous cells are theoretically possible with spherical cells, giving 3.6 times the dose contribution of the central cell nucleus for nitrogen, 1.5 times for 20 ppm boron-10, and 1.8 times for the overall contribution.

CAPILLARY SIMULATION

In this geometry, similar to that of Gavin[1], data for the capillaries in normal white brain matter were taken from Stewart et al[5]. Here the wall thickness was 1 um and the lumen diameter 6 um. The outside wall of the capillary was circular while the lumen was elliptical to allow the nucleus to be contained within the wall, as shown in figure 1. Calculations were made with a new program CAPIL6[6] for a neutron fluence of 6.0×10^{12} n cm^{-2} and uniform nitrogen and boron concentrations of 3.5% and 8 ppm respectively. Results are given in table 2 below for 10,000 cells scored, where "Events" is the average number of neutron capture reactions occurring in each section, and "Pass" is the number of passages of particles through the nucleus. The average dose to the nucleus is 5.69 Gy, resulting from an average of 2.52 nuclear passages per cell.

Table 2. Regional contributions to nucleus dose

Region	Nitrogen-14 Events	Pass	Dose(Gy)	Boron-10 Events	Pass	Dose(Gy)
Blood	11.4	0.26	0.22	7.7	0.27	0.83
Cytoplasm	10.2	0.19	0.23	6.9	0.21	1.03
Extracap	180.	0.85	0.99	122.	0.49	1.54
Nucleus	0.15	0.15	0.09	0.10	0.10	0.76
total	202	1.45	1.52	137	1.07	4.17

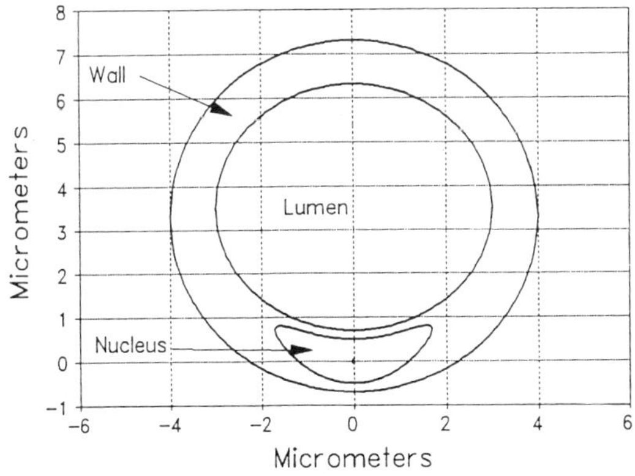

Figure 1. Cross section of the capillary model used in Monte Carlo calculations.

It is also of interest to obtain the spectrum or distribution of nucleus hits. This is given in table 3 where it can be seen that 8% of nuclei do not receive a single hit. The number of hits needed to kill a cell is not well known. However, if 4 or more hits are required, then 75% of cells survive, expanding as other cells suffer reproductive death, and so maintaining capillary integrity.

CAPILLARY DOSE FACTOR

The capillary dose factor (CDF) arises from the geometric reduction in dose to the nuclei of capillary endothelium cells from boron in the blood, relative to that for the cancer cells. The CDF is obtained by applying the expected concentrations of nitrogen and boron in the different compartments, as given in Table 4.

The blood borne boron concentrations for BSH and BPA are typical values required to achieve 20 ppm boron in the tumour. The extracapillary nitrogen concentration is assumed to be 1.7% for the brain and 2.8% for the tumour.

Table 3. Hit distribution for 10000 cells

Hits	0	1	2	3	4	5	6	7	8	9	10
Cells	816	1963	2576	2167	1365	703	258	111	33	6	2

Table 4. Boron and nitrogen concentrations in the brain for BSH, BPA

	Boron ppm BPA	BSH	Nitrogen % capillary
lumen	5	20	2.9
cytoplasm	5	(2)*	(1.4)
external	5	2*	2.9
nucleus	0.1#	0.4#	3.2

assume 2% nuclear conc,
* assume 10% into normal brain

Absorbed doses given in Table 5 are calculated for the nitrogen and boron values of Table 4 and the following parameters: fluence 10^{13} n cm^{-2}; nuclear radii 1.7,0.5,2.5; lumen radii 3.0 2.8 um. Endothelial nucleus doses are compared to the uniform boron distributions (u) to obtain the CDF for BSH and BPA.

Table 5 Dose to capillary endothelial nucleus (Gy)

Site	Nitrogen	BSH_u	BSH	BPA_u	BPA
Lumen	0.13	2.14	2.12	0.54	0.58
Cytoplasm	0.06	2.35	0.21	0.55	0.58
External	1.51	8.39	0.89	2.15	2.04
Nucleus	0.12	2.46	0.06	0.62	0.01
Total	1.82	15.34	3.28	3.86	3.22
CDF (N+B)			0.30		0.89

CONCLUSIONS

Contiguous cells make a major contribution to the dose to the nucleus of the central cell, such that the nitrogen dose can be more than doubled and the boron dose increased by half, depending on the cell density. This in turn depends on the types of cells being investigated. For endothelial cells a substantial dose sparing effect may arise relative to cells in a lesion, but not for isolated cancer cells in transit.

In the capillary model, comparable doses to nuclei of endothelial cells are received when the boron concentration in blood for BPA is one quarter of that for BSH. Under these conditions, comparable tumour doses are also delivered. The disparity arises because BSH does not pass the blood-brain barrier whereas BPA does. The capillary dose factor is CDF=0.3 for BSH and 0.9 for BPA, in agreement with the results of other studies.

ACKNOWLEDGEMENT

Work supported in part by the Government Employees Medical Research Fund.

REFERENCES

1 P R Gavin, F J Wheeler, R Huiskamp, C E deHaan, K H I Philipp: Large animal normal tissue tolerance with boron neutron capture: comparison of empirical results with a microdosimetric model, MIT Workshop on Macro and Microdosimetry, Cambridge, USA, Oct (1991).

2 D Gabel, S Foster, R G Fairchild: The Monte Carlo simulation of the biological effect of the $^{10}B(n,alpha)^{7}Li$ reaction in cells and tissue and its implication for boron neutron capture therapy, Radiat Res 111 14-25 (1987).

3 D E Charlton, Energy deposition in small ellipsoidal volumes by high - LET particles: application th thermal neutron dosimetry, Int J Radiat Biol 59 827-842 (1991).

4 J F Ziegler, TRIM, the transport of ions in matter, IBM-Research, Yorktown, NY (1992).

5 P A Stewart, C L Farrell, R F DelMaestro, The effect of cellular microenvironment on vessels in the brain. 1 Vessel structure in tumour, peritumour and brain from humans with malignant glioma. Int. J. Radiat. Biol. 60, 125-130 (1991).

6 D E Charlton, B J Allen, Monte Carlo calculations of ion interactions in the nuclei of capillary endothelial cells of the brain irradiated by thermal neutrons. To be published.

GADOLINIUM AS AN ELEMENT FOR NEUTRON CAPTURE THERAPY

Robert M. Brugger, Hungyuan B. Liu, Brenda H. Laster, Christopher R. Gordon, Dennis D. Greenberg, and Lynne S. Warkentien

Medical Dept., Brookhaven National Laboratory, Upton, NY 11973

Gadolinium is being studied as an alternate active element to B for Neutron Capture Therapy (NCT). The advantages of Gd-157 as compared to B-10 are that Gd-157 has a very large thermal neutron cross section (455,000 b), and Gd containing compounds that target tumors are being developed as MRI contrast agents. A disadvantage is that the products of the Gd + n reaction, gamma rays and Auger electrons, do not appear to be as well suited to NCT as the products of the B + n reaction, but these still may be effective in destroying tumors. Because of the advantages of Gd, an investigation of Gd as an element for NCT is proceeding at Brookhaven National Laboratory (BNL)[1].

The one Gd loaded MRI contrast agent, Gd-DTPA, that is now accepted for imaging does target brain tumors and will not be toxic at the levels of injection that will be needed for GdNCT[1]. Calculations and measurements of the physical dose show that with 150 ppm in a 50 cc tumor and a source of 5×10^{12} neutrons, greater than 30 Gy of dose from the prompt gamma rays will be delivered to the tumor[2]. The dose drops off rapidly away from the tumor similar to the dose distribution in brachytherapy. Measurements of this dose with a biological dosimeter agree with calculations. In addition, the biological dosimeter shows that the dose can be enhanced with the addition of IdUrd to the treatment. Thus far, for the compounds tried, the biological effect observed from Auger electrons has been modest. In another study, a different delivery system was examined. Calculations show that Gd metal seeds can be used with GdNCT in a combination of brachytherapy and NCT[3].

At BNL, preparations are being made to test in vitro compounds containing Gd and compare their response to the response of Gd-DTPA to determine if one or several compounds can be located that enter the cells and enhance the Auger effect. Figure 1 shows the shape of the two similar rotators with positions for cell vials that have been constructed for these tests. The first rotator is made of only paraffin which simulates healthy tissue and provides control curves. The second rotator has 135 ppm of Gd-157 in the paraffin to simulate a Gd loaded tumor. Cells are irradiated in vials in the paraffin rotator and in the Gd-paraffin rotator at the epithermal beam of the Brookhaven Medical Research Reactor (BMRR). This produces an irradiation similar to what a patient would receive in an actual treatment. A combination of irradiations are made with both rotators; with no Gd compound or IdUrd in the cell medium, with only Gd compound in the cell medium, with only IdUrd in the cell medium, and with both Gd compound and IdUrd in the cell medium. The first set shows the effects of gamma rays from the H(n,gamma) reaction and the prompt gamma rays from capture of neutrons by Gd. The second set shows if there is any effect of Gd being in the cell medium or inside the cells, i.e., an Auger effect. The third set shows the effect of enhancement by the IdUrd produced by the gamma rays from neutrons captured by either H or Gd. The fourth set

Advances in Neutron Capture Therapy, Edited by
A.H Soloway *et al.*, Plenum Press, New York, 1993

Table 1. Calculated and Measured Doses per MW-min in the Rotators.

Source	Calculation Physical	Calculation Biological	Measured Physical	Measured Biological
Paraffin Rotator				
D_γ	2.6 cGy	2.6 RBEcGy	2.2	
D_{N-14}	0.7 cGy	3.3 RBEcGy		
D_{fast}	0.5 cGy	2.4 RBEcGy		
Φ_{th} (n/cm^2/s)	7.1 x 10^8		6.6 x 10^8	
Irradiation to reach 10% cell survival (MW-min)		86.0	91.0	78±9
Gd Rotator				
D_γ	16.0	16.0	12.0	
D_{N-14}	0.4	1.8		
D_{fast}	0.5	2.4		
Φ_{th} (n/cm^2/s)	4.0 x 10^8		4.1 x 10^8	
Irradiation to reach 10% cell survival (MW-min)		35.9	44.8	39

The RBEs used; gamma rays = 1, fast neutrons = 4.5, N(n,p) = 4.5

combines all of the reactions and enhancements. Preliminary calculations and physical measurements of the doses that the cells will receive in these rotators have been made. Figure 2 shows preliminary cell survival measurements. The single target multihit model was used to fit sets of points. The curve labeled "para rotator control" is the cell survival experience of irradiation of cells in the paraffin rotator, no IdUrd or Gd added. The curve labeled "para rotator+IdUrd" shows the enhancement of the effects of the radiation when IdUrd is added to the cell medium. The four curves to the left are for irradiations with cells in the Gd loaded rotator; first with no Gd compound or IdUrd added to the cell medium, second with only Gd compound added, third with only Id-Urd added and fourth with both IdUrd and Gd compound added. Table 1 lists both measured and calculated results[2,3].

Iododeoxyuridine (IdUrd) incorporated in cellular DNA was used in conjunction with Gd. It was anticipated that the Auger electrons released in the Gd interaction with thermal neutrons might induce a photoelectric effect in iodine, thus contributing a high LET-type damage to the cells. Results in the preliminary studies suggest that the effect of IdUrd with Gd is additive rather than synergistic.

An MRI scan and a SPECT scan for DTPA of a boy with a brain tumor indicated that Gd-DTPA concentrates in a brain tumor while very little goes to brain, but that some Gd goes to the scalp[3]. Calculations have been made to see the effects of neutrons captured by the Gd in the skin. A tumor was placed about 1.25 cm below the surface on the side of a head model which simulated the boy's head. The skull and the brain areas of this model had typical skull and brain tissue compositions of elements. Guided by the SPECT scans, 188 ppm of Gd-157 were put in the tumor and 63 ppm in the skull and then the head was irradiated from the side with an epithermal beam representing the one at the BMRR. The dose components along the line through the center of the tumor were calculated and some of these are displayed in figure 3. The components of the background coming from capture of neutrons by tissue material are seen. In addition, the peaked dose to the tumor is seen at the location of the tumor. An additional dose, peaking at the front and back surfaces of the head, is present from neutrons captured by the Gd in the skin and producing gamma rays. However this dose is relatively small and does not exclude GdNCT as a feasible therapy. In addition, this surface effect can probably be reduced by determining the build up and wash out of the Gd in the tumor and in the scalp and selecting an optimum time after administration of the compound for irradiation. One set of compounds of particular interest at this time, for not only photodynamic therapy, but also BNCT, are porphyrins[4]. It has been shown that these compounds target tumors and that they enter the cells but possible not the nucleus of the cells. Porphyrins, since they are light sensitive, are being tested for photodynamic therapy. Porphyrins have been modified with two[4] and four[5] carborane cages on side chains giving porphyrin molecules with 20 and 40 B atoms per molecule of porphyrin[5]. These porphyrins are candidates for BNCT. In addition, a Gd atom can be added to the center of the ring giving one Gd in addition to the 20 or 40 atoms of B per molecule. For the 40 B atoms plus 1 Gd atom per molecule, the thermal neutron cross

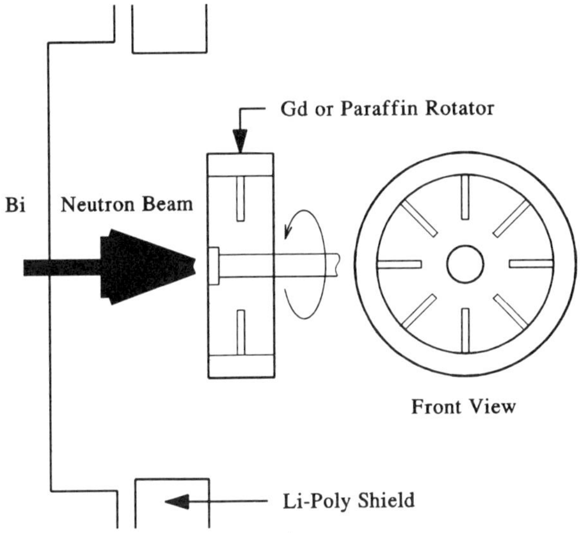

Figure 1. The Rotators at the BMRR Epithermal Beam Port. The left side is a side view with the neutron beam coming from the left. On the right side is a front view of the rotator as viewed along the neutron beam.

section/molecule is 154,000 b for the B, but 255,000 b/molecule for the Gd. The dose from the B fission is deposited at the location of the molecule with an RBE of 2.5, but the dose from the de-excitation of the Gd-158* is more broadly distributed by the gamma rays and a factor of about 0.06 must be applied for this distribution and an RBE of 1 for the gamma rays[2]. This indicates that the B is about 25 times more effective when only the gamma rays from Gd are compared to the fission products from B. However, for porphyrins, the Gd appears to be inside the cells and an Auger effect should be significant. The Auger electrons from Gd have energies as large as 50 keV and ranges up to 30 microns. Fifty percent more of the molecules will have a reaction of neutrons with Gd than with B. The effect of the Gd through the Auger reaction could be as great as the effect from the B. In vitro cell survival experiments should demonstrate this effect.

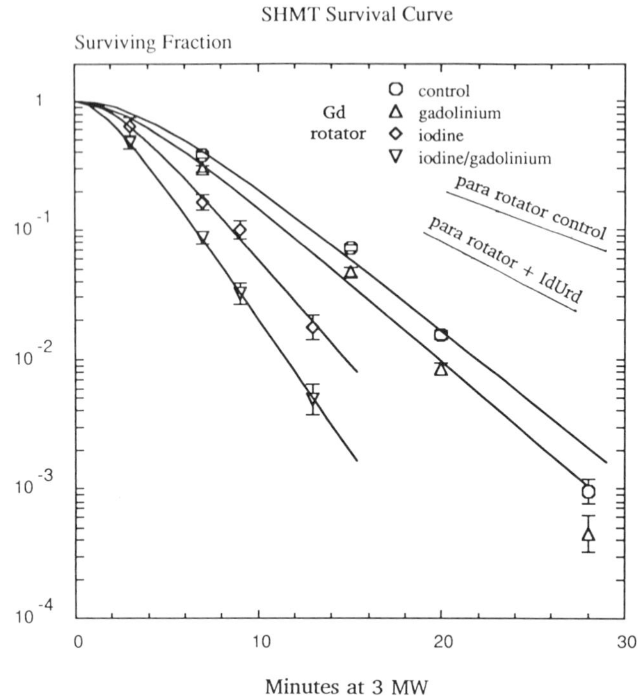

Figure 2. Preliminary Cell Survival Measurements in the Gd-Loaded and Paraffin Only Rotators.

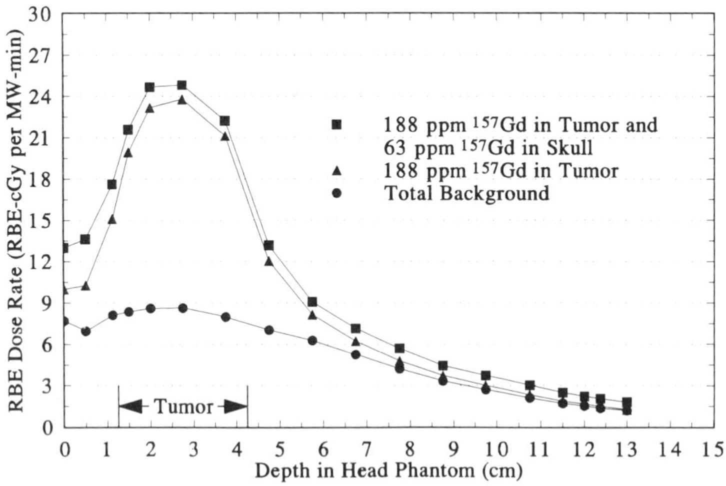

Figure 3. The RBE Dose Rate through a Head Model.

Table 2. Gd Brachytherapy Phantom at Li-Poly Face, at 35 mm Depth per MW-min.

	Calculated		Measured	
	Control	Seeds	Control	Seeds
Φ_{th} x 10^8 (n/cm^2/s)	7.96	6.35	---	5.89
D_γ (cGy)	4.22	10.37	---	8.06
D_{fast} (cGy)	0.35	0.35	---	---
$D_{N\text{-}14}$ (cGy)	0.15	0.12	---	---

Calculations have shown that Gd in the form of brachytherapy seeds and irradiated with thermal neutrons will deliver a significant dose to a tumor[3]. Seeds have been made of Gd metal and dose measurements made in a phantom. The seeds can be made of natural Gd since the cross section is so large that any neutron reaching the surface of a seed is captured. The magnitude of the dose produced is proportional to the surface area of the seeds and not their volume. The calculated and measured dose rates that the cells will experience when this phantom is placed in the epithermal beam of the BMRR are given in table 2.

In conclusion, recent studies at BNL of Gd as an element for NCT continue to indicate that GdNCT might be successful. Rotators for irradiating cells have been made so that in vitro screening can be made of Gd compounds. One compound that looks particularly interesting is porphyrin with both B and Gd added.

ACKNOWLEDGEMENTS

This research was supported by US Department of Energy contract DE-AC02-76CH00016.

REFERENCES

1. R. Brugger, S. Gibb, J-L. Shih, S. Pickup, X. Luo, J. Oro, and B. Laster. "Estimated Doses by Gadolinium Neutron Capture Therapy." Proceedings of the International Workshop on Marco and Micro Dosimetry and Treatment Planning for Neutron Capture Therapy, editors, O.H. Harling, R.G. Zamenhof, and G.R. Solares (to be published) July 1991.
2. J.A. Shih and R.M. Brugger. "Gadolinium as a Neutron Capture Therapy Agent," Medical Physics 19(3), (May/June 1992).
3. J.A. Shih and R.M. Brugger. "Neutron Induced Brachytherapy: A Combination of Neutron Capture Therapy and Brachytherapy," Medical Physics, 19, pp.369-376 (1992).
4. S.B. Kahl and M-S. Koo. "Synthesis and Properties of Tetrakis-Carborane-Carboxylate Esters of 2,4-Bis-(α,β-Dihydroxyethyl) Deuteroporphyrin IV." Progress in Neutron Capture Therapy for Cancer, editors, B.J. Allen, D.E. Moore, and B.V. Harrington, pp. 223-226 (1192).
5. M. Miura, P.L. Micca, J.C. Heinrichs, D. Gabel, R.G. Fairchild, and D.N. Slatkin. "Biodistribution and Toxicity of 2,4-Dioinyl Nido-o-Carbonanyldeuteroporphyrin IX in Mice," Biochemical Pharmacology 43: 467-476 (1992).

MONTE CARLO DOSE DISTRIBUTION OF GADOLINIUM NCT FOR BRAIN TUMORS OF DIFFERENT SITE AND LEVEL OF NECROSIS

Jerzy T. Masiakowski,† John L. Horton,† and Lester J. Peters§

†Department of Radiation Physics
§Department of Radiotherapy
The University of Texas M.D.Anderson Cancer Center
Houston, Texas 77030

INTRODUCTION

Almost all the energy released in the gadolinium-157 thermal neutron capture is emitted through prompt gamma rays with a spectrum up to 7.9 MeV. Low energy internal conversion electrons, mainly in the range 80-280 keV, are also emitted. An FDA approved carrier, Gd-DTPA, the gadolinium contrast agent for magnetic resonace imaging (MRI), is currently used clinically. Gadolinium uptake in brain tumors is provided by simple diffiusion through the disrupted blood-brain barrier. The dynamics of the uptake and washout phases are not identical because there are biological processes favoring the transport of gadolinium carriers into the extravascular space.[1] Doses of Gd-DTPA larger than those commonly used for MRI, or multiple injections can produce high gadolinium concentrations in tumor.

In our Monte Carlo calculations published elsewhere[2,3] we have shown the feasibility of gadolinium neutron capture therapy (GdNCT) for brain tumors, especially when the contribution of internal conversion electrons is taken into account. The inclusion of conversion electrons increases the average dose to tumor by about 45% compared to the dose from prompt gammas only. The paper by Shih and Brugger[4] reports experimental evidence confirming those calculations.

In this paper we report the dose distribution for GdNCT simulated for more realistic tumors with the gadolinium concentration decreasing towards the necrotic center.

METHODS

The three-dimensional dose distribution for GdNCT of brain tumors has been modeled in a semi-spherical head phantom 16 cm in diameter made of brain-equivalent material, using Monte Carlo and convolution techniques. The total dose was separated into two contributions: the background dose and the dose following the capture in ^{157}Gd. The background dose, due to all neutron-tissue interaction other than capture in gadolinium was calculated using Monte Carlo Neutron and Photon (MCNP) transport code converting the neutron and photon fluences calculated in a gadolinium loaded phantom into the dose to phantom with no gadolinium present. The dose following the capture in ^{157}Gd was obtained by convolving the dose distribution kernel for a small volume source with the capture probability distribution. The kernel was calculated using Electron and Gamma Shower (EGS) Monte Carlo code including gadolinium-158 prompt gammas and internal conversion electrons. The capture probability distribution was obtained by multiplying the MCNP calculated thermal neutron fluence times the distribution of macroscopic cross section for thermal neutron capture.

In this study we assumed a wide, monoenergetic 2 keV incident neutron beam. All our calculations were done under the assumption of two parallel opposed equally weighted wide neutron beams. While this simple field arrangement would not be optimal for most tumors, it does provide a "worst case" scenario for evaluating the utility of gadolinium capture therapy of primary tumors and allows modeling of the treatment of multiple brain metastases.

RESULTS AND DISCUSSION

The gadolinium NCT dose distribution is strongly dependent on tumor size, the value of gadolinium concentration and the distribution of the concentration throughout the tumor. An example of dose distributions for four different cases with the tumor at the center of the phantom is shown in Fig.1. The dose is shown as dose profiles in planes parallel to the central axis. The highest profile is the central axis profile

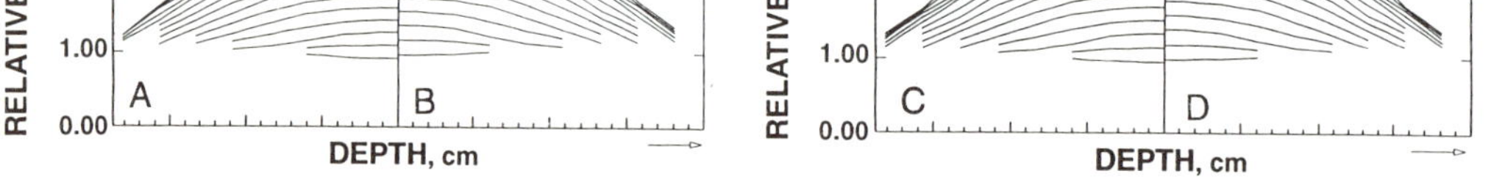

Fig.1. A comparison of GdNCT dose distribution profiles for tumors located at the center of the phantom with the homogeneous 150 μg/g ^{157}Gd concentration, 3 cm (A), and 4 cm in diameter (B), and with the ^{157}Gd concentration decreasing towards the center of 4 cm in diameter tumors; concentrations ranging from 150 to 60 μg/g (C), and from 200 to 80 μg/g (D).

followed by dose profiles in 0.5 cm increments away from the central axis. Cases A and B were calculated for a homogeoneus ^{157}Gd concentration of 150 μg/g throughout the tumors. The dose difference results from the difference in size, the tumor in A was 3 cm in diameter and in B was 4 cm in diameter. The background dose to the normal brain remains almost independent of the size of the tumor treated. All profiles in Fig.1 are symmetric so only half-phantom profiles are shown. In cases C and D the calculations were performed for a tumor 4 cm in diameter with a decreasing ^{157}Gd concentration towards the center of the tumor. Parabolic concentration profiles were assumed as shown in the inserts (Fig.1C and D). The maximum and minimum values of the concentration distributions were 150 and 60 μg/g (C), and 200 and 80 μg/g (D), respectively.

The final results of GdNCT dose distribution for two cases of large semispherical tumor and three cubic metastases in different locations are shown in Fig.2. The arrows

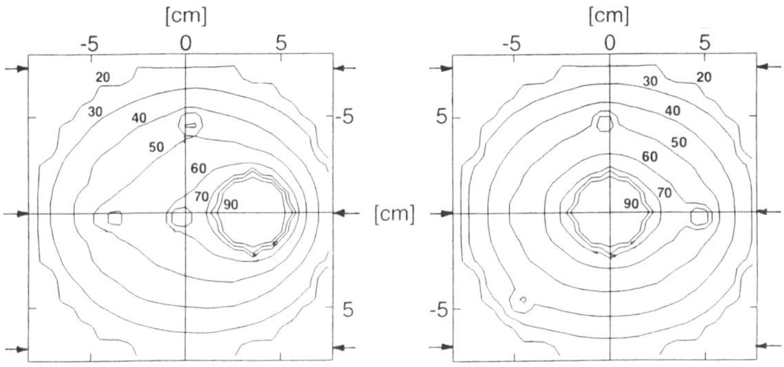

Fig.2. The %-isodose contours of GdNCT with two wide parallel opposed epithermal neutron beams (arrows) for two different cases of large tumor and three small nodes. The small metastases receive a 10-20% boost from ^{157}Gd neutron capture.

indicate the parallel opposed incident neutron beams covering the entire phantom. The %-isodose contours in the equatorial plane through the center of large tumor are drawn in 10% decrements starting with the 90% contour. The step-like shape of the outer 20% contour reflects the shape of the phantom used for the Monte Carlo calculations. The phantom was constructed of five coaxial cylinders to simulate a sphere. The parabolic gadolinium concentration profile through the large tumor was assumed, with the concentration of 150 and 60 μg/g at the boundary and at the center of tumor, respectively. A homogeneous gadolinium concentration of 150 μg/g was assumed for small, 1 cm cubic metastases. The large tumors receive well localized dose. The tumor dose enhancement is greater for shallow metastasis than for a deeper metastasis. This is due to the fact that the maximum of thermal neutron fluence occurs at a depth of about 3.5 cm under the surface. However, gadolinium uptake in tumor is an individual characteristic and may vary for the same type of tumor by as much as a factor of five.[5] As shown in Fig.1 (C) versus (D) an increase of gadolinium uptake by 1/3 causes a dramatic tumor dose enhancement. The decision whether the treatment could be beneficial or not would be made only on the basis of a precise map of gadolinium concentration for the particular tumor. This map can be provided by a quantitative MR measurement as it has been shown that both relaxation rates R_1 and R_2 are linearly dependent on Gd concentration up to 200 ppm[7].

ACKNOWLEDGEMENTS

This study was partialy supported by a grant from Cray Research, Inc. and the Unsworth Foundation.

REFERENCES

1. P.Gowland, P.Mansfield, P.Bullock, M.Stehling, B.Worthington, and J.Firth, Dynamic studies of gadolinium uptake in brain tumors using IR-EPI, *Magn. Reson. Med.* 26:241 (1992)
2. J.T.Masiakowski, J.L.Horton and L.J.Peters, Gadolinium neutron capture therapy for brain tumors: A computer study, *Med. Phys.* 19:1277 (1992)
3. J.T.Masiakowski, J.L.Horton and L.J.Peters, Contribution of conversion electrons to the dose in GdNCT for brain tumors: A computer study, *Med. Phys.* (submitted)
4. J.L.A.Shih and R.M.Brugger, Gadolinium as a neutron capture therapy agent, *Med. Phys.* 19:733 (1992)
5. S.C.Wang, D.L.White, J.M.Pope, and R.C.Brasch, Magnetic resonance imaging contrast enhancement versus tissue gadolinium concentration, *Invest. Radiol.* 25:S44 (1990)
6. J.Kenney, U.Schmiedl, K.Maravilla, F.Starr, M.Graham, A.Spence, and J.Nelson, Measurement of blood-brain permeability in a tumor model using magnetic resonance imaging with Gd-DTPA, *Magn. Reson. Med.* 27:68 (1992)
7. J. Hazle private communication.

EVALUATION OF DEPTH-DOSE DISTRIBUTIONS FOR GADOLINIUM NEUTRON CAPTURE THERAPY

Tetsuo Matsumoto

Atomic Energy Research Laboratory
Musashi Institute of Technology
Ozenji 971, Asao-ku, Kawasaki-shi
215 Japan

INTRODUCTION

In a previous communication, we reported the radiation effect of gadolinium (Gd) neutron capture reactions in cultured Chinese hamster cells.[1] Comparison of the results at 10 % survival level between cell cultures with and without Gd showed a 3.5-fold difference in the neutron fluence. In this paper, total depth-dose distributions have been calculated in a cylindrical phantom with a simulated tumor containing gadolinium.

METHODS

Distributions of neutron and capture gamma-ray fluxes were calculated using a two-dimensional coupled neutron photon transport code (DOT3.5). Coupled neutron photon cross-sections used for the calculations were based on the VITAMIN-C library.[2] The prompt gamma-ray and internal conversion electron spectra (figure 1a and 1b) of the $^{157}Gd(n,\gamma)$ reaction were taken from the published data.[3] The total photon and total internal conversion electron energies from the gadolinium capture reaction were 2.22 MeV and 0.054 MeV, respectively. The Auger and Coster-kronig electrons have not been considered in the calculation.

The model was R-Z geometry water phantom (20 cm in diameter x 20 cm in length) as shown in figure 2a. A cylindrical (3 cm x 3cm) tumor with 0 - 5,000 ppm Gd was placed near the surface and in a separate calculation at a depth of 3.5 cm. The thermal neutron beam spectrum, measured by time-of-flight method at the irradiation field of the Musashi reactor, was used (figure 2b). The source size was 10 cm in diameter and isotropic beam wa assumed.

Absorbed dose rates for neutrons and capture gamma-rays were calculated by multiplying fluxes by kerma factors[4] and dose conversion factors[5] in wet tissues (by weight 10%H, 12%C, 4%N, 73%O and 1% other elements), respectively. Absorbed dose rate for electrons was calculated by using reaction rates and intensities mainly for energies of 30, 70, 80, 130 and 180 keV (figure 1b).

Advances in Neutron Capture Therapy, Edited by
A.H Soloway *et al.*, Plenum Press, New York, 1993

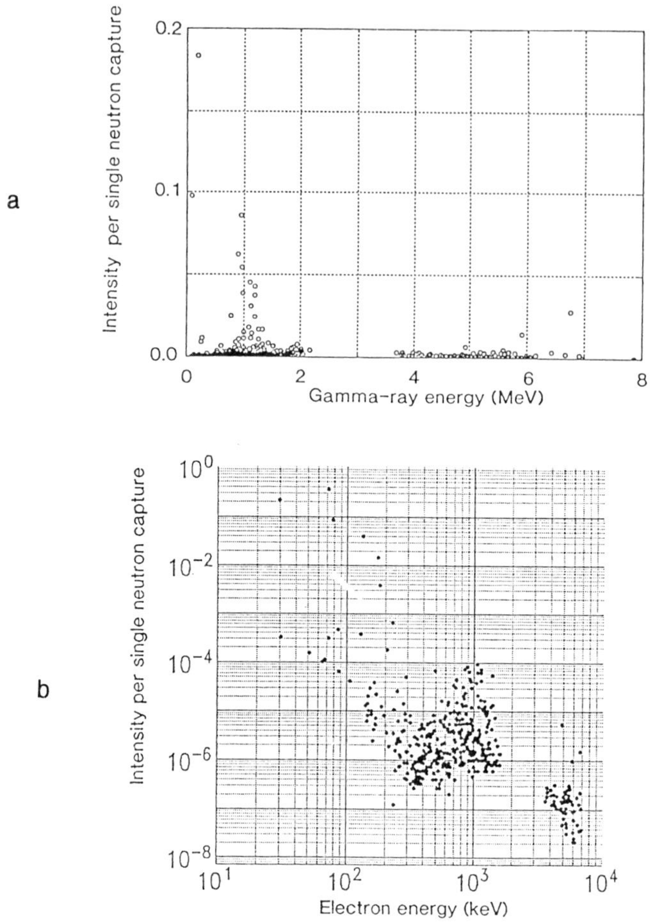

Figure 1. Yields of capture gamma-rays (a) and internal conversion electrons (b) taken from the data of Greenwood et al. (1978).

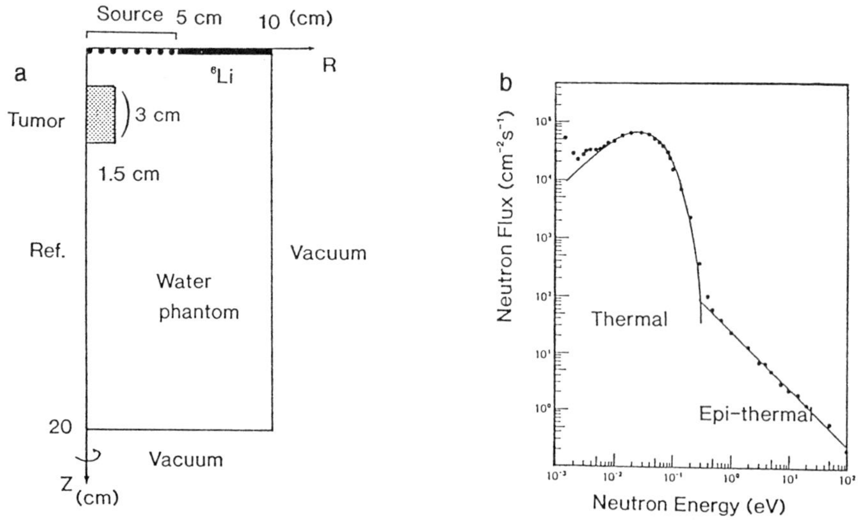

Figure 2. (a) Model for calculations. (b) The spectrum of incident neutrons (measured by time of flight method).

RESULTS AND DISCUSSION

Figure 3a and 3b show spatial distributions of thermal neutron flux and capture gamma-ray dose rate. A notable reduction in thermal neutron flux was observed across the simulated tumor which contained 5,000 ppm Gd. The capture gamma-ray dose rate increased across the tumor.

Figure 4a and 4b show the distributions of thermal neutron flux and capture gamma-ray dose rate along the central axis of the phantom which contains a simulated tumor at a depth of 0-3 cm. The thermal neutron flux was normalized to 1.5×10^9 $ncm^{-2}s^{-1}$ at the surface with the tumor containing no gadolinium. The capture gamma-ray dose is localized predominantly within the tumor because there is a high intensity of low energy photons from the $^{157}Gd(n,\gamma)$ reactions (figure 1a). However, the dose rate is lower than that reported by Shih and Brugger[6] which may cause the different data of prompt gamma-ray spectrum used for the calculations. A linear relation ship was not observed between the gadolinium concentration and gamma-ray dose rate in the tumor because of thermal neutron depression at higher gadolinium concentrations.

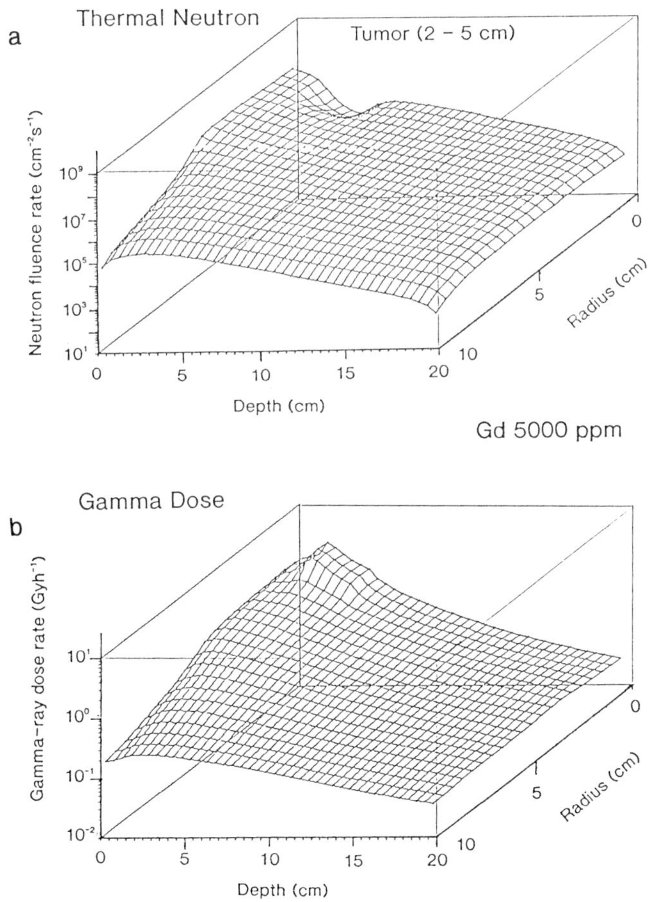

Figure 3. Spatial distributions of (a) thermal neutron flux and (b) capture gamma-ray dose rate in a water phantom with a simulated tumor containing 5,000 ppm Gd.

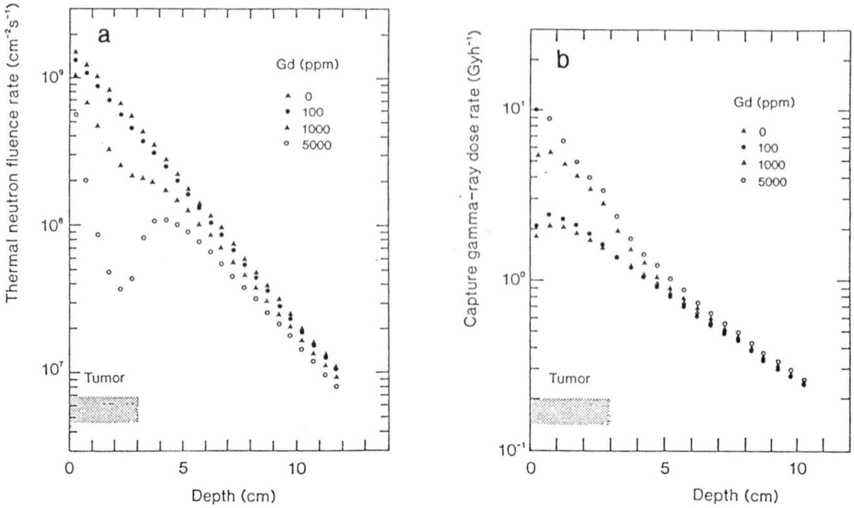

Figure 4. Distributions of (a) thermal neutron flux and (b) capture gamma-ray dose rate along the central axis in a water phantom with a 3 cm simulated tumor containing various Gd concentrations.

Figure 5a shows the total (neutron, gamma-ray including core gamma-ray and electron) depth-dose distributions along the central axis of the phantom. A tumor containing three different Gd concentrations was placed at a depth of 0-3 cm. A background dose rate at 0 ppm gadolinium was 5 Gy/h at the surface which was mainly due to the $H(n,\gamma)D$ and $^{14}N(n,p)^{14}C$ reactions. The data of core gamma-ray dose rate (0.7 Gy/h at the surface) in the phantom is shown elsewhere.[7] At 1,000 and 5,000 ppm gadolinium the total absorbed dose rate near the surface increased to 13 and 25 Gy/h, respectively, suggesting that gadolinium neutron capture therapy could be useful to treat a superficial tumor. The neutron, gamma-ray and electron contributions to total dose rate are 2.0, 5.5 amd 5 Gy/h, respectively, for the tumor containing 1,000 ppm gadolinium. It is interesting that the gamma-ray dose and the conversion electron dose for gadolinium are of comparable magnitude.

Figure 5b shows the total depth-dose distributions in a phantom containing a simulated tumor at a depth of 2-5 cm. Since the thermal neutron flux is generally decreased by one-half after traversing through 2 cm of the water phantom, the induced capture gamma-ray and internal conversion electron doses in the tumor are only a half of that for the tumor located near the surface.

CONCLUSION

A useful dose enhancement can be achieved in the tumor region by capture gamma-rays and internal conversion electrons from the $^{157}Gd(n,\gamma)^{158}Gd$ reactions but the dose is not selective at the cellular level.

ACKNOWLEDGEMENT

This research was partially suported by Grant-in-Aid from the Ministry of Education, Science and Culture.

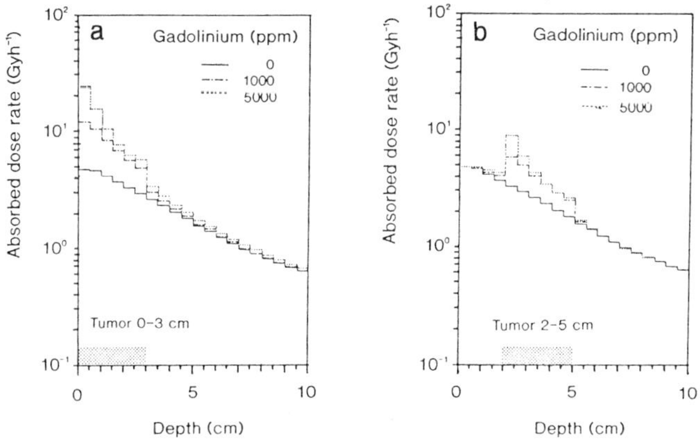

Figure 5. Total depth-dose distributions along the central axis in a water phantom with a 3 cm simulated tumor (a) at a depth of 0-3 cm and (b) at a depth of 2-5 cm containing three different Gd concentrations.

REFERENCES

1. Y.Akine, N.Tokita, T.Matsumoto, O.Oyama and O.Aizawa, "Gadolinium neutron capture reactions:A radiobiological assay", Progress in Neutron Capture Therapy for Cancer, B.J.Allen, D.E.Moore and B.V. Harrington eds., Plenum Press, New York (1992)
2. VITAMIN-C:The CTR processed multigroup cross-section library for neutronics studies:ORNL/RSIC-37, Oak Ridge National Laboratory (1980)
3. R.G.Greenwood, C.W.Reich, H.A.Baader, H.R.Koch and D.S.Bretig, "Collective and two-quasi-particle states in Gd-158 observed through study of radiative neutron capture in Gd-157", Nucl. Phys. A304, 327-437(1978)
4. International Commission on Radiation Units and Measurements, Report 13:"Neutron fluence, Neutron spectra and Kerma", ICRU Press, Washington (1969)
5. International Commission on Radiological Protection, Report 21: "Data for protection against ionising radiation from external sources" Pergamon Press, Oxford (1973)
6. J.A.Shih and R.M.Brugger, "Gadolinium as a neutron capture therapy agent", Progress in Neutron Capture Therapy for Cancer, B.J.Allen, D.E.Moore and B.V.Harrington eds., Plenum Press, New York (1992)
7. T.Matsumoto and O.Aizawa, "Head phantom experiment and calculation for boron neutron capture therapy", Phys. Med. Biol. 33, 671-86(1988)

GADOLINIUM NEUTRON CAPTURE THERAPY: GROWTH SUPPRESSION OF EHRLICH ASCITES AND SUBCUTANEOUSLY-IMPLANTED TUMOR CELLS IN MICE

Yasuyuki Akine[1], Nobuhiko Tokita[1*], Koichi Tokuuye[1], Michinao Satoh[1], Minako Sumi[1], Yoshinobu Fukumori[2], Tooru Kobayashi[3], and Keiji Kanda[3]

[1]The National Cancer Center Hospital, Tokyo, Japan
[2]Kobe-Gakuin University, Kobe, Japan
[3]Research Reactor Institute, Kyoto University, Osaka, Japan
*Visiting researcher

INTRODUCTION

A key to successful gadolinium neutron capture therapy (GNCT) is to sustain an appropriate concentration of ^{157}Gd in the tumor during neutron irradiation. Currently, experimental GNCT studies are being carried out using meglumine gadopentetate (MG), a contrast medium for magnetic resonance imaging (Akine et al., 1990). MG visualizes tumors when given intravenously, but its rapid clearance makes it less suitable for GNCT. Recently, MG-containing microcapsules have become available with an improved retention time *in vivo* (Fukumori et al., 1993). In this communication we report our preliminary results using these two forms of gadolinium carriers in two different models: one study involving MG given subcutaneously with tumor cells; the other with MG-containing microcapsules given intraperitoneally along with tumor cells.

SUBCUTANEOUS TUMOR MODEL

This study was aimed at evaluating dynamics of MG given subcutaneously and comparing GNCT to 3 MeV electron therapy for their abilities to suppress tumor formation in mice. Three-week-old ICR male mice were anesthetized, and Ehrlich ascites cells (3 x 10^6 cells/0.1 ml PBS) with 0.1 ml PBS or 0.1 ml MG (37.1 mg) were inoculated at two subcutaneous sites on the dorsum of each mouse. The mice were then individually exposed to thermal

Advances in Neutron Capture Therapy, Edited by
A.H Soloway *et al.*, Plenum Press, New York, 1993

neutrons (~1.1 x 10^{12} neutrons cm^{-2}) at the Kyoto University Reactor (KUR) for 8 minutes. The thermal neutron flux and gamma-ray dose rate at the irradiation port were 3 x 10^9 neutrons cm^{-2} s^{-1}, and 1 Gy/h, respectively. The biological endpoint was the time period required for mice to form tumors with an arbitrary size of 100 mm^2 (a product of two perpendicular axes). Once the margins were raised above the normal dermis, the tumor sizes were measured by a caliper with a margin of error less than 1 mm.

In a separate study three groups of ten mice each were inoculated with the same numbers of cells and were irradiated with 3 MeV electrons with a total dose of 5, 12 or 20 Gy at a dose rate of approximately 2 Gy/min. Changes in the amount of MG after subcutaneous injection were measured with ^{153}Gd-MG (0.5 m mol MG/50μCi/ml) by radioactivity counting.

RESULTS

It took 28.7 ± 2.3 days for mice treated with neutrons and MG (N+, Gd+) to form 100 mm^2 tumors; 15.2 ± 0.2 days for those treated with neutrons without MG.(N+, Gd-), and 7.3 ± 0.4 days for the mice unirradiated with (N-, Gd+) or without (N-, Gd-) MG. Time periods required for mice irradiated 5, 12 or 20 Gy to form 100 mm^2 tumors were 10.9 ± 2.4, 21.4 ± 5.4, and 35.0 ± 7.1 days, respectively. From these values, a time-dose relationship, $T = 4.3 + 1.5\,D$, has been derived, where T is the time periods in days, and D the electron dose in Gy. Thus, corresponding electron doses for the time periods required for groups (N+, Gd-) and (N+, Gd+) are 7.8 and 16.2 Gy, respectively. The presence of MG in this study enhanced the thermal neutron effect by 2.1 fold (16.2/7.8).

From the clearance data based on ^{153}Gd-MG, the residual concentration of MG at the inoculation site was estimated to be 78% (4.7 mg ^{157}Gd/ml) at the beginning and 61% (3.6 mg ^{157}Gd/ml) at the completion of neutron irradiation. Based on our *in vitro* biological dosimetry data, the 250 KVp x-ray equivalent dose for the all radiations (including electrons) resulting from 4 .7-3.6 mg ^{157}Gd/ml and 1.1 x 10^{12} neutrons cm^{-2} is predicted to be 21.6-24 Gy if the equation is extrapolated (Tokuuye et al 1992). Thus, a considerable discrepancy exists between the dose obtained in this study and the biological dosimetry data (16.2 vs 21.6-24 Gy).

ASCITES TUMOR MODEL

This study evaluates MG microcapsules (MG content 31% W/W) which release MG slowly (14% of the total MG released in 20 min) when suspended in saline (Fukumori et al. 1993). ICR mice were anesthetized and inoculated intraperitoneally (ip) with 10^7 Ehrlich ascites cells in 0.2 ml PBS. MG or placebo microcapsules (220 mg) were suspended in 0.5 ml dextran 40 just before use and were administered ip. The mice were individually exposed to thermal neutrons (~2.16 x 10^{12} neutrons cm^{-2}) at KUR. Nine mice received MG microcapsules and neutron irradiation (Gd+, N+), and ten were given placebo microcapsules and neutron irradiation (Gd-, N+). Control mice consisted of two groups of 14 mice each receiving 10^7 tumor cells and MG (Gd+, N-), or placebo (Gd-, N-) microcapsules without neutron irradiation. After treatment, tumor growths in the mice were followed to determine the survival (Akine et al.1992b).

RESULTS

Mice given MG microcapsules survived considerably longer than those given placebo microcapsules or control at 60 days ($p<0.0001$) with two of the nine mice surviving without disease at 180 days. All mice in other groups died within 17 days after treatment. Mice that had succumbed died as a result of progressive accumulation of tumor cells. None of the mice given MG or placebo microcapsules and neutron irradiation showed a sign of intestinal injury such as diarrhea. The estimated ^{157}Gd concentration in microcapsules was 2.5 mg/ml peritoneal fluid, and the dissolved ^{157}Gd in the peritoneal fluid was predicted to be 370 μg or less. At this concentration and the neutron fluence given, the radiation effect from the non-microcapsular ^{157}Gd dissolved in the peritoneal fluid was in the survival range of 20% levels. Hence, the cytocidal effect on the ascites tumor cells observed in this study was mainly due to radiations produced from the MG in the microcapsules rather than those from dissolved MG.

DISCUSSION AND CONCLUSIONS

In the subcutaneous tumor model the achieved enhancement of thermal neutron effects was disproportionately small when compared to that achieved *in vitro* under a similar condition. Non-uniform distributions and dynamics of subcutaneously-administered MG might have been responsible for the low enhancement effect. In the ascites tumor model slow-releasing microcapsules improved both the ^{157}Gd retention periods *in vivo*, and survival after GNCT. However, the effectiveness may have been reduced because of the inability for short-range electrons to reach the target molecules in the nucleus (Akine et al. 1992a). Although these two preliminary studies cannot be compared because Gd carriers and tumor model systems differ from each other, the results indicate the importance of sustaining appropriate amounts of ^{157}Gd in the tumor. Further studies are in progress to evaluate these ^{157}Gd carriers using the same experimental systems.

References

Akine, Y., Tokita, N., Matsumoto, H., Oyama, H., Egawa, S. and Aizawa, O., 1990, Radiation effect of gadolinium-neutron capture reactions on the survival of Chinese hamster cells. *Strahlenther Onkol* 166: 831.

Akine, Y., Tokita, N., Matsumoto, T., Oyama, H. and Aizawa, O., 1992a, Gadolinium-neutron capture reactions: a radiobiological assay, *in*: "Progress in Neutron Capture Therapy for Cancer", B. J. Allen, D. E. Moore, B. V. Harrington, eds., Plenum Press, New York, p361.

Akine, Y., Tokita, N., Tokuuye, K., Satoh, M., Fukumori, Y., Tokumitsu, H., Kanamori, R., Kobayashi, T. and Kanda, K., 1992b, Neutron-capture therapy of murine ascites tumor with gadolinium-containing microcapsules, *J Cancer Res Clin Oncol* 119: 71.

Fukumori, Y., Ichikawa, H., Tokumitsu, H., Miyamoto, M., Jono, K., Kamamori, R., Akine, Y. and Tokita, N., 1993, Design and preparation of ethyl cellulose microcapsules of gadopentetate dimeglumine for neutron capture therapy by means of Wurster process, *Chem Pharm Bull* in press.

Tokuuye, K., Tokita, N., Akine, Y., Kobayashi, T. and Kanda, K., 1992, Biological dosimetry for gadolinium neutron capture reaction (abstract), *In*; The fifth International Symposium on Neutron Capture Therapy, Columbus, Ohio, USA.

BIOLOGICAL DOSIMETRY FOR GADOLINIUM NEUTRON CAPTURE REACTION

Koichi Tokuuye[1], Nobuhiko Tokita[1*], Yasuyuki Akine[1], Tooru Kobayashi[2] and Keiji Kanda[2]

[1]National Cancer Center Hospital, Tokyo, Japan
[2]Research Reactor Institute, Kyoto University, Osaka, Japan
*Visiting researcher

INTRODUCTION

In gadolinium neutron capture reaction, photons and electrons of broad energy levels, from 20 or 30 KeV up to 7.9 MeV, are released. We have previously reported that electrons are the significant component of the over-all radiations involved in this reaction (Akine et al., 1992). Dose distributions for released photons have been reported previously (Brugger et al., 1989; Matsumoto, 1992), but dosimetry for electrons has yet to be elaborated (Masiakowski et al., 1992). This makes it impossible to estimate total absorbed doses in gadolinium neutron capture therapy. We have, therefore, carried out a biological dosimetry study using Chinese hamster cells as reported previously (Akine et al., 1990). Our intention was to derive an equation which enables us to estimate equivalent x-ray doses for any combination of gadolinium concentration and neutron fluence.

MATERIALS AND METHODS

Experiments were carried out using the standard radiobiological technique for cell survival assay. Exponentially-growing Chinese hamster (V79) cells were suspended in medium containing meglumine gadopentetate (MG, Magnevist, Schering AG, Berlin). The cell suspensions containing MG in Teflon tubes (6 cm long, 0.8 cm in outer diameter and 1

Advances in Neutron Capture Therapy, Edited by
A.H Soloway *et al.*, Plenum Press, New York, 1993

mm thick) were individually exposed to thermal neutrons generated by the Heavy Water Facility of the Kyoto University Reactor (KUR, 5MW, $3x10^8$ thermal neutrons/cm^2/sec.; a gamma dose rate of 1 Gy/hr). After exposure the cells were resuspended, diluted with fresh medium and aliquots of the cells were plated onto plastic petri dishes for colony formation. Colonies were fixed, stained and counted after 7 days. Neutron and Photon dosimetries were carried out with gold foils and thermoluminescent dosimetries (TLDs: BeO), respectively.

X-rays irradiation of cells was carried out using a 250 KVp x-ray unit (General Electric Maxitron with a filter of 2 mm Cu at 24 cm SSD) at a dose rate of 1.2-1.4 Gy/min at room temperature. After irradiation the cells were diluted and appropriate numbers of the cells were plated as described above. An ionization chamber was used for dosimetry.

RESULTS AND DISCUSSION

The x-ray survival curve, derived from 6 separate experiments, exhibits a shoulder at the low dose region. The curve was fit with a linear-quadratic model which was used to calculate the equivalent x-ray dose for any given survival level (Fig. 1a).

Survival curves for cells exposed to neutrons in the presence or absence of gadolinium (Gd) were derived from two separate experiments as shown in Fig.1b. The Gd-157 concentrations ranged from 0 to 3132 μg/ml. The survival curve for cells exposed to neutrons only was fit with a simple exponential function as expected for high LET thermal neutrons. On the other hand, survival curves for cells exposed to thermal neutrons and MG exhibit a shoulder in the low dose regions and were best fit with a linear-quadratic model. The presence of shoulders also suggest that the radiation effects are mainly of low LET. Based on the results, an x-ray equivalent dose for a given effect resulting from gadolinium neutron capture reaction can be estimated by matching the survival levels.

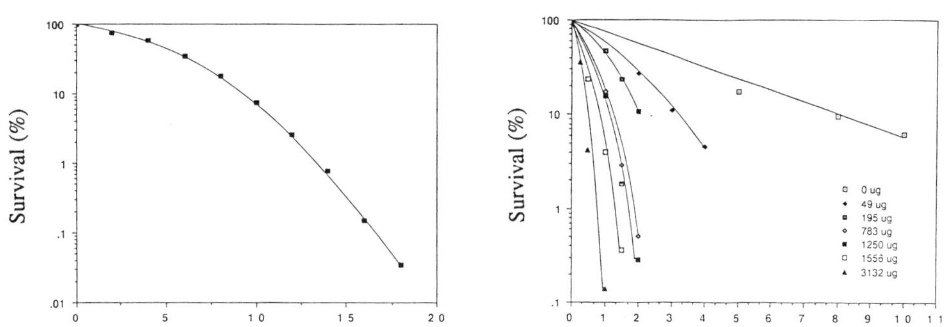

a. Dose (Gy) b. Fluence ($x10^{12}$neutrons/cm^2)

Fig.1 (a). Survival of cells exposed to 250 KVp x rays. **(b)** Survival curves for cells suspended with MG (0-3132 μg/ml Gd-157) and exposed to thermal neutrons. Note the presence of shoulders on the survival curves except for the one with 0 μg/ml Gd-157 (neutrons only).

For example, if gadolinium neutron capture therapy given under a condition resulted in 10% survival, this survival level corresponds to 9.1 Gy with x rays, hence, the dose from the gadolinium therapy was equivalent to 9.1 Gy x rays. After analysing the survival curves, an x-ray equivalent dose has been tentatively estimated as follows:

$$\text{Deq (Gy)}=0.56\ N^{1.36}\ G^{m}$$

where $m = 0.407N^{-0.267}$ and Deq is the x-ray equivalent dose in Gy; N is the value obtained by neutron fluence $/10^{12}/\text{cm}^2$; and G corresponds to the Gd-157 concentration (μg/ml).

It should be noted that the equation provides only approximate values and is valid within the data ranges shown in Fig. 1b.

To evaluate the extent of enhancement of the thermal neutron effect by MG, the thermal neutron fluence required for the 10% survival was plotted as a function of Gd-157 concentration (Fig. 2). The required fluence decreases rapidly between 0 to about 100 μg/ml, indicating a high rate of enhancement of the thermal neutron effect in this region. It is known that the probability of interacting with neutrons is linearly related to the Gd-157 concentration, therefore, if the Gd-157 concentration doubles, so does the radiation dose. In theory, Gd-157 at 783 μg/ml has a 16 times higher probability of interacting with neutrons than Gd-157 at 49 μg/ml, therefore, the dose for the former should be 16 times larger than the latter if the same neutron fluence was given. However, the dose increase obtained from this study was only 2.9 fold. The high enhancement effect seen between 0 to 100 μg/ml followed by a less significant enhancement beyond 100 μg/ml, therefore, may be explained by two possibilities: saturation effects, and reduced thermal neutron penetration. In the former, MG molecules would saturate the intracellular space perhaps at 100 μg/ml Gd-157. The intracellular Gd-157 is thought to be far more effective in cell killing because the target molecules in the nucleus are within the reach of short-range high LET electrons. Gd-157 located outside the cell, however, is less effective because

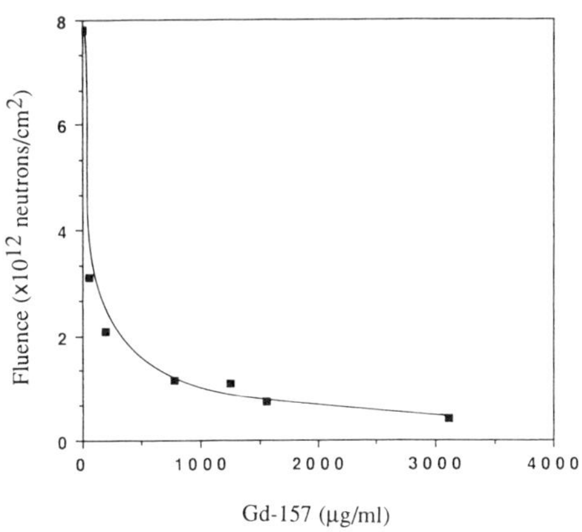

Fig. 2. Thermal neutron fluence required for 10% survival as a function of Gd-157 concentration.

only long-range electrons which are generally of low LET reach the target molecules in the nucleus. The high enhancement rate seen in 0-100 μg/ml Gd-157 may represent the effect of intracellular Gd-157. Another possibility for the high enhancement effect at 0-100μg/ml Gd-157 is that the penetration of thermal neutrons through the cell suspensions is minimally affected by the presence of Gd-157. At high concentrations, however, the degree of neutron penetration decreases rapidly, resulting in considerable loss in over-all radiations produced. Studies are needed on the intracellular distributions of MG.

CONCLUSIONS

Any survival level or dose may be achieved by a combination of a wide range of Gd-157 concentration and the neutron fluence. The results show that: (1) the radiations from gadolinium neutron capture reaction appear to be of low linear energy transfer(LET); (2) a high enhancement effect is observed between 0 and 100 μg/ml Gd-157, and (3) an equation has been derived tentatively to estimate equivalent x-ray doses for the effects produced by gadolinium neutron capture reaction under various combinations of Gd-157 concentrations and neutron fluences.

References

Akine, Y., Tokita, N., Matsumoto, H., Oyama, H., Egawa, S. and Aizawa, O., 1990, Radiation effect of gadolinium-neutron capture reactions on the survival of Chinese hamster cells. *Strahlenther Onkol* 166: 831.

Akine, Y., Tokita, N., Matsumoto, T., Oyama, H. and Aizawa, O., 1992, Gadolinium-neutron capture reactions: a radiobiological assay, *in*: "Progress in Neutron Capture Therapy for Cancer", B. J. Allen, D. E. Moore, B. V. Harrington, eds., Plenum Press, New York, p361.

Brugger, R. M. and Shih, J. A., 1988, Evaluation of gadolinium-157 as a neutron capture therapy agent. *Strahlenther Onkol* 165: 153.

Masiakowski, J. T., Horton, J. L. and Peters, L. J., 1992, Monte Carlo dose distribution of gadolinium NCT for brain tumors of different site and level of necrosis (Abstract), *in*: "The Fifth International Symposium on Neutron Capture Therapy for Cancer, Columbus, Ohio, USA., September 1992, p89.

Matsumoto, T., 1992, Transport calculations of depth-dose distributions for gadolinium neutron capture therapy. *Phys Med Biol* 37: 155.

CAPTURE OF p-BORONOPHENYLALANINE BY MELANIN-RELATED COMPOUNDS: COMPLEX FORMATION OF p-BORONOPHENYLALANINE WITH L-DOPA

K. Yoshino[1], T. Maruyama[1], Y. Mori[1], H. Kakihana[2], and Y. Mishima[3]

[1]Department of Chemistry, Faculty of Science.
Shinshu University, Matsumoto 390, Japan
[2]Department of Chemistry, Faculty of Science and Engineering
Sophia University, Chiyoda, Tokyo 102, Japan
[3]Department of Dermatology, School of Medicine
Kobe University, Kobe 650, Japan

INTRODUCTION

There are the biologically proven facts that first, ^{10}B-p-BPA accumulates ^{10}B in melanoma cells, and second, the accumulation is not permanent. After reaching a saturation concentration, melanoma cells release boron[1,2,3,4]. The reason for the accumulation has been considered as follows. 1. p-BPA resembles L-tyrosine or L-DOPA, which are precursors of melanin synthesis. 2. The melanin synthesis path is very active in melanoma cells. 3. Therefore, p-BPA might be incorporated to the melanin synthesis path.

However, there is the chemically proven fact that p-BPA cannot be oxidized by tyrosinase, which promotes the melanin synthesis[5]. This indicates that after p-BPA reaches melanoma cells, it is not oxidized to melanin. We have also looked for other reasons and have presupposed that melanin--related compounds capture p-BPA. Considering the fact that ^{10}B concentration in melanoma decreases gradually, we also assumed that the strength of the capture is not as strong as a peptide bonding, and the bond might be a labile complex formation between p-BPA and melanin-related compounds. Among the melanin-related compounds, L-DOPA was found to complex with p-BPA by ^{11}B-NMR measurement. In this paper, we describe the chemical findings on the complex formation.

EXPERIMENTAL METHODS

^{11}B-NMR measurements were carried out by JEOL FX90Q. The resonance frequency is 28.70 MHz. A quartz 10 mmϕ tube with a quartz D_2O capillary for lock was used. The samples were prepared under anaerobic condition in N_2 globe box. p-BPA was synthesized by a modified method of Snyder et al.[6] using methyl borate with natural abundance of boron.

RESULTS AND DISCUSSION

^{11}B-NMR spectra of 1:1 mole ratio solution of p-BPA and L-DOPA were measured at pH 5.1 - pH 11.5. The concentrations of each species are 0.05 M for the pH 5.4 - pH 11.5 samples, 0.025 M for the pH 5.1 - pH 6.1 samples, and 0.1 M for the pH 10.0 samples. Two signals were observed as shown in Figure 1. One signal changed the chemical shift from 4.1 ppm at pH 11.5 to 30.7 ppm at pH 5.1. This signal is assigned to free p-BPA[7]. The other signal at 10.7 ppm is independent of pH. This signal is attributed to the complex compound formed between the boronic acid group of p-BPA and two hydroxyl groups of L-DOPA[8,9]. The ^{11}B-NMR data of all spectra are summarized in Table 1. The formation constants are calculated using the signal area ratios and the initial concentrations of p-BPA and L-DOPA. The value decreases with decrease of pH. This is due to the fact that the chemical species which complex with L-DOPA is the boronate anion, $(HO)_3B^-C_6H_4CH_2CH(NH_3^+)COO^-$, which changes to boronic acid species, $(HO)_2B\text{-}C_6H_4CH_2CH(NH_3^+)COO^-$, with decrease of pH. At the pH of blood, about 7.4, more than 80% of p-BPA is clearly complexed with L-DOPA.

These results under anaerobic condition suggest that L-DOPA captures p-BPA by the complex formation at the place of the high L-DOPA concentration in a human body. But since oxygen exists in vivo and L-DOPA is oxidized to melanin in the melanoma cells, the experiments under aerobic condition should be carried out. Further, we need the p-BPA-fructose

Table 1. ^{11}B-NMR spectroscopic data of 1:1 solution of p-BPA and L-DOPA

p-BPA(M)	pH	δ_1	δ_2	δ_3	A_1	A_2	A_3	logK'
0.025	5.1	30.7	10.7	-	87	13	-	0.8
0.025	5.6	30.4	10.7	-	70	30	-	1.4
0.025	6.1	29.3	10.7	-	48	52	-	2.0
0.05	5.4	31.1	10.2	-	56	44	-	1.4
0.05	5.9	31.1	10.2	-	47	53	-	1.7
0.05	6.6	30.9	10.2	-	21	79	-	2.6
0.05	7.5	27.8	10.2	-	19	81	-	2.7
0.05	9.1	-	10.7	-	-	100	-	
0.1	10.0	-	10.5	6.9	-		-	
0.05	11.5	-	10.2	4.1	-	62	38	1.9

δ_n:chemical shift in ppm downfield of BF_3OEt_2

A_n: relative area of resonance (%)

K': equilibrium constant of complex formation

$$\text{p-BPA} + \text{L-DOPA} \overset{K'}{\rightleftharpoons} \text{BPA-L-DOPA COMPLEX}$$

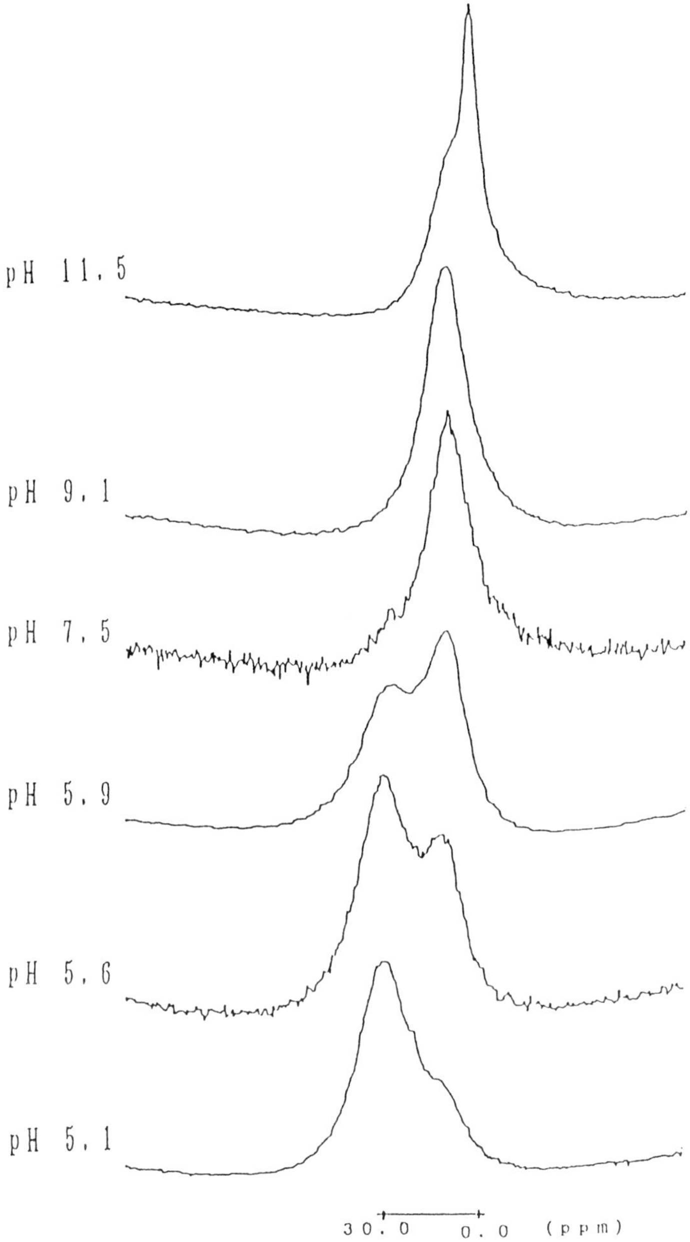

Figure 1. ^{11}B-NMR spectra of 1:1 solutions of p-BPA and L-DOPA. The concentrations are 0.025 M for pH 5.1, pH 5.6, and pH 5.9 samples, and 0.05 M for pH 7.5, pH 9.1, pH 11.5 samples. The reference signal of BF_3OEt_2 at δ =0.0 is not shown.

experiments because this complex solution is used in the trial of neutron capture therapy. In the second paper[10] we will describe the aerobic experiments modelling in vivo behavior of the BPA-fructose solution.

REFERENCES

1. K. Yoshino, M. Okamoto, H. Kakihana, Y. Mori, Y. Mishima, M. Ichihashi, M. Tsuji, and T. Nakanishi, "Chemical Assay of Boron and its Distribution in Melanoma-bearing Subjects" *in*: "Neutron Capture Therapy", H. Hatanaka ed., Nishimura Co Ltd., Niigata, Japan, 291-302 (1986).
2. J.A. Coderre, J.D. Glass, R.G. Fairchild, U. Roy, S. Cohen, and I. Fand, Selective targeting of boronophenylalanine to melanoma for neutron capture therapy, *Cancer Res.* 47:6377-6383 (1987).
3. C. Honda, M. Shiono, N. Wadabayashi, M. Ichihashi, Y. Mishima, T. Kobayashi, K. Kanda, Y. Hori, and K. Yoshino, Increased selective ^{10}B-uptake by malignant melanoma using systemic administration of $^{10}B_1$-BPA.fructose complex. *in*: "Progress in Neutron Capture Therapy for Cancer", B.J. Allen et al., ed., Plenum Press, New York 421-424 (1992).
4. Y. Mishima, M. Ichihashi, C. Honda, M. Shiono, T. Nakagawa, H. Obara, J. Shirakawa, J. Hiratsuka, K. Kanda, T. Kobayashi, T. Nozaki, O. Aizawa, T. Sato, H. Karashima, K. Yoshino and H. Fukuda, Advances in the control of human cutaneous primary and metastatic melanoma by thermal neutron capture therapy. *in*: "Progress in Neutron Capture Therapy for Cancer", B.J. Allen et al., ed., Plenum Press, New York, 577-583 (1992).
5. M. Tsuji, M. Ichihashi, and Y. Mishima, Selective affinity of $^{10}B_1$-p-boronophenylalanine.HCl to malignant melanoma: for thermal neutron capture therapy, *Jpn. J. Dermatol.* 93, 773-778 (1983) in Japanese.
6. H.R. Snyder, A.J. Reedy, and W. Lennarz, Synthesis of aromatic boronic acids. Aldehydo boronic acids and a boronic acid analog of tyrosine, *J. Am. Chem. Soc.* 80: 835-838 (1958).
7.. K. Yoshino, H. Kakihana, M. Okamoto, and Y. Mori, Chemical behaviour dopaborate and ^{10}B-p-boronophenylalanine. *in*:"Neutron Capture Therapy", H. Hatanaka ed., Nishimura Co Ltd., Niigata, Japan, 55-60 (1986).
8. K. Yoshino, M. Kotaka, M. Okamoto, and H. Kakihana. ^{11}B-NMR study of complex formation of borate with catechol and L-dopa, *Bull. Chem. Soc. Jpn.* 52, 3005-3009 (1979).
9. K. Yoshino, S. Oguchi, Y. Shimakata, and Y. Mori, M. Okamoto, M. Kotaka, and H.Kakihana, ^{11}B,^{13}C-NMR study of the complex formation of phenylboronate with catechol and L-dopa, *Kyoto Univ. Inst., TR* 260, 221-230 (1985).
10. K. Yoshino, H. Kakihana, Y. Mori, N. Torii, H. Takahashi, and Y. Mishima, Proposed mechanism of capture and release of p-boronophenylalanine through melanom cells: Chemical behaviour of p-boronophenylalanine-fructose complex in L-Dopa solution. *in*: "Advances in Neutron Capture Therapy", A.H. Soloway, et al. ed., This symposium proceedings.

PROPOSED MECHANISM OF CAPTURE AND RELEASE OF P-BORONO-PHENYLALANINE IN MELANOMA CELLS: CHEMICAL BEHAVIOR OF P-BORONOPHENYLALANINE-FRUCTOSE COMPLEX IN L-DOPA SOLUTION

K. Yoshino[1], H. Kakihana[2], Y. Mori[1], N. Torii[1], H. Takahashi[1], and Y. Mishima[3]

[1]Department of Chemistry, Faculty of Science
Shinshu University, Matsumoto 390, Japan
[2]Department of Chemistry, Faculty of Science and Engineering
Sophia University, Chiyoda, Tokyo 102, Japan
[3]Department of Dermatology, School of Medicine
Kobe University, Kobe 650, Japan

INTRODUCTION

In the previous paper[1], we have described the complex formation of p-Boronophenylalanine (p-BPA) with L-DOPA. Based on the observation we have proposed that the complex formation might be closely related to the selective accumulation of p-BPA in melanoma, where L-DOPA accumulates as a precursor of melanin synthesis. In this paper we have carried out a series of experiments, modelling conditions of the blood and using p-BPA-fructose complex[2] that has been used for melanoma patients.

EXPERIMENTAL METHODS

The apparatus and the reagents used were as the same as those used in the previous paper[1]. The samples were prepared in the aerobic condition.

RESULTS AND DISCUSSION

Model Experiment for In Vivo Behaviour of BPA-fructose Solution

To confirm further the capture of p-BPA by L-DOPA, we carried out a series of experiments modelling behaviours of p-BPA-fructose complex in blood and melanoma cells after injection. p-BPA-fructose complex solution of pH 7.4 was prepared by dissolving 100 mg p-BPA and 220 mg fructose in 3 mL water using 0.28 mL of 2 N NaOH. In this solution the concentration of p-BPA was 0.15 M. Our group has used this solution for Neutron Capture Therapy, administering it into melanoma patients by intravenous injection or peri-lesional injection. After injection, blood dilutes the p-BPA-fructose solution. Modelling this fact, we diluted the solution with pH 7.4 phosphate buffer solution. Figure 1 shows the ^{11}B-NMR spectra of 0.048M p-BPA and 0.122M fructose solution and those of its diluted solution to 1/2 and 1/8 concentrations. As shown in Figure 1, 0.048M p-BPA-fructose solution gave only one ^{11}B-NMR signal of p-BPA-fructose complex at 8 ppm. In the case diluted to half concentration, a shoulder peak of free p-BPA at 28 ppm appeared. In the case of 0.006 M, the shoulder peak of free p-BPA was clearly observed. The experimental data indicate that free p-BPA increases with dilution of p-BPA-fructose complex solution in blood.

After the p-BPA-fructose solution was diluted to 0.0048 M, we added L-DOPA to the solution. The addition of L-DOPA modells the situation when the p-BPA- fructose solution comes into melanoma cells, where L-DOPA is

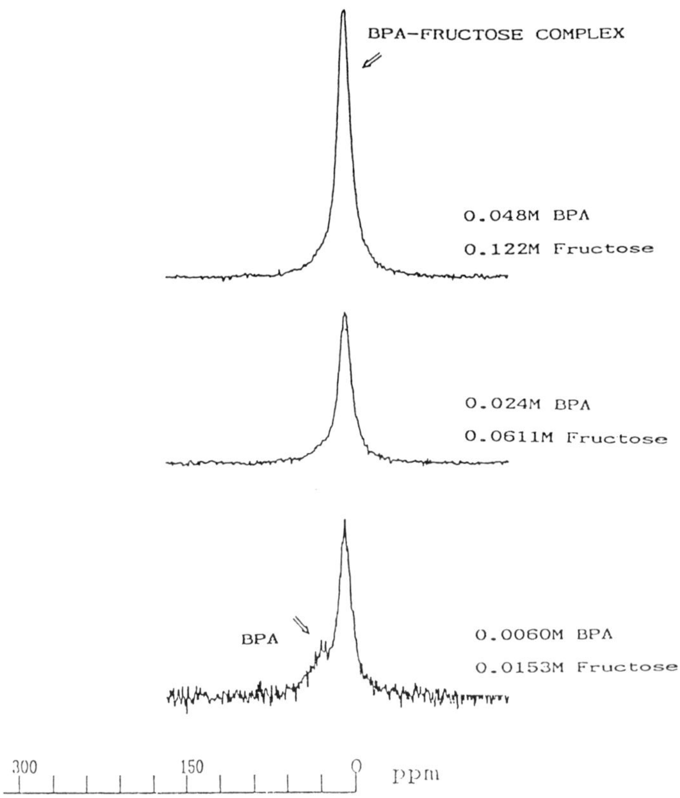

Figure 1. ^{11}B-NMR spectra of of p-BPA-Fructose solutions of pH 7.4. (a): 0.048 M p-BPA and 0.122 M fructose (b): 0.024 M p-BPA and 0.0611 M fructose (c): 0.0060 M p-BPA and 0.0153 M fructose. The reference signal of BF_3OEt_2 (= 0 ppm) is not shown.

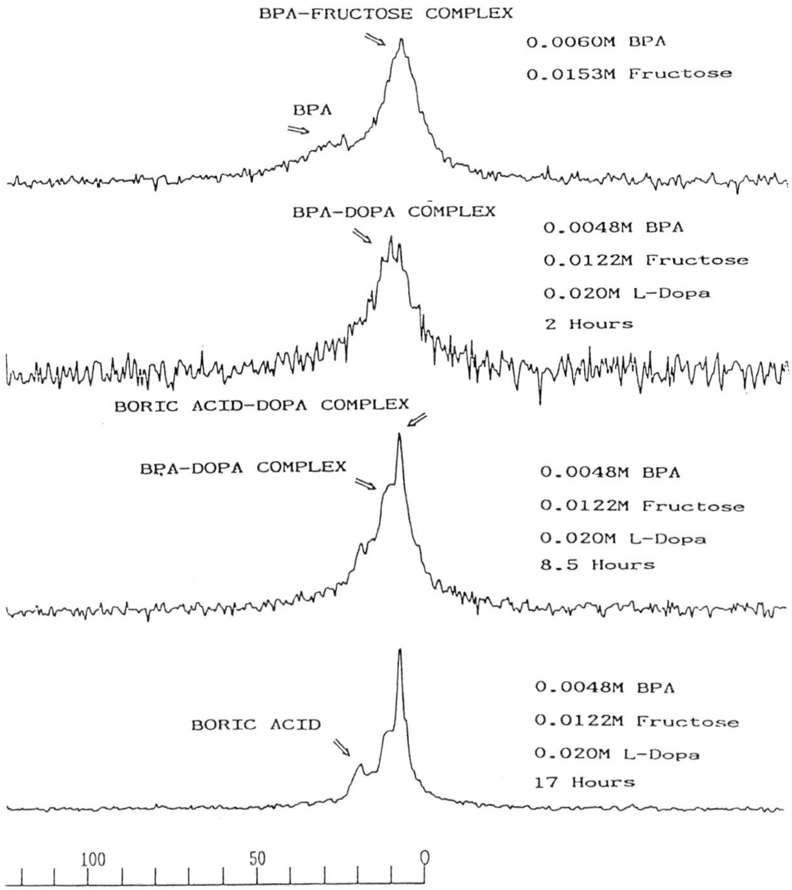

Figure 2. ^{11}B-NMR spectra of p-BPA-fructose-L-DOPA solutions
(a): 0.006M p-BPA and 0.015M fructose solution without L-DOPA
(b),(c),(d): 0.0048 M p-BPA, 0.0122 M fructose, and 0.020 L-DOPA solution at (b) 2 hours, (c) 8.5 hours, and (d) 17 hours after the start of the FT-measurement
The reference signal of BF_3OEt_2 (= 0 ppm) is not shown.

selectively accumulated[3] and exists in higher concentration than that in blood. After the addition of L-DOPA, ^{11}B-NMR measurement was started. As shown in Figure 2, two hours after the start of FT-accumulation, the signal of free p-BPA disappeared, and the most important point is that the main big signal of fructose complex in 8 ppm shifted to 11 ppm. The new signal is that of p-BPA-L-DOPA complex as shown in the previous paper[1]. This result indicates that nearly all of p-BPA-fructose complex is captured by L-DOPA to form p-BPA-L-DOPA complex. As shown in Figure 2 (c) and (d), the signal of this p-BPA-L-DOPA stepwisely decreased, and new signals clearly appeared. The new signals are assigned to boric acid-L-DOPA complexes and to 1:1 and 1:2 boric acid-fructose complexes. During the NMR measurement, the colour of the sample solution gradually changed to black. This indicates that L-DOPA was oxidized to quinone form by oxygen present in the sample solution. This observation implies that p-BPA-L-DOPA complex is decomposed into boric acid, phenylalanine, and dopaquinone as shown in Figure 3. Since tyrosinase promotes the oxidation of L-DOPA, this reaction may also take place in melanoma cells. Combining these chemically proven facts with the biologically proven facts, we can account for the behaviour of p-BPA around melanoma cells.

Proposed Mechanism of Capture and Release of p-BPA through Melanoma Cells

1. Soluble p-BPA-fructose complex compound comes around melanoma cells mainly because of the amino acid transport phenomenon. However, tyrosinase cannot oxidize p-BPA. This means p-BPA is excluded from the main path of melanin formation. 2. On the other hand, p-BPA forms a new complex compound with L-DOPA, which is selectively accumulated in melanoma cells and exists in higher concentration than in the blood. This means p-BPA moves and accumulates in melanoma cells. 3. In the melanoma cells, the p-BPA-L-DOPA complex is oxidized to form dopaquinone from the L-DOPA part of the complex, and to form boric acid and phenylalanine from the p-BPA part. 4. Boric acid is equilibrated with L-DOPA, but, since boric acid is not an amino acid, it has no selectivity to remain in melanoma cells and so moves into the blood. Then, the boric acid-L-DOPA equilibrium changes as the L-DOPA-borate complex decreases. 5. Since new blood with low concentrations of p-BPA comes contaneously around the melanoma cells, free p-BPA leaves the melanoma cells. The decrease of p-BPA in the p-BPA-L-DOPA equilibrium in the melanoma cells promotes the dissociation of p-BPA from the complex.

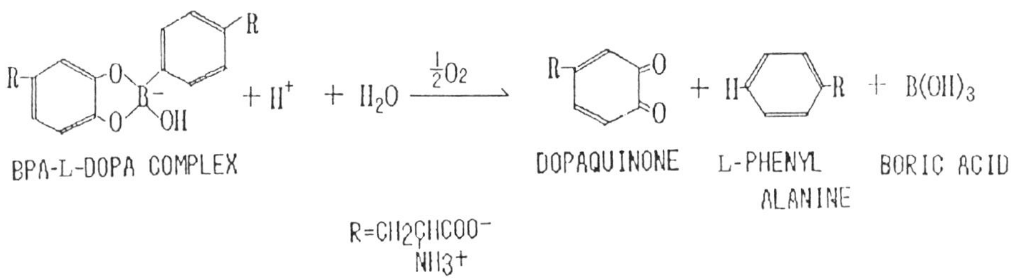

Figure 3. Oxidation reaction of p-BPA-L-DOPA complex.

REFERENCES

1. K. Yoshino, T. Maruyama, Y. Mori, Kakihana, and Y. Mishima, Capture of p-Boronophenylalanine by melanin-related compounds: Complex formation of p-Boronophenylalanine with L-Dopa. *in*: "Advances in Neutron Capture Therapy", A.H. Soloway, et al. ed., This symposium proceedings.
2. Y. Mori, A. Suzuki, K. Yoshino, and H. Kakihana, Complex formation of p-boronophenylalanine with some monosaccharides. *Pigment Cell Research* 2: 273-277 (1989).
3. M. Seiji, and S. Iwashita, Intracellular localization of tyrosinase and site of melanin formation in melanocyte, *J. Invest. Derm.* 45: 305-314 (1965).

RELATION BETWEEN STRUCTURES OF p-BORONOPHENYLALANINE RELATED COMPOUNDS AND THEIR USEFULNESS FOR BORON NEUTRON CAPTURE THERAPY

H. Kakihana[1], K. Yoshino[2], and Y. Mori[2]

[1]Department of Chemistry, Faculty of Science and Engineering, Sophia University. Chiyoda, Tokyo 102, Japan
[2]Department of Chemistry, Faculty of Science, Shinshu University, Matsumoto 390, Japan

^{10}B-compounds are taken up by cancer cells by two different mechanisms: 1) Biophysical uptake, and 2) Biochemical uptake. The actual uptake is a combination of these two.

The Blood-Brain Barrier (BBB) phenomenon is a typical example of biophysical uptake. Normal brain tissues possess the special characteristic of excluding certain compounds, while brain tumor cells will take them up. This means that in the case of brain tumor cells, we can concentrate our chemical efforts on the design and synthesis of compounds containing as many ^{10}B atoms as possible which will be excluded by normal brain. Borocaptate Sodium is a good example of such a special compound. We can conclude that the key chemistry for biophysical uptake is the "Synthetic Chemistry" especially of boron chemistry.

The mechanism of biochemical uptake is a much more complicated issue. To find effective and useful compounds, two additional expertises are required in addition to the competence in synthetic chemistry. Firstly, up to date knowledge on possible metabolic pathways: secondly an understanding of solution chemistry. In Figure 1, the melanin metabolic pathway is shown. It is clear that L-DOPA and its derivatives play key roles in this sequence. Our expectation is that L-DOPA and related compounds containing ^{10}B could be involved in some parts of the pathway and as a result, ^{10}B could be accumulated in these tumor cells.

Dopaborate

Based on this fundamental idea, we began investigating (sodium) dopaborate. Our experimental results proved that the bond between B and O of

Advances in Neutron Capture Therapy, Edited by
A.H Soloway *et al.*, Plenum Press, New York, 1993

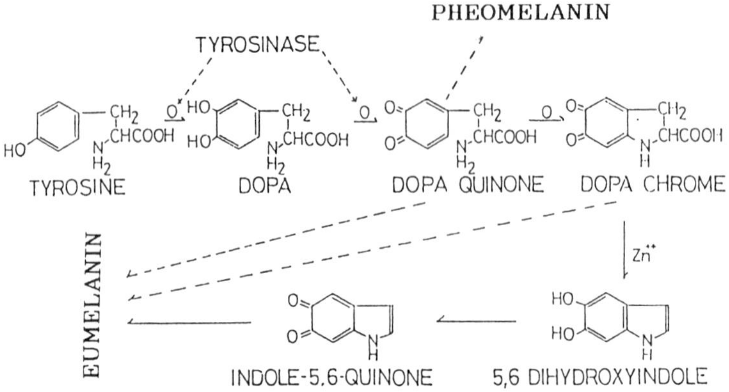

Figure 1. Melanin metabolic path.

the dopaborate molecule is a very weak one and by simple dilution with water, dopaborate is decomposed to boric acid and L-DOPA[1].

Dopaborate + H_2O = Boric Acid + L-DOPA

The chemical results have shown us that only L-DOPA enters tumor cells, but boric acid does not, because boric acid itself is not involved in the metabolic pathway.

p-Boronophenylalanine (p-BPA).

The second compound that we have intensively studied is p-BPA, and its salts and monosaccharide complexes. We selected these compounds based on our scientific knowledge that the bond between B and C in such molecules is much stronger than the bond between B and O of the dopaborate molecule. Our chemical studies have given quite important results. Firstly, as expected, the bond between B and C is strong and the molecules are never decomposed by dilution with water[1]. p-BPA itself has very low solubility, which limits its use to oral administration. p-BPA.HCl is readily dissolved in water. However, the pH of its aqueous solution is very low. This limits its usefulness to peri-lesional administration. p-BPA.NaOH is soluble in water, but the solution is basic, which limits its use again to peri-lesional administration.

p-BPA-monosaccharide complex.

p-BPA-monosaccharide complex is readily dissolved in water[2]. Moreover, pH of the aqueous solution can be easily adjusted to the phisiological pH of human blood. These chemical results demonstrates that p-BPA-monosaccharide complex has great versatility in its administration. It is available not only for peri-lesional and oral use but also by intravenous injections. Based on these facts, we can conclude that the p-BPA-monosaccharide complex is very close to the ideal compound for melanoma treatment. However, there still are two important problems that remain to be studied scientifically. Firstly, whether or not the complex remains soluble after extensive dilution,

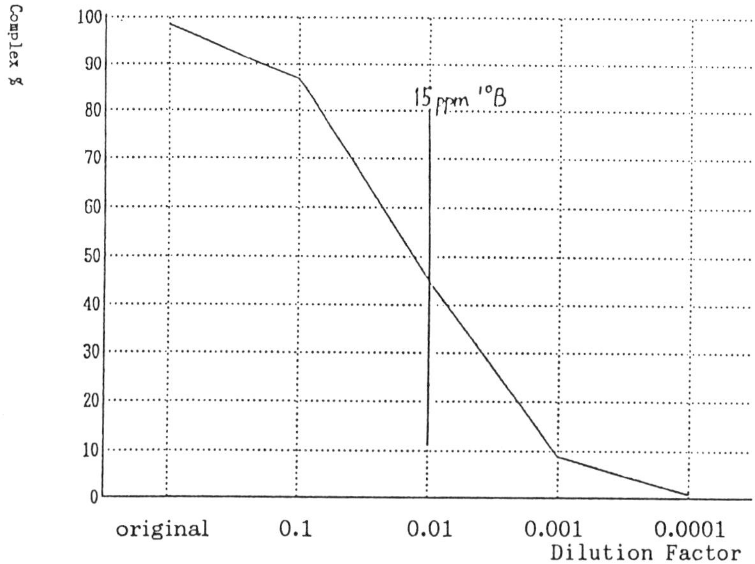

Figure 2. Calculated fraction of p-BPA-fructose complex against the total fraction of p-BPA species. The abscissa indicates the diluted concentration obtained by multiplying the original concentration by the dilution factor.

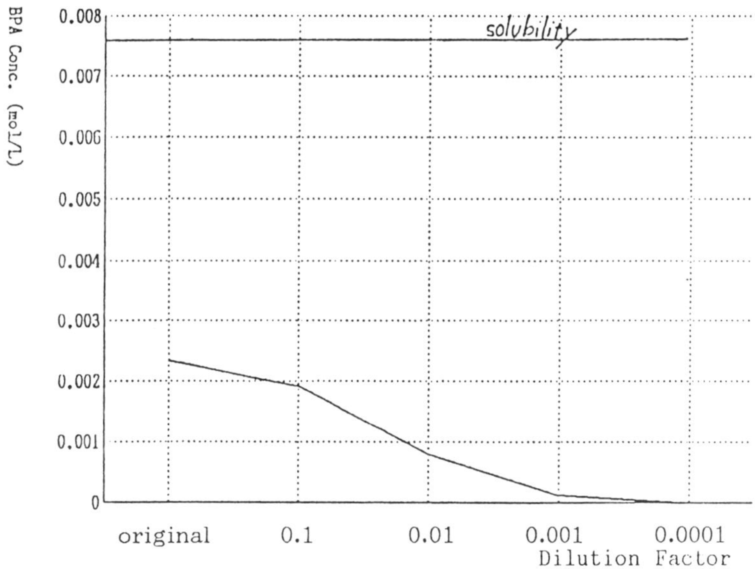

Figure 3. Calculated concentration of the uncomplexed p-BPA. The abscissa indicates the same concentration as shown in Figure 2.

which is the case, some hours after its clinical administration. Secondly, the reason why the uptake of the complex by tumor cells achieves a maximum concentration level. Several hours after administration, the p-BPA-monosaccharide complex is decomposed into two components: p-BPA and monosaccharide. This process occurs because the bonds between B atom of p-BPA and O atoms of monosaccharide are week as is the case of B and O of dopaborate. Since p-BPA has a low aqueous solubility, we were concerned that p-BPA might precipitate in human body before reaching tumor cells.

To probe the validity of this concern, we have designed and carried out a series of dilution experiments. An original solution is prepared containing 100 mg of p-BPA, and 220 mg of the monosaccharide, fructose, in 3 mL of H_2O. To adjust pH of the solution to that of blood, 0.28 mL of 2 N NaOH is added. This original solution is diluted stepwisely with water to 1/10000. No precipitation was observed in any of the diluted samples. Dissociation constants of p-BPA-Monosaccharide complex have been measured[3,4]. Using the values obtained we can calculate the percentage of p-BPA monosaccharide in each diluted solution. In Figure 2, the results obtained for the case of p-BPA-fructose are shown. Even after dilution, a high percentage of p-BPA remains as the soluble complex. 15 ppm is the maximum ^{10}B concentration in the blood of the patients under actual treatment conditions and 45% of the administrated p-BPA-fructose complex remains in this soluble form. Figure 3 presents the concentration of free p-BPA, and fortunately, its concentration is far lower than the solubility of free p-BPA. These experimental observations have confirmed that p-BPA will not precipitate in the vascular compartment before reaching the tumor cells.

The results of our research as to why the uptake of the complex by tumor cells achieves a maximum value, is described in other papers[5,6].

REFERENCES

1. K. Yoshino, H. Kakihana, M. Okamoto, and Y. Mori, Chemical behavior of dopaborate and ^{10}B-p-Boronophenylalanine, *in*:" Neutron Capture Therapy" H. Hatanaka ed., Nishimura Co., Ltd., Niigata, Japan (1986).

2. K. Yoshino, A. Suzuki, Y. Mori, H. Kakihana, C. Honda, Y. Mishima, T. Kobayashi, and K. Kanda, Improvement of solubility of p-boronophenylalanine by complex formation with monosaccharides, *Strahlenther. Onkol.* 165, 127-129 (1989).

3 Y. Mori, A. Suzuki, K. Yoshino, and H. Kakihana, Complex formation of p-boronophenylalanine with some monosaccharides, *Pigment Cell Research* 2, 273-277 (1989).

4 Y. Kinoshita, K. Yoshino, Y, Mori, H. Kakihana, and Y. Mishima, Complex formation of phenylboronic acid derivatives with aldopentoses and oligosaccharides, *in*:" Progress in Neutron Capture Therapy for Cancer", B.J. Allen et al., ed., Plenum Press, New York, 243-246 (1992).

5. K. Yoshino, T. Maruyama, Y. Mori, H. Kakihana, and Y. Mishima, Capture of p-Boronophenylalanine by melanin-related compounds: Complex formation of p-Boronophenylalanine with L-Dopa. *in*: "Advances in Neutron Capture Therapy", A.H. Soloway, et al. ed., This symposium proceedings.

6. K. Yoshino, H. Kakihana, Y. Mori, N. Torii, H. Takahashi, and Y. Mishima, Proposed mechanism of capture and release of p-boronophenylalanine through melanom cells: Chemical behaviour of p-boronophenylalanine-fructose complex in L-Dopa solution. *in*: "Advances in Neutron Capture Therapy", A.H. Soloway, et al. ed., This symposium proceedings.

SYNTHESIS OF *o*- AND *m*-BORONOPHENYLALANINE

K.Yoshino[1], T.Maruyama[1], H.Takahashi[1], Y,Mori[1], H.Kakihana[2], Y.Mishima[3], M.Ichihashi[3], C.Honda[3], M.Shiono[3]

[1]Department of Chemistry, Faculty of Science
Shinshu University, Asahi, Matsumoto 390, Japan
[2]Department of Chemistry, Faculty of Science and Engineering, Sophia University, Kioi-cho, Chiyoda-ku Tokyo 102, Japan
[3]Department of Dermatology, Kobe University School of Medicine, Kusunoki-cho, Chuo-ku, Kobe 650, Japan

INTRODUCTION

p-Boronophenylalanine (*p*-BPA) has been used for Boron Neutron Capture Therapy (BNCT). The synthetic route to *p*-BPA was developed by Snyder et al.[1]. *o*- and *m*-Boronophenylalanine (*o*- and *m*-BPA) are the isomers of *p*-BPA, and they may be incorporated into melanoma cells such as *p*-BPA and used for BNCT. Even if this incorporation dose not happen, this will give us informations about the relationship between structure and incorporation. Therefore, we have developed a way of synthesizing *o*-BPA and *m*-BPA.

EXPERIMENTAL METHODS AND RESULTS

We have developed a synthetic route to *o*-BPA (Figure 1). First, *o*-tolueneboronic acid 2 was prepared from *o*-bromotoluene 1 by a Grignard reaction with trimethyl borate[2,3]. The hydrolysis must be performed at low temperature to produce 2 in high efficiency. The bromination of product 2 by irradiation with a 200-watt unfrosted tungsten lamp prepared *o*-bromomethylbenzeneboronic acid 3. The alkylation of product 3 by acetaminomalonic ester anion prepared *o*-(2-carbethoxy-2-acetaminoethyl)-benzeneboronic acid 4[4]. Hydrolysis of the amide 4 was impossible under an alkaline condition, as reported by Kuszewski et al.[4]. We have carried out the hydrolysis and the decarboxylation of product 4 by boiling it in 2 M HCl_{aq}-MeOH and distilling

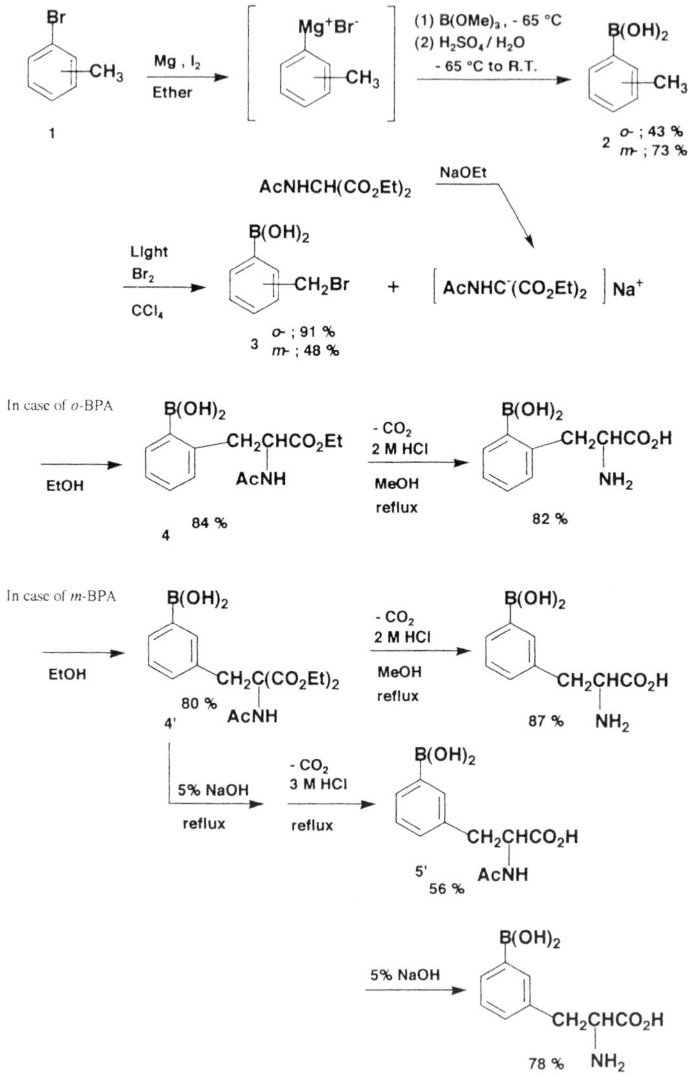

Figure 1. Synthetic schemes of *o*-BPA and *m*-BPA

Table 1. ^{13}C-NMR chemical sift data for *o*- and *m*-BPA.

Compound	δ (ppm)								
	C-1	C-2	C-3	C 4	C-5	C-6	carbonyl	α	β
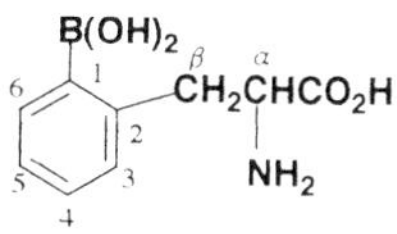	ND	140.7	132.6	133.3	129.9	136.9	174.2	57.4	38.2
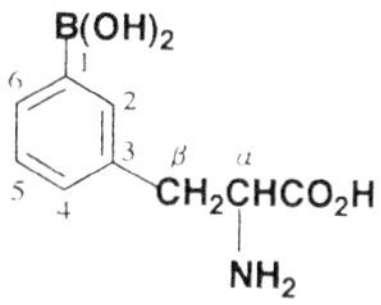	ND	137.2	136.1	134.5	131.4	135.7	173.9	56.9	38.4

^{13}C-NMR spectrum was recorded in 2 M HCl at 22.5 MHz at ambient temperature with TSP external reference. ND=Not detected.

Table 2. ^{11}B-NMR chemical shift and linewidth data for *o*- and *m*-BPA.

Compound	Condition	δ (ppm)	$W_{1/2}$ (Hz)
o-BPA	In 2 M HCl	28.3	515
	pH 12	2.04	116
m-BPA	In 2 M HCl	30.1	644
	pH 12	3.83	150

^{11}B-NMR spectrum was recorded at 28.7 MHz at ambient temperature with BF_3OEt_2 externel reference.

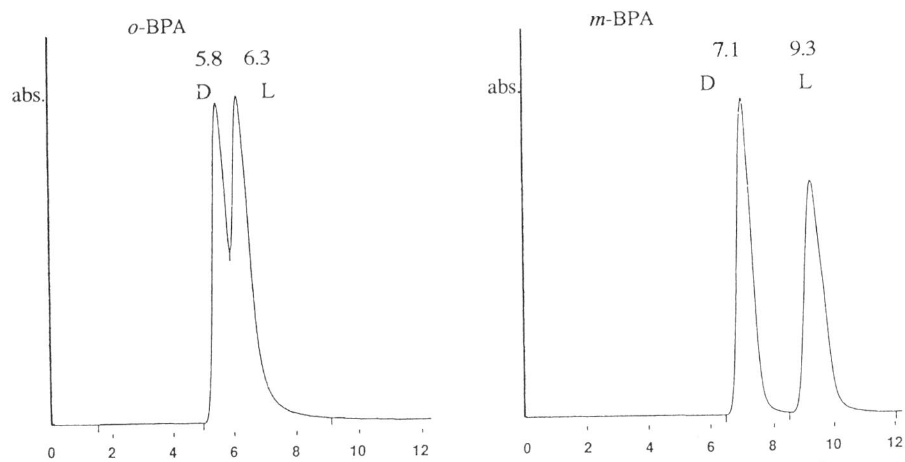

Figure 2. HPLC analysis results.

Cuolum : crowm ether-coated reversed phase packings type, CROWNPAK CR (+) (0.46×15 cm) (DAICEL CHEMICAL INDUSTRIES, LTD) ; Temp. : 25 ℃ ; Eluent : aqueous pH 1.5 $HClO_4$ solution ; Flow-rate : 1 mL/min ; Detection : 210 nm.

MeOH from the reaction solution. As a result, the desired product *o*-BPA was obtained in 82% yield.

We have also developed a synthetic route to *m*-BPA (Figure 1). The alkylation product *m*-(2,2-dicarbethoxy-2-acetaminoethyl)-benzeneboronic acid 4' was prepared similarly, using the same route as for *p*- and *o*-BPA. And the desired product *m*-BPA was obtained by working 2 M HCl on product 4' in 87% yield. *m*-BPA was also obtained by alkaline hydrolysis. Alkaline hydrolysis and decarboxylation of product 4' prepared *m*-(2-carboxy-2-asetaminoethyl)-benzeneboronic acid 5' in 56 % yield. And *m*-BPA was obtained by alkaline hydrolysis of product 5' in 78 % yield.

All compounds were identified by ^{13}C,^{11}B-NMR and by IR spectroscopy. ^{13}C and ^{11}B-NMR spectra were obtained on a JEOL FX 90 MHz spectrometer. The purities of *o*-BPA and *m*-BPA were confirmed by HPLC analysis (Figure 2). The ^{13}C and ^{11}B-NMR data of *o*-BPA and *m*-BPA are shown in Tables 1, 2.

DISCUSSION

On the preparation of *o*-BPA, the hydrolysis of the amides 4 was impossible under alkaline. We think this is due to the steric hindrance by intramolecular interaction between the carbonyl at the N-acetyl group and $-B(OH)_3^-$ under alkaline conditions. On the other hand, under acid conditions, $-B(OH)_3^-$ converts to $-B(OH)_2$, which does not seem to interact with carbonyl. Therefore, hydrolysis occurred under acid conditions. In this process, we could omit the synthetic route by alkaline hydrolysis, thus shortening the process (Figure 1).

REFERENCES

1. H. R. Snyder, Albert J. Reedy, and Wm. J. Lennarz, Synthesis of Aromatic Boronic Acids / Aldehydo Boronic Acids and Boronic Acid Analog of Tyrosine, J. Am. Chem. Soc. 80, 835 (1958).
2. K. Torssell, Acta. Chem. Scond., 8, 1779 (1954).
3. R. M. Washburn, E. Levenes, C. F. Albright, F. A. Billig and E. S. Cernak, Preparation, Properties, and Uses of Benzenboronic Acid, Advances in Chemistry Series, 23, 102 (1956).
4. J. R. Kuszewski, W. I. Lennarz, and H. R. Snyder, Aromatic Boronic Acids / Synthesis of *o*-Boronopenylalanine, COO-314-11 (1964).

SYNTHESIS OF CARBORANYLPHENYLALANINE FOR POTENTIAL USE IN NEUTRON CAPTURE THERAPY

Jognandan K. Prashar,[1] Douglas E. Moore,[1] J. Gerald Wilson[2] and Barry J. Allen[2]

[1]Department of Pharmacy, The University of Sydney, Sydney 2006,
[2]Australian Nuclear Science and Technology Organisation, Menai 2234, Australia

INTRODUCTION

The successful use of p-boronophenylalanine (BPA) as the agent for the delivery of boron to melanoma for NCT is based on it being incorporated with tyrosine as a precursor for melanin synthesis.[1] Ideally for the patient undergoing NCT, the boron concentration in the tumour should reach 20 to 30 ppm, and for BPA this level can be achieved by the administration of very high doses. However, the procedure would be improved by a compound with the same affinity for the tumour but with a larger boron content. Consequently a new generation of boronated amino acids are being synthesised to take advantage of the unique bonding characteristics of boron and its ability to form cluster compounds containing up to 12 boron atoms in a compact structure. Following this approach, we have synthesised a 1,2-dicarba-*closo*-dodecaboranyl derivative of phenylalanine and its *nido* analogue.

RESULTS AND DISCUSSION

One of the most useful general methods of synthesis of amino acids is the Sorensen process in which phthalimidomalonic ester is the central reagent. The desired amino acid side chain is introduced by alkylation at the alpha-carbon, and after the resulting product is subjected to hydrolysis and decarboxylation, the amino acid is obtained. Modifications to this procedure, involving the use of acetamidomalonic ester[2] or formamidomalonic ester[3] have been reported for the preparation of several amino acids.

Advances in Neutron Capture Therapy, Edited by
A.H Soloway *et al.*, Plenum Press, New York, 1993

The synthesis, as outlined in the Reaction Scheme, was commenced by preparation of methyl-*p*-bromo-benzoate (**1**) from *p*-bromobenzoic acid. Since addition of the boron cluster to an aromatic residue is most readily achieved by reaction of decaborane to an acetylene derivative,[4] the next step was the conversion of the bromo function of (**1**) to the acetylenic function. The common procedure for the introduction of the acetylenic substituent on to an aromatic nucleus is the Stephen-Castro coupling reaction between an aryl iodide and a protected cuprous acetylide in pyridine at reflux.[5] However, removal of the protecting groups often requires many steps and/or strongly alkaline conditions.[6] The Wittig reaction is possible, but requires more steps. Therefore a new procedure was used, involving the condensation of arylhalide with ethynyl trimethylsilane followed by removal of the silane group under very mild conditions.[7] Treatment of (**1**) with ethynyl trimethylsilane in the presence of triethylamine, triphenyl-phosphine and palladium acetate gave methyl 4-((trimethylsilyl)ethynyl)-benzoate (**2**) which on reaction with tetrabutyl ammonium fluoride furnished methyl 4-ethynyl-benzoate (**3**). Addition of decaborane to the solution of (**3**) in benzene in the presence of acetonitrile yielded methyl *p*-1,2-dicarba-*closo*-dodecaboranyl benzoate (**4**). Reduction of (**4**) with lithium aluminium hydride and treatment of the resulting benzyl alcohol derivative (**5**) with phosphorous tribromide in pyridine gave the benzyl bromide derivative (**6**).

Diethylphthalimidomalonate (**7**) was prepared by condensation of potassium phthalimide with ethyl bromomalonate at 110-120°C. Initial reaction of (**6**) with (**7**) in the presence of sodium hydride was unsuccessful. However, this condensation was finally achieved in good yield by refluxing the mixture of (**6**) and (**7**) in dimethylformamide and anhydrous potassium carbonate.[8] The hydrolysis and decarboxylation of (**8**) to form *p*-1,2-dicarba-*closo*-dodeca-boronophenylalanine (**11**) went smoothly. However the final product was contaminated with the side product phthalic acid, and therefore required chromatography. In order to eliminate chromatography from the final step, diethylformamidomalonate (**9**) was used instead of (**7**). This was readily prepared by nitrosation of diethylmalonate with sodium nitrite in acetic acid, followed by reduction with zinc and formic acid. Reaction of (**6**) with (**9**) in the presence of sodium ethoxide furnished (**10**) in good yield. On treatment of (**10**) with hydrochloric acid (6M) and acetic acid the final compound (**11**) was obtained in its pure form.

All products and intermediates in the synthesis were examined by high resolution mass spectrometry, ^{1}H-nuclear magnetic resonance and infra red spectroscopy, and their characteristics were as expected.

As is common for amino acids, the solubility of (**11**) in aqueous solutions of neutral pH was extremely low, in fact much lower than for BPA. The use of alkali was able to overcome the solubility problem, and for biodistribution experiments solutions were used having a pH of 8.8. In a further attempt to obtain a compound of improved solubility, the *closo* dodecaborane cage was converted to the more water soluble anionic *nido* form by reaction with sodium hydroxide in absolute ethanol.[9]

Biodistribution experiments with both the *closo* and *nido* compounds using the nude mouse with Harding-Passey murine melanoma xenograft were performed as previously described.[10] For both compounds, there was no preferential uptake by the tumour; the *closo* compound passed to the liver of the animal, while the *nido* compound was quickly cleared through the kidney and bladder.

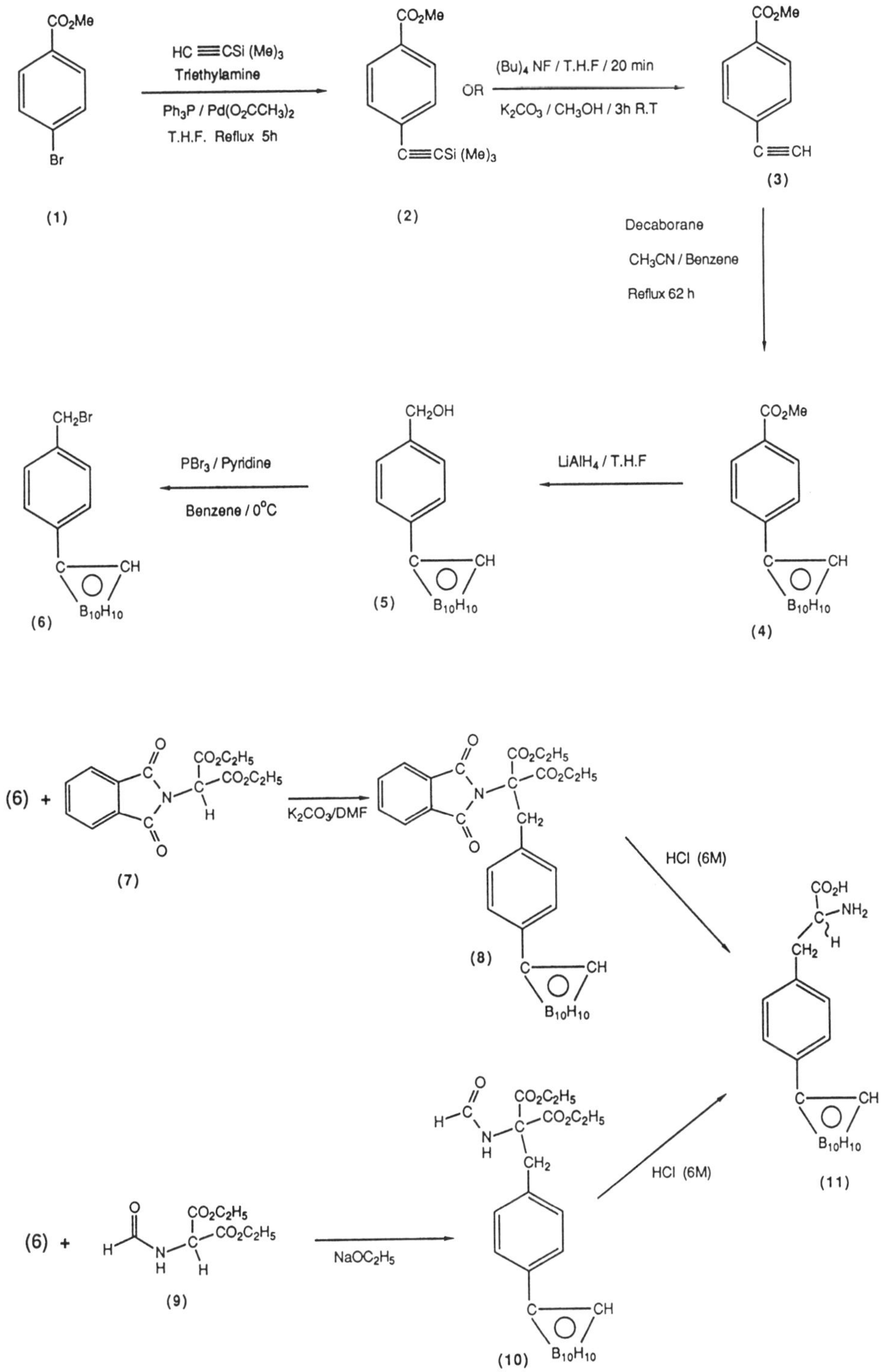

Fig. 1 **SCHEME**—Reaction Pathway for the Synthesis of Carboranylphenylalanine

ACKNOWLEDGMENT

This project is supported by The National Health and Medical Research Council of Australia.

REFERENCES

1. Y. Mishima, M. Ichihashi, S. Hatta, C. Honda, K. Yamamura, and T. Nakagawa, New neutron capture therapy for malignant melanoma - melanogenesis-seeking ^{10}B molecule. *Pigment Cell Res.* 2:226-234 (1989).
2. H.R. Snyder and C.W. Smith, A convenient synthesis of D,L-tryptophan. *J. Amer. Chem. Soc.* 66: 350-351 (1944).
3. A. Galat, Ethyl formylaminomalonate: a new intermediate in the synthesis of amino acids. *J. Amer. Chem. Soc.*, 69: 965 (1947).
4. T.L. Heying, J.W. Ager, S.L. Clark, D.J. Mangold, H.L. Goldstein, M. Hillman, R.J. Polak, and J.Y. Szymanski, A new series of organoboranes. I. Carboranes from the reaction of decaborane with acetylenic compounds. *Inorg. Chem.* 2: 1089-1092 (1963).
5. C.E. Castro and R.D. Stephens, The substitution of aryl iodides with cuprous acetylides. *J. Org. Chem.* 28: 3313-3315 (1963).
6. C.E. Castro, R. Havlin, V.K. Honwad, A. Malte, and S. Moje, Copper (I) substitutions. Scope and mechanism of cuprous acetylide substitutions. *J. Amer. Chem. Soc.*, 91: 6464-6470 (1969).
7. W.B. Austin, N. Bilow, W.J. Kelleghan, and K.S.Y. Lau, Facile synthesis of ethynylated benzoic acid derivatives and aromatic compounds via ethynyltrimethylsilane. *J. Org. Chem.* 46: 2280-2286 (1981).
8. A.K. Sen and S. Sarna, Alkylation in dimethylformamide. *J. Indian Chem. Soc.* 44: 44-645 (1967).
9. A. Varadarajan and M.F. Hawthorne, Novel carboranyl amino acids and peptides: reagents for antibody modification and subsequent neutron capture studies. *Bioconjugate Chem.* 2: 242-253 (1991).
10. B.J. Allen, S. Corderoy-Buck, D.E. Moore, Y. Mishima, and M. Ichihashi, Local control of murine melanoma xenografts in nude mice by neutron capture therapy, *in* "Progress in Neutron Capture Therapy for Cancer," B.J. Allen, D.E. Moore and B.J. Harrington, eds., Plenum Press, New York, (1992) pp. 425-428.

ASYMMETRIC SYNTHESIS OF CARBORANYL AMINO ACIDS WITH POTENTIAL USE IN BNCT

S. Sjöberg[1], M. F. Hawthorne[2], P. Lindström[1], J. Malmquist[1], J. Carlsson[3], A. Andersson[3], and O. Pettersson[3]

[1]Department of Chemistry, Uppsala University, Box 531, S-751 21 Uppsala, Sweden
[2]Department of Chemistry and Biochemistry, The University of California at Los Angeles, Los Angeles, California 90024, U.S.A.
[3]Department of Radiation Sciences, Uppsala University, Box 531, S-751 21 Uppsala, Sweden

Currently, much interest is being paid to the synthesis of boron compounds which could be delivered selectively to tumour cells either *per se* or by use of targeting strategies, including the use of monoclonal antibodies, growth factors and liposomes. α-Amino acids containing the *closo*-1,2-$C_2B_{10}H_{11}$-carborane cage, their corresponding *nido*-analogues and peptides derived from these amino acids are potential boron compounds for this purpose.

The synthesis of the *S*-form of carboranylalanine (**Ia**), via a nine step reaction starting with diethyl N-acetamidomalonate and involving an enzymatic resolution of N-acetylpropargylglycine, has been reported[1,2]. The configurational assignment was based on assumptions of the stereospecificity of the enzymatic step. It was suggested that carboranylalanine might be able to replace phenylalanine in biological systems, as the size of the icosahedral side chain is approximately that of a phenyl ring rotating about its 1,4 axis.[1] Later, these ideas gained experimental support.[2-4] Recently, the first asymmetric syntheses of both enantiomers of (**Ia**) have been reported.[5,6] The racemic acids **Ia**[7] and **Ic**[8] have also been described.

In this paper strategies for the asymmetric synthesis of both enantiomers of amino acids of type **I** are described and exemplified by the synthesis of **Ib - Id**.

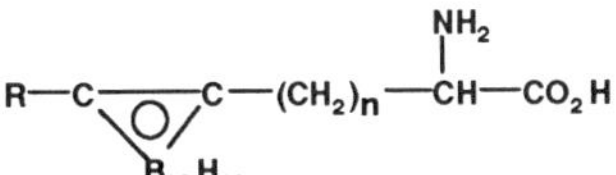

I

	R	n
Ia	H	1
Ib	H	3
Ic	CH_3	3
Id	CH_3	1

Advances in Neutron Capture Therapy, Edited by A.H Soloway *et al.*, Plenum Press, New York, 1993

The amino acids are obtained either via *a*) alkylation of chiral glycine enolate equivalents or *b*) electrophilic amination of a chiral enolate of a bornanesultam derivative of a carboranylalkanoic acid. In the former method the chirality is introduced at one of two different stages in the synthesis depending on the nature of the substituent R and the value of n: *a*)(*i*) a haloalkyl carborane (**II**) or *a*)(*ii*) an acetylenic halide (**III**) is allowed to react with the glycine enolate. In case *a*)(*ii*) the reaction product is converted to a carborane by reaction with the bis-acetonitrile complex of decaborane. The chiral glycine equivalents used are the imidazolidinone **IV** introduced by Fitzi and Seebach[9] and the bornanesultam derivative **V** introduced by Oppolzer[10]. Method *a)(ii)* has earlier been used in the synthesis of the enantiomers of carboranylalanine, **Ia**.[5,6]

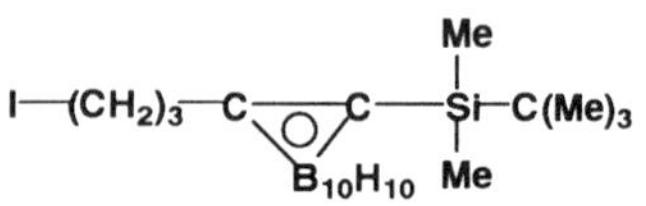

IIb

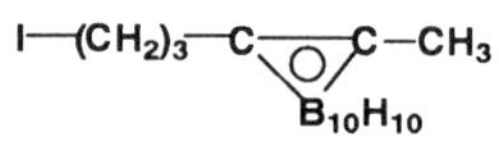

IIc

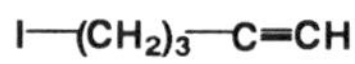

IIIb

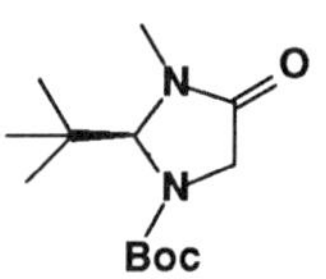

IV

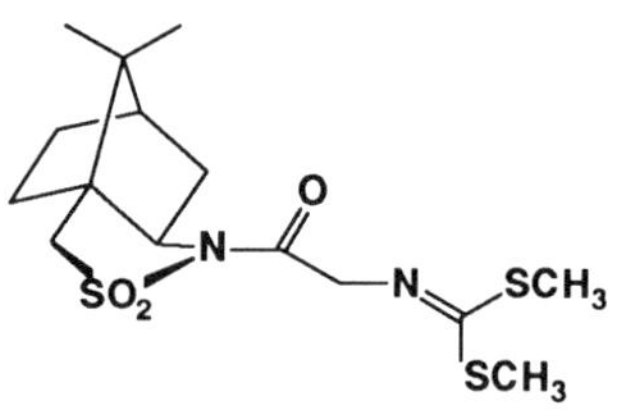

V

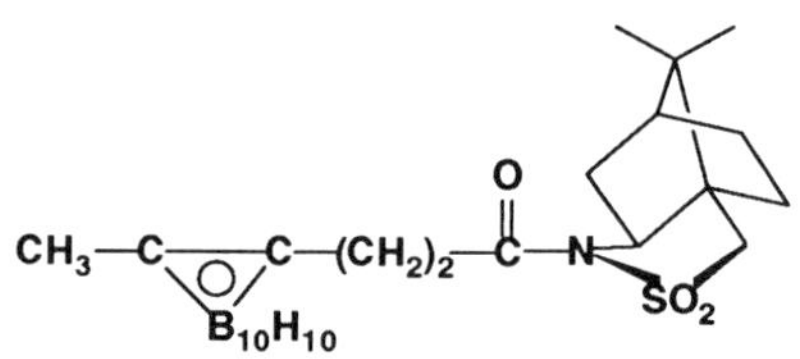

VI

Alkylation of **IV** with the protected haloalkyl carborane **IIb** according to route *a)(i)* gave after deprotection and acidic hydrolysis the amino acid (*S*)-**Ib**. The enantiomeric amino acid, (*R*)-**Ib,** was obtained via route *a)(ii)* starting with alkylation of the enantiomer of **IV** with the acetylenic halide **IIIb**. The amino acid (*S*)-**Ic** was prepared from both the imidazolidione **IV** and the sultam **V** via alkylation of the corresponding enolates with the haloalkyl carborane **IIc**.

The absolute configurations of the amino acids (*S*)-**Ia** , (*S*)-**Ib** and (*S*)-**Ic** were established by nuclear Overhauser measurements on the products obtained by alkylation of the imidazolidinone **IV** with propargylbromide, **IIb** and **IIc** respectively.

The amino acid (S)-**Id** was prepared according to route *b*) via electrophilic amination of the enolate derived from the chiral carboranylalkanoic acid derivative **VI** using the method described by Oppolzer[11]. The (S)-configuration suggested for this enantiomer of **Id**, rests solely on the stereochemical outcome of the amination reactions described by Oppolzer[11]. (*R*)-**Id** was obtained similarly from the enantiomer of **VI**.

The determination of the enantiomeric purities of the amino acids **Ia**[5,6] and **Ib-Id** was done via chromatographic separation (HPLC) of the diastereomeric derivatives formed with N-α-(2,4-dinitro-5-fluoro-phenyl)-(S)-alanineamide, according to the procedure described by Marfey[12]. For the separation, an HP 1084B LC with UV detector and a 250x4.6 mm C_{18} (Spherisorb ODS I, 10 μ) column was used. The solvents used were acetonitrile / water (50/7, v/v) and potassium dihydrogen phosphate (0.01 M in water) 25/75 to 50/50 during 0-10 min and then 20/80 for 10 min, flow 2 ml min^{-1}, column temp 40 °C, wave length 340 nm.

The separation of the diastereomeric derivatives, (S,S)-**VII** and (S,R)-**VII** obtained from an sample of (S)-**Ib** with EP=99% is shown on the next page. In all cases base line separation was obtained. The derivatives of the (S)-enantiomers of amino acids **Ia-Ic** all had the shortest retention times. The derivative of **Id** which was eluted first is the one which was obtained via amination of **VI**. A sample of (S)-**Ia** prepared according to the enzymatic resolution method[1,2] showed EP = 97 %.

NMR data and enantiomeric purities for the amino acids are summarized in the table below. Some racemization of (*S*)-**Ia** was observed during acidic hydrolysis of the corresponding imidazolidinone derivative.

Table. 1

Compound	NMR δ*	EP (%)
(S)-**Ia**	^{13}C NMR; 40.0 (CH_2), 55.0 (α-CH), 63.6 (BCH), 74.3 (BC), 171.9 (C=O); (75.4 MHz)	91-95
(S)-**Ib**	^{1}H NMR; 1.62 (m, 2H, β-CH_2), 1.76 (m, 2H, γ-CH_2), 2.30 (t, 2H, δ-CH_2), 3.50 (t, 1H, α-*CH*), 4.52 (s, 1H, cage-*CH*); (400 MHz)	99
(R)-**Ib**	^{1}H NMR; In accord with (S)-**Ib** ^{13}C NMR; 26.4 (γ-CH_2), 31.9 (β-CH_2), 38.4 (δ-CH_2) 55.7 (α-CH) 63.6 (BCH), 76.9 (BC), 174.8 (C=O); (400 MHz)	98-99
(S)-**Ic** (imidazolidinone route)	^{13}C NMR; 23.5 (CH_3), 26.7 (γ- CH_2), 31.8 (β-CH_2), 36.0 (δ-CH_2), 55.8 (α-CH), 76.9 (BC), 79.7 (BC), 173.7 (C=O); (100.6 MHz) ^{1}H NMR; 1.72 (m, 2H, β-CH_2), 1.83 (m, 2H, γ-CH_2), 2.06 (s, 3H, CH_3), 2.32 (t, 2H, δ-CH_2), 3.53 (t, 1H, α-*CH*); (400 MHz)	99
(S)-**Ic**-(sultam route)	^{1}H NMR as above	99
(S)-**Id**-HCl	^{1}H NMR; 2.16 (s, 3H, CH_3), 2.86 (dd 1H, β-*CH*), .3.22 (dd 1H, β-*CH*) 4.40 (dd, 1H,α-*CH*); (400 MHz)	99
Racemic **Ia**-HCl**	^{1}H NMR; 2.74 (dd, 1H, β-*CH*), 3.14 (dd 1H, β-*CH*), 4.15 (dd, 1H, α-*CH*), 4.72 (br. s, 1H, B*CH*); (400 MHz)	
Diastereomeric *nido*-compounds formed from racemic carboranylalanine (**Ia**)**	^{1}H NMR; 1.67 (br. s, 1H, BCH), 2.07 (dd 2H, β-CH_2), 3.88 (t 1H, α-CH), and 1.73 (br. s, 1H, BCH), 1.98 (dd 1H, β-CH), .2.32 (dd 1H, β-CH) 3.89 (dd, 1H, α-CH); (400 MHz)	

* NMR were recorded in CD_3OD solutions for **Ia** , **Ib Ic**, rac. **Ia**-HCl and the *nido*-compounds prepared from rac. **Ia**, and in D_2O solution for **Id**-HCl. Only C-bonded hydrogen atoms are shown.
** Racemic **Ia** was prepared from propargylbromide via the diethyl N-acetamidomalonate route.

Attention is drawn to the observation that carboranylalanine (**Ia**), in contrast to the amino acid **Ic,** easily undergoes degradation to a roughly equimolar mixture of the diastereomeric *nido*-analouges when the acid is heated in water The configurational assigments of the *nido*-compounds have not yet been accomplished. Crossover experiments with **Ia** and **Ic** resulted only in the degradation of carboranylalanine **Ia**.

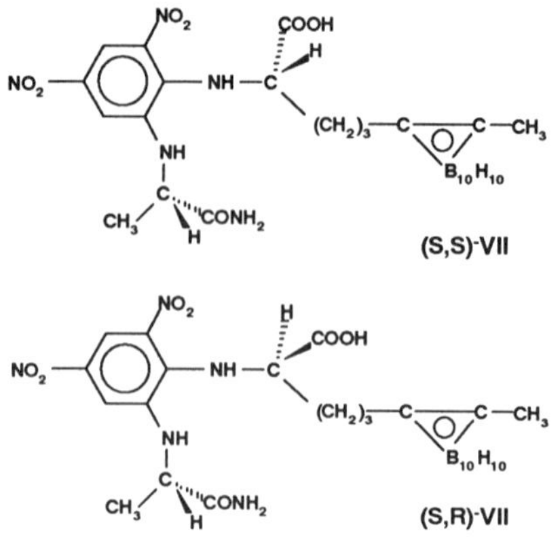

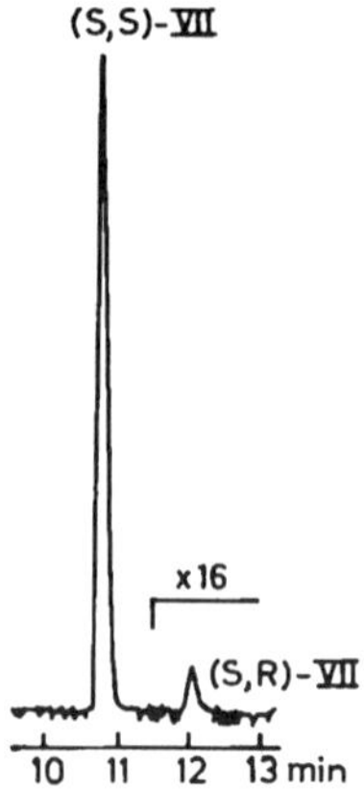

REFERENCES

1. O. Leukart, M. Caviezel, A. Eberle, E. Escher, A. Tun-Kyi and R. Schwyzer, L-o-Carboranylalanine, a boron analogue of phenylalanine, *Helv. Chim. Acta* 59:2184 (1976).
2. J.L. Fauchère, O. Leukart, A. Eberle and R. Schwyzer, The synthesis of [4-carboranylalanine, 5-leucine]-enkephalin, *Helv. Chim. Acta* 62:1385 (1979).
3. W. Fischli, O. Leukart and R. Schwyzer, Hormone-receptor interactions. Carboranylalanine (Car) as a phenylalanine analogue: Reactions with chymotrypsin, *Helv. Chim. Acta* 60:959 (1977).
4. O. Leukart, E. Escher and D. Regioli, Synthesis of angiotensins, bradykinins and substance P octapeptides in which the residue Phe and Thyr have been replaced by Car and of [Car[1], Leu[5]]-enkephalin, *Helv. Chim. Acta* 62:546 (1979).
5. A. Andersson, J. Andersson, J-O. Burgman, J. Capala, J. Carlsson, H. Conndé, J. Crawford, S. Graffmann, E. Grusell, A. Holmberg, E. Johansson, B.S. Larsson, B. Larsson, T. Liljefors, P. Lindström, J. Malmquist, L. Pellettieri, O. Pettersson, J. Pontén, A. Roberti, K. Russel, H. Reist, L. Salford, S. Sjöberg, B. Stenerlöw, P. Strömberg and B. Westermark, Programme for BNCT with accelerator-produced keV neutrons and related biological and chemical studies, *in*: "Progress in Neutron Capture Therapy for Cancer," B.J. Allen, D.E. Moore and B.V. Harrington, eds., Plenum Press, New York (1992).
6. S. Sjöberg and P. Lindström, Asymmetric synthesis of carboranyl amino acids. *Workshop on chemistry and Biology of BNCT*, May 1991 Uppsala.
7. I. M. Wyzlic and A. H. Soloway, A general convenient way to carborane-containing amino acids for boron neutron capture therapy, *Tetrahedron Lett.* 33;7489 (1992).
8. A. Varadarajan and M.F. Hawthorne, Novel carboranyl amino acids and peptides: Reagents for antibody modification and subsequent neutron-capture studies, *Bioconjugate Chem.* 2:242 (1991).
9. R. Fitzi and D. Seebach, Resolution and use in α-amino acids synthesis of imidazolidinone glycine derivatives, *Tetrahedron* 44:5277 (1988).
10. W. Oppolzer, R. Moretti and S. Thomi, Asymmetric alkylations of a sultam-derived glycinate equivalent: Practical preparation of enantiomeric pure α-amino acids, *Tetrahedron Lett.* 30:6009 (1989).
11. W. Oppolzer, O. Tamura and J. Deerberg, Asymmetric synthesis of α-amino acids and α-N-hydroxyamino acids via electrophilic amination of bornansultam-derived enolates with 1-chloro-1-nitrosylcyclohexane, *Tetrahedron Lett.* 31:991 (1990).
12. P. Marfey, Determination of -amino acids. II. Use of a bifunctional reagent, 1,5-Difluoro-2,4-dinitrobenzene. *Carlsberg Res. Commun.* 49:591 (1984).

NOVEL CARBORANYL AMINO ACIDS AND PEPTIDES

Robert R. Kane, Roger H. Pak, Lai-Ling Ng, and M. Frederick Hawthorne*

Department of Chemistry & Biochemistry
The University of California at Los Angeles
405 Hilgard Ave. Los Angeles, CA, 90024-1569

INTRODUCTION

Successful BNCT will require the selective accumulation of 5-30 ppm ^{10}B in tumor.[1] An attractive method for the delivery of boron to tumors is an antibody-based approach, which requires that $\sim 10^3$ boron atoms be delivered by each immunoprotein in order to achieve the necessary ^{10}B concentrations. This requirement presents a chemical challenge which has led us to an interest in the precise chemical synthesis of hydrophilic boron-rich 'trailer' molecules.[2] We previously described the synthesis of anionic *nido*-carborane[3] containing peptides **1** and **2** (derived from amino acid **3**, Figure 1), their attachment to monoclonal IgG antibodies (Mabs) against the CEA antigen, and the *in vivo* characterization of these immunoconjugates.[4] However, because we found the sequential coupling reactions leading to the synthesis of **1** and **2** to be prohibitively slow under solid-phase conditions, we have recently begun to explore the synthesis of similar ^{10}B-rich peptides in solution using a 'doubling' approach (segment synthesis), which is described herein.

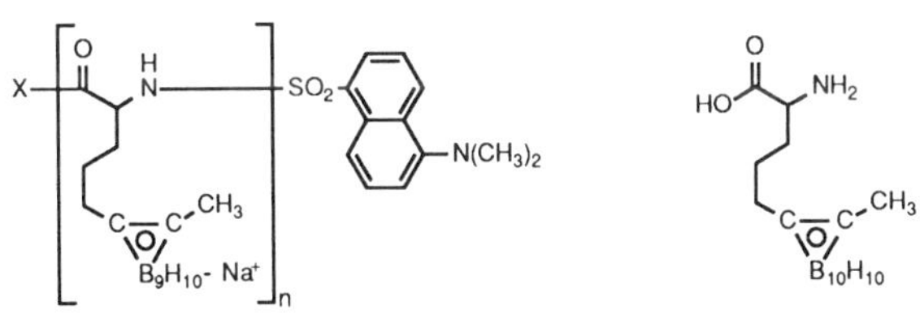

1 X=OH, n=2

2 X=(N$^{\varepsilon}$-Ac)lysine, n=10

3

Figure 1

Advances in Neutron Capture Therapy, Edited by
A.H Soloway *et al.*, Plenum Press, New York, 1993

RESULTS AND DISCUSSION

We began these studies by improving the synthesis of the *closo*-carborane amino acid monomer, as the synthesis of 3 had required nine steps from propargyl bromide (~30% overall). The TBDMS-protected carboranylpropyl iodide 4 efficiently monoalkylated imine anion 6 under standard conditions, affording amino ester•HCl 7 after mild hydrolysis (Scheme 1).[5] The silyl protected iodide 4 was utilized because a previous report from our laboratories had claimed that unprotected iodide 5 was unsuitable for imine-anion alkylation, presumably due to the relatively acidic carborane C-H vertex.[4a] We were surprised to discover that the readily available unprotected iodide 5 efficiently alkylates imine anion 6, affording amino ester 8 in good yield after mild hydrolysis (77% for two steps). Although the remainder of this communication describes the preparation of derivatives of protected amino ester 7, unprotected analogue 8 is currently employed since it is expected to exhibit very similar reactivity.

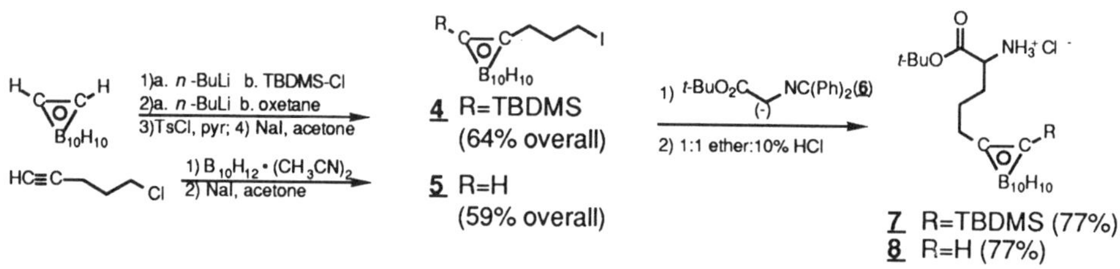

Scheme 1

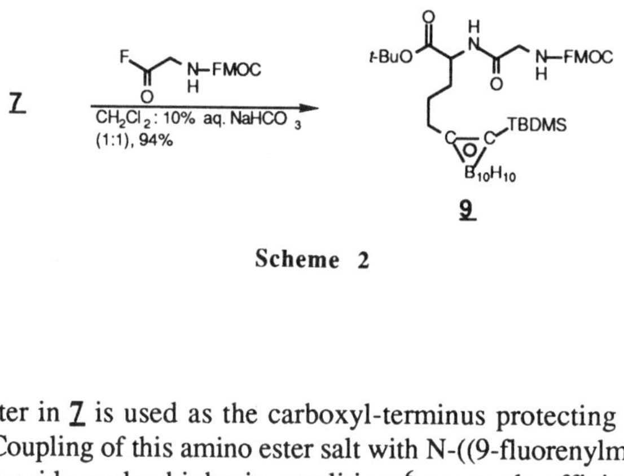

Scheme 2

The *t*-butyl ester in 7 is used as the carboxyl-terminus protecting group for segment peptide synthesis. Coupling of this amino ester salt with N-((9-fluorenylmethyl)oxy)carbonyl (FMOC) glycyl fluoride under biphasic conditions[6] proceeds efficiently, affording the orthogonally protected dipeptide 9 in good yield (Scheme 2). Glycine was incorporated into this peptide (and larger derived peptides) for a number of reasons. Glycine is an available source of a FMOC protected amino acid, allowing the ready conversion of 7 to an FMOC protected dipeptide with a 'spacer', hopefully reducing any unfavorable steric interactions. Additionally, the incorporation of glycine demonstrates the viability of incorporating natural amino acids into boron-rich peptides. We are presently investigating the incorporation of hydrophilic amino acids (or amino acid derivatives) into carborane containing peptides.

Dipeptide 9 is equipped to participate in repetitive segment condensations by virtue of the orthogonal conditions available for removal of the carboxyl and amino protecting groups. The *t*-butyl ester protecting group was selectively removed upon treatment of 9 with TFA, affording the N-FMOC amino acid **10** in 95% yield (Scheme 3). The amine protecting group could also be selectively removed. Initially, treatment of the N-FMOC amino ester **9** with DBU[7] afforded the amino ester **11** in 61% yield. These conditions were chosen to avoid the conversion of the *closo*-carborane to its anionic *nido* derivative,[8] a potential side reaction of FMOC deprotection using 1° or 2° amines.[9] Since our discovery that FMOC removal is sufficiently rapid that 1° or 2° amines can be utilized without *nido*-carborane formation, we have begun to use *tris*-(2-aminoethyl)amine (TAEA) since this reagent facilitates the removal of the dibenzofulvene formed in the deprotection reaction (CH_2Cl_2, room temperature, 30 min, 70%).[10]

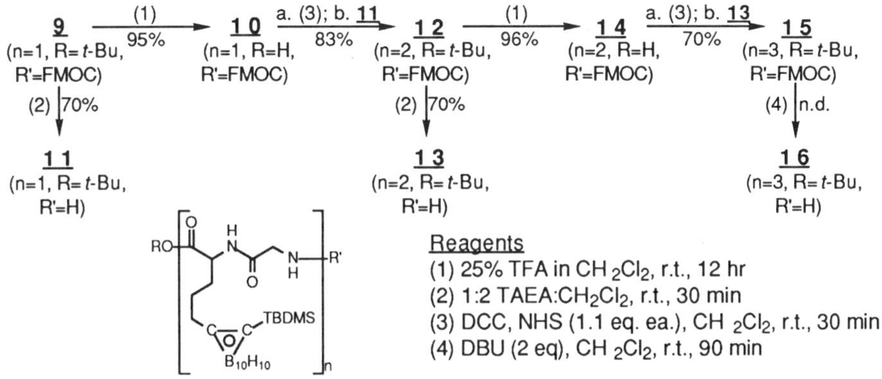

Scheme 3

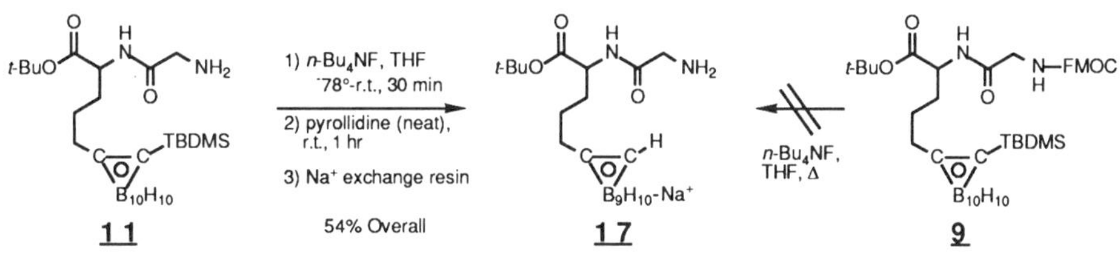

Scheme 4

The coupling of N-FMOC amino acid **10** with amino ester **11** was efficiently accomplished using *in situ* activation (DCC/NHS), thereby producing tetrapeptide **12** in 83% yield. This demonstrates one cycle of 'segment synthesis', effectively doubling the length of peptide **9** in three reactions. A second cycle of deprotection/coupling reactions again doubled the size of the peptide affording octapeptide **15** in reasonable yield (~50%). In order to facilitate characterization the amino terminus of **15** was deprotected.

After demonstrating the solution-phase segment synthesis, attention was turned to the manipulation of the carborane cages. Removal of the TBDMS carborane protecting group of **11**, using one equivalent of tetrabutylammonium fluoride in THF, was followed by conversion of the resulting *closo*-carborane-containing dipeptide to the anionic dipeptide **17** in 54% overall yield (Scheme 4). The general conversion of *closo*-carborane containing peptides to the corresponding anionic *nido*-derivatives had been described in a previous communication from our laboratories.[4a] An attempt to convert protected amino ester **9** directly to the *nido* amino ester **17** by extended treatment with tetrabutylammonium fluoride was less successful, and resulted in the formation of a complex mixture of products.[11]

CONCLUSIONS

Peptides derived from carborane-containing amino acids can be rapidly and efficiently synthesized in solution. The segment condensation strategy which was employed in this work is amenable to scale up, and provides a means for the incorporation of alternating 'natural' amino acids. These large homogeneous boron-rich compounds are important additions to the array of compounds available for use in BNCT.

ACKNOWLEDGEMENT

Financial support by the National Institutes of Health (Grants CA31753-09 and CA53870-01) is gratefully acknowledged. R.R.K. thanks the National Institutes of Health for a Postdoctoral Fellowship (CA09306-01). R.H.P. is a predoctoral trainee supported in part by USHHS National Institutional Research Service Award T32 GM08375.

REFERENCES

1. Fairchild, R.G.; Bond, V.P. (1985) *Int. J. Radiat. Oncol. Biol. Phys.* **11**, 831.
2. Hawthorne, M.F. (1991) *Pure and Appl. Chem.* **24**, 327-334.
3. Throughout this paper *closo*-carborane, *o*-carborane, or carboranyl refer to deriva tives of the *closo*-1,2-$C_2B_{10}H_{12}$ cage, while *nido*-carborane refers to derivatives of the [*nido*-7,8-$C_2B_9H_{11}]^-$ cage fragment.
4. a. Varadarajan, A.; Hawthorne, M.F. (1991) *Bioconjugate Chem.* **2(4)**, 242-253. b. Paxton, R.J.; Beatty, B.G.; Varadarajan, A.; Hawthorne, M.F. (1992) *Bioconjugate Chem.* **3(3)**, 241-247.
5. All new compounds reported in this communication have been appropriately characterized (HRMS, HPLC, multinuclear NMR, etc.). Yields are reported for material homogeneous by NMR, TLC and/or HPLC.
6. Carpino, L.A.; Mansour, E.-S.M.E.; Sadat-Aalee, D. (1991) *J. Am. Chem. Soc.* **56**, 2611-2614.
7. Kitas, E.A.; Wade, J.D.; Johns, R.B.; Perich, J.W.; Tregear, G.W. (1991) *J. Chem. Soc., Chem. Comm.*, 338-339.
8. Wiesboeck, R.A.; Hawthorne, M.F. (1964) *J. Am. Chem. Soc.* **86**, 1643-1644.
9. Hawthorne, M.F.; Wegner, P.A.; Stafford, R.C. (1965) *Inorg. Chem.* **4**, 1675-1678.
10. Carpino, L.A.; Sadat-Aalaee, D.; Beyermann, M. (1990) *J. Org. Chem.* **55(5)**, 1673-1675.
11. TBAF was expected to remove the FMOC protecting group (Happ, E., Scalfi-Happ, C., Chládek, Š. (1987) *J. Org. Chem.* **52(24)**, 5387-5391) and convert the *closo*-carborane to its *nido* derivative (Tomita, H.; Luu, H.; Onak, T. (1991) *Inorg. Chem.*, **30(4)**, 812-815), concurrently with removal of the silyl group.

ENANTIOSELECTIVE SYNTHESIS OF AN UNNATURAL AMINO ACID, L-CARBORANYLALANINE

Peggy A. Radel and Stephen B. Kahl

Department of Pharmaceutical Chemistry
University of California, San Francisco, CA 94143-0446

BACKGROUND AND TARGET SELECTION

Boron neutron capture therapy (BNCT) holds great promise as a radiochemotherapeutic treatment for various cancers, including malignant melanoma and glioblastoma multiforme. Its appeal lies in the potential for great selectivity in cell killing due to its binary nature. A noncytotoxic compound containing ^{10}B is first introduced to the tumor. Subsequent exposure to an epithermal neutron beam will cause a uniquely efficient neutron capture by ^{10}B. The ^{11}B intermediate undergoes rapid decomposition to an α–particle and recoil lithium atom. The sizeable kinetic energy of these particles (2.8 MeV) is deposited within 10-15 µm, roughly one cell diameter. The two independently manipulable requirements for this therapy allow great precision in treatment of neoplasms.

There is a need to develop compounds that would be more selective for tumor uptake and which would maximize the amount of ^{10}B introduced. Compounds incorporating carborane cages are attractive as these functionalities import ten boron atoms per cage, are very stable, and possess interesting physical characteristics. The initial target compound, L-carboranylalanine (**1**), was chosen as an isostere for phenylalanine.

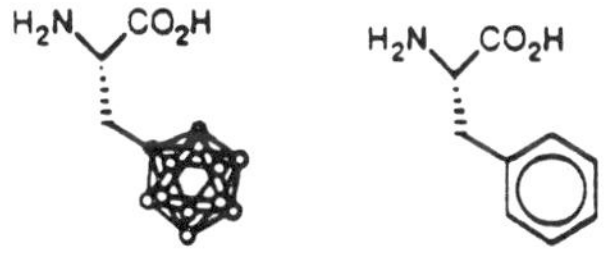

1

Figure 1. Structures of carboranylalanine and phenylalanine.

An initial area of application is in melanoma, as a false substrate for the abnormally high aromatic amino acid transport process involved in melanin biosynthesis. Carboranylalanine could provide much greater intracellular importation of ^{10}B than p-boronophenylalanine, which is currently in Phase I trials (1). We believe that carboranylalanine is a good biomimetic of phenylalanine based on its stability, hydrophobicity, and space filling characteristics. We conclude that carboranylalanine will

Advances in Neutron Capture Therapy, Edited by
A.H Soloway *et al.*, Plenum Press, New York, 1993

resemble phenylalanine (rotating freely about the 1,4 axis of its aromatic ring) and will be actively imported.

The use of peptide analogs in receptor mediated importation processes will be explored. We will be investigating several systems known to be upregulated in neoplastic cells relative to normal cells. Even with the increased bulk of the carborane cage, we believe highly active peptide analogs can be produced upon substitution with carboranylalanine. The increased hydrophobicity and increased area of contact with cell surface receptors may increase the binding of carboranylalanyl-substituted peptide analogs, leading to increased uptake and retention of these drugs. Design of peptide analogs bearing carboranylalanine will allow us to address specific cancers, such as prostate and breast, which presently have no well-localized drugs for BNCT. Use of peptide analogs does require that both L- and D- carboranylalanine be readily available.

SYNTHESIS

L-carboranylalanine has been synthesized by Schwyzer *et al*, in low yield (2). We wanted to find a higher yielding, enantioselective route that would also allow general access to other homologs. These improvements are required as the expensive, less preponderant natural isotope, ^{10}B, is the only species that will undergo neutron capture. In evaluating synthetic routes, brevity and late introduction of the carborane cage were attractive. Our attempts to utilize a strategy involving a stereoselective serine lactone ring opening with a nucleophilic metallocarborane in the last step were not successful (3). The little explored reactivity of that species was not sufficient to effect the transformation under the required conditions. A longer but potentially high yielding alternative route was then examined. Diastereoselective introduction of an azido group to imide systems had been reported by Evans and coworkers; the azidoimide was then converted to the free amino acid (4). Our investigation was performed with the natural isotope distribution (80% ^{11}B and 20% ^{10}B), to be repeated with ^{10}B enriched materials , if warranted.

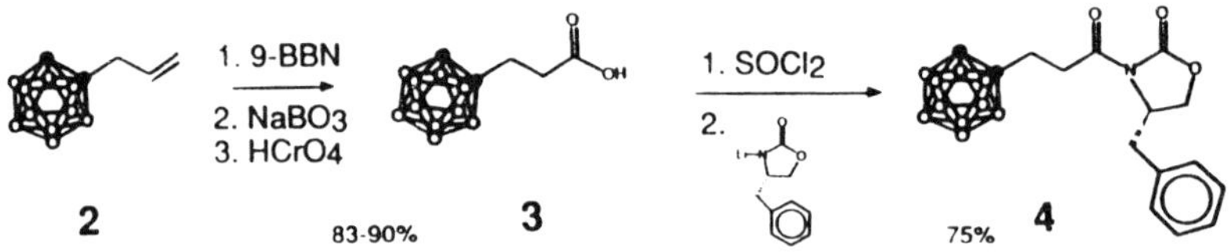

Scheme 1

The required imide was synthesized starting from the commercially available allyl carborane(**2**) (Scheme 1). Hydroboration of this molecule required very mild alkaline peroxidic workup to prevent carborane cage destruction (5). The results illustrate the interesting balance between steric and electronic directing effects in this reaction. Use of diborane in THF led to isolation of 35% of the secondary alcohol and 65% of the desired primary alcohol. The inductive directing effect could be overcome by steric effects when a bulky hydroborating agent, 9-BBN, was used. Using this reagent, none of the secondary alcohol was isolated by chromatography nor detected in the ^{13}C NMR of the crude reaction product. The crude alcohol was oxidized with Jones reagent to yield the acid **3** in

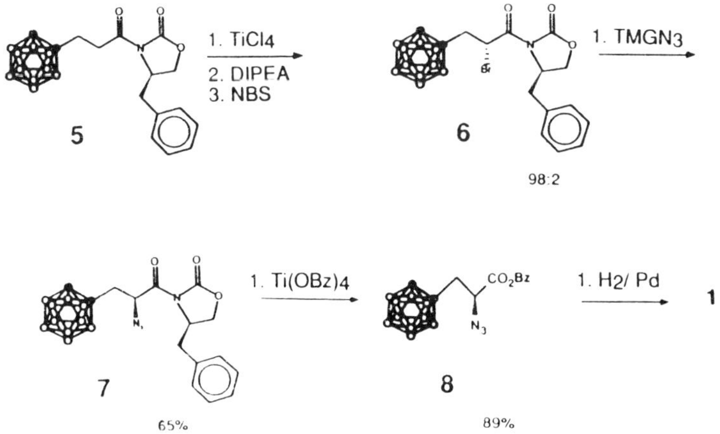

Scheme 2

90% yield from **2**. The acid was converted to the acid chloride which was reacted with the lithium salt of R- or S-4-benzyl-2-oxazolidone to yield imide **4** or **5** in 75% yield.

Direct azidization was attempted, but it was shown that the imide **4** did not survive exposure to potassium amide bases, even for very short periods at low temperature. Indirect, two step introduction of the α-azido group was therefore examined. Stereoselective bromination of imide **5** (Scheme 2) was attempted with di-n-butylborontriflate, diisopropyl-amine and N-bromosuccinimide. Irresolvably incomplete enolization was observed with commercially available triflate. We decided to apply another enolization not previously reported in asymmetric brominations, which utilized titanium tetrachloride to chelate the imide system (6). This allowed highly diastereoselective bromination, yielding a 98:2 ratio of the desired isomer **6**. This reaction still requires optimization; at present 10-20% may not enolize. Azide displacement, utilizing tetramethylguanidinium azide, proceeded with inversion at -23°C and was completely stereoselective. Highly purified (>99:1 by HPLC analysis) azidoimide **7** could be isolated in 65% over-all yield from **5**.

Removal of the oxazolidone "template" was accomplished under mild conditions by transesterification of **7** with titanium tetrabenzyloxide in 89% yield (7). The more usual peroxidic alkaline hydrolysis would destroy the carborane cage. Hydrogenolysis of azidobenzyl ester **8** proceeds without difficulty to the desired amino acid in 89% yield. Similar yields and diastereoselectivites were obtained in using imide **5** as a starting material to obtain D-carboranylalanine.

Analysis to quantify the ratio of enantiomers present in the final product has not been straight-forward. The published optical rotation was not reproducible. Conversion of the amino acid to the α-N-Mosher's amides, with methyl or benzyl ester of the carboxylic acid, did not yield base line separation of the ^{19}F NMR signals of the diastereomeric amides. GLC separation of the azidobenzylester **8** on a chiral cyclodextrin column was not successful and HPLC separation on a Nucleosil Chiral-2 or a Chiralcel OB column did not yield resolved enantiomers. However, use of a Chiralcel OJ column did permit baseline resolution of enantiomers. From the reaction sequence utilizing carefully chromatographed intermediates, a 98:2 ratio of isomers for the R-azido-benzylester 8 was found. Less careful chromatography of the corresponding intermediates led to a 5:95 ratio in the S-azidoester.

PROPOSED BIOLOGICAL TESTING

Once the stereochemical analysis has been satisfactorily completed, cell culture studies will be conducted with L-carboranylalanine to ascertain the extent of uptake in B16 melanoma cells. Assay will be by the ICP-AAS method for boron content. Cell survival studies upon thermal neutron irradiation can then be performed.

Peptides will be synthesized on supported systems, purified, then studied in similar fashion for tumor uptake in appropriate cell lines. If significant uptake has occurred, appropriate binding assays may be used to ascertain the mode of entry.

CONCLUSION

We have synthesized L-carboranylalanine enantioselectively and in good yields. The route lends itself to eventual gram scale synthesis. D-carboranylalanine has also been synthesized, in similar yields and diastereoselectivities. The introduction of the α-amino group is accomplished via highly diastereoselective bromination (98:2) of an imide system with subsequent azide displacement to yield an azidoimide, which can be converted without significant erosion of stereointegrity to the desired amino acid.

L-carboranylalanine may find an immediate application as a selective tumor localizing drug for melanoma in BNCT. Both L- and D-carboranylalanine will be utilized by incorporation into peptide analogs targeted for receptor mediated processes, to develop new tissue specific localizing drugs for various cancers in BNCT treatment. L- and D-carboranylalanine will also prove to be useful structure/activity probes in general peptide design/ function elucidation work.

ACKNOWLEDGEMENTS

The authors wish to acknowledge support from the Department of Energy, in the form of grant DE-FG03-89ER6073. Partial funding was also made available from E.G. and G., Idaho, C91-103476. P. A. Radel wishes to thank Larry Wainschel and David Pate for their assistance.

REFERENCES

1. Mishima, Y., Ichihashi, M., Honda, C., Shono, M., Nakagawa, T., Obara, H., Shirakawa, J., Hiratsuka, J., Kanda, K., Kobayashi, T., Nozaki, T., Aizawa, O., Sato, T., Korashima, H., Yoshino, K.,& Fukuda, H. (1992) in *Proc. of the Fourth International Symposium on Neutron Capture Therapy for Cancer,* (4th: 1990: Sydney, Australia) ed. B. J. Allen, D. E. Moore, and B. V. Harrington, Plenum Press.

2. Leukart, O., Caviezel, M., Eberle, A., Escher, E., Tun-Kyi, A. & Schwyzer, R. (1976) *Helv. Chim. Acta* . **59**, 2184-2187.

3. Arnold, L. D., Drover, J. C. G., Vederas, J. C. (1987) *J. Am Chem. Soc.* **109**,4649-4659.

4. Evans, D. A., Britton, T. C., Ellman, J. A., & Dorow, R. L., (1990) *J. Am. Chem Soc.* **112**, 4011-4030.

5. Kabalka, G. W., Shoup, T. M., & Goudgaon, N. M., (1989) *J. Org. Chem.* **54**, 5930-5933.

6. Evans, D. A., Urpi, F., Somers, T. C., Clark, J. S., & Bilodeau, M. T. (1990) *J. Am. Chem. Soc.* **112**, 8215-8216 .

7. Evans, D. A., Ellman, J. A., & Dorow, R. L. (1987) *Tetrahedron Let.* **28**, 1123-1126 .

SYNTHESIS AND EVALUATION OF CARBORANE-CONTAINING AMINO ACIDS FOR BORON NEUTRON CAPTURE THERAPY

Iwona M. Wyzlic[1], Albert H. Soloway[1], Rolf F. Barth[2] and Joan Rotaru[2]

[1] College of Pharmacy
[2] Department of Pathology
The Ohio State University
Columbus, OH 43210

INTRODUCTION

Malignant melanoma, a neoplasm derived from neuroectodermal cells, is usually cutaneous in origin but can arise in a wide variety of extracutaneous sites including the oral, rectal and vaginal mucosa and the eye. It has a high propensity to metastasize to the brain, is usually multicentric, and for all intents and purposes is totally refractory to any current therapeutic modalities. For this reason it is a possible candidate for Boron Neutron Capture Therapy. Although melanoma cells are variably resistant to photon irradiation, they are highly sensitive to alpha-particles.

p-Boronophenylalanine (BPA)[1] is one of several compounds that has been evaluated as a potential capture agent for the treatment of melanoma. The rationale for its use was the avidity of melanomas for aromatic amino acids and their subsequent incorporation into melanin. The success in using BPA as a means of delivering boron to melanoma is now well documented in animals and in man.[2-4] However, this compound is more than 30 years old and contains only one boron atom per molecule. It is precisely for this reason that there is an important need for the synthesis of other boron-containing amino acids, possesing higher percentages of boron and a greater specificity for and persistence within melanoma.

The objective of the present research is to synthesize carborane-containing amino acids (Figure 1) that might meet these requirements and be superior to BPA for the treatment of melanoma by means of BNCT.

CHEMISTRY

We have developed a new, convenient method for the synthesis of carborane-containing amino acids.[5] It yielded two new boron-containing amino acids, *p*-(*o*-carboran-1-yl)-phenylalanine (CBPA, 1) and *O*-(*o*-carboran-1-ylmethyl)-tyrosine (CBT, 2), and previously synthesized *o*-carboranylalanine[6-8] (CBA, 3) (Figure 1) in good yields (70-77%).

Advances in Neutron Capture Therapy, Edited by
A.H Soloway *et al.*, Plenum Press, New York, 1993

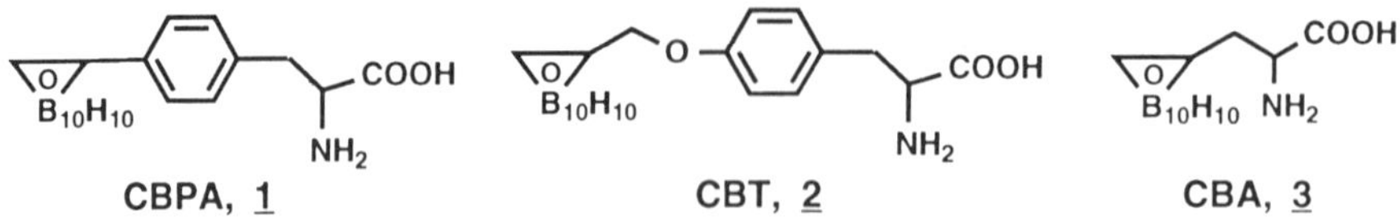

Figure 1

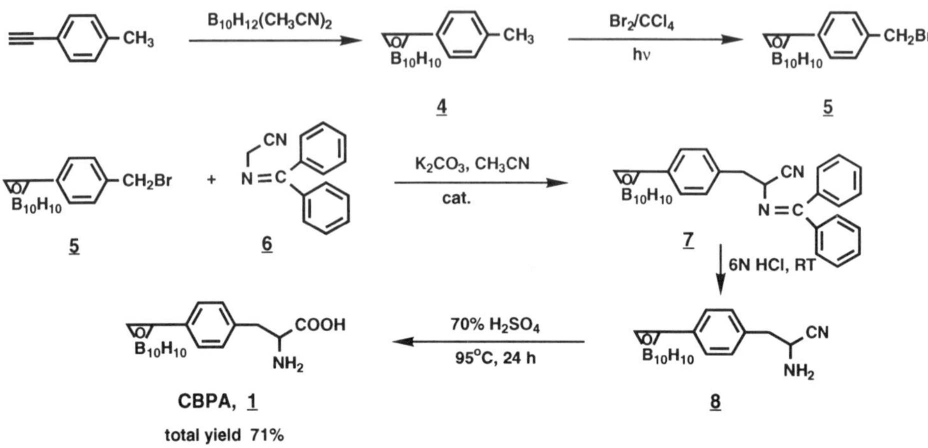

Scheme 1

p-(*o*-Carboran-1-yl)-phenylalanine (CBPA, 1) is an analogue of *p*-boronophenylalanine (BPA). This new amino acid possesses 10 boron atoms and is highly stable to strongly acid conditions, and might be transported into tumor cell by the same transport system that is operative for BPA.

The synthetic scheme for the synthesis of CBPA, 1 is shown in Scheme 1. Boronation of 4-ethynyltoluene with decaborane-acetonitrile complex followed by bromination afforded *p*-(*o*-carboran-1-yl)-benzyl bromide 5. The bromination conditions (Br_2, CCl_4, hν) that were used did not effect the aromatic or carborane nuclei. The synthesis of 1 (Scheme 1) involved a phase transfer alkylation[9,10] of commercially available *N*-(diphenylmethylene)aminoacetonitrile, 6 , with 5 to yield the mono-alkylated product 7. The partial hydrolysis of the alkylated Schiff's base, 7, with 6N HCl afforded the aminonitrile 8. In order to hydrolyze the cyano function to carboxy function stronger acidic conditions (70% H_2SO_4, 95°C, 24h) were required to obtain *p*-(*o*-carboran-1-yl)-phenylalanine (CBPA, 1) with total yield 71% (from 4-ethynyltoluene).

The same procedure has been applied for the synthesis of *O*-(*o*-carboran-1-ylmethyl)-tyrosine (CBT, 2) (Figure 1), a carborane-containing analogue of tyrosine. The *O*-alkylation of *p*-cresol with propargyl bromide afforded *O*-(3-prop-1-ynyl)-*p*-cresol. Boronation of this compound with decaborane-acetonitrile complex followed by bromination gave *p*-[*O*-(*o*-carboran-1-ylmethyl)]-benzyl bromide that has been used directly for alkylation of N-(diphenylmethylene)aminoacetonitrile, 6. The two step hydrolysis of the alkylation product afforded *O*-(*o*-carboran-1-ylmethyl)-tyrosine (CBT, 2) with total yield 77% (from *p*-cresol).

o-Carboranylalanine (CBA, 3) (Figure 1) is a potential analogue of phenylalanine. The dimensions of the cage are only slightly larger than space occupied by a benzene ring rotating about its C(1)-C(4) axis, and its two carbon atoms participate in the delocalized bonding. Our synthesis of CBA involved a phase transfer alkylation of compound 6 with propargyl bromide to yield mono-alkylated product. Boronation of this product with a decaborane-acetonitrile complex followed by the hydrolysis of the alkylated Schiff's base afforded o-carboranylalanine 1 (CBA, 3) with total yield 70%.

Carborane-containing amino acids 1 (CBPA), 2 (CBT) and 3 (CBA) are highly lipophilic. The hydrophilicity of these structures may be increased by conversion of their hydrophobic neutral carborane cages to anionic [*nido*-7,8-$C_2B_9H_{11}$]- moieties.[11,12]

The initial base degradation studies were carried out with *o*-carboranylalanine 3 (Scheme 2). A mixture of amino acid 3 and sodium hydroxide in absolute alcohol was refluxed under argon for 24h. The excess base was neutralized by passing CO_2 through the reaction mixture and separation of Na_2CO_3 by filtration. The solvent was removed under vacuum and the sodium salt of *nido*-carboranylalanine, 9 was obtained in 90% yield. This anion was isolated as the tetraphenylphosphonium salt, 10.

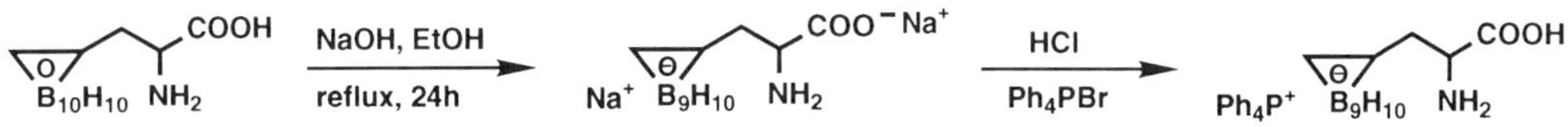

CBA, 3   9   10

Scheme 2

BIOLOGICAL STUDIES

Toxicity studies

The cell line MRA27 is a human melanotic melanoma derived from a 60 year old male Caucasian and obtained, in passage number 150-160 *in vitro*, from the laboratory of E.K. Rofstad, The Norwegian Radium Hospital, Oslo, Norway. It is propagated in McCoy's media containing 10%FBS, L-glutamine, penicillin, streptomycin, non-essential amino acids and a mixture of insulin, sodium pyruvate and oxalacetic acid.

Toxicity of vehicle and compounds was determined by a (^{3}H)-TdR uptake assay. Between 5-6000 cells/well were seeded into 96 well plates and grown for 24 hours at 37°C in a 5% CO_2, humidified incubator. Growth medium was replaced with media containing the test compound of interest and incubated an additional 24 hours at which time 0.1 μCi (^{3}H)-TdR was added. After an additional 24 h incubation period, the cells were harvested by a Skatron Cell Harvester and radioactivity measured by a Beckman LS 7000 Scintillation Counter to quantify cell growth.

Toxicity of hydroxypropyl-β-cyclodextrin, used to solubilize compounds, was determined and only nontoxic concentrations were used. Initial toxicity assays of all compounds were performed using a final concentration of 1.75% β-cyclodextrin in the test media. The compounds were initially solubilized in 56% β-cyclodextrin, sterilized through 0.22μ syringe filters and diluted to 1.75% in culture media. Boron levels in test media were determined by direct current plasma - atomic emission spectroscopy.[13]

The results of toxicity studies for amino acids: CBPA, 1, CBT, 2, CBA, 3 and aminonitriles: CBACN (precursor of CBA) and CBTCN (precursor of CBT) indicated that all three amino acids are non-toxic, whereas their precursors are highly toxic (Figures 2 and 3).

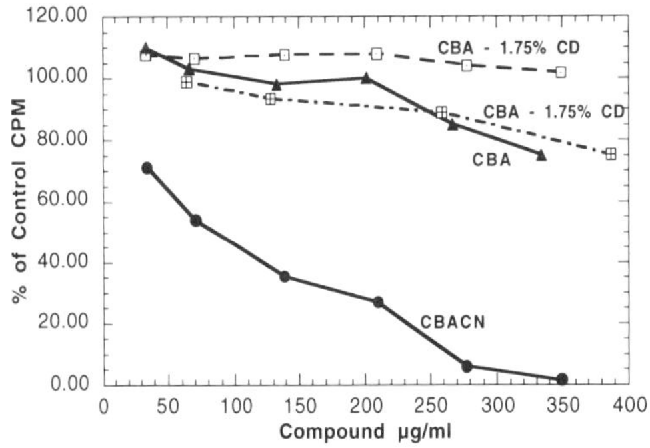

Figure 2

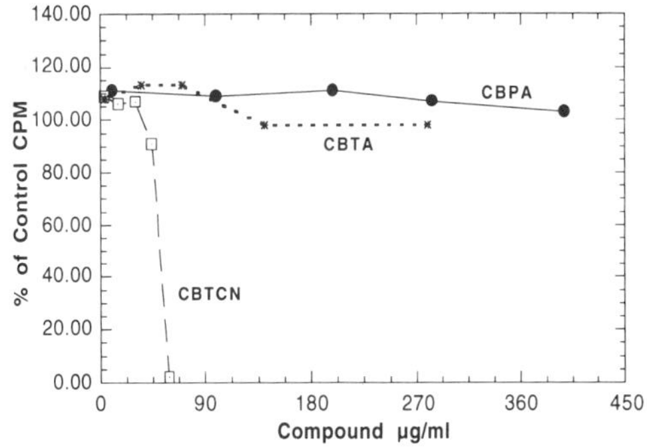

Figure 3

CONCLUSIONS

A general, convenient procedure for the synthesis of carborane-containing amino acids has been developed. Two new boron-containing amino acids, *p*-(*o*-carboran-1-yl)-phenylalanine (CBPA), 1, and *O*-(*o*-carboran-1-ylmethyl)-tyrosine (CBT), 2 and previously known *o*-carboranylalanine (CBA), 3 have been synthesized in good yields (70-77%). CBA, 3 has been converted to sodium salt of its *nido*-structure, 9 and subsequently isolated as tetraphenylphosphonium salt, 10.

Toxicity studies for amino acids: CBPA, 1, CBT, 2 and CBA, 3 indicated:

CBPA - non-toxic up to 398 μg/ml
CBT - non-toxic up to 280 μg/ml
CBA - non-toxic up to 196 μg/ml

and for two aminonitriles - amino acids precursors: CBACN and CBTCN indicated:

CBACN - toxic at 35 μg/ml
CBTCN - toxic at 30-40 μg/ml

Cellular uptake studies are currently in progress.

ACKNOWLEDGEMENT - This research has been supported by the U.S. Department of Energy grant DE-FG02-90ER60972 and contract DE-AC02-76CH000616 and the National Cancer Institute of the U.S. Public Health Service R01 CA 53896.

REFERENCES

1. H.R. Snyder, A.J. Reedy and W.J. Lennarz, Synthesis of aromatic boronic acids. Aldehydo boronic acids and a boronic acid analog of tyrosine, *J. Am. Chem. Soc.* 80:835 (1958).
2. J.A. Coderre, J.A. Kalef-Ezra, R.G. Fairchild, P.L. Micca, L.E. Reinstein and J.D. Glass, Boron neutron capture therapy of a murine melanoma, *Cancer Res.* 48:6313 (1988).
3. Y. Mishima, M. Ichihashi, M. Tsui, H. Sasumu, M. Ueda, C. Honda and T. Susuki, *J. Invest. Dermatol.* 92 (Suppl.):3215 (1989).
4. Y. Mishima, M. Ichihashi, M. Tsui, S. Hatta, M. Ueda, C. Honda and T. Susuki, Treatment of malignant melanoma by single thermal neutron capture therapy with melanoma-seeking ^{10}B-compound, *Lancet* 2:388 (1989).
5. I.M. Wyzlic and A.H. Soloway, A general, convenient way to carborane-containing amino acids for boron neutron capture therapy, *Tetrahedron Lett.* 33:7489 (1992).
6. O. Leukart, M. Caviezel, A. Eberle, E. Escher, A. Tun-Kyi and R. Schwyzer, L-*o*-Carboranylalanine, a boron anlogue of phenylalanine, *Helv. Chim. Acta* 59: 2184 (1976).
7. S. Sjoberg, *Workshop on Chemistry and Biology on BNCT*, 1991, Uppsala.
8. P. Radel and S. Kahl, Enantioselective synthesis of an unnatural amino acid, L-carboranylalanine, *BUSA III*, 1992, Pullman.
9. M.J. O'Donnell and T.M. Eckrich, The synthesis of amino acid derivatives by catalytic phase-transfer alkylations, *Tetrahedron Lett.* No.47:4625 (1978).
10. M.J. O'Donnell and R.L. Polt, A mild and efficient route to Schiff base derivatives of amino acids, *J. Org. Chem.* 47:2663 (1982).
11. A. Varadarajan and M.F. Hawthorne, Novel carboranyl amino acids and peptides: Reagents for antibody modification and subsequent neutron-capture studies., *Bioconjugate Chem.* 2:242 (1991).
12. M.F. Hawthorne, D.C. Young, P.M. Garrett, D.A. Owen, S.G. Schwerin, F.N. Tebbe, and P.A. Wegner, The Preparation and characterization of the (3)-1,2- and (3)-1,7-dicarbadodecahydroundecaborate(-1) ions, *J. Am.Chem. Soc.* 90:862 (1968).
13. R.F. Barth, D.M. Adams, A.H. Soloway, E.B. Mechetner, F. Alam and A.K.M. Anisuzzaman, Determination of boron in tissues and cells using direct-current plasma atomic emission spectroscopy, *Anal. Chem.* 63:890 (1991).

SYNTHESIS, ANTIVIRAL ACTIVITY, CYTOTOXICITY, AND CELLULAR PHARMACOLOGY OF 5-CARBORANYL-PYRIMIDINE NUCLEOSIDES

Raymond F. Schinazi,[1,2] Naganna Goudgaon,[1,2] José Soria,[3] and Dennis C. Liotta[3]

Emory University, Departments of Pediatrics[1] and Chemistry,[3] and Veterans Affairs Medical Center,[2] Atlanta, Georgia, 30033

INTRODUCTION

Our laboratories have been involved in the synthesis and evaluation of nucleosides for the treatment of viral infections and cancer for more than a decade. Our group was the first to synthesize 5-dihydroxyboryl-2'-deoxyuridine (DBDU), a novel compound that was shown by us to destroy hamster V-79 cells when irradiated with low energy neutrons, thus producing cell destruction by a boron neutron capture reaction.[1] This reaction is a consequence of the $^{10}B(n,\alpha)^{7}Li$ reaction itself, as well as a concomitant self-sensitization to those radiations provided by the presence of the nucleoside analogue in DNA. DBDU is related to the natural nucleoside thymidine and may be useful for the treatment of gliomas and other human tumors. The aim of this work was to synthesize additional boron containing nucleosides that are hydrolytically stable and that are substrates for nucleoside kinases found in tumor cells. The design of the ideal compound for BNCT requires the following criteria:

(1) The new agents must exhibit very low toxicity *in vitro* and *in vivo*. If this condition is met, adverse effects that result from non-selective uptake of these agents can be minimized; (2) Since the primary emphasis of our work is on the treatment of gliomas, the materials in question must be able to effectively permeate the blood-brain-barrier (BBB) even if disruption of the BBB by the tumor occurs. Thus, our efforts have focused on modifications that enhance the lipophilicity of nucleosides; (3) In order to minimize the amount of time required for an effective dose of the drug to infuse into the tumor, high boron content carboranyl nucleosides were chosen; (4) In addition to providing a high boron content, substitution of carboranyl groups on nucleosides causes a significant increase in their lipophilicity; (5) In order to maximize the destructive power of these agents in tumor tissue during the neutron capture reaction, we have also sought boron-containing nucleosides that will be transported into the nuclei of the tumor cells. This requires that the materials be substrates for both cellular kinases and, after triphosphate formation, DNA polymerases; and (6) In addition to the anabolic requirements stated above (*i.e.*, cell permeation, phosphorylation and incorporation into the cellular DNA), it is also desirable that the agent in question be resistant to catabolism. In order to accomplish these goals, nucleosides that are known to be phosphorylated intracellularly at sites that tolerate moderately sized substituents (*e.g.*, the 5-position of pyrimidine nucleosides) were modified.

RESULTS AND DISCUSSION

A. Carboranyl-Containing 3'-Thia-Nucleoside Analogues. We have recently demonstrated the viability of 3'-heteranucleosides, such as racemic 2',3'-dideoxy-3'-thiacytidine

Advances in Neutron Capture Therapy, Edited by
A.H Soloway *et al.*, Plenum Press, New York, 1993

BCH-189 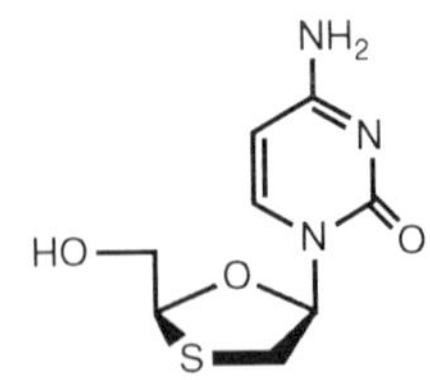

FTC 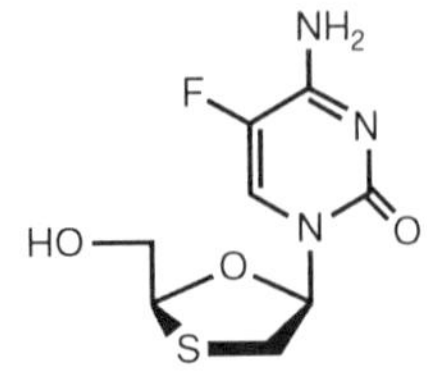

(BCH-189) and 2',3'-dideoxy-5-fluoro-3'-thiacytidine (FTC), as both anti-human immunodeficiency virus (HIV) and anti-hepatitis B virus (HBV) agents.[2-5]

These compounds not only show good activity against these viruses, but also exhibit extremely low toxicity. In addition, they readily permeate cell membranes, are excellent substrates for cellular kinases, and resist many other catabolic processes that are commonly observed with other 2'-deoxycytidine nucleosides.[5] As a consequence, we believe that these compounds represent a good starting point for further elaboration into potential sensitizing agents for BNCT. Therefore, the first targets we have pursued are **1**, **2**, **3** and **4**.

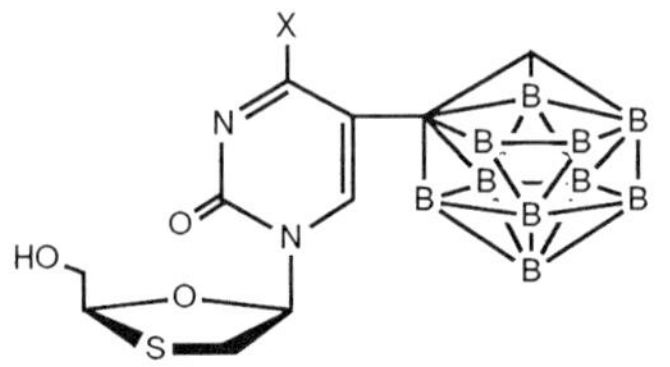

1: X = OH
2: X = NH_2

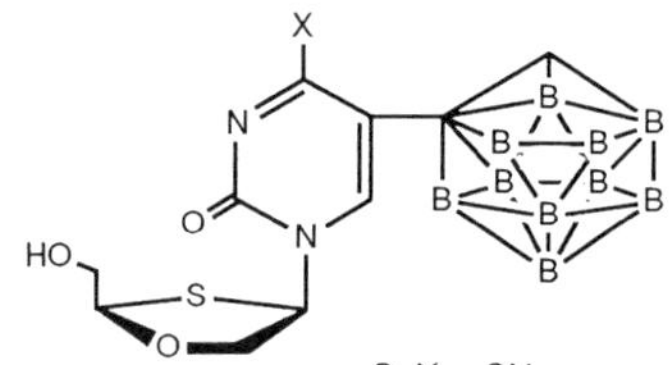

3: X = OH
4: X = NH_2

The key synthetic intermediate **6** is easily prepared in racemic form using the synthetic protocol recently developed by Liotta and co-workers.[2] The conversion of racemic **6** to the corresponding uracil carborane (*i. e.*, **1** and **3**) has already been accomplished using the general route shown below. Work on the cytidine derivatives is currently ongoing using these well-established synthetic protocols.

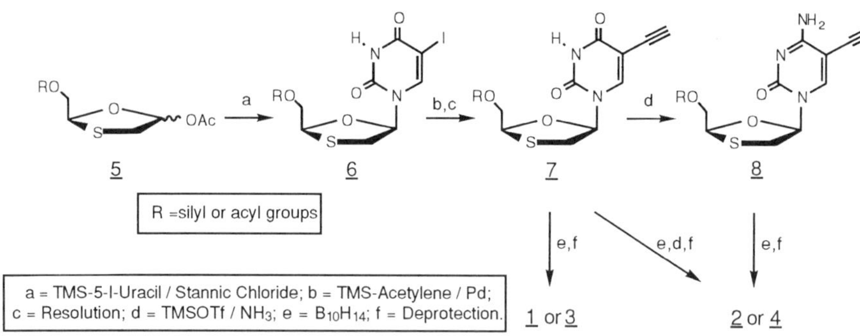

Recently, we have found that racemates of virtually all of the 3'-thia- and 3'-oxanucleosides can be resolved using lipase-mediated hydrolyses of appropriate 5'-acyl derivatives.[4] The enantiomers of cytidine analogues can also be resolved by cytidine/2'-deoxycytidine deaminase since we have found that only the (+)-enantiomer are substrates for this enzyme. Having easy access to both enantiomers of these materials could prove to be quite important. Specifically, we attribute the extremely low toxicity of (-)-BCH-189 (3TC) and (-)-FTC to the fact that these compounds are "unnatural" L-enantiomers. Thus, these materials are apparently accepted as substrates by cellular kinases, but are not recognized by most other key cellular enzymes and, therefore, may not interfere with normal cell functions. The use of "unnatural" enantiomers could represent an important new approach for designing a variety of chemotherapeutic agents that exhibit low toxicity and that could be useful for treating gliomas and other tumors.

B. Synthesis of 5-Carboranyl-2'-Deoxyuridine (CDU). A major objective was to extend the work already performed with DBDU[1] by synthesizing compounds that have a high boron content and an intact 3'- and 5'-hydroxyl function. The rationale was that these compounds would be lipophylic, be readily phosphorylated by cellular kinases, and also incorporate into DNA as analogues of natural 2'-deoxypyrimidine nucleosides. We have successfully synthesized 5-carboranyl-2'-deoxyuridine (CDU) from 5-iodo-2'-deoxyuridine **1** and hydroxymethyl-CDU. These compounds had similar chemical characteristics to authentic samples obtained from Dr. Y. Yamamoto.[6]

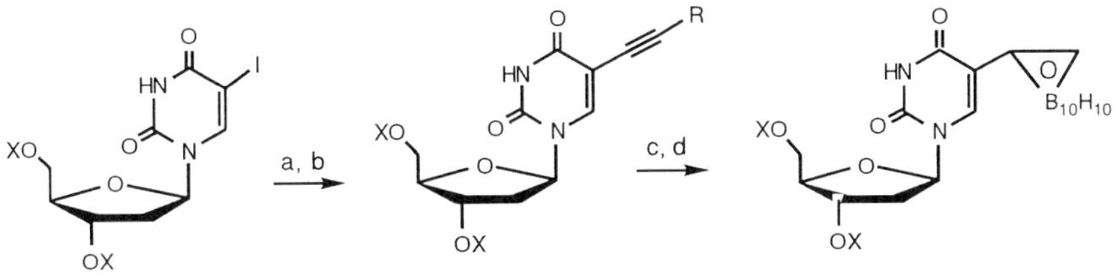

1: X = H
2: X = Bz

3: R = SiMe3, X = Bz
4: R = H, X = Bz

5: X = Bz
6: X = H

a = HC≡C-SiMe$_3$, (Ph$_3$P)$_2$PdCl$_2$/CuI, Et$_3$N, 50°C; b = Bu$_4$NF/THF;
c = $B_{10}H_{12}(CH_3CN)_2$, toluene,Δ; d = NaOMe/MeOH

C. Cellular uptake and phosphorylation of CDU. Studies were undertaken using cold CDU in order to follow the intracellular profiles of the parent drug and metabolites detected within CEM cell. CEM cells were suspended in RPMI 1640 medium containing 10% fetal calf serum and antibiotics (1.8 x 10^6 cells/ml, 10 ml per time point) and incubated with addition of 100 μM CDU. Cells were exposed to the drug for 12 h. At that time point the medium was removed and the cells were washed 2 times with cold Hank's balanced salt solution. Extraction was performed with addition of 0.2 ml of 80% cold acetonitrile/water and storage overnight at -20°C. The following morning, the suspensions were centrifuged and extractions were repeated 2 times for 0.5 hours at -20°C. The total supernatants (0.6 ml) were filtered through Acrodisc® LC PVDF 0.45 μm and then lyophilized to dryness. The residues were resuspended in 205 μl of water and aliquots of 88 μl were analyzed by HPLC. Quantitation of intracellular parent drug and metabolic derivatives are conducted by HPLC methods developed in our laboratories. Significant amounts of cold CDU were found intracellularly at 12 h after treatment which we estimate to be 137.5 ± 9.3 μM (average of 3 different determinations). *A new more polar metabolite than CDU was noted, which was not seen in the control untreated cell extract.* When this new metabolite was collected and pooled and then treated with affinity purified bovine kidney alkaline phosphatase (EC3.1.3.1) it disappeared and a new peak corresponding to CDU appeared (as determined by HPLC). Furthermore, snake venom phosphodiesterase I (EC3.1.4.1) had no effect on the new metabolite, even when the cell-treated extract was incubated for 4 h with this enzyme. Interestingly, the new metabolite was not found in the medium in cells that were treated with CDU, suggesting that it is formed intracellularly. Based on these experiments, we conclude that the new metabolite is probably the 5'-monophosphate of CDU. In addition, we have demonstrated that CDU and 5-carboranyl-2',3'-dideoxy-3'-thiauridine are not substrates for commercially available *E. Coli* thymidine phosphorylase (dThdPase).

D. Cytotoxicity and other biological activity. An indirect way to determine if the nucleosides synthesized by our group are phosphorylated is to determine if they exhibit even at high concentration some biological effect such as antiviral activity or cytotoxicity. CDU and related analogues and 5-carboranyl-2',3'-dideoxy-3'-thiauridine were evaluated in CEM, in human peripheral blood mononuclear (PBM) cells and in rapidly dividing Vero cells for cytotoxicity using a trypan blue exclusion method.[5] These compounds and related analogues were also evaluated against HIV-1 (strain LAV) in human lymphocytes (Table 1). It is of interest that the least toxic compound evaluated in all 3 cell systems was 5-hydroxymethylcarboranyluridine. These results suggest that other 5-hydroxymethylcarboranyl-derivatives such as 5-hydroxymethylcarboranyl-2'-deoxyuridine should be synthesized. Efforts towards this end are ongoing by synthesizing first the protected 5-propargyl-2'-deoxyuridine that can easily be converted to the desired compound. CDU and 5-carboranyluridine (CU) exhibited significant toxicity in Vero cells. 5-Carboranyl-2',3'-dideoxy-3'-thiauridine was a weak, but selective inhibitor of HIV-1 in human PBM cells. It is conceivable that the (-)-form of 5-carboranyl-2',3'-dideoxy-3'-thiauridine may be less toxic than the racemic compounds. Studies aimed at resolving the enantiomers are ongoing.

The finding that some of these novel nucleosides have biological activity strongly suggests that they are phosphorylated in various cells. The ability of these compounds to sensitize cells to neutrons is currently under investigation at Brookhaven National Laboratory.

Table 1. Biological activity of various boron-containing compounds in cell culture

Compound	CEM cells IC_{50}, μM On day 6	Vero cells IC_{50}, μM On day 3	PBM cells IC_{50}, μM On day 6	Anti-HIV-1[1] EC_{50}, μM On day 6
5-Carboranyl-2'-deoxyuridine (CDU)	70.9	17.1	> 100	73.2
5-Carboranyluridine (CU)	76.1	26.4	> 100	> 100
5-Hydroxymethylcarboranyluridine (HMCU)	> 100	> 100	> 100	> 100
5-Carboranyl-2',3'-dideoxy-3'-thiauridine	37.9	34.2	> 100	12.1
5'-O-Butyryl-5-carboranyl-3'-thia-2',3'-dideoxyuridine	18.8	4.6	> 100	6.4

[1] In PBM cells.

ACKNOWLEDGMENTS

This work is supported by the National Institutes of Health Grant RO1-CA-53892. We thank Dr. Yoshinori Yamamoto (Tohoku University, Sendai, Japan) for providing authentic samples of CDU, HMCDU, and CU. We also thank Ms. Annette Peck for excellent technical assistance.

REFERENCES

1. R.F. Schinazi and W.H. Prusoff, Synthesis of 5-dihydroxyboryl-2'-deoxyuridine and related boron-containing pyrimidines, *J Org Chem.* 50:841-847 (1985).
2. W.-B. Choi, L.J. Wilson, S. Yeola, D. Liotta, and R.F. Schinazi, *In situ* complexation directs the stereochemistry of N-glycosylation in the synthesis of thyalanyl and dioxolanyl nucleoside analogues, *J Amer Chem Soc.* 113:9377-9379 (1991).
3. S.-L. Doong, C.-H. Tsai, R.F. Schinazi, D.C. Liotta, and Y.-C. Cheng, Inhibition of the replication of hepatitis B virus *in vitro* by 2',3'-dideoxy-5-fluoro-3'-thiacytidine and related analogues, *Proc Natl Acad Sci.* 88:8495-8499 (1991).
4. L.K. Hoong, L.E. Strange, D.C. Liotta, G.W. Koszalka, C.L. Burns, and R.F. Schinazi, Enzyme-mediated enantioselective preparation of the antiviral agent 2',3'-dideoxy-5-fluoro-3'-thiacytidine [(-)-FTC] and related compounds, *J Org Chem.* 57:5563-5585 (1992).
5. R.F. Schinazi, A. McMillan, D. Cannon, R. Mathis, R.M. Lloyd, A. Peck, J.-P. Sommadossi, M. St. Clair, J. Wilson, P.A. Furman, G. Painter, W.-B. Choi, and D.C. Liotta, Selective inhibition of human immunodeficiency viruses by racemates and enantiomers of cis-5-fluoro-1-[2-(hydroxymethyl)-1,3-oxathiolane-5-yl]cytosine, *Antimicrob. Agents Chemother.* 36:2423-2431 (1992).
6. Y. Yamamoto, T. Seko, H. Nakamura, H. Nemeto, H. Hojo, N. Mukai, and Y. Hashimoto. Synthesis of carboranes containing nucleoside bases. Unexpected high cytostatic and cytocidal toxicity towards cancer cells. *J Chem Soc Commun.* 157-158 (1992).

SYNTHESIS OF CARBORANE-CONTAINING NUCLEOSIDES, NUCLEOTIDES, AND CARBOHYDRATES FOR BNCT

Werner Tjarks[1], Abul K. M. Anisuzzaman[1], Albert H. Soloway[1], Liang Liu[2], and Rolf F. Barth[2]

[1] College of Pharmacy and [2] Department of Pathology,
The Ohio State University, Columbus, Ohio 43210, U.S.A.

INTRODUCTION

The radiobiological effectiveness of the neutron capture reaction is 2-5 times greater when it occurs in the cell nucleus rather than in the cytoplasm.[1,2] Boronated nucleic acid precursors and components may be suitable agents to achieve high selective boron concentrations in the nucleus of tumor cells.

Boron-containing nucleosides are one type of compound in this category. They could be taken up more selectively by rapidly proliferating malignant cells than by mitotically less active normal cells, intracellularly entrapped because of phosphorylation, and, ideally, incorporated into cellular DNA or RNA.

A second class of compounds are the 5'-mono-, 5'-di-, and 5'-triphosphates of boron-containing nucleosides.The nucleotides are usually unable to penetrate cell membranes because they carry at least two negative charges under physiological conditions. In addition, they are rapidly cleaved *in vivo* to the parent nucleosides by serum phosphatases. Therefore, it seems that 5'-phosphates of boronated nucleosides are not suitable candidates for BNCT *per se*. However, a highly lipophilic boron moiety, such as the carborane cage, linked to the base or sugar part of a 5'-phosphate could increase the molecule's stability and may compensate for the ionic character of the nucleotide. An additional decrease in ionic charge and increase in lipid solubility could be achieved by administering such a carborane-containing nucleotide in the form of an appropriate phosphate diester . On entry into tumor cells, phosphodiesterase hydrolysis could produce the free carboranyl nucleotide. Boronated 5'-phosphates can also be used as reference compounds to quantify the intracellular phosphorylation of the parent nucleoside by modern HPLC procedures after incubating this nucleoside with tumor cells *in vitro*.

A third group of compounds are boronated "antisense-type" oligonucleotides. However, antisense oligonucleotide research is in a very preliminary state and many problems (e.g. *in vivo* tumor selectivity, *in vivo* stability, and large-scale production) have not been resolved as yet.[3] It remains to be determined whether or not antisense oligonucleotides can be useful agents in the treatment of cancer and AIDS.

RESULTS AND DISCUSSION

2'-*O*-(*o*-carboran-1-ylmethyl)uridine (9), the first carborane-containing nucleoside, was synthesized in our laboratories.[4,5] This compound was originally designed as an intermediate in the synthesis of oligonucleotides. Initial molecular modeling studies have shown that a carborane cage in 2'-position of one or more nucleoside units of an oligonucleotide would not significantly compromise its ability to hybridize with cellular RNA or DNA.[6] However, a nucleoside such as compound (9), possessing a bulky pendant group on the ribose moiety, should not show any tumor selectivity, since it is an unsuitable substrate for kinases. Detailed *in vitro* and *in vivo* studies with compound (9) confirmed this assumption (see "Cellular uptake, subcellular distribution and toxicity of carboranyl uridines" by L.Liu, R.F. Barth, W.Tjarks, A.H. Soloway, and David H. Ives in this proceedings).

The previously described synthesis of 2'-*O*-(*o*-carboran-1-ylmethyl)uridine[5] has two major disadvantages: 1) the methodology is limited to the preparation of certain carboranyl pyrimidine nucleosides but cannot be adapted to the synthesis of purine nucleosides or the direct preparation of pyrimidine nucleosides possessing free amino functions;[7] 2) the chromatographic separation of isomeric intermediates was very difficult and time-consuming, even for 2.5 g preparation of the target compound. The synthetic route for the synthesis of 2'-*O*-(*o*-carboran-1-ylmethyl)uridine (9), which is described below, does not involve extensive separation procedures and should be suitable for the synthesis of various carboranyl nucleosides. In addition, a generally applicable method for the synthesis of 5'-monophosphates of carborane-containing nucleosides is presented using disodium 2'-*O*-(*o*-carboran-1-ylmethyl)uridine-5'-phosphate (11) as an example.

Diacetone-D-allose[8] (1) was alkylated with propargyl bromide to yield compound (2), which was subsequently converted to the carboranyl derivative (3) as described previously for the synthesis of 3-*O*-(*o*-carboran-1-ylmethyl)-1,2:5,6-di-*O*-isopropylidene-D-glucose from diacetone-D-glucose.[5] The overall yield for both reactions was 64%. Removal of the 5,6-*O*-isopropylidene protective group of (3) was achieved by controlled acidic hydrolysis in H_2O/AcOH (3:2, v/v) for 7 h at 60 °C to afford compound (4) in 87% yield. Selective benzoylation of the hydroxy group in position 6 of allofuranose (4) was accomplished by the action of benzoyl chloride in dichloromethane at -25 °C in the presence of equimolar amounts of pyridine. The yield was 91%. At higher temperatures, increased formation of the 5,6-bis benzoylated was observed. The allose derivative (6) was obtained in 95% yield by refluxing compound (5) in H_2O/AcOH (3:2, v/v) for 1 h. Proton and carbon-13 NMR indicated ~80% β-allopyranose, ~10% α-allopyranose, and ~10% allofuranose configuration for compound (6). Periodate-oxidation ($NaIO_4$) in H_2O/EtOH (1:1, v/v) for 1 h at ambient temperature afforded 5-*O*-benzoyl-2-*O*-(*o*-carboran-1-ylmethyl)-3-*O*-formyl-D-ribofuranose, which is not shown in the scheme. The 3-formyl group appeared to be rather unstable even under described reaction conditions and was completely removed by stirring the crude product for 72 h in 0.1 M HCl/acetone (1:1, v/v) at room temperature. The resulting 5-*O*-benzoyl-2-*O*-(*o*-carboran-1-ylmethyl)-D-ribofuranose was acetylated without purification (acetic anhydride in pyridine at 25 °C for 12 h) to yield ribofuranose derivative (7). The overall yield for the three reactions was 82%. NMR data indicated 95% β-configuration for (7).

Condensation of compound (7) with trimethylsilyl protected uracil in the presence of trimethylsilyl trifluoromethane as the catalyst (acetonitrile, 4 h, room temperature) affords the uridine derivative (8) in 71% yield. This compound was deprotected in the same way as described previously for 3',5'-Di-*O*-acetyl-2'-*O*-(*o*-carboran-1-ylmethyl)-uridine[5] to obtain 2'-*O*-(*o*-carboran-1-ylmethyl)uridine (9) in 94% yield. Spectroscopic and chromatographic data indicated 100% β-configuration for the purified material. This result demonstrates a strong *trans*-directing influence of the carborane moiety in the 2'-position on substituents at the anomeric carbon. It can be expected that the condensation of other silylated pyrimidines or purines with ribofuranose (7) might also lead to a high β to α ratio for the resulting nucleosides.

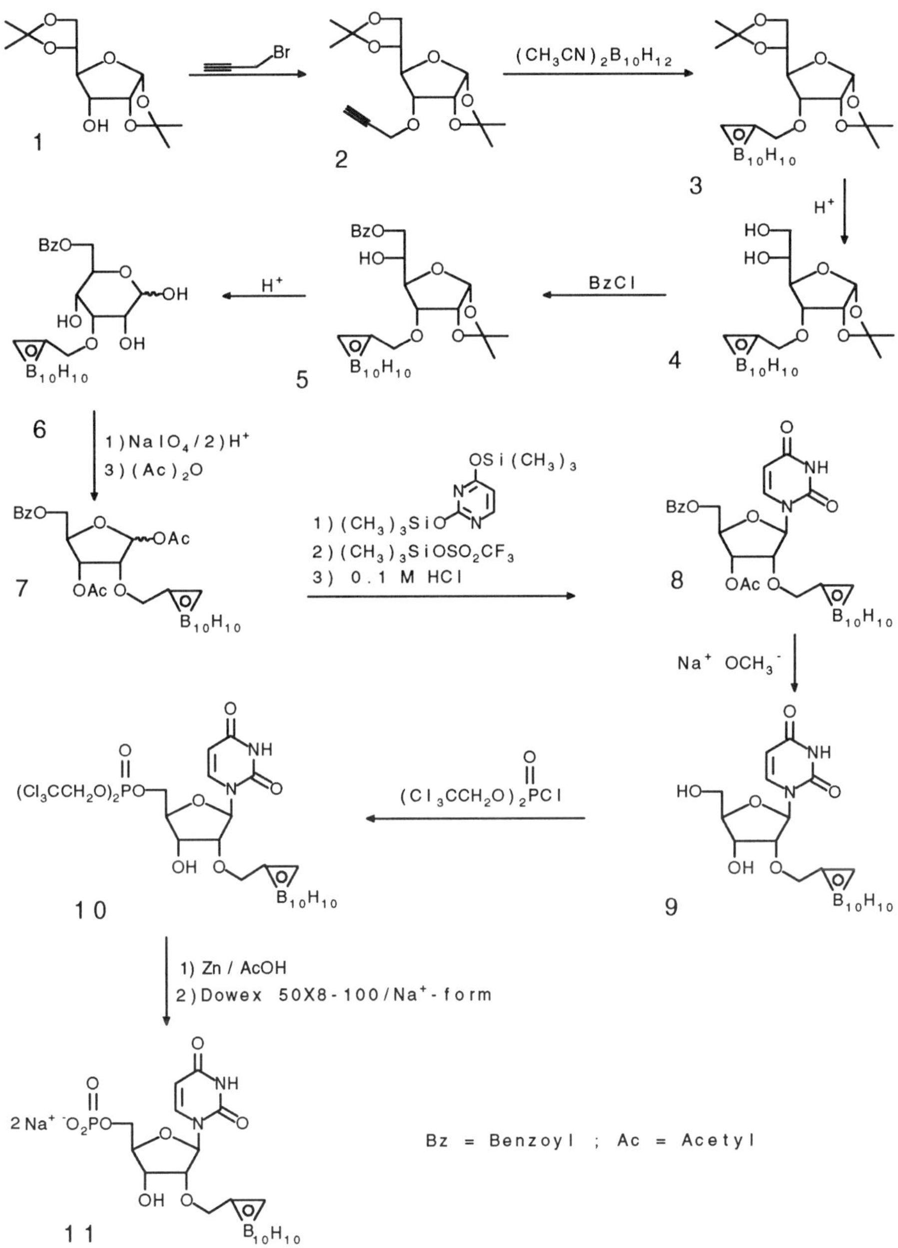

SCHEME

Compound (10) was obtained in 85% yield by reacting the carboranyl uridine (9) with bis(2,2,2-trichloroethyl)phosphorochloridate in acetonitrile in the presence of pyridine for 48 h at 4 °C. Reductive removal of the trichloroethyl groups of (10) was accomplished by stirring the material for 5 h at ambient temperature in H_2O/AcOH (1:9, v/v) with zinc powder (yield: 69%). The resulting zinc phosphate was converted to the disodium salt (11) using a strongly acidic ion-exchange resin and was purified by reversed phase flash chromatography (C_8).

CONCLUSIONS

The synthesis of 2'-*O*-(*o*-carboran-1-ylmethyl)uridine (9) in 39% overall yield from diacetone-D-allose[8] (1) is suitable for large-scale preparation. The described method is potentially applicable to the synthesis of other carborane-containing pyrimidine and purine nucleosides. The preparation of disodium 2'-*O*-(*o*-carboran-1-ylmethyl)uridine-5'-phosphate (11) from the carboranyl nucleoside is readily feasible and presents a general way for the synthesis of 5'-monophosphates of carborane-containing nucleosides. Compound (11) is the first described carborane-containing nucleotide. It remains to be determined whether or not this particular nucleotide possesses any tumor selectivity in contrast to its parent nucleoside.

ACKNOWLEDGEMENT

This research has been supported by the U.S. Department of Energy grant DE-FG02-90ER60972 and contract DE-AC02-76CH000616, the N.I.H. grant 1R01 CA53896, and a grant from The Ohio State University Comprehensive Cancer Center.

REFERENCES

1. T. Kobayashi and K. Kanda, Analytical calculations of boron-10 dosage in cell nucleus for neutron capture therapy, *Radiat. Res.* 91:77 (1982).
2. D. Gabel, S. Foster, and R.G. Fairchild, The monte-carlo simulation of the biological effect of the $^{10}B(n,\alpha)^{7}Li$ reaction in cells and tissues and its application for boron neutron capture therapy, *Radiat. Res.* 111:14 (1987).
3. V.V. Vlassov and L.A. Yakubov, Oligonucleotides in cells and in organisms: pharmacological considerations, *in*: " Prospects for Antisense Nucleic Acid Therapy of Cancer and AIDS ", E. Wickstrom, ed., Wiley-Liss, Inc., New York (1991).
4. A.K.M. Anisuzzaman, F. Alam, and A.H. Soloway, Synthesis of a carboranyl nucleoside for potential use in neutron capture therapy, *Polyhedron* 9:891 (1990).
5. W. Tjarks, A.K.M. Anisuzzaman, L. Liu, A.H. Soloway, R.F. Barth, D. J. Perkins, and D.M. Adams, Synthesis and *in vitro* evaluation of boronated uridine and glucose derivatives for boron neutron capture therapy, *J. Med. Chem.* 35:1628 (1992).
6. S. Black, personnel communication.
7. D. Wagner, J.P.H. Verheyden, and J.G. Moffat, Preparation and synthetic utility of some organotin derivatives of nucleosides, *J.Org. Chem.* 39:24 (1974).
8. D.C. Baker, D. Horton, and C.G.Tindal, D-Allose. From 1,2:5,6-di-*O*-isopropylidene-α-D-glucofuranose by way of 1,2:5,6-di-*O*-isopropylidene-α-D-ribo-hexofuranose-3-ulose hydrate, *Methods Carbohydr. Chem. 7:3* (1972).

NOVEL CARBORANYL DIOLS AND THEIR DERIVED PHOSPHATE ESTERS

Robert R. Kane, Christine S. Lee, Cindy L. Coe, Melissa A. St. Rose, Karin Drechsel†, and M. Frederick Hawthorne*

University of California at Los Angeles, Los Angeles, CA, 90024-1569

† On leave from Institut für Anorganische Chemie der Technischen Hochschule Aachen Templergraben 55, W-5100 Aachen, Germany

INTRODUCTION

Tumor–directed antibodies or their immunoreactive fragments are attractive candidates for the selective delivery of ^{10}B for BNCT, provided that ~1000 ^{10}B atoms can be attached to each immunoreactive protein without significantly altering its biological properties.[1] Previous studies have revealed problems associated with randomly conjugating whole monoclonal antibodies (MAbs) with large numbers of small boron-containing compounds[2] or with limited numbers of heterogeneous[3] or homogeneous boron-rich polymers.[4] Recently, we have described an approach to this problem based on the synthesis of a homogeneous boron-rich 'trailer' compound and it's conjugation to a specific site of a tumor-directed antibody fragment (Fab-SH).[5] The success of this approach rests upon the ability to precisely synthesize a hydrophilic "trailer" molecule containing ~1000 ^{10}B atoms. An oligophosphate–based boronated "trailer" is an attractive target, as the requisite coupling chemistry is well developed (in the context of DNA synthesis[6]) and the polyanionic oligomers would be expected to be inherently hydrophilic. We report herein preliminary results concerning the solution-phase synthesis of hydrophilic boron-rich oligophosphates, novel compounds that may find utility as intermediates in the assembly of tumor-localizing boron-rich compounds .[7]

RESULTS & DISCUSSION

The use of derivatives of *o*-carborane[8] (**1**) is pervasive in BNCT research, as these relatively stable boron-rich compounds can be readily functionalized.[9] Additionally, lipophilic *closo* carborane derivatives can be converted under mild conditions to stable anionic *nido* carborane derivatives (**2**, Scheme 1) which exhibit enhanced hydrophilicity.[10] The oligophosphates described herein are derived from the *o*-carborane diol **3**, which is prepared by the condensation of dilithio-*o*-carborane with an excess of trimethylene oxide (90%).[11,12] Treatment of diol **3** with one equivalent of TBDMSOTf affords **4** (48%) after silica gel

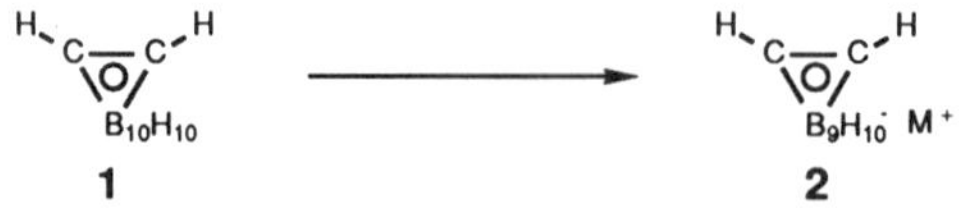

Scheme 1

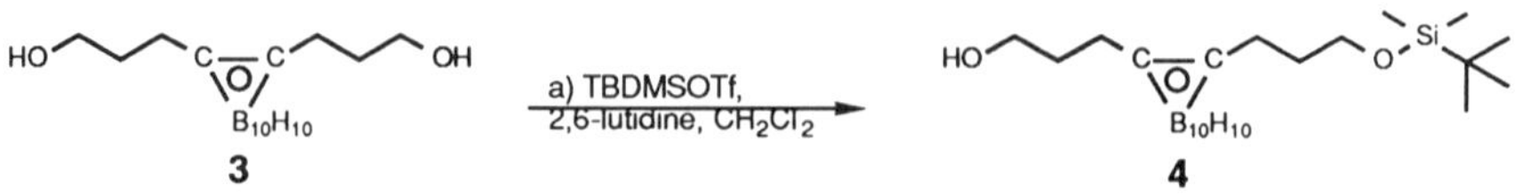

Scheme 2

chromatography of the reaction mixture (starting material and its diprotected analog are the two major contaminants). The coupling of monoprotected *o*-carboranyl diol **4** with isobutanol was then examined under a variety of conditions (Scheme 3). The results of these coupling reactions are summarized in Table 1. The simplicity, speed, economy and efficiency of the dichlorophosphite coupling reaction (Entry 4, Table I) prompted us to use this method in subsequent studies.

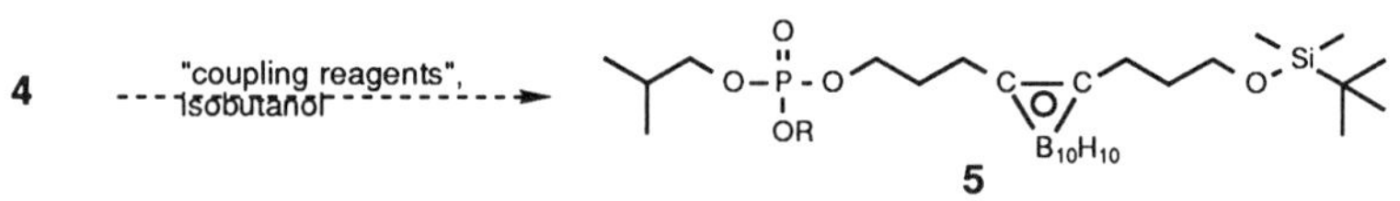

Scheme 3

Table 1. Yields for the synthesis of compound **5** under various conditions.

Entry	Coupling reagent	R	Yield
1	$Cl_2P(O)OR$	2-ClC_6H_4-	25%
2	$(BTO)_2P(O)OR^a$	"	47%
3	Cl_2POR^b	"	68%
4	Cl_2POR^c	"	88%
5	$ClP(OR)N(i\text{-}Pr)_2{}^{d,c}$	$NC(CH_2)_2$-	54%^{e}

Notes: (a) BT = benzotriazole; (b) The initially formed phosphite triester was oxidized *in situ* with aqueous iodine (0.1 M); (c) The initially formed phosphite triester was oxidized *in situ* with 0.1 M iodine in THF/H_2O/2,6-Lutidine (40/1/10); (d) The intermediate phosphoramidite was isolated in 90% yield, and was coupled with isobutanol in the presence of tetrazole; (e) Yield from two steps.

Accordingly, phosphotriester **5** was converted under acidolytic conditions to alcohol **6**.[13] Coupling of the monophosphate alcohol **6** with another portion of alcohol **4** under the 'optimized' coupling conditions afforded diphosphate **7** in moderate yield (35% from two steps, Scheme 4). A second iteration of deprotection and coupling provided triphosphate **8** in a low but reproducible yield (18% from two steps). Current work is focused upon improving yields, especially in the deprotection step, which has consistently been disappointingly low

(~60%). It is likely that our final strategy will employ a hydroxyl protecting group other than the hardy *t*-butyldimethylsilyl group.

At this point we decided to explore the removal of the phosphate protecting groups as well as the conversion of the *closo*-carboranes to the anionic *nido*-derivatives. Examining the

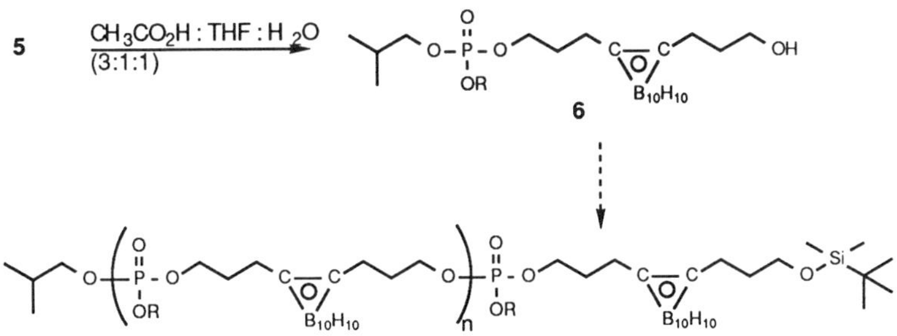

7 n=1, R=2-ClC$_6$H$_5$-
8 n=2, R=2-ClC$_6$H$_5$-

Scheme 4

reactions of monophosphate **5** in a model study, we found that the 2-chlorophenyl phosphate protecting group could be efficiently removed under standard conditions.[14] Accordingly, treatment of compound **5** with *syn*-pyridine-2-aldoxime and tetramethylguanidine in THF at room temperature followed by a cation exchange (Na^+ form cation exchange resin) afforded sodium salt **9** in 87% yield (Scheme 5). When this anion was suspended in neat pyrrolidine at room temperature (1 hr) the *closo*-carborane was quantitatively converted to the *nido*-cage (**10**, Scheme 5). Unfortunately, these conditions also led to substantial elimination, and alkene **11** was also isolated. Milder conditions for this important transformation are being explored. In any case, the isolation of the extremely hydrophilic phosphate **10** demonstrates the viability of the phosphate approach to the synthesis of boron-rich macromolecules.

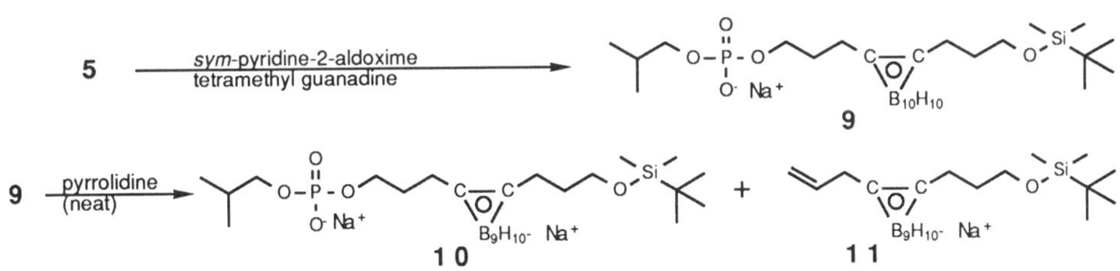

Scheme 5

CONCLUSIONS

In summary, we have demonstrated the viability of synthesizing short boron-rich oligophosphates in solution. However, the combination of steadily decreasing yields and increasingly difficult purifications has precluded the use of this methodology for the synthesis of lengthy oligomers. The solution-phase synthesis noted herein is useful for the synthesis of large quantities of short boron–rich oligophosphates that can be further derivatized, and as such describes a noteworthy advance in the synthesis of extremely hydrophilic boron-rich compounds for use in labeling biomolecules for boron neutron-capture therapy of cancer.

ACKNOWLEDGEMENT

Financial support by the National Institutes of Health (Grants CA31753-09 and CA53870-01) is gratefully acknowledged. R.R.K. thanks the National Institutes of Health for a Postdoctoral Fellowship (CA09306-01).

REFERENCES

1. It can be calculated that antibodies with this load of ^{10}B would deliver ~30 ppm ^{10}B in tumor.
2. Mizusawa, E.; Dahlman, H.L.; Bennett, S.J.; Goldenberg, D.M.; Hawthorne, M.F. (1982) *Proc, Nat. Acad. Sci. U.S.A.* **79**, 3011-3014. Goldenberg, D.M.; Sharkey, R.M.; Primius, F.J.; Mizusawa, E.; Hawthorne, M.F. (1984) *Proc, Nat. Acad. Sci. U.S.A.* **81**, 560-563. Mizusawa, E.; Thompson, M.R.; Hawthorne, M.F. (1985) *Inorg. Chem.* **24**, 1911–1916.
3. For example, see: Alam, F.; Soloway, A.H.; Barth, R.F.; Mafune, N.; Adams, D.M.; Knoth, W.H. (1989) *J. Med. Chem.* **32**, 2326–2330. Pettersson, M.L.; Courel, M.-N.; Girard, N.; Gabel, D.; Delpech, B. (1989) *Strahlenther Oncol.* **165(2/3)**, 151-152.
4. Varadarajan, A.; Hawthorne, M.F. (1981) *Bioconjugate Chem.* **2(4)**, 242-253. Paxton, R.J.; Beatty, B.G.; Varadarajan, A.; Hawthorne, M.F. (1992) *Bioconjugate Chem.* **3(3)**, 241-247.
5. Hawthorne, M.F. (1991) *Pure and Appl. Chem.* **24**, 327-334.
6. Gait, M.J. (ed.) *Oligonucleotide Synthesis: A Practical Approach*; IRL, Ltd.: Oxford, 1984.
7. Some of these results were presented at the 203rd National Meeting of the American Chemical Society (ORGN#242).
8. Throughout this paper *closo*-carborane, *o*-carborane, or carboranyl refer to derivatives of the *closo*-1,2-$C_2B_{10}H_{12}$ cage, while *nido*-carborane refers to derivatives of the [*nido*-7,8-$C_2B_9H_{11}]^-$ cage fragment.
9. For example, see Grimes, R.N. Carboranes, Academic Press, New York: 1970.
10. Hawthorne, M.F.; Wegner, P.A.; Stafford, R.C. (1965) *Inorg. Chem.* **4**, 1675. Wiesboeck, R.A.; Hawthorne, M.F. (1964) *J. Am. Chem. Soc.* **86**, 1643-1644.
11. Yields are reported for material homogeneous by NMR, TLC and/or HPLC. All new compounds reported in this communication have been appropriately characterized (HRMS, HPLC, multinuclear NMR, etc.).
12. Key NMR data for compound **3**: ^1H (200 MHz, CD_3OD) δ 3.56 (triplet, J=5.9 Hz, 4 H), 2.37 (aa'bb' pattern, 4 H), 1.83-1.69 (multiplet, 4 H); ^{13}C (50 MHz, CD_3OD) δ 81.35, 61.48, 33.62, 32.60; ^{11}B (160 MHz, CD_3OD) δ -4.44 (doublet, $J_{B\text{-}H}$=146.0, 2 B), -8.91(doublet, $J_{B\text{-}H}$=153.7, 2 B), -10.09(doublet, $J_{B\text{-}H}$=161.4, 6 B).
13. A 3:1:1 mixture of acetic acid : tetrahydrofuran : water was added to the silyl-protected alcohol and the mixture was stirred at room temperature until all the starting material was dissolved (~4 hours). Corey, E.J. (1972) *J. Am. Chem. Soc.* **94**, 6190-6192. Attempted deprotection of **4** with TBAF resulted in the formation of an intractable mixture of products.
14. Reese, C.B.; Titmas, R.C.; Yau, L. (1978) *Tetrahedron Lett.* **(30)**, 2727-2730.

ETHER LIPIDS AS POTENTIAL BORON CARRIERS FOR BORON NEUTRON CAPTURE THERAPY: SYNTHESIS OF *rac*-1-(9-*o*-CARBORANYL)NONYL-2-METHYL-GLYCERO-3-PHOSPHOCHOLINE (B-Et-11-OMe)

Peter Lemmen, Birgit Werner, and Benedikt Streicher

Organisch-Chemisches Institut der Technischen Universität München
Lichtenbergstr. 4, D-8046 Garching / Germany

INTRODUCTION

Prerequisite for the fascinating concept of boron neutron capture therapy (BNCT) is the availability of boron compounds which accumulate selectively and with a high ratio tumor/healthy tissue. A number of concepts towards the synthesis of such nontoxic boron compounds are followed at present in this context. They include boron loaded antibodies[1], thioureas[2], porphyrins[3], and low density lipoproteins[4].

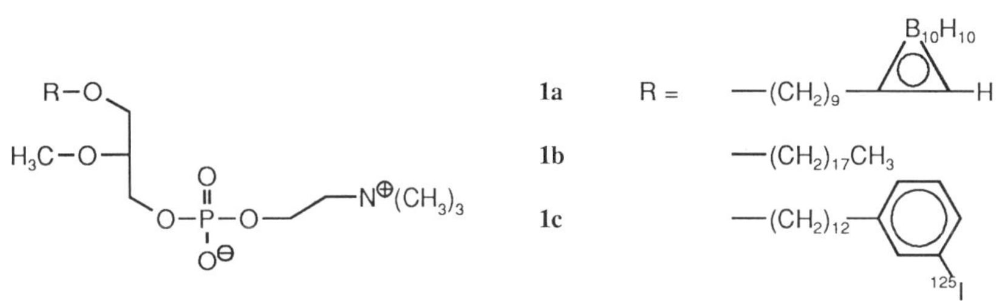

For various reasons, boron containing ether lipids[5] of type **1a** appear to be promising compounds for this purpose: Natural ether lipids and their artificial analog Et-18-OMe **1b** are accumulated selectively in various tumors[6,7]. This accumulation can be understood partly by the absence of the catabolic enzyme O-alkyl glycerol mono oxidase (AGMO)[8] in a great number of natural and experimental tumors[9]. The artificial ether lipid **1b** is not a substrate for AGMO itself. However, it seems to be cleaved by AGMO after a preceeding dephosphorylation by phospholipases[10]. By this way cells can accumulate ether lipids to a significat extent: **1b** has been accumulated to 8 - 18% of the total phospholipid content of the plasma membrane and 3 - 7% of the nuclear membrane[11] in fibrosarcoma and carcinoma cells. As shown for **1b**, ether lipids can pass the blood brain barrier[12].

The radioiodinated analog **1c**, proposed for tumor imaging by γ-scintigraphy[13], is accumulated with a tumor/blood ratio of 12; 16% of the administered dose were found in

Advances in Neutron Capture Therapy, Edited by
R.F. Barth and A.H. Soloway, Plenum Press, New York, 1993

the tumor after 24h. If these properties also apply to boronated analogues of ether lipids like **1a**, these are very promising boron carriers for BNCT. We propose their utilisation in this context. In this communication we report the synthesis of *rac*-1-(9-*o*-carboranyl)nonyl-2-methyl-glycero-3-phosphocholine (B-Et-11-OMe) (**1a**) as a first representative.

The lability of the carborane moiety towards base necessitates a specially adapted strategy of their synthesis, standard procedures can not be used in this case. The formation of ether bonds under strongly basic Williamson conditions has to be conducted prior to the introduction of the carborane.

Introduction of the phosphocholine-moiety under standard conditions via the bromoethyl phosphate[14] or via the dioxaphospholane[15] would destroy the carborane moiety by amine mediated degradation. Phosphorylation by oxazaphospholane[16] circumvents drastically basic conditions. By this reagent, the essential atoms of the phosphocholine head group of phospholipids can be introduced simultaneously and under mild conditions.

SYNTHESIS

Base catylyzed ring opening of allyl-glycidyl ether **2** by benzyl alcohol gave the 1-allyl-3-benzyl diether of glycerol **3**. This was converted into it's 2-methyl-ether **4** by alkylation with dimethylsulfate in DMF solution at 100°C. Under these conditions the allyl group rearranged to give a 1-(1-propenyl)ether. This group could be cleaved by acid. Reaction with 10-undecin-1-ol-mesylate under phase transfer conditions yielded 1-undecinyl-2-methyl-3-benzyl-glycerol **6**.

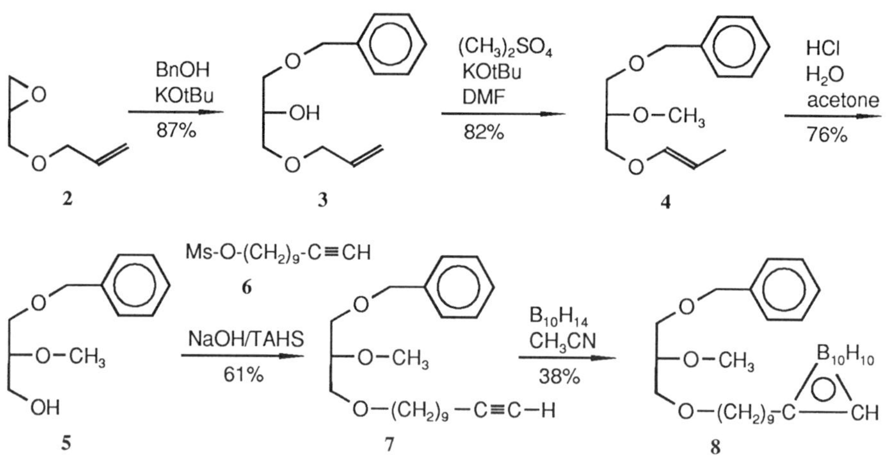

Reaction of **6** with the acetonitrile adduct of decaborane (CAUTION: toxic, volatile) in toluene[17] yielded *rac*-1-(9-*o*-carboranyl)nonyl-2-methyl-3-benzyl-glycerol **8** in 38% yield after chromatography.

After hydrogenolytic cleavage of the benzyl protective group, the resulting **9** was subjected to phosphorylation: **9** was caused to react with 2-chloro-3-methyl-1,3,2-λ^3-oxazaphospholane (**10**)/ triethylamine at 0°C in dichloromethane for one hour. The carborane moiety is stable under these conditions, as was proven by ^{11}B-NMR-spectroscopy. An excess of **10** was removed by a brine wash. The λ^3-oxazaphospholane **11** was oxidized by tert.-butyl hydroperoxide to yield the λ^5-oxazaphospholane **12**. The latter was ringopened by hydrolysis under N-methylating conditions to yield **1a**. The final product was purified by chromatography, precipitated from ether and characterized by ^{1}H, ^{13}C, ^{31}P and ^{11}B-NMR-spectroscopy. According to ^{11}B-NMR, the *o*-carborane moiety is not changed.

Work is under way to extend this sequence to related boron containing ether lipids for BNCT.

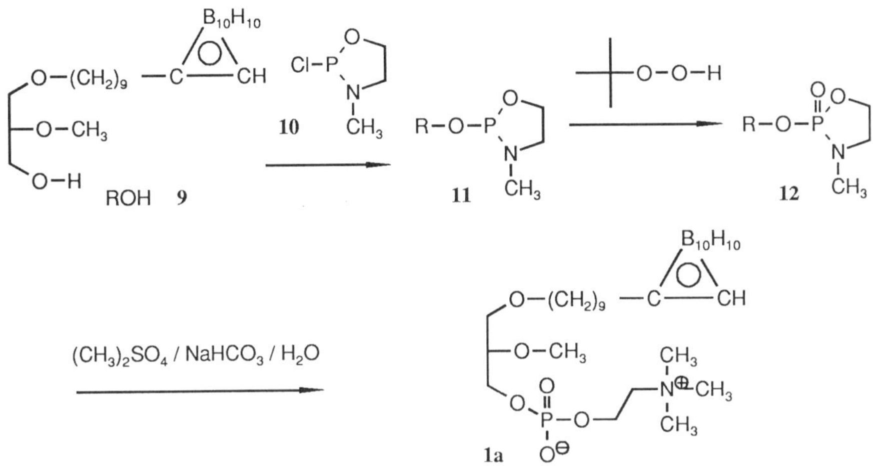

EXPERIMENTAL

1-Benzyloxy-2-methoxy-3-(9-*o*-carboranyl)nonyloxy-propane (8)

The reaction is carried out under a blanket of dry nitrogen.

Acetonitrile (2.6mL, 50mmol) is added to a solution of decaborane (3.1g, 25mmol) in toluene (25mL) and the mixture heated to 80°C for 1.5h. After cooling, a solution of **7** (5.3g, 15mmol) in toluene (25mL) is added carefully and the mixture warmed to 80°C. After 24h, ^{1}H-NMR-examination of a sample shows complete disappearance of the acetylenic H-atoms. To the cooled mixture are added carefully methanol (80mL), water (10mL) and conc. hydrochloric acid (10mL) and the mixture is warmed for another hour. After cooling, ether and water are added and the organic phase is washed with sodium bicarbonate and sodium chloride solutions. The raw product is distributed between hexane and methanol (4 stages). The hexane fraction is subjected to chromatography (silica gel, 10*7cm, eluent: hexane/ 5-20% ethyl acetate); pure fractions were pooled to yield **8** (2.65g, 38%). ^{11}B-NMR: δ= -2.37 (J_{BH}= 150 Hz, 1B, B-9 or B12), -5.81 (J_{BH}= 127 Hz, 1B, B-9 or B12), -9.29 (J_{BH}= 155 Hz, 2B), -11.36, -12.04, -12.88ppm (2B each).

MS (chemical ionisation) (exptl. intensity; calc. int.): 461 (3; 2), 463 (33; 26), 464 (63; 62), 465 (100; 100), 466 (92; 99), 467 (50; 53), 468 (10; 11), 469 (2; 2).

rac-1-(9-*o*-Carboranyl)nonyl-2-methyl-glycero-3-phosphocholine (B-Et-11-OMe) (1a)

Under nitrogen a solution of **9** (2.71g, 7.23mmol) and triethyl amine (3mL, 22mmol) in dichloromethane (30mL) is dropped at 0°C into a solution of 2-chloro-3-methyl-1,3,2-λ^3-oxazaphospholane (**10**) (2.0g, 14.5mmol) in dichloromethane (20mL). After one hour the reaction is stopped by extraction with brine. The residue is oxidized with tert.-butyl-

hydroperoxide (2.7mL, 3M in pentane) at -50°C (1h, then r.t.). (Intermediates **11** (δ= 137) and **12** (δ= 21 ppm) are monitored by their ^{31}P-NMR-spectra).
Dimethylsulfate (2.2mL, 11.8mmol) and sodium hydrogen carbonate (conc. solution, 3mL) are added. After stirring overnight, ammonia (5mL, conc. solution) is added and the product isolated by distribution in the system chloroform / methanol / water 1:1:1 and precipitation from acetone at -70°C. Chromatography (silica gel, 5*5.5cm, eluent: chloroform / methanol / water 250/250/25mL, later 250/250/50mL) yielded **1a** (2.12g, 54%). (DC: silica gel, eluent: chloroform/methanol/water 5/5/1, RF: 0.2). ^{31}P-NMR: δ= -0.71 ppm. ^{11}B-NMR: δ= -2.61, -5.97, -9.52, -11.67 ppm (shoulder to high field).
^{1}H-NMR: δ = 4.29 (m, O-$\mathbf{CH_2}$$CH_2$-N$^{\oplus}$), 3.95 (m, sn-3-**H**), 3.7 (br. s, $CB_{10}H_{10}\mathbf{CH}$), 3.55 (m, O-$CH_2\mathbf{CH_2}$-N$^{\oplus}$), 3.55 (m, sn-2-**H**), 3.5 (m, sn-1-**H**), 3.45 (s, O-$\mathbf{CH_3}$), 3.4 (m, O-$\mathbf{CH_2}$-$(CH_2)_8$), 3.20 (s, N$^{\oplus}$$(\mathbf{CH_3})_3$), 3 - 1 (br. m, $B_{10}\mathbf{H}_{10}$), 2.20 (AA'part of AA'XX'.., -CH_2-CH_2-$C_2B_{10}H_{11}$), 1.51 (m, O-CH_2-$\mathbf{CH_2}$-$(CH_2)_7$), 1.45 (m, $\mathbf{CH_2}$-CH_2-$C_2B_{10}H_{11}$), 1.26 ppm (m, $(\mathbf{CH_2})_5$). ^{13}C-NMR: δ = 79.66 (J_{CP} = 8.0 Hz, sn-2-**C**), 75.45 ($(CH_2)_9\mathbf{C}B_{10}H_{10}CH$), 71.68 (O-$\mathbf{C}H_2$-$(CH_2)_8$), 70.24 (sn-1-**C**), 66.21 (J_{CP} = 5.2 Hz, O-$\mathbf{C}H_2CH_2$-N$^{\oplus}$), 64.93 (J_{CP} = 5.2 Hz, sn-3-**C**), 61.13 ($(CH_2)_9CB_{10}H_{10}\mathbf{C}H$), 59.23 ($J_{CP}$ small, O-$CH_2\mathbf{C}H_2$-N$^{\oplus}$), 57.76 (O-$\mathbf{C}H_3$), 54.25 (N$^{\oplus}$$(\mathbf{C}H_3)_3$), 38.01 (-$CH_2$-$\mathbf{C}H_2$-$C_2B_{10}H_{11}$), 29.62, 29.28, 29.19, 29.02, 28.86 ($(\mathbf{C}H_2)_5$), 25.96 ppm (O-CH_2-$\mathbf{C}H_2$-$(CH_2)_7$). (Assignments are proven by 135°DEPT, ^{13}C^{1}H-correlated and ^{1}H^{1}H-COSY-experiments.)

ACKNOWLEDGEMENTS

The expert technical assistance of Mrs. M. Koenig is gratefully acknowledged. We are indebted to Deutsche Forschungsgemeinschaft and Fonds der Chemischen Industrie for financial support and to the CEC for support of the "European Collaboration on BNCT".

REFERENCES

1. A. Varadarajan, R.M. Sharkey, D.M. Goldenberg, and M.F. Hawthorne, *Bioconjugate Chem.* 2:102 (1991).
2. H. Ketz, W. Tjarks, and D. Gabel, *Tet. Lett.* 31:4003 (1990).
3. M. Miura, D. Gabel, G. Oenbrink, and R.G. Fairchild, *Tet. Lett.* 31:2247 (1990).
4. B.H. Laster, S.B. Kahl, E.A. Popenoe, D.W. Pate, and R.G. Fairchild, *Cancer Research* 51:4588 (1991).
5. P. Lemmen and B.Werner, *Chem. Phys. Lipids*, in the press.
6. F. Snyder and R. Wood, *Cancer Res.* 29:251 (1969).
7. B. Arnold, R. Reuter, and H.U. Weltzien, *Biochim. Biophys. Acta* 530:47 (1978).
8. A. Tietz, M. Lindberg, and E.P. Kennedy, *J. Biol. Chem.* 239:4081 (1964).
9. J.F. Soodsma, C. Piantadosi, and F. Snyder, *Cancer Res.* 30:309 (1970).
10. J. Kötting, C. Unger, and H. Eibl, *Lipids* 22:831 (1987).
11. W.J. van Blitterswijk, H. Hilkmann, and G.A. Storme, *Lipids* 22:820 (1987).
12. H. Eibl, *Angew. Chem.* 96:247 (1984) *Angew. Chem. Int. Ed. Engl.* 23:257 (1984).
13. R.E. Counsell, S.W. Schwendner, K.L. Meyer, T. Haradahira, and M.D. Gross, *J. Nuc. Med.* 31:332 (1990).
14. R. Berchtold, *Chem. Phys. Lip.* 30:389 (1982).
15. T.T. Nguyen and P. Chabrier, *Bull. Soc. Chim. Fr.* 667 (1974).
16. R. Stumpf and P. Lemmen, *Z. Naturforsch.* 45B:1729 (1990).
17. M.M. Fein, D. Grafstein, J.E. Paustian, J. Bobinski, B.M. Lichstein, N. Mayes, N.N. Schwartz, and M.S. Cohen, *Inorg. Chem.* 2:1115 (1963).

A BORONATED METALLOPORPHYRIN FOR USE IN NEUTRON CAPTURE THERAPY AND MAGNETIC RESONANCE CONTRAST ENHANCEMENT OF GLIOMA

Stephen B. Kahl,[1] Myoung-Seo Koo,[1] Robert M. Straubinger,[2] Leryang R. Huang,[2] Robert J. Fiel,[3] Richard Mazurchuk[3] and James J. Alletto[3]

[1]University of California, San Francisco
[2]State University of New York at Buffalo
[3]Roswell Park Cancer Institute

INTRODUCTION

Porphyrins are a unique class of metal chelating agents that have shown selective affinity for a variety of neoplasms. The highly water-soluble boronated porphyrin known as BOPP has previously been reported to show highly selective uptake and retention in a murine glioma model, and to produce tumor-boron levels well in excess of that needed for therapy.(1) We wish to report the synthesis and testing of a Mn(III) chelate of BOPP which has the potential to serve simultaneously as a tumor-selective ^{10}B delivery agent and as a proton MRI contrast enhancement agent for glioma. The compound is found to selectively localize in a rat 9L gliosarcoma in a fashion very similar to the free base BOPP and to preferentially enhance the tumor/normal brain contrast of T_1-weighted images for at least 92 hours. Based on the behavior of BOPP at the corresponding dose, it can be estimated that the tumor boron content of the MnBOPP-treated animals is in the 30 μg/g range at 24 hours.

MATERIALS AND METHODS

The preparation of MnBOPP was carried using the basic method of Loach and Calvin.(2) The BOPP diacid was stirred with manganese diacetate in glacial acetic acid until the free base porphyrin had been totally consumed as indicated by lack of any remaining red fluorescence. The acetic acid was removed *in vacuo* and the residue repeatedly partitioned between ether and water. Removal of the ether gave the metallated porphyrin diacid. Conversion to the water-soluble dipotassium salt was by ion exchange, and gave the product in essentially quatitative yield.

The rat brain tumor cell line 9L was obtained from Dr. D. Deen of the Brain Tumor Research Center of UCSF. Cells were maintained in Dulbecco-Modified Minimum Essential Medium (DMEM) containing 5% fetal bovine serum and 5% calf serum. Cultures were renewed from frozen stocks approximately monthly. Fischer 344 rats weighing ~300 gm were anesthetized and placed in a stereotoxic frame. Approximately 4 x 10^4 9L cells in 1.5 μl saline were injected into the caudate nucleus through a burr hole to a depth of 4.5 mm. Rats with 14-day old tumors were anesthetized and the heads positioned using a stereotactic head holder. Rats were imaged on the coronal plane before and immediately after the administration of MnBOPP. The metalloporphyrin was given as a single tail vein bolus at a dose of 30 mg/kg. Subsequent images were obtained at various times over a period of 1-4 days. Animals were anesthetized for acquisition of each image.

The magnetic resonance imaging was performed on a GE Sigma Advantage 1.5 Tesla whole body clinical imager with a custom-built transceiver rat head coil. All images were obtained using a T_1-weighted spin-echo sequence, with TE = 20 msec and TR = 400 msec, 3 mm slice thickness, an 8 cm field of view, and a 256 x 256 matrix.

The GE Signa imager is equipped with software which provides values of mean signal intensity (SI) and the standard deviation measured from any user-defined region of interest. Signal intensities of the tumor region and the corresponding normal brain region in the right hemisphere were measured on all images. The difference in signal intensity between the tumor and normal brain was expressed as $(SI_{tumor} - SI_{normal})/SI_{normal} \times 100$ and thus represents the percent contrast enhancement of the tumor.

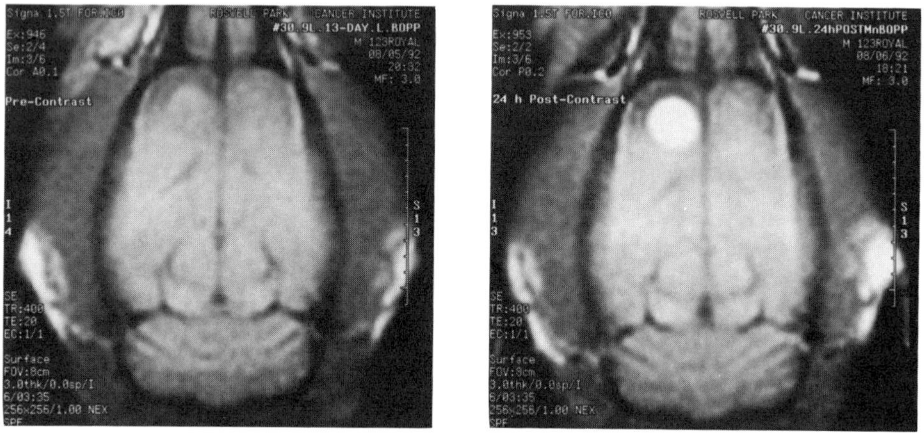

(a) (b)

Figure 1. Coronal spin-echo images of rat brain with 14-day old tumor in the frontal region of the left hemisphere. Pre-contrast image (a) and subsequent post-contrast image at 24 hours (b) after intravenous administration of MnBOPP.

RESULTS

Rats were imaged pre- and post-administration of MnBOPP. The glioma is apparent in the pre-contrast control image as a slightly hyper-intense area semi-circumscribed on the anterior by a hypo-intense crescent in the frontal region of the left hemisphere (Fig. 1a). The posterior margin of the tumor is nearly iso-intense with normal tissue. The immediate post-contrast image (not shown) taken at 3 minutes after injection shows the tumor as slightly enhanced, most notably along the midline, the medial margin of the tumor and extending anteriorly between the hypo-intense border of the tumor and posteriorly to the olfactory bulb. The image obtained at 24 hours (Fig. 1b) shows that MnBOPP is selectively retained in the tumor and that the tumor mass is homogeneously enhanced. The MnBOPP is able to delineate clearly an apparent boundary of the tumor and to enhance significantly the tumor/normal brain contrast. The maximal contrast enhancement of the tumor/normal brain by MnBOPP was observed at 24 hours post-injection reaching 62% at that time. At all times from 24 hours to 92 hours the degree of enhancement of tumor over normal brain was 53% or greater, indicating that MnBOPP is retained by the tumor for at least 4 days.

DISCUSSION

Many brain tumors appear to be iso-intense, or nearly so, with normal brain tissue in conventional T_1-weighted MR images. At the same time, the T_2-weighted images often fail to delineate tumor from hemorrhage, CSF, edema and necrosis. In our 9L model, the pre-contrast T_1-weighted images show the tumor as heterogeneous and poorly delineated. When hemorrhage is present, the pre-contrast control image may show a slight hyper-intense region in T_1-weighted images as in Figure 1a. However, the tumors remain relatively obscure in non-contrast enhanced images.

Several studies have demonstrated that water-soluble paramagnetic metalloporphyrins can be used as tumor-selective MR contrast enhancement agents. In particular, $MnTPPS_4$ has been shown to be an effective contrast enhancement agent for brain tumors,[(2)] as well as other solid tumors.[(3-6)] This compound shows a very different time-course of uptake and enhancement from MnBOPP with a maximal enhancement of 38% taking place at just five minutes after injection of a dose equivalent to 30 mg/kg of MnBOPP.[(7)] At 24 hours the tumor remains enhanced (22%), but to a much lesser degree than MnBOPP.

Results obtained by various investigators with $MnTPPS_4$ suggested that the manganese chelate BOPP might be an effective tumor-specific contrast enhancement agent since we had recently reported that BOPP itself was highly selective for C6 glioma cells in a murine model.[(1)] Moreover, tumor porphyrin levels equivalent to 65 ppm boron at non-toxic doses could readily be achieved with maximal tumor uptake at 24 hours. The Mn(III) derivative of BOPP was readily prepared in essentially quantitative yield from BOPP diacid. When administered to rats bearing the 9L gliosarcoma at a dose of 30 mg/kg, MnBOPP appears to follow a kinetic distribution pattern very similar to the free BOPP. As with BOPP, maximal tumor uptake occurs at 24 hours while blood level drops rapidly after an early peak. Very little, if any, compound is seen in normal brain. Compared with $MnTPPS_4$ at a similar dose, the degree of tumor enhancement with MnBOPP is greater (62% vs. 38%) and more prolonged (up to 96 hours). In spite of its ability to localize and enhance the MRI signal in tumors, the limitation of $MnTPPS_4$ appears to be its relatively poor ratio of effective to toxic dose. Although little is known about the toxicity of MnBOPP, the lack of toxicity of BOPP itself at reasonable doses suggests that MnBOPP will be similarly non-toxic.

The present work demonstrates the MR contrast enhancing capabilities of MnBOPP. Additionally it may serve to open the door to a dual use of the compound. The dose-response curve for BOPP in the C6 glioma model is approximately linear at 30 mg/kg[(1)] and may be extrapolated to a value of ~30 µg/g B at 24 hours post-injection. Since the time of maximum contrast enhancement of MnBOPP is comparable to the time of maximum tumor uptake of BOPP, we estimate that the tumor boron content of MnBOPP-treated rats at 24 hours is in the 25-30 ppm range needed for neutron capture therapy. MnBOPP is thus unique in that it represents the first example of a single agent that can enhance contrast between tumor and normal tissue and be potentially effective as an agent for boron neutron capture therapy.

REFERENCES

1. Hill, J.S., Kahl, S.B., Kaye, A.H., Stylli, S.S., Koo, M.-S., Gonzales, M.F., Vardaxis, N.J., and Johnson, C.I., *Proc. Natl. Acad. Sci. USA* **89**:1785-1789 (1992).
2. Bockhorst, K., Höhn-Berlage, M. Kocher, M., and Hossmann, K.-A., **8**:499-504 (1990).
3. Chen, C.-W., Cohen, J.S., Myers, C.E., and Sohn, M., *Fed. Eur. Bio. Soc.* **168**:70-74 (1984).
4. Fiel, R.J., Button, R.M., Gilani, S., Mark, E.H., Musser, D.A., Henkelman, R.M., Bronskill, M.J., and van Heteren, J.G., *Magn. Reson. Imaging* **5**:149-156 (1987).
5. Ogan, M.D., Revel, D., and Brasch, R.C., *Invest. Radiol.* **22**: 822 (1987).
6. Fiel, R.J., Musser, D.A., Monk, E.H., Mazurchuk, R., and Alletto, J.J., *Magn. Reson. Imaging*, **8**:255-259 (1990).
7. Huang, L.R, Straubinger, R.M., Kahl, S.B., Koo, M.-S., Alletto, J.J., Mazurchuk, R., Chan, R.I., Thamer, S.L., and Fiel, R.J., *Magn. Reson. Imaging* (1992) (in press).

SYNTHESIS AND BIOLOGICAL PROPERTIES OF CARBORANYLAZIRIDINE

Yoshinori Yamamoto, Hiroyuki Nakamura, and Hisao Nemoto

Department of Chemistry
Faculty of Science
Tohoku University
Sendai 980, Japan

INTRODUCTION

Cellular and molecular studies have indicated that under aerobic conditions RSU-1069 and RSU 1164 alkylate DNA at the phosphate and purine bases via the aziridine group, a process that leads to DNA strand breakage.[1] Accordingly, it occurred to us that a carborane containing aziridine group may become a potentially useful boron carrier for boron neutron capture therapy.[2] We wish to report that the first synthesis of carboranylaziridine **1**, its selective uptake by B-16 Melanoma, and its selective cytotoxicity toward cancer cells; the last observation has not been made with the previously known ^{10}B carriers.

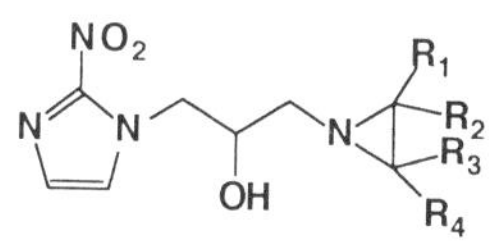

RSU 1069: $R_1=R_2=R_3=R_4=H$
RSU 1164: $R_1=CH_3$, $R_2=R_3=R_4=H$

SYNTHESIS OF CARBORANYLAZIRIDINE

1-Carboranyl-3-methylaziridino-2-propanol **1** (MACB) contains both a carborane framework unstable under basic conditions and an aziridine group labile under acidic conditions. In fact, the thermal reaction[3] of **2** with the methylaziridine **3a** resulted in the decomposition of the carborane framework. Next, we examined the low temperature reaction of **2** with several metallated aziridines (M=Li,[4] $AlEt_2$,[5] MgBr,[6] $PbBu_3$,[7] and CuLn[8]). Among these reagents, the lead and copper aziridines gave a promising result; **3b** and **3c** afforded **1** in ca.50% yield along with the recovered **2**. Although **3b** gave **1** in a good yield, there was a problem on purification of **1**. For a large scale preparation, we used the copper method. To a THF solution (3 mL) of 2-methylaziridine (0.07 mL, 10 mmol) was added nBuLi in hexane (0.8 mL × 1.6 M, 1.2 mmol) at -78°C under Ar. The mixture was stirred for 30 min. Copper cyanide (53.7 mg, 0.6 mmol) was added and the mixture was allowed to warm to -38°C. At this temperature, CuCN dissolved and the color of the solution turned clear dark yellow. The mixture was again cooled to -78°C. Epoxy-1-allylcarborane[9] (50 mg, 0.25 mmol) in THF (2 mL) was added, and the mixture was allowed to warm to r.t. over 1 hr. The reaction was quenched with a 1:1 mixture of aqueous saturated NH_4Cl and 28% NH_4OH solution (5 mL). The usual work-up and purification by silica gel column chromatography gave **1** in 50% yield : mp 103°C.

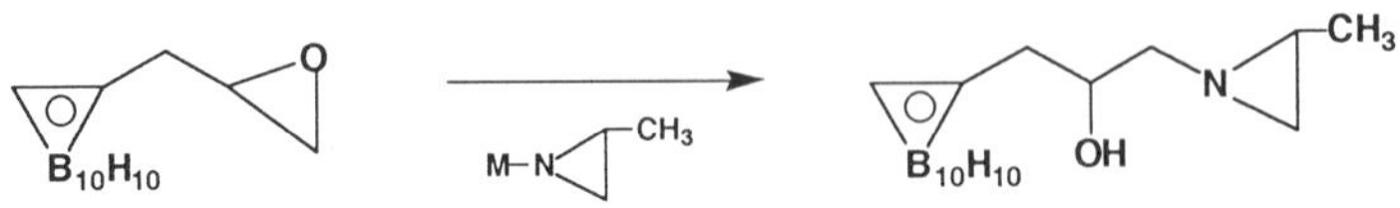

2

3 a; M=H
b; M=$PbBu_3$
c; $[Cu(CN)Li_2]_{1/2}$

1, MACB

CYTOTOXICITY TOWARD CANCER CELLS AND A NORMAL CELL (FIBROBLAST)

The growth inhibition of several human tumor cell lines and normal cell (TIG-1-20) with MACB and with BPA is summarized in Table 1. The cells were suspended in Eagle-MEM medium supplemented with 10% fetal calf serum, and cultured with different doses of each compound in Falcon 60 mm ϕ dishes in a CO_2-incubator at 37°C for 3 days. The cells were trypsinized and counted, and results are presented as the concentration of agent that resulted in 50% growth inhibition of untreated cultures (For kinetics of the growth inhibition, see supplementary material).

Table 1. Growth inhibition of human tumor and normal cell lines

Compound	Cell line	IC_{50}	
		ppm (10^{-6}g.dm^{-3})	M (mol dm^{-3})
MACB	B-16	1.40	5.45×10^{-6}
	HepG 2	2.13	8.28×10^{-6}
	TIG-1-20	4.70	1.83×10^{-5}
	IMR 32	8.20	3.19×10^{-5}
	1-87	15.5	6.02×10^{-5}
BPA	B-16	2100	8.55×10^{-3}
	TIG-1-20	2050	8.35×10^{-3}

MACB exhibited relatively high growth inhibition towards B-16 Melanoma and Hep G 2 (liver cancer cell), with IC_{50} values of the order of 10^{-6} M. Very interestingly, these values were significantly lower than those of TIG-1-20 (human fetal lung normal cell) (ca. the order of 10^{-5}M), indicating that MACB possesses selective cytotoxicity toward certain cancer cells. Such selective cytotoxicity was not observed with the boron carriers reported previously. For example, BPA (p-boronophenylalanine) which has been clinically used to treat a patient for skin cancer[10] exhibited essentially the same cytotoxicity toward B-16 and TIG-1-20 (Table 1). Accordingly, there is a possibility that MACB may serve as an anti-cancer agent. The cytotoxicity of MACB toward 1-87 (lung cancer) and IMR 32 (brain tumor) were lower than that toward TIG-1-20.

UPTAKE BY B-16 MELANOMA AND FIBROBLAST CELLS

Boron accumulation in B-16 and TIG-1-20 cells was measured by using the ICP (induced coupled plasma) method. The cells (4.5-5.0×10^6) were incubated for 1-24 h with Eagle's MEM containing MACB (1.5 ppm) or BPA•HCl (200 ppm). The concentrations of the carriers were adjusted to those of their IC_{50} values. At 1, 2, 3, 6, and 24 h, the cells were washed three times with PBS (-) (5 mL), and processed for boron measurement by ICP (Table 2 and supplementary material). Regardless of MACB or BPA, the boron incorporation into B-16 cells at 24 h increased slightly in comparison with that at 2 h. Very interestingly, the incorporation into TIG-1-20 after 24 h decreased by using MACB, although it increased with BPA. As shown in the last column of Table 2, the accumulation ratio of the tumor to normal cells was 1.32 with MACB and 0.83 with BPA. It is clear that the selective uptake of boron by the tumor cells has been realized with MACB.

Table 2. The comparison of boron incorporation into tumor and normal cells pre-incubated with MACB and BPA[a]

	B-16 (Tumor Cell)			TIG-1-20 (Normal Cell)			Accumulation
Compounds	2 h	24 h	Increment [b]	2 h	24 h	Increment [b]	ratio[c]
MACB	0.25	0.26	1.04	0.14	0.11	0.79	1.32
BPA	0.22	0.28	1.27	0.17	0.26	1.53	0.83

[a] Boron incorporated into the cells is shown in $\mu g/10^6$ cells. [b] The boron incorporation at 24 h is divided by that at 2 h. [c] The increment in B-16 is divided by that in TIG-1-20.

ACKNOWLEDGMENT

We acknowledge Professors Y. Fujii and K. Sogawa of our Department for valuable technical assistance to the cell experiment, and Professors Y. Mishima and M. Ichihashi of Kobe University for providing us with B-16 cells.

REFERENCES

1. A. R. J. Silver and P. O'Neill, *Biochem. Pharm.*, 35:1107 (1986). A. R. J. Silver, P. O'Neill, T. C. Jenkins, and S. S. McNeil, *Int. J. Rad. Oncol. Biol. Phys.*, 12:1203 (1986). G. E. Adams, I. Ahmed, P. W. Sheldon, and I. J. Stratford, *Brit. J. Cancer*, 49:571 (1984). G. E. Adams, I. Ahmed, P. W. Sheldon, and I. J. Stratford, *Int. J. Rad. Oncol. Biol. Phys.*, 10:1653 (1984). I. J. Stratford, P. O'Neill, P. W. Sheldon, A. R. J. Silver, J. M. Walling, and G. E. Adams, *Biochem. Pharm.*, 35:105 (1986).
2. (a) Y.Yamamoto, *Pure Appl. Chem.*, 63:423 (1991). Y. Yamamoto, T. Seko, H. Nakamura, H. Nemoto, H. Hojo, N. Mukai, and Y. Hashimoto, *J. Chem. Soc., Chem. Commun.*, 157 (1992). H. Nemoto, J. G. Wilson, H. Nakamura, and Y. Yamamoto, *J. Org. Chem.*, 57:435 (1992) and references cited therein. (b) A. K. M. Anisuzzaman, F. Alam, and A. H. Soloway, *Polyhedron*, 9:891 (1990). (c)

S. B. Kahl and M. S. Koo, *J. Chem. Soc., Chem. Commun.*, 1769 (1990). (d) J. L. Maurer, F. Berchier, A. J. Serino, C. B. Knobler, and M. F. Hawthorne, *J. Org. Chem.*, 55:838 (1990). J. L. Maurer, A. J. Serino, and M. F. Hawthorne, *Organometallics*, 7:2519 (1988). (e) R. F. Schinazi and W. H. Prusoff, *J. Org. Chem.*, 50:841 (1985). (f) M. Miura, D. Gabel, G. Oenbrick, and R. G. Fairchild, *Tetrahedron Lett.*, 31:2247 (1990).

3. P. O'Neill, T. C. Jenkins, I. J. Stratford, A. R. J. Silver, I. Ahmed, S. S. McNeil, E. M. Fielden, and G. E. Adams, *Anti-Canc. Drug Design.*, 1:271 (1987).
4. C. L. Kissel and B. Rickborn, *J. Org. Chem.*, 37:2060 (1972).
5. G. H. Posner, *Angew. Chem. Int. Ed. Engl.*, 17:487 (1978). L. E. Overman, L. A. Flippin, *Tetrahedron Lett.*, 22:195 (1981).
6. M. C. Carre, J.-P. Houmounou, and P. Caubere, *Tetrahedron Lett.*, 26:3107 (1985).
7. J. Yamada, M. Yumoto, and Y. Yamamoto, *Tetrahedron Lett.*, 30:4255 (1989).
8. Y.Yamamoto, N. Asao, and T. Uyehara, *J. Am. Chem. Soc.*, 114:5427 (1992). The use of $(R_2N)_2CuLi$ led to decomposition of the carborane framework. $(R_2N)Cu(CN)Li$ and $(R_2N)_2Cu(CN)Li_2$ (**3c**) gave **1** in good yields.
9. T. L. Heying, J. W. Ager, S. L. Clark, R. P. Alexander, S. Papetti, J. A. Reid, and S. I. Trotz, *Inorg. Chem.*, 2:1097 (1963).
10. Y.Mishima, C. Honda, M. Ichihashi, H. Obara, J. Hiratsuka, H. Fukuda, H. Karashima, T. Kobayashi, K. Kanda, and K. Yoshino, *The Lancet*, 11:388 (1989).

PREPARATION OF *META*-CARBORANYLPROPYL-2-NITROIMIDAZOLE DERIVATIVES FOR APPLICATION TO BNCT

D.S. Wilbur, D.K. Hamlin, J.C. Livesey, R.R. Srivastava, G.E. Laramore, and T.W. Griffin

Department of Radiation Oncology
University of Washington
Seattle, WA 98195

INTRODUCTION

Many types of tumor localizing boronated compounds have been studied for application to boron neutron capture therapy (BNCT) of cancer (1-4). However, compounds which are designed to localize in hypoxic regions of tumors (5) have not been studied. This is understandable as the focus of much of the previous research in this area has been directed towards use of thermal or epithermal neutron beams to treat the cancer, where the radiation dose from the heavy particles produced by neutron capture (He-4 & Li-7) must be lethal to all cancer cells for efficacious therapy. In contrast to this, our research focus has been to evaluate augmentation (boost) of the tumor dose obtained from a fast neutron beam (which also contains epithermal neutrons) when boron rich compounds are administered. Importantly, in an application that is directed at augmentation of fast neutron therapy (FNT) it is not necessary that all tumor cells contain boron-10 to obtain an efficacious therapy. Indeed, application of agents that are preferentially retained in hypoxic tissues in a combination therapy involving FNT and BNCT seems ideal as it targets high LET particles to a region of the cancer that is particularly resistant to radiation damage (6).

A large number of compounds have been studied (often as radiation sensitizers) which appear to be activated and retained in hypoxic tissue (7-9). The majority of these agents fall in a category termed bioreductive alkylation agents (10), where either a quinone (11) or aromatic nitro group (12) is enzymatically reduced in the cell to activate the compound for alkylation. Alkylation can selectively occur in hypoxic tissues as rapid reoxidation in normoxic tissues deactivates the alkylation process. Of the many possible compounds that might be studied as boron-10 carrier molecules, 2-nitroimidazole derivatives are attractive as many are tolerated in relatively large doses (13). This is important as the administered doses will have to be high to obtain concentrations necessary for neutron capture of boron-10.

Initial studies to evaluate hypoxia localization of boronated compounds have focused on the synthesis of carboranyl derivatives of 2-nitroimidazole that are similar in structure to misonidazole, **1**. Although there are issues of neurotoxicity with **1**, simple carboranyl derivatives of misonidazole were chosen for study since there is a large amount of information available on its *in vivo* characteristics, including distribution in animals, tumor uptake and retention, and concentrations achievable in patients. Additionally, the carboranyl derivatives that were chosen appeared to be reasonable initial compounds in a series of boronated 2-nitroimidazole derivatives under consideration, as their chemistry appeared to be relatively

Advances in Neutron Capture Therapy, Edited by
A.H Soloway *et al.*, Plenum Press, New York, 1993

straight forward. Thus, we have synthesized carboranyl-2-nitroimidazole derivatives **2** and **3**. The synthesis of the *ortho*-carborane derivative **2** has been previously presented (14). A preliminary report of the synthesis of the *meta*-carborane derivative **3** is presented here.

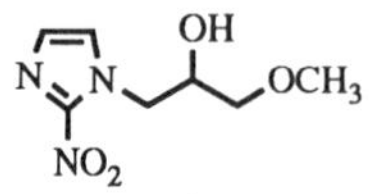

1

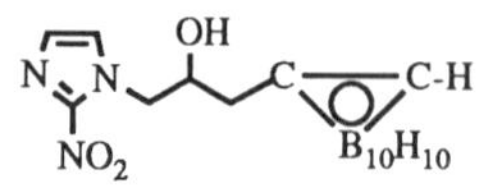

2

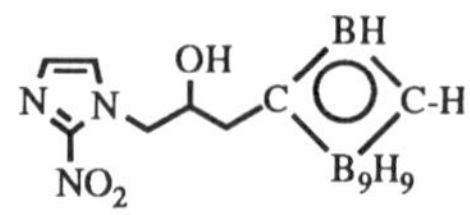

3

EXPERIMENTAL

General

^{1}H, ^{13}C, and ^{11}B NMR were obtained on either a Varian VXR-300 (300 MHz) or a Bruker AC-200 (200 MHz). IR data were obtained on a Perkin-Elmer 1420 IR spectrophotometer. Spectral data were consistent with the structures shown in Scheme I. HPLC-MS was conducted on a HP 5989 using negative chemical ionization.

HPLC analyses were conducted on a gradient system which consisted a Hewlett-Packard quaternary 1050 pump, multiple wavelength UV detector, and a H.P. computer workstation. HPLC evaluations were obtained on a C-18 column eluting with gradients composed of MeOH/1% HOAc in H_2O solution at 1 mL/min. Evaluation of *nido* carborane derivatives was conducted using a gradient containing MeOH / 5 mM tetrabutylammonium phosphate (TBAP) in H_2O.

meta-Carborane, **4a**, was obtained as a gift from Prof. M.F. Hawthorne (UCLA) or purchased from Consumer Health Research USA (Los Angeles). *meta*-Trimethylsilyl-carborane, **4b**, was prepared from *ortho*-trimethylsilylcarborane by thermal rearrangement (15). All compounds were purified by either kugelrohr sublimation or column chromatography prior to use. Purity of compounds was determined by HPLC analysis and spectral data. Except where noted, reported yields are those of isolated, purified products.

Syntheses of *meta*-Carboranylpropyl Epoxide, 5a

Initially, the synthesis of **5a** was accomplished by reaction of *meta*-carborane, **4a**, in benzene with *n*BuLi at ice/H_2O temperature, followed by addition of epichlorohydrin. This mixture was refluxed for 5 h and the reaction was stopped with addition of 10% aqueous $NaHCO_3$. A 60% crude yield was obtained. Purification of the crude product by silica gel column chromatography eluting with 20% EtOAc/hexanes containing 2% Et_3N gave 33% yield of pure **5a**.

Later syntheses of **5a** began with the synthesis of **5b**, which was conducted similarly to that described above. In the synthesis, compound **4b** was reacted in anyhd. ether with *n*BuLi at ice/H_20 temperature, followed by reaction of the lithio intermediate with epichlorohydrin at reflux for 5 h. A 79% crude yield of **5b** was obtained. This crude product (one compound by HPLC) was dissolved in anhyd. THF and was reacted with nBu_4NF at 0°C. The crude **5a** was isolated to yield a nearly quantitative conversion. Purification could be accomplished as described above.

Coupling Reactions of 5a with 2-Nitroimidazole to Prepare 3

Reaction of **5a** with 2-nitroimidazole was conducted at reflux in absolute EtOH containing K_2CO_3 for 24 h. The reaction mixture was then allowed to cool and the EtOH was removed under vacuum. The crude material was taken up in EtOAc and washed with $NaHCO_3$ solution. The EtOAc was dried and evaporated to yield a faint yellow solid.

Purification was accomplished by preparative TLC eluting with 100% EtOAc. When impure **5a** (three components by HPLC) was used, a crude yield of 49% was obtained, and a purified yield of 20% was obtained. When purified **5a** was used, a 62% yield was obtained of a product which contained only the desired **3** (one compound by HPLC). This crude product could be recrystallized for purification.

Attempted Synthesis of *nido* Carboranyl Derivative 6

A 10% solution of **3** in pyrrolidine was heated to 80°C. A vigorous bubbling occurred within 10 min and the reaction mixture turned very dark. Aliquots of the reaction mixture were placed on the HPLC for analysis. A transient peak was observed with a similar retention time to that of the *nido* derivative of the corresponding *ortho*-carborane derivatized 2-nitroimidazole (14), however, its appearance was accompanied by two other major peaks with many minor peaks also being present.

RESULTS AND DISCUSSIONS

The synthetic steps for preparation of *meta*-carboranylpropyl-2-nitroimidazole **3** are shown in Scheme I. The initial synthetic step to prepare **5a** was first accomplished by reaction of *meta*-carborane, **4a**, with *n*BuLi and epichlorohydrin in anhyd. THF. Difficulties in purifying **5a** led to the use of crude **5a** in the coupling reactions with 2-nitroimidazole. From this reaction **3** was obtained and purified by preparative silica TLC (100% EtOAc). High resolution mass spectral data confirmed the presence of **3** (mass calc. 314.2507; found 314.2508: 3 isotopic peaks were matched), however, NMR data could not be interpreted as

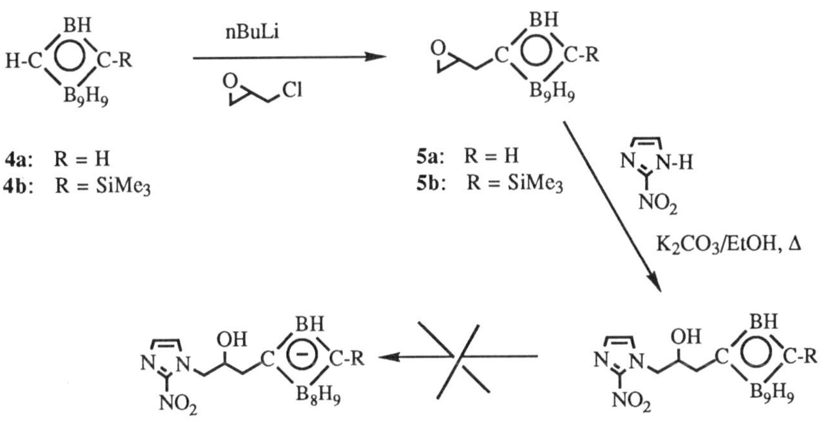

6 **3**

Scheme I: Synthesis of 1-(3'-*meta*-carboranyl-2'-hydroxy)propyl-2-nitroimidazole, **3**.

being exclusively compound **3**. Re-evaluation of HPLC conditions, which had previously indicated a single compound had been obtained, resulted in separation of (a closely eluting) second compound. HPLC-MS analysis of the second, minor peak indicated that it was similar to structure **3**, but had an additional 57 mass units. This additional mass could have been from either a propyl epoxide (C_3H_5O) or *n*-butyl (C_4H_9) group, and indeed, ^{1}H-NMR data indicated that both adducts were present in the crude reaction mixture. However, after purification by preparative TLC ^{1}H-NMR data suggested that the product from reaction with the di-propyl epoxide (minor peaks in 1.8-2.8 ppm region of spectrum) had been removed, leaving some unassigned peaks at 0.8-0.9 ppm and 1.2-1.3 ppm, which may be that of a butyl group. Similarly, previous studies with *ortho*-carborane had indicated that a phenyl group had been incorporated in the carborane moiety when phenyl lithium was used. Due to the formation of the (difficult to separate) impurity, a method for purification of epoxide **5a** was sought. Purification of **5a** was accomplished on silica gel eluting with EtOAc/Et_3N, however the overall yield of the reaction was only 33%.

An alternative synthetic approach was taken in the synthesis of **3** to improve the yield. Studies of the thermal rearrangement of *ortho*-carboranes had provided trimethylsilyl-*meta*-carborane **4b**, which appeared to be a useful intermediate for protecting one carborane cage carbon from alkylation (10). It was found that **4b** could be monoalkylated using *n*BuLi and epichlorohydrin in THF to give **5b**. Removal of trimethylsilyl protecting group was facile, providing **5a** in a high overall yield (75%). Coupling yields of **5b** after silica column purification were improved (61%) over those obtained with the crude epoxide (36-49%).

Water solubility of the carboranyl-2-nitroimidazoles **2** and **3** is very low. The *ortho*-carboranyl derivative **2** readily formed the anionic *nido* derivative at room temperature in pyrrolidine (16), which made it quite water soluble (14). It was expected that harsher conditions would be necessary for *nido* carborane formation of the *meta*-carborane derivative **3**. Thus, we studied the reaction of *meta*-carborane in pyrrolidine to evaluate the formation of *nido* derivative and found that it required elevated temperature (80°C) over a 35 min period. Unfortunately, no isolable *nido* derivative, **6**, was obtained when **3** was subjected to these conditions.

SUMMARY

The synthesis of a *meta*-carborane derivative of 2-nitroimidazole has been achieved. Alkylation of *meta*-carborane with epichlorohydrin, to make the requisite propyl epoxide intermediate, was complicated by the addition of a second propyl epoxide and (possibly) a butyl group. An alternate synthetic route using the trimethylsilyl substituted carbon derivative of *meta*-carborane gave the desired epichlorohydrin alkylation product in high yield, and after deprotection resulted in an improved yield of the desired product, **3**. An attempt to make the *meta*-carboranyl derivative **3** more water soluble by forming the anionic *nido* derivative **6** was unsuccessful. Continuing studies will be directed at obtaining new carboranyl derivatives of 2-nitroimidazoles which have water solubilizing groups attached.

ACKNOWLEDGMENT

We are grateful to Hewlett-Packard for use of their HPLC-MS in these studies, and to Steve Harnos of Hewlett-Packard Instruments for operation of the HPLC-MS. We also thank Prof. M.F. Hawthorne (UCLA) for his generous gift of *meta*-carborane which was used in these studies.

REFERENCES

1. Soloway, A.H. (1964) Boron Compounds in Cancer Therapy. In *Progress in Boron Chemistry*, eds., Steinberg, H and A.L. McCloskey, A.L. (The Macmillan Co, N.Y.) Chapt. 4, pp 203-234.
2. Fairchild, R.G. & Bond. V.P. (1984) New compounds for neutron capture therapy (NCT) and their significance. *Strahlentherapie* **160**, 764-770.
3. Barth, R.F., Soloway, A.H. & Fairchild, R.G. (1990) Boron Neutron Capture Therapy of Cancer. *Cancer Res.* **50**, 1061-1070.
4. Hawthorne, M.F. (1991) Biochemical applications of boron cluster chemistry. *Pure and Appl. Chem.* **63**, 327-334.

5. Coleman, C. N. (1988) Hypoxia in Tumors: A Paradigm for the Approach to Biochemical and Physiological Heterogeneity. *J. Natl. Cancer Inst.* **80**, 310-317.
6. Adams, G.E. (1981) Hypoxia-Mediated Drugs for Radiation and Chemotherapy. *Cancer* **48**, 696-707.
7. Kennedy, K.A. (1987) Hypoxic cells as specific drug targets for chemotherapy. *Anti-Cancer Drug Design* **2**,181-194.
8. Kirkpatrick, D.L. (1989) The Development of Hypoxic Tumor Cell Cytotoxic Agents. *Pharmac. Ther.* **40**, 383-399.
9. Kennedy, K.A., Teicher, B.A., Rockwell, S. & Sartorelli, A.C. (1990) The Hypoxic Tumor Cell: A Target for Selective Cancer Chemotherapy. *Biochem. Pharmacol.* **29**, 1-8.
10. Moore, H.W. (1977) Bioactivation as a Model for Drug Design Bioreductive Alkylation. *Science* **197**, 527-532.
11. Powis, G. (1989) Free Radical Formation by Antitumor Quinones. *Free Radical Biol. & Med.* **6**,63-101.
12. Brown, J.M. (1982) The Mechanisms of Cytotoxicity and Chemosensitization by Misonidazole and Other Nitroimidazoles. *Int. J. Radiation Oncology Biol. Phys.* **8**, 675-682.
13. Jenkins, T.C. (1990) Hypoxia-selective agents: radiosensitizers and cytotoxins, in *The Chemistry of Antitumor Agents*, ed., Wilman, D.E.V. (Chapman and Hall, New York) Chapt. 13, pp 342-369.
14. Wilbur, D.S., Hamlin, D.K., Livesey, J.C., Laramore, G.E., & Griffin, T.W (1991) Synthesis of Carborane Containing Nitroimidazole Compounds for Application to BNCT. Proceedings of the 4th International Symposium on *The Synthesis and Applications of Isotopes and Isotopically Labelled Compounds.* Toronto, Canada, Sept. 3-7.
15. Wilbur, D.S., Hamlin, D.K., Laramore, G.E. & Griffin, T.W. (1993) Preparation of *meta*-Carborane Derivatives by Thermal Rearrangement in a Small Scale Laboratory Apparatus. In *Proceeding of 5th International Symposium on Neutron Capture Therapy* (this volume).
16. Varadarajan, A. & Hawthorne, M.F. (1991) Novel Carboranyl Amino Acids and Peptides: Reagents for Antibody Modification and Subsequent Neutron-Capture Studies. *Bioconjugate Chem.* **2**, 242-253.

HYPOXIA-SELECTIVE CELLULAR TOXICITY OF CARBORANYL-NITROIMIDAZOLE

J. C. Livesey, L. W. Wiens, D. S. Wilbur, D. K. Hamlin, G. E. Laramore

Department of Radiation Oncology, RC-09
University of Washington Medical Center
Seattle, Washington 98195

INTRODUCTION

The binary nature of neutron capture therapy requires the coincident presence of neutrons of appropriate energy and an atomic nucleus of a sufficient capture cross-section. The successful application of boron neutron capture therapy will require novel delivery schemes which exploit characteristics unique to the tumor environment. Some of these tumor-selective characteristics currently under investigation for boron targeting (with their corresponding drug class exploiting this characteristic) include tumor specific antigens (monoclonal antibodies), hydrophobicity (porphyrins), a high growth rate compared to surrounding normal tissue (boronated amino acids and nucleotides), and leaky vascular networks (lipid vesicles) (Barth *et al.,* 1991).

Many solid tumors are thought to contain appreciable areas existing at low oxygen tensions; these hypoxic areas may contain radiation-resistant clonogenic cells capable of repopulating the tumor mass after incomplete tumor therapy (Gray *et al.,* 1953). Studies have been performed on the elimination of hypoxic regions in tumors during radiotherapy by such manipulations as hyperbaric oxygen(Henk, 1981), high LET particles (Parker, 1988; Suit *et al.,* 1990), and the pharmacologic modification of oxygen delivery (Coleman, 1988). The carriage of boron to tumors as 2-nitroimidazole derivatives is based on the presence of hypoxic regions. This approach may be considered an extension of a large effort to develop oxygen-mimetic radiosensitizing drugs. In our approach, the retention of the drug in hypoxic tissue, rather than directly increasing photon radiation damage, is used as the selective delivery characteristic which, when coupled with moderated fast neutron irradiation, provides augmentation to neutron therapy.

Our first efforts to prepare modified 2-nitroimidazole drugs carrying boron have been reported (Wilbur *et al.,* 1991). Initial biological characterization of this drug, 1-(3'-*ortho*-[*nido*]carboranyl-2'-hydroxy)propyl-2-nitroimidazole (CNI-1) is described here.

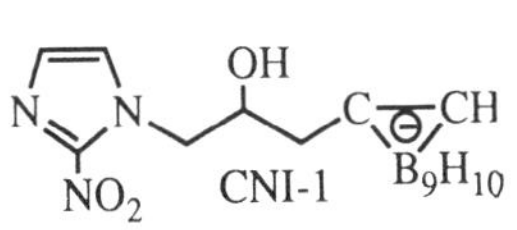

Figure 1. Structure of *nido*-3-Carboranyl-2-hydroxypropyl-2'-nitroimidazole (CNI-1).

Advances in Neutron Capture Therapy, Edited by
A.H Soloway *et al.*, Plenum Press, New York, 1993

Table 1. Octanol-Water Partition Coefficients of some 2-Nitroimidazole Drugs.

Drug	*P*	Reference
Misonidazole	0.43	Brown and Workman, 1980
Desmethyl-misonidazole	0.13	Coleman *et al.*, 1982
Etanidazole (SR-2508)	0.046	Brown *et al.*, 1981
Pimonidazole (Ro-03-8799)	8.5	Smithen *et al.*, 1980
CNI-1	0.05	This work

METHODS

The cesium salt of 1-(3'-*ortho*-[*nido*]carboranyl-2'-hydroxy)propyl-2-nitroimidazole (CNI-1) was synthesized as described (Wilbur *et al.*, 1991; Wilbur *et al.*, 1992). Spectroscopic data collected on this product were consistent with the proposed structure. Just prior to use, the sodium salt was prepared by passing a solution of the cesium salt over an ion-exchange column (Alltech Maxi-Clean) followed by elution with a H_2O/MeCN/MeOH (5:4:1) mixture.

The octanol-water partition coefficient was determined by the method of Purcell *et al.* (1973). For this determination, CNI-1 was measured by ultraviolet spectroscopy using a molar extinction coefficient of 6825 $mol^{-1}cm^{-1}$ at 322 nm.

The toxicity of drugs was tested in KHTn cells, a variant of the KHT tumor (Kallman *et al.*, 1967). These cells were grown in Sigma 1:1 medium (equal parts Ham's F12 with Dulbecco's MEM) supplemented with antibiotics and 10% (v/v) fetal calf serum. For hypoxia experiments [following the general method of Rasey *et al.* (1990)], cells were grown in the center of glass Petri dishes having notched edges to allow the free exchange of gases. After two days growth, the cells were treated with CNI-1 or misonidazole dissolved in medium and the dishes were placed in aluminum chambers from which the gas was repeatedly evacuated and refilled with gas of a defined oxygen tension. After five evacuation/refilling cycles, the chambers were sealed and incubated at 37°C for the required time. At the end of the incubation period, the chambers were opened and an included solution of sodium dithionite tested for the maintenance of hypoxic conditions. Cells were dissociated with trypsin, counted and diluted to achieve the required number of cells per plate. After 10 days growth, each fixed and stained colony of ≥ 50 cells was taken as evidence of one surviving cell in the drug treated cell suspension.

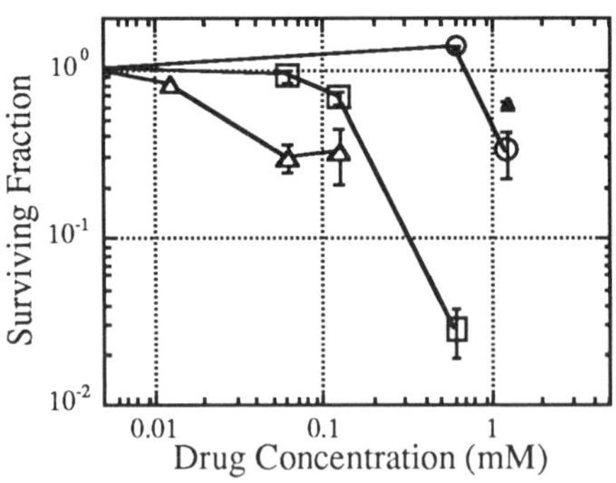

Figure 1. Toxicity of CNI-1 under hypoxic (open symbols) and oxic (filled symbol) conditions. CNI-1 was incubated with KNTn cells for one (○), four (□), or twelve (△) hours. Results for the corresponding oxic incubation condition (1 mM for 12 hours) is shown for comparison. Mean ± S.D. (N≥ 4 over 2 separate experiments).

RESULTS AND DISCUSSION

CNI-1 was found to have a octanol/water partition coefficient of 0.05, indicating a substantial partitioning into the aqueous phase with low lipid solubility for the *nido* anion. This value should be compared to other 2-nitroimidazoles previously studied as photon radiosensitizers (Table 1). Note the similarity with that of etanidazole, currently in clinical trials as a radiosensitizing drug showing lower toxicity than misonidazole (Coleman, 1990).

Initial estimates of the *in vitro* toxicity of CNI-1 in KHTn cells indicated that the compound is toxic only under hypoxic conditions. Figure 1 illustrates that when present at 1 mM concentration in culture medium for a four hour incubation, the plating efficiency of oxic KHTn cells was not substantially diminished. By comparison, cells incubated under hypoxic conditions for either one or four hours showed a dose-dependent reduction in plating efficiency. This reduction in plating efficiency under hypoxia was of a magnitude similar to that demonstrated by similar incubations of cells with misonidazole (Figure 2) Alterations in cell survival after hypoxic incubation with misonidazole are similar to literature reports of the effects of this drug in other cell lines (Hall and Roizin-Towle, 1974).

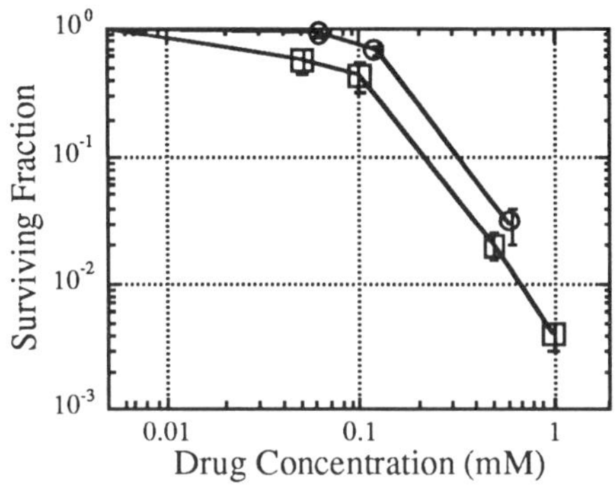

Figure 2. Comparison of the toxicity of CNI-1 (○) and misonidazole (□) incubated with KHTn cells for four hours under hypoxic conditions. Shown are the mean ± S.D. (N≥ 3).

SUMMARY

1-(3'-*ortho*-[*nido*]carboranyl-2'-hydroxy)propyl-2-nitroimidazole (CNI-1) has an octanol/water partition coefficient of 0.05. It exhibits hypoxia-selective toxicity *in vitro* in a rodent tumor model, with a magnitude similar to that of misonidazole. We interpret these data as indicating that the compound is metabolized selectively under hypoxia to species toxic to tumor cells in a manner similar to previously characterized radiosensitizing 2-nitroimidazole compounds. This data is encouraging for the development of compounds to deliver boron to hypoxic tumor cells.

ACKNOWLEDGEMENT

We appreciate the generous provision of the KHTn cell line by Drs. Janet Rasey and Norma Nelson which was used in these studies.

LITERATURE CITED

Barth, R.F., Soloway, A.H., Fairchild, R.G., 1991, Boron neutron capture therapy of cancer. *Cancer Res* 50: 1061-1070.

Brown, J.M. Workman, P., 1980, Partition coefficient as a guide to the development of radiosensitizers which are less toxic than misonidazole. *Radiat. Res.* 82: 171-190.

Brown, J.M., Yu, N.Y., Brown, D.M., Lee, W.W., 1981, SR-2508: A 2-nitroimidazole amide which shoudl be superior to misonidazole as a radiosensitizer for clinical use. *Int. J.of Radiat. Oncol, Biol. Phys.* 7: 695-703.

Coleman, C.N., 1988, Hypoxia in tumors: a paradigm for the approach to biochemical and physiologic heterogeneity. *J. Natl. Cancer Inst.* 80: 310-317.

Coleman, C.N., Wasserman, T.H., Phillips, T.L., Strong, J.M., Urtasun, R.C., Schwade, J.G., Johnson, R.J., Zagars, G., 1982, Initial pharmacology and toxicology of intravenous desmethylmisonidazole. *Int. J.Radiat. Oncol. Biol. Phys.* 8: 371-375.

Gray, L.H., Conger, A.D., Ebert, M., Hornsey, S., Scott, O.C.A., 1953, The concentration of oxygen dissolved in tissues at the time of irradiation as a factor in radiotherapy. *Br. J. Radiol.* 26: 638-648.

Hall, E.J. Roizin-Towle, L., 1974, Hypoxic sensitizers: radiobiological studies at the cellular level. *Radiol.* 117: 453-457.

Henk, J.M., 1981, Does hyperbaric oxygen have a future in radiation therapy? *Int. J. Radiat. Oncol. Biol. Phys.* 7: 1125-1128.

Kallman, R.F., Silini, G., Van Putten, L.M., 1967, Factors influencing the quantitative estimation of the in vivo survival of cells from solid tumors. *J. Natl. Cancer Inst.* 39: 539-549.

Parker, R.G., 1988, An appraisal of particle radiation therapy research. *Int. J. Radiat. Oncol. Biol. Phys.* 15: 1435-1439.

Purcell, W.P., Bass, G.E., J.M., C., 1973, Experimental determination of partition coefficients. *Strategy of drug design: A guide to biological activity*. New York, Wiley-Interscience. 126-143.

Rasey, J.S., Nelson, N.J., Chin, L., Evans, M.L., Grunbaum, Z., 1990, Characteristics of the binding of labeled fluoromisonidazole in cells in vitro. *Radiat. Res.* 122: 301-308.

Smithen, C.E., Clarke, E.D., Dale, J.A., Jacobs, R.S., Wardman, P., Watts, M.E., Woodcock, M., 1980, Novel, nitro-1-imidazolyl)-alkanolamines as potential radiosensitizers with improved therapeutic properties. *Radiation Sensitizers, Their Use in the Clinical Management of Cancer*. New York, Masson. 22-33.

Suit, H.D., Goitein, M., Munzenrider, J., Verhey, L., Urie, M., Gragoudas, E., Koehler, A., Gottschalk, B., Sisterson, J., Tatsuzaki, H., 1990, Increased efficacy of radiation therapy by use of proton beam. *Strahlenther. Onkol.* 166: 40-44.

Wilbur, D.S., Hamlin, D.K., Livesey, J.C., Srivastava, R.R., Laramore, G.E., Griffin, T.W., 1993, Preparation of *meta*-carboranylpropyl-2-nitroimidazole derivatives for application to BNCT. *Neutron Capture Therapy*. New York, Plenum, in press.

Wilbur, D.S., Hamlin, D.K., Livesey, J.C., Laramore, G.E., Griffin, T.W., 1991, Synthesis of carborane containing nitroimidazole compounds for application to BNCT. *4th International Symposium on the Synthesis and Applications of Isotopes and Isotopically Labelled Compounds*, Toronto, Ontario, Canada.

PREPARATION OF *META*-CARBORANE DERIVATIVES BY THERMAL REARRANGEMENT IN A SMALL SCALE LABORATORY APPARATUS.

D.S. Wilbur, D.K. Hamlin, G.E. Laramore, and T.W. Griffin

Dept. of Radiation Oncology
University of Washington
Seattle, WA 98195

INTRODUCTION

Carboranes, principally C_2-*ortho*-carboranes, **1**, have been of interest for modification of compounds being developed for boron neutron capture therapy (BNCT) of cancer (1-7). The other two possible regioisomers, *meta*-carborane, **2**, and *para*-carborane, **3**, which can be produced through thermal rearrangement reactions (8-10) of **1** have not been utilized in the studies, in part due to the perceived difficulties in their preparation. However, it should be noted that some significant differences in the chemical properties of the *meta* and *para* isomers exist, which make them attractive alternatives in synthesis of compounds for BNCT. Importantly, *ortho*-carboranes are reactive with amines, except under controlled conditions, resulting in the removal of one boron atom from the cage forming *nido*-carboranes. Contrary to this, the *meta*- and *para*- isomers do not react readily with amines under normal reaction conditions. Another difference in the regioisomeric cage structures are their lipophilicities, which increase (i.e. *para* > *meta* > *ortho*) as the dipolar nature of the ring decreases.

We have been interested in attaching large numbers (i.e. >100) of carborane cage moieties to polymeric materials for coupling to monoclonal antibodies or similar protein carrier molecules. We have been particlularly interested in polyamines as these compounds can be modified through reactions of activated groups such as carboxylate active esters, imidates, epoxides, and other functionalities which contain leaving groups (e.g. alkyl halides and tosylates). Due to the limitations of the use of *ortho*-carboranes in reactions involving amines, an investigation to develop a simple apparatus which could be used in the laboratory to produce *meta*-carboranes was conducted.

BACKGROUND

The $R_1R_2C_2B_{10}H_{10}$ carborane moiety can be obtained in three regioisomeric forms (with respect to carbon atoms) as shown below. The *ortho*-isomer, **1**, can be converted to the *meta*-isomer, **2**, through thermal rearrangement in an autoclave heated at 465-500°C under an inert atmosphere (argon) for 24 hours (8) or similarly at 500-510°C for 8 hours (9). *meta*-Carborane, **2**, can be converted (partially) to the *para*-isomer, **3**, by elevating the temperature to 615°C for 17 hours (10). Conversion of **1** to **2** and **3** via a flow-through tube heated to 550-600°C has also been described (11).

In general, with the exception of the methyl (R = CH_3) and phenyl (R = C_6H_5) derivatives, the rearrangement of derivatives of *ortho*-carboranes to produce derivatized *meta*-carboranes has not been practical due to the thermal decomposition of the appended group. However, it has been found that bulky R- groups such as $(CH_3)_3Si$-, $C_6H_5(CH_3)_2Si$-, $Cl(C_6H_5)_2Si$- and $CH_3(C_6H_5)_2Si$- when disubstituted could be rearranged at 260-300°C in a high temperature/pressure bomb without thermal decomposition (12). Unfortunately, the rearrangements take 2-10 days to complete, and there is no easy method of following the reaction progress.

Although *meta*-carborane can be purchased, rearrangement of *ortho*-carboranes was of interest for the following reasons: *1*) incorporation of boron-10 into the *meta*-carborane derivatives is important for further synthetic studies, and *2*) rearrangement of easily prepared mono-substituted *ortho*-derivatives could produce monosubstituted *meta*-derivatives for further modification.

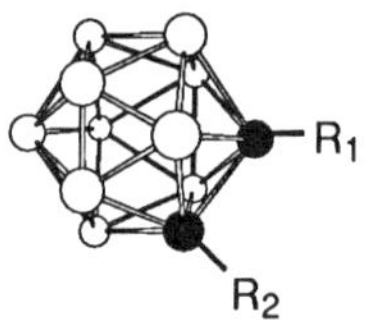

(*ortho*-) 1,2-carborane, **1**

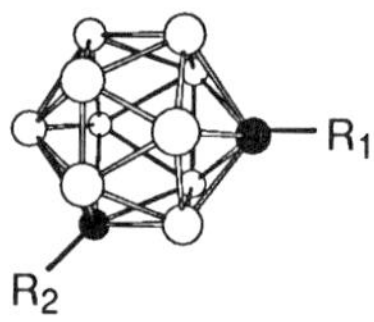

(*meta*-) 1,7-carborane, **2**

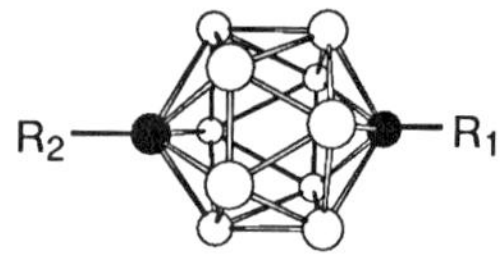

(*para*-) 1,12-carborane, **3**

EXPERIMENTAL

General

All reagents were used as obtained unless otherwise noted. ^1H, ^{13}C, and ^{11}B NMR were obtained on either a Varian VXR-300 (300 MHz) or a Bruker AC-200 (200 MHz) for all products. IR data were obtained on a Perkin-Elmer 1420 IR spectrophotometer. Spectral data were consistent with conversion of *ortho*-carborane to *meta*-carborane. Melting points were obtained on a Melt-Temp II and are uncorrected.

HPLC equipment consisted of a Hewlett-Packard 1050 isocratic pump, variable wavelength UV detector, H.P. 1047 RI detector, and a H.P. computer workstation. HPLC evaluations were obtained on a C-18 column eluting with 80% MeOH/20% of a 1% HOAc in H_2O solution at 1 mL/min; retention times for the *meta*-carboranes were roughly twice that of the starting *ortho*-carborane (note retention times in Table I). Thus, the results from the rearrangement were easily evaluated by HPLC analysis.

Carboranes Studied

1,2-Carborane, **1**, *1,7-Carborane*, **2**, used in the study were obtained from Aldrich Chemical Co. or from Consumer Health Research USA, and were used after sublimation. *1-Methyl-1,2-carborane*, **4**. was prepared by the method of Hawthorne et al. (13). *1-Trimethylsilyl-1,2-carborane*, **6** was prepared by a method similar to that previously reported (14). The crude **6** was purified by kugelrohr sublimation at 100°C / 0.4 torr to yield (83%) of a colorless solid, mp 97.5-98.5°C (sealed tube), reported mp 94-95°C. *1-tert-Butyldimethylsilyl-1,2-carborane*, **8** was prepared by the method described by Hawthorne et al. (15). The crude product was purified by kugelrohr distillation at 110°C / 0.4 torr to give 70% yield, mp 65-66°C. *1-Methyl-2-Trimethylsilyl-1,2-carborane*, **10** was prepared in a manner analogous to preparing **4**, using refluxing trimethylsilyl chloride to quench the Grignard reagent. The crude product was purified by silica column chromatography eluting with 90:10 hexane/EtOAc.

Thermal Rearrangement Apparatus

The apparatus consisted of a 2.8 cm x 46.4 cm quartz thermolysis tube packed with quartz helices, thermolysis oven (Thermolyne 21100 tube oven), borosilicate glass sample introduction chamber and sample boat, heat tape and temperature controller for heating sample chamber, condensation chamber made of a 500 mL flask packed with 5 mm glass beads, cooling bath, dry-ice condenser (filled with ice-H_2O), gas flow meter, mercury bubbler, and an oil bubbler. A schematic of the apparatus is shown in Figure 1.

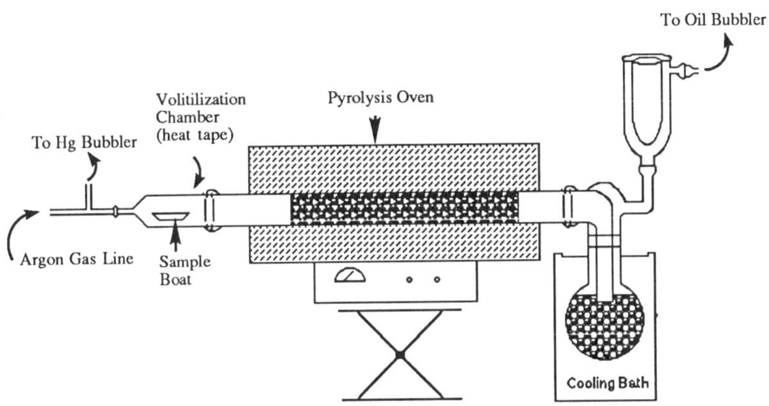

Figure 1: Apparatus for Thermal Rearrangement of Carboranes

Thermal Rearrangement Procedure

Column Preparation : The thermolysis apparatus was prepared for a rearrangement experiment by baking out the oven at 1200°C under a stream of argon for 1 h (total time allotted 3 h for heating and returning to a stable 550°C).

Rearrangement Experiment: In a typical experiment, the oven was heated to 550-560°C under a flow of argon, the condensation chamber was filled with 230 mL of CCl_4 and cooled with ice-water, the apparatus was charged with 1-2 g of *ortho*-carborane in a glass boat inserted in the sample chamber, and the sample chamber is then heated slowly to 150°C under an argon flow of 400 mL/min. The progress of the thermolysis was followed visually by disappearance of the *ortho*-carborane. For the less volatile carboranes **1** and **4**, a heat gun was employed to remove the *ortho*-carborane from cooler glass surfaces. The thermolysis oven was turned off and allowed to cool. The product was collected by rinsing the glassware/beads with $CHCl_3$ and evaporating the CCl_4 / $CHCl_3$ solution to dryness.

RESULTS AND DISCUSSION

A simple quartz glass thermolysis apparatus was constructed based on the studies of Papetti et al. (11). The temperature of the oven was maintained in all experiments at 550-560°C for comparison purposes, but other (lower?) temperatures may work on some derivatives. We found that our initial dry-ice/IPA bath did not condense all of the rearranged product effectively. Thus, the condensing chamber was altered and glass beads were placed in the flask to give more surface area. Even this design did not condense all of the product (which was a fine fog), so the trap was filled with CCl_4(cooled with ice/H_2O) and the hot gasses were bubbled through. In most experiments 1-2% of the unrearranged starting material was noted by HPLC analysis. Indeed, HPLC analysis was found to be a very quick method of determining the product purity as the *meta*-carboranes were much more lipophilic in their nature (see retention times reported in Table 1).

Table 1. Thermal Rearrangement Yields for Some *ortho*-Carboranes

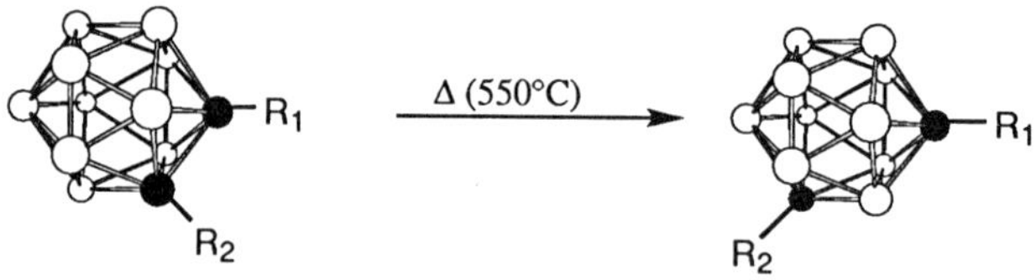

o-Carborane	R_1	R_2	R.T.(min)	*m*-Carborane	R.T.(min)	Yield
1	H	H	2.8	**2**	3.6	69%
4	Me	H	3.5	**5**	5.6	48-80%
6	Me_3Si	H	7.7	**7**	14.6	74-99%
8	tBu(Me_2)Si	H	19.3	**9**	38.6	30%*
10	Me	Me_3Si	11.8	**11**	26.6	64%

R.T. = HPLC retention times on a reverse phase C-18 column at 1 mL/min, eluting with 80% $MeOH/H_2O$;
* Non-purified yield; Product was 58% of collected rearranged product(s)

Rearrangement of unsubstituted *ortho*-carborane, **1**, in the pyrolysis apparatus was difficult (slow) due to its low volatility. The sublimation of *ortho*-methylcarborane was more facile, however it also was relatively slow (taking hours) to volatilize and pass through the oven. Thus, we prepared 1-trimethylsilyl-1,2-carborane, **6**, and 1-methyl-2-trimethylsilyl-1,2-carborane, **10**, and evaluated their conversion to *meta*-carboranes **7** and **11** in the apparatus. These compounds were more easily volatilized and gave good isolated yields, with >98% conversion of *ortho* to *meta* . Contrary to these results, rearrangement of the bulkier *tert*-butyldimethylsilyl carborane **8** resulted in a mixture of three products, with none of the unrearranged **8** being present. While there was approx. 58% of the desired product present (by HPLC) in the mixture, two other unidentified products were also present. Yields of *meta*-carboranes from rearrangement experiments are given in Table I.

SUMMARY

meta-Carborane derivatives are different from *ortho*-carborane derivatives as they are more stable towards attack by nucleophiles such as amines, and the chemical reactions at carbons alpha to the carborane cage are more characteristic of other organic reactions. These factors make the *meta*-carboranyl derivatives attractive intermediates for their use in preparing various boron-10 containing compounds for application to BNCT. Although *meta*-carborane is now available commercially, having a quick and efficient method of producing it, and some of its' functionalized derivatives, will allow application of ^{10}B containing *meta*-carboranes in a number of different studies.

The thermal rearrangement reactions are easy to conduct and result in moderate to high yields of nearly isomerically pure *meta*-carboranes. Additional studies need to be conducted to determine if the oven temperature can be decreased to improve the isolated yields of *ortho*-carboranes containing *tert*-butyldimethylsilyl and trimethylsilyl groups. A further modification of the condensing flask to allow introduction of additional CCl_4 is planned.

LITERATURE CITED

1. Fairchild, R.G. & Bond, V.P. (1984) New Compounds for Neutron Capture Therapy (NCT) and their Significance. *Strahlentherapie* **160**, 764-770.
2. Barth, R.F., Soloway, A.H. & Fairchild, R.G. (1990) Boron Neutron Capture Therapy of Cancer. *Cancer Res.* **50**, 1061-1070.
3. Wellmann, F., Abraham, R., Muller, R. & Gabel, D. (1991) Synthesis and Biological Behavior of a Boronated Analogue of the Antiestrogen U23,469-M. *Z. Naturforsch.* **46c**, 252-256.
4. Yamamoto, Y. (1991) Molecular design and synthesis of B-10 carriers for neutron capture therapy. *Pure & Appl. Chem.* **63**, 423-426.
5. Soloway, A.H., Anixuzzaman, A.K.M., Alam, F., Barth, R.F. & Liu, L. (1991) The development of carboranyl nucleic acid precursors for use in neutron therapy of tumors. *Pure & Appl. Chem.* **63**, 411-413.
6. Miura, M., Gabel, D., Oenbrink, G. & Fairchild, R.G. (1990) Preparation of Carboranyl Porphyrins for Boron Neutron Capture Therapy. *Tetrahedron Lett.* **31**, 2247-2250.
7. Kahl, S.B., Joel, D.D., Nawrocky, M.M., Micca, P.L., Tran, K.P., Finkel, G.C. & Slatkin, D.N. (1990) Uptake of a *nido*-carboranylporphyrin by human glioma xenografts in athymic nude mice and by syngeneic ovarian carcinomas in immunocompetent mice. *Proc. Natl. Acad. Sci. USA* **87**, 7265-7269.
8. Grafstein, D. & Dvorak, J. (1963) Neocarboranes, a new family of stable organoboranes isomeric with the carboranes. *Inorg. Chem.* **2**, 1128-1133.
9. Stanko V.I. & Klimova A.I. (1966) Neobarene Family. *J. Gen. Chem. USSR* **36**, 450-453.
10. Papetti, S. & Heying, T.L. (1964) *p*-Carborane [1,12-Dicarbaclovododecaborane (12)] *J. Am. Chem. Soc.* **86**, 2295.
11. Papetti, S., Obenland, C. & Heying, T.L. (1966) Vapor Phase Isomerization of *o*-Carborane. *Ind. Eng. Chem., Prod. Res. Dev.* **5**, 334-337.
12. Salinger, R.M. & Frye, C.L. (1965) Facile Polyhedral Rearrangement of Icosahedral Silylcarboranes. *Inorg. Chem.* **4**, 1815-1816.
13. Hawthorne, M.F., Andrews, T.D., Garrett, P.M., Olsen, F.P., Reintjes, M., Tebbe, F.N., Warren, L.F., Wegner, P.A. & Young, D.C. (1967) Icosahedral Carboranes and Intermediates Leading to the Preparation of Carbametallic Boron Hydride Derivatives. *Inorg. Synth.* **10**, 91-118.
14. Heying, T.L., Ager, J.W., Clark, S.L., Alexander, R.P., Papetti, S., Reid, J.A. & Trotz, S.I. (1963) A New Series of Organoboranes. III. Some Reactions of 1,2-Dicarbaclovododecaborane(12) and its Derivatives. *Inorg. Chem.* **2**, 1097-1105.
15. Gomez, F.A., Johnson, S.E. & Hawthorne, M.F. (1991) A Versatile Protecting Group for 1,2-Dicarba-*closo*-dodecaborane(12) and the Structure of a Silylcarborane Derivative. *J. Am. Chem. Soc.* **113**, 5915-5917.

CARBONYL NONAHYDRODECABORATE(1-): A VERSATILE SYNTHON FOR FUNCTIONALIZED BORANE ANIONS

Kenneth Shelly, M. Frederick Hawthorne, and Carolyn B. Knobler

Department of Chemistry and Biochemistry
University of California at Los Angeles
Los Angeles, CA 90024 USA

INTRODUCTION

Despite the surging interest in neutron capture therapy, the development and testing of new boron containing compounds has not received as much attention as other areas of BNCT. The most intensively studied boron agents are the mercaptoundecahydro-*closo*-dodecaborate dianion ($[B_{12}H_{11}SH]^{2-}$, "BSH") and 4-(dihydroxyboryl)phenylalanine (BPA), yet both of these species have been known for decades.[1,2] The development of new boron agents for BNCT requires convenient high-yield methods for the introduction of boron moieties into suitable chemical structures.

Common strategies employed in the synthesis of new materials for BNCT include the attachment of boron-containing residues to existing compounds known to accumulate in tumor tissues or polymeric carrier delivery agents. The most popular large boron-rich moiety used to accomplish this is the carboranyl group derived from *ortho*-carborane, $C_2B_{10}H_{12}$.[3] This molecule offers the advantages of relative stability, high boron content, and the versatility of its derivative chemistry. The use of the carboranyl moiety may be limited, however, due to its extreme hydrophobicity, although in many cases it may be transformed to the more hydrophillic $-C_2B_9H_{10}^-$ group.

The development of water-soluble boron carriers would be facilitated by a readily available boron-rich hydrophilic group capable of derivatization and conjugation. Polyhedral boranes such as the $[B_{10}H_{10}]^{2-}$ ion[4] and the $[B_{12}H_{12}]^{2-}$ ion[5] are good candidates because of their hydrolytic stability and high boron content, and they are easily synthesized from decaborane in good yield. Their rich derivative chemistry, however, is usually of limited utility because of low yields or the production of multiple products or isomers that may be difficult to separate.[1,6-8]

Carbon monoxide is one of the most versatile derivatives known in polyhedral borane chemistry, capable of a wide variety of functional group transformations.[8-10] Unfortunately the synthesis of the carbonyl boranes themselves is usually difficult.[8,11] The direct substitution with carbon monoxide on $[B_{10}H_{10}]^{2-}$ and $[B_{12}H_{12}]^{2-}$ is shown in Eqn. 1 and 2. These reactions require severe conditions (high temperature and pressure) and produce

Advances in Neutron Capture Therapy, Edited by
A.H Soloway *et al.*, Plenum Press, New York, 1993

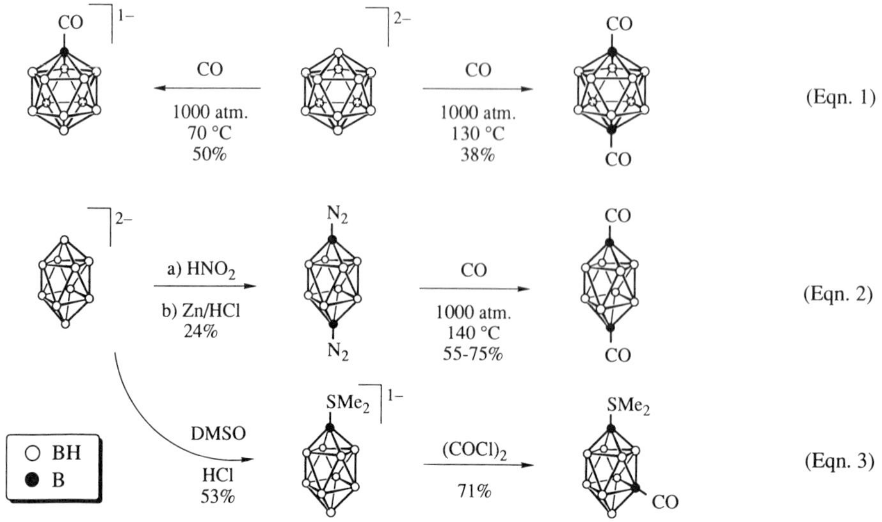

relatively low yields of product. A more convenient method using oxalyl chloride[12] (Eqn. 3) to produce monocarbonyl derivatives of ligand substituted $[B_{10}H_{10}]^{2-}$ ions proceeds in more reasonable yield; however the substituted $[B_{10}H_9L]^-$ precursors are available in only low to moderate yields.[6,7]

Our interest in the production of a variety of water-soluble boron species for encapsulation in liposomes[13,14] prompted us to reexamine the carbonylation of polyhedral borane anions. This has led to the development of a convenient method to produce monofunctionalized derivatives of $[B_{10}H_{10}]^{2-}$ in high yield.

DISCUSSION

The reaction of oxalyl chloride with $[B_{10}H_{10}]^{2-}$ proceeds rapidly and essentially quantitatively at room temperature with the evolution of carbon monoxide to produce $[2\text{-}B_{10}H_9CO]^-$. The cleanest reaction is observed employing $[Ph_3PMe][B_{10}H_{10}]$ in CH_2Cl_2, but the reaction has been used with several salts of $[B_{10}H_{10}]^{2-}$ (cesium, tetramethylammonium, triethylammonium) and with other solvents such as acetonitrile and tetrahydrofuran. The reaction product exhibits a strong infrared peak at 2129 cm^{-1}. This absorption lies in the frequency range typical of carbon monoxide coordinated to boron, which is higher than that of metal carbonyls due to the lack of d orbital back-bonding. The $^{11}B\{^1H\}$ spectrum of the reaction mixture in CH_2Cl_2 exhibits seven signals, consistent with equatorial substitution, and shows only traces of byproducts. The substituted boron is indicated by a high field resosance at -43.8 ppm (relative to $BF_3{\bullet}Et_2O$) which is a singlet in the proton coupled spectrum. $[Ph_3PMe][2\text{-}B_{10}H_9CO]$ may be isolated as a light tan solid in 85% yield, or the reaction mixture may be used directly in further reactions. The structure of $[2\text{-}B_{10}H_9CO]^-$ has been determined by X-ray crystallography[15] and shows a linear B-C-O array with a C-O distance of 1.13 Å.

The utility of the $[2\text{-}B_{10}H_9CO]^-$ anion results from its ease of conversion to a variety of other species as depicted in Figure 1. Although $[2\text{-}B_{10}H_9CO]^-$ is relatively insensitive to moisture in the solid state or in solution, it is easily hydrated in aqueous solvent mixtures to form $[2\text{-}B_{10}H_9CO_2H]^{2-}$ in accordance with Equation 4, which lies predominantly to the

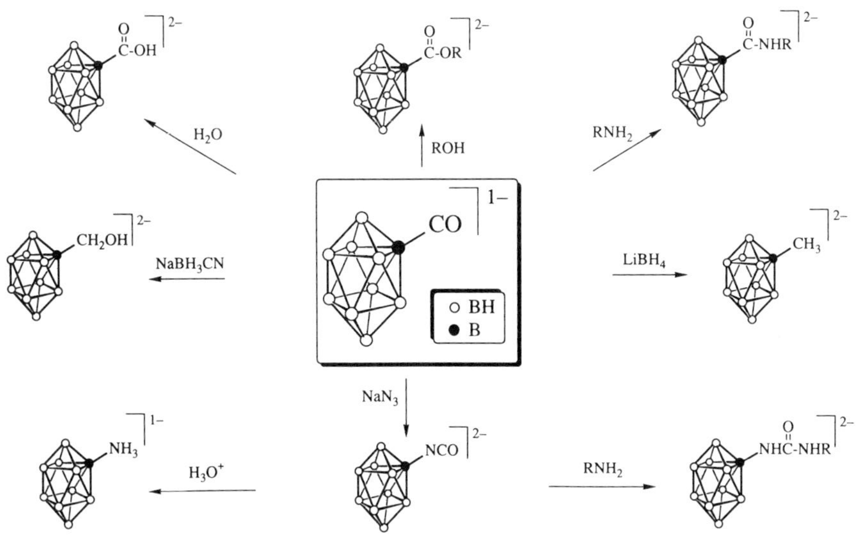

Figure 1. Transformations of the $B_{10}H_9CO^-$ ion.

$$[2\text{-}B_{10}H_9CO]^- \;+\; H_2O \;\rightleftharpoons\; [2\text{-}B_{10}H_9CO_2H]^{2-} \;\; H^+ \qquad \text{(Eqn. 4)}$$

right in aqueous solutions. The transformation of the carbonyl group from -C≡O to >C=O is indicated by the appearance of an infrared stretch at 1690 cm^{-1} as well as a -C–O stretch at 1256 cm^{-1}. The identity of $[Ph_3PMe][2\text{-}B_{10}H_9CO_2H]$ has been confirmed by crystallographic analysis.

The equilibrium in Equation 4 suggests the formation of other acyl derivatives of the decahydrodecaborate anion. The reaction of $[2\text{-}B_{10}H_9CO]^-$ with an amine (RNH_2) results in the formation of an amide ($[2\text{-}B_{10}H_9CONHR]^{2-}$) as long as an excess of the amine is present to absorb the acid formed by the reaction. Esters are produced by the combination of $[2\text{-}B_{10}H_9CO]^-$ with alcohols; in this case, the reaction is slow or incomplete unless an auxiliary base is added to absorb the acid formed.

The reduction of $[2\text{-}B_{10}H_9CO]^-$ with borohydride ion (BH_4^-) in refluxing THF produces $[2\text{-}B_{10}H_9CH_3]^{2-}$ (confirmed by X-ray crystallography). Although this is a novel species, as an easily synthesized single alkyl isomer of substituted $[B_{10}H_{10}]^{2-}$, it has little utility as a route to other derivatives. Milder reaction conditions employing cyanoborohydride (BH_3CN^-) as reducing agent result in a mixture of products including $[2\text{-}B_{10}H_9CH_2OH]^{2-}$, an ion with the potential of derivatization or conjugation with many other moieties. The optimized production and isolation of this species is in progress.

A very useful transformation of $[2\text{-}B_{10}H_9CO]^-$ occurs from its reaction with azide ion in acetonitrile. The carbonyl undergoes a Curtius-type rearrangement to form the isocyanate $[2\text{-}B_{10}H_9NCO]^{2-}$, indicated by an infrared absorption at 2304 cm^{-1}. This ion, whose structure has been determined by X-ray crystallography, is stable in neutral aqueous solution but is rapidly hydrolyzed in acidic media. The hydrolysis product, $[2\text{-}B_{10}H_9NH_3]^-$, has been previously synthesized by other more direct methods but in reduced yield compared to the present route. The isocyanate ion is also reactive with alcohols and amines, enabling the synthesis of a variety of amine derivatives of $[B_{10}H_{10}]^{2-}$.

The $[2\text{-}B_{10}H_9CO]^-$ is conveniently prepared in high yield from the decahydrodecaborate anion. This carbonyl species may be conjugated to existing species or derivatized to a variety of other soluble $[B_{10}H_{10}]^{2-}$ derivatives. The $[2\text{-}B_{10}H_9CO]^-$ ion therefore offers a high-yield route to the production of many boron-rich species for boron neutron capture therapy.

ACKNOWLEDGMENTS

This research was conducted as part of the BNCT program of the Idaho National Engineering Laboratory (INEL) and was performed under the auspices of the U.S. Department of Energy, Office of Energy Research, under DOE Field Office, Idaho, Contract No. DE-AC07-76ID01570.

REFERENCES

1. Knoth, W.H. (1964) *J. Am. Chem. Soc.* **86**, 3973-3983.
2. Snyder, H.R., Reedy, A.J. & Lennarz, W.J. (1958) *J. Am. Chem. Soc.* **80**, 835-838.
3. Hawthorne, M.F. (1991) *Pure & Appl. Chem.* **63**, 327-334.
4. Hawthorne, M.F. & Pitochelli, A.R. (1959) *J. Am. Chem. Soc.* **81**, 5519.
5. Pitochelli, A.R. & Hawthorne, M.F. (1960) *J. Am. Chem. Soc.* **82**, 3228-29.
6. Hertler, W.R. & Raasch, M.S. (1964) *J. Am. Chem. Soc.* **86**, 3661-3668.
7. Hertler, W.R., Knoth, W.H. & Muetterties, E.L. (1965) *Inorg. Chem.* **4**,280-287.
8. Knoth, W.H., Sauer, J.C., Balthis, J.H., Miller, H.C. & Muetterties, E.L. (1967) *J. Am. Chem. Soc.* **89**, 4842-4850.
9. Haslinger, F., Soloway, A.H. & Butler, D.N. (1966) *J. Med. Chem.* **9**, 581-584.
10. Alam, F., Soloway, A.H., Barth, R.F., Mafune, N., Adams, D.M. & Knoth, W.H. (1989) *J. Med. Chem.* **32,** 2326-2330.
11. Knoth, W.H., Sauer, J.C., England, D.C., Hertler, W.R. & Muetterties, E.L. (1966) *J. Am. Chem. Soc.* **88**, 935-939.
12. Hertler, W.R., Knoth, W.H. & Muetterties, E.L. (1965) *Inorg. Chem.* **4**, 288-293).
13. Shelly, K., Hawthorne, M.F. & Schmidt, P.G. (1992) in *Progress in Neutron Capture Therapy for Cancer*, eds. Allen, B.J., Moore, D.E. & Harrington, B.V. (Plenum Press, New York), p. 259-264.
14. Shelly, K., Feakes, D.A., Hawthorne, M.F., Schmidt, P.G., Krisch, T.A. & Bauer, W.F. (1992) *Proc. Natl. Acad. Sci. USA* **89**, 9039-9043.
15. Shelly, K., Hawthorne, M.F. & Knobler, C.B. (1992) *Inorg. Chem.* **31**,2889-2892.

SYNTHESIS OF MONOCARBON CARBORANES WITH POTENTIAL FOR APPLICATION IN ^{10}B NEUTRON CAPTURE THERAPY

J.H.Morris,[1] G.S.Peters,[1] F.E.Rayment,[1] M.D.Spicer,[1] F.S.Mair[2] and A.Martin[2]

[1]Department of Pure and Applied Chemistry, Strathclyde University, Glasgow, U.K.
[2]University Chemical Laboratory, Lensfield Road, Cambridge, U.K.

INTRODUCTION

The pseudoicosahedral carborane cluster anion [*closo*-1-$CB_{11}H_{12}$]$^-$ and its neutral derivative 1-Me_3N-1-$CB_{11}H_{11}$ are potential boron carriers in ^{10}B NCT. The alternatives of an anionic derivative or a neutral (zwitterionic) species with a tumour-seeking functional group at a boron atom offer scope for modification of biological properties and solubility. The available range of substituted derivatives of these monocarbon carboranes have previously been limited by the methods available to introduce substituents.[1,2] In this work we have achieved controlled substitution of the cluster using low-temperature functionalised boron insertion reactions on *nido*-monocarbon carborane precursors.[3]

RESULTS AND DISCUSSION

The *nido* carboranes [7-$CB_{10}H_{13}$]$^-$ and 7-Me_3N-7-$CB_{10}H_{12}$ were readily deprotonated by butyllithium to the anions [7-$CB_{10}H_{11}$]$^{3-}$ and [7-Me_3N-7-$CB_{10}H_{10}$]$^{2-}$ respectively. These anions undergo functionalised boron insertion reactions with compounds "$RBCl_2$" according to scheme 1.

[7-X-*nido*-7-$CB_{10}H_{10}$]$^{2-}$ + $RBCl_2$ $\rightarrow$ [1-X-2-R-*closo*-1-$CB_{11}H_{10}$]

X = H^-, R = Ph (**1**), MeC_6H_4 (**2**), Me_2N (**3**), $^iPr_2N$ (**4**), $Cl(CH_2)_4O$ (**5**), and F (**6**)
X = Me_3N, R = Ph (**7**), Me_2N (**8**), $^iPr_2N$ (**9**), $Cl(CH_2)_4O$ (**10**), $Br(CH_2)_4O$ (**11**), and F (**12**).

SCHEME 1

The structures of the compounds were established on the basis of their ^{1}H, ^{11}B and ^{11}B-^{11}B COSY NMR spectra, and by X-ray crystallography on selected compounds (for example, product (**5**)[4], Fig.1), the details of which will be reported elsewhere. All the products were identified as pseudoicosahedral monocarbon carboranes with substituents in the 2-position, but in several cases further structural complications occurred. Product (**3**) was isolated as the neutral 2-Me_3N-substituted derivative after methylation. Product (**4**) was isolated directly as the protonated neutral 2-$^{i}Pr_2NH$-derivative, and X-ray crystallography established its heavy atom positions (Fig.2).

In compounds (**8**) and (**9**), more substantial rearrangements occurred. Compound (**8**) was identified as 1-Me_2NH-2-CH_2Cl-*closo*-1-$CB_{11}H_{10}$ (Fig.3), and it has not yet been established whether the chlorine in the substituent originates from the $RBCl_2$ reagent or from the use of dichloromethane in the work-up procedure. The reaction to produce product (**9**) results in at least three closely related products. Two of these comprised different crystal modifications of 1-Me_2N-2-$H(^{i}Pr)_2NCH_2$-*closo*-1-$CB_{11}H_{10}$ (Fig.4), one of which incorporated 0.5 CH_2ClCH_2Cl.

Products with functionalised substituents in the 2-position offer scope for introduction of tumour-seeking groups and are therefore potential boron carriers for BNCT.

EXPERIMENTAL METHOD

General method for the synthesis of [2-R-*closo*-1-$CB_{11}H_{11}$]$^-$ salts

A solution of $[Me_3NH][CB_{10}H_{13}]$ (0.19 g, 1.0 mmol)in THF (6cm^3) was treated with a solution of $^{n}BuLi$ (2.0 cm^3, 1.6M in hexanes), and butane and Me_3N were removed *in vacuo*. The resultant clear yellow solution of "$Li_3CB_{10}H_{11}$" was cooled to 0°C, and excess $RBCl_2$ (see table 1) added by syringe under inert atmosphere (Ar or N_2), and the mixture stirred at room temperature for ca 1 hr. Volatiles were removed *in vacuo*, and organic and monoboron impurities were removed by digestion in hot 10% NaOH, filtration, and precipitation with a solution of a large cation halide (1.2 mmol). The products were purified by column chromatography on silica.

General method for the synthesis of 1-Me_3N-2-R-*closo*-1-$CB_{11}H_{10}$ and related derivatives

A solution of 7-Me_3N-7-$CB_{10}H_{12}$ (ca. 1.0 mmol) in THF was treated with a solution of $^{n}BuLi$ (3.0 cm^3, 1.6M in hexanes) and volatile products removed *in vacuo*. The resultant solution was treated with excess "$RBCl_2$" (see table 2) added under inert atmosphere. The solution was stirred for ca. 30 min. and solvent and volatiles were removed *in vacuo*. Methanol (10 cm^3) and conc. HCl (2 cm^3) were added and the mixture refluxed for 6 hr. The solution was cooled to 0°C and treated with NaOH and Me_2SO_4 (5 cm^3)to remethylate any demethylated byproducts. The solvent was removed *in vacuo*, and the product extracted into CH_2Cl_2, and purified by column chromatography on silica eluted with CH_2Cl_2.

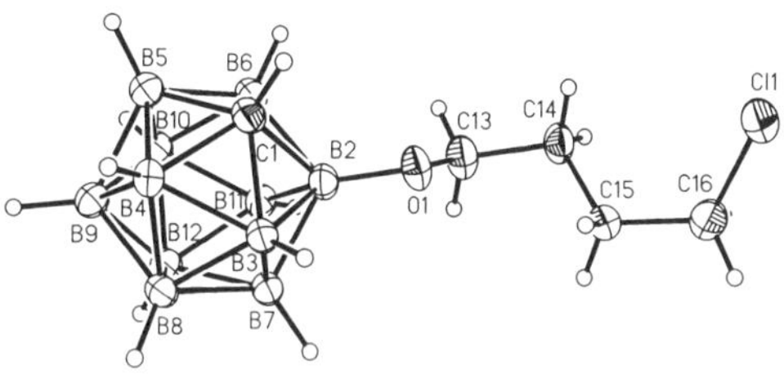

Figure 1. $[2\text{-Cl(CH}_2)_4\text{O-}closo\text{-1-CB}_{11}\text{H}_{11}]^-$

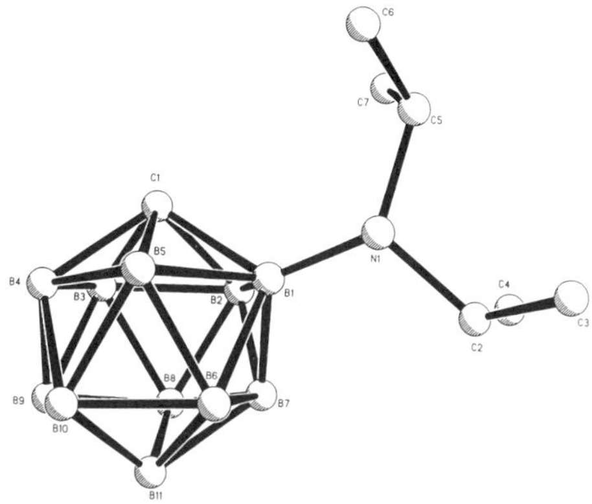

Figure 2. $2\text{-}^i\text{Pr}_2\text{NH-}closo\text{-1-CB}_{11}\text{H}_{11}$

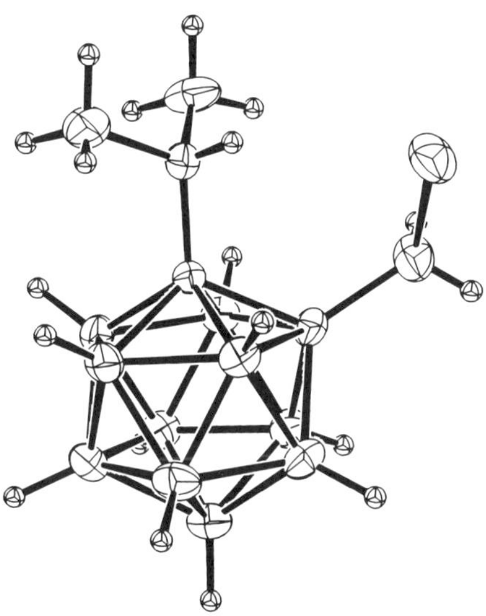

Figure 3. $1\text{-Me}_2\text{NH-2-CH}_2\text{Cl-}closo\text{-1-CB}_{11}\text{H}_{10}$

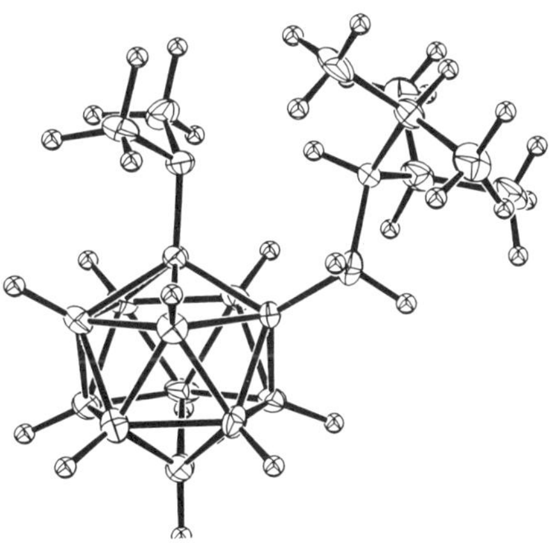

Figure 4. $1\text{-Me}_2\text{N-2-H(}^i\text{Pr)}_2\text{NCH}_2\text{-}closo\text{-1-CB}_{11}\text{H}_{10}$

TABLE 1. Experimental details for insertion products of the $[CB_{10}H_{11}]^{3-}$ anion.

Compound	$Li_3CB_{10}H_{11}$ (mmol)	Insertion reagent (mmol)		Cation	Found % C H N(P) Cl	Calc. % C H N(P) Cl
1	1.0	$PhBCl_2$	7.6	$[^nBu_4N]^+$	59.8 11.8 3.1	59.9 11.4 3.0
2	5.7	$TolBCl_2$	20.0	$[Ph_4P]^+$	61.2 5.9 4.8	60.3 6.1 4.7
3*	1.58	Me_2NBCl_2	11.9		24.2 10.2 6.7	23.9 10.0 7.0
4+	1.31	$^iPr_2NBCl_2$	20.9		34.6 10.9 5.7	34.6 10.8 5.8
5!	1.11	BCl_3SMe_2	11.6	$[Ph_4P]^+$	59.3 6.8 5.2 6.0	59.1 6.7 5.3 6.0
6	0.64	BF_3OEt_2	9.7	$[BzMe_3N]^+$	42.3 9.0 4.5	42.5 8.7 4.5

* The product was methylated with Me_2SO_4 and purified by chromatography.

+ The reaction was carried out in CH_2Cl_2.

! The reagent cleaved THF to give $Cl(CH_2)_4OBCl_2$.

TABLE 2. Experimental details for insertion products of $[7\text{-}Me_3N\text{-}7\text{-}CB_{10}H_{10}]^{2-}$.

Compound	$Li_2[Me_3NCB_{10}H_{10}]$ (mmol)	Insertion reagent (mmol)		Yield %	Found % C H N	Calc. % C H N
7	1.72	$PhBCl_2$	7.0	21.9	43.2 8.5 5.4	43.3 8.7 5.1
8	1.07	Me_2NBCl_2	15.0	25.6	22.7 8.5 5.7	20.4 8.1 5.6
9	1.22	$^iPr_2NBCl_2$	11.0	19.3	41.3 10.8 9.7	40.0 11.1 9.3
10	1.12	BCl_3SMe_2	9.6	25.0	31.2 8.9 4.5	31.2 8.8 4.6
11	0.65	BBr_3SMe_2	3.9	37.7	28.0 8.2 4.0	27.3 7.8 3.4
12	1.45	BF_3OEt_2	20.3	35.0	23.7 9.0 5.9	21.2 8.7 6.4

REFERENCES

1. S.A.Khan, J.H.Morris and S.Siddiqui, *J. Chem. Soc., Dalton Trans.*, 1990, 2053.
2. S.A.Khan, J.H.Morris, M.Harman and M.B.Hursthouse, *J. Chem. Soc., Dalton Trans.*, 1992, 119.
3. J.H.Morris, S.A.Khan, F.Mair and G.Peters, *Proceedings 4th Int. Symp. on NCT for Cancer Dec 1990*, B.J.Allen, D.E.Moore and B.V.Harrington Eds., Plenum Press, 1992, 289.
4. F.S.Mair, J.H.Morris, D.F.Gaines and D.Powell, *J. Chem. Soc., Dalton Trans.*, 1993, 135.

SYNTHESIS AND EVALUATION OF AMINE REACTIVE CARBORANYL DERIVATIVES FOR APPLICATION TO BORON-10 CONTAINING POLYMERS

D.S. Wilbur, D.K. Hamlin, R.R. Srivastava, G.E. Laramore, and T.W. Griffin

Dept. of Radiation Oncology
Univ. of Washington
Seattle, WA 98195

INTRODUCTION

The use of monoclonal antibodies to selectively deliver boron-10 to cancer cells for boron neutron capture therapy (BNCT) has been under active investigation for over 20 years (1). A major difficulty in the application of boronated antibodies is the high concentrations of boron-10 required in the tumor tissue (e.g. 35 mg B-10/g tissue) (2). Simple calculations have shown that each antibody molecule must have a minimum of 600-1000 boron atoms to achieve the concentrations of boron-10 required (1-4). Seemingly, the only way to achieve such a large number of boron atoms per antibody molecule is to incorporate the boron into a polymer, then attach the boronated polymer to the antibody. Further, the coupling of boron containing polymer with the antibody must be accomplished in a manner that does not adversely affect its antigen binding or dramatically change the physical properties of the antibody (e.g. net ionic charge). Thus, it seems that a reasonable objective would be to prepare a neutral boronated polymer which could be site-specifically attached to the antibody.

One approach to producing boronated polymers for coupling with antibodies is to react functionalized icosahedral $B_{12}H_{12}^{2-}$ boranes (5) or $C_2B_{10}H_{10}$ carboranes (6) with preformed polyamines, such as polylysine (3,7). In contrast to anionic icosahedral boranes, carboranes are neutral (in *closo* form), and can be made water soluble with appropriate substituents, making these structures attractive for modification of polyamine compounds. However, some *closo*-carboranes can be degraded to their anionic *nido* counterparts in the presence of amines (6). Importantly, two competing reactions can occur when reacting polyamines with *closo*-carboranes, those being reaction with functionalities of the side chain (desired reactions) and reaction with boron in the carborane cage (undesired reactions).

Our goal to prepare a neutral boronated polylysine polymer for modification of antibodies led to an investigation which was designed to evaluate reactions of functionalized carboranes with polyamines in a model system. In the investigation several carboranes containing amine reactive functionalities were synthesized and their reactions with 1° amines were studied. A preliminary report of that investigation is presented herein.

Advances in Neutron Capture Therapy, Edited by
A.H Soloway *et al.*, Plenum Press, New York, 1993

EXPERIMENTAL

General

^1H, ^{13}C, and ^{11}B NMR were obtained on either a Varian VXR-300 (300 MHz) or a Bruker AC-200 (200 MHz). IR data were obtained on a Perkin-Elmer 1420 IR spectrophotometer. Spectral data were consistent with the structures shown in Table I. GC-MS data were obtained on a HP 5890 Series II GC (J&W DB-1 column - 30 m x 0.25 mm) interfaced with a Kratos Profile HV-3 mass spectrometer (Mach 3 software).

Thin layer chromatography was conducted on silica gel using EtOAc/hexane solvent mixtures. Visualization of the developed TLC plates was accomplished by I_2, phosphomolybdic acid (10%) or $AgNO_3$ (15% solution).

HPLC equipment consisted of a Hewlett-Packard 1050 isocratic system with RI detection, and a H.P. 1050 gradient system with UV detection (220 & 254 nm). HPLC evaluations were conducted at a flow rate of 1 mL/min on C-18 columns eluting with 80% MeOH / 20% of a 1% HOAc in H_2O mixture on the isocratic system, and varying quantities of MeOH / 5 mM TBAP in H_20 as gradients (for detection of *nido* carboranes). Reaction mixtures were analyzed on both systems to detect *closo* and *nido* products

Carborane Syntheses

All compounds were purified by either kugelrohr sublimation or column chromatography prior to use. Purity of compounds was determined by HPLC analysis and spectral data. Reported yields are those of isolated, purified products.

Synthesis of *ortho*-carboranylcarboxylate tetrafluorophenyl (TFP) esters **1a, 1b**, and **1c** were accomplished in two synthetic steps. Initally, the TFP esters of the corresponding alkynes were prepared by reaction with tetrafluorophenol and DCC in anhyd. THF. The alkynyl carboxylate TFP esters were then reacted with decaborane in refluxing acetonitrile to give **1a** (22%), **1b** (57%), or **1c** (35%). The *meta*-carboranylcarboxylate TFP ester, **2**, was prepared from carbonylation of methyl-1,7-carborane using nBuLi/CO_2 in anhyd. ether (56%), followed by esterification with an excess of DCC and tetrafluorophenol in anhyd. THF (65%). Synthesis of the carboranylbutylimidate, **3**, was prepared as described by Wong, Tolpin, and Lipscomb (8). Synthesis of the *ortho*-carboranylalkyl halides **4a** (24%) and **4c** (42%) were accomplished by reaction of the corresponding alkyne with decaborane in acetonitrile as previously described (8,9). Synthesis of the *ortho*-carboranylalkyl tosylates were accomplished by reaction of the corresponding alkynyl tosylates with decaborane in refluxing acetonitrile to give **4b** (73%) and **4d** (55%).

Synthesis of *ortho*-carboranylpropyl epoxide **5** was most efficiently accomplished by reaction of *tert*-butyldimethylsilyl-1,2-carborane (10) with *n*BuLi and epichlorohydrin (88%), followed by removal of the silyl group (85%). Synthesis of *meta*-carboranylpropyl epoxide **6** was best accomplished by reaction of trimethylsilyl-1,7-carborane (11) with *n*BuLi and epichlorohydrin (79%), followed by removal of the silyl group (95+%).

Reactions with 1° Amines

Reactions were conducted by dissolving 10 mg of the starting functionalized carborane, **1a**-**6**, in 1 mL of CH_3CN (or EtOH). To this solution was added one equivalent of the amine (phenethylamine, glycine benzyl ester, or aminoacetonitrile). In those examples where an amine salt was used, one equivalent of triethylamine was added. The reaction was followed over 24 h at room temperature by HPLC analysis or TLC analysis. Reactions in which starting material remained after 24 h were prepared again and reaction product composition was followed while refluxing for 24 h. Reactions of epoxides **5** and **6** were also examined under the conditions employed to couple these reagents with 2-nitroimidazole (12), by reaction of the epoxide with the amine in K_2CO_3/EtOH at reflux for 24 h. Evaluation of the product composition was done by HPLC, GC-MS, and NMR. The major products obtained are outlined in Table 1.

RESULTS AND DISCUSSION

Carboranes containing carboxylate active esters, an alkyl imidate ester, alkyl halides/ tosylates, and propyl epoxides were synthesized as these functional groups can undergo reaction with amines. The amines chosen for reaction with the functionalized carboranes were 1° amines to mimic polylysine. Phenethylamine was chosen because it afforded a UV active product to aid in the monitoring of reaction progress. Glycine benzyl ester and amino-acetonitrile were chosen for study as these could be used in further modification of the carborane (i.e. glycine spacer) if reactivity with a large polymer were found to be a problem in later studies. All reactions were conducted with one equivalent of amine in an attempt to prepare exclusively the adduct.

1,2-Carboranes substituted with carboxylate tetrafluorophenyl (TFP) esters (**1a**, **1b**) and propionate TFP ester (**1c**) were synthesized to evaluate the aminolysis reaction versus reaction with the boron cage. Reactions of the amines with **1a**-**1c** in MeOH resulted in a mixture of amide adduct (Table I) and methyl ester. Reactions in acetonitrile gave exclusively the desired product. The TFP ester of 1,7-carborane **2** also gave exclusively the desired adduct. In contrast to the TFP esters, reaction of the carborane butylimidate ester **3** was found to be slow at room temperature, but provided the desired adduct as the major product. Reactions of the carboranyl alkyl halides **4a** and **4c**, and the carboranyl alkyl tosylates **4b** and **4d**, only resulted in formation of the *nido* carborane, with no evidence of adduct formation. Reactions of the propyl epoxides **5** and **6** with amines in acetonitrile resulted in *nido* formation from **5** and no reaction with **6**. Evaluation of the reactions of **5** and **6** under conditions used for coupling them with 2-nitroimidazole (i.e. K_2CO_3/EtOH/D) (12) again resulted in ring degradation of **5**, whereas reaction of **6** appeared to give an intermediate product (by HPLC) which may have been the desired adduct, then reacted further resulting in a product consistent with the *nido* form of the amine adduct (Table 1). This result was surprising as the 1,7-carboranes are fairly resistant to ring degradation. However, intramolecular amine degradations may be more facile than degradation by amines in intermolecular reactions.

SUMMARY

Some general observations can be made from this limited study. These are:

1. Reactions of 1° amines with TFP esters in acetonitrile are rapid, yielding amide adducts within a few minutes at room temperature.
2. Reactions of 1° amines with *ortho*-carboranylbutylimidate are slow in basic solution at room temperature, but do yield amidine bonded carboranyl derivatives over time.
3. Reactions of 1° amines with carboranylpropyl epoxides are slow, requiring elevated temperatures (e.g. 60 °C). *ortho*-Carborane derivatives form *nido*-carboranes before epoxide reaction. The *meta*-carborane derivatives appear to initially react with the epoxide in preference to the boron cage, but ultimately give *nido*-carborane products under the reaction conditions studied.
4. Reactions of 1° amines with carboranyl alkyl halides and alkyl tosylates are slow, requiring elevated temperatures, and result in cage degradation prior to displacement (or elimination) of the leaving group.

Based on the results obtained, the carboxylate TFP esters will be studied further for modification of polyamine molecules. To avoid formation of the nido compound, it is likely that the polyamine will need to be added to a large excess of the TFP ester. While avoiding the formation of *nido*-carborane may be possible in reaction mixtures, one must wonder if

Table 1. Reactions of Functionalized Carboranes with 1° Amines

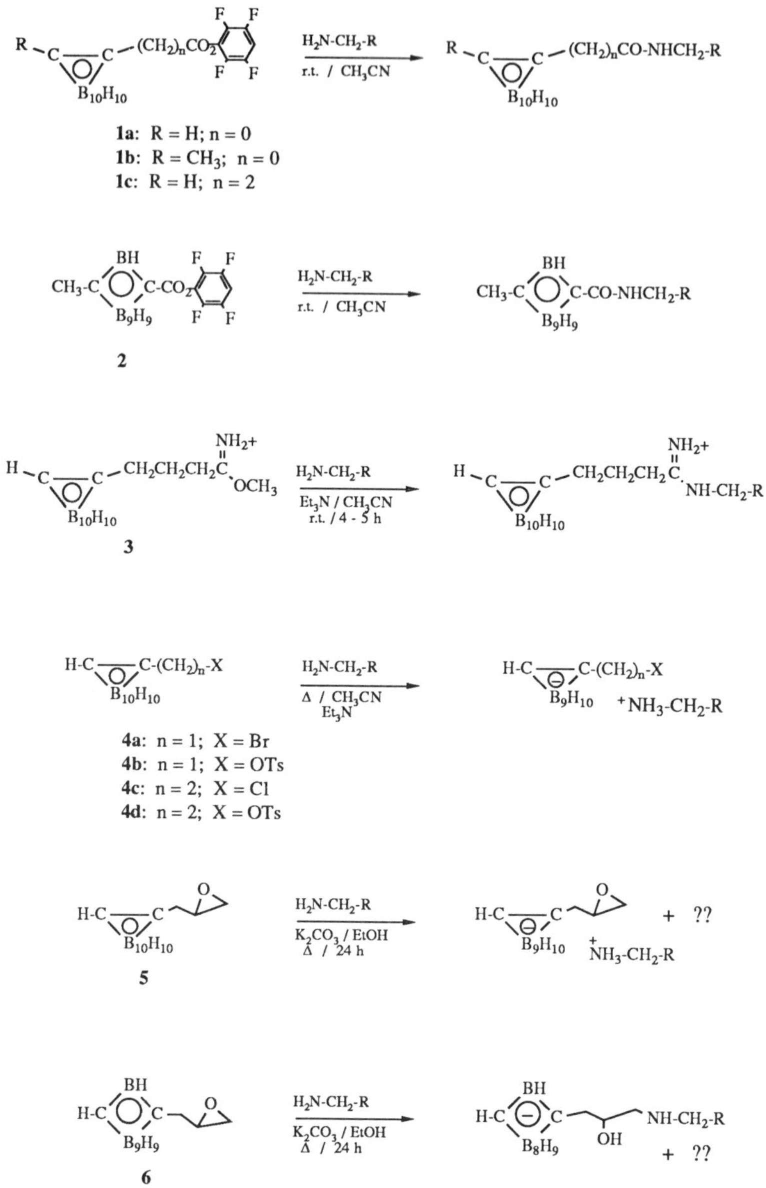

R = Peptide; R = CH_2-C_6H_5 (Phenethylamine); R = CO_2-CH_2-C_6H_5 (Glycine Benzyl Ester); R = CN (Aminoacetonitrile)

proximity to amines *in vivo* will ultimately result in its formation. Therefore, our primary focus will be on the use of the more amine resistant 1,7-carboranyl carboxylate TFP esters in future studies.

LITERATURE CITED

1. Alam, F., Barth, R.F. & Soloway, A.H. (1989) Boron Containing Immunoconjugates for Neutron Capture Therapy of Cancer and for Immunocytochemistry. *Antibody, Immunoconj. Radiopharm.* **2**, 145-163.
2. Barth, R.F., Soloway, A.H. & Fairchild, R.G. (1990) Boron Neutron Capture Therapy of Cancer. *Cancer Res.* **50**, 1061-1070.
3. Barth, R.F., Alam, F., Soloway, A.H., Adams, D.M. & Steplewski, Z. (1986) Boronated monoclonal antibody 17-1A for potential neutron capture therapy of colorectal cancer. *Hybridoma* **5**[Suppl 1], S43-S50.
4. Hawthorne, M.F. (1991) Biochemical applications of boron cluster chemistry. *Pure & Appl. Chem.* **63**, 327-334.
5. Middaugh, R.L. (1975) *Closo*-Boron Hydrides, in *Boron Hydride Chemistry*, ed. Muetterties, E.L. (Academic Press, New York) Chapt. 8, pp 273-300, .
6. Grimes, R.N. (1970) *Carboranes* (Academic Press, New York).
7. Alam, F., Soloway, A.H., Barth, R.F., Mafune, N., Adams, D.M. & Knoth, W.H. (1989) Boron neutron capture therapy: linkage of a boronated macromolecule to monoclonal antibodies directed against tumor associated antigens. *J. Med. Chem.* **32**, 2326-2330.
8. Wong, H.S., Tolpin, E.I. & Lipscomb, W.N. (1974) Boron Hydride Derivatives for Neutron Capture Therapy. Antibody Approach. *J. Med. Chem.* **17**, 785-791.
9. Fein, M.M., Grafstein, D., Paustian, J.E., Bobinski, J., Lichstein, B.M., Mayes, N., Schwartz, N.N. & Cohen, M.S. (1963) Carboranes. II. The Preparation of 1- and 1,2-Substituted Carboranes. *Inorg. Chem.* **2**, 1115-1119.
10. Gomez, F.A., Johnson, S.E. & Hawthorne, M.F. (1991) A Versatile Protecting Group for 1,2-Dicarba-*closo*-dodecaborane(12) and the Structure of a Silylcarborane Derivative. *J. Am. Chem. Soc.* **113**, 5915-5917.
11. Wilbur, D.S., Hamlin, D.K., Laramore, G.E. & Griffin, T.W. (1993) Preparation of *meta*-Carborane Derivatives by Thermal Rearrangement in a Small Scale Laboratory Apparatus. Proceedings of 5th Int. Symposium on Neutron Capture Therapy (this volume).
12. Wilbur, D.S., Hamlin, D.K., Livesey, J.C., Srivastava, R.R., Laramore, G.E. & Griffin, T.W. (1993) Preparation of *meta*-Carboranylpropyl-2-nitroimidazole Derivatives for Application to BNCT. Proceedings of 5th Int. Symposium on Neutron Capture Therapy (this volume).

SYNTHESIS AND PRELIMINARY *IN VIVO* TOXICITY EVALUATION OF AN IODINATED SULFIDOBORATE

M. Miura,[1] P.L. Micca,[2] J.C. Heinrichs[2] and D.N. Slatkin[2]

[1]Department of Applied Science
[2]Medical Department
Brookhaven National Laboratory,
Upton, New York 11973

INTRODUCTION

Methods have been proposed to localize compounds for neutron-capture therapy[1-4] noninvasively using nuclides such as ^{99m}Tc, ^{111}In, ^{55}Mn and ^{157}Gd, which generally form ionic rather than covalent bonds. It might also be desirable to have an imaging nuclide bound covalently to a neutron-capture agent to assure correspondence between the localization of the imaging nuclide and the distribution of the capture agent. Because imaging nuclides ^{123}I and ^{127}I do form covalent bonds, we sought to label a ^{10}B-containing agent for boron neutron-capture therapy [BNCT] with iodine. An iodine-labeled ^{10}B-carrier would enable the biodistribution of ^{10}B to be imaged noninvasively in patients so as to optimize the timing of neutron exposure for BNCT.

While $Na_2B_{12}H_{11}SH$ [BSH] is in clinical use for BNCT in Japan, its disulfido dimer $Na_4B_{24}H_{22}S_2$ [BSSB][5] is also under consideration for use in BNCT. We now describe the synthesis of an iodinated analogue of the dimer for possible use in BNCT and compare its toxicity in normal mice to that of BSSB.

METHODS

Chemistry

The sulfidohydroborate dimer $K_4H_{11}B_{12}SSB_{12}H_{11}$ [K_4BSSB], which was ion-exchanged from the cesium salt (95 atom% ^{10}B-enriched, Callery Chemical Co., Callery, Pennsylvannia), was treated with iodine and iodine monochloride in tetrachloroethane and allowed to stir at reflux under nitrogen overnight (Figure 1) as described previously[6] for periodination of $B_{12}H_{12}^{-2}$. A product was precipitated under these oxidizing conditions using tetramethylammonium chloride and was purified by recrystallization in 24% molar yield from BSSB. The product was characterized by elemental analyses (Galbraith Laboratories Inc, Knoxville, Tennessee) and infrared [IR] spectroscopy and shown to be, most likely, the periododimer $(Me_4N)_4I_{11}B_{12}SSB_{12}I_{11}$ [$(Me_4N)_4$IBSSB]. There was no BSSB identified in the recrystallized product by thin layer chromatography. The beige-colored tetramethylammonium salt was ion-exchanged to the sodium salt for an initial toxicity study in mice.

Advances in Neutron Capture Therapy, Edited by
A.H Soloway *et al.*, Plenum Press, New York, 1993

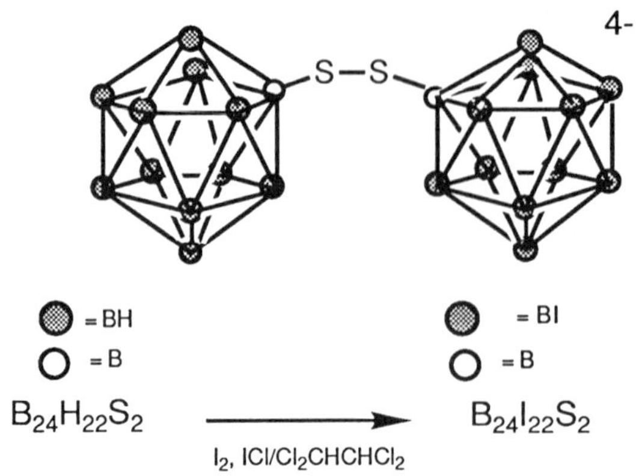

Figure 1. Synthesis of IBSSB from BSSB.

Table I. Hematologic analyses of the blood of mice given BSSB (Group A), IBSSB (Group B) or saline (Group C) in doses described. Mouse group and number [MS], leukocytes [WBC] ($10^3/\mu l$), erythrocytes [RBC] ($10^6/\mu l$), hemoglobin [HGB] (g/dl), hematocrit [HCT] (%), mean erythrocyte volume [MCV] (μm^3), mean erythrocyte hemoglobin [MCH] (pg), mean erythrocyte hemoglobin concentration [MCHC] (100·MCH/MCV %), erythrocyte volume distribution width [RDW] (100·σ/MCV %), platelets [PLT] ($10^3/\mu l$), mean platelet volume [MPV] (μm^3), granulocytes [GRN] ($10^3/\mu l$), lymphocytes [LYM] ($10^3/\mu l$), midrange-size leukocytes [MID] ($10^3/\mu l$).

MS	WBC	RBC	HGB	HCT	MCV	MCH	MCHC	RDW	PLT	MPV	GRN	LYM	MID
A1	3.4	7.76	14.8	42.2	54.4	19.1	35.0	24.8	1090	4.80	0.10	3.2	0.10
A2	5.1	8.81	17.1	49.2	55.8	19.4	34.8	24.9	1382	4.80	0.40	4.2	0.50
A3	4.0	7.74	15.5	44.1	57.0	20.0	35.1	23.9	1101	4.90	0.30	3.4	0.30
A4	4.9	8.81	17.0	47.9	54.4	19.3	35.5	23.7	1391	4.80	0.30	4.5	0.10
A5	5.6	8.46	16.7	47.7	56.4	19.7	35.0	24.2	1316	4.50	0.70	3.9	1.0
B1	2.6	8.05	15.6	44.8	55.6	19.4	34.8	24.2	1405	4.70	0.20	2.2	0.20
B2	4.8	8.28	15.2	45.7	55.2	18.4	33.3	23.6	1395	5.10	0.20	4.5	0.10
B3	4.4	7.92	15.5	43.4	54.8	19.6	35.7	23.9	1536	4.80	0.20	4.0	0.20
C1	4.7	7.88	15.6	44.5	56.5	19.8	35.1	23.9	1384	4.50	0.30	4.2	0.20
C2	4.4	8.72	17.5	48.7	55.9	20.1	35.9	23.8	1223	4.80	0.20	3.9	0.30
C3	3.7	8.66	16.9	48.2	55.7	19.5	35.1	24.3	1334	5.00	0.50	2.6	0.60
C4	3.3	8.64	16.7	48.3	55.9	19.3	34.6	24.5	1249	4.80	0.30	2.8	0.20
C5	4.4	8.85	17.4	50.1	56.6	19.7	34.7	23.9	1222	4.80	0.40	3.6	0.40

Toxicity

Three groups of female mice (BNL Hale-Stoner strain) were each given either a single intraperitoneal [i.p.] injection of A) 17 μg B per gram of body weight [gbw] as Na_4BSSB, B) 50 μg B/gbw as Na_4IBSSB; or C) 0.9% NaCl in water [saline]. Injection volumes were all 0.015 ml/gbw. Five days after the injection, mice were euthanized by deep halothane inhalation anesthesia, thoracotomy, and removal of ≈0.7-0.9 ml blood through the right ventricle; hematological and chemical analyses were performed on blood specimens from each mouse.

Table II. Chemical analyses of blood plasma from mice given BSSB (Group A), IBSSB (Group B) or saline (Group C) in doses described. Mouse group and number [MS], glucose [GLU] (mg/dl), blood urea nitrogen [BUN] (mg/dl), blood creatinine [BCR] (mg/dl), aspartate transaminase [AST] (U/l), alanine transaminase [ALT] (U/l), alkaline phosphatase [ALP] (U/l), creatine phosphokinase [CPK] (U/l), total protein [TPR] (g/dl), albumin [ALB] (g/dl), calcium [CA] (mg/dl), chloride [CHL] (meq/l), inorganic phosphorus [P] (mg/dl), potassium [K] (meq/l), sodium [SOD] (meq/l).

MS	GLU	BUN	BCR	AST	ALT	ALP	CPK	TPR	ALB	CA	CHL	P	K	SOD
A1	311	16	0.4	46	28	128	50	5.7	3.4	11.7	119	11.7	10.0	156
A2	217	26	0.4	41	21	115	37	5.3	3.3	10.9	118	8.8	6.9	154
A3	315	23	0.4	56	36	130	64	6.0	3.6	11.1	117	9.6	9.2	154
A4	268	23	0.5	70	34	103	141	5.3	3.2	11.3	119	10.9	9.9	153
A5	249	23	0.6	80	28	132	215	6.0	3.6	11.6	121	9.0	6.5	161
A6	269	30	0.5	40	19	143	58	5.8	3.4	11.0	120	9.4	7.2	158
B1	332	15	0.4	80	33	108	89	5.5	3.0	11.2	119	10.1	7.9	155
B2	206	27	0.4	77	47	123	148	5.8	3.3	11.9	119	10.8	7.8	156
B3	252	26	0.4	89	66	123	75	5.8	3.2	11.3	117	9.8	6.0	155
B4*	278	28	0.5	44	22	110	35	5.9	3.3	11.5	116	9.9	7.1	154
C1	307	24	0.5	63	26	149	134	5.6	3.0	11.6	121	11.7	11.3	153
C2	298	17	0.4	58	26	156	113	5.3	3.2	11.1	121	11.0	11.0	154
C3	286	31	0.5	49	26	120	36	5.5	3.2	11.3	115	8.8	6.3	154
C4	295	28	0.5	41	28	112	35	5.5	3.3	11.1	119	9.0	6.6	155
C5**	234	32	0.5	112	54	126	245	5.9	3.5	11.4	123	9.7	9.5	158
C6	226	33	0.5	50	19	136	99	6.1	3.6	11.6	121	10.1	7.9	159
C7	410	29	0.6	30	10	164	58	5.8	3.3	12.0	121	11.2	10.5	157

*Mouse not given entire dose of IBSSB; excluded from analyses.
**Mouse appears to have had liver dysfunction by its abnormal liver enzyme values; excluded from analyses.

RESULTS AND DISCUSSION

The infrared spectrum of IBSSB shows a marked reduction in intensity of the strong absorbance at ≈2500 cm^{-1} (corresponding to BH stretching) from that of BSSB. This could be due to a minor contamination by BSSB or to incomplete iodination. Two absorbances appeared at 968 and 944 cm^{-1} in IBSSB which were not in the spectrum of BSSB. These correlate with the BI stretching frequencies in $B_{12}I_{12}^{-2}$ (940 and 925 cm^{-1}).[6] The higher spectral frequencies of the BI stretch in IBSSB than in $B_{12}I_{12}^{-2}$ may be due to inductive effects, which suggests that sulfur exerts a greater electron-withdrawing influence than does iodine on ortho-, meta- and para- BI bonds.

A single i.p. injection of IBSSB at a dose corresponding to 50 µg B/gbw appears to have been minimally more toxic than was a similar injection of three-fold less BSSB-boron, as judged by hematological and chemical tests of the blood 5 days later. Inadvertently, the BSSB group received 17 µg ^{10}B/gbw, only about one-third of the intended boron dose, 50 µg/gbw. Nevertheless, it is useful to compare the subacute toxicity of a relatively high dose of a substance, IBSSB, which had never tested been in an animal, with that of its parent compound, BSSB, which is known to be non-lethal at ≈35 µg B/gbw,[7] twice the lower of the two boron doses used. Fourteen chemical and thirteen hematological tests of blood from mice given either BSSB (Group A) or IBSSB (Group B) showed no significant differences between the two groups using the Wilcoxon Two-Sample

Table III. Ranks and rank sums of the ALT levels in blood plasma from IBSSB- and saline-treated mice.

Rank	1	2	3	4	5	6	7	8	9	*Rank sum*
IBSSB	--	--	--	--	--	--	33	47	66	*24*
Saline	10	19	26	26	26	28	--	--	--	*21*

test.[8] Alanine transaminase [ALT] levels in the mice given IBSSB were significantly higher than those in mice given physiological saline (Group C) (Table III). No other differences were noted between mice in groups B and C.

CONCLUSIONS

Periodination of BSSB can be carried out under reaction conditions similar to those used for $B_{12}H_{12}^{-2}$. We surmise from this preliminary *in vivo* experiment that IBSSB may be slightly hepatotoxic to mice. Since the calculated weight ratio of iodine to ^{10}B in IBSSB is 11.5, the presence of a therapeutically adequate concentration of ^{10}B in tumor, for example 30 μg ^{10}B per gram, would be associated with 345 μg I per gram tumor. This would be suitable for imaging a tumor in the brain and for quantifying the iodine in the image using computed tomography [CT] without using a conventional contrast enhancement agent. Contrast enhancement in malignant gliomas is typically ≈2-20 EMI scale units; 1 EMI scale unit corresponds to 77 μg I/ml tissue.[9] It follows that each added EMI scale contrast enhancement unit observed in a tumor image pixel by CT after 95% ^{10}B-enriched IBSSB administration would correspond to an additional 6.7 μg ^{10}B/ml in the tumor zone corresponding to that pixel. Thus, ^{10}B concentrations useful for BNCT of malignant gliomas could be quantified by conventional CT using nonradioactive ^{10}B-IBSSB as a combined radiographic contrast agent and ^{10}B-carrier. Alternatively, iodine-radiolabeled IBSSB would allow visualization and quantification of iodine in the tumor by single photon emission computed tomography [SPECT].

The microdistribution of boron delivered by ^{125}I-labeled IBSSB can be assessed using ^{125}I-mediated silver-grain autoradiography [ARG] in a contiguous photographic emulsion. A grain/disintegration efficiency approaching 50% should be attainable and spatial resolution would be comparable to that feasible with tritium-based ARG.[10]

Because boron and iodine are linked covalently in IBSSB, noninvasive quantification of iodine in a clinical tumor image would thereby also quantify boron noninvasively. Quantitative imaging could be performed using either radiolabeled or nonradiolabeled IBSSB in SPECT or CT, respectively, provided that *in vivo* deiodination of IBSSB, if it occurs at all, is sufficiently slow. In view of the potential clinical advantages of an iodinated boron carrier over an unlabeled boron carrier for BNCT, further studies of the toxicology of IBSSB are underway and experiments on the pharmacokinetics, tissue microdistribution, noninvasive imaging and radiotherapeutic efficacy of IBSSB are planned.

ACKNOWLEDGEMENTS

This work was supported by U.S. Department of Energy Contract No. DE AC02-76CH00016 and by National Institutes of Health Grant 1R01CA38107-01A2.

REFERENCES

1. Slatkin, D., A history of boron neutron capture therapy of brain tumours, *Brain*, 114:1609 (1991).

2. Hainfeld, J.F., Uranium-loaded apoferritin with antibodies attached: Molecular design for uranium neutron-capture therapy, *Proc. Natl. Acad. Sci. USA*, 89:11064 (1992).

3. Martin, R.F., d'Cunha, G., Pardee, M., Allen, B.J., Induction of double-strand breaks following neutron capture by DNA-bound Gd-157, *Int. J. Rad. Biol. Phys. Chem. Med.* ,54:205 (1987).

4. Shih, J.-L. A., Brugger, R.M., Gadolinium as a neutron capture therapy agent, *Med. Phys.*, 19:733 (1992).

5. Joel, D.D., Fairchild, R.G., Laissue, J.A., Saraf, S.K., Kalef-Ezra, J.A., Slatkin, D.N., Boron neutron capture therapy of intracerebral rat gliosarcomas, *Proc. Natl. Acac. Sci., USA*, 87:9808 (1990).

6. Knoth, W.H., Miller, H.C., Sauer, J.C., Balthis, J.H., Chia, Y.T., Muetterties, E.L., Chemistry of boranes. IX. Halogenation of $B_{10}H_{10}^{-2}$ and $B_{12}H_{12}^{-2}$, *Inorg. Chem.*, 3:159 (1964).

7. Slatkin, D., Micca, P., Forman, A., Gabel, D., Wielopolski, L., Fairchild, R., Boron uptake in melanoma, cerebrum and blood from $Na_2B_{12}H_{11}SH$ and $Na_4B_{24}H_{22}S_2$ administered to mice, *Biochem. Pharm.*, 35:1771 (1986) [cf. mouse groups E and F in Table 1 therein.]

8. Miura, M., Micca, P.L., Heinrich., J.C., Gabel, D., Fairchild, R.G., Slatkin, D.N., Biodistribution and toxicity of 2,4-divinyl-nido-o-carboranyldeuteroporphyrin IX in mice, *Biochem. Pharm.*, 43:467 (1992).

9. Gado, M.H., Phelps, M.E., Coleman, R.E., An extravascular component of contrast enhancement in cranial computed tomography, *Radiology*, 117:589 (1975).

10. Rogers, A.W., "Techniques of Autoradiography." Third Edition. Elsevier/North-Holland Biomedical Press, Amsterdam (1979).

BORON DELIVERY TO TUMORS MEDIATED BY MACROMOLECULES AND VESICLES

M. Frederick Hawthorne

University of California at Los Angeles, Los Angeles, CA

INTRODUCTION

During the past decade many types of macromolecular species potentially capable of selectively transporting large quantities of ^{10}B to tumor have been examined for the purposes of BNCT. These large boron carriers have either been obtained by the conjugation of high-molecular weight biomolecules, such as antibodies, with small boron-containing reagents or the purposeful synthesis of large boron-rich oligomers which are then attached as reagents to antibodies or other peptides targeted for tumor cell receptor sites. In addition to the use of large molecules as boron carriers, recent research has demonstrated the utility of synthetic unilamellar liposomes as tumor cell-selective vesicles capable of the *in vivo* delivery of therapeutic quantities of boron to tumor with low injected doses of boron. A group of receptor-mediated vesicular delivery systems are under preliminary examination which are comprised of low density lipoproteins filled with hydrophobic carborane derivatives. The characteristic common to all of these macromolecular and vesicular tumor targeting boron delivery systems is their use of receptor sites that are either unique to tumor cells or grossly overexpressed by tumor cells relative to the cells which comprise normal tissues.

BORON DELIVERY USING IMMUNOPROTEINS

Bale and Spar[1] in 1967 proposed the use of tumor associated antibodies as means to deliver ^{10}B to malignant cells for BNCT. At the time, the only antibodies available for this purpose were polyclonal, structurally heterogeneous and not available in large quantities. Nonetheless, we[2] and others[3] successfully conjugated available antibodies in a random manner using small boron-rich reagents such as the 4-carboranyldiazonium cation[2] and functionalized polyhedral borane anions[3]. These earliest attempts to produce viable boron-containing polyclonal (and monoclonal[4] at a later date) conjugates were generally unsuccessful since only a few conjugation reagents could be attached to the antibody before immunoreactivity was lost. In addition, the boron contents of the conjugates were much too low for further consideration. Water-solublizing groups[5] including sugars[6] were incorporated into boron conjugation reagent structures with little avail. Interest in the use of

Advances in Neutron Capture Therapy, Edited by
A.H Soloway *et al.*, Plenum Press, New York, 1993

immunoproteins as boron delivery vehicles waned until the discovery of hybridoma procedures by Millstein and Köhler which made monoclonal antibodies (Mabs) plentiful[7].

Simple arithmetical considerations led to the conclusion that each of the 10^6 antigenic sites assumed to be located on each tumor cell would require approximately 10^3 ^{10}B-atoms in order to achieve the 10^9 ^{10}B-atoms per cell estimated for effective BNCT. This result could only be achieved if each Mab molecule were bonded to at least 10^3 ^{10}B-atoms. The obvious method for achieving such a high boron loading of protein was to attach boron-rich oligomers to only a few random points in the Mab structure, usually indigenous lysine ε-amino groups. The first efforts in this direction are exemplified by the work of Barth and Soloway[8] in which the ε-amino groups of poly-L-lysine were reacted with boron-rich conjugation reagents derived from the $B_{10}H_{10}^{2-}$ polyhedral ion. The conjugated oligomer was then coupled to Mab to achieve a product with 2 x 10^3 B-atoms per Mab molecule, but with greatly reduced immunoreactivity. In this instance the nonuniformity of the poly-L-lysine oligomer, the randomness of the conjugation of these species with boron reagent and the similar randomness of the conjugation of the oligomeric reagent with Mab lysine ε-amino groups yields a highly heterogeneous product which would be difficult to reproduce if desired and imposible to structurally modify in a controlled manner. Very recent work[9] with a "starburst" dendrimer carrying 48 primary amino groups provided an oligomer of uniform molecular weight, but subsequent conjugation of the polyamine to boron reagent and coupling of the product to Mab introduced structural uncertainties. However, the product contained over 2 x 10^3 B-atoms per mab molecule while the immunoreactivity was about 80% of that presented by the native antibody. Animal experiments[10] with these dendrimer-containing conjugates gave excessive liver uptake relative to tumor; behavior characteristic of every highly substituted Mab examined thus far.

Rather than attack pristine Mab molecules with a randomized array of oligomeric conjugation reagents we have synthesized designed oligomeric structures for this purpose.[11,12] These precisely synthesized oligomers were peptides comprised of boron-rich α-amino acids and amino-terminated with a fluorescent group. The carboxyl terminus was employed for active ester mediated conjugation with Mab lysine residues in a random fashion. The peptides themselves were trivially designated as "boron trailer reagents" or simply, "trailers" and synthesized from the amino acids illustrated in Figure 1 using Merrifield solid phase syntheses. Figure 2 depicts the structure of an undecapeptide which was conjugated with T84.66 anti-CEA Mab and studied with nude mice bearing ten-day-old LS 174T tumor xenografts. A fluorescent conjugate bearing 600 B-atoms per Mab was obtained [12,13] and shown to have retained 95% of the original Mab immunoreactivity. In mice,[13] 43 and 4.3% of the injected dose of conjugate per gram of tissue localized in the liver and tumor tissues, respectively, 48 hours post injection (27 μg/mouse). Blood contained only 1.2% ID/g. These partial biodistribution results were in agrement with data obtained by others[8-10] and the dramatic uptake of Mab conjugate by liver is identified as the single largest problem to be

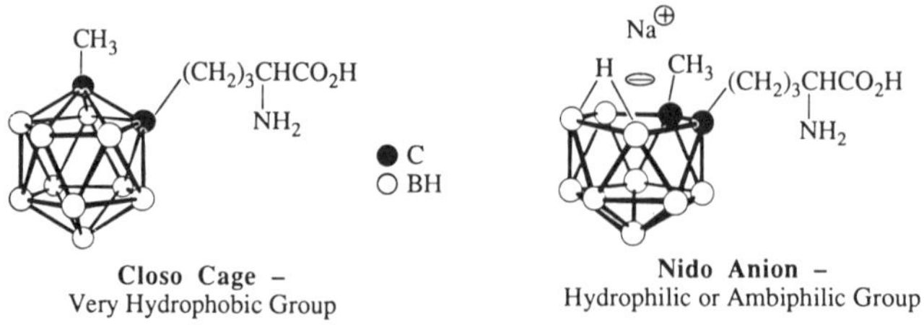

Figure 1. Two boron-rich α-amino acids employed in peptide synthesis as their *t*-BOC derivatives.

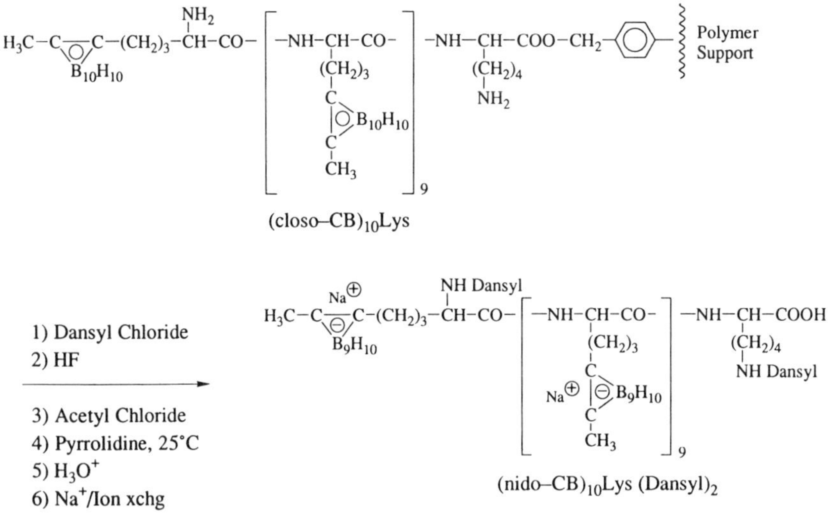

Acetyl may occasionally replace dansyl in these species.

Figure 2. An undecapeptide trailer reagent synthesized using the Merrifield method.

surmounted if immunoproteins are to ever succeed as boron carriers for BNCT. While Mabs coupled with trailers provide a less disordered immunoconjugate than those previously available, it is clear that the random attachment of trailers to Mab molecules is not amenable to complete structural control. Also, the use of a large IgG Mab as the basic molecular structure to be modified may contribute to excessive liver accretion. Accordingly, we have begun a new approach to this endeavor which makes use of an engineered FabSH anti-CEA fragment conjugated via its thiol group to a manifold peptide which, in turn, collects a fixed number of trailer oligomers. The latter species could be based upon either peptide or diphosphate ester structures. Figure 3 illustrates the construction of the complete FabSH conjugate. The total array could have a molecular weight of approximately 50kD. At this time, work is progressing on the FabSH (and dithiol Fab') molecules and new hydrophilic trailers and manifold structures. Oligomeric diphosphate esters of carboranyl diols appear to be attractive candidates for trailer applications.

It has been pointed out by us[12] that bispecific antibodies might serve to initially attach large boron-rich arrays to tumor cells by serving as a bridge between tumor cell wall antigen and hapten groups present in free trailer species. Figure 4 presents a pictorial representation

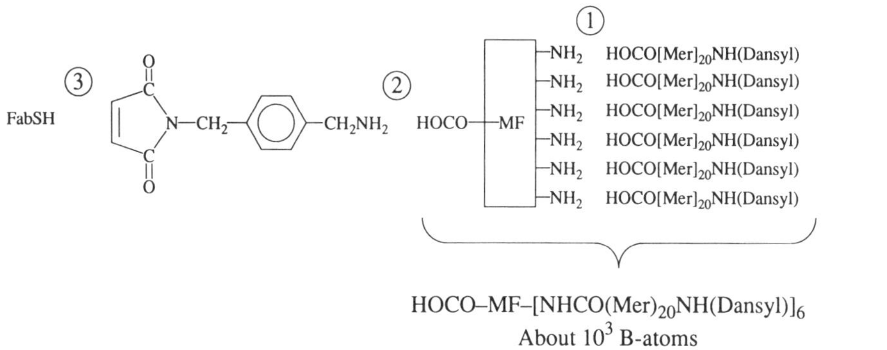

Figure 3. Sequence of reactions leading to immunodirected boron-rich oligomers. The HOCO-$(Mer)_{20}NH_2$ precursor is a peptide or a polyamide with 20 boron-rich repeating units. The HOCO-$(MF)(NH_2)_6$ species is a peptide manifold containing five lysine residues.

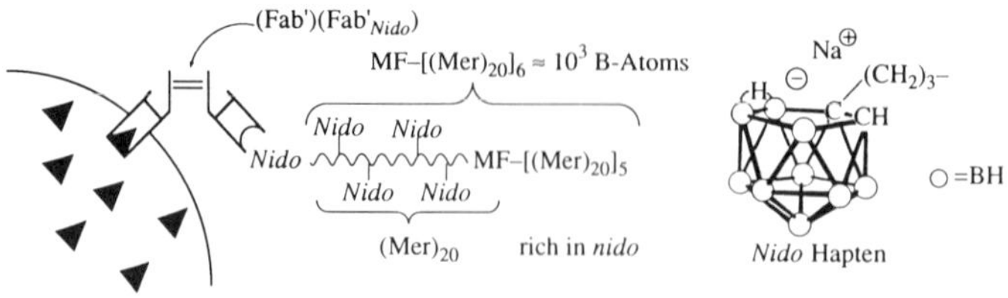

Figure 4. The bispecific antibody binding hapten-containing boron-rich molecule to cell wall CEA. The synthetic bispecific antibody is initially injected and binds to tumor cell wall CEA. The manifold trailer MF-$[(Mer)_{20}]_6$ is injected later and the *Nido* hapten targets its specific antibody binding site now held to the cell wall by CEA antigen/antibody interaction.

of this concept which would hopefully attach 10^3 B-atoms to each bispecific antibody which is complexed with a cell wall antigen without the use of covalent chemical bonds. The bispecific antibody will be synthesized by linking a Fab'_{CEA} to a Fab'_{nido}. The required Fab' fragments are obtained from their respective monoclonal IgG by enzymatic cleavage. While the Fab'_{CEA} is available from well known Mabs, the Fab'nido (which recognizes the [*nido*-7,8-$C_2B_9H_{11}$-]$^-$ hapten) is derived from an IgG_{nido} available from a newly cloned, but productive hybridoma. The hapten employed in the trailer molecule also serves as the source of boron and could be of either the peptide or diol diphosphate type.

While one considers bridging, non-covalent bonding interactions which may secure a boron reservoir to a tumor cell wall antigen prior to endocytosis of the combined structure, one should not overlook the possible application of the classical four-fold streptavidin $(biotin)_4$ host-guest complex having an equilibrium constant for biotin binding of approximately 10^{15}. The desired bispecific system could be constructed from the FabSH antibody fragment described above. Bonding of the thiol function of FabSH to streptavidin with a linker would provide an immunoprotein capable of both cell wall antigen complexation and binding with up to four biotin-terminated trailer molecules which are also fluorescence labeled for *in vitro* analytical purposes. Work has begun with component prototypes of this system.

GROWTH FACTOR RECEPTORS AS TARGETS FOR BORON DELIVERY

Although the cell wall antigens associated with a variety of tumor cell lines have been investigated as Mab conjugate receptor sites, it is possible to employ the receptors of growth and regulatory peptides, such as epidermal growth factor (EGF), as targets for Mabs[14] conjugated with boron trailer molecules.

Since receptors such as those associated with EGF are extravagantly expressed by many types of tumor cells compared to nonmalignant cells, it is also possible that the parent peptide such as the 53-amino acid EGF can be conjugated with boron-rich trailer reagents and employed for boron delivery through the normal EGF pathway.[15] Since the EGF and related growth factor molecules are relatively small, the size of the attached trailer molecules would be critical. A small trailer would be less perturbing to the biological function of the EGF, but the boron-delivery capability of such a conjugate would be compromised. A large trailer would provide more boron per conjugated EGF molecule, but receptor binding and cell entry of the conjugate could be impaired. Cultured glioma cells have been employed to demonstrate the cytotoxicity of EGF-^{131}I-dextran conjugates.[16]

NATURAL AND SYNTHETIC VESICLES AS BORON DELIVERY VEHICLES

Vesicles, such as the naturally occurring low density lipoproteins[17] (LDL's) and synthetic species such as uni-and multilamellar liposomes[18] have the potential to deliver relatively huge quantities of boron-containing species to tumor cells. The LDL delivery mechanism is well understood and is mediated by apoprotein B-100, imbedded in the phospholipid shell of the LDL, which binds to the tumor cell LDL receptor site. Such receptor sites are overexpressed by rapidly growing tumor cells which require LDL-delivered cholesterol for their continued growth. Replacement of the cholesteryl fatty acid esters, which fill the interior of the phospholipid shell of the LDL, with hydrophobic fatty alcohol esters of carborane carboxylic acids provides an LDL which is rich in boron and apparently unchanged in its ability to bind LDL receptor sites. Following complexation of the boron-containing LDL with the tumor cell wall receptor, the receptor-LDL array is internalized, stripped of the LDL and the receptor returned to the cell wall surface. This pumping action will proceed as long as LDL's are available and the cells internal regulatory mechanism perceives a need for additional cholesterol. Since the boron-rich LDL's contain little available cholesterol, active LDL uptake may proceed in an essentially unfettered manner until terminated by another means. This LDL delivery method was devised by Kahl[17] and recently shown to provide as much as 240 ppm of intracellular boron in tumor cells grown in culture. Although the required LDL's must be harvested from a compatible donor and converted to the desired product, this inconvenience is mitigated by the excellent tumor cell boron levels achieved.

Figure 5 depicts the cutaway structure of a synthetic liposome and the structure of the phospholipid which, when mixed with an equal weight of cholesterol, provides the lipid bilayer. Such unilamellar liposomes have been under investigation in our laboratories for the past five years. The selective delivery of therapeutic quantities of boron to tumor (EMT 6 murine adenocarcinoma and P1798 murine lymphosarcoma) in mice with high tumor/blood ratios and low injected doses (5-15 mg/kg body wt.) has been demonstrated[18]. The selective accretion of boron in tumors is thought to be the result of liposome leakage from immature *ad hoc* tumor vasculature into the tumor bed. The migration of the free liposomes to tumor cells followed by endocytosis has been demonstrated using fluorescence microscopy with fluorescent liposomes. The boron-containing species introduced as a solute in the aqueous core of the liposome is the sodium salt of a polyhedral borane anion. Considerable resources have been devoted to the discovery of efficacious borane derivatives and this effort continues along with a search for suitable boron-containing compounds which may be added to the lipid bilayer without compromise of the structural integrity of the liposome. Borane derivatives capable of reaction with indigenous intracellular protein amino groups have been designed

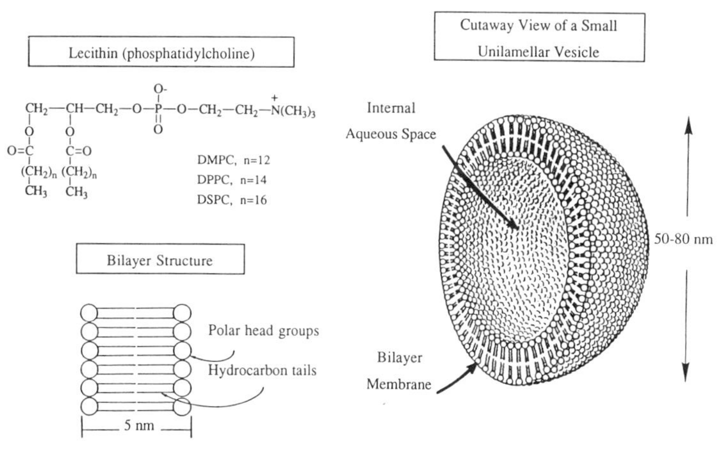

Figure 5. Cutaway cross-sectional view of a typical liposome, its bilayer construction and the phospholipid used in the bilayer.

and synthesized. Such reactivity insures retention of the boron species within the cell while blood, liver and spleen are cleared. Figure 6 presents a time point biodistribution experiment. Only tumor, blood, liver and spleen data are presented because all other tissues do not take up appreciable quantities of boron. The ease with which liposomes can be prepared coupled with the freedom they provide for the development of reactive boron derivatives make them among the most attractive known candidates for the realization of BNCT.

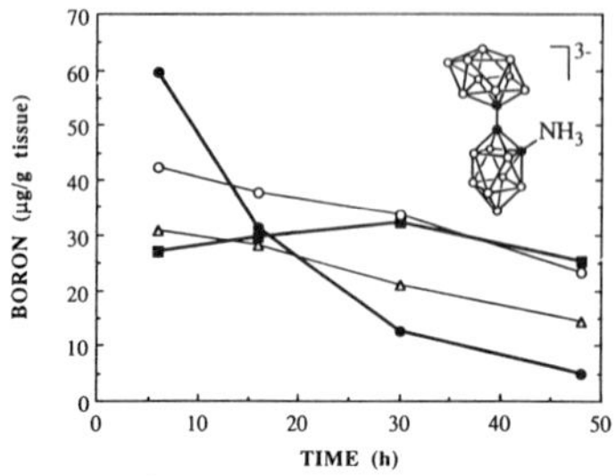

Figure 6. Biodistribution of liposomal $Na_3B_{20}H_{17}NH_3$ in Balb/c Mice bearing EMT 6 tumors. Injected dose 11mg/kg body weight. ●, Blood; ■, tumor; ○, liver and Δ, spleen. After 48 hrs. tumor contains 25.4ppm B and tumor/blood = 5.3.

ACKNOWLEDGEMENTS

The author wishes to thank his coworkers and colleagues in the BNCT community for their direct and indirect contributions to this paper. He also thanks the National Cancer Institute for continuing support through grant R01-CA31753 and the Idaho Engineering Laboratory of the U.S. Department of Energy under DOE field Office, Idaho, Contract DE-AC07-76IDO1570.

REFERENCES

1. Bale, W.F. & Spar, I.L. (1967). *Adv. Biol. Med. Phys.* **5**, 286-356.
2. Hawthorne, M.F., Wiersema, R.J. & Takasugi, M. (1972). *J. Medicinal Chem.* **15**, 449-452.
3. Barth, R.F., Johnson, C.W., Wei, W-Z., Carey, W.E., Soloway, A.H. & McGuire, J. (1982). *Cancer Detect. Prevent.* **5**, 315-323.
4. Mizusawa, E., Dahlman, H.L., Bennett, S.J., Goldenberg, D.M. & Hawthorne, M.F. (1982).*Proc. Natl. Acad. Sci. USA*, **79**, 3011-3014.
5. Sneath, R.L., Wright, J.E., Soloway, A.H., O'Keefe, S.M., Dey, A.S. & Smolnycki, W.D. (1976). *J. Medicinal Chem.* **19**, 1290-1294.
6. Maurer, J.L., Berchier, F.B., Serino, A.J., Knobler, C.B. & Hawthorne, M.F. (1990). *J. Org. Chem.*, **55**, 838-842.
7. Köhler, G. & Milstein, C. (1975). *Nature*, **256**, 495-497.
8. Alam, F., Soloway, A.H., Barth, R.F., Mafune, N., Adams, D.M. & Knoth, W.H. (1989). *J. Medicinal Chem.*, **32**, 2326-2330.
9. Barth, R.F., Soloway, A.H., Adams, D.M. and Alam, F. Delivery of boron-10 for neutron capture therapy by means of monoclonal antibody-starburst dendrimer immunoconjugates in: "Progress in Neutron Capture Therapy for Cancer", Allen, B.J., Moore, D.E. and Harrington, B.V. eds., Plenum Publishing Corp., New York (1992).
10. Soloway, A.H., Private communication.
11. Varadarjan, A., & Hawthorne, M.F. (1991). *Bioconjugate Chemistry*, **2**, 242-247.
12. Hawthorne, M.F. (1991). *Pure Appl. Chem.* **63**(3), 327.
13. Paxton, R.J., Beatty, B.G., Hawthorne, M.F., Varadarajan, A., Williams, L.E., Curtis, F.L., Knobler, C.B., Beatty, J.D., & Shively, J.E. (1991). *Proc. Natl. Acad. Sci. USA*, **88**, 3387-3391.
14. Mendelsohn, J. (1990). *Seminars in Cancer Biology*, **1**, 339-344.
15. Malmquist, J. & Sjöberg, S. (1992). *Inorg. Chem.*, **31**, 2534-2537.
16. Andersson, A., Capola, J., Carlsson, J. (1992). *Neuro-Oncology*, in press.
17. Laster, B.H., Kahl, S.B., Popenoe, E.A., Pate, D.A., & Fairchild, R.G. (1991). *Cancer Res.* **51**, 4588-4593.
18. Shelly, K., Feakes, D.A., Hawthorne, M.F., Schmidt, P.G., Krisch, T.A., & Bauer, W.F. (1992). *Proc. Natl. Acad. Sci. USA*, **89**, 9030-9043.

IN VIVO DISTRIBUTION OF BORONATED MONOCLONAL ANTIBODIES AND STARBURST DENDRIMERS

Rolf F. Barth[1], Dianne M. Adams[1], Albert H. Soloway[2] and Michael V. Darby[2]

Department of Pathology and College of Pharmacy[2], The Ohio State University, Columbus, Ohio 43210, U.S.A.

INTRODUCTION

Monoclonal antibodies (MoAbs) are theoretically among the most specific of all tumor targeting agents. In the span of ten years a voluminous literature has developed on their potential use for the diagnosis and treatment of cancer. Drugs, radionuclides, and toxins have been linked to MoAbs, but, unfortunately, their therapeutic efficacy has been limited due to a variety of problems[1]. Attempts to use MoAbs to target ^{10}B to tumors for BNCT have been unsuccessful for many of these same reasons. In order to sustain a lethal $^{10}B\ (n,\alpha)Li^7$ reaction at the cellular level, approximately $\approx 10^9$ ^{10}B atoms must be delivered to each tumor cell[2]. This requirement has led to the use of macromolecules such as boronated poly-D,L-lysine (BPL) or boronated starburst dendrimers (BSD), which can be linked to MoAbs by means of heterobifunctional reagents[3,4]. The purpose of the present study was to evaluate the *in vivo* distribution of a MoAb directed against the murine B16 melanoma[5] that has been boronated using a starburst dendrimer, and to compare this to the distribution patterns of native MoAb and unmodified dendrimers of varying molecular weights.

MATERIALS AND METHODS

Iodination of MoAbs and starburst dendrimers. MoAb IB16-6, boronated starburst dendrimers (BSD) and boronated IB16-6 were radiolabeled with ^{125}I using the iodogen oxidation method. Twenty μg of iodogen (1,3,4,6 - tetrachloro -3α, 6α diphenylglycoluril, Pierce Chemical Co., Rockford, IL) in chloroform were evaporated to dryness in 10 x 75 mm soda lime glass tubes. One mg of dendrimer or MoAb was dissolved in 200 μl of sodium phosphate buffer, pH 7.2, and added to the reaction tube. This was followed by the addition of 1 mCi of ^{125}I sodium iodide (Dupont-NEN Boston, MA). The reaction was allowed to proceed for 10 minutes, after which unbound ^{125}I was removed from the radiolabeled substrate by Sephadex G-25 column chromatography using PD-10 disposable columns (Pharmacia-LKB Technology Piscataway, NJ). The percentage of ^{125}I bound to the protein was determined by instant thin layer chromatography and trichloracetic acid precipitation. If the level of binding was <85%, the sample was further purified using a Centricon 10 micro-concentrator (Amicon, Danvers, MA) in order to eliminate the remaining unbound ^{125}I. Starburst dendrimers (SD) are precision macromolecules composed of repetitive polyamido amino groups arranged in a starburst pattern[6]. The number of reactive terminal amino groups range from 3 (1° generation) to 6144 (12° generation), and molecular weights range from 3.6 x 10^2 to 1.40 x 10^6 D. In the present study 1°, 2°, 3°, 4° and 5° generation dendrimers (Polysciences Inc., Warrington,PA) having 3, 6, 12, 24 and 48 reactive terminal amino groups respectively, and molecular weights ranging from 360 to 10633 D were radiolabeled using Bolton-Hunter reagent (ICN Biomedicals Inc., Costa Mesa, CA) with minor modifications of the procedure described in the product insert.

Conjugation of antibody. Fifth generation starburst dendrimers were linked to MoAb IB16-6 using the heterobifunctional reagent m-maleimidobenzoyl-N-hydroxysulfosuccinimide ester (sMBS) to derivatize the dendrimer, and N-succinimidyl 3-(2-pyridyldithio) propionate (SPDP) to derivatize the MoAb, as previously described[4]. Dendrimers were boronated using $Na(CH_3)_3NB_{10}H_8NCO$ as the boronating species, and sMBS and SPDP as the linkers to IB16-6.

Biodistribution of ^{125}I radiolabeled SD, BSD and immunoconjugates. C57Bl/6 mice were placed on drinking water supplemented with 0.1% Lugol's iodine solution 48 hours prior to the initiation of and then for the duration of the distribution studies. Groups of four mice were used for each time point, and were injected intraperitoneally (i.p.) with the different ^{125}I labeled test agents. Animals were bled via the retro-oribital sinus at varying time intervals, ranging from 1 to 120 hours post injection, and then were killed by cervical dislocation. Liver, spleen and muscle samples were obtained, and radioactivity was determined by counting in a Tracor Analytic gamma scintillation counter. Distribution was determined as percent of the total dose of radioactivity administered, and expressed as percent uptake per gram tissue. For tumor localization studies animals were implanted subcutaneously (s.c.) with Bl6 melanoma cells into the right flank, and the tumor was allowed to grow to ≈ 4 mm in diameter at which time animals were injected i.p. with the test agents.

RESULTS

The distribution data for ^{125}I labeled MoAb IB16-6, ^{125}I-boronated BSD and ^{125}I-BSD-IB16-6 immunoconjugate following i.p. administration to C57B1/6 mice carrying the B16 melanoma are shown in Fig 1. Less than 0.6% of the native antibody localized in the liver at 1 day (Fig 1A) and this rapidly decreased to <0.2% by 5 days. In contrast, 20% of the injected dose of BSD (Fig 1B) and 11% of the ^{125}I-BSD-IB16-6 (Fig 1C) were present in the liver at 1 day and this decreased to 13% and (Fig 1B) 8% (Fig 1C) respectively at 3 days. The amount ^{125}I-IB16-6 localized in the tumor (Fig 1A) was 0.4% at 1 day and 0.2% at 5 days, indicating that although this MoAb was highly specific *in vitro*[5], it had poor *in vivo* tumor localizing activity.

A fourfold increase in the number of boron atoms per molecule of dendrimer (Fig 1D) did not produce any biologically significant alterations in distribution compared to that seen in Fig 1C. These observations raised questions concerning the suitability of starburst dendrimers as boron carriers and led to the next series of experiments summarized in Fig. 2. The distributions of ^{125}I labeled 1°, 2°, 3°, 4° and 5° generation, non-boronated starburst dendrimers (SD-1, SD-2, SD-3, SD-4 and SD-5) were studied in non-tumor bearing C57B1/6 mice. Animals were killed at 1, 6, 24 and

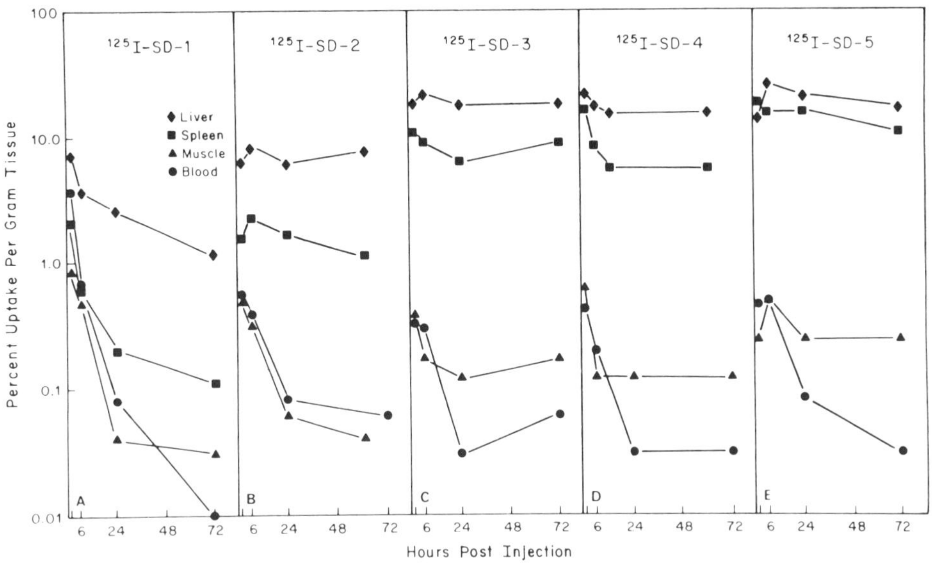

Figure 1. Distribution of ^{125}I labeled MoAb immunoconjugates and dendrimer in B16 melanoma bearing mice. All test agents were iodinated using iodogen. Boronated starburst dendrimer BSD (Fig 1B) was prepared using $Na(CH_3)_3NB_{10}H_8NCO$ as the boronating species and a fifth generation dendrimer. BSD was conjugated to MoAb IB16-6 using s-MBS and SPDP as linkers (Fig 1C)[4]. BSD'-IB16-6 (Fig 1D) had 4x more boron per molecule of antibody than BSD-IB16-6 shown in Fig 1C. Test agents were injected i.p. into C57B1/6 mice carrying s.c. implants of the B16 melanoma, and the animals were killed 1, 3 or 5 days later.

72 hours following i.p. administration of the iodinated dendrimers. Six hours following injection, hepatic localization was 3.7% for SD-1, 8.5% for SD-2, and ranged from 17.6 to 26.8% for SD-3, SD-4, and SD-5. By 72 hours this had decreased to 1.2% for SD-1, 7.9% for SD-2, and ranged from 15.3 to 20% for SD-3, SD-4 and SD-5. The amounts of SD-1 detected in the spleen decreased from 0.6% at 6 hours to 0.1% at 72 hours, and remained constant at 1% for SD-2. In contrast, splenic values for SD-3, SD-4 and SD-5 were 9, 10, and 16% respectively at 6 hours, and had not fallen from these levels at 72 hours.

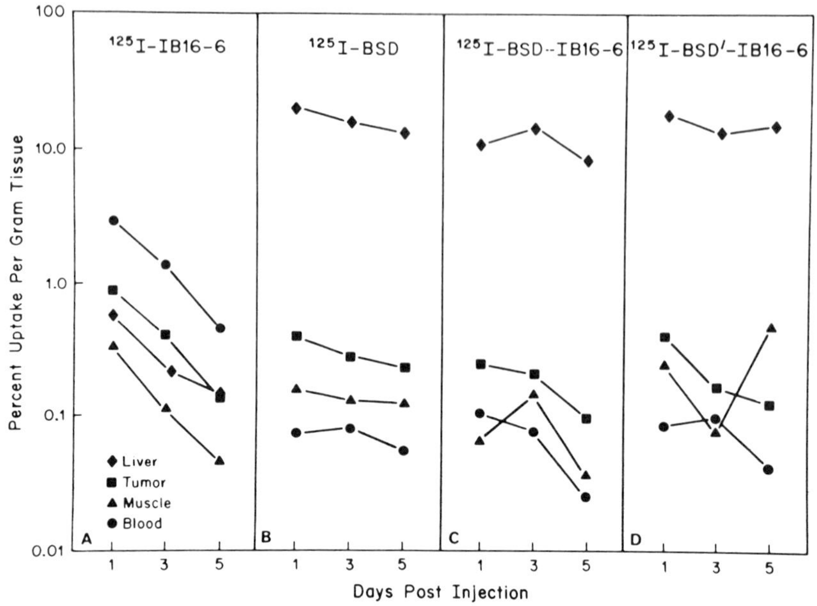

Fig 2. Distribution of ^{125}I labeled 1°, 2°, 3°, 4° and 5° generation starburst dendrimers. All dendrimers were iodinated using the Bolton-Hunter reagent. Test agents were injected i.p. into non-tumor bearing C57B1/6 mice, and the animals were killed 1, 6, 24 and 72 hours later.

DISCUSSION

The data obtained in the present series of experiments demonstrate that starburst dendrimers have an intrinsic propensity to localize in the liver and spleen, the two major organs of the reticulo-endothelial system. The absolute amounts localizing in the liver appeared to be directly related to the generation of the dendrimer, i.e. molecular weight and the number of reactive terminal amino groups. The distribution of BSD-5 (Fig 1B) and MoAb IB16-6 linked to BSD-5 (BSD-IB16-6) (Fig 1C) were identical, indicating that the boronated dendrimer probably was the major determinant for increasing hepatic localization. Increasing the number of boron atoms per molecule of antibody did not result in any further increase in hepatic localization (Fig 1D). The amounts of IB16-6 localizing in subcutaneous implants of the B16 melanoma decreased from 1% for the native, non-boronated antibody to 0.2% for the immunoconjugate BSD-IB16-6 (Fig 1C). The small quantities of native IB16-6 that localized in the tumor preclude the use of starburst dendrimers of the molecular size necessary to heavily boronate this MoAb, so that it could deliver 10^3 boron atoms per molecule. Although hepatic localization of lower molecular weight dendrimers, such as SD-1 and SD-2 was decreased, the amount of boron that each of these could provide to an antibody molecule makes them unsuitable. In conclusion, starburst dendrimers do not seem to be well suited for the linkage of boron to MoAbs for immunotargeting, if they have to be administered systemically. On the other hand, they may be very useful for linking boron to antibodies and other ligands for use in immuno-electron microscopy, where the boron atoms can be detected by means of electron energy loss spectroscopy/electron spectroscopic imaging[7]. MoAbs still remain attractive targeting agents for the delivery of ^{10}B to tumors, but new approaches are necessary. One possibility is to use bispecific antibodies, one combining site of which recognizes a tumor associated antigen, and the other a boronated macromolecule, such as a starburst dendrimer. Studies currently in progress in several laboratories should determine whether or not this approach is feasible.

ACKNOWLEDGEMENTS

We thank Polysciences,Inc. Warrington, PA for generously providing us with starburst dendrimers and Stephanie Walters for secretarial assistance. This work was supported by Department of Energy grant DE-ACO2-76CH000016.

REFERENCES

1. Waldmann, T.A. (1991) Science 252, 1657-1662.
2. Barth, R.F., Soloway, A.H. and Fairchild, R.G. (1990) Cancer Res. 50, 1061-1070.
3. Alam, F., Barth, R.F. and Soloway, A.H. (1989) Antibody, Immunoconjugates and Radiopharmaceut. 2, 145-163.
4. Barth, R.F., Soloway, A.H., Adams, D.M. and Alam F. (1992) in Prog. in Neutron capture Therapy, Plenum Press. pp 265-268.
5. Johnson, C.W., Barth, R.F., Adams, D., Holman, B., Price, J.E. and Sautins, I. (1987) Cancer Res 47:1111-1117.
6. Roberts, J.C., Adams, Y.E., Tomalia, D.E., Mercer-Smith, J.A. and Lavellee, D.K. (1990) Bioconjugate Chem 1 305-308.
7. Bendayan, M., Barth, R.F., Gingras, D., Londono, I., Robinson, P.T., F., Alam F., Adams, D.M., and Mattiazzi, L. (1989), J. Histochem and Cytochem 37:573-580.

BINDING OF BORONATED POLYLYSINE TO IMMUNOGLOBULIN BY WAY OF GLYCOSIDE MOIETIES: IMMUNOREACTIVITY AND BORON CONTENT

S. Novick[1], M.R. Quastel[1], S. Marcus[2], D. Chipman[2], G. Shani[3], R. F. Barth[4] and A. Soloway[5]

Depts of Nuclear Medicine[1], Biology[2] and Nuclear Engineering[3] of the Soroka Medical Center and Ben Gurion University of the Negev, Beer-Sheva, Israel; Depts of Pathology[4] and Pharmacy[5], Ohio State University, Columbus, Ohio, U.S.A.

INTRODUCTION

This report shows a new approach for linking the boronated polylysine(BPL) at specific sites on the antibody that are distant from the antigen binding site, that is, to the group made functional by oxidation of the carbohydrate residues present at the CH_2 region of the antibody. By this technique we estimate that over 10^4 boron atoms are bound per antibody molecule without demonstrable loss of immunoreactivity.

MATERIALS AND METHODS

A. Binding of BPL to IgG

Polylysine was heavily boronated with isocyanate borane derivative $NaMe_3NB_{10}H_8NCO$ as described by Alam et al[1].

SPDP [{3-(2-pyridyldithio)proprionic acid}-N-Hydroxy succinimide ester] was added to BPL and incubated at room temperature for 1 hour. The complex was separated by passage through a Sephadex G25 column. Disulfide bonds were reduced using Cleland's reagent (DL-Dithiothreitol) for 1 hour. The product was separated on Sephadex G25 and was termed 'A'.

Immunoglobulin was oxidized using sodium m-periodate for 2 hours in the dark on ice and then separated on a Sephadex G25 column. Hydrazine was added to MBS-Sulfo (Maleimidobenzoyl-N-Hydroxysulfosuccinimide ester) and MB-hydrazide was extracted with diethyl ether. The MB hydrazide was added to the oxidized IgG for an hour, followed by sodium cyanoborohydride for 3 hours in the dark. The mixture was separated on a Sephadex G25 column. The resulting product was termed 'B'.

The treated polylysine (A) was added to the MB-hydrazine treated-IgG (B). Coincubation of A and B for 24 hrs at room temp. gave the antibody conjugated with boronated polylysine:

Advances in Neutron Capture Therapy, Edited by
A.H Soloway *et al.*, Plenum Press, New York, 1993

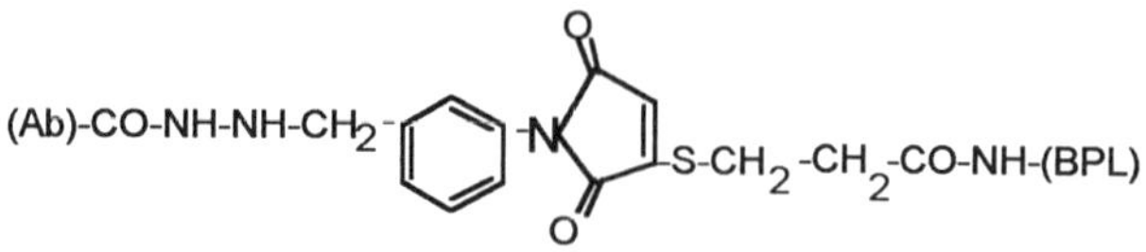

Final separation of products was done on a Sephacryl S-300 column, and aliquots were collected with a fraction collector. Fractions were eluted with 80% Trisma buffer at pH=8.5/20% DMF. 70-75 fractions were collected. Protein content per fraction was measured by the Lowry method or by absorbance at 280 nm. Fluorescence was measured at the maximum excitation of 340 nm and maximum emission at 510 nm.

B. Immunoreactivity of the IgG complex

Immunoreactivity of the IgG-BPL complex was measured quantitatively by modifying an IRMA (immunoradiometric) test. The basis for this test is the competition of two antibodies for an antigen (report in preparation). We used for this purpose a double antibody IRMA test designed for the measurement of serum thyroid stimulating hormone (TSH). The antibodies used were (a) ^{125}I-radiolabeled monoclonal anti-TSH and (b) the BPL-bound polyclonal anti-TSH prepared according to the technique already described.

The procedure is summarized as follows: A given amount of high TSH serum obtained from hypothyroid patients was placed in a series of test tubes containing the first antibody of the TSH test. To these tubes were added aliquots of fractions containing the BPL-IgG complex following passage through the Sephacryl S300 column, followed immediately by ^{125}I-anti-TSH monoclonal antibody. The degree of competition of the polyclonal BPL-IgG complex with the radiolabeled monoclonal was easily measured using an automated gamma counter, and provided a simple and reproducible measure of antibody immunoreactivity.

C. Measurement of boron concentrations

Measurements of boron concentrations were carried out by direct current plasma atomic emission spectroscopy[2] .

RESULTS

1. Binding of BPL to IgG

Figure 1 shows the passage of a dansylated polylysine-IgG complex through the Sephacryl-S300 column. Three peaks are observed. The first peak demonstrates the complex with the highest molecular weight, which is also labeled with dansyl chloride, evidence that the polylysine is bound to the antibody. The second peak which is not fluorescent, demonstrates free IgG. The last peak belongs to the lower molecular weight polylysine, and is labeled with fluorescent dansyl chloride.

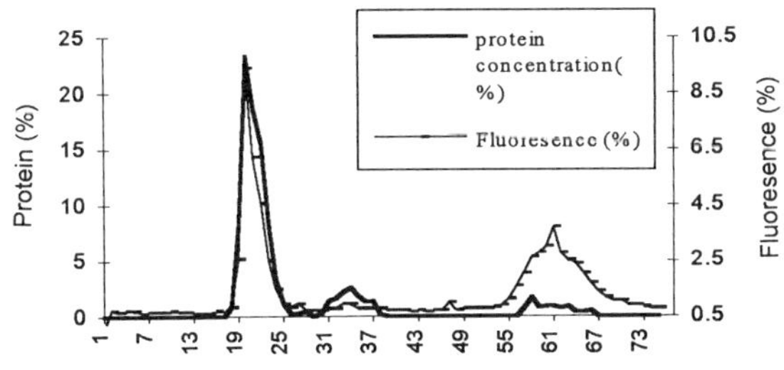

Figure 1. Elution pattern of IgG on sephacryl S-300 gel column chromatography.

2. Immunoreactivity of the IgG-BPL complex

Figure 2 shows the immunoreactivity of the BPL-IgG complex (using the modified double antibody IRMA procedure for serum TSH). The ordinate shows the counts per minute(cpm) of the TSH-antibody complex. Aliquots from the fractions of the BPL-IgG passed through the Sephacryl-S300 column were added to the reaction. In the absence of added BPL-IgG, the radiolabeled monoclonal anti-TSH bound maximally. The fall in the cpm over the region of the BPL-IgG indicates interference by BPL-IgG with binding of the radiolabeled second antibody. Unbound polylysine had no effect on the IRMA test. Denatured anti-TSH polyclonal IgG, which normally strongly interferes with the IRMA reaction, completely loses this property when heated. Both BPL-IgG (anti TSH) and IgG (anti-TSH) were equally and effectively competitive with the radiolabeled second monoclonal antibody of the IRMA, indicating that binding of the boronated polylysine by the method herein reported did not affect the measured immunoreactivity of the antibody.

3. Boron content of the BPL-IgG complex

Figure 2 also shows the relationship between immunoreactivity and boron content. There was a close to linear relationship between the total BPL-IgG concentration and the boron content (Fig. 3). This relationship indicates that there are about 10^4 atoms of boron per antibody -BPL complex.

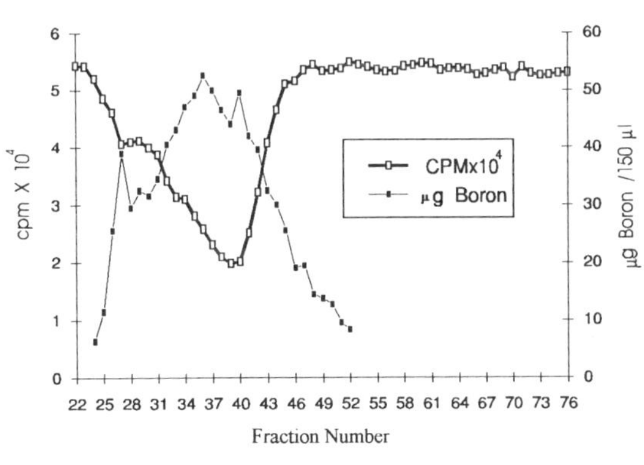

Figure 2. Immunoreactivity and boron concentration

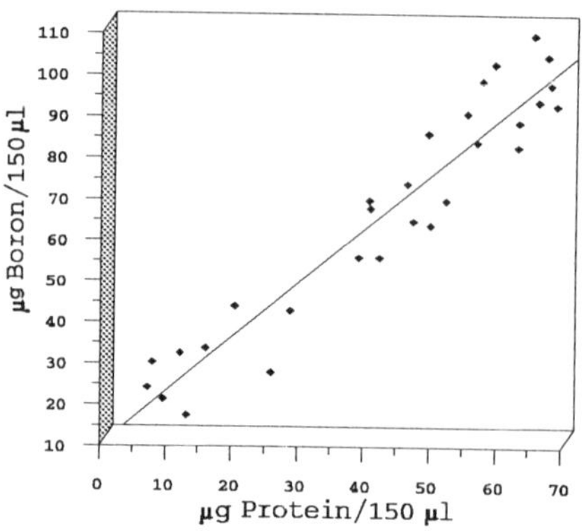

Figure 3. Boron concentration relative to the protein content

DISCUSSION

There are 2 major considerations of regarding the use of antibodies as carriers of boron to cells. The first is that sufficient boron is brought to the cells in order to render boron-neutron activation events highly likely. The second is that the binding of the boronated carrier to the antibody should not affect it's immunoreactivity. Prior studies have shown that when boronated polymers are bound to antibody, a significant deterioration of immunoreactivity may occur[3].

In this communication, we demonstrate a technique for the binding of BPL to IgG by way of the carbohydrate moieties of the antibody molecule, modified from that described by Rodwell et al[4]. When BPL is bound to antibody by the technique herein described, boron can be measured in the elution fractions proportional to the amount of IgG-BPL complex present. The amount of boron present was about 10^4 atoms per antibody molecule. If it is assumed that the number of binding sites on a cell is of the order of 10^5, the resultant total binding to the cell should be about 10^9. This is of the magnitude required to support an effective boron-neutron effect.

Our data suggest that the modified IRMA can serve as a simple quantitative technique for measuring changes in the immunoreactivity of an antibody. Using the TSH IRMA system, we show that binding BPL to anti-TSH polyclonal antibody by the technique herein described does not affect it s immunoreactivity.

REFERENCES

1. Alam F., Soloway A.H., Barth R.F., Mafune N., Adams D.M. & Knoth W.H. (1989) J. Med. Chem. **32**, 2326-2330.

2. Barth R.F., Adams D.M., Soloway A.H., Mechetner E.B., Alam F. & Anisuzzaman ABK (1991) Anal. Chem. **63**, 890-892.

3. Pettersson M.L., Courel M.N. , Girard N. , Abraham R. , Gabel D. , Thellier M. & Delpech B. (1990) J. Immunol. Meth. **126**, 95-102.

4. Rodwell J.D. , Alvarez V.L. , Lee C. , Lopes A.D. , Goers J.W.F , King H.D. , Posner H.J. & McKearn T.J (1986) Proc. Nat. Acad. Sc. (US) **82,** 2632-2636.

BORONATED PEPTIDES AND NUCLEIC ACID COMPONENTS FOR NCT

B.F. Spielvogel,[1] A. Sood,[1] J. Tomasz,[1,2] B.R. Shaw,[2] S. Karthikeyan,[1] W. Powell,[1] B. Laster,[3] R.M. Brugger,[3] and J. Coderre[3]

[1]Boron Biologicals, 533 Pylon Drive, Raleigh, NC 27606
[2]Dept. of Chemistry, Duke University, Durham, NC 27706
[3]Brookhaven National Laboratory, Upton, NY 11973

INTRODUCTION

We are interested in the synthesis, pharmacological, and neutron capture properties of boron-containing intermediate antimetabolites. These include analogues of amino acids,[1] peptides,[2] choline[3] derivatives, phosphonoacetates[4] or boranophosphates,[5] and more recently, the boron analogs of nucleosides[6] and nucleotides.[7]

Modified nucleic acid components are of increasing interest because of their potent antitumor and antiviral properties. In the case of boronated nucleosides and nucleotides, of considerable added importance is their potential for use in Boron Neutron Capture Therapy (BNCT). Boronated nucleic acid components may localize in the nucleus more efficiently than many of the other BNCT compounds in development today.

Various boronated amino acids and peptides also show promise for use in BNCT. We have prepared a number of dipeptides using normal amino acids (in particular, the aromatic amino acids tyrosine and phenylalanine) and amine-carboxyboranes (boron analogues of amino acids) and have examined their suitability for BNCT. We also report here data on several boronated nucleic acid components.

Chemistry

Although dipeptides containing an amine-carboxyborane and a normal amino acid can be prepared by general condensation methods, we have found that the PPh_3/CCl_4 coupling method is best.[2] The boronated nucleosides used in this report are all nucleosides containing a cyanoborane group on the N^7 position of the guanine base and can be prepared as previously reported.[6] The compounds used in this study are shown in Figure 1.

Advances in Neutron Capture Therapy, Edited by
A.H Soloway *et al.*, Plenum Press, New York, 1993

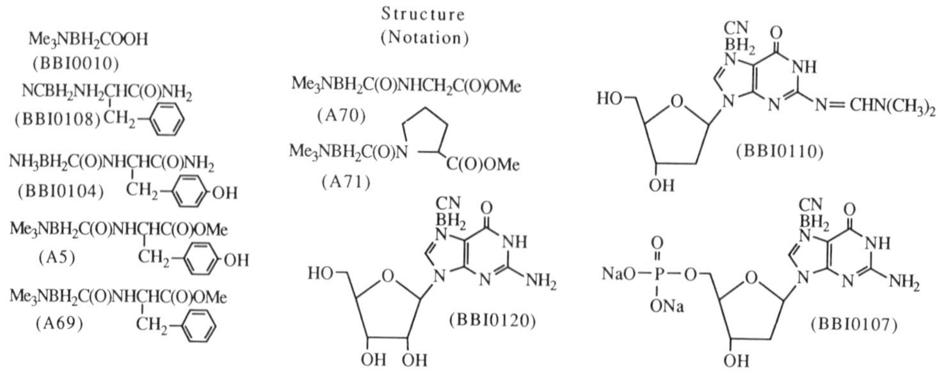

Figure 1. Structures of compounds used in the current study.

METHODS

Cell Lines

B16 Melanoma: Since the dipeptides contain amino acids that are precursors of the pigment melanin, these compounds might be expected to be taken up by melanoma cells through peptide transport systems to supply the demand for melanin precursors. The subclone B16G3.12 has been well characterized and has been used for several experiments with BPA. This subclone was maintained in alpha MEM to which 10% (v/v) fetal bovine serum, 1% (v/v) antibiotics (penicillin, 10,000 IU/ml and streptomycin 100 μg/ml) and 1% (v/v) L-glutamine (200mM) were added.

GS-9L: This cell line serves as a non-melanoma control to determine whether the dipeptides have a broader utility than just for melanoma. This cell line originated from an N-nitrosomethyl-urea-induced neoplasm in a Fischer 344 rat. Cells were maintained in DMEM medium supplemented with inactivated 10% fetal bovine serum.

V79 cells: V79 Chinese hamster lung fibroblastic cell line is a widely used normal cell line and is maintained in the same medium as the GS-9L cell line.

Cell Irradiations

Irradiation of cells in vitro was carried out as previously described. (Gabel et al, Radiat. Res. 43, 534, 1984) Before irradiation, cells were preincubated for 18 hours with the test boron compound. The same concentration of boron compound was maintained in the medium during trypsinization and harvesting, as well as during the reactor irradiation. For irradiation, 10^5-10^6 cells in 1.5 ml of growth medium were transferred into 1.5 ml polyethylene centrifuge tubes and inserted into a cell-vessel rotator. The tubes were rotated at 2 rpm in the thermal neutron beam of the Brookhaven Medical Research Reactor at a distance of 25 cm from the port for exposures (integrated time x reactor power) of 2, 4, 6, 9, 12, 16, 18, 20, or 24 MW.min. Following the irradiation, the cells were diluted with ^{10}B-free medium and plated into petri dishes for colony-forming survival assay. Two or three different dilutions were used per tube with five replicate dishes per dilution. After two weeks, the plates were washed with PBS, fixed with 4% formaldehyde in PBS and Giemsa-stained. Colonies with >50 cells each were counted. Survival fraction was expressed as percentage of the colony counts on control plates. The data were fit by a linear-quadratic line that was forced to pass through the origin or by a single-hit, multi-target fitting program.

For irradiations under washed conditions, the cells were incubated with the boron compound for 18 hours, followed by removal of the medium, three washes with cold phosphate buffered saline, trypsinization and resuspension in boron-free medium as described above.

Data Analysis

D_0 values were obtained by measuring the difference in irradiation time which reduced cell survival by a factor of 1/e (i.e. by 63%) on the linear portion of the curve. Results were analyzed by comparison to the effects of boric acid on the particular cell line. Calibration curves for each cell line were constructed by plotting $1/D_0$ versus boron (as boric acid) concentration. For a particular test compound, the $1/D_0$ value read-off of the calibration curve would indicate a "boric acid equivalent" concentration, or the concentration of boric acid (assumed to be uniformly distributed inside and outside the cell) required to produce the same cell kill effect. The boric acid equivalent value divided by the actual boron concentration in the incubation medium is termed the "Radiation Effect." A Radiation Effect of 1.0 indicates the same effectiveness as boric acid; a value of 2.0 indicates that the test compound is twice as effective as the same amount of boric acid; a value of 0.5 indicates that the test compound is only half as effective as the same concentration of boric acid. Radiation Effect values greater than unity can be interpreted as boron accumulation inside the cell to a greater degree than found in the incubation medium. Radiation Effect values less than unity could imply exclusion from the cell, or association with the cell membrane.

Toxicity

The plating efficiency (PE) of the cells exposed to the test compound divided by the PE of the untreated control cells. A toxicity value of 1.0 signifies no toxicity, whereas a value of 0 indicates complete lethality.

Results and Discussion

In the V-79 Chinese Hamster Lung Fibroblast cell experiments (Table I), the experiments with A5, washed, indicate a high degree of retention in the cells. A Radiation effect of 1.09 after washing was obtained. Further washing experiments and time course of washout are warranted. Both nucleosides BBI-0107 and 0110 are not permanently bound within the cell. Both wash out quickly (and completely for 0107).

In the B16 Murine Melanoma screen (Table 2), both A5 and A69 show a significant Radiation Effect; more effective than BPA in this particular cell line. BBI-0120, a nucleoside, shows significant binding and cell kill after washing. Further washed experiments and a time course of washout are warranted.

In the 9L-Rat Gliosarcoma cell line (Table 3), both A5 and A69 show very high Radiation Effects, equal to or greater than BPA. BBI-0010 is boro-betaine, the boron containing half of the dipeptides A5 and A69. BBI-0010 is not as effective as the dipeptides and only slightly less effective than boric acid.

Table 1. V-79 Chinese Hamster Lung Fibroblast Cells

Compound	Ambient/ Washed	Tox.	Rad. Effect	n
A69	Ambient	1.16	1.59	1
	Washed	1.13	0.35	1
A70	Washed	1.15	0.45	1
A71	Washed	1.11	0.00	1
A5	Washed	1.05	1.09	3
BBI-0107	Ambient	1.07	0.63	1
	Washed	1.05	0.00	1
BBI-0110	Ambient	0.78	0.54	1
	Washed	0.78	0.14	1

Table 2. B16 Murine Melanoma

Compound	Ambient/ Washed	Tox.	Rad. Effect	n
A5	Ambient	.82	1.80	2
	Washed		0.48	2
A69	Ambient	.89	1.33	2
	Washed		0.83	2
BBI-0108	Ambient	.38	0.95	1
	Washed		0.61	1
BBI-0120	Ambient	.71	0.91	1
	Washed		1.12	1
BBI-0104	Ambient	1.60	0.71	1
	Washed		0.38	1
Boric Acid	Ambient	1.00	1.00	1
BPA	Ambient	1.05	1.05	2
BSH	Ambient	0.66	0.50	2

Table 3. 9L - Rat Gliosarcoma

Compound	Ambient/ Washed	Tox.	Rad. Effect	n
A5	Ambient	1.11	3.85	3
	Washed	1.11	0.56	3
A69	Ambient	0.82	2.45	5
	Washed	0.82	0.46	3
BBI-0010	Ambient	1.19	0.89	2
BPA	Ambient	0.84	2.5	5
	Washed	0.90	1.3	
BSH	Ambient	0.83	0.72	5

Cell Survival Curves

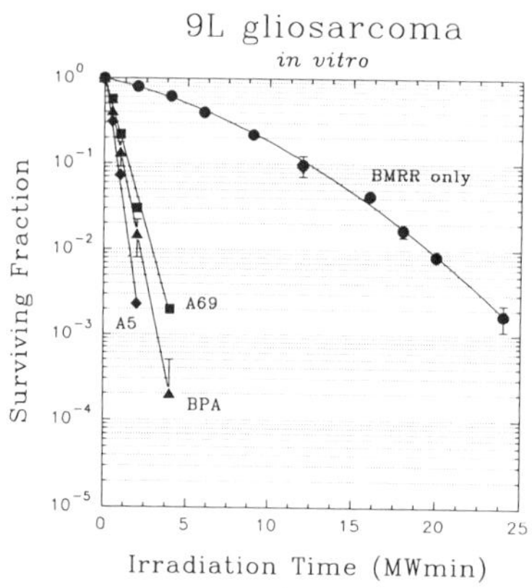

Figure 2. Effects of dipeptides A5 and A69 on 9L rat gliosarcoma cell survival under ambient conditions.

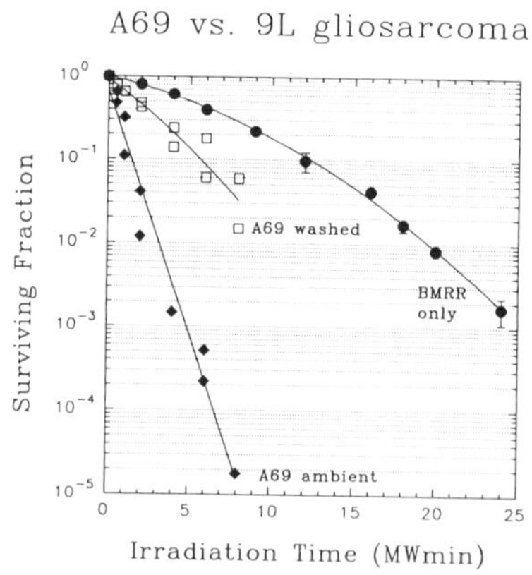

Figure 3. Effects of dipeptide A69 on 9L rat gliosarcoma cell survival under ambient and washed conditions.

Figure 2 shows the effectiveness of the dipeptides A5 and A69 against the 9L rat gliosarcoma; the effect of BPA is included for reference. The ^{10}B concentrations present during the incubation and the irradiation were: 22.2 μg ^{10}B/ml for A5; 21.9 μg ^{10}B/ml for A69; and 26.8 μg ^{10}B/ml for BPA. See Table 3 for a normalization of the Radiation Effects to the effect produced by boric acid.

Note that both dipeptides A5 and A69 are equally as effective as (or more effective than) BPA.

Figure 3 shows the effect of washing the cells after incubation and prior to irradiation. The effectiveness of A69 is greatly reduced by washing. This indicates that the compound is not firmly fixed within the cells. The same effect was observed with A5. See Tables 1, 2, and 3 for quantification of the effects of washing.

Biodistribution Study

A biodistribution study was carried out with the dipeptide A5 in C57B 1/6 mice bearing subcutaneous B16 G3.12 melanomas. Mice were given 20 mg (35 μg ^{10}B/g body weight) A5 orally (intragastric intubation) as an aqueous slurry in 0.5 ml water and sacrificed at 2, 4, 6, 8, (n=2 mice per point) and 10 (n=1) hours. Boron content in tissues was determined by the prompt gamma technique. This is the same boron dose (in μg ^{10}B/g) used with BPA; under these conditions BPA produces 20 μg ^{10}B/g tumor at 4-6 hrs post administration. At the 4 hr time point there is a 2:1 differential concentration between tumor and normal tissues, but the boron concentration was about 7μg ^{10}B/g.

CONCLUSIONS

All of the derivatives or precursors of the dipeptides A5 and A69 tested (BBI-0104, 0010, A70, A71) were not significantly better than the parent dipeptides.

In vitro, A5 and A69 were more effective than boric acid and equal to or better than BPA.

In vivo, given orally, tumor uptake of A5 was low. Perhaps the oral absorption route (GI tract) is not as effective for these compounds as for BPA. If the solubility of these dipeptides could be improved, i.v. administration may prove effective. Conversion of the dipeptide esters to the sodium salts or free acids is underway. These derivatives are highly soluble. Tests are underway.

The boronated nucleoside, BBI-0120, showed significant binding and cell kill after washing. We have shown that boronated nucleosides can be incorporated into DNA. Further work is clearly warranted.

ACKNOWLEDGEMENTS

This research was supported by the Department of Energy (Grants DE-FG05-90ER81012 and DE-FG05-90ER81065) and by the American Cancer Society (FRA301 and NP741).

REFERENCES

1. Spielvogel, B.F. in *Boron Chemistry-4*, Parry, R.W.; Kodama, G., eds., Pergamon Press, NY, 1980, 119-129.
2. Sood, A., Sood, C.K., Spielvogel, B.F., Hall, I.H. *Eur. J. Med. Chem.* 1990, 25, 301.
3. Sood, A., Sood, C.K., Spielvogel, B.F., Hall, I.H., Wong, O.T. *J. Pharm. Sci.* 1992, 81, 458.
4. Sood, A., Sood, C.K., Hall, I.H., Spielvogel, B.F. *Tetrahedron* 1991, 47, 6915.
5. Tomasz, J., Shaw, B.R., Porter, K.W., Spielvogel, B.F., Sood, A. *Angew. Chem. Int. ed.* 1992, 31, 1373..
6. Sood, A., Spielvogel, B.F., Shaw, B.R. *J. Am. Chem. Soc.* 1989, 111, 9234.
7. Sood, A., Shaw, B.R., Spielvogel, B.F. *J. Am. Chem. Soc.* 1990, 112, 9000.

TARGETING OF BORONATED ANTI-AFP MoAB TO HEPATOMA CELLS: APPLICATIONS OF BORONATED MoAB FOR BNCT

H. Yanagiё,[1] T. Takahashi,[1] Y. Fujii,[1] M. Sekiguchi,[1]
H. Nariuchi,[2] T. Kobayashi,[3] and K. Kanda [3]

[1] Department of Clinical Oncology,
[2] Department of Alergology, Institute of Medical Science,
University of Tokyo, 4-6-1 Shiroganedai, Minato-ku,
108 Tokyo, Japan
[3] Research Reactor Institute, Kyoto University, Kumatori,
Sennan-gun,
590-04 Osaka, Japan

INTRODUCTION

The cytotoxic effects of boron neutron capture on cancer cells were first reported by Locher [1] and Kruger [2]. Cell destruction in boron neutron capture therapy (BNCT) is due to the nuclear reaction between ^{10}B and thermal neutrons that proceeds as follows:

$$^{10}B + {}^{1}n \longrightarrow {}^{7}Li + {}^{4}He + 2.4\ MeV$$

The produced alpha-particles (^{4}He) killed cells in the range of 10 μm from the site of the capture reaction. It is thus theoretically possible to kill tumor cells without affecting adjacent normal cells, provided efficient ^{10}B atoms can be made to selectively accumulate in tumor cells. BNCT has been applied clinically to malignant brain tumors[3] or malignant melanoma[4]. In malignant brain tumors, the blood-brain barrier is destroyed, thus making it possible for the ^{10}B compound to easily penetrate the tumor cells. In malignant melanoma, the ^{10}B conjugated precursor of melanin pigments is easily taken up by melanoma cells. This may possibly lead to extended application of

Advances in Neutron Capture Therapy, Edited by
A.H Soloway *et al.*, Plenum Press, New York, 1993

BNCT for treating various tumors, provided a sufficient amount of ^{10}B compound is delivered selectively by the monoclonal antibody that has reacted with the tumor cells. Takahashi et al.[5] prepared anti-alpha fetoprotein (AFP) monoclonal antibody conjugated with ^{10}B-compound. This boronated anti-AFP MoAb exerted cytotoxic effect on hepatoma cells in BNCT *in vitro* .

MATERIALS AND METHODS

Chemicals

The cesium salt of undecahydro-mercaptocloso-dodecaborate ($Cs_2\,^{10}B_{12}H_{11}SH$) was kindly supplied by Shionogi Research Laboratories Co. Ltd. (Osaka, Japan). N-hydroxy-succinidiml 3-(2-pyridyldithio) propionate (SPDP) was purchased from Pharmacia Fine Chemicals (Uppsala, Sweden).

Mice and Target Tumor Cells

Male BALB/c nu/nu mice were obtained from Nihon SLC (Shizuoka, Japan) and used at 6 to 7 weeks of age. Rat alpha-fetoprotein (AFP) producing hepatic carcinoma cell line AH66 was kindly provided by Dr. H. Sato, Sasaki Institute, Sasaki Foundation, Tokyo, and maintained in RPMI 1640 medium (Hazleton Biologics, Inc, Kansas, USA) supplemented with 10 % fetal calf serum (Cell Culture Laboratories, Ohio, USA) and 100 µg/ml kanamycin.

Anti-rat AFP Monoclonal Antibody

The anti-rat AFP monoclonal antibody (Ig G 1) was obtained in the ascite form from pristane (2, 6, 10, 14 - tetramethylpentadecane ; Aldrich Chemical Co., Milwaukee, WI) treated BALB/c mice previously i.p. injected with the hybridoma clone established by Takahashi et al.[5] It was purified by passage through protein A-MAPS-II Kits column (Bio-Rad Laboratories, Richmond, CA, USA).

Boronation of Anti-AFP MoAb

After being dialized against 0.01 M HEPES buffer (pH 7.45) supplemented 0.15 M NaCl, one ml of monoclonal antibody (2.0 mg/ml) was incubated with 20 mM SPDP for 30 min. at room temperature. Free SPDP was removed by passage through a Sephadex G25 column. 50 mM ^{10}B compound was suspended in the antibody solution. Following incubation at 4 C overnight, the boronated monoclonal antibodies were passed through a Sephadex G25 for removal of the free ^{10}B compound.

Determination of ^{10}B Compound Concentration Conjugated with Anti-AFP MoAb

The amount of ^{10}B compound conjugated with MoAb was determinated by the prompt gamma-ray spectrometry at the Research Reactor Institute, Kyoto University [6]. Boron concentration was determined by measuring the gamma ray at the peak corresponding to a 478 keV area .

Measurement of Accumulated ^{10}B Concentration in Each Organ *in vivo*

1 x 10 7 AH66 cells were injected subcutaneously into the backs of male BALB/c nu/nu mice. About 10 - 14 days after injection, when tumor weight reached 100-300 mg, 1.5 mg boronated anti-AFP MoAb or control 1.5 mg boronated anti-dinitrophenol (DNP) MoAb were injected intravenously from the tail veins. ^{10}B concentration the of tumor, blood or several organs at 12, 24, 72 and 120 hrs after injection of ^{10}B-MoAb conjugates was measured by prompt gamma-ray spectrometry. Six mice were used in each part of this study.

RESULTS AND DISCUSSION

Boronated anti-AFP MoAb was obtained by mixing 2 mg of anti-AFP MoAb-SPDP solution with 50 mM ^{10}B compound. 14.7 ppm of ^{10}B compound was conjugated to 2 mg anti-AFP MoAb, as confirmed by prompt gamma-ray spectrometry. The number of ^{10}B atoms conjugated directly to MoAb was estimated as 442 per antibody.

Boron concentrations in tumor tissue obtained at 12, 24, 72 and 120 hrs after injection of 1.5 mg ^{10}B conjugated anti-AFP MoAb were 1.27 ± 0.30 (SD, n=6), 2.38 ± 0.76, 2.59 ± 0.18 and 2.20 ± 0.06 ppm, respectively. For control ^{10}B conjugated anti-DNP MoAb, the values were 0.89 ± 0.01, 0.92 ± 0.27, 1.20 ± 0.07 and 1.16 ± 0.18 ppm, respectively. The concentrations in blood were less than 0.40 ± 0.10 ppm using anti-AFP MoAb and less than 0.51 ± 0.14 ppm using anti-DNP MoAb at each sampling time (12, 24, 72 and 120 hrs). ^{10}B concentration in organs (liver, spleen, lung, kidney, heart and brain) after injection of ^{10}B-anti-AFP MoAb was less than 0.70 ppm at each sampling time. The number of ^{10}B atoms delivered to tumor cells was calculated as 7.28×10^{8}, 1.10×10^{9}, 2.39×10^{9} and 9.46×10^{8} at each sampling time following injection of ^{10}B-anti-AFP MoAb. ^{10}B atoms were made to selectively accumulate in AH66 tumor using ^{10}B-anti-AFP MoAb conjugate *in vivo*.

According to Alam et al., 10^{9} ^{10}B atoms are required to destroy one tumor cell in BNCT for effective cytotoxic effect [7]. In this experiment, 2.39×10^{9} ^{10}B atoms accumulated in AH66 tumor cells at 72 hrs after injection of ^{10}B-anti-AFP MoAb conjugates. Alam et al. reported 1300 boron atoms to conjugate in a molecule of monoclonal antibody using SPDP [8]. Barth et al. [9] and Takahashi et al. [5] reported the boronation of a monoclonal antibody using more than 10^3 atoms to markedly reduce antibody reactivity. In this experiment, we conjugated 442 ^{10}B atoms to a molecule of anti-AFP monoclonal antibody. Boronated MoAb showed no reduction in antibody reactivity.

In the present experiments, the selective accumulation of ^{10}B atoms in tumors was achieved using ^{10}B-anti-AFP MoAb conjugate *in vivo*.. *In vivo* BNCT is still attended with problems. Firstly, the number of ^{10}B atoms conjugated to a molecule of MoAb should be increased to about 1000 ^{10}B atoms per MoAb, without reduction in MoAb reactivity.

Secondly, ^{10}B accumulation in AH66 tumor cells was highest at 72 hrs after injection of ^{10}B-anti-AFP MoAb conjugate, the time when the cytotoxic effect must be assessed with thermal neutron irradiation in *in vivo* BNCT.

REFERENCES

1. Locher, G. L., Biological effects and therapeutic possibilities of Neutrons., Amer. J. Roentgenol. Radium. Ther., 36: 1 (1936).
2. Kruger, P. G., Some biological effects of nuclear disintegration products on neoplastic tissue., Proc. Natl. Acad. Sci. USA, 26: 181 (1940).
3. Hatanaka, H., " Boron-Neutron Capture Therapy for Tumors .", Nishimura Co. Ltd., Niigata.
4. Mishima, Y., Honda, C., Ichihashi, M., Treatment of malignant melanoma by single thermal neutron capture therapy with melanoma-seeking ^{10}B-compound., Lancet, 11: 388 (1989).
5. Takahashi, T., Fujii, Y., Fujii, G. and Nariuchi, H., Preliminary study for application of anti- alpha-fetoprotein monoclonal antibody to boron-neutron capture therapy., Jpn. J. Exp. Med., 57: 83 (1987).
6. Kobayashi, T. and Kanda, K., Microanalysis system of ppm-order ^{10}B concentrations in tissue for neutron capture therapy by prompt gamma-ray spectrometry., Nucl. Instr. Methods., 204: 525 (1983).
7. Alam, F., Soloway. A. H., Barth, R.F., Johnson, C.W. and Carey, W.E., Boronation of polyclonal and monoclonal antibodies for neutron capture., Brookhaven Natl Lab Report., 51730: 229 (1984).
8. Alam, F., Soloway, A. H., McGuire, J. E., Bart, R.F., Carey, W.E. and Adams, D., Dicesium N-succinimidyl 3-(undecahydro-closo-dodecaboranyl-dithio) propionate, a novel heterobifunctional boronating agent., J. Med. Chem., 28: 522 (1985).
9. Barth, R. F., Soloway, A. H. and Fairchild, R. G., Boron neutron capture therapy of cancer., Cancer Res., 50: 1061 (1990).

EPIDERMAL GROWTH FACTOR AS A POTENTIAL TARGETING AGENT FOR DELIVERY OF ^{10}B TO MALIGNANT GLIOMAS

Jacek Capala[1], Rolf F. Barth[1], Dianne M. Adams[1], Albert H. Soloway[2], and Jörgen Carlsson[3]

[1]Department of Pathology and [2]Department of Medicinal Chemistry,
The Ohio State University, Columbus, OH
[3]Department of Radiation Sciences,Uppsala University, Uppsala, Sweden

INTRODUCTION

Epidermal growth factor (EGF) is a single-chain, 53-residue polypeptide that binds to a transmembrane glycoprotein with tyrosine kinase activity[1]. After binding, the receptor-ligand complex is promptly internalized and then degraded in lysosomes. The EGF receptor (EGFR) gene is homologous to the c-*erb* B oncogene[2] and is often amplified in human glioblastomas. Studies by Bigner *et al.*[3] revealed that in a series of 33 human glioma biopsies, 15 showed amplification of of the EGFR gene. Similiar or even higher frequencies of amplification have been observed by others[4,5], and this often is associated with an increase in cell surface receptor expression[6,7]. The distribution of EGFR in high grade gliomas is variable, which probably reflects the cellular heterogenity of these tumors. However, in 25 of 42 human gliomas studied by Torp *et al.*[6], more than 50% of the tumor cells bound biotinylated EGF. Attempts have been made to use EGFRs as targets for radioimmunotherapy, and in one series of patients with recurrent glioblastomas, 50% showed some evidence of a clinical response with no significant toxicity[8,9], although there was no followup report indicating improved survival. Nevertheless, the EGFR still remains one of the more attractive targets on high grade gliomas[10]. Targeting the EGFR by means of boronated EGF may be easier than using boronated monoclonal antibodies (MoAbs) directed against EGFR since EGF has a relatively low molecular weight ($\sim$6 kD) compared to MoAbs, and it should be possible to construct relatively small conjugates, which would diffuse more rapidly into poorly vascularized regions of the tumor. The rapid internalization and degradation of EGF-EGFR complexes, as well as their high affinity constant, are further advantages, if the boron persists intracellulary. The purpose of the present study was to develop methodology for the boronation of EGF and to study its *in vitro* affinity for glioma cells.

MATERIALS AND METHODS

EGF Binding Studies

The human malignant glioma U-343MGaCl2:6 cell line expressing 2.5 x 10^5 high affinity and 8.6 x 10^5 low affinity receptors per cell were used in all experiments. The cells were incubated at 37 °C in Ham's F-10 medium supplemented with 10% FCS, L-glutamine (2 mM), streptomycin (100 μg/ml) and penicillin (100 IU/ml). Receptor grade EGF (Sigma Chemical Co., St. Louis, MD or Janssen Biochimica, Beese, Belgium), derived from mouse submaxillary glands, was used in experiments with radiolabeled EGF. Recombinant EGF (Upstate Biotechnology, Inc., Lake Placid, NY) was used as a substrate for conjugation with boronated "starburst" dendrimer. EGF was radiolabeled with ^{125}I-sodium iodide by means of the Chloramine-T method. The labeled protein and free radioactivity were separated by means of column chromatography on G-25 Sephadex (Pharmacia, Uppsala, Sweden). Binding of ^{125}I-EGF as a function of time was determined as follows. Petri dishes, 3.5 cm in diameter, were seeded with 2.8 x 10^5 glioma cells and then incubated at 37 °C with 120 nM of ^{125}I-EGF (0.1 μCi/ml medium) for varying time periods ranging from 15 to 180 min. Then the cells were washed 6 times with cold PBS (pH=7.4) and the cell associated radioactivity was determined with a 1214 RackBeta 'Excel' liquid scintillation counter (LKB Wallac, Turku, Finland). Intracellular degradation of ^{125}I-EGF was determined following 30 min incubation with ^{125}I-EGF (1.11 μCi/ml). The cells were washed 6 times with cold serum-free medium and further incubated in medium containing 30 μM chloroquine or 5 mM lidocaine, both of which are inhibitors of lysosomal proteolytic enzymes[11]. Cell associated radioactivity was measured at the end of incubation in the presence or absence of lysosomal inhibitors. Binding of boronated EGF was determined by means of a competitive binding assay in which 4.5 x 10^5 cells/petri dish were incubated at 4 °C for 90 min in medium containing 1.6 nM of ^{125}I-EGF and various concentrations of native or boronated EGF in the range 0.016 to 166 nM. Then, the cells were washed 6 times in cold PBS and the cell associated activity was determined. Affinity constants (K_A) were calculated using nonlinear, least square regression analysis of the binding data according to the method of Scatchard.

Conjugation of EGF with Boronated "Starburst" Dendrimers

"Starburst" dendrimers are composed of repetitive polyamido amino groups arranged in a starburst pattern (Polysciences, Inc., Warrington, PA). They were boronated with $Na(CH_3)_3NB_{10}NCO$ as previously described[12]. For conjugation, EGF was incubated overnight at 4 °C with a 10-molar excess of boronated "third generation" (MW = 2414) dendrimer in the presence of 0.1 M 1-ethyl-3-(3-dimethylaminopropyl) carbodiimide hydrochloride (EDC, Pierce, Rockford, IL), following which the mixture was passed through a Sephadex G-25 column (1.2 x 19 cm) and eluted with PBS (pH=7.4). One ml fractions were collected and protein concentrations were determined spectrophotometrically by measuring absorbance at 280 nm. Boron was quantified by direct current plasma atomic emission spectroscopy (DCP-AES) using an ARL Spectran VB spectrometer (Applied Research Laboratories, La Brea, CA). Fractions containing peak concentrations of both protein and boron were pooled and used in further experiments.

RESULTS AND DISCUSSION

As shown in Figure 1, maximum cellular binding was observed after 30 min incubation with ^{125}I-EGF, following which cell associated radioactivity declined to 73%, suggesting downregulation of EGF membrane receptors. If the medium was replaced with medium lacking ^{125}I-EGF, cell associated radioactivity decreased rapidly within 60 min to 10% due to degradation of EGF. Radioactivity released into the incubation medium was in a low molecular weight form, as determined by means of gel filtration. These results confirmed that EGF bound with high affinity and specificity to its receptors on glioma U-343MGaCl2:6 cells and that the receptor-ligand complex was promptly internalized.

Incubation in medium containing chloroquine or lidocaine reversibly inhibited cellular degradation of EGF and ~70 % of the initial radioactivity remained cell associated for up to 6 hours (Figure 2), suggesting that lysosomal proteolytic enzymes were responsible for intracellular degradation of EGF. Cell associated radioactivity rapidly decreased after the medium had been replaced with drug-free medium. Again, all of the radioactivity released from the cells was in a low molecular weight form.

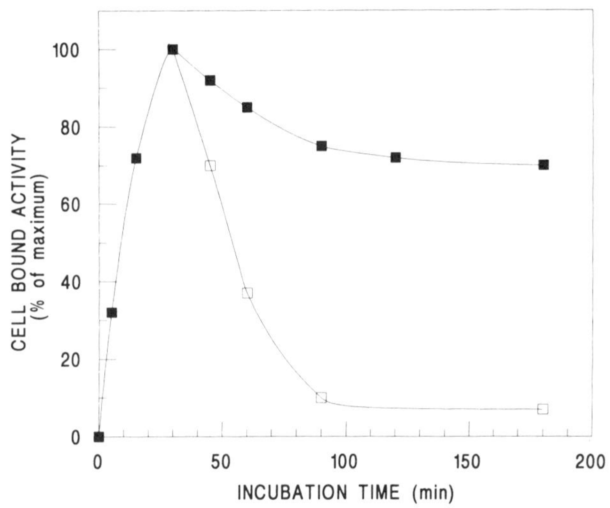

Figure 1. Cell bound radioactivity as a function of time. (■) continuous incubation in ^{125}I-EGF containing medium; (□) incubation in medium devoid of ^{125}I-EGF.

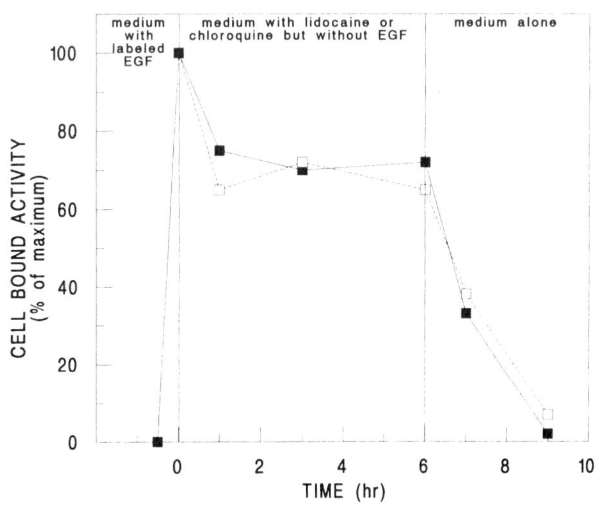

Figure 2. Cell bound radioactivity as a function of time with or without lidocaine (■) or chloroquine (□).

Using 1-ethyl-3-(3-dimethylpropyl) carbodiimide hydrochloride (EDC) as a coupling reagent, we have conjugated a boronated "starburst" dendrimers to EGF. Based on protein and boron determinations, it was calculated that the conjugates had > 1000 atoms of boron per EGF molecule. According to previous reports[7], the U-343 glioma expresses 2 x 10^5 EGF receptors per cell, and if these could be saturated with boronated EGF, then 2 x 10^8 boron atoms would be delivered per cell. Binding of the conjugate and native EGF to EGF-receptors was compared by competitive binding assays. In these experiments 166 nM of native or boronated EGF produced a 92% or 88% decrease in ^{125}I-EGF binding, respectively (Figure 3). In agreement with previously reported data[7], the K_A obtained for native EGF was 2.1 x 10^9 M^{-1}, while that obtained for B-EGF was decreased to 2.3 x 10^8 M^{-1}, indicating that linkage of the boronated "starburst" dendrimer to EGF resulted in a 10-fold decrease in the K_A of the conjugate. This reduction in K_A may have been due to steric hindrance caused by attachment of the dendrimers to carboxyl groups of asparagine and glutamine, which are located at sites involved in the binding of EGF to its receptor. Nevertheless, the K_A of the boronated conjugate compares favorably to the K_A of MoAbs that have been used for *in vivo* targeting of radionuclides and toxins[13].

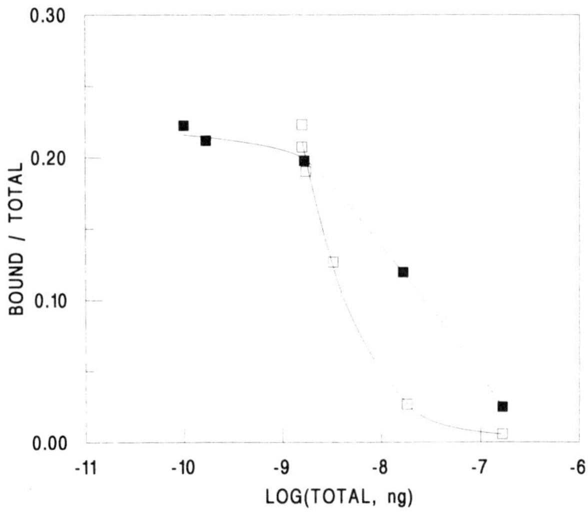

Figure 3. Displacement curves obtained in a competitive binding assay when increasing concentrations of native (□) or boronated (■) EGF were used as a competitor.

If boronated EGF is to be used for BNCT of gliomas, the following criteria must be met: 1) the conjugates must retain their high affinity for EGF receptors *in vivo*; 2) following either systemic or local administration, sufficient amounts must reach glioma cells dispersed within normal brain; 3) the boron must persist intracellulary within glioma cells for a longer time than that required to clear from normal brain. Studies are planned to investigate each of these requiements. Although the upregulated expression of EGFR on human glioma cells suggests that this may be a possible target for ^{10}B-EGF conjugates, within any given tumor there will be considerable heterogenity in EGFR expression. Furthermore, the expression of EGFR on some normal neurons and on reactive astrocytes[6] is a problem that will be addressed in animal studies that currently are in progress in our laboratory.

ACKNOWLEDGEMENTS

This work was supported by U.S. Dept. of Energy contract #DE-AC02-76CH000016. J. Capala has been supported in part by the Swedish Cancer Society.

REFERENCES

1. Carpenter, G. (1979) *Ann. Rev. Biochem.* **48**,193-216.
2. Downward, J., Yarden, Y., Mayers, E., Scare, G., Totty, P., Stockwell, A., Ullrich, A., Schlessinger, J., & Waterfield, M.D. (1984) *Nature* **307**,521-527.
3. Bigner, S.H., Wong, A.J., Mark, J., Kinzler, K.W., Vogelstein, B. & Bigner, D.D. (1987) *Cancer Genetics and Cytogenetics* **29**,165-170.
4. Tuzi, N.L., Venter, D.J., Kumar, S., Staddon, S.L., Lemoine, N.R. & Gullick W.J. (1991) *Br. J. Cancer* **63**,227-233.
5. Ekstrand, A.J., James, C.D., Cavenee, W.K., Seliger B., Petterson, R.E. & Collins, V.P. (1991) *Cancer Res.* **51**,2164-2172.
6. Torp, S.H., Helseth, E., Dalen, A. & Unsgaard, G. (1991) *Cancer Immunol. Immunother.* **33**,61-64.
7. Bigner, S.H., Humphrey, P.A., Wong, A.J., Vogelsein, B., Mark, J., Friedman, H.S. & Bigner, D.D. (1990) *Cancer Res.* **50**,8017-8022.
8. Epenetos, A.A., Courtney-Luck N., Pickering, D., Hooker, G., Durbin, H., Lavender, J.P & McKenzie, C.G. (1985) *Br. Med. J.* **290**,1463-1466.
9. Kalofonos, H.P., Pawlikowska, T.R., Hemingway A., Courtenay-Luck, N., Dhokia B., Snook, D., Sivolapenko, G.B., Hooker, G.R., McKenzie, C.G., Lavender P.J., Thomas, D.G.T. & Epenetos, A.A. (1989) *J. Nucl. Med.* **30**,1636-1645.
10. Harris, A.L. (1990) *Cancer Cells* **2**:321-323.
11. Carpenter, G. & Cohen S. (1976) *J. Cell Biol.* **71**,159-171.
12. Barth, R.F., Soloway, A.H., Adams, D.M. & Fazlul, A. (1992) in *Progress in Neutron Capture Therapy For Cancer,* eds. Allen, B.J., Moore, D.E. & Harrington, B.V. (Plenum Press, Yew York), pp. 265-268.
13. Oredipe, O.A., Barth, R.F., Rotaru, J.H. & Steplewski, Z. (1992) *Hybridoma* **11**,607-615.

EVALUATION OF A ^{10}B-LABELLED DNA LIGAND

A.Corder[1], A. Whittaker[2], D.P. Kelly[2], H. Meriaty[3], B.J. Allen[3] and R.F. Martin[1]

[1]Molecular Sciences Group, Peter MacCallum Cancer Institute
Melbourne, Australia
[2]Department of Chemistry, University of Melbourne
[3]ANSTO, NSW, Australia

INTRODUCTION

The basis of Boron Neutron Capture Therapy (BNCT) is the phenomenon of thermal neutron capture by ^{10}B, resulting in the release of fission products. The range of the fission products is such that the radiochemial damage is largely confined to the cell containing the neutron capture event. We have been investigating the use of ^{10}B-labelled DNA ligands as a means of targeting the BNC reaction to nuclear DNA in the cell. There are two main advantages of this strategy. Firstly, DNA is the critical target for radiobiological effects and the use of ^{10}B-DNA ligands locates the damage in the DNA[1,2]. Secondly, the genome represents a large reservoir of binding sites for DNA ligands. Taking a figure of about 6.8 x 10^9 base pairs per genome, and assuming that a DNA ligand can be accommodated at a binding ratio of say, one ligand per 10bp without seriously compromising cell viability, then the number of available ligand binding sites is more than 6 x10^8 per cell. For a cell nucleus 9μm in diameter, this represents more than 27 ppm of ^{10}B, for a DNA ligand with one ^{10}B atom per molecule.

Bibenzimidazoles, exemplified by Hoechst 33342, bind in the minor groove of DNA, and are widely used as fluorescent DNA stains. Moreover, Hoechst 33342 is used in fluorescence activated cell sorting in which viable cells are recovered after staining and sorting[3]. We therefore chose a bibenzimidazole as the DNA-binding moiety for a ^{10}B-labelled DNA ligand.

EXPERIMENTAL

Two ^{10}B-labelled DNA ligands have been synthesised; details of the synthesis and characterisation will be reported elsewhere. The first compound is ^{10}B-*para*-Hoechstboronic acid (pHBA - 4-[15''-(4'''-methylpiperazin-1'''-yl)-2'', 5'-bi-1*H*-benzimidazol-2'-yl]phenyl boronic acid)[4] and the second is its cyclic ester derivative. It was thought that the general reactivity of boronic acids with 1,2-diols, in particular carbohydrates and RNA, might compromise delivery of the ligands to the cell nucleus, but the uptake of the ester did not prove to be any better than that of the boronic acid. Results reported in this paper refer to uptake experiments and cell survival studies using p-HBA only. The details of the synthesis and characterization of pHBA will be reported elsewhere; characterization included ^{1}H- and ^{13}C-nmr and FAB ms.

For both the uptake studies and the cell survival studies, V79 cells were grown in α-media with 10% FCS. The cells were suspended in growth medium at a population density of 1 x 10^6 cells/ml. pHBA acid was added to a concentration of 32 μM or 64 μM. In the case of the cell survival experiments, tubes were also set-up containing a ^{10}B-borate buffer to a final concentration of 1, 0.5, or 0.25 mM. The borate buffer samples were essentially boron neutron capture references used in a similar way as described by the Brookhaven group[5,6]. These references enabled us to compare our results with those described previously for other ^{10}B compounds. After the appropriate additions, the cells were incubated for 60 minutes at 37°C, prior to irradiation or nuclear isolation.

For the uptake studies, nuclei were isolated using Triton X-100 and then the pHBA extracted from the nuclei using 0.5M triethylamine in ethanol/water (1 vol:1 vol) and measured spectrophotometrically. The spectrophometric measurements of pHBA stock solutions, and cell and nuclear extracts, were based on a molar extinction coefficient of 2.6 X 10^4 in a solution 0.1% trifluoroacetic acid in methanol: water (45:55; v/v). The extinction coefficient was determined using a stock solution that had been assayed for Boron content. The uptake results assume that all the pHBA is extracted from the nuclei. The small amount of fluorescence detected in the pellet after extraction supports this assumption.

In the case of the cell survival experiments, the cells were irradiated for varying periods which equate to 1, 2, 3, and 4 Gray. Irradiation was carried out in the epithermal beam at the MOATA reactor at ANSTO[7]. The cells were then plated for clonogenic assay[8], and the fraction of cells surviving calculated after seven days. For the toxicity studies (Table 1), cells were treated for 2 hours prior to plating for clonogenic assay. Cell survival data was analysed using programme 1R, version 1990, from BMDP Statistical Software Inc. (Los Angeles, California).

RESULTS AND DISCUSSION

Toxicity

Toxicity was measured as survival relative to the plating efficiency of controls without any added ligand(Table 1). The ^{10}B-borate was not toxic at any of the three concentrations, however some cytotoxicity was evident for 64 μM of ^{10}B-pHBA. The extent to which such cytotoxicity might be a limiting factor for clinical BNCT will only become evident after *in vivo* studies.

Table 1. Toxicity *In Vitro*

Compound		n*	%Survival ± SD
^{10}B-borate	1 mM	2	92 ± 14
	0.5 mM	5	104 ± 31
	0.25 mM	5	107 ± 9
^{10}B-pHBA	32 μM	20	85 ± 32
	64 μM	5	64 ± 18
Hoe 33342	20 μM	8	80 ± 4.3

* Number of experiments

Cell Survival

Results of cell survival studies are shown in figure 1. The irradiated control shows that there is some cell kill from thermal neutron irradiation alone. However, after irradiation in the presence of the pBHA, cell kill is significantly enhanced; the dose-modification factor is about 2. Irradiation of cells in the presence of 0.25 mM ^{10}B-borate, which is assumed to be randomly distributed throughout the media and the cells, resulted in a similar enhancement of cell kill to that occuring in the presence of 64 μM pHBA. This difference in the amount of ^{10}B required for cell kill (namely 0.25 mM versus 64 μM) reflects the preferential localization of the DNA ligand form of ^{10}B relative to the critical target, namely the cell nucleus. The morc curvilinear shape of the 64 μM pHBA survival curve, compared to that for 0.25 mM ^{10}B-borate, might be a consequence of the radioprotective properties of bibenzimidazole DNA ligands[8], which would certainly be a contraindication for the potential of pHBA for BNCT.

Table 2 attempts to compare our pHBA results with other ^{10}B-pharmaceuticals described by the Brookhaven group and their collaborators[5,6], who have compared different compounds with reference to experiments using ^{10}B-borate. Boric Acid Equivalents (BAE) is defined as the ^{10}B-borate concentration that yields equivalent survival to that obtained for a particular ^{10}B concentration of a particular ^{10}B-pharmaceutical. The ratio of BAE to actual boron concentration of the test compound in the medium is then calculated; the higher ratios indicating increasing "effectiveness".

Assuming that ^{10}B-borate is uniformly distributed between medium and cells, a BAE: ^{10}B-concentration ratio of less than 1 would suggest that the compound was excluded from the cell, whereas a high ratio would suggest a preferential concentration of the ^{10}B-compound relative to the ultimate radiobiological target, presumably nuclear DNA. Of the first four entries in Table 2, BOPP is clearly "better" than BSH and BSSB, and the ^{10}B-LDL is outstanding with a ratio of 10.6.

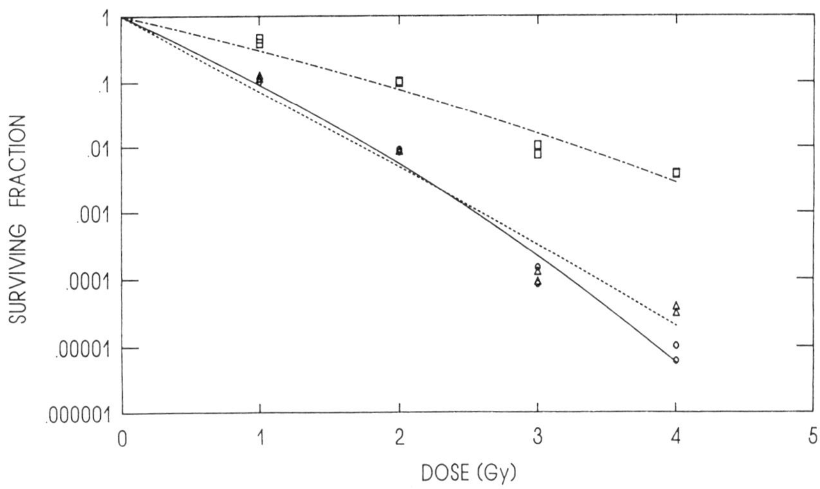

Figure 1. Three survival curves for thermal neutron irradiation of V79 cells; irradiation-only controls([]), 0.25 mM ^{10}B-borate reference(Δ), and 64μM ^{10}B-HBA(o).

Table 2. Comparison of Boric Acid Equivalences

Compound	BAE μg/g	^{10}B in MEDIUM μg/g	BAE/^{10}B
BSH	9.5	30	0.32
BSSB	12.0	30	0.40
BOPP	41.0	30	1.4
B-LDL	37.0	3.5	10.6
^{10}B-pHoeBA	250 μM	64 μM	3.9

A further aspect of the data in Table 2 is the distinction between "ambient" and "washed" conditions. In the ambient case, the cells are irradiated in the presence of the medium containing the ^{10}B-compound. In contrast, for the washed conditions, after incubation with the ^{10}B compound in the nominated concentration cell culture medium, the cells are washed and irradiated in ^{10}B-free medium. Thus a high BAE: ^{10}B-concentration ratio for washed conditions indicates not only favourable distribution of the ^{10}B-pharmaceutical during incubation, but also (relative) retention during washing and incubation (during irradiation) in ^{10}B-free medium. This consideration makes the results for ^{10}B-LDL particularly notable. All our experiments with ^{10}B-pBHA were done under ambient conditions.

Calculation of the BAE:^{10}B-concentraton ratios involves comparison of survival curves for ^{10}B-borate and the ^{10}B-compound in question. For the Brookhaven group data, this comparison involved analysis of survival curves using the multi-hit formula. In our case, we have compared ^{10}B borate with the ^{10}B-ligand results in a more semi-quantitative fashion. That is, the survival curve for 250 μM ^{10}B borate is similar to that of 64 μM pHBA. In spite of this difference in comparing the data, we suggest that it is reasonable to compare the BAE/^{10}B ratios in Table 2.

Such a comparison suggests the following sequence in order to decreasing "effectiveness":

^{10}B-LDL > ^{10}B-pHBA > ^{10}B-BOPP.

However, it is important to emphasize that comparing all of these compounds in this way merely reflects the relative distribution of ^{10}B between the medium, the cells, and the subcellular compartments, and the relative efficiency of the BNC event in each of these compartments.

Nuclear Uptake

Fluoresence microscopy studies have indicated that the pHBA is taken-up into the cell nucleus (data not shown). This method of assessing uptake is not very quantitative, as Hoechst flouresence is markedly enhanced in the DNA-bound form. Therefore to quantitate the uptake we used the spectrophotometric method described. Preliminary results have indicated that for an initial concentration of 64 μM, 1.2 nmoles of the pHBA acid is taken up per 10^6 nuclei. From this uptake measurement and our cell survival results we can calculate the number of boron neutron capture events occuring in the nucleus at the thermal neutron fluence which corresponds to a lethal event.

From the survival curves for the irradiation-only controls and for the 64 μM ^{10}B-pHBA we have calculated that the thermal neutron fluence which yields one lethal event due to BNC (ie. 0.84 Gy which corresponds to 37% survival for 64 μM ^{10}B-HBA relative to the radiation-only control) is approximately 3×10^{12} ncm^{-2}. For a concentration of 1.2 nmoles/10^6 nuclei, there are 7×10^8 ^{10}B atoms per nucleus. Using these values and the thermal neutron capture cross section of ^{10}B, there is an estimated 8 Boron neutron capture events per nucleus per lethal event.

This value is somewhat higher than expected, especially since BNC events in other parts of the cell would further contribute to cell kill. The measured value of ^{10}B-HBA in the nuclei (viz 1.2 nmole per 10^6 nuclei) also seems high from another standpoint; it corresponds to about 30 ppm of ^{10}B (for a nuclear diameter of 9 μ), which would be expected to yield more cell kill than that observed.

The calculated neutron fluence is a possible source of error; it is based on activation of gold foils placed on the near (beam side) and far edges of the tubes containing the cells. Subsequent experiments will involve dosimetry with gold solutions in replicate tubes.

CONCLUSIONS

We have shown that ^{10}B-labelled pHBA significantly enhances cell kill. At a concentration of 64 μM the dose modifying factor is ~2 at 10% survival. The utility of such ^{10}B-pharmaceuticals for BNCT will depend on their distribution between tumour and normal tissues in irradiated field. We are now embarking on distribution studies in a mouse glioma model.

ACKNOWLEGEMENTS

This project was supported by the Anti Cancer Council of Victoria, including an ACCV Cancer Research Scholarship awarded to A. Whittaker. Access to the MOATA reactor was supported by a grant from the Australian Institute of Nuclear Science and Engineering. The authors thank Dr. Jane Matthews and Jenny Smith from the PMCI Statistical Centre for the statistical analysis of the survival curve data.

REFERENCES

1. T. Kobayashi and K. Kanda. Analytical calculation of boron-10 dosage in cell nucleus for neutron capture therapy, *Radiat. Res.* 91: 77 (1982).
2. D.E. Charlton. Energy deposition in small ellipsoidal volumes by high-LET particles: application to thermal neutron dosimetry, *Int. J. Radiat. Biol.* 59: 827 (1991).
3. R.E. Durand. Use of Hoechst 33342 cell separation from multicell systems, *J Histochem. Cytochem.* 30: 117 (1982).
4. A.D. Whittaker, D.P. Kelly, M. Pardee and R.F. Martin, Synthesis of ^{10}B- and ^{157}Gd-labelled DNA ligands, for neutron capture therapy, *in*: "Progress in Neutron Capture Therapy for Cancer, B.J. Allen, D.E. Moore and B.V. Harrington, Eds. Plenum Press, New York and London (1992).
5. R.G. Fairchild, S.B. Kahl, B.H. Laster, J. Kalef-Ezra, and E.A. Popenoe. *In Vitro* determination of uptake, retention, distribution, biological efficacy, and toxicity of boronated compounds for neutron capture therapy: A comparison of porphyrins with sulfhydryl boron hydrides, *Cancer Res.* 50: 4860 (1990).
6. B.H. Laster, S.B. Kahl, E.A. Popenoe, D.W. Pate and R.G. Fairchild. Biological efficacy of boronated low-density lipoprotein for boron neutron capture therapy as measured in cell culture, *Cancer Res.* 51: 4588 (1991).
7. R.F. Martin, G. D'Cunha, M. Pardee and B.J. Allen. Induction of double-strand breaks following neutron capture by DNA-bound ^{157}Gd, *Int. J. Radiat. Biol* 54: 205 (1988).
8. L. Denison, A. Haigh, G. D'Cunha and R.F. Martin, DNA ligands as radioprotectors: molecular studies with Hoechst 33342 and Hoechst 33258, *Int. J. Radiat. Biol.* 61: 69 (1992).

SYNTHESIS AND EVALUATION OF ^{157}Gd-LABELLED DNA LIGANDS

A. Whittaker[2], D.P. Kelly[2], M. Pardee[1], A. Corder[1], H. Meriaty[3], B.J. Allen[3] and R.F. Martin[1]

[1] Peter MacCallum Cancer Institute, Molecular Sciences Group
481 Lt. Lonsdale Street, Melbourne, Victoria, Australia
[2] University of Melbourne, Chemistry Department
Grattan Street, Parkville, Victoria, Australia
[3] Australian Nuclear Science and Technology
Lucas Heights Research Laboratories, Menai, NSW, Australia

INTRODUCTION

The interest in the potential role of ^{157}Gd in neutron capture therapy[1,2,3] obviously stems from its very high thermal neutron capture cross section. It is about 60 times higher than that of ^{10}B. However, unlike the fission reaction associated with boron neutron capture, ^{157}Gd undergoes an n-gamma reaction. The very high energy of the gamma photons is in marked contrast to the limited range of the fission products of boron neutron capture. That limited range, of the order of a cell diameter, is a crucial feature of the rationale of boron neutron capture therapy. However, some of the gamma energy from the ^{157}Gd capture reaction is converted, with the emission of conversion electrons[4]. More importantly, the vacancies left by conversion electrons result in the emission of Auger electrons. The Auger electron emission, exemplified by the Auger emitting isotope ^{125}I, is a focus of intense radiochemical damage, and the range of damage is limited to molecular dimensions[5-8].

If the gadolinium is in the form of a tripositive cation (Gd^{3+}), and bound electrostatically to plasmid DNA, then neutron capture induces DNA double stranded breaks, presumably due to the Auger electron effect[1]. If the Gd^{3+} is sequestered from the DNA by chelating agents such as EDTA or DTPA, then the level of DNA double-strand-breaks is much reduced[1]. The idea of trying to incorporate Gd into a DNA-binding compound is therefore an obvious one; it will be necessary to produce such ligands in order to exploit the Auger electron aspect of ^{157}Gd neutron capture[9]. Also, there are other good reasons for employing DNA ligands for neutron capture. In particular, the large reservoir of DNA binding sites that are represented in the cell, provide an opportunity to maximise ^{157}Gd (or ^{10}B) uptake[10].

We have synthesised three Gd-labelled DNA ligands (Fig. 1) All three are minor groove binding bibenzimidazoles with a two carbon chain to a nitrogen which is conjugated to a metal chelating group namely, diethylenetriaminepentaaceticacid (DTPA)[11]. In one case (Gd Hoe monomer), of the five carboxylic acid groups in the DTPA, one is involved in the amide bond to the DNA ligand, leaving 4 negative charges. The gadolinium is in a tripositive state, so the gadolinium chelate region has an overall charge of -1. A similar ligand has a second amide group, so the chelate region is neutral. The third ligand (Gd Hoe dimer) actually has two DNA ligand moieties, so again the gadolinium chelate region is neutral.

Spectrophotometric studies indicate that the Gd-ligands bind to DNA[9], but unfortunately the uptake into cultured cells is rather poor. Delivery systems such as liposomes or protein conjugates will have to be developed in order to get substantial uptake. In the meantime, we have to ask the question, that if these ligands are bound to DNA, then does neutron capture by the Gd induce DNA double-strand breaks? This question is the focus of this paper.

Advances in Neutron Capture Therapy, Edited by
A.H Soloway *et al.*, Plenum Press, New York, 1993

EXPERIMENTAL

The experimental system used to assay DNA strand breaks exploits the properties of plasmid DNA - pBR322 - a double stranded circular molecule comprised of a little more than 4000 base pairs. The intact molecule is supercoiled and compacted. Introduction of a single-strand break into either of the two strands relaxes the molecule into an open circular form, and double stranded breaks

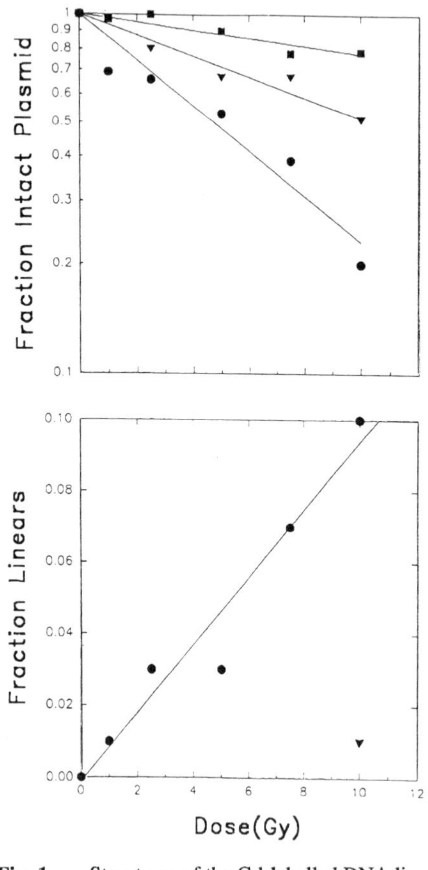

Fig. 1. Structures of the Gd-labelled DNA ligands

yield linear molecules. All three of these species can be separated by agarose gel electrophoresis. Single-strand-breaks (ssbs) are assayed by disappearance of intact plasmid DNA. Double-strand-breaks (dsbs) are assayed by the appearance of linear species. The experimental details have been described[12].

Experiments with $^{157}GdCl_3$ were similar to those described previously[1], except that DNA strand breakage was quantitated. The final concentrations of the components in the irradiated solutions (typically 100µl, in Ependorf tubes) were as follows:

- pBR322 DNA (from Boehringer Mannheim); 50µg/ml (75µM bp)
- Tris HCl buffer (pH7); 40mM,
- EDTA; 0.2mM,
- $^{157}GdCl_3$; 2.5mM (omitted from some controls) and,
- DTPA; 10mM (only added to one set of samples).

After irradiation in the thermal neutron beam in the MOATA reactor at Ansto, for periods corresponding to the indicated total doses, samples containing 0.5µg of pBR322 DNA were analysed by agarose gel electrophoresis.

For experiments with the ^{157}Gd-DNA ligands, M13 RF DNA was used instead of pBR322; the larger size conferring increased sensitivity. All the experiments described involved the ^{157}Gd-Hoe monomer, but preliminary experiments with the other two ligands have yielded similar results. The final concentrations of the components in the irradiated solutions were as follows:-

- M13 mp9 RF DNA (obtained from Boehringer Mannheim and dialysed against citrate buffer); 50µg/ml,
- Tris HCl buffer (pH7); 20mM,
- sodium citrate buffer (pH 7); 50mM,
- ^{157}Gd-Hoe monomer; 25µM,
- methanol; 2.5% (v/v), and
- DTPA; 10mM (only added to some samples).

After thermal neutron irradiation for periods corresponding to the indicated total doses, samples containing 0.5µg of M13 DNA were analysed by agarose gel electrophoresis. The DNA ssb yields were calculated by extrapolation of the 120Gy data points back to zero dose; a linear dose response was then assumed.

RESULTS AND DISCUSSION

Experiments with $^{157}GdCl_3$

The results are shown in Fig. 2. The upper panel shows the ssb data. Each of the lines represents the loss of intact plasmid with increasing radiation dose, so the steeper the line the higher the yield of single strand breaks. The control shows the effect of radiation only, without $^{157}GdCl_3$. The inclusion of gadolinium clearly increases the yield of DNA ssbs. If excess chelating agent (DTPA) is present, so the Gd^{3+} is sequestered off the DNA, there is much less single-strand breakage, but some damage is still evident.

The lower panel of Fig. 2. is perhaps the most interesting in that it shows induction of DNA dsbs. There are no DNA dsbs in the radiation-only control. In contrast, the inclusion of $^{157}Gd^{3+}$ results in obvious dose-dependent double-strand DNA breakage. In the presence of DTPA, which takes the Gd^{3+} off the DNA, DNA dsbs are only (just) detected at the highest radiation dose. Clearly the $^{157}Gd^{3+}$ must be on the DNA in order for neutron capture to result in DNA dsbs. We therefore attribute the DNA dsb induction to Auger electrons. The appreciable DNA single-strand breakage that occurs in the presence of DTPA is likely to be due to the other components of the ^{157}GdNC reaction, probably conversion electrons.

It is interesting to note the relative yields of DNA ssbs and DNA dsbs. The radiation dose that yields an average of 1 DNA ssb per plasmid molecule, (ie. when 37% of the plasmid DNA is intact), results in about 6% of linear plasmid molecules. In other words, there about 6 DNA dsbs per 100 DNAssbs. For low LET radiation, this sort of experiment yields the ratio of around 1 DNA dsb per 100 DNA ssbs[13]. On the other hand, the high LET products generated by ^{10}B-neutron capture yield about 12 DNA dsbs per 100 DNA ssbs in the same sort of experiment[14].

Experiments with the ^{157}Gd-labelled DNA ligand

At the outset it is important to note that the ^{157}Gd-DNA ligand is at a much lower concentration (25µM) than the $^{157}GdCl_3$ in the previous experiments. Consequently, much higher radiation doses are required to produce detectable damage, even with the larger (M13 RF) DNA target molecule.

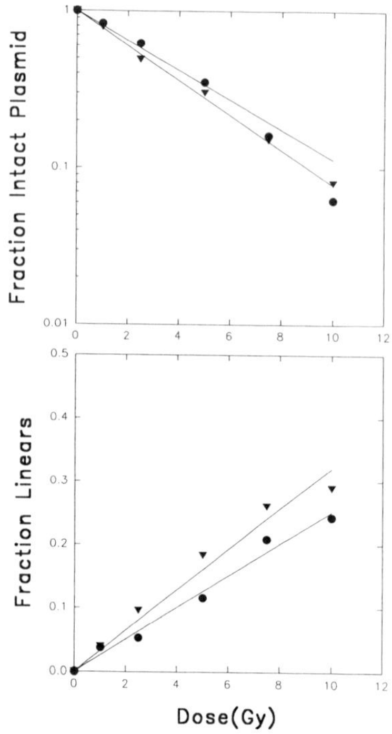

Fig. 2. The upper panel shows induction of DNA ssbs in pBR322 DNA in radiation-only controls ([]), test samples with $^{157}GdCl_3$ (•), and samples with both $^{157}GdCl_3$ and DTPA (∇). The lower panel shows the DNA dsbs results, with the same symbol key as for DNA ssbs.

Table 1 DNA strand breakage induced in M13 RF DNA by ^{157}GdNC of DNA-bound ^{157}GdHoe monomer. Number of DNA strand breaks per DNA molecule (relative to non-irradiated control) after 120Gy MOATA irradiation:

	DNA ssbs	DNA dsbs*
Irradiation only	0.12	-
DNA/GdHoe monomer	2.2	-
DNA/Gd Hoe monomer + DTPA	0.34	-

*DNAdsbs were not detectable in any of the samples

The radiation-only control shows a low yield of DNA ssbs after what is quite a high dose in the thermal neutron beam, but when the Gd-DNA ligand is present there is a much higher yield of DNA ssbs. Inclusion of excess DTPA, which takes the Gd out of the DNA ligand clearly reduces the yield of DNA ssbs. The most important aspect of the result is that there are no detectable DNA dsbs. The experiments with Gd^{3+} yielded 0.06 DNA dsbs per DNAssb; this level of DNAdsbs induction would have been easily detected in the experiments shown in Table 1.

Since the yield of DNA ssbs is decreased in the presence of DTPA, it would appear that a considerable portion of the observed DNA breakage is due to the Auger electron effect, even though this damage does not include DNA dsbs.

CONCLUSIONS

In contrast to $^{157}Gd^{3+}$ which yields DNA dsbs upon neutron capture, DNA-bound GdHoe monomer does not yield detectable DNA dsbs, only DNA ssbs. Our interpretation of this is that in the case of Gd^{3+}, the cation is presumably bound to the phosphate and therefore close to the DNA the helix. In the DNA/Gd ligand complex the Gd atom is apparently much further from the DNA helix, taking the DNA out of range of the high LET component (ie. that yields DNA dsbs) of the Auger electron shower associated with neutron capture by the ^{157}Gd atom.

We are now trying to design DNA ligands in which the Gd is forced into a position closer to the DNA helix.

ACKNOWLEGEMENTS

This project was supported by the Anti Cancer Council of Victoria, including an ACCV Cancer Research Studentship awarded to A. Whittaker. Access to the MOATA reactor was enabled by a grant from the Australian Institute of Nuclear Science and Engineering.

REFERENCES

1. R.F. Martin, G. D'Cunha and M. Pardee and B.J. Allen. Induction of double-stranded breaks following neutron capture by DNA-bound ^{157}Gd, *Int. J. Radiat. Biol.* 54: 205 (1988).
2. R.M. Brugger and J.A. Shih. Evaluation of gadolinium - 157 as a neutron capture therapy agent, *Strahlenther. Onkol.* 165: 153 (1989).
3. Y. Akine, N. Tokita, T. Matsumoto, H. Oyama and O. Aizawa. Gadolinium-neutron capture reactions: a radiobiological assay, *in*: "Progress in Neutron Capture Therapy for Cancer", B.J. Allen, D.E. Moore and B.V. Harrington, Eds., Plenum Press, New York and London (1992).
4. R.C. Greenwood *etal.* Collective and two-quasiparticle states in ^{158}Gd observed through radiation neutron capture in ^{157}Gd, *Nuclear Physics* A304: 327 (1978).
5. D.E. Charlton and J.L. Humm. A method of calculating intial DNA strand breakage following decay of incorporated ^{125}I, *Int. J. Radiat. Biol.* 53: 353 (1988)
6. D.E. Charlton. The range of high LET from ^{125}I decays, *Radiat. Res.* 107: 163 (1986).
7. R.F. Martin and W.A. Haseltine. Range of radiochemical damage to DNA with decay of iodine-125, *Science* 213: 896 (1981).
8. R.F. Martin and N. Holmes. Use of an ^{125}I-labelled DNA ligand to probe DNA structure, *Nature* 302: 452 (1983).
9. R.F. Martin, A. H. Haigh, C. Monger, M. Pardee, A.D. Whittaker, D.P. Kelly, and B.J. Allen. ^{157}Gd-neutron capture: Potential of ^{157}Gd-labelled DNA ligands for neutron capture therapy, *in*: "Progress in Neutron Capture Therapy for Cancer, B.J. Allen, D.E. Moore and B.V. Harrington, Eds., Plenum Press, New York and London (1992).
10. A. Corder, A. Whittaker, D.P. Kelly, H. Meriaty, B.J. Allen and R.F. Martin. Synthesis and Evaluation of ^{10}B-labelled DNA ligands, **These proceedings.**
11. A.D. Whittaker, D.P. Kelly, M. Pardee and R.F. Martin. Synthesis of ^{10}B-and ^{157}Gd-labelled DNA ligands for neutron capture therapy, *in*: "Progress in Neutron Capture Therapy for Cancer", B.J. Allen, D.E. Moore and B.V. Harrington, Eds. Plenum Press New York and London (1992).
12. L. Denison A. Haigh, G. D'Cunha and R.F. Martin. DNA ligands as radioprotectors: molecular studies with Hoechst 33342 and Hoechst 33258, *Int. J. Rad. Biol.* 61: 69 (1992).
13. R.E. Krisch, M.B. Flick and C.N. Trumbore. Radiation chemical mechanisms of Single- and Double-strand break formation in irradiated SV40 DNA, *Radiation Research* 126, 251 (1991).
14. A. Corder and R.F. Martin. Bibenzimidazole as radioprotectors against BNC-induced damage. **These Proceedings.**

CHEMICAL AND ENZYMATIC INCORPORATION OF BORON INTO DNA

B.F. Spielvogel,[1] A. Sood,[1] W. Powell,[1] J. Tomasz,[1,2] K. Porter,[1,2] and B.R. Shaw[2]

[1]Boron Biologicals, 533 Pylon Drive, Raleigh, NC 27606
[2]Dept. of Chemistry, Duke University, Durham, NC 27706

INTRODUCTION

Neutron capture events occurring in the cell nucleus associated with genetic material should be highly lethal.[1,2] We and others have prepared a variety of boronated nucleic acid derivatives or DNA binding boronated ligands toward achieving this goal.[3] Approaches we have used include synthesis of nucleosides with N-boronated bases[4] and (oligo)nucleotides with boronation on the phosphodiester backbone.[5] Previously, we reported that the radiolabeled [2-^{14}C]2'-deoxycytidine-N^3-cyanoborane was taken up much faster in leukemic Tmolt3 cells than in rapidly dividing non-tumorous human Bg-9 fibroblasts.[6] Evidence was also reported for a greater percentage of binding or incorporation into the DNA and RNA of cancer cells than in non-cancer cells.

In view of the above, we initiated two studies: (1) To obtain definitive information on the incorporation of boronated species into DNA, boronated nucleoside triphosphates were prepared to determine (via molecular biology techniques) if they could act as substrates for DNA polymerase. (2) To provide new agents to target genetic material, highly boronated oligonucleotides were prepared chemically.

BORONATED NUCLEOSIDE TRIPHOSPHATES

Synthesis

Two triphosphate species, **1** and **2**, were prepared for this study (Figure 1). **1** was synthesized by reaction of the N^7-cyanoborane adduct of 2'deoxyguanosine with $P(O)Cl_3$ followed by reaction with tri-n-butylammonium pyrophosphate in DMF. Subsequent hydrolysis produced **1** in 40% yield. Compound **2** was prepared as previously described[7] and exists as a 3:2 mixture of diastereomers. Both **1** and **2** possess sufficient stability to carry out the experiments described below.

DNA Incorporation Studies: Compound 1

Compound **1** was found to be a substrate for DNA polymerases and was readily incorporated into M13mp2 DNA duplex. The presence of cyanoborane

Advances in Neutron Capture Therapy, Edited by
A.H Soloway *et al.*, Plenum Press, New York, 1993

on the N^7 of guanines in this newly synthesized DNA was determined by the following assays: 1) Normal DNA is readily methylated at N^7 of guanine by dimethylsulfate and the methylated DNA is easily cleaved upon heating. The boronated DNA was found to be resistant to this cleavage as it could not be methylated at the N^7 position of guanine. 2) There are numerous restriction enzymes that cleave normal DNA at specified sites. Some of these enzymes, such as *HaeIII, PvuI, PuuII* and *Sau3AI,* require a guanine in their binding sites and bind to the N^7 position of guanine. Since, in boronated DNA, this position is blocked, no binding to or cleavage of DNA occurs. Details of these studies will be published separately.

Boronated Nucleoside Triphosphates

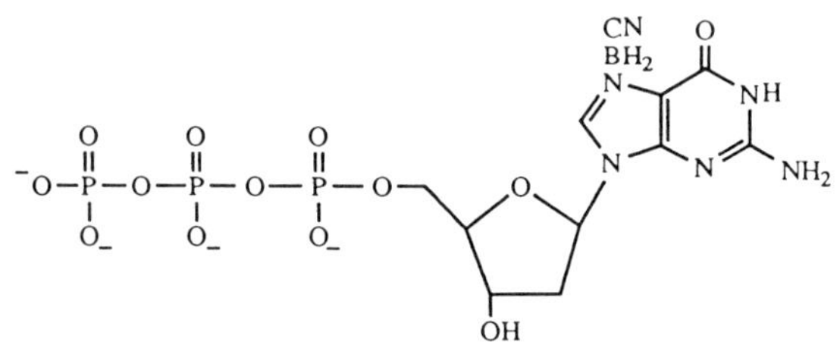

1

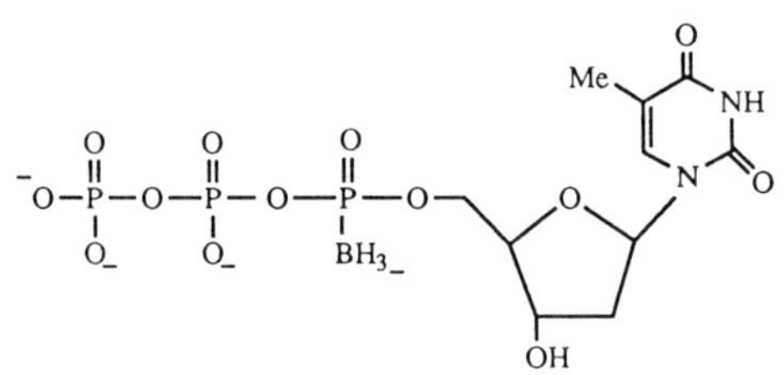

2

Figure 1. N^7-Cyanoborano-2'-deoxyguanosine 5'-triphosphate (**1**), 5'-αP-borane thymidine triphosphate (**2**)

DNA Incorporation Studies: Compound 2

Since replacement of one of the non-bridging oxygens of the α-phosphate group with a BH_3 group in a nucleoside triphosphate results in the formation of two diastereomers, we wanted to see which diasteromers, if either, was the substrate for DNA polymerase. We previously determined that the mixture of the two (P-BH_3) diastereoisomers could substitute (without pause) for deoxythymidine 5'-triphosphate in the extension of a deoxyribo 17-mer primer by Sequenase using a 25-mer template containing one 2'-deoxyadenosine residue.[7] The two diasteromers were separated by HPLC and, surprisingly, both were found to be substrates for the Sequenase (polymerase). The rate of incorporation, however, may be different.

SYNTHESIS OF CARBORANYLPHOSPHORAMIDATE OLIGONUCLEOTIDES

Oligonucleotides may be expected to have significant selectivity and affinity for nucleic acid targets.[8] In addition to binding single stranded RNA sequences, other nucleic acid targets, including double stranded DNA, are possible. Boronated oligonucleotides, in addition to inhibitory activity of tumor or viral genes, may also be used for neutron capture. Indeed, there is evidence that, as RNA is bound by any "antisense" oligonucleotide, additional RNA is expressed. Thus, a means for concentration of boronated oligomers in the cell may be useful for BNCT.

In addition to constructing backbone boronated oligomers of the type previously prepared, we devised a route that could deliver large numbers of boron atoms to nucleic acid targets. One type shown in Figure 2 is the oligonucleotide phosphoramidate with either a charged or uncharged backbone.

To prepare such oligos, several carboranyl primary or secondary amines, where the amino group is separated from the cage by one, two, or three methylene groups, were prepared. Our preparative route was designed to use the H-phosphonate method for preparing DNA using a Biosearch/Milligen 8750 synthesizer. Preliminary model reactions using diethyl phosphite were carried out to establish the general synthesis methodology (Figure 3).

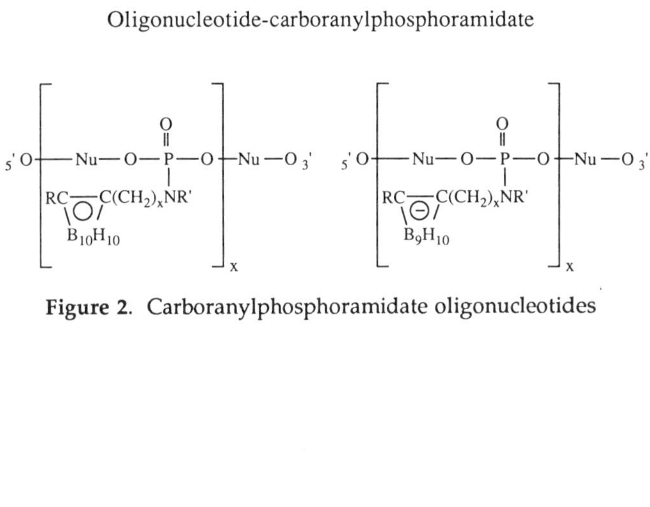

Figure 2. Carboranylphosphoramidate oligonucleotides

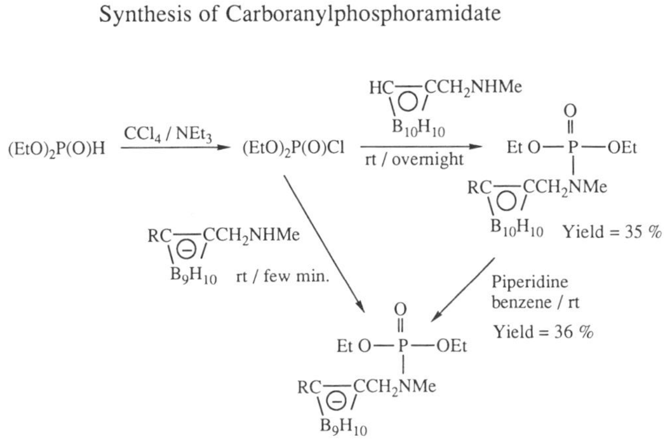

Figure 3. Model studies on the synthesis of carboranylphosphoramidates

After successful model reactions, two oligonucleotides, a 12-mer with only thymine bases and a 14-mer with a mixed sequence were synthesized according to Figure 4. Both carboranyl phosphoramidates were considerably nonpolar as expected. The 12-mer was converted as shown to $(dT)_{12}$ and identified by comparison with an authentic sample of $(dT)_{12}$. The backbone structure of 14-mer was confirmed by 31-P nmr which showed a broad peak at 10.1 ppm corresponding to phosphoramidate structure. The peak is broad due to the overlap of several signals. No peak was observed near 0 ppm where normal phosphate is expected. The ^{11}B nmr of the crude mixture showed signals corresponding to the nido-carborane structure indicating that 3 day NH_4OH treatment at room temperature is sufficient to open up the closo-carborane in this particular case.

Synthesis of Dodecathymidylyl Carboranylphosphoramidate

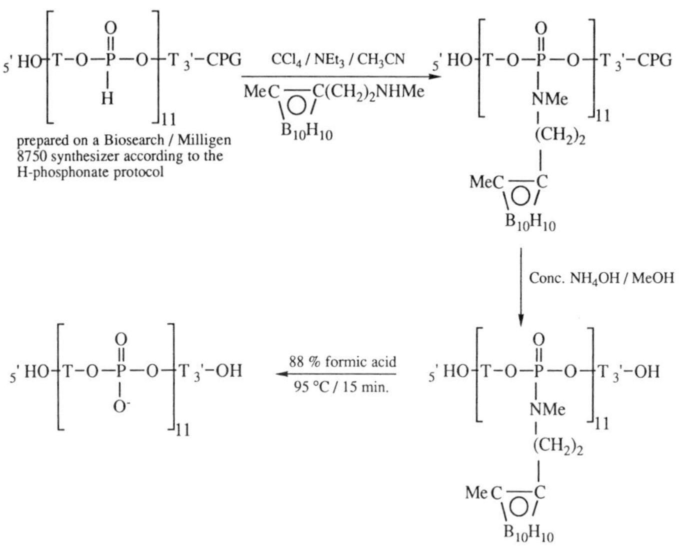

Figure 4. Preparation method for synthesis of oligonucleotides.

CONCLUSIONS

This work has established, for the first time, that boronated nucleoside triphosphates can serve as substrates for DNA polymerases and be incorporated into DNA. The feasibility of chemically preparing boronated DNA, specifically oligonucleotides with a carboranyl phosphoramidate backbone, has been demonstrated. This work shows that suitably designed boronated nucleic acid derivatives can be prepared to target genetic material.

ACKNOWLEDGMENTS

This research was supported by the Department of Energy (Grant DE-FG05-90ER81065) and in part by the NIH (Grant 1-R43-AI30887-01) and by the American Cancer Society (NP-741).

REFERENCES

1. T. Kobayashi and K. Kanda, Analytical calculation of boron-10 dosage in cell nucleus for neutron capture therapy, *Radiat. Res.* 91:71 (1982).
2. D. Gabel, S. Foster, and R.G. Fairchild, The Monte-Carlo simulation of the biological effect of the $^{10}B(n\alpha)^{7}Li$ reaction in cells and tissue and its implication for boron neutron capture therapy, *Radiat. Res.* 111:14 (1987).
3. B.J. Allen, D.E. Moore, and B.V. Harrington, Eds., "Progress in Neutron Capture Therapy for Cancer" Plenum Press. New York (1992). See articles on pps 207, 211, 219 and 231 and references therein.
4. A. Sood, B.R. Shaw, and B.F. Spielvogel, Boron containing nucleic acids: Synthesis of cyanoborane adducts of 2'-deoxynucleosides *J. Amer. Chem. Soc.* 111:9234 (1989).
5. A. Sood, B.R. Shaw, and B.F. Spielvogel, Boron containing nucleic acids 2: Synthesis of oligodeoxynucleoside boranophosphates, *J. Amer. Chem. Soc.* 112:9000 (1990).
6. A. Sood, B.F. Spielvogel, B.R. Shaw, L.D. Carlton, B.S. Burnham, E.S. Hall, and I.H. Hall, The synthesis and antineoplastic activity of 2'-deoxynucleoside-cyanoboranes in murine and human culture cells, *Anti Cancer Res.* 12:335 (1992).
7. J. Tomasz, B.R. Shaw, K. Porter, B.F. Spielvogel, and A. Sood, 5'-P-Borane-substituted thymidine monophosphate and triphosphate. Angewandte Chemie. Int. Ed. 31:1373 (1992).
8. For a recent review, see S.T. Crooke, Therapeutic Applications of Oligonucleotides, *Bio Technol.* 10:882 (1992).

LIPOSOMAL DELIVERY OF BORON TO MURINE TUMORS FOR BORON NEUTRON CAPTURE THERAPY

D.A. Feakes,[1] K.J. Shelly,[1] M.F. Hawthorne,[1] P.G. Schmidt,[2] C.A. Elstad,[2] G.G. Meadows[3] and W.F. Bauer[4]

[1]University of California at Los Angeles, Los Angeles, CA
[2]Vestar, Inc., San Dimas, CA
[3]Washington State University, Pullman, WA
[4]Idaho National Engineering Laboratory, Idaho Falls, ID

Liposomes, of a specific size and composition, have demonstrated unique ability to selectively deliver significant quantities of boron to murine tumors. Aqueous solutions of water soluble compounds with a high boron content have been encapsulated into the liposomes and studied *in vivo*. Although the liposomes deliver their contents to the interior of the tumor cells, the characteristics of the boron compound determine whether boron will be retained by the tumor cells or simply cleared from the system. Investigation of a series of compounds which vary in characteristics such as size, charge, and reactivity have allowed us to develop a working hypothesis which predicts *a priori* , with some degree of success, the ability of these boron compounds to be retained by tumor cells.

MATERIALS AND METHODS

The polyhedral borane anion derivatives used were prepared by published methods or slight modifications thereof.[1-5] Distearoyl phosphatidylcholine was obtained from Avanti Polar Lipids, and cholesterol was supplied by Calbiochem or Sigma.

Liposome emulsions were prepared by probe sonication of a dried film composed of equimolar amounts of the phospholipid and cholesterol with the hydrating solution (typically 5 mL, 250-300 mM in the boron-containing salt) at 65°C for 15-30 min. The vesicles were separated from the remaining free borane salt by eluting through a column of Sephadex G-25 (medium) with isotonic phosphate-buffered saline or lactose. Liposomal preparations were diluted with the appropriate buffer to a lipid concentration of 23-24 mg/mL and sterilized by filtration through a 0.22 μm Millipore membrane. The integrity of the encapsulated borane salt was confirmed by ^{11}B NMR at 162 MHz.

All murine biodistribution studies utilized female Balb/c mice (16-20 g), with EMT6 tumors implanted in the right flank 7-10 days prior to the experiment. Tumor mass at the time of sacrifice was 125-350 mg. Injections of liposome emulsions (200 μL) were made in the tail vein. Boron analyses were perfomed by inductively coupled plasma atomic emission spectroscopy (ICP-AES)[6] at the Idaho National Engineering Laboratory. Each data point represents the average of five mice. For clarity, error bars are not shown in the graphical data; standard deviations were typically 5-15% of the average values.

Advances in Neutron Capture Therapy, Edited by
A.H Soloway *et al.*, Plenum Press, New York, 1993

BIODISTRIBUTION STUDIES IN MICE

Murine biodistribution results for liposomes containing $Na_2B_{10}H_{10}$ and $Na_2B_{12}H_{11}SH$ have been reported.[7] They are characterized by the delivery of the boron species to the tumor (approximately 10 μg boron/g tissue at 6 hours) followed by rapid clearance. Although neither species provides therapeutically useful boron accumulation in tumor, these initial experiments verified that liposomes could be used to deliver their contents to tumor cells although the compounds themselves were not suitable for *in vivo* retention of the boron and BNCT.

The murine biodistributions of the isomers, $Na_2(n\text{-}B_{20}H_{18})$ and $Na_2(i\text{-}B_{20}H_{18})$, are quite similar to each other.[7] Six hour tumor boron values are approximately 20 μg boron/g tissue or twice that of the previously studied ten-boron species. The boron tumor concentration is essentially maintained over the 48 hour time period, especially in the case of the $Na_2(i\text{-}B_{20}H_{18})$ species.

$Na_2(i\text{-}B_{20}H_{18})$ in buffer solution was injected in concentrations equivalent to the encapsulated species (Table 1). Although the unencapsulated species had some affinity for liver, accumulation was not as high as that observed with the encapsulated species. The remaining tissues examined contained only trace amounts of boron 6 hours post-injection.

The murine biodistribution of $K_4B_{20}H_{17}OH$ is shown in Figure 1A. The blood loses 80% of its boron content between 6 and 30 hours, resulting in a 48 hour tumor to blood ratio of 3.4. Although the liver and spleen rapidly clear, the tumor loses boron as well. There is a 57% decrease in the tumor boron concentration over the 48 hour time period resulting in a final boron concentration of 7.3 μg boron/g tissue.

The double injection experiment (Figure 1B) with $Na_2(i\text{-}B_{20}H_{18})$ was performed by injecting liposomes at 0 and 24 hours (280 μg boron each injection, mice receiving both injections received 560 μg boron). The first 24 hour time period is analogous to the single injection experiment.[7] At 24 hours, the second injection was administered, increasing the boron concentration in all of the tissues. Boron cleared from the various tissues at the usual rates, resulting in a 48 hour tumor boron concentration of 26.6 μg boron/g tissue and a tumor to blood ratio of 2.6.

Liposomes loaded with $Na_2B_{10}H_9NCO$ displayed the biodistribution shown in Figure 1C. The 6 hour tumor boron concentration (19.2 μg boron/g tissue) only slowly decreases with time (20% loss between 6 and 30 hours) as compared to the rapid clearance of $Na_2B_{10}H_{10}$ (46% loss between 6 and 24 hours). Blood boron levels drop rapidly, 79% between 6 and 30 hours, resulting in a 48 hour tumor to blood ratio of 5.0.

The murine biodistribution of $Na_3B_{20}H_{17}NH_3$ is shown in Figure 1D. The tumor boron concentration increases over a period of approximately 30 hours and then slowly decreases, resulting in a final tumor boron concentration which is still 94% of the initial tumor boron concentration. All other tissues clear steadily over the 48 hour time period. The low blood boron concentration at 48 hours yields a tumor to blood boron ratio of 5.3.

Table 1. Murine tissue boron concentrations after injection of free $Na_2(i\text{-}B_{20}H_{18})$ (200 μg of boron, ≈11 mg/kg of body weight) in buffered solution.

Tissue	Boron concentration, μg/g of tissue ± SD 6 h	24 h
Blood	0.9 ± 0.1	1.9 ± 0.7
Liver	18.1 ± 0.8	8.9 ± 1.5
Spleen	0.9 ± 0.1	1.5 ± 0.5
Tumor	1.2 ± 0.3	1.9 ± 0.6

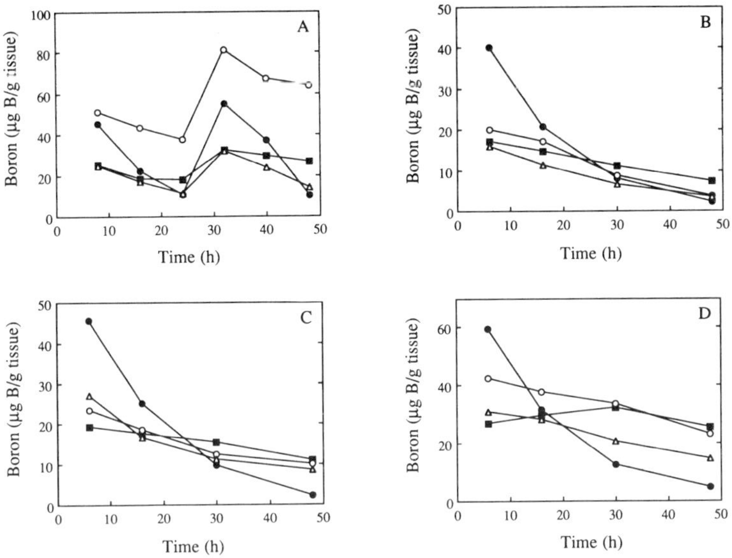

Figure 1. Murine tissue boron concentrations from delivery of borane salts by liposomes. ●, Blood; ■, tumor; ○, liver; Δ, spleen; Borane salts, with the injected doses were $Na_2(i\text{-}B_{20}H_{18})$, 280 µg of boron each injection (≈15 mg/kg of body weight) (*A*), $K_4B_{20}H_{17}OH$, 200 µg of boron (≈11 mg/kg of body weight) (*B*), $Na_2B_{10}H_9NCO$, 166 µg boron (≈9 mg/kg of body weight) (*C*), $Na_3B_{20}H_{17}NH_3$, 198 µg of boron (≈11 mg/kg of body weight) (*D*).

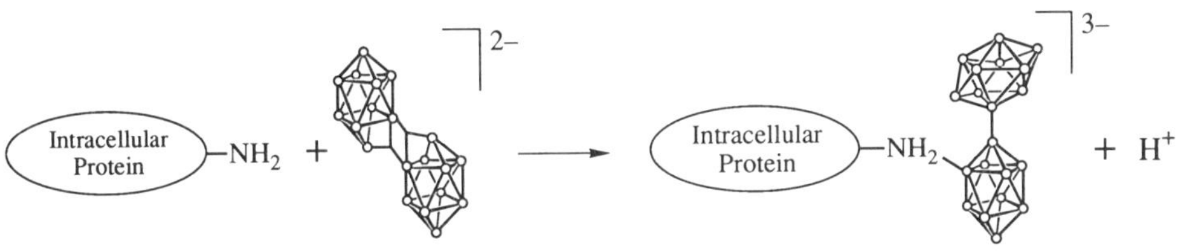

Figure 2. Proposed reaction of $Na_2(n\text{-}B_{20}H_{18})$ with the amino group of an intracellular protein moiety.

DISCUSSION

Initial investigations of the liposomal delivery of boron to murine tumors utilized $Na_2B_{10}H_{10}$ and $Na_2B_{12}H_{11}SH$ because of their high boron content, ease of synthesis, water solubility, hydrolytic stability, and lack of toxicity. These compounds were successfully delivered to the tumor, but quickly cleared due to their inability to be retained by the tumor cells.[7] Emphasis changed to the product resulting from the oxidative coupling of $Na_2B_{10}H_{10}$, $Na_2(n\text{-}B_{20}H_{18})$, and its photoisomer, $Na_2(i\text{-}B_{20}H_{18})$, since these compounds are known to undergo nucleophilic attack.[3,4] Additionally, the osmotic stress upon the liposomes encapsulating the $Na_2B_{20}H_{18}$ species is equivalent to those encapsulating $Na_2B_{10}H_{10}$ even though twice the concentration of boron is delivered. These compounds possess the potential to react with intracellular protein moieties as illustrated in Figure 2.

Biodistribution studies of both the normal and the photoisomer demonstrated retention of the delivered boron by the tumor cells over the 48 hour period, suggesting an intracellular reactivity suitable for tumor retention.[7] Further investigations, including the determination of the biodistribution of unencapsulated $Na_2(i\text{-}B_{20}H_{18})$ and its encapsulated hydrolysis product, $Na_4B_{20}H_{17}OH$, have established that liposomal $Na_2(i\text{-}B_{20}H_{18})$ is the active species *in vivo*. Neither the unencapsulated $Na_2(i\text{-}B_{20}H_{18})$ nor encapsulated $Na_4B_{20}H_{17}OH$ are retained by the tumor.

Double injection experiments are of interest because they enable a second dose of the boron compound to be administered once the initial blood boron has cleared. In the case of $Na_2(i\text{-}B_{20}H_{18})$, the second dose was administered at 24 hours. All normal tissues then cleared at their usual rates. However, because the tumor retains boron, the tumor boron concentration is further increased by the second dose, resulting in a final tumor boron concentration of 26.6 μg boron/g tissue.

The biodistribution of two other compounds believed to possess the potential for intracellular protein reaction, $Na_2B_{10}H_9NCO$ and $Na_3B_{20}H_{17}NH_3$, were examined. Compared to its unsubstituted precursor, $Na_2B_{10}H_{10}$, $Na_2B_{10}H_9NCO$ retained the boron delivered to the tumor (80% versus 54% over the first 24 hours). Since the only difference between the two compounds is the amine reactive isocyanate moiety, the proposed intracellular protein reaction appears to be substantiated.

The reduced amine derivative of $Na_2(n\text{-}B_{20}H_{18})$, $Na_3B_{20}H_{17}NH_3$, is the only compound studied to date which acccumulates in tumor over approximately 30 hours. This easily oxidized compound probably reverts to the more reactive $B_{20}H_{17}NH_3^{1-}$ species *in vivo*. The tumor boron concentration after 48 hours is essentially equal to the boron concentration after 6 hours.

Both the normal isomer, $Na_2(n\text{-}B_{20}H_{18})$, and the reduced amine, $Na_3B_{20}H_{17}NH_3$, have recently been investigated in Balb/c mice bearing P1798 murine lymphosarcoma. The biodistributions are similar to the EMT6 results.

CONCLUSION

Based on the data accumulated to date, a working hypothesis has been proposed which allows the prediction, *a priori* and with some degree of success, of which boron compounds will be retained by the tumor when delivered by liposomes. This hypothesis is that compounds possessing the potential to react with intracellular protein moieties will be retained within the tumor cells. This hypothesis is substantiated by the biodistributions of $Na_2(n\text{-}B_{20}H_{18})$, $Na_2(i\text{-}B_{20}H_{18})$, $Na_2B_{10}H_9NCO$, and $Na_3B_{20}H_{17}NH_3$. The retention in tumor of the reduced amine species, $Na_3B_{20}H_{17}NH_3$, is attributed to its ability to be oxidized to a more reactive $Na_3B_{20}H_{17}NH_3$ species. Compounds which do not possess either a reactive functional group or the ability to be oxidized intracellularly to a more reactive species, such as $K_4B_{20}H_{17}OH$, are not retained by the tumor. Double injection experiments, such as the one reported for $Na_2(i\text{-}B_{20}H_{18})$, can be used to further increase the boron concentration retained by the tumor once the initial boron has cleared from normal tissues. The diversity and utility of liposomal delivery of boron compounds to tumors is one of the most promising systems known to date.

ACKNOWLEDGEMENTS

This research was conducted as part of the BNCT program of the Idaho National Engineering Laboratory and was performed under the auspices of the U.S. Department of Energy (DOE), Office of Energy Research, under DOE Field Office, Idaho, Contract DE-AC07-76ID01570.

REFERENCES

1. Hawthorne, M.F. & Pilling, R.L. (1967) in *Inorganic Syntheses Vol. IX*, ed. Tyree, Jr., S.Y. (McGraw-Hill Publishing Co. Inc., New York), pp. 16-19.
2. Hawthorne, M.F. & Pilling, R.L. (1966) *J. Am. Chem. Soc.* **88**, 3873.
3. Hawthorne, M.F., Pilling,R.L. & Garrett, P.M. (1965) *J. Am. Chem. Soc.* **87**, 4740-4746.
4. Hawthorne, M.F., Pilling, R.L., Stokely, P.F. & Garrett, P.M. (1963) *J. Am. Chem. Soc.* **85**, 3704.
5. Shelly, K., Knobler, C.B. & Hawthorne, M.F. (1992) *Inorg. Chem.* **31**, 2889-2892.
6. Tama, S.R., Moore, D.E. & Allen, B.J. (1987) *Anal. Chem.* **59**, 2161-2164.
7. Shelly, K., Hawthorne, M.F. & Schmidt, P.G. (1992) in *Progress in Neutron Capture Therapy for Cancer*, ed. Allen, B.J., Moore, D.E. and Harrington, B.V. (Plenum Publishing Corp., New York), pp. 259-263.

LOW DENSITY LIPOPROTEIN RECONSTITUTIONS WITH ALKYL AND ARYL CARBORANES

Stephen B. Kahl, David W. Pate, and Larry A. Wainschel

Department of Pharmaceutical Chemistry
University of California, San Francisco
San Francisco, CA 94143-0446

INTRODUCTION

Low density lipoproteins (LDL's) are among the most attractive of possible boron delivery methodologies by virtue of their high carrying capacity for hydrophobic boron compounds and their cellular internalization through receptor-mediated processes. Previous work in our laboratory has demonstrated that long chain unsaturated fatty alcohol esters of carborane carboxylic acid can be very efficiently exchanged for the cholesterol ester core of human LDL. In collaboration with coworkers at Brookhaven National Laboratory using *in vitro* techniques, we have found that: (a) boron concentrations exceeding 10 times that needed for therapy were easily obtained; (b) boron remained firmly bound despite repeated washings and suspension in boron-free medium; (c) boron distribution was intracellular, with a biological efficacy indicative of a cytoplasmic location; (d) *in vitro* uptake and efficacy studies were consistent with a receptor-mediated binding mechanism; (e) boron uptake, retention, and efficacy were better by far than those obtained with any other boron compound previously studied in our *in vitro* screening system.(1) The present study was undertaken to determine whether simple alkyl and aryl carboranes could be similarly exchanged for LDL cholesterol and, if so, whether any identifiable structure activity relationships among the compounds could be identified. We also wished to determine whether reconstituted LDL's suffered serious aggregation and to demonstrate that ICP-AES and prompt gamma spectrometric boron assay procedures produced coincident numbers.

MATERIALS AND METHODS

Allyl, methyl, phenyl and butenyl orthocarborane and metacarborane were purchased from Dexsil Chemical Corp. and required further purification (either vacuum distillation or recrystallization) before use. Pure hexyl carborane was a generous gift from Callery Chemical Co. Elaidyl carborane carboxylate (ECC) was synthesized by the method Kahl(2) from elaidyl alcohol (Nu-Chek Prep) and carborane carboxylic acid chloride. Low density lipoprotein was isolated from an overweight but otherwise healthy volunteer by density gradient ultracentrifugation and reconstituted with boron compounds by the method of Kahl and Callaway.(3) Protein content of reconstituted solutions was determined by the micro-Lowry method. Boron analyses were carried out at Brookhaven National Laboratory (prompt gamma) and at Idaho National Engineering Laboratory (ICP-AES).

Advances in Neutron Capture Therapy, Edited by
A.H Soloway *et al.*, Plenum Press, New York, 1993

RESULTS

Table 1 shows the results of one LDL reconstitution. Each entry represents the average ± standard deviation for three reconstitutions of each compound. The boron analyses were carried out at BNL using the prompt gamma technique. All samples were made using unenriched material. The boron content is reported here as natural boron so the errors associated with the phenyl, methyl and meta carborane points are substantial.

Table 2 displays the boron/protein ratios obtained for a series of samples either singly or doubly filtered through a 0.45 micron filter. Each number represents the average of four boron/protein ratios with each of these ratios in turn obtained using prompt gamma boron values.

Table 3 presents a comparison of boron analytical data obtained from prompt gamma and ICP-AES analyses of identical reconstituted LDL samples. The ICP-AES data were obtained using a longer and more forceful bomb digestion technique than is normally used. Prompt gamma data are again based on ^{10}B determinations of natural abundance samples.

DISCUSSION

Low density lipoproteins are attractive targets as tumor-selective boron carriers for a number of reasons. Their oily core provides a favorable domain for lipophilic drugs and prodrugs, and it has been demonstrated that the natural core components can be replaced, in part or *in toto*, with lipophilic drugs.(1,4) They are constituted of natural components able to survive in tissue and plasma for significant time periods. Their small particle size (17-25 nm) allows them to diffuse from vascular to extravascular compartments. Interaction of the surface protein, Apo B-100, with receptors on the target surface results in endocytosis of the LDL particle and delivery of its core contents to the

Table 1. LDL reconstitutions with alkyl and aryl carboranes

Compound	Ave. μg B/gm	Ave. μg protein/gm	Ave. μg B/μg protein
1-hexyl carborane	472 ± 8	660 ± 15	0.71 ± .05
1-allyl carborane	379 ± 125	673 ± 100	0.55 ± .11
1-butenyl carborane	67 ± 20	508 ± 100	0.13 ± .04
1-phenyl carborane	59 ± 35	492 ± 60	0.11 ± .06
1-methyl carborane	39 ± 10	224 ± 13	0.11 ± .06
meta carborane	22 ± 5	511 ± 60	0.04 ± .02

Table 2. Boron/protein ratios for singly vs. doubly filtered samples

	μg B/μg Protein	
Compound	single filtered	double filtered
1-hexyl carborane	0.71 ± 0.05	0.68 ± 0.04
1-allyl carborane	0.34 ± 0.10	0.30 ± 0.04
ECC	0.25 ± 0.02	0.22 ± 0.02
1-methyl carborane	0.21 ± 0.05	0.39 ± 0.15
1-phenyl carborane	0.15 ± 0.05	0.14 ± 0.04
1-butenyl carborane	0.06 ± 0.02	0.09 ± 0.03

Table 3. Comparison of prompt gamma and ICP-AES measurements of boron content in identical samples.

sample #	prompt gamma	ICP
1	512.5	590.0
2	484.5	526.4
3	418.5	480.6
4	30.5	39.2
5	49.0	47.3
6	38.0	42.1
7	68.0	92.7
8	82.0	125.1
9	49.5	46.9
10	113.0	68.3
11	30.0	39.6
12	34.0	40.0
13	22.5	33.7
14	27.0	32.4
15	17.0	43.8
16	520.5	547.6
17	370.5	400.1
18	246.0	314.2

peri-nuclear region of the cell. This receptor-mediated process is significantly enhanced in certain pathophysiologic states including many cancers and the atherosclerotic plaques characteristic of arteriosclerosis. We have previously demonstrated that hydrophobic boron compounds, particularly ortho-carborane derivatives of unsaturated long chain fatty acids, can be efficiently introduced into the LDL core and that the resultant boronated LDL particles behave in cell culture in a manner consistent with receptor-mediated uptake.(1) In fact, the very high degree of core substitution obtained with these derivatives appears to negate the normal negative feedback regulation by cholesterol and results in a non-saturable uptake process.

In a preliminary report several years ago, we noted that several alkyl and aryl carboranes could also serve as reconstitution agents.(3) The present study was undertaken to confirm these findings and to seek evidence of any relationship between the nature of the side chain substitutent and the reconstituting ability of the compound. Table 1 presents results from a representative set of these reconstitutions. These quantitative results support our previous findings that n-hexyl and allyl carborane reconstitute into LDL far more efficiently than do other substituted and unsubstituted carboranes. With the exception of meta carborane, the trend in measured protein values is mimicked by overall reconstitution efficiency as expressed in μg boron/μg protein, i.e., high measured protein in a reconstituted sample generally reflects high boron content. Moreover, the variance of both boron and protein values within a given set of reconstitutions as expressed as a percentage of the mean value is invariably low for "good" reconstituters and high for "poor" ones. These data suggest that "poor" reconstituters disrupt the Apo B-100 protein which makes up 40% of the spherical particulate surface. The molecular causes of this disruption are certainly not clear, but may be related to the physicochemical nature of the carboranes. Those carboranes which are crystalline at room temperature (ortho, meta, methyl, butenyl and phenyl) do not reconstitute well, whereas those which are oils (hexyl and allyl) do. All of the carborane esters of fatty alcohols which were previously found to be "good" reconstituters were oils. Whatever the mechanism, it is clear that only hexyl and allyl carborane are viable candidates among this series for LDL-mediated tumor boron delivery.

It was also of interest to determine whether the single filtration following reconstitution to remove the starch support was sufficient to remove LDL aggregates. There had been some suggestion in the literature that reconstituted LDL aggregated and that these aggregates were rapidly removed by the reticuloendothelial system when injected into animals. Table 2 presents the results of our study of this phenomenon. The singly filtered samples all have boron/protein ratios within experimental error of similar determinations from Table 1 with the exception of allyl carborane which usually is associated with a large standard deviation. The elaidyl carborane carboxylate has been our standard for all previous *in vitro* work, and the ratio reported here is consistent with values for a number of previous samples. It is clear that the second filtration step has a relatively minor effect on the over-all boron/protein ratio of the boronated LDL. Indeed, in each of the pairs of ratios both values are within experimental error of each other. However, the raw data (not shown) indicate an interesting effect. For the better compounds, i.e. hexyl, allyl and ECC, both the absolute boron content and the absolute protein content are reduced by approximately 10% during the second filtration. In contrast, in the poorer compounds the drop in boron content varies widely while the protein content falls predictably by 10-15%. This is further confirmation that such compounds are unsuitable for LDL incorporation and disrupt the LDL in an as yet unknown fashion.

The third goal of this study was to assess the comparability of the two most commonly used boron assays, prompt gamma and ICP-AES spectrometry. Since the tissue boron content of a series of planned animal biodistribution experiments using boronated LDL is to rely on the ICP-AES method, it is important to demonstrate that both methods give comparable numbers. An early attempt at this comparison gave wildly different values for the same sample with significant carry over from one analysis to the next. This was attributed to the very "greasy" nature of the LDL matrix and high thermal stability of the carborane cage. The results presented in Table 3 were generated after increasing the severity and length of the sample bomb digestion procedure. For samples containing up to ~600 ppm boron, the ICP technique gives boron concentration values that are generally higher than with prompt gamma analysis. Interestingly, the degree of agreement is greatest at the highest values, e.g. samples 1-3 and 16-18. Perhaps not coincidentally, these samples contained hexyl and allyl carborane, respectively. The samples with greatest divergence, i.e. samples 13-15, contained meta carborane. This may indicate that the new sample digestion procedure is still not adequate to overcome the greater chemical stability of meta carborane compared to ortho carborane.

Although the present study does suggest that n-hexyl or allyl carboranes could be used for small animal studies of boronated LDL, we believe it more prudent to continue using ECC. Animal distribution studies with this compound encapsulated in LDL have begun and will be reported upon at a later date.

ACKNOWLEDGMENTS

The authors wish to thank Dr. Brenda H. Laster of BNL and Dr. William F. Bauer of INEL for carrying out the boron analyses. The submitted paper has been authorized by a contractor of the US government under DOE Contract No. DE-ACO7-76IDO1570. Accordingly, the US government retains a non-exclusive, royalty-free license to publish or reproduce the published form of this contribution, or allow others to do so, for U.S. government purposes.

REFERENCES

1. Laster, B.H., Kahl, S.B., Popenoe, E.A., Pate, D.W., and Fairchild, R.G., *Cancer Res.* **51**:4588-4593 (1991).
2. Kahl, S.B., *Tetrahedron Lett.* **31**:1517-1520 (1990).
3. Kahl, S.B. and Callaway, J.C., *Strahlenther. Onkol.* **165**:137-139 (1989).
4. Firestone, R.A., Pisano, J.M., Falek, J.R., McPhaul, M.M. and Krieger, M., *J. Med. Chem.* **27**:1037-1043 (1984).

A RAPID METHOD FOR THE DIRECT ANALYSIS OF BORON IN WHOLE BLOOD BY ATOMIC EMISSION SPECTROSCOPY

W.F. Bauer[1*], P.L. Micca[2] and B.M White[1]

[1]Idaho National Engineering Laboratory, EG&G Idaho, Inc.
Idaho Falls, ID 83402
[2]Medical Department, Brookhaven National Laboratory
Upton, NY 11973

INTRODUCTION

The INEL BNCT program has been involved in normal tissue dose tolerance studies using dogs as the large animal model[1]. In order to accurately predict when to begin an irradiation and then determine the length of the radiation required to achieve the desired radiological dose at a specified mean boron concentration, a rapid technique for the determination of boron is required. It has previously been shown that after administration and distribution of $Na_2B_{12}H_{11}SH$ (BSH), the elimination phase has a $T_{1/2}$ of ~300 minutes[2] and one should be able to roughly predict blood B concentrations at specified times during the elimination. However, the physiology can be expected to vary somewhat from animal to animal and the blood B concentrations are expected to vary as well. These variations mean that target B concentrations can easily be missed.

There are numerous methods available for B analysis in biological samples, however, most require some form of time consuming digestion procedure (20 minutes minimum) to eliminate the matrix and, in many cases, to convert the B to a single form. Rapid analysis of animal tissues by atomic emission spectroscopy has been performed by suspending homogenized tissue in a deionized water slurry[3]. This analysis procedure, including tissue homogenization, could be performed in ~10 minutes. Since a heparinized blood sample would not require homogenization, it should readily be suspendable in solution.

This paper describes the analysis of heparinized whole blood using slurry sample introduction into both an inductively-coupled plasma and a direct current plasma atomic emission spectrometer (ICP-AES and DCP-AES, respectively). A non-ionic surfactant (Triton X-100™) is used to form a stable slurry of suspended blood cells. The non-ionic surfactant was selected to minimize cell destruction and ICP or DCP problems caused by excess alkali metal content.

Advances in Neutron Capture Therapy, Edited by
A.H Soloway *et al.*, Plenum Press, New York, 1993

Table 1. Results of an initial study comparing slurry nebulization and acid digestion of whole blood spiked with either BSH or tetraborate. Analysis was by ICP-AES.

	Slurry µg B/g	Digestion µg B/g	Difference[a]	RPD[b]
Blood	< 0.42	< 0.40		
15 µL BSH	9.50	9.57	-0.08	-0.8
30 µL BSH	18.43	17.96	0.47	2.6
45 µL BSH	30.16	29.76	0.40	1.3
60 µL BSH	39.16	38.02	1.14	2.9
75 µL BSH	49.20	48.38	0.82	1.7
60 µL $B_4O_7^{2-}$	53.02	50.68	2.34	4.5
40 µL $B_4O_7^{2-}$	36.27	34.24	2.02	5.7
30 µL $B_4O_7^{2-}$	27.70	26.31	1.40	5.2
20 µL $B_4O_7^{2-}$	19.39	18.25	1.14	6.1
10 µL $B_4O_7^{2-}$	9.69	9.60	0.09	1.0
Mean			0.97	3.0
Standard Deviation			0.79	2.3

[a] difference = digestion - slurry
[b] Relative % difference = difference × 100/mean

EXPERIMENTAL

Samples were prepared as slurries simply by weighing a 300 mg aliquot into a 10 mL volumetric flask, adding 0.2 mL of a 2.5% solution of Triton X-100™, and diluting to the mark with deionized, boron free water. Standards were prepared in a similar 0.05% solution of Triton X-100™ from commercial standard solutions (SPEX Industries and NIST). Salts of BSH were from Callery Chemical Company and Centronics Limited and p-boronophenylalanine (BPA) was obtained from Callery Chemical Company. The $Na_4B_{24}H_{22}S_2$ (BSSB) was prepared from BSH by the method described in reference 4. Standard solutions containing ^{10}B were prepared from dried 95% ^{10}B enriched $B(OH)_3$ obtained from NIST. Boron analysis was performed by ICP-AES at both the 208.959 and 249.678 nm emission lines while the DCP-AES analysis used the 208.959 and 249.778 nm emission lines of boron.

RESULTS AND DISCUSSION

Initial experimentation was performed using blood obtained from a healthy labrador retriever. Aliquots (1 mL) of the heparinized whole blood were spiked with differing quantities of a BSH stock solution (691 µg B/mL) and a commercial standard containing B (1000 µg/mL) as ammonium tetraborate. The resulting solutions were analyzed by standard ICP-AES procedures after digestion of 200 mg with nitric acid[5]. Slurries for ICP-AES consisted of 200 mg aliquots of the spiked blood that had been suspended in 0.05% Triton X-100™. The results of this experiment are shown in Table 1. Plotting the ICP-AES analysis results for slurry nebulization versus the standard digestion procedure results in the regression line $C_{slurry}=(1.03_8 \pm 0.01_4)C_{digest}-(0.09 \pm 0.43)$. The results from the two sample preparation techniques agree very well, however, a slight (~3%) positive bias may be present in the slurry results. This slight bias was not a great concern because it most likely was the result of having mismatched standard and sample matrices. This initial study

Table 2. Effect of blood volume on the recovery of B by DCP-AES with slurry sample introduction.

mg Blood	0	109	219	300	400	500
% Recovery	100	97	101	103	101	101

Table 3. Comparison of typical results for B analysis by prompt γ, DCP-AES and ICP-AES.

	Concentrations in μg $^{10}B/g$			
Sample	Prompt γ[a]	Prompt γ[b]	DCP-AES	ICP-AES[c]
BPA Solution	26.9		27.1	
9.5 μg $^{10}B/mL$ Standard	9.6	9.5	9.7	
Blank Blood Spike (Borate)	12.5		12.2	
BSSB Solution	122	125	114	
A 300 min Blood	48.5	47.3	43.2	50.5
A 360 min Blood	43.6	41.8	37.7	42.2
A 420 min Blood	37.5	38.9	33.3	38.0
A 489 min Blood	32.3	32.2	28.8	32.5
A 558 min Blood	29.0	29.5	26.3	26.6
A 627 min Blood	24.3		22.6	26.4
B 360 min Blood	44.2		40.0	45.7
B 420 min Blood	40.3		34.8	40.5
B 496 min Blood	36.0		31.7	37.1
B 562 min Blood	33.3		29.1	33.6
B 622 min Blood	30.2		26.5	30.8
C 240 min Blood	60.7		54.0	62.7
C 300 min Blood	54.1		47.2	56.0
C 360 min Blood	45.9		42.0	48.4
C 420 min Blood	39.3		34.7	39.4
C 506 min Blood	33.2		28.1	32.3
C 571 min Blood	29.9		26.3	30.7
C 639 min Blood	26.8		24.4	27.6

[a] Using multipoint calibration at a later date
[b] Using only a 2 point calibration on the day of the infusion
[c] Same slurries as for DCP but analysis was several days later

utilized calibration standards that were prepared in only a deionized water matrix. To avoid this potential problem, all subsequent standards were prepared in 0.05% Triton X-100™.

Since the aspiration and nebulization of the blood slurries for boron analysis by ICP-AES worked well for both BSH and a simple borate, the technique was taken to BNL to be used with the DCP-AES in direct support of the large animal irradiations. Because the detection limits for the 208.959 nm B emission on the DCP-AES were only ~0.023 μg/mL, it was decided to use 300 mg of heparinized whole blood instead of the 200 mg used in the ICP-AES experiments. To be sure that there were no significant physical or spectral interferences associated with this change, increasing quantities of B free whole dog blood

Table 4. Mean % errors for whole blood B analysis by prompt γ, ICP-AES and DCP-AES.

Sample Type	Mean % Errors[a]			
	Prompt γ[b]	ICP-AES	DCP-1[c]	DCP-2[c]
Overall Error	0.1 ± 2.8	1.6 ± 3.7	-9.5 ± 4.4	-3.9 ± 6.4
Cage Compounds	0.2 ± 3.0	1.6 ± 3.7	-11.0 ± 2.3	-5.8 ± 4.1
Borate Compounds	-0.8		0.0 ± 1.9	

[a] Assumes prompt γ with multipoint calibration is true value
[b] Prompt γ with 2 point calibration
[c] DCP-AES analysis prior to (-1) and after (-2) viewing area adjustments

were spiked with B from a commercial standard. The results of this experiment are shown in Table 2 and indicate that no significant interferences exist with as much as 5% blood in the slurry when using DCP-AES.

Some typical analysis results for the same blood samples by prompt γ[6] and slurry nebulization with DCP-AES and ICP-AES are shown in Table 3. The ICP-AES analyses were performed at the INEL on the same slurry preparations that were used for DCP-AES at BNL. Prompt γ was in some cases done twice, once the day the sample was taken using only a blank and a single standard for calibration and a second time when a multipoint calibration curve could be obtained. Examination of Table 3 indicates that the DCP-AES results are always lower for blood and the BSSB compound and near normal for BPA and boric acid. Assuming that the multipoint prompt γ results are near the true values, the % errors were calculated and averaged as presented in Table 4. The precisions for all methods and B compounds are quite comparable, however the first DCP-AES analysis shows a mean -11% error for boron cage compounds. This discrepancy was not noted in other DCP-AES analyses when the samples were digested[7]. Viewing a cooler region of the DCP appeared to give an enhanced recovery for B as BSSB so some blood slurry samples were rerun viewing this region. The results of this run are presented in the last column of Table 4. Even though the recovery of the cage compounds was somewhat better, the mean error was still ~-6% for B in cage compounds. The reason for the lower DCP-AES

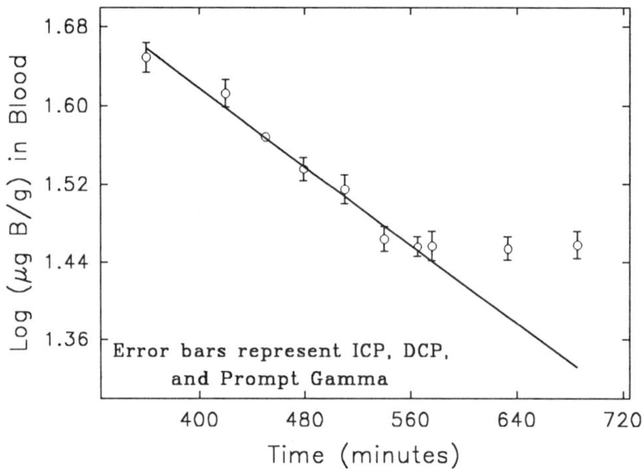

Figure 1. Elimination of B in an atypical dog following administration of BSH. Each point represents the mean of the DCP-AES analysis and ICP-AES and prompt γ performed at a later date.

results is not known but believed to be related to the fact that the sample never reaches the core of the DCP and is not contained within the plasma. All subsequent B analysis by DCP-AES were corrected for this negative bias after the analysis of a reference standard or sample.

Figure 1 indicates the utility of the rapid analytical technique in treatment planning. In this dog, the B elimination was quite normal until ~560 minutes. At this point the dog was taken to the reactor and anaesthetized. This, in some way, disrupted the normal elimination and during the course of the irradiation there was no net change in the blood B level. By using the rapid DCP-AES analysis of the blood slurries, adjustments could be made in the irradiation schedule knowing that the elimination had ceased.

CONCLUSIONS

Blood samples can be prepared and analyzed for B as slurries within 5-15 minutes of receipt. The analysis does not appear to be affected by blood slurry concentrations of up to 5% and the slurries are stable for several days. When B is present as a cage type compound, DCP-AES results are typically 5-15% low. This negative error can easily be corrected for if a suitable reference standard or a reference analysis of a sample is available. Boron analysis in the slurries by ICP-AES is unaffected by the compound type, however, caution should be used in order to assure that positive errors do not result from volatile B compounds.

ACKNOWLEDGEMENTS

This work was performed under the auspices of the U.S. Department of Energy, Office of Energy Research, under DOE Field Office, Idaho, Contract No. DE-AC07-761D01570.

REFERENCES

1. P.R. Gavin, C.E. DeHaan, S.L. Kraft, M.P. Moore, L.R. Wendling, and R.V. Dorn, Acute and late reactions following boron neutron capture epithermal neutron therapy in dogs with spontaneous brain tumors, (1992) in *Progress in Neutron Capture Therapy for Cancer*, eds B.J. Allen et. al. (Plenum Press, NY) pp. 507-514.
2. P.R. Gavin, S.L. Kraft, L.R. Wendling, and D.L. Miller, Canine spontaneous brain tumors - a large animal model for BNCT (1989) *Strahlenther. Onkol.* **165**, 225.
3. M.D. Wichman, R.C. Fry, and M.K. Hoffman, Slurry method for rapid DCP atomic emission determination of residual bone fragments in mechanically separated turkey (1986) *Appl. Spectr.* **40**, 351.
4. G.R. Wellum, E.I. Tolpin, A.H. Soloway, and A. Kaczmarczyk, Synthesis of μ-disulfido-bis(undecahydro-*closo*-dodecaborate)(4-) and of a derived free radical (1977) *Inorg. Chem.* **16**, 2120.
5. W.F. Bauer, D.A. Johnson, S.M. Steele, K. Messick, D.L. Miller, and W.A. Propp, Gross boron determination in biological samples by inductively coupled plasma-atomic emission spectroscopy (1989) *Strahlenther. Onkol.* **165**, 176.
6. R.G. Fairchild, D. Gabel, B.H. Laster, D. Greenberg, W. Kiszenick, and P.L. Micca, Microanalytical techniques for boron analysis using the $^{10}B(n,\alpha)^{7}Li$ reaction (1986) *Med. Phys.* **13**, 50.
7. R.F. Barth, D.M. Adams, A.H. Soloway, E.B. Mechetner, F. Alam, and A.K.M. Anisuzzaman, Determination of boron in tissues and cells using direct-current plasma atomic emission spectroscopy (1991) *Anal. Chem.* **63**, 890.

QUANTIFICATION OF BORON IN BIOLOGICAL TISSUE BY SECONDARY ION MASS SPECTROMETRY

V.K.F. Chia,[1] R.J. Blieler,[1] D.B. Sams,[1] C.M. Jones,[1] R.W. Odom,[1] W.F. Bauer,[2] A.K. Gianotto,[2] C.D. Swartz,[3] and J.P. Weidner[3]

[1]Charles Evans & Associates, Redwood City, CA 94063
[2]Idaho National Engineering Laboratory, Idaho Falls, ID 83415-3519
[3]Washington State University, Pullman, WA 99163

INTRODUCTION

The combination of boron-containing drugs and neutron irradiation to combat certain brain cancers has proven to be a viable method for cancer treatment. The effectiveness of this treatment depends on the selectivity of the carrier compound delivering boron to the tumor cells. To date, the most effective brain tumor treatment drug is borocaptate sodium ($Na_2B_{12}H_{11}SH$ or BSH). As new drugs are constantly being synthesized to improve selective delivery of boron to cancer cells any analytical technique capable of quantifying boron at the subcellular level will play an important role in determining the drug's potency for boron neutron capture therapy (BNCT).

Electron probe X-ray microanalysis (EPXMA) is an established technique for obtaining quantitative information on the distribution of elements at the ultrastructural level.[1] However, EPXMA is unable to detect boron because of X-ray absorption in the boron region by the beryllium window of the detector. The complimentary technique of electron energy loss spectroscopy (EELS) is less often used primarily because analyses must be performed on extremely thin (<60 nm) uniform sections of tissue. Secondary ion mass spectrometry (SIMS), although universally recognized as a sensitive quantitative technique by the semiconductor industry, is often not exploited by the biological and medical community even though ion microscopy has been utilized sucessfully for a variety of biological applications.[2]

This paper describes the application of SIMS using a Cameca mass spectrometer to localize and quantify boron in rat tissue after BSH intravenous delivery. The methodology for preparing tissue samples suitable for SIMS analysis is described and quantitative elemental images from the liver tissue are presented. It should be noted that this is our first SIMS experiment with infused rats and these boron localization results should be treated as preliminary.

Advances in Neutron Capture Therapy, Edited by
A.H Soloway *et al.*, Plenum Press, New York, 1993

EXPERIMENTS AND RESULTS

A laboratory rat was intravenously infused with ^{11}BSH (100 mg B/kg body weight) and the boron allowed to distribute to the tissues for about one hour. This rat and a control rat (not administered with boron) were then euthanized and the liver and three regions of the brain (cerebral hemispheres and brainstem) removed and immediately cryopreserved using a liquid nitrogen metal mirror fixation unit.[1] At the time of writing this paper, only the liver tissues from the infused and control rats were analyzed by SIMS and no inductively coupled plasma atomic emission spectrocopy (ICP-AES) data were available for comparison.

Cryofixation, as opposed to chemical fixation, is the *only* method by which both ultracellular and intracellular element contents can be simultaneously preserved. The frozen hydrated tissues were then cryoultramicrotomed at about -140°C using no embedding medium, including the "low temperature" types or freeze substitution technique, which may inadvertently cause some form of chemical change to occur in the tissues. Freeze-drying was performed on the cryosections and the freeze-dried sections placed onto high-purity silicon wafers for SIMS analysis. Although not performed during this test experiment, it is imperative that conventional transmission electron microscopy (CTEM) is performed in parallel to the SIMS analysis to evaluate the quality of the crysections with respect to induced structural artifacts from the crypreparation procedures.

We have used brain homogenate as the matrix standard in this study because our primary goal is to accurately quantify boron in brain tumors. Figure 1 shows boron calibration curves of ^{11}B-boric acid and ^{11}BSH in brain homogenate and ^{11}BSH in gelatin. The absolute boron concentrations were quantified by ICP-AES. A similar experiment using the isotopically enriched form of these compounds was also performed. Irrespective of the boron isotope used, the slope of the calibration plots for BSH in brain homogenate was consistently lower than with boric acid. However, similar slopes are observed for the plots of ^{11}B-boric acid in brain homogenate and ^{11}BSH in gelatin suggesting that the sampled homogenate analysed was more aqeous containing unbound boron.[3] Similar experiments in aqueous media are being performed to confirm this hypothesis.

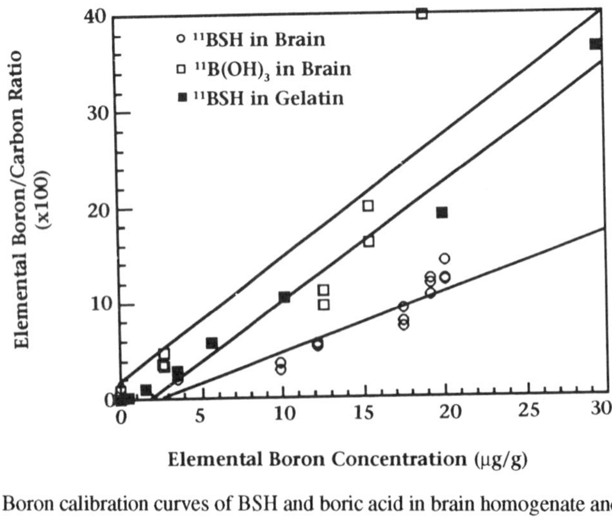

Figure 1. Boron calibration curves of BSH and boric acid in brain homogenate and gelatin.

For the purpose of this study, an average value of the brain homogenate and gelatin calibration results was used for SIMS quantification in the liver tissue. Dynamic SIMS profiles acquired from the livers of the diffused and control rat are shown in Figure 2. Boron concentrations of 53 μg B/g in the diffused rat and 13 μg B/g in the control rat were observed. An optical micrograph of the freeze-dried section and a boron secondary ion image of the infused rat's liver are shown in Figure 3. Boron is homogeneously distributed and is not observed in the tissue voids indicating that no diffusible boron was lost during sample preparation. The potassium secondary ion image in Figure 4 shows clusters of high intensity spots probably corresponding to cells. The observed potassium/sodium ratio of about eight suggests that the tissue was viable at the time of crypreservation.

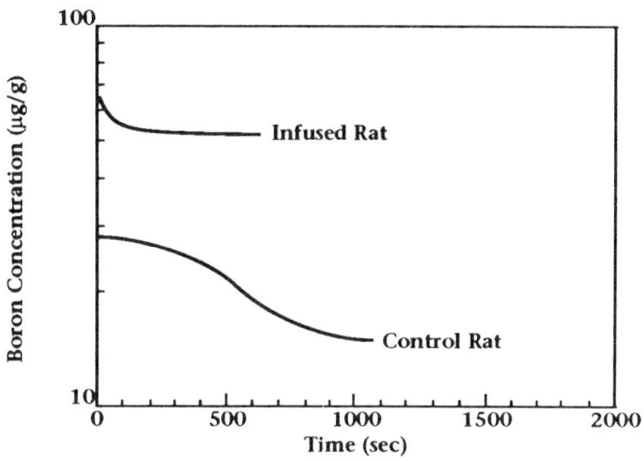

Figure 2. SIMS depth profiles of boron in the liver of an infused BSH rat and control rat.

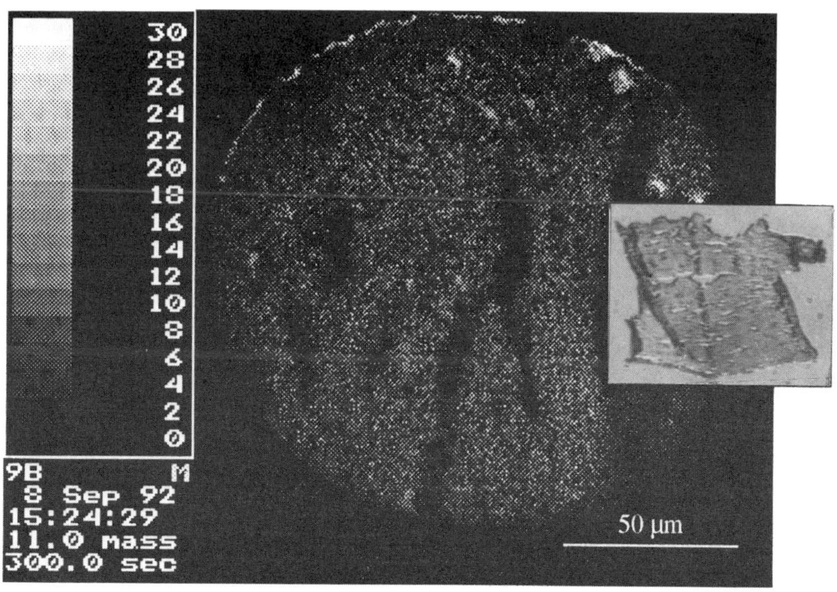

Figure 3: Boron secondary ion image of liver tissue from the infused rat.
Insert shows an optical micrograph of the freeze-dried liver section prepared at Duke University Medical Center.

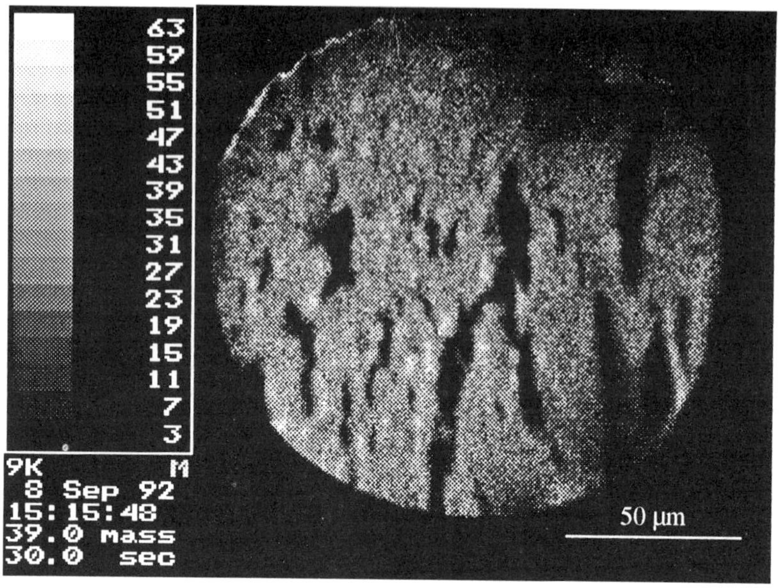

Figure 4. Potassium secondary ion image of liver tissue from the infused rat.

CONCLUSIONS

These preliminary results illustrate that with appropriate cryopreservation of tissue quantitative imaging is possible by SIMS. Quantification using brain homogenate gave reasonable boron concentration values in the liver; a comparison with ICP-AES results will, confirm the accuracy. Boron was homogenously distributed in the freeze-dried liver section of the infused rat. The viability of the tissue at the time of cryopreservation was confirmed by the high potassium to sodium ratio. The high intensity potassium spots observed probably correspond to individual cells. Further improvements in the cryopreservation methodology is required to obtain the higher resolution images needed for this localization study.

REFERENCES

1. A. LeFurgey, M. Bond and P. Ingram, Frontiers in Electron Probe Microanalysis: Applications to Cell Physiology, *Ultramicroscopy*, 24: 185 (1988)
2. R.W. Linton, I.H. Musselman, S.R. Bryan, Laser and Ion Microprobe Mass Spectrometry - Applications to Human Tissues, in: "Microprobe Analysis in Medicine," P. Ingram, J.D. Shelburne and V.L. Roggli, eds., Hemisphere Publishing Co., NY (1989)
3. W.F. Bauer, K.M. Bradshaw and T.L. Richards, Interaction Between Boron Containing Compounds and Serum Albumin Observed by Nuclear Magnetic Resonance, in: "Progress in Neutron Capture Therapy for Cancer, " B.J. Allen et al, eds., Plenum Press, NY (1992)

FTIR SPECTROMETRY FOR THE ASSAY OF POLYHEDRAL BORON COMPOUNDS IN PLASMA

Douglas E. Moore,[1] Yohanes Setiawan,[1] Ned Blagojevic[2] and Barry J. Allen[2]

[1]Department of Pharmacy, The University of Sydney, Sydney 2006
[2]Australian Nuclear Science and Technology Organization, Menai 2234, Australia

INTRODUCTION

Many of the new compounds being developed for neutron capture therapy contain the icosahedral boron cage structure with up to 12 boron atoms. The infrared spectrum of each compound provides a fingerprint which can be used as an aid in their identification.[1] In particular, the most intense and identifiable band is the boron-hydrogen stretching vibration at about 2500 cm^{-1}. This high energy region is relatively free from overlap due to absorption by solvent and other molecules in the system. Additionally, the B-H stretch band is not significantly influenced by interaction from the surrounding atoms. The advent of Fourier Transform infrared instrumentation (FTIR) has enhanced the capabilities of this spectroscopic technique, especially in the area of sensitivity and speed of assay. Thus FTIR is used routinely, for example to confirm the introduction of the boron cluster in the synthesis of new polyhedral boron compounds,[2] and in characterising reconstituted boronated low density lipoproteins.[3] Very recently, FTIR was applied to the determination of sodium borocaptate (BSH) in aqueous solution and urine.[4]

The FTIR microscope spectrometer is a relatively recent innovation which is useful as a detector of molecular probes in intact tissue.[5] The spatial resolution that microspectroscopy is able to achieve approaches the diffraction limit of one-half of the wavelength being used. For boron compounds with absorption around 2500 cm^{-1} (4 μm), the minimum dimension of the sample for spectral measurement is 2 to 4 μm. Therefore, single cell microscopic imaging by infra-red spectroscopy is feasible, and has been used for the study of a human red cell.[6]

As a first stage in the application of FTIR spectroscopy to the study of boron compound distribution in tissue, we describe here the sensitivity, accuracy and precision of the technique when used for the assay of boron in blood samples.

EXPERIMENTAL

Chemicals. Decaborane and dicesium mercaptoundecahydro-*closo*-dodecaborate (BSH) were from Callery Chemical Co., USA), and p-(1,2 dicarba-*closo*-dodecaborane) benzyl alcohol (DBBA) was synthesised in our laboratory. Bovine serum album was obtained from Sigma Chemical Co., USA. Carbon tetrachloride was of analytical grade (Ajax Chemicals, Sydney). All water used was doubly distilled in an all glass system. Human plasma was obtained from the Sydney Blood Bank.

FTIR Spectroscopy. Infrared spectra were recorded with a Bio-Rad FTIR spectrometer model FTS® 20/80, equipped with a Bio-Rad 3200 Data Station and a wide-range mercury cadmium telluride detector. A resolution of 8 cm^{-1} with 6 mm aperture were used, and 256 scans were collected and apodized with a triangular function for each sample. A variable pathlength liquid cell with CaF_2 windows was used to record spectra of all liquid samples. The pathlength of 1 mm was used for CCl_4 solutions and 100 μm for aqueous solutions.

Calibration Solutions. Standard solutions of decaborane and DBBA were prepared containing 5 mg/ml of boron in CCl_4. The solutions for calibration were derived from a series of dilutions ranging in concentration from 0 - 1.5 mg/ml. BSH standard solution was prepared containing 10 mg/ml of boron in water, and dilutions were made to provide solutions containing 0 to 5 mg/ml. The stock solutions were stored at room temperature.

Extraction of Boron Compounds from Aqueous Samples. The DBBA stock solution was added to (i) phosphate buffer solutions pH 7.4 (PBS), (ii) PBS containing human serum albumin, and (iii) human plasma samples, to give boron concentrations between 0 to 200 ppm. Each sample was extracted twice with 25 ml of CCl_4. The CCl_4 extracts were combined and evaporated to dryness at 35^{0}C under vacuum. The residue was reconstituted with 5 ml of fresh CCl_4, and examined by FTIR. The experiments were performed in triplicate.

RESULTS AND DISCUSSION

To obtain the maximum sensitivity for the detection of the B-H absorption band at 2600 cm^{-1}, various solvents were examined for their transmittance in this region. As pointed out by Lu and Munro[4], the use of water as solvent means that the strong water absorption bands at 1640 and 3300 cm^{-1} and the weaker band at 2100 cm^{-1} interfere strongly with the relatively weak B-H absorption. Thus they used a very short pathlength apparatus (15 μm) for aqueous samples with background subtraction processing and achieved a sensitivity of 5 ppm boron for BSH in urine. We have confirmed the difficulty of BSH determination in aqueous solutions, and report a lower limit of detection of 50 ppm boron using a 100 μm pathlength and simple peak height quantification, or 5 ppm boron by background subtraction techniques.

Blood samples demand some pre-treatment to remove blood components with interfering absorptions. Solvent extraction is the usual means of achieving this, and carbon tetrachloride was chosen as it is the most transparent solvent in the IR spectral region of interest and is a good solvent for many of the hydrophobic carborane compounds. The study was performed using DBBA as an example of a polyhedral boron compound which is readily extractable from blood.

The IR spectrum in a 1 mm pathlength cell of DBBA in CCl_4 at a concentration of 20 ppm boron (Figure 1) displays the characteristic B-H absorption band at 2600 cm^{-1} suitable for quantification. Figure 2 shows the calibration curve for DBBA in CCl_4 solution under these conditions. Directly from the absorbance values of the major B-H stretching band a straight line was found with a correlation coefficient (R^2) equal to 0.9987. The detection limit of boron for DBBA (or other polyhedral boron compounds) in CCl_4 was found to be about 1 ppm (signal/noise = 2.5).

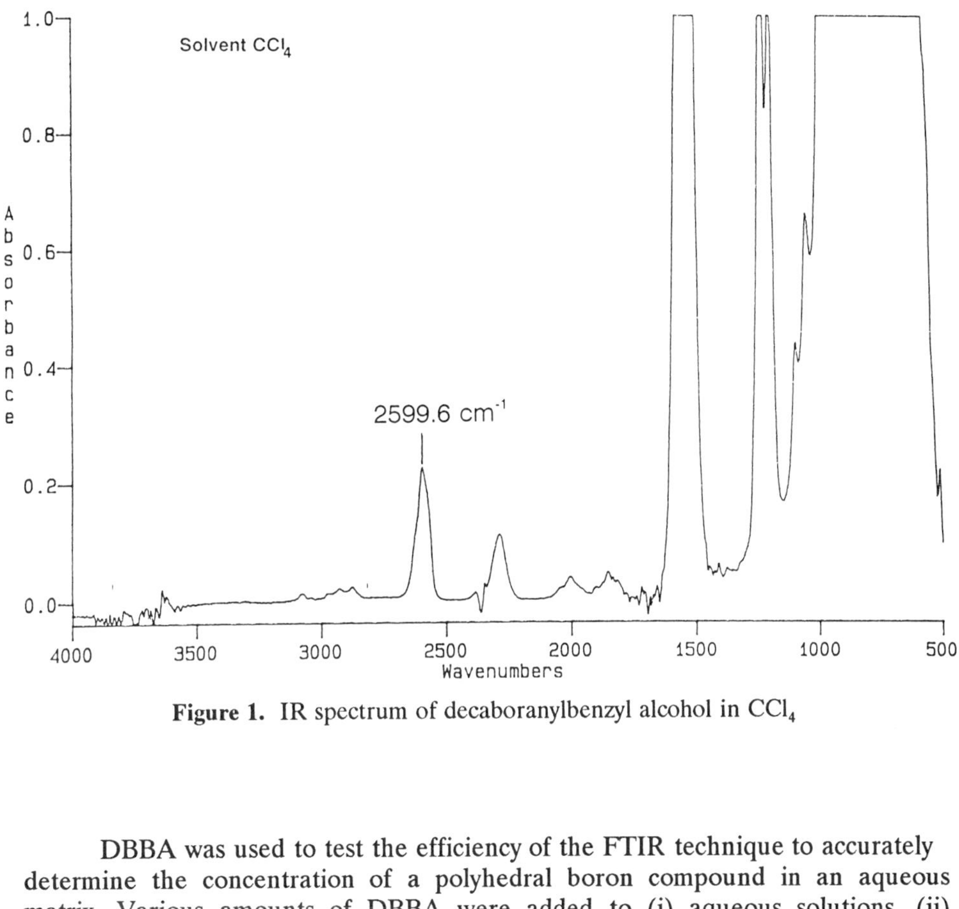

Figure 1. IR spectrum of decaboranylbenzyl alcohol in CCl_4

DBBA was used to test the efficiency of the FTIR technique to accurately determine the concentration of a polyhedral boron compound in an aqueous matrix. Various amounts of DBBA were added to (i) aqueous solutions, (ii) phosphate buffered saline containing human serum albumin, and (iii) human plasma, and the aqueous phase was extracted twice with CCl_4. The quantification limit for boron in the three aqueous matrices was determined to be 3.0 ppm, being the concentration which could be detected with a coefficient of variation less than 10% in all three solutions. Except at the lowest boron concentrations in plasma, the recovery of boron was above 90%. The results for DBBA in human plasma are shown in Table 1. In the extraction of plasma with CCl_4, no protein precipitant (such as phosphotungstic acid) was used, in an attempt to minimise the possibility that DBBA could be precipitated and trapped in the protein pellet. Protein binding of a small amount of DBBA could explain the slightly lower recovery of the lowest concentration samples in plasma.

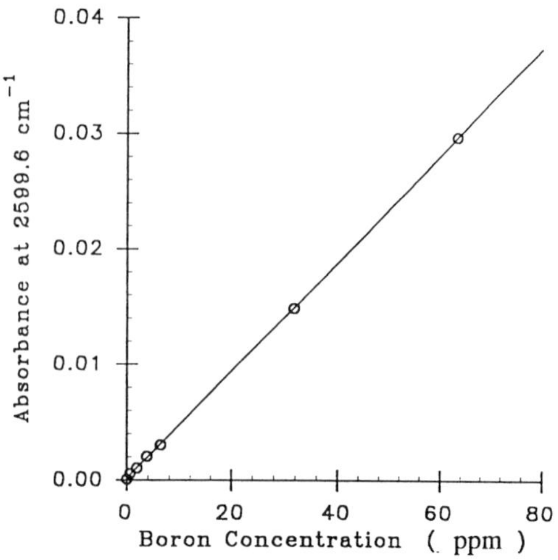

Figure 2. Calibration curve for DBBA in CCl_4 using absorbance at 2599.6 cm^{-1}

The FTIR technique as described above may be extended to the assay of other polyhedral boron compounds with suitable solubility in organic solvents such as CCl_4. Where the compound is a weak acid or base, the unionised fraction may be extracted quantitatively with CCl_4 after adjusting the pH of the aqueous phase to an appropriate pH at which the compound is predominantly unionized.

The procedure developed here can be carried out within 30 minutes. When 5 mL of plasma is used, at a sensitivity level of 5 ppm, 25 μg of boron is detected. This quantity is determined by the volume requirements of the liquid cell. The use of a micro sample cuvette and smaller sample volumes could potentially improve the minimum detectable amount to below 5 μg boron in 5 mL of plasma.

In an NCT protocol it is essential that a rapid method be available to monitor the boron concentration in the blood of patients so that the appropriate neutron dose can be calculated. It is suggested that FTIR could fulfil that purpose for some polyhedral boron compounds.

Table 1 . Recovery of Boron added as DBBA to 5 ml of plasma

Boron concentration expected (μg/ml)	Boron concentration found ± SD* (μg/ml)	CV (%)	Recovery of Added Boron ± SD* (%)
2.69	2.22 ± 0.13	6.06	82.53 ± 5.00
5.37	4.50 ± 0.36	7.91	83.72 ± 6.62
10.75	9.92 ± 0.36	3.59	92.28 ± 3.31
26.87	24.47 ± 1.20	4.88	91.04 ± 4.45
53.74	51.47 ± 2.82	5.49	95.76 ± 5.25
107.49	105.75 ± 2.93	2.77	98.39 ± 2.73
188.11	180.39 ± 4.31	2.39	95.90 ± 2.29

SD* = standard deviation of three replicates

ACKNOWLEDGMENT

Financial assistance was provided by the Australian Research Council and the Government Employees Assistance to Medical Research Fund.

REFERENCES

1. L.J. Bellamy, W. Gerrard, M.F. Lappert, and R.L. Williams, Infrared spectra of boron compounds, *J. Chem. Soc.* 2412-2415 (1958).
2. A. Varadarajan and M.F. Hawthorne, Novel carboranyl amino acids and peptides: reagents for antibody modification and subsequent neutron capture studies. *Bioconjugate Chem.* 2: 242-253 (1991).
3. S.B. Kahl and J.C. Callaway, New tumor localizers: advances in the use of low density lipoproteins. *Strahlenther. Onkol.* 165: 137-139 (1989).
4. D.R. Lu and C. Munro, Quantitative analysis of mercaptoundecahydrododecaborate by Fourier transform infrared spectroscopy, *Pharm. Res.* 9: 1199-1202 (1992).
5. J.A. Reffner and W.T. Wihlborg, FT-IR molecular microanalysis system, *Proceedings of the XIIth International Congress for Electron Microscopy*. 270-271 (1990).
6. A. Dong, R.G. Messerschmidt, J.A. Reffner, and W.S. Caughey, Infrared spectroscopy of a single cell - the human erythrocyte, *Biochem. Biophys. Res. Comm.* 156:752-756 (1988).

PROMPT GAMMA-RAY ANALYSIS TO DETERMINE ^{10}B CONCENTRATIONS

M.W. Konijnenberg[1], C.P.J. Raaijmakers[1], G. Constantine[2],
L.G.H. Dewit[1], B.J. Mijnheer[1], R.L. Moss[2], and
F. Stecher-Rasmussen[3]
[1]The Netherlands Cancer Institute, Amsterdam, the Netherlands
[2]Institute for Advanced Materials JRC, Petten, the Netherlands
[3]Netherlands Energy Research Foundation ECN, Petten, the Netherlands

INTRODUCTION

For a successful application of BNCT knowledge of the boron pharmacokinetics is of great importance. For the future clinical trials at Petten Prompt Gamma-Ray Analysis (PGRA) will be applied to determine ^{10}B concentrations in patients' blood samples. The boron concentration within the tumour volume can then be deduced from pharmacokinetic studies. With the rapidly obtained results from PGRA it will be possible to tailor the irradiation of each patient according to their individual boron pharmacokinetics.

At the High Flux Reactor (HFR) in Petten a Prompt Gamma-Ray Analysis (PGRA) setup for on-line measurements of ^{10}B concentrations in various sample types has been

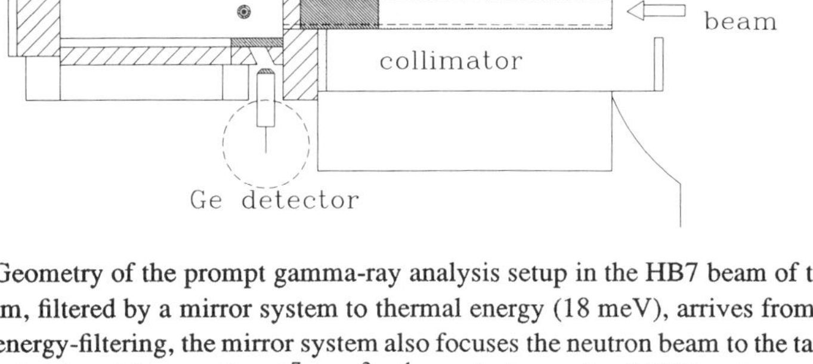

Figure 1: Geometry of the prompt gamma-ray analysis setup in the HB7 beam of the HFR. The neutron beam, filtered by a mirror system to thermal energy (18 meV), arrives from the right. In addition to energy-filtering, the mirror system also focuses the neutron beam to the target position, giving a flux density of $\Phi_{th} = 2 \times 10^7\ cm^{-2}s^{-1}$, a cadmium ratio of 1000 and a γ-ray component of 0.01 Gy/h.

developed. It is located at the thermal beam port HB7, close to the therapy beam HB11. It is the aim of this study to investigate whether during the healthy tissue tolerance studies of dogs at Petten [1, 2] immediate determinations of the boron levels in blood can be performed by this PGRA method. This rapid assay greatly improved the accuracy in the dose delivery, as the start and stop of the irradiations was determined by the PGRA data.

PROMPT GAMMA-RAY ANALYSIS

PGRA boron-determination for BNCT has been used in Japan [3] and the US [4, 5] for several years. The intensity of 478 keV prompt gamma-rays formed in the $^{10}B(n,\alpha\gamma)^7Li$ reaction, during irradiation of the sample, is directly related with its ^{10}B- concentration. This γ-ray is emitted in 93.5% of the decays of the excited ^7Li state. The incident neutron fluence is monitored by the intensity of the 2223 keV $H(n,\gamma)$ line.

The geometry of our set-up has changed only slightly from the situation reported before [6]. As shown in the schematic lay-out (fig. 1) the viewing angle of the high purity germanium detector to the target position is directed 60° backward from the beam. This direction reduces the background level due to the slight decrease of scattered neutrons at backward angles. A sample changer is applied which not only enabled automatic data acquisition but showed a better positioning of the samples in the beam as well.

For the determination of the ^{10}B-concentration in a sample the ratio of the intensities of the boron and the hydrogen line (or PGRA signal) is used in combination with a calibration curve. This calibration curve, depicted in fig. 2, was measured before with a set of samples with well-known ^{10}B-concentrations in the range from 1 to 180 ppm and determines the relation between the PGRA signal, and the ^{10}B-concentration.

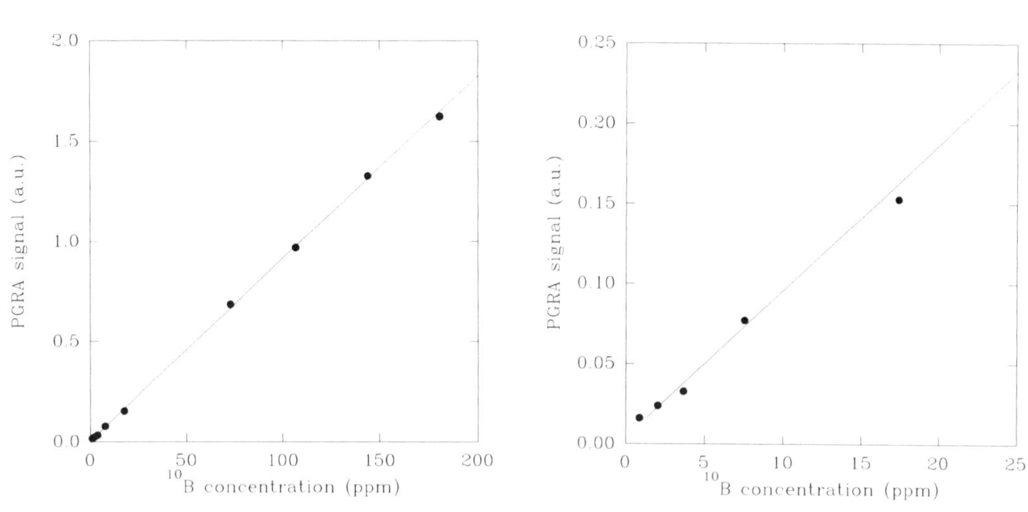

Figure 2: Calibration curve defining the relation between the PGRA signal and the ^{10}B-concentration for 10 samples with boric acid solutions. The PGRA signal is calculated by the ratio of the intensities in the 478 keV boron and the 2223 keV hydrogen line. The straight line fits the PGRA signals perfectly with a χ^2 = 1.6.

RESULTS AND DISCUSSION

Canine blood samples for the healthy tissue tolerance studies performed at the BNCT facility have been analysed using PGRA. These studies revealed a great inter-animal variation in the boron level and the decision of the start of the irradiation was therefore completely based on individual PGRA concentration measurements. For instance the times between the administration of BSH and the start of the irradiation for the 50 ppm ^{10}B concentration varied between 90 and 230 minutes. All samples were afterwards checked with the Inductively Coupled Plasma/Atomic Emission Spectroscopy (ICP/AES) method also available at ECN [6]. The comparison showed a good agreement, within 5%.

Table 1: Comparison of the results of independent analysis by the PGRA and the ICP/AES method of the same samples of BSH. Values are expressed as percentage total boron per gram of BSH, not corrected for the water content. For the PGRA results an enrichment factor of 95.8 % of ^{10}B was assumed, as stated by the manufacturer (Centronics).

Sample	PGRA (% B)	ICP/AES (% B)
1	57.2	58.7
2	53.0	58.0
3	59.1	57.7
4	55.1	43.9
5	61.3	56.5

Further comparisons were made between PGRA and ICP/AES with a set of BSH samples. The concentrations of the BSH samples as determined by both methods are almost identical (table 1), apart from sample 4 which showed an unexplained 22 % discrepancy. The BSH examined in the above comparison was intended to use in patients for a human pharmacokinetic study performed at the Bremen St.-Jürgen Straβe Hospital [7]. In the near future more comparisons will be made between the outcome of PGRA and ICP/AES for various types of samples.

CONCLUSIONS

Accurate determinations, $1\sigma = 3$ %, of the relevant ^{10}B concentration in 1 ml blood samples are possible within 10 minutes. The detection limit of our setup is 1 ppm ^{10}B, with an accuracy of about 10 %. Good agreement was obtained with ICP/AES for blood and BSH samples at clinical relevant concentrations. The method proved to be extremely useful for the determination of start of the irradiation of dogs, which is based on these boron concentration measurements using PGRA.

ACKNOWLEDGEMENTS

The idea of directing our detector at $-60°$ was a wonderful advice P. Micca of Brookhaven National Lab. gave us to improve our set-up. We would like to thank R. Huiskamp of

ECN for providing us the ICP-data and helping us with the manufacture of boron samples. This work was financially supported by the Netherlands Cancer Society (NKB grant NKI 90-03).

References

[1] K. Philipp, R. Huiskamp, J. Casado, A. Siefert, P. Gavin and R. Moss,Developments in the healthy tissue tolerance studies as a prerequisite for clinical application of BNCT with borocaptate sodium at the HFR Petten, contribution to this symposium.

[2] A. Siefert, J. Casado, K. Philipp, R. Huiskamp, P. Gavin E. Dühmke and R. Moss,Brain effects observed in the canine healthy tissue tolerance studies for BNCT with borocaptate sodium at the epithermal neutron beam of the HFR, Petten, contribution to this symposium.

[3] T. Kobayashi and K. Kanda,Microanalysis system of ppm-order ^{10}B concentrations in tissue for neutron capture therapy by prompt gamma-ray spectrometry,*Nucl. Inst. Meth. 204(1983)525–531*

[4] R.G. Fairchild, D. Gabel, B.H. Laster, D. Greenberg, W. Kiszenick and P.L. Micca,Microanalytical techniques for boron analysis using the $^{10}B(n, \alpha)^{7}Li$ reaction,*Med. Phys. 13(1986)50–56*

[5] R. Rogus, O.K. Harling, I. Olmez and S. Wirdzek,Boron-10 prompt gamma analysis using a diffracted neutron beam, in*Progress in Neutron Capture Therapy for Cancer*, edited by B.J. Allen et al., Plenum Press, New York, 1992

[6] M. Konijnenberg, F. Stecher-Rasmussen, C.P.J. Raaijmakers, R. Huiskamp, R.L. Moss, A.C. Begg, L. Dewit, V.G.A. Gregoire and B.J. Mijnheer,Boron concentration measurements for the Petten BNCT project, in*Progress in Neutron Capture Therapy for Cancer*, edited by B.J. Allen et al., Plenum Press, New York, 1992

[7] D. Haritz and D. Gabel,Results of continued clinical investigations of BSH in patients with malignant glioma, contribution to this symposium.

PROBING BORONATED NUCLEOSIDE UTILIZATION BY GLIOMA CELLS USING ION MICROSCOPY

Brian D. Bennett*, Jonathan Mumford-Zisk*, Rolf F. Barth†, Albert H. Soloway•, and George H. Morrison*

* Department of Chemistry, Cornell University, Ithaca, NY 14853
† Department of Pathology, Ohio State University, Columbus, OH 43210
• College of Pharmacy, Ohio State University, Columbus, OH 43210

INTRODUCTION

Boron Neutron Capture Therapy for Cancer will be successful only if agents are found that target tumors selectively and provide ^{10}B concentrations ≥ 30 µg/g (tumor)[1]. Further, these agents produce 2-5 times the therapeutic effect if they localize in the nucleus rather than uniformly throughout the cell.[2] Nucleosides can be used to target tumor cells, since transporter and kinase activities are often elevated compared to normal tissue.[3]

We are now utilizing halogenated nucleosides to probe the intracellular fate of nucleosides in tumor cells. Such probes serve as reference, or "lead" compounds in the search for boronated nucleic acid precursors. The boronated nucleoside 2'-O-(o-Carboran-1-ylmethyl)-uridine (CBU-2') has been shown to preferentially localize in the cytoplasm of F-98 rat glioma cells.[4] This localization may be consistent with incorporation into ribosomal RNA, which is ubiquitous in the cytoplasm. However, rRNA precursors should also localize in the nucleolus, since this structure functions to produce ribosomes for transport to the cytoplasm.

The subcellular distribution of CBU-2' is compared to the RNA precursor 5-Fluorouridine (FUrd) and the DNA precursor 5-Bromodeoxyuridine (BrdUrd) in this study. The intracellular distribution of FUrd was studied since it is known to selectively accumulate into nucleolar RNA. BrdUrd, a deoxythymidine analog, gives distinct nuclear labeling of cells actively replicating DNA.

METHODOLOGY

Ion microscopy (IM) is an imaging technique based on secondary ion mass spectrometry (SIMS). Samples are placed in high vacuum and bombarded with a beam of energetic primary ions. The secondary ions, which are generated by the impact of the primary ion beam on the surface, are accelerated into a double-focusing mass spectrometer. The stigmatic optics maintain the lateral image of the surface. After mass filtration, the secondary ions are projected onto an array detector so that their final positions correspond to their original positions on the surface. The final output of the instrument is a two dimensional image showing the distribution of a specific element on the sample surface. Ion microscopy is capable of localizing any element or isotope in the specimen, most with a sensitivity in the low ppm range, with a spatial resolution of about 0.5 μm. Three-dimensional imaging is easily obtained due to the fact that the primary beam is continually removing surface layers. Thus, a series of image acquisitions forms a multidimensional image stack that reveals three-dimensional element maps.

The GS-9L rat glioma cells used in this work are freeze fractured-freeze dried tissue culture. This cryopreparation technique provides samples for ion microscopy which are free from redistribution and matrix effects. Cryopreparation of cells in this work was performed as previously described.[5] Freeze fractured-freeze dried cell cultures have been shown to give rise to valid elemental maps.[6] Thus quantitative concentration information is available at each resolvable point inside the cell. Cells damaged by either a physiological process or the cryopreparation itself, are screened out by comparing the potassium, sodium and calcium levels to physiologically normal levels, i.e. high intracellular potassium relative to sodium (K/Na ratio >> 1) and a higher concentration of calcium in the cytoplasm relative to the nucleus.[5] The cytoplasmic localization of calcium allows easy identification of the nucleus in healthy cells.

Halogenated nucleosides serve as excellent reference compounds for boronated nucleoside evaluation, since the halogen tag is readily detected as a negative ion.[7] Fluorine from FUrd and bromine from BrdUrd were imaged as negative secondary ions at m/z 19 and 81, respectively. Mass interference, from polyatomic species, was exhaustively characterized and will be reported elsewhere. The polyatomic mass interferences can be described briefly as relatively low. For the FUrd and BrdUrd additions given in the figure captions below, polyatomic mass interference accounted for 5% and 2% of the ion current at m/z 19 and 81. Further, this background is homogeneously distributed within freeze-fractured cells. Thus, selective subcellular localization of fluorine and bromine are accurately assessed. Boron was imaged as a positive secondary ion at m/z 11 and was not detected in controls. In all images concentration increases with brightness (i.e. "hot spots" are brightest).

RESULTS AND DISCUSSION

GS-9L cells were pulsed simultaneously with CBU-2', FUrd, and BrdUrd. Figure 1 shows the localization of F^-, Br^-, and B^+ from the same "spot" of fractured cells. Overall, the localizations are cytoplasmic for boron and nuclear for fluorine and bromine. In separate experiments cells were treated with FUrd, BrdUrd, and CBU-2' by the themselves. The localizations were essentially the same as in the simultaneous administration.

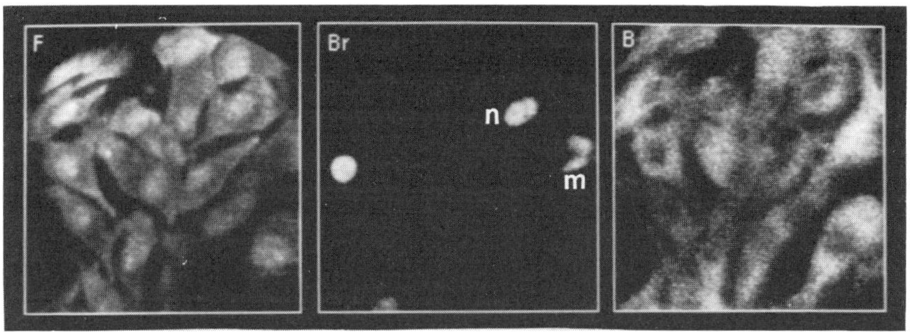

Figure1. GS-9L cells were treated with FUrd (10 μM), BrdUrd (40 μM), and CBU-2' (10 μg/mL) for 3 hours in DMEM (Gibco) supplemented with 10% FBS (Gibco) @ 37⁰ C, 5% CO_2. The fluorine, bromine, and boron image are given left to right. The S-phase cell described in the text is labeled with an "n" and the mitotic cell is assigned an "m".

One cell, assigned "n" in the Br image, shows distinct nuclear labeling. This cell was in S-phase (synthesizing DNA) sometime during the pulse period and incorporated BrdUrd into its DNA. Boron from CBU-2' is concentrated in the cytoplasm of this cell. This indicates that CBU-2' is not incorporated into actively replicating DNA.

Another cell appears to have been caught in mitosis at the end of the pulse period (Figure 1, cell labeled "m" in the Br image). Note the mitotic figure in the Br image. The F image of the corresponding cell does not display accumulation in the nucleus as seen in the surrounding cells. This is consistent with FUrd's role as a RNA precursor, since RNA synthesis is limited during mitosis. The nuclear localization of F from FUrd may be due to the rapid utilization and turnover of heterogeneous nuclear RNA (hnRNA gives rise to mRNA). Boron from CBU-2' is relatively homogeneous in this cell. Exclusion of boron from the nucleus may be limited due to the dispersion of the nuclear envelope, which is not reassembled around the two groups of daughter chromosomes until telophase.

Subcellular fractionation studies have shown that 5-fluorouracil from FUrd preferentially incorporates into nucleolar RNA,[8] consistent with ribosomal RNA incorporation. In ion microscopy preferential accumulation of fluorine from FUrd in the nucleolus is made clear by three-dimensional imaging.

Figure 2 shows the same cell referred to in figure 1 (labeled "n"). This cell was isolated in the ion microscope for higher magnification and three-dimensional imaging to reveal nucleoli. The image represents one image "slice" within the three-dimensional image stack. The ion images are of the same nucleus with F on the left and Br to the right. This nucleus shows high fluorine and low bromine in several discrete regions. These regions most likely correspond to nucleoli since subcellular fractionation studies have shown FUrd to selectively accumulate in nucleolar RNA (nucleolus/nucleoplasm ratio > 20).[9] The empty pockets in the Br image are consistent with nucleolar localization, since DNA replication is limited in this structure. Also, in a recent study, we have correlated the ion microscopic localization of F from FUrd to nucleoli stained with Pyronine Y and localized by laser scanning confocal microscopy.[10] We have not found CBU-2' to give a nucleolar localization in this simultaneous pulsing scheme or with cells treated with CBU-2' alone. Since rRNA within the ribosome is transported from the nucleus to the cytoplasm, it is unlikely that incorporation into rRNA would lead to an exclusively cytoplasmic localization. If CBU-2' was indeed an RNA precursor the cytoplasmic localization would be accompanied by selective nucleolar accumulation.

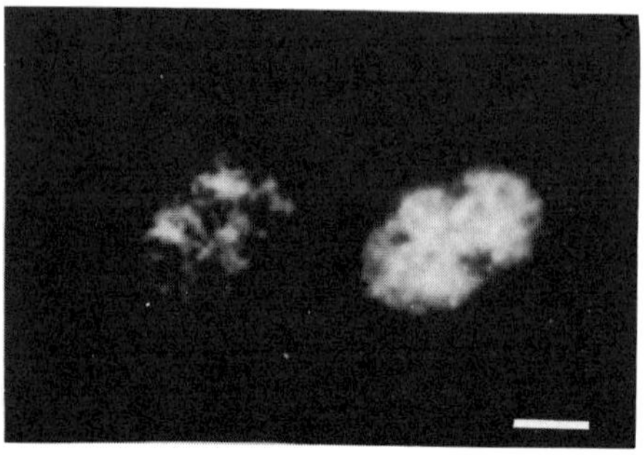

Figure 2. The same cell labeled "n" in figure 1 has been isolated and magnified further by the ion microscope. The fluorine and bromine ion images of the cell's nucleus are given, with F on the left and Br to the right. Note the high F^- intensity in the nucleoli, which correspond to the low intensity spots in the Br image. Bar = 6μm.

CONCLUSIONS

Simultaneous pulsing of the boronated nucleoside CBU-2' with FUrd and BrdUrd was utilized to assess whether CBU-2' was incorporated into RNA or DNA. FUrd, which incorporates selectively in nucleolar RNA, and BrdUrd, which incorporates selectively in DNA show intracellular distribution patterns which are not similar to CBU-2'. This information leads one to consider the synthesis of other nucleoside analogs as targets for nucleic acid incorporation. Thus, the subcellular localization of FUrd, BrdUrd, and the boronated nucleoside by ion microscopy appears to provide an effective methodology for screening new compounds.

Acknowledgment
This work was supported by DOE DE-FG02-91ER61138 grant to G.H.M. and a NSF Graduate Fellowship to I. G. The Cornell NIH/NSF Developmental Resource for Biophysical Imaging and Optoelectronics was used for cell culture work.

REFERENCES

1. Barth, R.F., Soloway, A.H., & Fairchild, R.G. (1990) *Cancer Res.* **50,** 1061.
2a. Gabel, D., Foster, S., & Fairchild, R.G. (1987) *Radiation Res.* **111,** 14.
2b. Kobayashi, T. & Kanda, K. (1982) *Radiation Res.* **91,** 77.
3. Weber, G. (1983) *Cancer Res.* **43,** 3466.
4. Bennett, B.D., Zha, X., Gay, I., & Morrison ,G.H. (1992) *Biol. Cell* **74,** 105.
5. For a comprehensive review see Chandra, S. & Morrison, G.H. (1992) *Biol. Cell.* **74,** 31.
6. Ausserer, W.A., Ling, Y-C., Chandra, S., & Morrison, G.H. (1989) *Anal. Chem.* **61,** 2690.
7. Hallegot, P., Girod, C., LeBeau, M.M., & Levi-Setti, R. in*Secondary Ion Mass Spectrometry, SIMS VII.*, eds. Benninghoven, A., Evans C., McKeegan, K.D., & Storms, H.A. (John Wiley & Sons, New York), p. 327.

8. Takimoto, C.H., Cadman, E.C., & Armstrong, R.D. (1986) *Mol. Pharmacol.,* **29,** 637.
9. ibid
10. unpublished results

SUBCELLULAR LOCALIZATION STUDY OF BORONATED URIDINE AND GLUCOSE DERIVATIVES IN CELL CULTURES USING ION MICROSCOPY

Isabelle Gay[1], Brian D. Bennett[1], Jonathan Mumford-Zisk[1], Rolf F. Barth[2], Albert H. Soloway[3], and George H. Morrison[1]

[1]Department of Chemistry
Cornell University, Ithaca, N.Y. 14853

[2]Department of Pathology
[3]College of Pharmacy
The Ohio State University, Columbus, Ohio 43210

The success of boron neutron capture therapy depends on the uptake and retention of boron by tumor cells. The precise intracellular localization and concentration of boron must be known to be able to estimate the radiobiologic effects of boronated drugs. Therefore, it is critical to have analytical methods for evaluating the uptake, localization, and retention of boronated compounds in cell cultures and animal models. Ion microscopy is one such method.

In this study, ion microscopy was used to determine the localization of boronated uridine and glucose derivatives in F98 glioma cells. The four boronated compounds studied were 2'-O-(o-Carboran-1-ylmethyl)uridine (CBU-2'), 3'-O-(o-Carboran-1-ylmethyl)uridine (CBU-3'), 5'-O-(o-Carboran-1-ylmethyl)uridine (CBU-5'), and 3-O-(o-Carboran-1-yl methyl)-D-glucose (CA). These compounds were designed with the hope that they would be taken up through the action of the nucleoside transport system and be converted to the phosphate thereby becoming entrapped within the cells.[1] It was further hoped that these compounds would be triphosphorylated and then be incorporated into the nucleic acids of the tumor cells. Using ion microscopy we were able to determine the amount of boron localized in the cytoplasm and nucleus of the cells.

Ion microscopy is an imaging technique based on secondary ion mass spectrometry (SIMS). It can be used to quantitatively localize any element or isotope (H to U) at submicron spatial resolution.[2-4] The sample is bombarded with a primary ion beam which causes the sputtering of ions from the sample. It is these secondary ions (either positive or negative) which are directly imaged. The ions are dispersed according to their energies, filtered, and then dispersed according to their mass-to-charge ratio. The transmitted mass can then be imaged using a microchannel plate/ fluorescents screen assembly or detected in various other modes. The optics of the ion microscope are such that there is a one-to-one correspondence between the relative position of atoms in the sample and the final position of ions as they strike the detector. The ion images represent the surface chemistry of the sample. Continuous sputtering provides three-dimensional information. Ion images of several different elements for a given field of view allow various experimental comparisons. These images allow the experimenter to correlate elemental distribution with cell morphology with a spatial resolution of .5μm. Therefore, ion microscopy can be used to do drug uptake and retention studies, cell physiological studies, and studies of the kinetics and transport within cells.[5]

Advances in Neutron Capture Therapy, Edited by
A.H Soloway *et al.*, Plenum Press, New York, 1993

METHODS

Sample preparation is one of the key steps in any ion microscopic study. It is crucial that the native elemental distributions of the cells be preserved while at the same time satisfying the instrumental requirements. The sample should be compatible with high vacuum, and it should be conducting since it will be held at 4500V in the ion microscope. A sample preparation technique has been developed in our lab which satisfies all of these requirements .[6] The cells are grown and treated on silicon wafers, a conducting substrate which does not harm the cells in any way. Once the cells have undergone their treatments, they are sandwiched using a clean smooth silicon wafer. The sandwich is then fast frozen in cryogenic fluids and fractured by prying the two halves apart under liquid nitrogen. The bottom chips (the ones on which the cells were grown) are freeze-dried over night. Fast-freezing and prying the two halves apart produces large areas containg many cells grouped together which have been fractured in a plane through the apical cell membrane. This exposes the cell interiors free from any contamination although it makes it impossible to analyze the plasma membrane.

Figure 1 shows several freeze-fractured and freeze-dried LLC-PK_1 cells studied using ion microscopy and scanning electron microscopy.[5] From the SEM images it is apparent that the two of the cells in the top right corner show freezing damage from the sample preparation. The same information is also apparent using ion microscopy. The potassium level in these cells is very low, while the sodium level is high. There is also

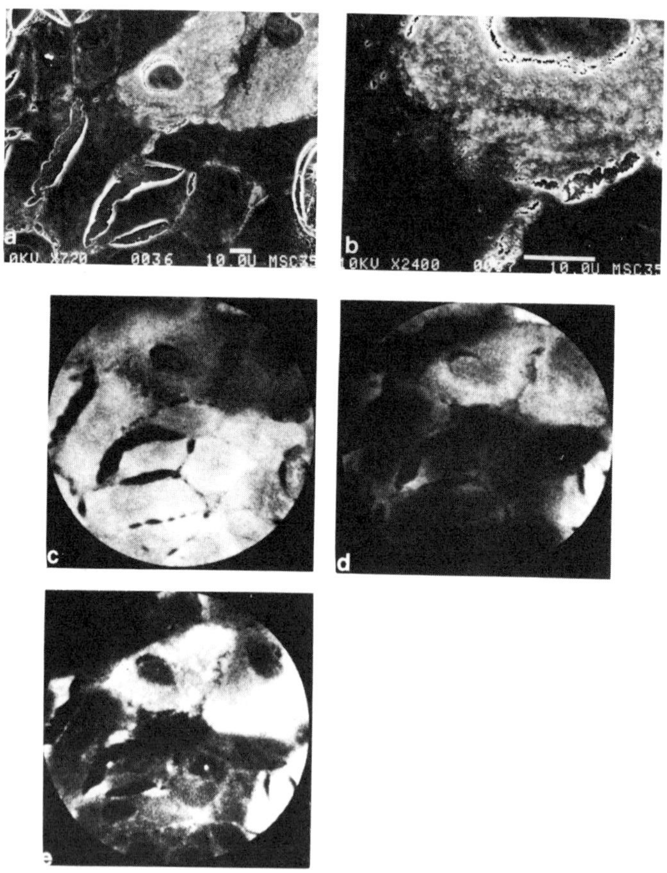

Figure 1. **(a)** A SEM image showing freeze-fractured freeze-dried LLC-PK_1 cells. The two cells in the top right corner are damaged. **(b)** A higher magnification SEM image clearly shows the damage. SIMS images of the same cells reveal intracellular distributions of **(c)** ^{39}K, **(d)** ^{23}Na, and **(e)** ^{40}Ca. (Photographs reproduced from S. Chandra and G.H. Morrison, *Biol. Cell* (1992) 74, 31-42. By copyright permission of Editions Scientifiques Elsevier.)

excessive loading of calcium in the cell cytoplasm. The potassium/sodium ratio for a typical healthy mammalian cell should be around 10, and it is known that the mitochondria of damaged cells accumulate massive quantities of calcium.[6,7] Thus, ion microscopy can easily be used to recognize and discard injured or damaged cells. It is imperative that these ion microscopic studies focus on healthy cells so as to accurately determine the physiological response of the cells to boronated compounds.

Once ion images are obtained they can be digitally processed so as to determine ion concentrations. A method has been developed in our lab for using relative sensitivity factors for converting ion intensity information to local dry weight elemental concentrations.[9]

Experiment

In this study F98 glioma cells were grown on silicon chips for 24 hours in regular media (DMEM + 10% FBS). This media was removed after 24 hours and the cells were incubated for six hours in fresh media containing the compounds studied. Concentations of 10μg/mL were used for each compound. The cells were then frozen, fractured, and freeze-dried. Finally, they were analyzed using ion microscopy.

The boron distribution of CBU-2' and CA in F98 cells was first compared to determine the effect of the base moiety on boron localization. Then the study was expanded to include CBU-3' and CBU-5' to examine the effect of the position of the carborane cage on boron distribution.

A fifth compound, 5-iodo-2'-deoxyuridine, was used as a morphological marker to label the nuclei of cells. This compound is known to be a DNA precursor.[10] By looking at the distribution of iododeoxyuridine in F98 glioma cells, it became possible to predict the distribution of an ideal boronated compound localized in the nucleus of the cells.

RESULTS AND DISCUSSION

All four boronated compounds gave similar results: the cytoplasm of the F98 glioma cells contained twice as much boron as the nucleus. Figure 2 shows the distribution of boron in the cytoplasm and nucleus of the cells. Using the quantitative imaging method developed in our lab we were able to determine the boron dry weight concentration and estimate the wet weight concentration in the nucleus and cytoplasm of the cells by assuming 85% cell water. The F98 cells show a large accumulation of boron. The fact that the boron concentration in the media was 3μg/mL B and that the intracellular boron concentration ranged from 50μg/mL to 200μg/mL shows that boron accumulates extensively in these cells.

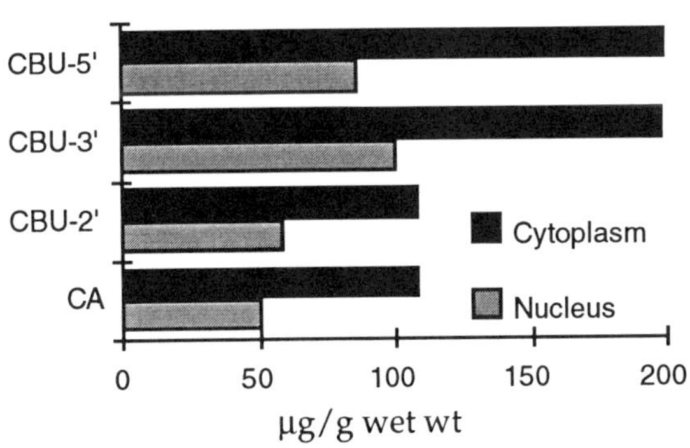

Figure 2. The boron distribution in F98 glioma cells treated with the boronated uridine and glucose derivatives. (Wet weight calculated assuming 85% cell water.)

Figure 3 shows images obtained for cells treated with CA. The K, Na, and Ca images were used to determine which cells were healthy , and to distinguish between the cytoplasm and nucleus of the cells. (The smaller cell at the bottom right corner of the images has been injured as is evidenced by the low potassium, high sodium, and high calcium levels.) Figure 4 shows the iodine localization for the cells treated with iododeoxyuridine. The iodine is located primarily in the nucleus (with the exception of the nucleoli) of the cells, as was expected. Had the boronated compounds been incorporated into the nucleus or the DNA of the cells this is the type of image we would have seen. If they had been incorporated into the RNA of the cells we would have seen a nucleolar localization.[11,12]

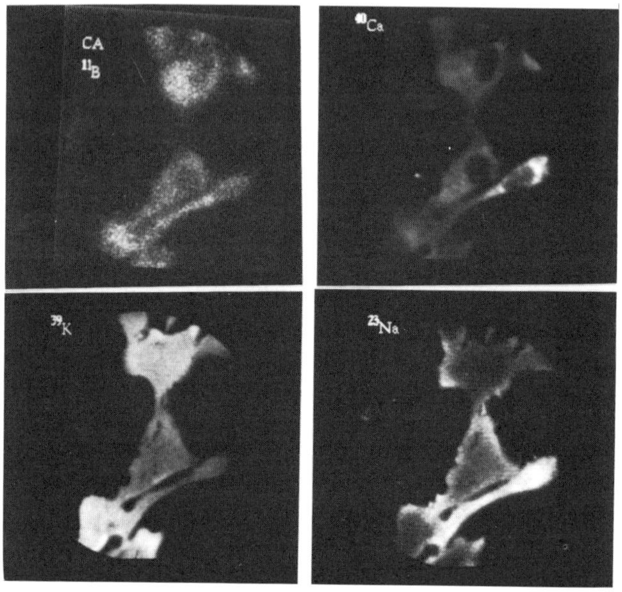

Figure 3. ^{11}B, ^{40}Ca, ^{39}K, and ^{23}Na images of F98 glioma cells treated with 10μg/mL CA for six hours..

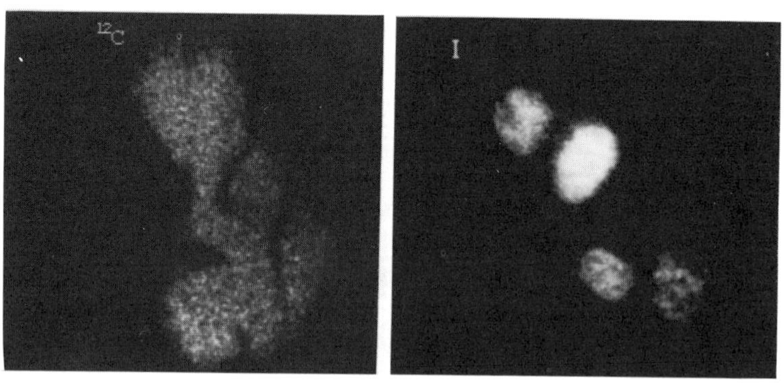

Figure 4. ^{127}I and ^{12}C images of F98 glioma cells treated with 10μg/mL iododeoxyuridine for six hours.

CONCLUSION

Ion microscopy is an efficient and sensitive method for determining the uptake, distribution, and concentration of boronated compounds at a subcellular resolution. Using ion microscopy we were able to determine that the four boronated uridine and glucose derivatives accumulate preferentially and extensively in the cytoplasm of F98 glioma cells. The position of the carborane cage and the existence of the base moiety did not influence the boron localization in that all four compounds showed twice as much boron in the cytoplasm as in the nucleus. The distribution of these boronated compounds seems to be more a function of the carborane cage itself (possibly due to the lipophilicity of the cage) rather than the carborane cage-nucleoside conjugation, and it is inconsistent with incorporation into the nucleic acids of the cells.

ACKNOWLEDGEMENTS

This work supported by DOE DE-FG02-91ER61138 grant to GHM and a NSF Graduate Fellowship to IG. The Cornell NIH/NSF Developmental Resource for Biophysical Imaging and Optoelectronics was used for cell culture work.

REFERENCES

1. W. Tjarks, A.K.M. Anisuzzaman, L. Liu, A.H. Soloway, R.F. Barth, D.J. Perkins, and D.M. Adams, Synthesis and in vitro evaluation of boronated uridine and glucose derivatives for boron neutron capture therapy, *J. Med. Chem.* 35:1628 (1992).
2. S. Chandra and G.H. Morrison, Ion microscopy in biology and medicine, *Methods Enzymol.* 158:157 (1988).
3. G.H. Morrison and G. Slodzian, Ion microscopy, *Anal. Chem.* 47:932A (1975).
4. M.S. Burns, Application of secondary ion mass spectrometry (SIMS) in biological research: a review, *J. Microsc. (Oxf)*, 127:237 (1982).
5. S. Chandra and G.H. Morrison, Sample preparation of animal tissues and cell cultures for secondary ion mass spectrometry (SIMS) microscopy, *Biol. Cell,* 74: 31 (1992).
6. S. Chandra, G.H. Morrison, and C.C. Wolcott, Imaging intracellular elemental distribution and ion fluxes in cultured cells using ion microscopy: a freeze-fracture methodology, *J. Microsc. (Oxf)* 144:15 (1986).
7. B. Alberts, D. Bray, J. Lewis, M. Raff, K. Roberts, and J.D. Watson. "Molecular Biology of the Cell," Garland Publishing, Inc., New York (1989).
8. A. P. Somlyo, A. V. Somlyo, and H. Shuman, Electron probe analysis of vascular smooth muscle: composition of mitochondria, nuclei, and cytoplasm, *J. Cell Biol.* 81:316 (1979).
9. W.A. Ausserer, Y.C. Ling, S. Chandra, and G.H. Morrison, Quantitative imaging of boron, calcium, magnesium, potassium, and sodium distributions in cultured cells with ion microscopy, *Anal. Chem.* 61:2690 (1989).
10. J.A. Aten, P.J.M. Bakker, J. Stap, G.A. Boschman, C.H.N. Veenhof, DNA double labelling with IdUrd and CldUrd for spatial and temporal analysis of cell proliferation and DNA replication, *Histochem. J.* 24: 251 (1992).
11. J.P. Berry, G. Lespinats, and F. Escaig, Ion microscopy of various pyrimidic nucleotides and analogs in the nuclei of cultured tumor cells, *Scanning,* 14:57 (1992).
12. B.D. Bennett, J. Mumford-Zisk, R.F. Barth, A.H. Soloway, and G.H. Morrison, Probing boronated nucleoside utilization by glioma cells using ion microscopy , in these *Proceedings*.

THE ANALYSIS OF BORON IN DEXTRAN-SULFHYDRYLBORANE CONJUGATES

Anders Holmberg [1], Lennart Meurling [2], Beat Wernli [3], Regin Weinreich [4]

1. Kabi Pharmacia Diagnostics AB, 75182 Uppsala, Sweden
2. Mercodia AB, 75228 Uppsala, Sweden
3. Paul Scherrer Institute, CH-5232 Villigen-PSI, Switzerland
4. Institute of Medical Radiobiology, University of Zurich and PSI, CH- 8029 Zurich Switzerland

INTRODUCTION

In the present study the sodium salt of sulfhydrylborane (BSH) was covalently linked to dextran derivatives. This reaction was accomplished in two ways, activation of dextran with 1-cyano-4-dimethylaminopyridine (1) with subsequent reaction with 2-aminoethylpyridyl-disulphide which underwent a disulphide exchange reaction with BSH. In the second way, dextran was allylated to a multiallyl derivative which in good yield reacted with BSH in a free radical initiated addition reaction (2). For the assessment of boron content of the derivatives elemental analysis of sulphur (3) was compared with Inductive Coupled Plasma Mass Spectrometry, ICP-MS of boron (4), the reference method in the analysis of the boron dextran derivatives. This was done under the assumption that the ratio of sulphur to boron would be constant during all the synthesis steps. The study was carried out to investigate a possibility to simplify the analysis of boron and for other clinical purposes. Our immediate interest in boronated dextran is its practical use for the treatment of arteriovenous malformations by BNCT in line with the ideas of Larsson (5) and Hatanaka (6).

Advances in Neutron Capture Therapy, Edited by
A.H Soloway *et al.*, Plenum Press, New York, 1993

MATERIALS AND METHODS

The Elemental Analysis of Sulphur

The samples were burned in a Carlo Erba 1106 elemental analyser over copper oxide with oxygen injected into the carrier gas. The analyser has an electron capture detector, ECP. One determination takes 5 min. The detection limit is 0.002 μg. or about 0.0004 % in a normal 0.5 mg sample.

The ICP-MS of Boron

The ICP-MS was carried out in a VG-PQ2 Plus equipment from FISONS Instruments with flow injection system. ^{10}B and ^{11}B were determined simultaneously. Thus calibration samples can consist of any isotopic composition. One determination takes 10 min., 5 measurements were carried out, each in peak-jump modus. Because of the high sensitivity of the method, the external boron contamination (e.g., glass vessels) had to be controlled and studied seriously.

For ICP-MS, the samples were dissolved in deionized water, diluted to a boron concentration of about 100 μg /l and generally handled in plastic vessels. In order to correct for the contamination of the glass parts of the ICP equipment, all determinations were carried out between two background measurements.

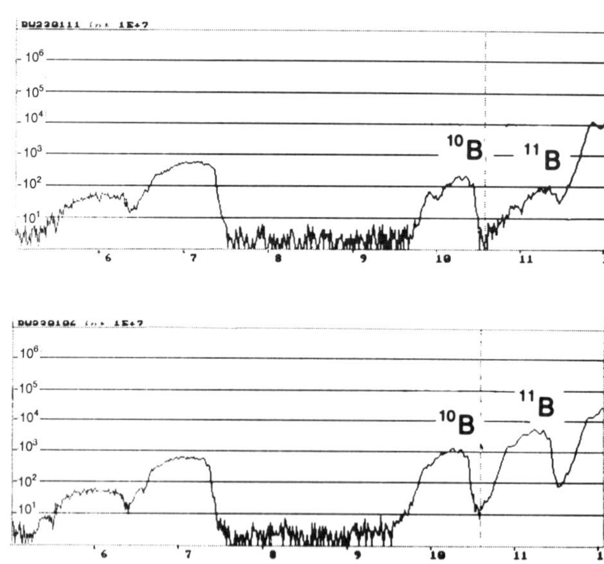

Figure 1. Logarithmic mass spectrogram of a 10B enriched dextran conjugate (above) and a calibration standard of boric acid of natural isotopic composition (below).

RESULTS

The boron-dextran derivatives were handled identically after the synthesis, in both analytical methods. The elemental analysis gave figures from 0.25% to 0.45% sulphur, corresponding to a boron content from 10 to 18 µg/g conjugate. With the ICP-MS however, the result of the measurements was 8 to 17 µg/g conjugate. This boron content corresponds to 10 to 20 cages per dextran chain. The assumption that the ratio between sulphur and boron is constant seems to be valid.

DISCUSSION AND CONCLUSIONS

A strategy for the design of efficacious BNCT agents is to couple a carrier molecule with ligands containing boronated compounds. To construct such BNCT agents many factors have to be considered, such as the activation degree of the carrier, the ratio between carrier and ligand in the preparatory step of the coupling reaction, and the resulting substitution degree of the ligand. To estimate the result of the synthesis repeated analysis of the boron has to be performed. In general the methods for the analysis of boron are complicated, demanding special instrumentation (7). Thus a possibility to simplify this analysis was investigated. This is a first attempt to use elemental analysis of sulphur as an indirect method to determine boron in BSH-dextran derivatives. Provided that other sources of sulphur can be controlled, it seems possible to use elemental analysis of sulphur as a simple indirect method to determine boron during the development of sulfhydrylborane conjugates. The 10-20% lower boron content obtained with ICP-MS may be corrected by an empirical factor. This uncertainty may reflect systematic errors in both or either of the analytical techniques. The BSH-dextran conjugates studied were shown to contain boron at levels that should be suitable for treatment of AVM by BNCT.

REFERENCES

1. Kohn J. & Wilchek M. (1983) *FEBS Letters* **154**, 1, 209-210.

2. Stacey &Harris (1963) in *Org. Reactions*, **13**, 150-376.

3. Kirsten W. & Nordenmark B. (1987) *Analyt. Chim. Acta*, **196**, 59-68.

4. (1989) *Applications of Inductively Coupled Plasma Mass Spectrometry*, Eds. Date A R, & Gray A L. (Blackie , Glasgow and London).

5. Larsson B. (1988) in *Clinical Aspects of Neutron Capture Therapy*, 21-26, Eds. Fairchild R. G. , Bond V. P. , and A. V. Woodhead, Plenum New York (1989).

6. Hatanaka H., Nakagawa Y. and Sakaki S. (1992) *5th International Symposium on Neutron Capture Therapy.* **122**, 62.

7. Moore, D.E. (1990) *J. Pharm. Biomed. Analysis*. **8**, 547-553.

DETERMINATION OF STRONGLY PROTEIN-BOUND BOROCAPTATE SPECIES BY HIGH PERFORMANCE LIQUID CHROMATOGRAPHY WITH ON-LINE INDUCTIVELY COUPLED PLASMA ATOMIC EMISSION SPECTROSCOPY DETECTION OF BORON

N.J. Hotz and W.F. Bauer*

Idaho National Engineering Laboratory, EG&G Idaho, Inc.
Idaho Falls, ID 83415

INTRODUCTION

Borocaptate compounds have been studied for many years as a means of delivering ^{10}B to brain tumors prior to neutron exposure for BNCT. The actual nature of the observed binding of mercapto-undecahydrododecaborate ($B_{12}H_{11}SH^{2-}$, BSH) anion and it's dimer ($B_{24}H_{22}S_2^{4-}$, BSSB) to plasma proteins has been of interest as this will affect both the toxicity and the biodistribution of these anions throughout the body. The toxicity of the borocaptate species may be related to the ability to bind to proteins and compete for sites which normally bind metabolic by-products.

Early work concerning the protein binding of BSH indicated that covalent binding via a disulfide linkage may occur between BSH and serum proteins[1,2]. If this is true, a minimum amount of BSH would be required to fill the strong covalent binding sites and only BSH in excess of this level would be available for tumor uptake. However, nuclear magnetic resonance (NMR) spectroscopy does not show evidence of covalent binding of BSH to albumins via a covalent disulfide bond[3], therefore indicating that binding is primarily ionic. In an experiment performed in our laboratory, phosphate buffered saline rinses performed on bovine serum albumin/BSH retained above ultrafiltration membranes were successful in removing some of the borocaptate that was considered "bound" to the protein. The dilution of the BSH and BSA, which is believed to shift the equilibrium and result in the release of the ionically bound BSH, could be performed by gel filtration chromatography (GPC). Since GPC is not an equilibrium method, the drug-protein complex will dissociate and the drug separated from the protein[4]. In the case of BSH binding to protein, only strongly associated or covalently bound compounds would be expected to co-elute with the high molecular weight proteins.

The primary disadvantage with GPC to determine the protein binding of BSH and related compounds is the typical UV-detectors used for liquid chromatography cannot distinguish between protein and protein with bound boron. To determine the B containing proteins, fractions can be collected and B determined by a separate technique such as

Advances in Neutron Capture Therapy, Edited by
A.H Soloway *et al.*, Plenum Press, New York, 1993

inductively-coupled plasma atomic emission spectroscopy (ICP-AES). Two modes of on line HPLC detection are clearly needed, one to detect the protein and one to detect the boron. To perform this dual detection, an ICP-AES has been coupled to a high performance liquid chromatograph with a UV detector to allow serial detection of the protein (UV at 280 nm) and boron (ICP-AES at 249.77 nm).

MATERIALS AND METHODS

Solutions of bovine serum albumin (BSA, Sigma A-6793) and human serum albumin (HSA, Sigma A-1667) were made at 4% w/w in pH 7.3, 0.10 M phosphate buffer. Varying amounts of $Li_2B_{12}H_{12}$ (BH), $Na_2B_{12}H_{11}SH$ (BSH), $Cs_4B_{24}H_{22}S_2$ (BSSB) and $B(OH)_3$ were added to the albumin solutions and allowed to incubate for 24 hours before GPC. The binding of the boron compounds was also performed at different incubation temperatures and incubation times. Twenty microliter volumes were injected into a two column series of Macrosphere GPC 300 columns (Altech & Associates) with a flow rate of 0.5 mL/min of pH 7.3, 0.10 M phosphate buffer. Deionized water containing 0.1% Triton X-100 (Sigma X-100R-S) at 2.0 mL/min was added to the UV detector effluent in order to stabilize the ICP and to maintain the optimum delivery flow to the nebulizer on the ARL 3520 ICP-AES. The conventional ICP-AES sample introduction system with a concentric glass nebulizer and spray chamber were utilized. Protein peaks were monitored at 280 nm and the B emission line at 249.77 nm was monitored to detect any eluting boron compounds. The diagram in Figure 1 shows a general schematic of the instrumental setup. Ultrafiltration and equilibrium dialysis experiments were done with 30000 molecular weight cutoff membranes.

RESULTS AND DISCUSSION

A decrease of the baseline noise in the ICP-AES signal during elution of the protein peak was observed during early experiments. In order to maintain the lower noise level, a 0.1% aqueous solution of Triton X-100 was added to the carrier flow immediately after

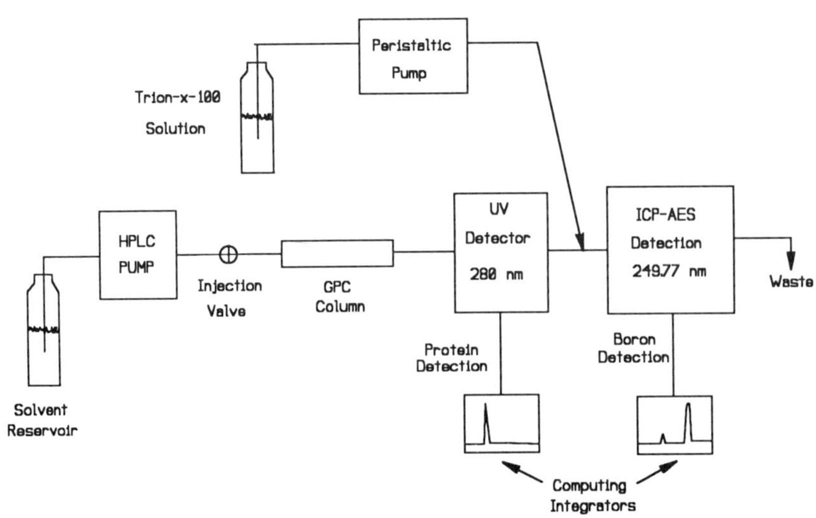

Figure 1. HPLC apparatus for the separation and detection of boron containing proteins.

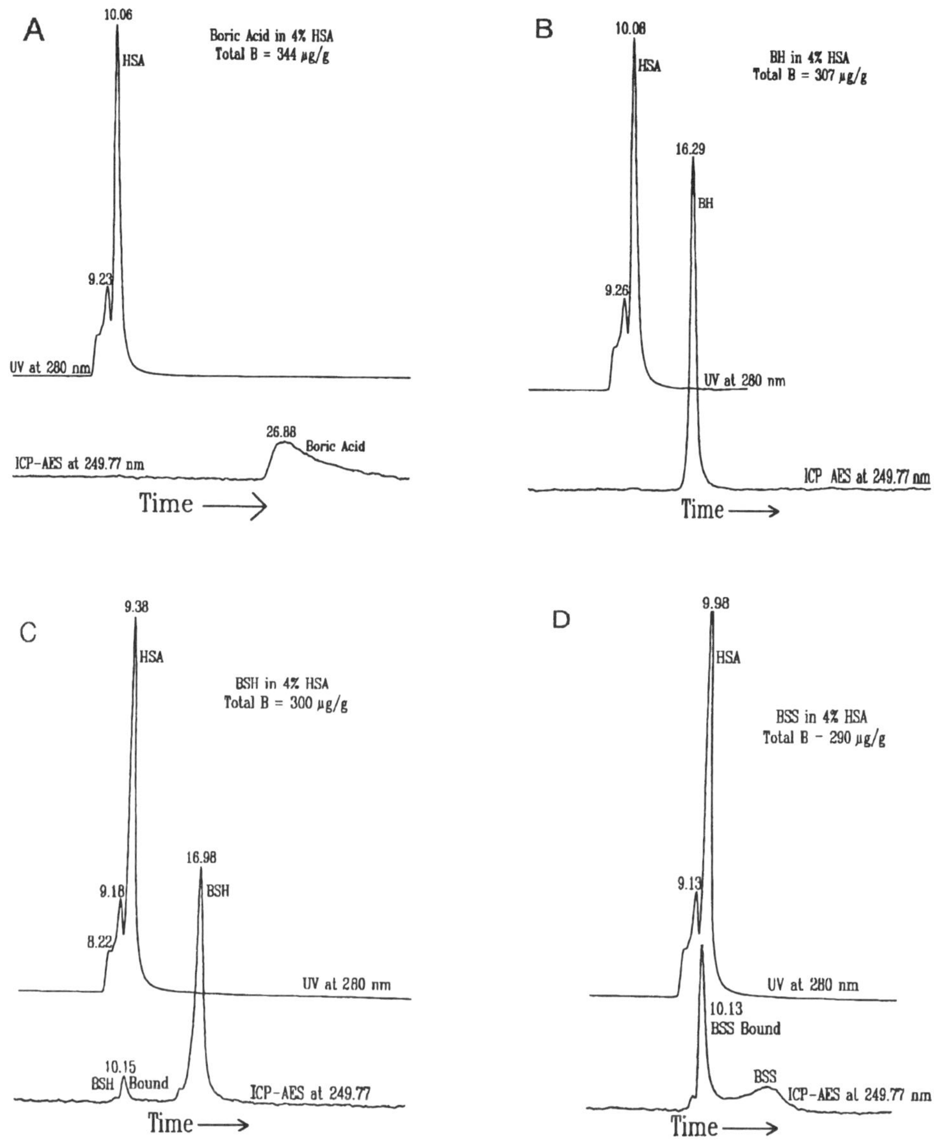

Figure 2. Chromatograms showing the nature of the interaction of A) $B(OH)_3$, B) $B_{12}H_{12}^{-2}$, C) BSH, and D) BSSB with HSA. Detection is via UV (top) and ICP-AES (bottom).

the UV-detector. This "makeup" flow was necessary to prevent significant band broadening caused by the dead volume of the spray chamber and the addition of the surfactant served to noticeably reduce the noise levels.

Figures 2A through 2D demonstrate the binding differences found between the various B compounds with HSA after a 24 hour incubation period at 37°C. Boric acid (Figure 2A) and BH (Figure 2B) are completely separated from any association with the albumin molecule. Binding of BH to HSA has been observed during ultrafiltration experiments, however, no binding is observed during GPC. The binding of these compounds appears to be solely ionic as previously reported[1,2]. The BSH anion, however, shows only a small quantity of B that is tightly bound to the albumin molecule (Figure 2C), most likely due to the BSSB and BSSOB already present in the BSH stock solution. This conclusion is supported by the fact that the peak areas indicated ~7% of the B was bound to the protein at all total B concentration levels and by Figure 2D which shows almost complete binding of the BSSB boron to the albumin. The tailing and poor peak shape of the "free BSSB" may indicate that the binding is via a covalent disulfide exchange type reaction. This strong binding of BSSB to albumin has been previously noted[5] and the reactivity may be partially explained by the ability of BSSB to form free radicals under certain conditions[6] and the importance of such free radicals in the formation of disulfide bonds[7]. The binding of BSH to albumin is essentially complete within the time of mixing (~5 minutes) therefore, BSH would not be expected to react significantly with serum proteins during the course of a normal infusion and elimination period. The observed levels of BSH binding to HSA are at significantly lower levels than those determined by ultrafiltration or equilibrium dialysis (65-85% bound). The ICP-AES (boron) chromatographs in Figure 3 demonstrate that only a limited number of sites are available for covalent binding and that subsequent concentration doublings of BSSB do not lead to an appreciable increase in "covalently" bound boron.

CONCLUSIONS

A method for determining protein bound boron has been described and demonstrated. The BSH anion is easily separated from its association with albumin molecules indicating that the binding is primarily of an ionic nature. The BSH dimer, BSSB, is quite strongly associated with the albumin molecule due to the ability to attack sulfhydryl and disulfide groups on the parent protein and create a series of disulfide exchange reactions. The results presented here suggest that no significant binding of BSH to albumin would be expected during the course of an infusion and BNCT treatment. Future work will examine BSH binding in real animal systems after infusion of the compound, i.e. *in situ* incubation of BSH with blood proteins.

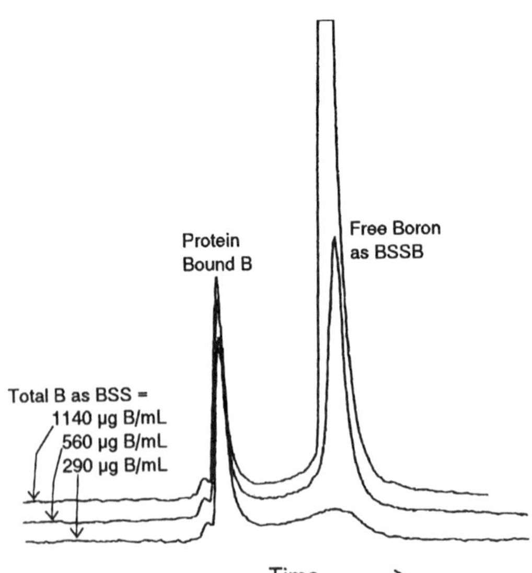

Figure 3. GPC chromatograms with ICP-AES detection of BSSB incubated with 4% HSA. The albumin molecule becomes saturated with covalently bound B at ~1.3 BSSB per albumin.

ACKNOWLEDGEMENTS

This work was performed under the auspices of the U.S. Department of Energy, Office of Energy Research, under DOE Field Office, Idaho, Contract No. DE-AC07-761D01570.

REFERENCES

1. A. H. Soloway, H. Hatanaka and M. A. Davis (1967) *J. Med. Chem.* **10**, 714-717.
2. T. Nakagawa, T. Nagai. Parts I, II, and III (1976) *Chem. Pharm. Bull.* **24**, 2934-2954.
3. W. F. Bauer, K. M. Bradshaw, and T. L. Richards (1992) in *Progress in Neutron Capture Therapy for Cancer*, ed. ed. B.J. Allen et al. (Plenum Press, New York), pp. 339 -343.
4. T. C. Pinkerton and K. A. Koeplinger (1990) *Anal. Chem.* **62**, 2114-2122.
5. D. Slatkin, P. Micca, A. Forman, D. Gabel, L. Wielopolski, and R. Fairchild (1986) *Biochemical Pharmacology*, **35**, 1771-1776.
6. G. R. Wellum, E. I. Tolpin, A. H. Soloway, and A. Kaczmarczyk (1977) *Inorganic Chemistry* **16**, 2120-2122.
7. J. March (1977) Advanced Orgainic Chemistry, 2nd ed. (McGraw-Hill Book Company, New York) p. 1115.

OPTIMISATION OF NEUTRON CAPTURE RADIOGRAPHY FOR ANALYSIS OF BORON-10 IN TISSUES

Haider A. Meriaty and Barry J. Allen

Biomedicine and Health Programme, Ansto
PMB 1, Menai, NSW 2234, Australia

INTRODUCTION

The $^{14}N(n,p)^{14}C$ reaction concomitantly occurs with the $^{10}B(n,\alpha)^{7}Li$ reaction in tissue. Although the cross section of nitrogen-14 is much lower than that for boron-10, the abundance of the former in tissue is much higher. Consequently, the rates of both reactions could be similar in normal tissue. CR-39 is a solid state nuclear track detector and is sensitive to the products of both reactions, therefore distinguishing between the proton and both alpha and lithium-7 ions is essential for meaningful boron-10 imaging.
The proton has a longer etch-induction time (t_i) than the other ions i.e. α and ^7Li. Using an optimum etching time (t_o) reveals only the pits of the heavy ions. The pit diameter rates for proton and heavier ions are calculated theoretically to determine (t_o) and this is verified experimentally.

THEORY

The etching rate ratio is the ratio of the etching velocity along the ion track (V_T), to the etching velocity of the undamaged surface (bulk etching rate, V_B) of the detector. It is postulated that the etching rate ratio (V) is proportional to the following function[1]:

$$V = h\,(z^* / \beta)^2\ \{1 + 0.05\ \mathrm{Ln}[\beta^2 / (1 - \beta^2) - 0.05\ \beta^2\} \qquad (1)$$

where:

z^* : effective ion charge and it is given by equation (2).
β : ratio of particle velocity to the speed of light.
h : constant.

The effective charge (z^*) of a moving ion is given by an empirical equation[2]:

$$z^* = z\ [1 - \exp(-130\ \beta / z^{2/3})] \qquad (2)$$

where z is atomic number of the ion.

For an incident ion normal on the detector surface and non-excessive etching time (t_e), the pit diameter rate is expressed[2] by:

$$d / t_e = 2\ V_B\ [(V - 1) / (V + 1)]^{1/2} \qquad (3)$$

The bulk etching rate V_B for CR-39 was reported to be proportional to the exponential of the temperature T (Kelvin) and concentration C (mol/l) of NaOH etchant[4]:

$$V_B = 1.276\ \exp[2.3\ (0.828\ C + 0.049\ T - 0.002\ C\ T - 17.624)] \qquad (4)$$

Advances in Neutron Capture Therapy, Edited by
A.H Soloway *et al.*, Plenum Press, New York, 1993

The energies of the alpha, lithium-7 ion and proton are calculated from the Q values of the capture reactions. Thus, the corresponding ion velocity (m s^{-1}) of mass m (kg) and energy E (J) can be calculated by:

$$\vartheta = [(2\,E) / m]^{1/2} \qquad (5)$$

The critical angle (θ_c) and etching efficiency (η) are a function of etching velocity[2]

$$\theta_c = \text{Arcsin}(1/V) \qquad (6)$$

$$\eta = 1 - \text{Sin}(\theta_c) \qquad \text{(For a thin external ion source)} \qquad (7)$$

METHODS

To utilise Eq.1 to calculate the etching parameters, the proportionality constant h must be determined. Its value was derived from a published experimental value of the rate of increase of pit diameter ($d / t_e = 2.5$ µm h^{-1}) of alpha particles from the $B^{10}(n,\alpha)^7Li$ reaction[5] using the parameters: T = 343.15 °K, C = 6.25 M and t_e = 1 hour and calculated values of V, β and z^*. Then Eq.1 was applied to determine V values of the proton and Li-7. The β and z^* of the latter ions were calculated from Eq.5 and Eq.2 respectively.

To calculate the optimum etching times for ^7Li and proton ions, their pit diameter rates were calculated from Eq.3. A light microscope was used to determine the track diameters of alpha and ^7Li ions. Other etching parameters were also determined, namely, the critical angles (θ_c) and the etching efficiencies (η) of the ions using Eq.6 and Eq.7.

To measure the pit diameter of the heavier ions, solid samples of boron-10 were used. Solutions of 1.25 and 40 ppm of methyl methacrylate polymer and decaborane ($B_{10}H_{14}$) were prepared. A drop of the solution formed a film of 150x150x0.001 mm on a perspex slide of 1 mm thickness. The film was then covered tightly with a 20x20 mm CR-39 foil (Pershore 1% DOP, 250 µm thickness and 2 years old) for later exposure to thermal neutrons and pit diameter measurements.

Harding Passey melanoma cells (1-2 x 10^6) were implanted on the right thigh of two female nude mice. After 10 days, the tumour sizes were 0.6 mm diameter. The mouse of weight 22.6 g was intraperitoneally injected with 5 mg boronophenylalanine (BPA.HCl) in 0.5 ml saline and the second mouse (19.08 g) was used as a control. Two hours post injection, both mice were sacrificed and microtome-sections of 8 ± 2 µm were prepared. The microtome procedures involved fixation in 10% formalin/phosphate buffer, dehydration in ethanol and then xylene, embedding in paraffin wax. The sections were then mounted on perspex slides of 1mm thickness and covered with a 20x20 mm foils of CR-39. The foils were firmly attached to perspex slides.

To ensure a good contact between the samples and the detectors, they were sandwiched between two plates of aluminium (140x145x1.5 mm) and bolted together. The detectors were then exposed to a thermal neutron fluence of 2.49×10^{10} n cm^{-2} s^{-1} (at low flux) from the Moata reactor at Ansto. The neutron fluence was measured by gold foils.

All detectors were exposed to day light for eighteen hours prior to the etching process to enhance their etching sensitivity[2]. Then, they were etched in a stirred NaOH solution (6.25 M) at 70 ± 1 °C in a glass flask with a reflux column.

RESULTS

The following conditions were determined:
1. h = 1.96x10-3 and the optimum etching time, t_o = 1 hour.
2. T_i = 20 minutes for alpha and ^7Li particles.
3. Average pit diameter of heavier ions d = 3.95 ± 0.22 µm for t_e = 1.667 hour.
The etching parameters are given below:

Ion	V	d/t_e ($\mu m\ h^{-1}$)	θ_c (degree)	η%
α	5.18	2.5†	11.1	81
^7Li	16.24	2.9	3.5	93.9
p	1.06	0.52	70.6	5.7

†: Published experimental value[5].

DISCUSSION

The diameters of 10 alpha pits were measured from a photomicrograph and their average diameter was calculated. For elliptical shaped pits, the average of their minor and major axes were taken. The actual etching time was 1.667 hours and the pit diameter was corrected to one hour. The average d/t_e was found to be $2.37 \pm 0.13\ \mu m\ h^{-1}$. For a non-excessive t_e, the growth of a pit diameter is linear with t_e (Eq.3) thus, the scaling to one hour is valid. The percentage difference between the reported and the measured pit diameters of alpha and ^7Li is 5.2%. The estimated proton diameter from Eq.1 agrees with a published result[6] (d = 0.517) with CV = 0.6%.

The percentage difference in V between the calculated and the experimental values is 20%. This difference may be due to the calculated value of V_B or brand of CR-39 (Hungarian MA-ND/α) or/and to the difference in η of alpha and ^7Li ions ($\Delta\eta$ = 12.9%). However, V varies about 13% per degree of the etchant temperature[3].

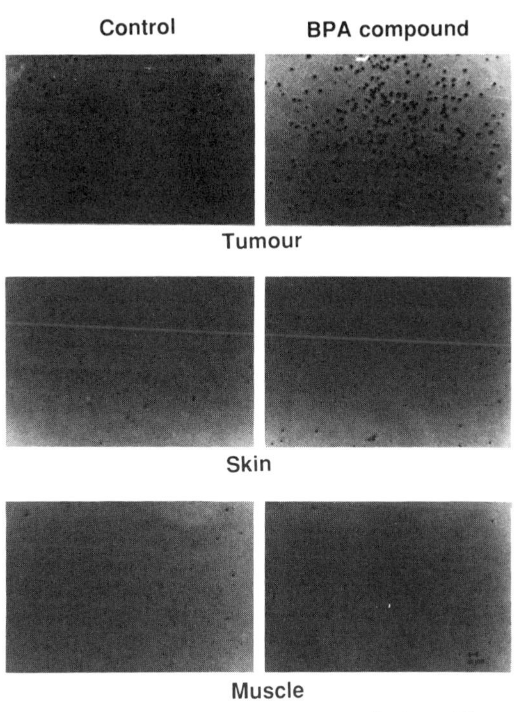

Figure 1: Boron-10 distributions in BPA.HCl-injected tissues and control. Dark dots correspond to alpha and ^7Li tracks on the CR-39 detector.

Two different etching times were performed on the detectors of the boron calibration samples. At t_e = 20 minutes, the pits were barely observable and they become obvious at t_e = 30 minutes. The etch-induction time (t_i) of alpha and ^{7}Li was therefore adopted to be 20 minutes.

The pit density of the background and a prepared sample from 1.25 ppm boron calibration solution was found to be almost the same. Higher boron-10 concentrations showed an increase in the pit densities. Hence, the 1.25 ppm concentration was considered to be the minimum detection sensitivity of CR-39 for boron-10.

Pit distributions in tumour, skin and muscle tissues for the control and the BPA.HCl injected mice were obtained. A high pit density is observed in the tumour with BPA.HCl in contrast to the control. The pit densities of the control tissues as well as of the skin and muscle of the injected mouse are close to background pit density. Therefore, tumour tissue takes up BPA.HCl selectively. The derived etching and irradiating conditions are therefore valid to selectively detect alpha and Li-7 ions but not protons.

CONCLUSIONS

The developed semi-empirical equation (Eq.1) provides an estimation of the etching rate ratios (V) for the ions of interest. Thereby, the critical angles, the pit diameters and the etching efficiencies of alpha, ^{7}Li ions and protons were determined. The efficiency ratios of alpha and ^{7}Li ions to protons are 14.2 and 16.5 respectively. The uncertainty in the experimental alpha diameter could be 13% due to the higher etching efficiency of the ^{7}Li ion.

To avoid revealing proton pits, the optimum etching time is 60 minutes for T = 70 °C and C = 6.25 M NaOH. Under these conditions, the highest pit density were observed in the tumour of the BPA.HCl-injected mouse relative to the other tissues including the control.

ACKNOWLEDGMENT

We are grateful to J. Mallesch and B. Izard for in vivo support and tissue and tumour sections.

REFERENCES

[1] B. G. Cartwright and E. K. Shirk, A Nuclear-Track-Recording Polymer of Unique Sensitivity and Resolution, Nuclear Instruments & Methods, 153:457-460, 1978.

[2] R. L. Fleischer, P. B. Price and R. M. Walker, Nuclear Tracks in solids: Principles and Applications, Berkeley University of California Press, 1975.

[3] J. H. Adams, A Precision Etching Bath for Plastic Track detectors, In: Solid State Nuclear Track Detectors, eds. François H., Kurtz N., Massué J. P., Monnin M, Schmitt R. and Durani S. A., (1980), Oxford, Pergamon.

[4] M. Fromm, F. Membrey, A. Chambaudet, R. Saouli and A. El-Rahamany, A Study of CR-39 Bulk Etch Properties under Various Temperature and Concentration Conditions to Modelize the Dissolution Rate, Nucl. Tracks Radiat Meas. Vol.19, Nos 1-4, pp. 169-170, 1991.

[5] K. Freyer, H. C. Treutler, K. Dietze, I. Hunyadi and I. Csige, Vacuum Treatment of CR-39 for the Reduction of Backgrounding in Neutron Induced Autoradiography of Boron, Nucl. Tracks Radiat. Meas., Vol.18, No.3, pp.309-313, 1991.

[6] A. Khan Hameed and R. Brandt, Track-Registration-and Development Characteristics of CR-39 Plastic Track Detector, Nucl. Tracks, Vol.7, No.3, pp.129-139, 1983.

THE APPLICATION OF AN ALPHA AUTORADIOGRAPHIC TECHNIQUE FOR DETERMINATION OF BORON-10 CONCENTRATIONS IN A STUDY OF INTRATUMORAL INJECTION OF BSH AND CBU-2'

Yirun Jiang[1], Sherry Ng[1], Thomas E. Blue[1], Joan Rotaru[2], and Rolf F. Barth[2]

[1]Nuclear Engineering Program
[2]Department of Pathology
The Ohio State University
Columbus, Ohio 43210

INTRODUCTION

The accurate measurement of ^{10}B distributions in biological samples with a sensitivity in the ppm range is essential for evaluating the potential usefulness of various boron-containing compounds for BNCT[1]. The purpose of the present study was to develop an alpha track autoradiographic technique to qualitatively and quantitatively determine the spatial distribution of ^{10}B in tissue samples. The technique uses the Solid State Nuclear Track Detector, CR-39, which is composed of a polycarbonate resin, to record the charged products resulting from the $^{10}B(n,\alpha,\gamma)^{7}Li$ reaction. It provides ^{10}B concentration information on a group of images: a black and white alpha track radiographic image, a color densitometric image, and a color histological tissue image.

EXPERIMENTAL PROCEDURE

F98 glioma cells were implanted[2] stereotactically into the right caudate nucleus of Fisher rats and ~16 days later the animals were injected intratumorally with 10 μl of either 142 mM sodium borocaptate (BSH, total 162 mg ^{10}B) or 95 mM 2'-O-(o-carboran-l-ylmethyl) uridine (CBU-2', total 90.3 mg ^{10}B) using the same stereotactic coordinates. Control rats were carried out using the identical procedure except that the compounds were injected into the non-tumor bearing brains. The animals were sacrificed 3, 6, 12, and 24 hours after the compounds were injected, and the brains were removed and frozen at -80°C. Subsequently, the frozen brains were cut coronally into 20μm thick sections, placed on slides and air dried. Boron-coated tape pointers were then affixed to the slides to provide a means for aligning the tissue section images and the alpha track images[3].

Next the CR-39 detector slides were placed on top of the tissue sections and secured with tape. The CR-39/tissue sections/slide sandwiches and a set of ^{10}B in Tissue Equivalent Liquid standards were then irradiated in the thermal column of The Ohio State University Research Reactor to a high fluence of 10^{12} neutrons/cm^2. Following the first irradiation, the slides were separated, and the tissue sections were covered by additional CR-39 detector slides to be ready for a second irradiation. The CR-39 detector slides and the standards from the first irradiation were etched in a 6.25 M NaOH solution at 70°C

Advances in Neutron Capture Therapy, Edited by
A.H Soloway *et al.*, Plenum Press, New York, 1993

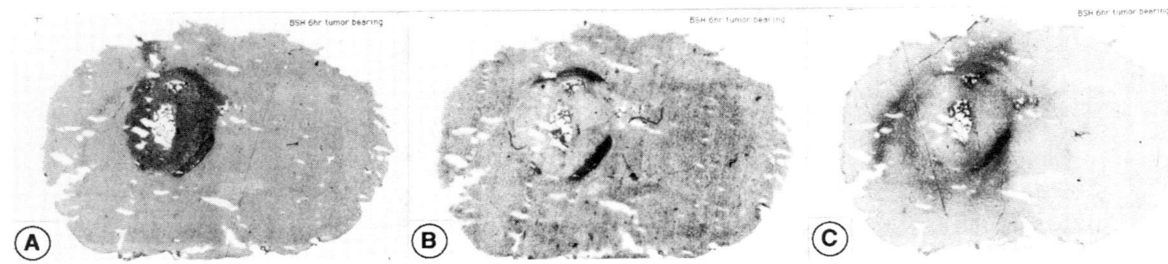

Figure 1. BSH at 6 hour sacrifice time for a tumor-bearing rat: A) histological tissue image; B) unstained tissue image, dark part specified the bleeding in the brain; C) alpha track image.

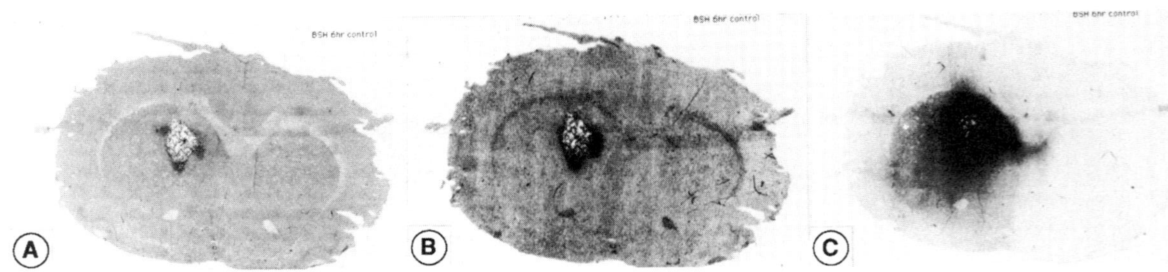

Figure 2. BSH at 6 hour sacrifice time for a non-tumor bearing rat: A) histological tissue image; B) unstained tissue image, dark part specified the bleeding in the brain; C) alpha track image.

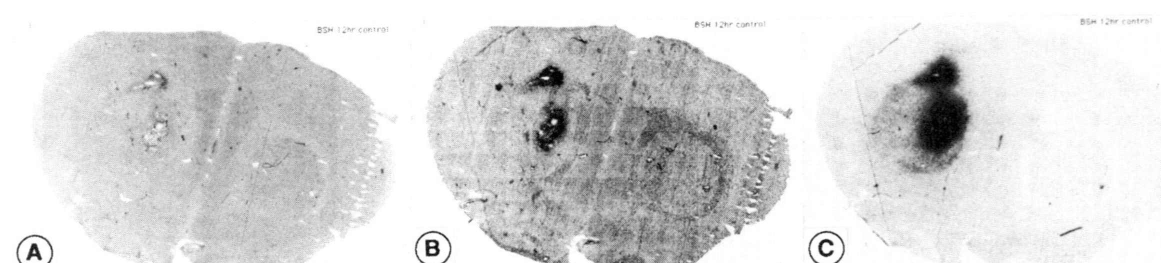

Figure 3. BSH at 12 hour sacrifice time for a non-tumor bearing rat: A) histological tissue image; B) unstained tissue image, dark part specified the bleeding in the brain; C) alpha track image.

for 100 minutes to produce tracks that were visible by light microscopy. The CR-39 detector slides from the first irradiation provide alpha track radiographic images that qualitatively give distributions of the boron concentrations within the tissue sections, as well as an estimation of boron concentrations by comparison with the standards. Then the second set of CR-39/tissue sections/slide sandwiches along with another set of standards were irradiated to a low fluence. The fluence which was estimated from the results of the first irradiation was used to produce about 20 tracks per counting field (80 μm X 80 μm) in order to avoid track overlapping during track counting. Then the CR-39 detector slides and the standards from the second irradiation were etched using the same procedure described above. The tissue sections on the slides were then stained with hematoxylin and eosin and mounted with a coverslip.

Imaging of the tissue section slide and the alpha tracks on the CR-39 detector slide were performed using our Macintosh Imaging System (MIS)[3]. Alignment between the tissue images and the alpha track images was established using the boron tape alignment marks. The high fluence black & white alpha track image was then converted to a color densitometric image by setting the gray levels recorded with MIS to colors that higher densities were converted to longer wavelengths. All images were then printed using a Tektronix color printer.

RESULTS AND DISCUSSION

The boron concentration within the tissue section was proportional to track density. However, because of overlapping tracks, the darkness of the track images for the high fluence irradiation is not linearly related to track density. Even so, colors were assigned to the darkness (gray levels) to better visualize boron concentrations. The accurate boron concentrations within the tissues are given by counting the CR-39 detector slides from the low fluence irradiation using our Track Counting System. Unfortunately, the low fluence data was not available at the time of the meeting.

Due to space limitations only examples of our results in black and white are presented in Fig. 1-3. The results indicated that both BSH and CBU-2' diffused out of the tumors eventually. No significant tumor cell intake of either compounds has been seen with this injection method. However, BSH appears to bind strongly to blood (Fig. 3 shows this most clearly). Both BSH and CBU-2' diffused faster when injected intratumorally than when injected in non-tumor bearing brains, as can be seen for BSH by comparing Fig. 1-c and Fig. 2-c. Also, all results indicated that BSH diffused faster and cleared more rapidly than CBU-2', for both the tumor bearing and the control groups.

ACKNOWLEDGMENT

This work has been supported by grant DE-AC02-76CH00016 from the Department of Energy.

REFERENCES

1. R. F. Barth, A. H. Soloway, and R. G. Fairchild: Boron neutron capture therapy of cancer. *Cancer Res.* **50**: 1061-1070(1990).
2. N. R. Clendenon, R. F. Barth, W. A. Gordon, J. H. Goodman, F. Alam, A. E. Staubus, C. P. Boesel, A. J. Yates, M. L. Moeschberger, R. G. Fairchild, and J. A. Kalef-Ezra: Boron neutron capture therapy of a rat glioma. *Neurosurgery* **26**: 47-55(1990).
3. J. E. Woollard, T. E. Blue, J. F. Curran, M. C. Dobelbower, H. R. Busby and R. F. Barth: An alpha autoradiographic technique for spatial quantification of boron-10 concentrations in tissue. *Nucl. Sci. Eng.* **110**: 96-103(1992).

NMR STUDIES OF THE INTERACTION OF BOROCAPTATE SODIUM WITH SERUM ALBUMIN

Peng-Peng Z. Tang[1], Martin P. Schweizer[1], William F. Bauer[3] and Kenneth M. Bradshaw[1,2]

1. Dept. of Medicinal Chemistry and Dept. of Radiology, University of Utah, Salt Lake City, UT 84112
2. Neutron Technology, Boise, ID 83702
3. Idaho National Engineering Laboratory, Idaho Falls, ID 83415-3519

INTRODUCTION

The interaction between BSH and serum albumin is of interest because it is related to the pharmacokinetics of BSH. Early research[1] suggested that covalent disulfide bridge might be involved in the interaction, but the idea was contradicted by subsequent studies[2,3,4] using ^{11}B Nuclear Magnetic Resonance (NMR) spectroscopy. NMR has been proved to be a very powerful technique to study the protein binding of a small molecule, such as BSH[5], and boron NMR is typically suitable to study the binding effect of the boronated agent. Both ^{11}B and ^{10}B have NMR activity, but ^{11}B has better NMR sensitivity (16.5% vs. 2% relative to ^{1}H) and a larger natural abundance (81% vs. 19%)[6]. Even though ^{10}B is the neutron active nucleus for BNCT, the more sensitive nucleus ^{11}B is more appropriate for NMR studies. Since the isotopic difference does not make any differences in the structure and binding effect of a compound, the result obtained from the ^{11}B NMR of BSH can hold for the ^{10}B enriched agent.

The chemical shift of boron NMR is very sensitive to the chemical environment. According to chemical differences, the 12 boron nuclei of BSH can be classified into 4 groups, as indicated in Fig. 1. Previous research indicated that the formation of a covalent disulfide bridge moves the ^{11}B chemical shift of sulfhydryl boron (B1) 3 ppm downfield[2] since a covalent binding to BSH changes the chemical environment around the boron nuclei.

The relaxation rates of a quadrupole nucleus are also very sensitive to protein binding. The binding can produce large changes in the quadrupole relaxation rate as a result of changes in the electric field gradients and the correlation times. In the interaction of albumin with BSH , the ^{11}B linewidth, $W^{b,e}$, of protein bound boron is obtained by subtracting the ^{11}B linewidth, W^{u}, of unbound BSH (obtained from free BSH solution) from the observed ^{11}B linewidth, W^{o}, of BSH in albumin solution ($W^{b,e} = W^{o} - W^{u}$). The ^{11}B spin-spin relaxation rate of protein bound BSH and the rate of exchange between bound and unbound BSH contribute to $W^{b,e}$. When the bound and unbound BSH are under fast exchange, the linewidth $W^{b,e}$ decreases as the temperature increases[5].

The relaxation rate of a nucleus is also related to its correlation time (τ_c). The correlation time, according to Stoke's Law, is proportional to the solvent viscosity[5]. When more than 1% (w/v) aqueous solution of protein is used, the effect of viscosity (η) on relaxation rate cannot be neglected. Under the so called 'extreme narrowing condition', the quadrupole relaxation rate (R) is linearly related to the correlation time (τ_c)[5]. Assuming that this condition applies to our system, the ^{11}B longitudinal relaxation rate (R_1) is proportional to the viscosity. Since the flow times of water (F^*_t) and that of the aqueous solutions of serum albumin (F_t) measured with an Otswald viscometer are proportional to their viscosities, the ratio f^v is proportional to the ratio of the viscosities of the solutions (eq. 1), and therefor can be applied to the measured ^{11}B longitudinal relaxation rate (R_1) as a correction coefficient to account for the viscosity effect (eq. 2).

$$F_t = C\frac{\eta}{\xi} \Longrightarrow \frac{\eta^*}{\eta} = \frac{\xi^* F_t^*}{\xi\ F_t} = f^v \tag{1}$$

$$R_1^* = f^v R_1 \tag{2}$$

where F_t, R_1 and f^v are defined as above, C is an apparatus constant for the viscometer, ξ is the density of the solution, and * corresponds to values for water.

After correction for the viscosity effect, the observed ^{11}B longitudinal relaxation rate of BSH upon titration with protein reflects the interaction between BSH and the protein, having contributions from both bound and unbound BSH as follows:

$$R_1^{\circ} = R_1^{b} f + R_1^{u}(1-f) \tag{3}$$

where R^b and R^u are the relaxation rates for bound and unbound BSH, respectively, R° is the observed relaxation rate with albumin present after correction for viscosity, and f is the fraction of bound BSH at a chosen concentration of protein. If the relaxation rate of bound BSH can be obtained, the fraction f can be derived from this equation and the concentrations of bound and unbound BSH ($[B]^b$ and $[B]^u$) can also be obtained .

With known concentrations of bound and unbound BSH, one can evaluate the binding constant K ($mole^{-1}$) and the number of binding sites, n, on serum albumin according to the binding equation[7], assuming there is only one type of binding site:

$$\frac{[B]^b}{[B]^u[SA]_0} = Kn - K\frac{[B]^b}{[SA]_0} \tag{4}$$

(where $[SA]_0$ is the concentration of serum albumin).

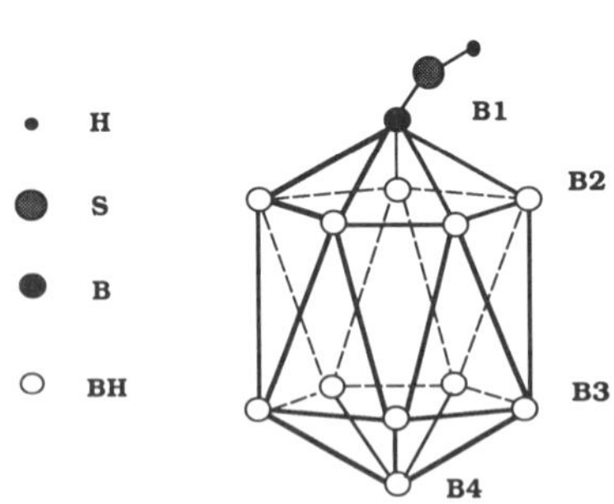

Figure 1. Structure of BSH Anion

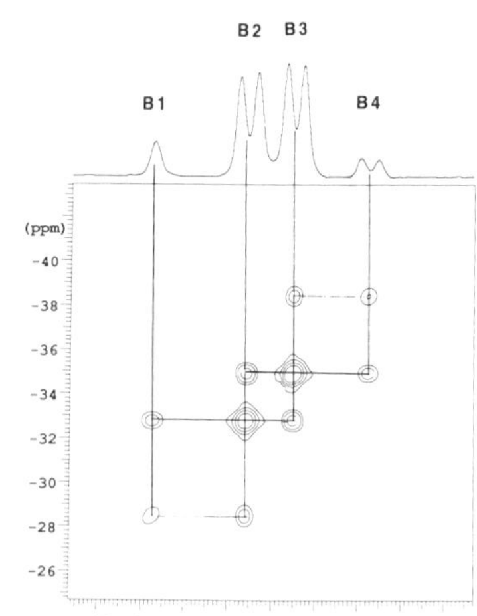

Figure 2. ^{11}B COSY spectrum. The connections among the boron nuclei are indicated.

EXPERIMENTAL PROCEDURES

All NMR studies were carried out on a Varian UNITY spectrometer (11.7 Tesla, 160 MHz for ^{11}B). The chemical shifts of four types of ^{11}B in BSH (see Fig. 1). were measured with variation of temperature and concentration of serum albumin. Three types of serum albumin - Human (HSA), Dog (DSA) and Bovine (BSA)- were used at concentrations of 1% to 5% (w/v). The proteins were purchased from Sigma and used without further purification. Saturated boric acid solution was used as an external reference for the chemical shift. The boron connectivities were determined by a ^{11}B chemical shift correlation spectrum (Fig. 2). The ^{11}B and proton coupling constant was measured. The concentration of BSH was 200ugB/mL D_2O, at boron natural abundance, in phosphate buffer at pD = 7.4. A 3000 Hz sweep width and 8k data points were used. Quartz NMR sample tubes and baseline correction were used to exclude the background interference of the borosilicate glass. The ^{11}B linewidths and longitudinal relaxation rates of BSH were studied at 295K and 310K with protein titration from 1% to 5% (w/v). Two to 3 samples were used for each protein concentration. The ^{11}B linewidth and R_1 were measured 2 to 3 times for each sample, and calculated 2 -3 times with different phase and baseline correction. The calculated values were then analyzed statistically. The ^{11}B linewidths, $W^{b,e}$, were obtained by subtraction ($W^{b,e} = W^o - W^u$) as discussed above. The viscosity effect on the longitudinal relaxation rates was corrected as follows. The flow times of water and of 1% - 4% aqueous solution of BSA were measured using an Otswald viscometer immersed in a water bath at 295K and 310K. Two mL of each sample was transferred into the viscometer. The flow time of each solution were averaged from 5 to 6 measurements. The densities of water and protein solutions were taken from literature[8]. Since the molecular weights of the three proteins are similar, the flow times measured for BSA were also used for HSA and DSA. The viscosity correction coefficient, f^V, and the longitudinal relaxation rates, R^*_1, were then calculated using equations (1) and (2).

RESULTS AND DISCUSSION

The ^{11}B chemical shifts of BSH are listed in table 1. The chemical shifts remain the same for various temperatures (295 to 310K) and varying concentrations of BSA, HSA and DSA (1% to 5%). The ^{11}B chemical shift of B1 moves 3 ppm downfield when a disulfide bridge is formed due to BSH reacting with cysteine[2]. However, the observation here indicates that there is a non-covalent binding, with no disulfide bond formed between BSH and these proteins. The observed $^1J_{^{11}B-^1H}$ is 126 Hz for all the proton bonded borons and does not change with protein titration and temperature variation which is consistent with non-covalent binding.

Table 1. The Chemical Shift* of BSH

B1	B2	B3	B4
-28.41 ± 0.08	-32.80 ± 0.04	-34.92 ± 0.03	-38.50 ± 0.00

* Saturated H_3BO_3 =0 as external reference.

Table 2 shows the subtracted linewidths, $W^{b,e}$, for B2 and B3 of BSH in the presence of 1% to 3% concentrations of HSA. The fact that $W^{b,e}$ decreases as temperature increases from 295K to 310K indicates a fast exchange between the unbound and bound BSH as discussed above. A similar behavior was observed during the titration of all three types of serum albumin, which indicates a fast exchange in the BSH binding with all these proteins.

Table 2. The Line width $W^{b,e}$ (Hz) of BSH *vs.* temperature change

	B2 1%HSA	B2 2%HSA	B2 3%HSA	B3 1%HSA	B3 2%HSA	B3 3%HSA
295K	19.83 ± 1.77	46.03 ± 2.82	61.33 ± 3.42	15.57 ± 0.55	37.97 ± 2.37	56.92 ± 2.77
310K	4.33 ± 3.22	16.33 ± 1.59	26.33 ± 7.35	1.60 ± 3.61	14.50 ± 2.51	20.47 ± 2.75

The longitudinal relaxation times of boron were measured at 295 and 310K with the titration of serum albumin. After correction for the viscosity, the longitudinal relaxation rate, R^*_1, of the four types of boron are plotted *vs.* the concentration of protein (Fig. 3). Similar patterns are observed from the plots of all three types of protein. The R^*_1 of B1 is less affected by the

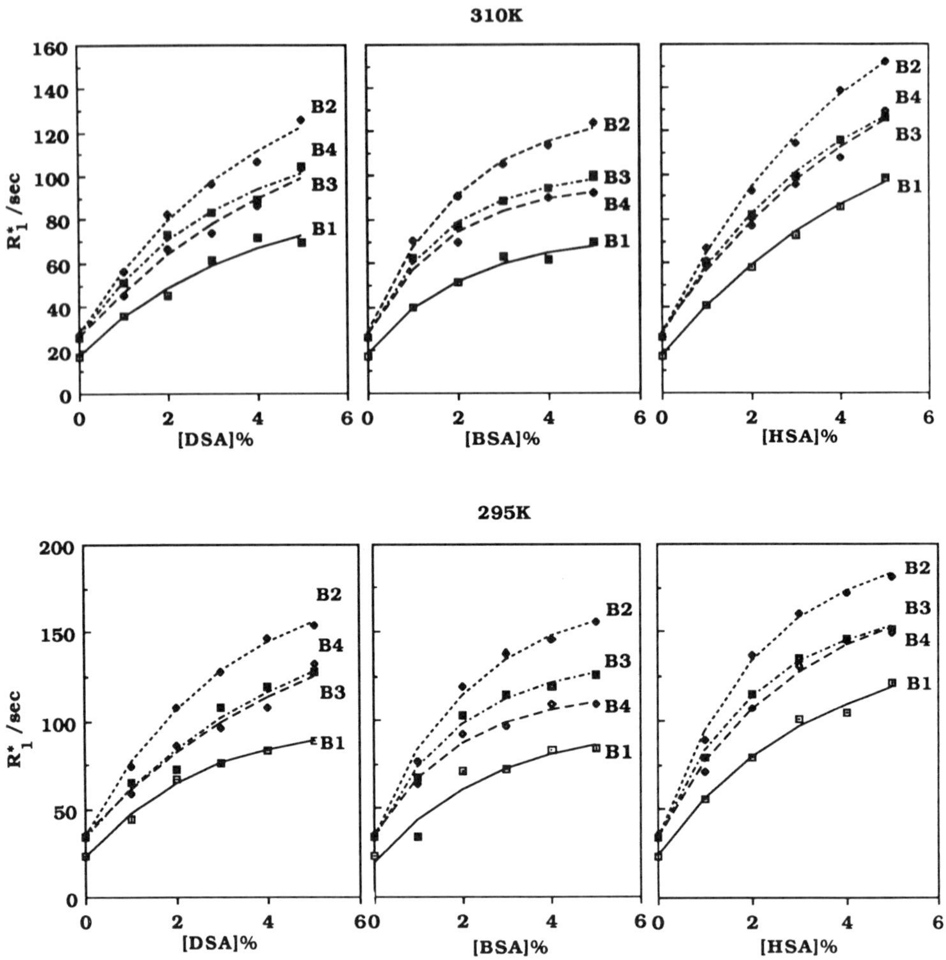

Fig 3. The longitudinal relaxation rate *vs* the concentration of the serum albumin. The points are the measured rate after viscosity correction and the lines are that calculated using the fitting function.

titration, as indicated by the lowest position in these plots. This also indicates that there is no formation of a disulfide bridge, since B1 would be most affected by such a event.

These plots also show that the longitudinal relaxation rate linearly increases with the concentration of serum albumin up to about 3%, then reaches a maximum value. This behavior may indicate that the interaction of BSH and serum albumin is saturated at about 3% protein, with the limited amount of BSH available. To prove that saturation occurs, a saturation function was used to fit the observed data:

$$Y = Y_0 - ae^{-bX} \tag{5}$$

where Y is the observed relaxation rate at albumin concentration X, Y_o is the relaxation rate as X approaches infinity, and a and b are the fitting coefficients. According to the binding equilibrium, as the concentration of serum albumin approaches infinity, the BSH can be considered as all bound to albumin, therefore Y_o corresponds to the longitudinal relaxation rate of bound BSH. With the measured relaxation rate of unbound BSH and the calculated rate of bound BSH, the binding fraction is calculated using equation 3, and the binding constant K and the number of binding sites n are evaluated using equation 4. Table 3 lists the binding parameters.

Table 3. The Binding Parameters

A. The binding fraction (f) of BSH at various concentration of serum albumin

	295K			310K		
	DSA	HSA	BSA	DSA	HSA	BSA
1%	.26 ±.01	.31 ±.04	.34 ±.03	.23 ±.02	.21 ±.02	.42 ±.03
2%	.46 ±.03	.55 ±.07	.63 ±.03	.41 ±.03	.35 ±.03	.63 ±.02
3%	.60 ±.03	.70 ±.06	.74 ±.05	.54 ±.05	.48 ±.03	.79 ± .03
4%	.71 ±.03	.79 ±.03	.83 ±.03	.64 ±.05	.59 ±.03	.87 ±.03
5%	.79 ±.02	.85 ±.05	.89 ±.03	.74 ±.02	.68 ±.03	.93 ±.01

B. The binding constant (K) and the number of binding sites (n) on albumin

	295K			310K		
	DSA	HSA	BSA	DSA	HSA	BSA
K($mole^{-1}$)	2.89	3.11	3.69	1.80	1.27	4.87
n	3.40	4.15	4.70	3.48	3.42	5.14

The obtained number of binding sites is consistent with that obtained using other methods[9].

CONCLUSIONS

1) There is non-covalent binding and no formation of a disulfide bridge between BSH and the three types of albumin.
2) The bound and unbound BSH are under fast exchange.
3) The number of binding sites on albumin is about 3 to 5 .
4) The binding constant is about 3 ($mole^{-1}$) at 295K and 1.5 ($mole^{-1}$) at 310K , which means that the binding effect decreases as temperature increases. The exception was observed for BSA, which gives a binding constant 4.87 ($mole^{-1}$) at 310K . The inconsistency may result from the purity and water content of the sample.

ACKNOWLEDGMENT

This project was supported in part by the US. Dept. of Energy, contract # DE-AC07-76ID01570.

REFERENCES

1. T. Nakagawa and T. Nagai, Interaction between serum albumin and mercaptoundecahydrododecaborate ion (an agent for boron-neutron capture therapy of brain tumor), *Chem. Pharm. Bull.* 24:2934 (1976).
2. E. G. Samsel and D. L. Miller, High resolution ^{10}B and ^{11}B nuclear magnetic resonance (NMR) spectroscopy of $Na_2B_{12}H_{11}SH$ impurities and metabolites, *Strah. Onkol.* 165:140 (1989).
3. W. F. Bauer, K. M. Bradshaw and T. L. Richards, Interaction between boron containing compounds and serum albumin observed by nuclear magnetic resonance, in 'Progress in Neutron Capture Therapy for Cancer' Ed: B. J. Allan, D. E. Moore and B. V. Harrington. Plenum Press, p 339, 1990.
4. K. M. Bradshaw, W. F. Bauer and T. L. Richards, Observation of the $Na_2B_{12}H_{11}SH$ binding to serum albumin species, presented at *the boron USA II workshop,* Research Triangle Park, NC (1990).
5. R. A. Dwek, *in* 'Nuclear Magnetic Resonance in Bio-Chemistry', Clarendon Press Oxford, (1975).
6. F. A. Bovey, *in* 'Nuclear Magnetic Resonance Spectroscopy' 2nd Ed. p3, Acad. Press, (1988).
7. A. Z. Khan and L. Aarons, Design and analysis of protein binding experiments, *J. Theo. Biol.* 140:145 (1989).
8. CRC Handbook of Chemistry and Physics, Ed. R. C. Weast, CRC Press, 60: F11 and D258, (1980).
9. W. F. Bauer, INEL Internal Report, (1991).

PURITY AND QUALITY DETERMINATIONS OF BOROCAPTATE SODIUM

A. K. Gianotto and W. F. Bauer*

Idaho National Engineering Laboratory, EG&G Idaho, Inc.
Idaho Falls, ID 83415

INTRODUCTION

The chemical purity or "quality" of any drug can influence it's action in the body. Knowledge of the chemical purity of the boron delivery agent borocaptate sodium ($Na_2B_{12}H_{11}SH$ or BSH) can have a direct effect upon it's toxicity[1], biodistribution[2] and the proper interpretation of experimental data[3]. The FDA requires that drugs seeking Orphan Drug status or an investigational new drug (IND) be chemically characterized[4]. The purity of BSH is most often determined by liquid chromatographic procedures[1,4]. The purity and "quality" of any chemical compound is only as good as the techniques used to characterize it and the number of potential impurities actively looked for in the sample. For these reasons, the somewhat more rigorous approach to purity determinations of BSH described here has been adopted by our laboratory.

EXPERIMENTAL

High performance liquid chromatography (HPLC) was performed with 20-50 μL injections onto a 300 or 250 x 8 x 4 or 3 mm column packed with 5 or 10 μm particles of Nucleosil® C_{18} (octadecyl) from Machery-Nagel. The mobile phase was a 50:50 $CH_3OH:H_2O$ solution with 5 mM TBAS ion pairing reagent (Alltech Associates, Inc.) adjusted to pH 8.1 and flowing at 1.0 mL/min with the 4 mm inner diameter columns. Detection was performed from 200-300 nm with a Waters 991 Photodiode Array Detector.

Elemental determinations of boron, sulfur, and sodium were performed by inductively coupled plasma-atomic emission spectroscopy (ICP-AES). Aliquots (150 μL) of samples and standards previously prepared for HPLC were either simply diluted or were digested with 3 mL of HNO_3 prior to the final dilutionto 10 mL. Analyses of carbon, hydrogen and nitrogen were performed on a Model EA 1108, Carlo Erba CHNS Analyzer. A nitrogen atmosphere was utilized as much as possible for preparation of the samples used in the combustion analyses to avoid additional contamination with water.

The ^{10}B and ^{11}B nuclear magnetic resonance (NMR) spectra were obtained on a Bruker Model AC300 NMR. The sample was simply diluted with water and spiked with D_2O to provide a lock signal.

Advances in Neutron Capture Therapy, Edited by
A.H Soloway *et al.*, Plenum Press, New York, 1993

Samples were prepared and handled for Fourier transform infrared (FTIR) spectroscopic analysis entirely under nitrogen as potassium bromide (KBr) pellets. The actual sample mass in each pellet was not accurately determined, consequently, the spectra were normalized to offset sample mass differences. All spectra were collected on a Digilab Model 65 FTIR.

DISCUSSION

Normally, HPLC is used to determine the "purity" of BSH salts. The primary analytes are usually the dimer (BSS) and oxidized dimer (BSSO). Typical chromatograms showing the separation of the oxidized species from the BSH parent peak are shown in Figure 1 and some actual BSS and BSSO concentrations are listed in Table 1. A variety of problems with the chromatography and it's use in determining contamination levels are also demonstrated in Figure 1. Sample AA has an obvious interference on the leading edge of the BSS peak which can be expected to bias the results high. Sample AA also has significant levels of some contaminant that elutes very late (>25 minutes) in the chromatogram. These late eluting components have not yet been identified but show up consistently in this sample and in many others to varying degrees. There are also additional contaminants that elute in and around the BSS and BSSO peak and some possibly even before the BSH parent peak. These components have not been positively identified but are believed to be byproducts of the reaction to produce BSH (see for example ref. 5 and 6).

Because of the possibility of serious contamination by other chemicals going undetected by HPLC, ICP-AES was used to determine elemental concentrations of S, B and Na in the BSH samples. Table 2 gives some typical results from these determinations. The utility of this analysis is demonstrated by sample FF which shows a noticeable problem with an excess of Na (Na/S=2.47 and B/Na=5.23). The HPLC chromatograms of FF were unremarkable, however this sample obviously contains a nonboron containing sodium salt.

The B/S ratios in Table 2 should be at or near 12, however these ratio's are typically between 12 and 13, possibly partially due to a slight bias (1-2%) in the B

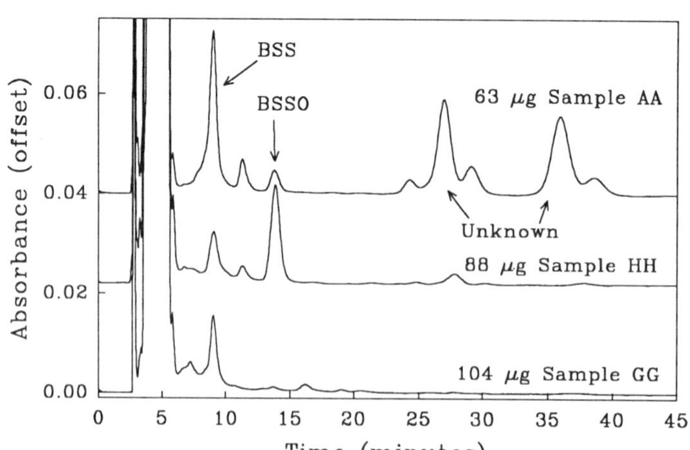

Figure 1. Chromatograms of BSH samples. 20 μL injection into 5 mM TBAS in 50:50 $CH_3OH:H_2O$ flowing at 1 mL/min. Chromatogram plotted at 215 nm.

Table 1. Typical results of HPLC analyses for BSS and BSSO contamination of BSH salts.

Sample	Compound[a]	BSS^{4-} (wt%)	$BSSO^{4-}$ (wt%)	$B_{12}H_{12}^{2-}$ (wt%)[b]
AA	$Na_2B_{12}H_{11}SH$	0.9 ± 0.1	0.097 ± 0.004	ND
BB	$Na_2{}^{10}B_{12}H_{11}SH$	0.07 ± 0.02	0.277 ± 0.009	ND
CC	$Na_2{}^{10}B_{12}H_{11}SH$	1.1 ± 0.1	0.34 ± 0.01	~37
DD	$Cs_2B_{12}H_{11}SH$	0.35 ± 0.05	0.29 ± 0.02	~ 3
EE	$Cs_2B_{12}H_{11}SH$	0.40 ± 0.06	0.14 ± 0.02	< 1
FF	$Na_2{}^{10}B_{12}H_{11}SH$	0.09 ± 0.02	ND[c]	ND
GG	$Na_2{}^{10}B_{12}H_{11}SH$	0.23 ± 0.05	ND[c]	ND
HH	$Na_2{}^{10}B_{12}H_{11}SH$	0.20 ± 0.07	0.35 ± 0.02	

ND = Not detected
[a] All concentrations have been corrected for isotopic abundance.
[b] Single chromatograms with a single $B_{12}H_{12}^{2-}$ standard.
[c] Not detected at less than ~0.05 weight percent.

determinations. The B/S and Na/S ratios of sample CC are excessively high even though the B/Na ratio is near normal. This is an obvious indication of a boron containing sodium salt, most likely to be $Na_2B_{12}H_{12}$. A closer look at the HPLC chromatogram results indicated that the $B_{12}H_{12}^{2-}$ anion may have been present but eluted just slightly before, but was not completely resolved from the BSH peak. Since the HPLC conditions used to determine BSS and BSSO were unable to separate $B_{12}H_{12}^{2-}$ from BSH, the conditions were adjusted and the chromatograms of CC in Figure 2 were obtained. The presence of $B_{12}H_{12}^{2-}$ was confirmed. The utility of the photodiode array detection is also demonstrated in Figure 2. Both chromatograms were from a single run and the increased detectibility of $B_{12}H_{12}^{2-}$ at 205 nm could be exploited. The NMR spectra in Figure 3 also indicates the significant $B_{12}H_{12}^{2-}$ contamination of sample CC. Most NMR spectra of BSH show a slightly larger peak than expected peak at δ -15 ppm. Could this peak height enhancement be due in part to non BSH contaminants?

Table 2. Mole ratios determined from elemental analyses by ICP-AES.

Sample	Compound	Mole Ratios		
		B/S	B/Na	Na/S
AA	$Na_2B_{12}H_{11}SH$	13.3	6.4	2.1
BB	$Na_2{}^{10}B_{12}H_{11}SH$	12.7	6.5	2.0
CC	$Na_2{}^{10}B_{12}H_{11}SH$	19.1	5.2	3.7
DD	$Cs_2B_{12}H_{11}SH$	13.2 ± 0.2		
EE	$Cs_2B_{12}H_{11}SH$	12.6 ± 0.3		
FF	$Na_2{}^{10}B_{12}H_{11}SH$	12.83 ± 0.06	5.23 ± 0.06	2.47 ± 0.06
GG	$Na_2{}^{10}B_{12}H_{11}SH$	12.8 ± 0.6	6.4 ± 0.1	2.0 ± 0.1
HH	$Na_2{}^{10}B_{12}H_{11}SH$	13	6.5	2.0

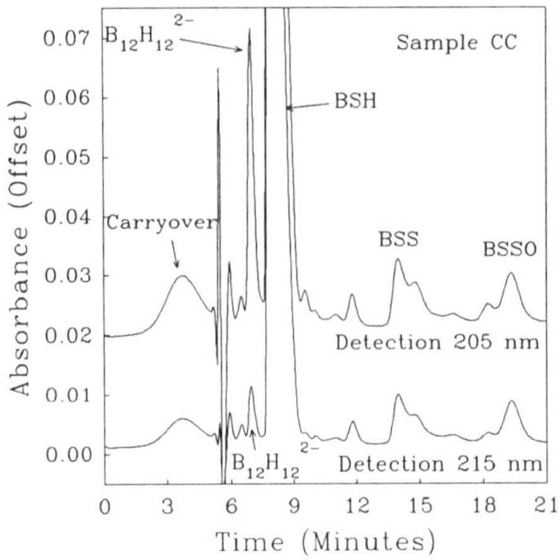

Figure 2. HPLC showing major contamination of CC with $B_{12}H_{12}^{2-}$. Conditions same as in Figure 1 but flowing at 0.5 mL/min.

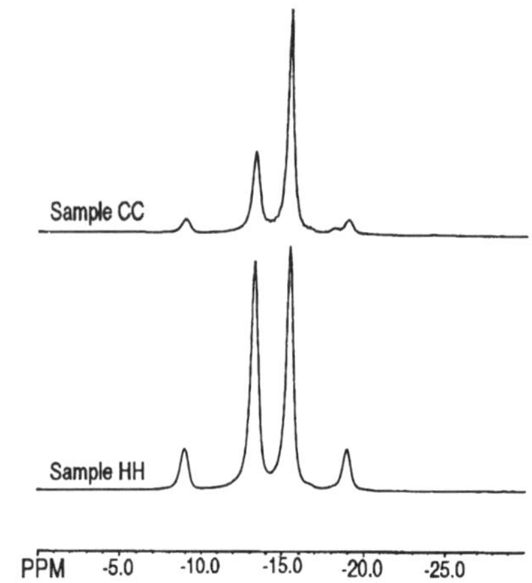

Figure 3. ^1H decoupled ^{10}B NMR spectra for ^{10}BSH (HH) and ^{10}BSH with significant $^{10}B_{12}H_{12}^{2-}$ (CC).

Many of the potential contaminants are organic in nature such as reagents used in the synthesis and some intermediate products. Elemental CHN analysis should point out significant contamination of BSH with these materials. The CHN analysis results indicated that sample FF had a relatively high level of C contamination (~ 1.7 wt% C) and sample EE had excessive contamination by an organic material (~5.89 wt% C). Subsequent Fourier Transform Infrared (FTIR) spectroscopy (Figure 4) confirmed the presence of an organic solvent in sample EE that was not identifible by HPLC and indicated that an unidentified hydrocarbon was present in FF. Also notable in the IR spectra are the obvious presence of significant water in samples FF and AA while water is absent in sample EE.

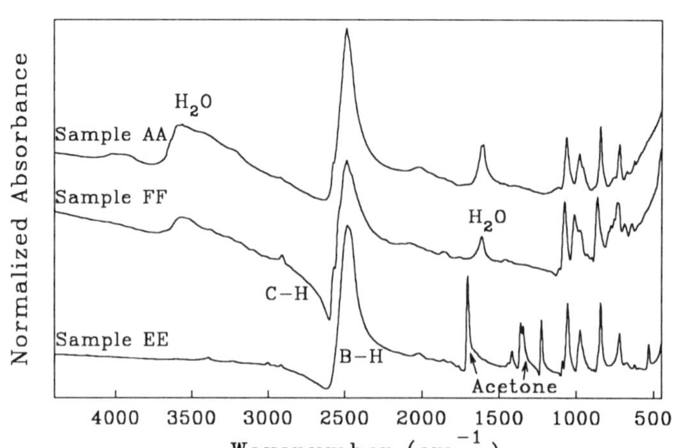

Figure 4. Infrared spectra of BSH samples obtained from KBr pellets.

CONCLUSIONS

The value of a multitechnique approach to the chemical characterization of BSH salts has been demonstrated. Liquid chromatography may be used to quantitatively determine BSS and BSSO as well as to identify the presence of other unknown compounds. Elemental analysis may be used to confirm stoichiometry and can ultimately identify some impurities such as other salts containing B, S, or Na. Spectroscopic such as NMR and FTIR techniques can not only identify the presence of major contaminants but will also provide confirmation of the chemical nature of the contaminant. With the use of all of the analytical techniques described here, some unidentified impurities have been discovered and these impurities are known to exist in much of the Na_2BSH used for BNCT research.

ACKNOWLEDGEMENTS

This work was performed under the auspices of the U.S. Department of Energy, Office of Energy Research, under DOE Field Office, Idaho, Contract No. DE-AC07-761D01570.

REFERENCES

1. T. Yoshizaki, I. Ikechi, and K. Nagasawa (1986) in *Neutron Capture Therapy*, ed. H. Hatanaka (Nishimura Co., Ltd., Niigata, Japan), pp. 394-400.
2. D. Slatkin, P. Micca, A. Forman, D. Gabel, L. Wielopolski, and R. Fairchild (1986) *Biochemical Pharmacology* **35**, 1771-1776.
3. A.H. Soloway, F. Alam, and R.F. Barth in *Workshop on Neutron Capture Therapy*, eds. R.G. Fairchild and B.P. Bond, (Brookhaven National Laboratory, Upton), BNL-51994, pp162-172.
4. J.M. Shremmer, D.J. Noonan, J.L. Russell, Jr., D.C. Johnson and T.Z. Polta (1986) in *Neutron Capture Therapy*, ed. H. Hatnaka (Nishimura Co., Ltd., Niigata, Japan) 388-393.
5. E. I. Tolpin, G.R. Wellum and S.A. Berley (1978) *Inorg. Chem.* **17**, 2867-2873.
6. B. Grüner, Z. Plzák, and I.Vinš (1991) *J. Chrom.* **588**, 201-210.

DETERMINATION OF p-BORONOPHENYLALANINE IN BIOLOGICAL TISSUES WITH HIGH PERFORMANCE LIQUID CHROMATOGRAPHY

K. Yoshino[1], T. Takasoh[1], H. Hatanaka[2], F. Komada[3], K. Okumura[3], Y. Mishima[4], C. Honda[4], and Y. Mori[1]

[1]Department of Chemistry, Faculty of Science, Shinshu University, Matsumoto 390, Japan
[2]Hyogo Prefectural Institute of Public Health, Kobe 652, Japan
[3]Department of Hospital Pharmacy, Kobe University, Kobe 650, Japan
[4]Department of Dermatology, School of Medicine, Kobe University, Kobe 650, Japan

INTRODUCTION

Many ways of determining ^{10}B in biological tissues have been developed[1-4]. These methods are adequate for the trial of neutron capture therapy, because ^{10}B concentration and tumor/blood ^{10}B ratio are key points for success. But, we want to know the concentration of ^{10}B-compound itself in tissues to answer following questions.

1. Whether ^{10}B-compounds are stable after being administered into the human body? Do the ^{10}B compounds decompose in human blood?
2. After reaching malignant tissue, does the ^{10}B-compound remain unchanged or does it change to another compound? For example, if p-Boronophenylalanine (p-BPA) is incorporated in the protein of melanoma cells, the concentration of p-BPA will be lower than the total ^{10}B concentration.

In this paper, we will describe an analytical method of determining p-BPA in tissues using High Performance Liquid Chromatography (HPLC).

EXPERIMENTAL METHODS

Tissues (1 g) were homogenized 0.4 M perchloric acid (3 mL) solution and chloroform (3 mL) by shaking for 1 min. The mixture was centrifuged. The aqueous phase was applied to a SEP-PAK C_{18} cartridge (Waters). The cartridge was washed with water (2 mL) and p-BPA was eluted with 2 M ammonia solution (2 mL). The eluate was evaporated till dry. The residue was dissolved in 0.1 M HCl solution (1 mL) with phenylalanine as an internal reference. The obtained solution (20μL) was applied to a HPLC column ($5C_{18}$-AR, 0.46X25cm, NAKARAI), and phosphate buffer (pH 3.1) served as the mobile phase. Boron in the organic phase after the homogenization was analyzed by the alkali fusion-methylene blue method[4].

RESULTS AND DISCUSSION

The pretreatment procedure for tissues has been developed by Hatanaka et al. and described in a literature[5]. In this process, there seemed to be a question that whether p-BPA remained in aqueous perchloric acid solution without being extracted to the organic phase. We have checked this point. First we have carried out the homogenization procedure without tissue sample and measured the UV spectrum of both aqueous and organic phases. As a result, the UV peaks of p-BPA were observed in the aqueous phase and no peaks of p-bpa were observed in the organic phases.

Another possibility is that p-BPA is attached to the decomposed tissues during homogenization and extracted into the organic phase with them. To obtain an answer we have analyzed boron in the organic phase by the alkali fusion methylene blue method. In the experiments, the organic phase was washed with water for three times, and collected by dissolving it with methanol. As shown in Figure 1, the absorbances obtained for the 1 mL blood samples with different amounts of p-BPA up to 12 ppm boron were as same as those for the reagent blanks, which means no boron was found in the organic phase.

Next, we have examined an effect of the standing time after mixing p-BPA with blood on the extraction of p-BPA into the organic phase. By this experiment, we can check whether p-BPA changes to another compound that

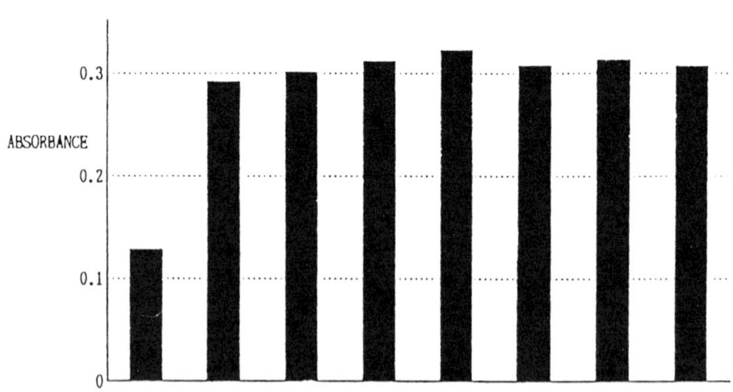

	BG1	BG2	BG2	0ppm	3ppm	6ppm	9ppm	12ppm
	0.129	0.292	0.302	0.312	0.323	0.308	0.313	0.307

Figure 1. Alkali fusion-methylene blue assay results for organic phase after homogenization of 1 mL blood sample with different amounts of p-BPA. BG1 is the reagent blank of this method and BG2 is the reagent blank including the reagents for homogenization.

Table 1. Standing time effect after elution by ammonia on the peak area of p-BPA

Run	Blood(cm3)	p-BPA(Bμg)	time (min.)	Area
1	1	4	0	17017
2	1	4	20	15177
3	1	4	40	13401
4	1	4	60	12665
5	1	4	80	10164

can be extracted into the organic phase. The absorbances obtained for the 1 mL blood samples with the different mixing times up to 5 hours were as same as those of the reagent blanks, which means that p-BPA does not change to the extractable compound.

The second check point for the pretreatment is the effect of ammonia on the decomposition of p-BPA. Ammonia was used as an elution solution for the mini column (SEP-PAK). The effect of the standing time after the elution on the peak area of p-BPA was investigated. As shown in Table 1, the peak area in HPLC decreased with the standing time. Snyder et al. reported[6] that refluxing of p-BPA (0.5 g) in 5 % sodium hydoroxide solution (20 mL) resulted in the formation of phenylalanine. Therefore, the decreasing peak area of p-BPA in HPLC indicates that the B-C bond of p-BPA was hydrolysed in ammonia solution.

Calibration Curve and Standard Deviation: A typical example of the HPLC chart for blood sample was shown in Figure 2. The calibration curve for 1 mL blood was shown in Figure 3. The peak areas are corrected for the

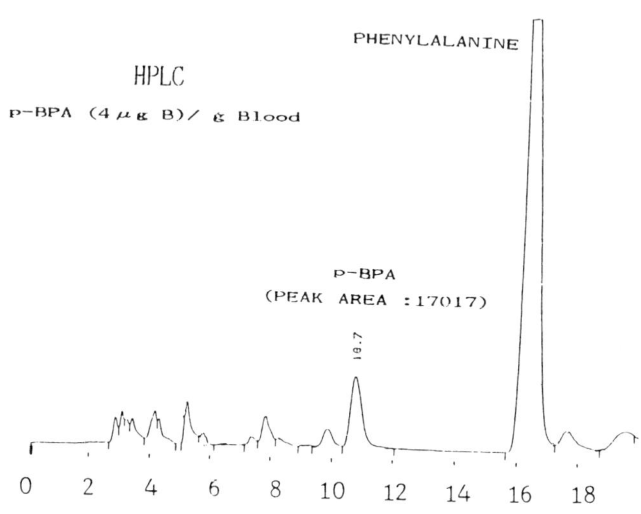

Figure 2. A typical example of the HPLC chart for blood sample. Flow rate: 1 mL/min., Temp.: 25°C, Detection: 210nm. The peak at 10.7 min.: p-BPA and that at 16.4 min.: phenylalanine (internal reference).

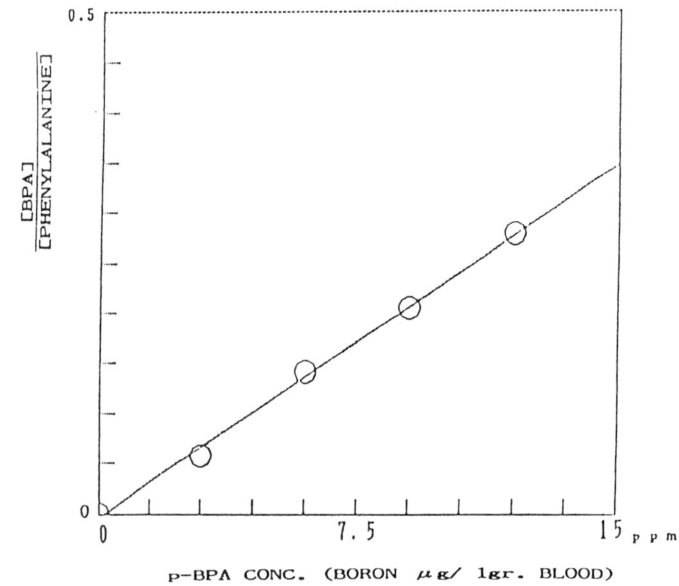

Figure 3. Calibration curve for HPLC analysis of p-BPA in blood.

cleavage of p-BPA in ammonia solution. The vertical line corresponds to the p-BPA/phenylalanine peak ratio. The calibration curve was a straight line with the correlation factor of 0.9989. The standard deviation for 5 samples of p-BPA (4μg B) in 1mL blood was 3.6%.

The HPLC method has enough sensibility for the tissue samples after administration of p-BPA into a patient. It will be possible to investigate whether p-BPA itself or different boron species accumulates in the malignant cells by determining p-BPA with the HPLC method and analyzing boron with other boron-measurement methods.

REFERENCES

1. I. Ikeuchi, and T. Amano, A colorimetric determination of boron in biological materials, *Chem. Pharm. Bull.* 26: 2619-2623 (1978).
2. T. Kobayashi, and K. Kanda, Microanalysis system of ppm-order ^{10}B in tissue for neutron capture therapy by prompt gamma-ray spectrometry, *Nuclear Instruments and Methods* 204: 525-531 (1983).
3. R.F. Barth, D. M. Adams, A.H. Soloway, E.B. Mechetner, F. Alam, and A.K.M. Anisuzzaman, Determination of boron in tissues and cells using direct-current plasma atomic emission spectroscopy, *Anal. Chem.* 63: 890-893 (1991).
4. K. Yoshino, C. Nishio, E. Ubukata, T. Maruyama, Y. Mori, Y. Mishima, C. Honda, M. Shiono, N. Wadabayashi, M. Ichihashi, T. Kobayashi, and K. Kanda, Rapid spectrophotometric determination of boron in biological tissue with alkali fusion decomposition. in:"Progress in Neutron Capture Therapy for Cancer", B.J. Allen, et al., ed., Plenum Press, New York, 309-312,(1992).
5. H. Hatanaka, F. Komada, M. Shiono, Y. Mishima, and K. Okumura, "Tissue distribution of p-boronophenylalanine administered orally as a cyclodextrin inclusion complex to melanoma-bearing hamsters, *Pigment Cell Research* 5: 38-40 (1992).
6. H.R. Snyder, A.J. Reedy and W.J. Lennarz, Synthesis of aromatic boronic acids. Aldehydo boronic acids and a boronic acid analog of tyrosine, *J. Am. Chem. Soc.* 80: 835-838 (1958).

LARGE ANIMAL STUDIES ON THE USE OF BNCT

FOR THE TREATMENT OF BRAIN TUMORS

P.R. Gavin, C.E. DeHaan, M.P. Moore, J.P. Weidner, C.D. Swartz,[1] S.L. Kraft,[2] C.A. Atkinson, C.R. Amaro, W.F. Bauer,[3] and A. Siefert[4]

[1]Washington State University, Pullman, WA, [2]Kansas State University, Manhattan, KS, [3]Idaho National Engineering Laboratory, Idaho Falls, ID, [4]Joint Research Centre, Petten, The Netherlands

INTRODUCTION

Animal models are useful to study the relative effects of BNCT on normal tissues and tumors. True efficacy studies of the modality on glioblastoma or other human malignancies requires human clinical trials. The use of large animals, primarily dogs, to study the effects of BNCT has been and continues to be of major interest. The use of large animals permits the study of normal tissue tolerance of the tissues of the head at an acceptable total body dose. These studies have been extended from normal dogs to dogs with induced and spontaneous tumors. While induced tumors in rodents are often used to study tumor response, the total body dose accompanying BNCT severely limits their use for normal tissue tolerance studies, especially with the epithermal-neutron beams.

LITERATURE REVIEW

Previous large animal model studies were done primarily with dogs, BSH, and thermal-neutron beams.[1-4] The reported tolerance of the normal canine brain was in the 35-90 Gy range, far exceeding the known tolerance of the canine brain of 15 Gy from a single fraction of megavoltage x-rays.[5] The authors reported the geometric sparing effects to the brain capillary endothelial cells may be responsible for the tolerance observed at these high physical radiation doses. However, the "contaminant" gamma and neutron irradiation should have exceeded tolerance. Several compounding factors must be considered. Some of the studies were done through the intact skin and overlying musculature, while some had the superficial tissues reflected to allow the thermal-neutron beam better penetration into the brain. Even with the reflected studies, the volume of the brain irradiated to these high doses would be small due to the extremely rapid fall-off of the thermal-neutron beam. Boron measurement accuracy and neutron dosimetry have also improved considerably since these irradiations were performed. In addition, the pharmacokinetics of the BSH compound have since been well defined following a single administration of intravenous BSH.[6] It should also be recognized that the peak physical dose assumes this dose is delivered to a volume large enough to obtain charged particle equilibrium. Therefore, BNCT doses as quoted could only be achieved within the lumen of vessels measuring 20 μ in diameter or greater. The capillary (5-7 μ diameter) dose would be substantially less.[7]

Studies using thermal- and fast-neutron beams and BNCT following BSH administration have been performed using swine.[8] Swine skin is an excellent model for human skin. However, the authors reported difficulty in obtaining a relative biological effectiveness (RBE) for the boron capture reactions due to insufficient knowledge of the BSH distribution in the skin of swine at that time.

Advances in Neutron Capture Therapy, Edited by
A.H Soloway *et al.*, Plenum Press, New York, 1993

An interesting study was performed in neonatal beagles bearing an implanted, viral-induced, intracerebral tumor.[9] There was a suggestion from the results that the BNCT was successful in eradicating some of these tumors. Unfortunately, this is a highly immunogenetic tumor and spontaneous regression was noted in the control group. Numbers of animals treated versus control were insufficient to establish statistical significance. In the treated group, there were 5 animals without residual tumor at necropsy and subsequent histopathologic examination. Two of the animals lived beyond 250 days and were sacrificed with no apparent complications. Two animals died 20 and 33 days post-treatment with no residual tumor but a suggestion of normal tissue damage. One animal that was not normal, and was sacrificed at 23 days post-treatment, also showed no residual tumor. This experiment would indicate that the treatment was successful in eradicating the tumors from these animals. The radiation dose to the normal tissues may have been excessive in some animals. The use of induced immunogenetic tumors in dogs is most appropriate for short-term studies. Due to the problem with spontaneous remission in some animals, the model is not well suited to look at tumor control and delayed normal tissue reactions that generally occur 5-6 months following treatment.

CURRENT STUDIES

Currently there are 2 active large animal BNCT programs--one, the Idaho National Engineering Laboratory (INEL) sponsored Washington State University (WSU) group using the Brookhaven Medical Research Reactor (BMRR) epithermal beam, and the other, a European community group using the Petten epithermal beam. The INEL group has irradiated 60 normal laboratory dogs to peak physical doses of 0-64.5 Gy with epithermal neutrons alone or after the intravenous infusion of BSH. The European group has irradiated 15 dogs to date with blood-boron concentrations of 0-62 μg/gm and total peak physical doses of 6.5-23 Gy including a 4-fraction study.

Prior to the normal tissue tolerance studies, a large study involving the pharmacokinetics of BSH in normal dogs and dogs with spontaneous brain tumors has been done.[10] BSH acts in a predictable, 2-compartmental model during the first 10 hours post-administration. Boron concentrations in the normal tissues can be accurately predicted from blood-boron measurements.[11] Most tissues of the head have similar concentrations on a μg basis with the exception of the normal brain. BSH appears to be excluded by the normal blood-brain barrier, and the boron within the normal brain can be largely accounted for by 4% blood volume.

Tumor concentrations in dogs with spontaneous tumors have been measured primarily at 3 post-administration time intervals.[12] The time intervals studied were 2, 6, and 12 hours following the end of infusion. All animals received 50 mg boron per kg body weight BSH intravenously at a rate of 1 mg boron per kg body weight per minute. The mean tumor concentration was generally less than the blood concentration. Improvements in tissue boron analysis allowed tumor samples as small as 10 mg to be accurately determined. A large degree of heterogeneity of the tumor blood-boron concentrations was detected.

The pharmacokinetic data allowed normal tissue concentrations to be readily determined from blood-boron concentrations, sampled following intravenous BSH administration, prior, during, and after irradiation. The INEL group utilizes a 5x10 cm portal centered over the right hemisphere of large retriever-type dogs. The mean target blood-boron concentrations during irradiation were 0, 25, 50, or 100 μg/g. The European group utilizes an 8 cm diameter incident beam centered over the brain of beagles. The dogs irradiated Petten have a mean target of blood-boron concentration of 0, 25, or 50 μg/g. Both studies utilize a single dorsal portal. As mentioned above, the Petten group is also looking at the effect of fractionation utilizing 4 fractions in some animals; all other animals were treated with a single fraction.

To date, the WSU dogs have been monitored for up to one year post-irradiation. The total body radiation effect can be detected by changes in the complete blood count.[13] The changes in the circulating white blood cells could be due to the blood-pool irradiation that would occur during these lengthy irradiations. However, the depression of the platelets would indicate significant total-body bone marrow irradiation. Platelet counts as low as 40,000/mm^3 have been observed. All changes in the blood count monitored to date have been responsible for no significant clinical problems, and have returned to normal values within 40 days post-irradiation. The Petten group has noted similar changes in the complete blood counts.[14]

Skin tolerance has been studied by both groups. The WSU dogs have reactions that have varied from mild epilation and hair color change to dermal necrosis.[15] The group at Petten, to date, has only observed epilation, depigmentation, and dry desquamation.[14] The higher range of doses used in the WSU dogs revealed some animals with a biphasic skin reaction. The initial epithelial reaction peaked at 3 weeks post-irradiation. In the highest dose groups, this reaction consisted of total epilation, depigmentation, and a moist desquamation. These areas healed and became re-epithelialized over the next several weeks. At 3 months post-irradiation, dermal necrosis ensued. The nature of the epithermal beams prevented surgical resection of the necrotic tissue and even prevented successful grafting techniques containing their own vascular supply. The BMRR reactor is an isotropic beam and the margins of the irradiation field become indistinct after scattering in the patient. While the moist desquamation in the early phase and the dermal necrosis in the later phase occur within the incident beam, significant irradiation spreads from the incident beam and easily-recognized histologic damage to the dermis was detected throughout the dorsal region of the head. This loss of sharp beam edges is readily seen in hair color changes.[15] In contrast, the Petten beam is a collimated beam and the incident 8 cm field had the majority of the change and there was an additional 1 cm rim with milder change. The remaining edge is rather sharply defined.[14] While this 1 cm rim increases the easily recognized change by 50% in volume, the remainder of the head and skin appear unaffected. At this time, it should be emphasized that dermal necrosis, while especially devastating in BNCT, should not be a problem when patients are irradiated with blood-boron concentrations between 25 and 100 ppm. At those blood-boron concentrations, brain necrosis occurs at a dose significantly below that required for dermal necrosis. The dermal necrosis observed was so severe that the animals could not be maintained long enough to ensure changes would occur within the CNS. However, groups with physical doses well below that seen in the dermal necrosis animals had the CNS effects mentioned below.

The irradiation effects on the central nervous system, in both WSU and Petten dogs, ranged from an elevated cerebral spinal fluid protein at 3 weeks post-irradiation to ischemic necrosis as early as 19 weeks post-irradiation. The ischemic necrosis seen at 5-6 months post-irradiation was rapidly progressive, could not be alleviated with supportive medical care, and resulted in the animal being humanely terminated.[14,16]

There is an inverse relationship between the blood-boron concentration and the peak physical dose required for an isoeffect.[17] Evidence to date clearly indicates that there is significant "dose sparing" to the capillary endothelium from the boron fission fragments following neutron capture. The results from epithermal neutron irradiation alone, and with boron capture, have been used to arrive at RBE estimates for the fast-neutron component, and the compound effect (the product of the microdosimetry and RBE) for BSH. The low gamma dose rate may result in a dose reduction factor for the gamma contamination.[18] These factors will require further study for better understanding. An excellent example of the dose sparing effects is illustrated in Table 1. The 5 dogs irradiated at 50 ppm blood-boron concentration at 27 Gy showed no magnetic resonance changes at 6 weeks post-irradiation. Four of 4 dogs irradiated at the same peak physical dose at 25 ppm blood-boron concentration had readily demonstrated magnetic resonance changes at 6 months.

Table 1. Six month post-irradiation MRI

Treatment Parameters	Dog Number	MR Brain Changes T1	T2	T1+Gd
	2314	0	0	0
50 ug/g Boron @ 27 Gy	2452	0	0	0
	2746	0	0	0
	2763	0	0	0
	3290	0	0	0
	2311	2	2	2
25 ug/g Boron @ 27 Gy	3300	1	2	1
	2320	2	2	1
	2465	1	2	2

T_1 = TR 400 TE 30
T_2 = TR 2000 TE 80
Gd - Gadolium-DTPA contrast agent
Relative MR change: 0 = no change, 1 = mild, 2 = moderate, 3 = severe
Treatment parameters: boron - mean blood boron-concentration
Gy = total physical radiation dose at the thermal peak

The currently used RBEs and compound factors for BSH appear to be useful in predicting the outcome from experimentation as indicated in Table 1. In addition, these factors have been used to arrive at the dosage groups utilizing the Petten reactor in their dog irradiation studies. The peak dose takes into account the different beam parameters, and size of the dog, which will influence the depth of the peak radiation dose. Due to the complexity and number of factors involved, it should be emphasized that the totality of the factors involved appear to be predictive. Any individual item is not necessarily correct, and a change in any individual factor would necessitate changes in the other components. Figure 1 indicates the various equivalent doses used to predict the response from 4 dogs. Two of these dogs (#997 & 946) were irradiated at BMRR and 2 dogs (#976 & 504) at Petten. These factors would have predicted that 3 dogs (#976, 504 & 946) would have exceeded the threshold dose of 14.8-15 Gy that would result in a 50% lethality following gamma or megavoltage x-ray irradiation. It would have predicted that one animal was slightly beneath that level. The experimental evidence confirms that the same 3 of 4 animals were affected with a lethal ischemic brain necrosis, and the animal predicted to be spared showed only sublethal, clinically quiescent change that was only detected on routine magnetic resonance examination. Figure 2 illustrates the MRI of ischemic necrosis seen that is fatal in dogs at 5-6 months post-irradiation and Figure 3 is representative of the MRI of sublethal clinically quiescent change seen in some animals at the 6 month post-irradiation routine examination.

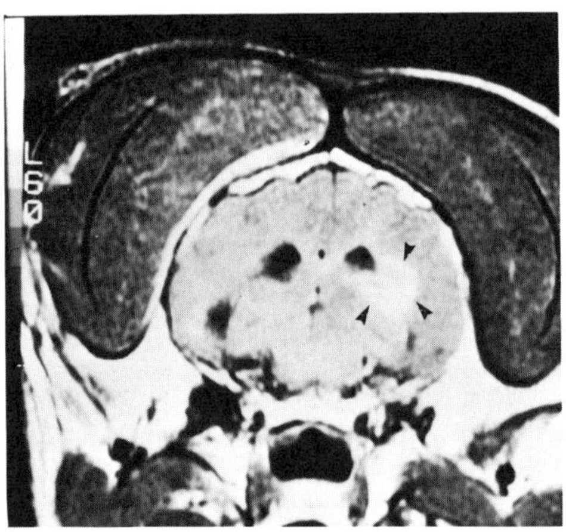

Figure 1. Equivalent Peak Radiation Dose. Dogs 976 and 504 were irradiated at Petten and dogs 997 and 946 at BMRR. The RBEs or compound factors used were: total gamma and other = 1; N (n,p) = 2.7; fast neutron = 4.5; and ^{10}B (n,α) = 0.33. The reference line at 14.8 eq Gy represents the LD_{50} for brain necrosis from a single exposure of x-rays.[5]

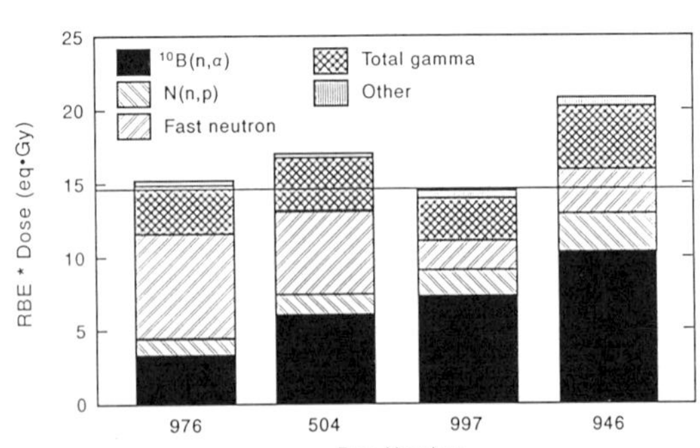

Figure 2. Lethal brain necrosis. Transverse section T_1 weighted MRI following contrast administration of a dog that developed fatal radiation necrosis 6 months post irradiation. Arrow indicates the contrast enhanced area that was histopathologically confirmed to be a hemorrhagic necrotic area.

SPONTANEOUS TUMOR DOG RESULTS

Twenty-one (21) dogs with spontaneous brain tumors have also been irradiated with the BMRR epithermal beam following intravenous infusion of BSH (Table 2). These studies were an extension of the normal tissue tolerance study to detect any change in normal tissue tolerance that might occur in animals with spontaneous brain tumors. The study also measured objective changes within the tumor at acceptable normal tissue injury levels. All dogs were treated with a single, dorsal, 10x10 portal regardless of tumor location. The total peak physical dose varied from 16-26 Gy and the mean blood-boron concentration during irradiation varied from 14-85 μg/g. The post-irradiation followup period varied from hours to greater than 3 years. No known acute toxicities have been observed. The 2 animals that survived only hours following administration of the compound and epithermal irradiation might indicate an acute toxicity. However, no histopathologic evidence of acute toxicity was detected and it is felt that the deaths

Table 2. Canine Tumor Treatment Summary

DOG	SEX/ AGE	DIAGNOSIS	SURVIVAL DAYS	CAUSE OF DEATH	^{10}B AVG. (ppm) TARGET	MEAS.	DOSE (cGy) TARGET	CALC.	TOLERANCE EXCEEDED
50 mg ^{10}B/kg/body weight (BSH administered IV)									
- 79	F/7	Oligodendroglioma	348	Died-Urinary obstruction	25	18.5	1900	1923	
- 80	M/4	Astrocytoma	12	Died-Anesthesia	25	21.6	1900	1722	
- 83	M/11	Meningioma	170	Died-Focal encephalomalacia	25	21.4	1900	1829	+
- 87	M/?	Meningioma	140	Euthanasia-PND	25	20.6	1900	1653	
- 93	F/6	Oligodendroglioma	300	Died-PND	25	24.9	1900	1912	+
- 94	F/5	Meningioma	>1022		25	14.3	1900	1887	+
- 95	M/6	Meningioma	0.5	Died-AND following irradiation	25	24.5	1900	1743	
-100	F/7	Meningioma	43	Euthanasia-PND	25	28.9	2300	2185	
-108	F/4	Choroid plexus papilloma	115	Unknown	50	51.9	1900		
-118	F/9	Astrocytoma	158	Died-PND	50	85.3	2300		
-129	F/3	Choroid plexus papilloma	386	Euthanasia-PND	50	56.9	2300	2141	
-132	M/13	Meningioma	54	Died-Surgery	50	34.5	2300		
-134	F/8	Meningioma	33	Died-PND	50	20.6	1900	1729	
-139	M/11	Meningioma	13	Died-Leg infection	50	57.5	2300		
-140	F/3	Choroid plexus papilloma or Meningioma	532	Euthanasia-PND	50	40.3	2300		
-142	F/6	Meningioma	181	Euthanasia-PND	50	54.2	2300	2708	+
-147	M/9	Malignant meningioma	102	Euthanasia-PND	50	37.1	2300		
-151	M/6	Intra-axial enhancing mass	>531		50	67.6	2300		
-152	M/10	Meningioma	99	Euthanasia-Unrelated tumors	50	60.8	2300		
100 mg ^{10}B/kg/body weight (BSH administered IV)									
-155	F/11	Meningioma	3	Died-Unidentified	100	101.6	2600		
-161	M/8	Choroid plexus papilloma	22	Euthanasia-PND	100	102.2	2600		

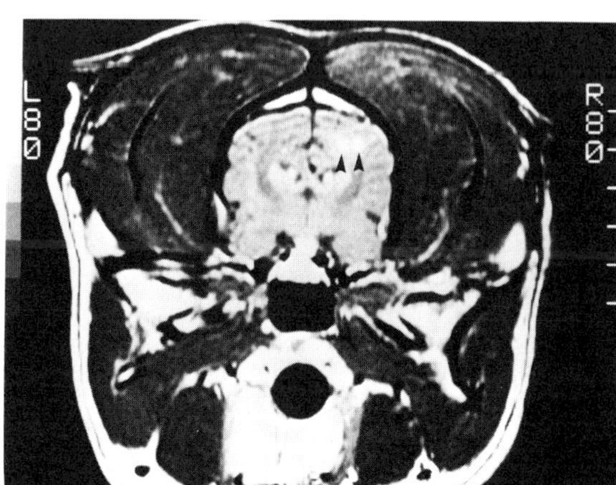

Figure 3. Sublethal brain necrosis. Transverse section of a T_1 weighted MRI following contrast administration. The areas (arrow) were detected on the 6 month post radiation exposure examination. The areas were not detected on the 12 month post exposure examination. Histopathologic examination at that time revealed a glial scar.

were attributed to the combined brain tumor and/or anesthesia required for treatment. Several animals did have changes in the peripheral blood similar to that described above, as well as elevation of the cerebral spinal fluid protein 3 weeks post-irradiation. In addition, most of the animals had epilation from partial to complete with pigmentation reversal in the hair that regrew. The most severe skin change seen was a moist desquamation in the center of the incident field. No dermal necrosis was encountered. Ischemic necrosis of the normal brain occurred in 2 animals (#83 and 142). Animal #83 had considerable peritumor edema at the time of treatment that could not be controlled or could not be eliminated with corticosteroids. It is felt that the edematous brain contained significant amounts of boron at the time of irradiation and led to ischemic necrosis in that region.[12] Dog #142 had ischemic necrosis similar to that seen in the normal dogs and would be indicative of an apparent excessive dose to the normal capillary endothelium and the calculated dose received was significantly higher than planned. Dogs #93 and 94 had magnetic resonance visible change at the 6 month post-treatment evaluation similar to that described in the sublethal changes in the normal dogs. These changes subsequently were not visible as was seen in the normal dogs at 1 year post-irradiation.

It can be seen that the BNCT as delivered in a single dose through a dorsal portal was not effective in controlling all of the tumors. The response of the tumors was at least equal to that available from conventional therapy in the dog recently reported.[19] The response to intraaxial glial tumors appeared better than that reported for the aggressive conventional therapy and included surgery, irradiation, and chemotherapy. The current studies should be viewed in a phase-I/phase-II trial perspective. Subsequent irradiation of dogs with spontaneous brain tumors will tailor the portal size and location and may involve multiple ports to achieve a better physical dose distribution than the initial studies reported above. While proper controls are difficult in spontaneous tumor work of large animals, certainly objective tumor response and long term survival would be noteworthy. It should be emphasized that of the dogs in the initial study of spontaneous tumors, only one animal was lost to followup in eventual histopathologic examination following treatment, thus providing a wealth of material to study the effect of BNCT on the tumor and normal tissues.

The dog has proven to be an excellent model to study the normal tissue tolerance to BNCT with BSH. Purpose-bred dogs and dogs with spontaneous tumors will be used to study other boron compounds with potential for BNCT. The results from the large animal study form a basis for eventual human clinical trials. The same clinical examination and treatment planning tools will be directly adapted to the human trials.

ACKNOWLEDGEMENT Work supported in part by the U.S. Department of Energy, Office of Energy Research, under DOE Field Office, Idaho, Contract No. DE-AC07-76IDO1570.

REFERENCES

1. R.G. Zamenhof, A.H. Madoc-Jones, and O.K. Harling, A clinical trial of neutron capture therapy for brain tumors at New England Medical Center and the Massachusetts Institute of Technology, Proceedings of Workshop on Neutron Capture Therapy, Cambridge, MA, January 22-23 (1986).
2. J.E. Kirsch, G.L. Brownell, R.G. Zamenhof, W.H. Sweet, J.R. Messer, E.P. Richardson, A. Tateuchi, and H. Hatanaka, Irradiation of the normal canine brain after craniectomy with the current boron neutron capture therapy regimen at MIT-reactor, in: Boron-Neutron Capture Therapy for Tumors, H. Hatanaka, ed., Nishimura Co., Ltd., Niigata (1986).
3. Y. Ushio, T. Hayakawa, K. Morimoto, H. Mogami, T. Takeuchi, K.Amano, T. Miobe, and H. Hatanaka, Pathologic study of the effect of boron-neutron capture radiation on normal brains of dogs, in: Boron-Neutron Capture Therapy for Tumors, H. Hatanaka, ed., Nishimura Co., Ltd., Niigata (1986).
4. A. Takeuchi, and H. Hatanaka, Tolerance of canine brain to the revised boron neutron capture therapy, in: Boron-Neutron Capture Therapy for Tumors, H. Hatanaka, ed., Nishimura Co., Ltd. (1986).
5. J.R. Fike, C.E. Cann, R.L. Davis, J.K. Borcick, T.L. Phillips, and L.B. Russel, Computed tomographic analysis of the canine brain: Effects of hemobrain x-irradiation, Rad Res 99:294-310 (1984).
6. S.L. Kraft, P.R. Gavin, C.E. DeHaan, W.F. Bauer, and T.E. Ary, The biodistribution of boron in normal canine tissues following borocaptate sodium administration and the effect of plasma exchange, in: Progress in Neutron Capture Therapy for Cancer, Allen, Moore, and Harrington, eds., Plenum Press, New York (1992).

7. P.R. Gavin, C.D. DeHaan, S.L. Kraft, M.P. Moore, F.J. Wheeler, and D.L. Miller, Dosimetric considerations: Radiation tolerance of the normal canine brain following boron neutron capture therapy, Proceedings of the 38th Annual Scientific Meeting, Radiation Research Society, New Orleans, Poster Abstract (1990).
8. J.O. Archambeau, R.G. Fairchild, and H.J. Brenneis, The response of the skin of swine to increasing absorbed doses of radiation from a thermal neutron beam, a degraded fission neutron beam, and the $^{10}B(n,\alpha)^7Li$ reactor, Rad Res 45:145-164 (1971)
9. C.L. Brownell, J.E. Kirsch, J.C. Murphy, M. Ashtari, W.C. Schoene, C. Rumbaugh, and G.R. Wellum, Pre-clinical neutron capture therapy trials at MIT using $Na_2B_{12}H_{11}SH$, Proceedings of First International Symposium on BNCT, Boston, MA, October 12-14, Brookhaven National Laboratory, Long Island, NY (1983).
10. S.L. Kraft, P.R. Gavin, C.E. DeHaan, C.W. Leathers, W.F. Bauer, D.L. Miller, and R.V. Dorn III, Borocaptate sodium: A potential delivery compound for boron neutron capture therapy evaluated in dogs with spontaneous intracranial tumors, Proceedings National Academy of Science, in press, (1992).
11. S.L. Kraft, P.R. Gavin, W.F. Bauer, D.L. Miller, and M.L. Griebenow, Biodistribution of boron in dogs with spontaneous intracranial tumors following borocaptate sodium administration Part II: Boron concentrations in normal canine tissues. Submitted to Rad Res.
12. S.L. Kraft, P.R. Gavin, C.W. Leathers, C.E. DeHaan, W.F. Bauer, D.L. Miller, and R.V. Dorn III, Biodistribution of boron in dogs with spontaneous intracranial tumors following borocaptate sodium administration Part I: Tumor boron concentrations. Submitted to Rad Res.
13. P.R. Gavin, C.E. DeHaan, S.L. Kraft, Y.D. Harker, P.D. Randolf, and F.J. Wheeler, Regional and total body dose following BNCT epithermal head irradiation: Biologic and dosimetric evaluation, in: Progress in Neutron Capture Therapy for Cancer, Allen, Moore, and Harrington, eds., Plenum Press, New York (1992).
14. K.H.I. Philipp, R. Huiskamp, J. Casado, A.S. Siefert, P.R. Gavin, and H.L. Moss, Developments in healthy tissue tolerance studies as a prerequisite for clinical application of BNCT with borocaptate sodium at the high flux reactor in Petten. In these proceedings.
15. C.E. DeHaan, P.R. Gavin, S.L. Kraft, F.J. Wheeler, and C.A. Atkinson, Qualitative dose response of the normal canine head to peithermal irradiation with or without boron capture, in: Progress in Neutron Capture Therapy for Cancer, Allen, Moore, and Harrington, eds., Plenum Press, New York (1992).
16. A. Siefert, J. Casado, K. Philipp, R. Huiskamp, P. Gavin, E. Duhmke, and R.L. Moss, Brain effects observed in the canine healthy tissue tolerance studies for BNCT with borocaptate sodium at the epithermal neutron beam of the HFR Petten. In these proceedings.
17. P.R. Gavin, S.L. Kraft, C.E. DeHaan, M.L. Griebenow, and M.P. Moore, A large animal model for boron neutron capture therapy, in: Progress in Neutron Capture Therapy for Cancer, Allen, Moore, and Harrington, eds., Plenum Press, New York (1992).
18. R. Huiskamp, P.R. Gavin, F.J. Wheeler, A.S. Siefert, and K.H.I. Philipp, Dose effect comparisons between HFR and BMRR irradiated dogs with respect to healthy tissue tolerance. In these proceedings.
19. G.L. Heidner, J.N. Kornegay, R.L. Page, R.K. Dodge, and D.E. Thrall, Analysis of survival in a retrospective study, J Vet Intern Med 5:219-226 (1991).

LESSONS LEARNED FROM HYPOXIC SENSITIZER AND PHOTODYNAMIC THERAPY RESEARCH OF POTENTIAL USE TO BNCT

J. Donald Chapman

Tumor Biology and Biophysics
Department of Radiation Oncology
Fox Chase Cancer Center
Philadelphia, PA 19111

INTRODUCTION

As the limits of single modality treatments of human cancer are realized, research efforts have been directed towards more-complex experimental therapies, some of which require the administration of both a chemical and a physical agent. Of the binary cancer therapies listed in Table 1, those involving hypoxic cell radiosensitizers, photodynamic therapy (PDT) and boron neutron capture therapy (BNCT) have much in common. They all require a combination of drug and physical agent in target cells for efficacy. In fact, the individual components of PDT and BNCT have little or no independent efficacy. Significant developments in all three fields have occurred over the past 20 years as evidenced by the several theme workshops and conferences devoted to each research modality.

Research in my laboratory has addressed several facets of both hypoxic radiosensitizer and PDT research. In this manuscript, lessons learned from these two cancer treatment modalities of potential interest to BNCT are presented. In particular, lessons learned about the selective drug delivery to target cells within tumors and the delivery of physical activating agents to tumor volumes are discussed. Tumor response to binary therapies can result from mechanisms which, on first analysis, might not be apparent. In particular, it is important to establish for any new cancer therapy the contribution of direct and indirect tumor cell kill for producing tumor response.

Advances in Neutron Capture Therapy, Edited by
A.H Soloway *et al.*, Plenum Press, New York, 1993

TABLE I

BINARY CANCER THERAPIES

- Hypoxic cell radiosensitizers
- Combination chemo/radiotherapy
- Photodynamic therapy (PDT)
- Immunoradiotherapy
- Boron neutron capture therapy (BNCT)

DELIVERY OF DRUGS TO TARGET CELLS IN VIVO

Research with mammalian cells identified nitroimidazoles as effective hypoxic radiosensitizers (1). The drugs tested in clinical trials to date, metronidazole, misonidazole, desmethylmisonidazole and SR-2508, are relatively low molecular weight compounds (~200 Da) with octanol/water partition coefficients (P) between 1.0 and 0.04. The limiting neurotoxicity of these drugs was less for SR-2508 compared to misonidazole resulting from the selective exclusion of the more hydrophilic compound by neurons and brain tissue (2). The intracellular concentration of delivered drug to cells depends somewhat upon the P value of the radiosensitizer (3). The pharmacokinetics of radiosensitizer in blood, tumor and normal tissues of rodents and humans has been well characterized in several studies. To date, clinical research has utilized systemic routes for radiosensitizer delivery to hypoxic cells within tumors. Unfortunately, very few of these clinical trials with misonidazole have shown radiosensitizing efficacy (4). Final analysis of the RTOG clinical trial with head and neck cancer and SR-2508 has yet to be reported. In all studies, radiosensitizer dosage was limited by toxicity and drug levels delivered to tumor cells would be expected to mimic only a small fraction of the radiobiological oxygen effect (5). As well, new studies reveal that only a fraction of human tumors of specific site and/or histology contain significant fractions of hypoxic cells (6). These and other factors probably contribute to the lack of a positive clinical demonstration of radiosensitizing effectiveness by the hypoxic sensitizers tested to date.

PDT utilizes photosensitizing compounds with quite different chemical properties. Many of the preclinical investigations and most clinical studies have been performed with Photofrin, a mixture of monomer, dimer, polymer and oligomer esters and ethers of hematoporphyrin (7). Several of the second-generation photosensitizers also contain porphyrin-related structures. These compounds have molecular weights of 550-800 Da, contain multiple sites at which to attach substituent sidechains and are, in general, extremely lipophilic. Photosensitizers have been administered both ip and iv for delivery to target cells in solid tumors. The pathways by which a specific photosensitizing drug is transported in vivo are multiple, can involve albumin, LDL and HDL as transporters and can utilize cell surface receptors for specific delivery and internalization (8). Initially it was thought that hematoporphyrin derivatives were selectively taken up by neoplastic tissue relative to normal tissues, thus providing a pharmacological selectivity for those cells requiring inactivation (9). Recent studies show that the delivery of most photosensitizers to tumor tissue is not selectively enhanced but that their efflux and clearance from solid tumors may be slower than that from normal tissue (10). The net result is that at 24-72 hr after photosensitizer administration, the relative amount of photosensitizer in tumor compared to normal tissue can be greater than 1.0 as measured by fluorescence techniques. PDT trials with cancers of the bladder and pulmonary tract deliver light, unidirectionally, to diseased organ surfaces and rely upon increased levels of photosensitizer in tumor relative to normal tissue for selectivity. Several preliminary clinical studies have shown encouraging response rates for the treatment of superficial lesions (11). One current goal of PDT research is the development of a single-structure photosensitizing compound which can be activated by longer wavelength light (670-800 nm). Research with novel photosensitizers has yet to determine which chemical properties, blood transporters and cell receptors are most important for the delivery of these lipophilic compounds to target cells in solid tumors.

As regards BNCT, lessons learned from hypoxic radiosensitizer and PDT research about drug delivery suggest that the following principles are important:

1) lipid/water solubility,
2) ionization status at physiological pH,
3) mechanism(s) of diffusion and/or transport into cells,
4) electron affinity (redox reactivity),
5) stereochemistry and shielding of reactive site(s), and
6) pro-drug formulations.

The first three properties are probably essential information for the understanding of the pharmacology and biodistribution of any class of drug, with boron-containing compounds as no exception. Chemical considerations 4-6 (above) could play a role in optimizing the delivery of some, but not all, classes of drug. The chemistry and the pharmacology of compounds containing dodecacarborate substituents are uncharted fields which demand experimental rigor.

DELIVERY OF ACTIVATING (PHYSICAL) AGENTS

Hypoxic cell sensitizers are administered sometime prior to each dose fraction prescribed by conventional radiotherapy and the physical activating agent is delivered by the various portals, fields and dose fractions to a total tumor dose as optimized by prior clinical practice. The development of novel radiation procedures was not required for the clinical testing of sensitizers.

Most experimental PDT of human tumors has utilized light delivered by lenses or diffusing tips on optical fibers. These techniques result in the unidirectional exposure of superficial tumors by specific wavelength light, specifically ~ 630 nm for activating Photofrin. The attenuation of light delivered to any tissue, including tumor, is strongly dependent upon wavelength and has been described by several mathematic models (12). The delivery of activating light to target cells within a solid human tumor poses a considerable problem. My laboratory has researched the delivery of light at various wavelengths to solid tumors by laterally diffusing optical fibers implanted in Iridium-192 brachytherapy needles. Attenuation of various wavelength light with distance from a single laterally diffusing fiber has been measured and input to a 2-dimensional photodosimetry treatment planning code to predict the homogeneity of light delivery to rat prostate carcinomas (13). The spacing of parallelly implanted fibers to produce relative light intensities of ≤15% of the mean were determined (Table 2). For Dunning R3327-AT prostate tumors, needle spacings increased from 0.7 cm to 1.3 cm for light wavelengths of 630 to 830 nm. In fact, for wavelengths of 730 nm and longer, needle spacings can be greater than 1.0 cm, the distance commonly used for current cancer treatment by interstitial brachytherapy. Thus, longer wavelength light can be delivered to tumor volumes with relative uniformity. Novel photosensitizing drugs which can be activated by these longer wavelengths might be suitable for the interstitial treatment of accessible tumors by PDT.

TABLE II

MAXIMUM INTERSTITIAL NEEDLE SPACING FOR A 7-FIBER APPLICATOR REQUIRED TO PRODUCE LIGHT DOSES WITH UNIFORMITY OF $\leq$ 15% IN DUNNING R3327-AT TUMORS (13)

Wavelength	Spacing (cm)
630	0.7
670	0.8
700	0.9
740	1.05
750	1.1
790	1.2
830	1.3

The physical agent required for BNCT is a "thermal" neutron beam. To date, these have been derived from fast neutrons generated within nuclear reactors and slowed down or moderated by appropriate scattering material. Unfortunately, most reactor channels constructed to deliver "thermal" and/or epithermal neutron beams also deliver some γ-radiation. The treatment planning of tumor volumes for neutron exposures required in BNCT can, consequently, be complex. This is a most important area for research and development in support of clinical BNCT which will benefit from some techniques developed for conventional photon therapy. There is little information in the photodosimetry principles associated with PDT of relevance to BNCT. The use of radiation physics phantoms, biological phantoms and rodent tumors will be important for the preclinical testing of various boron-containing compounds activated by various neutron sources.

CELLULAR TARGETS OF TUMOR RESPONSE IN VIVO

The mechanism of tumor response potentiation by hypoxic cell radiosensitizers has been well established by studies with cells and rodent tumors. These sensitizers must be present within mammalian cells at the time of irradiation to have activity. Upon their systemic administration to animals, hypoxic sensitizers are transported to solid tumors in the blood and diffuse, with time, into hypoxic tumor cells. Their rate of metabolism is much lower than that of molecular oxygen, which facilitates delivery to the viable hypoxic tumor cells. Upon irradiation with X- or γ-rays, these electron-affinic compounds mimic, in part, free radical reactions in which O_2 is involved (14). Radiation-induced molecular damage to cellular DNA and cell killing are enhanced (15). The mechanism of tumor cell death after irradiation with hypoxic sensitizers is indistinguishable from that of aerobic and hypoxic cells. If viable hypoxic cells in human tumors contribute to tumor radioresistance, hypoxic sensitizer therapy should provide some improvement in tumor response. Since the oxygenation status of individual human tumors is not routinely determined, appropriate clinical trials with hypoxic radiosensitizers may not, as yet, have been performed.

Mechanistic studies of cell and tumor response to PDT have been most instructive. When tumor or normal (endothelial) cells are illuminated in vitro in the presence of photosensitizing compounds, cell death is prompt (within 30 min) and appears to be mediated by membrane damage (16-18). The lipophilic nature of most photosensitizing compounds dictates their partitioning into plasma, mitochondrial and lysosomal membranes (19). Rapid damage to several membrane functions have been observed, including loss of intracellular/extracellular ion balance (20), reduction in oxidative phosphorylation (21), inactivation of several enzyme functions located in membranes (22) and damage to lysosomes (23). At somewhat longer times after treatment, cellular DNA is degraded into nucleosomal fragments, characteristic of apoptotic cell death (24). These evidences describe a mechanism of cell inactivation by PDT whose time-course and properties are qualitatively different from those of cells undergoing proliferative cell death after treatment by ionizing radiation and most chemotherapeutic drugs. Consequently, PDT research has defined another molecular target within cells by which tumor eradication might be effected.

Studies designed to characterize tumor response to PDT in vivo showed little or no direct cell inactivation after treatments which produced significant tumor response (25). Several studies from different laboratories suggest that vascular endothelial cells are the most sensitive targets within solid tumors (26) and that vascular damage, blood stasis and perfusion shutdown occur rapidly after tumor illumination (27-30). Tumor cell inactivation results from secondary ischemic death, which can be potentiated by the administration of bioreductive chemotherapy (31,32). PDT is possibly the first cancer treatment which can produce significant tumor response and cures by selectively damaging the tumor vasculature. Its research should provide insight into the potential role of tumor vasculature as a target for tumor treatment either alone or in combination with other modalities. The optimization of tumor treatment procedures by PDT will be strongly dependent upon a clear understanding of the target(s) responsible for tumor response. The illumination of neoplastic tissue 24-72 hr after the systemic administration of a photosensitizing drug may not be an optimum procedure if endothelial cell and vascular damage should be maximized for effective tumor cure.

As regards BNCT research, it is important that early research establish whether or not tumor response results from direct radiation inactivation of tumor cells or by some other secondary mechanism(s). The delivery of boron-containing compounds to solid tumors for their activation by "thermal" neutrons will require a knowledge of which cellular targets are most important for effecting tumor response.

Hopefully some information learned from hypoxic sensitizer and PDT research about drug delivery to specific cellular targets within tumor tissue and of experimental systems utilized to investigate tumor response might be of use to BNCT research. The delivery of "thermal" neutrons to target cells in solid tumors poses unique dosimetry problems which will not benefit significantly from the photodosimetry of PDT.

ACKNOWLEDGEMENTS

The light attenuation studies which generated the data of Table II were performed with Matthew Fenning and Darrell Brown. The assistance of Barbara Haagen and Anne Carson in preparing this manuscript is appreciated.

REFERENCES

1. Chapman, J.D., Reuvers, A.P. and Borsa, J. (1973) *Brit. J. Radiol.* **46**, 623-630.
2. Brown, J.M. and Lee, W.W. (1980) in *Radiation Sensitizers: Their Use in the Clinical Management*, eds. Brady, L.W. (Mason Publishing, USA) pp. 2-13.
3. Koch, C.J., Stobbe, C.C. and Baer, K.A. (1986) *Int. J. Radiat. Oncol. Biol. Phys.* **12**, 1151-1155.
4. Urtasun, R.C., Coleman, C.N., Wasserman, T.H. and Phillips, T.L. (1984) *Int. J. Radiat. Oncol. Biol. Phys.* **9**, 1691-1696.
5. Brown, J.M. (1989) *Int. J. Radiat. Oncol. Biol. Phys.* **16**, 987-993.
6. Chapman, J.D. (1991) *Radiother. Oncol. Suppl.* **20**, 13-19.
7. Kessel, D. (1990) in *Phototherapy of Cancer*, eds. Morstyn, G. and Kaye, A.H. (Harwood Acad. Publ., London) pp. 23-25.
8. Reyftmann, J.P., Morliere, P., Goldstein, S., Santus, R., Dubertret, L. and Lagrange, D. (1984) *Photochem. Photobiol.* **40**, 721-729.
9. Kessel, D. and Chou, T.H. (1983) *Cancer Res.* **43**, 1994-1999.
10. Henderson, B.W. and Dougherty, T.J. (1992) *Photochem. Photobiol.* **55**, 145-152.
11. Dougherty, T.J. and Marcus, S.L. (1992) *Eur. J. Cancer* **28A**, 1734-1742.
12. Arnfield, M.R., Chapman, J.D., Tulip, J., Fenning, M.C. and McPhee, M.S. (in press) *Photochem. Photobiol.*
13. Chapman, J.D., Fenning, M.C., Stobbe, C.C., Gatt, S., Brown, D.Q. and Curran, Jr., W.J. (in press) *Int. J. Radiat. Oncol. Biol. Phys.*
14. Chapman, J.D., Reuvers, A.P., Borsa, J. and Greenstock, C.L. (1973) *Radiat. Res.* **56**, 291-306.
15. Dugle, D.L., Chapman, J.D., Gillespie, C.J., Borsa, J., Webb, R.G., Meeker, B.E. and Reuvers, A.P. (1972) *Int. J. Radiat. Biol.* **22**, 545-555.
16. Kessel, D. (1977) *Biochemistry* **16**, 3443-3449.
17. Santus, R., Kohen, C., Kohen E. & 4 others (1983) *Photochem. Photobiol.* **38**, 71-77.
18. Reyftmann, J.P., Kohen, E., Morliere, P., Santus, R., Kohan, C., Mangel, W.F., Dubertret, L. and Kirschberg, J.G. (1986) *Photochem. Photobiol.* **44**, 461-469.
19. Santus, R. and Morliere, R. (1991) *Photochem. Photobiol.* **54**, 1071-1077.
20. Dubbelman, T.M.A.R. and VanSterenick, J. (1984) *B.B.A.* **771**, 201-207.
21. Hilf, R., Murant, R.S., Narayanan, U. and Gibson, S.L. (1986) *Cancer Res.* **46**, 211-217.
22. Hilf, R., Smail, D.B., Murant, R.S., Leakey, P.B. and Gibson, S.L. (1984) *Cancer Res.* **44**, 1483-1488.
23. Morliere, P., Santus, R., Maziere, J.C., Bazin, M., Kohen, E., Smith, K.M. and Dubertret, L. (1991) *J. Cell. Pharmacol.* **2**, 11-19.
24. Agarwal, M.L., Clay, M.E., Harvey, E.J., Evans, H.H., Antunez, A.R. and Oleinick, N.L. (1991) *Cancer Res.* **51**, 5993-5996.
25. Henderson, B.W., Waldow, S.M., Wang, T.S., Potter, W.R., Malone, P.B. and Dougherty, T.J. (1985) *Cancer Res.* **45**, 572-576.

26. Fingar, V.H., Wiemann, T.J.M. and Doak, K.W. (1990) *Cancer Res.* **50**, 2599-2603.
27. Castellani, A., Pace, G.P. and Concioli, M. (1963) *J. Pathol. Bacteriol.* **86**, 99-102.
28. Selman, S.H., Kreimer-Birnbaum, M., Klaunig, J.E., Goldblatt, P.J., Keck, R.W. and Britton, S.L. (1984) *Cancer Res.* **44**, 1924-1927.
29. Star, W.M., Marijnissen, H.P.A., Van Den Berg-block, A.E., Versteeg, J.A.C., Franken, K.A.P. and Reinhold, H.A. (1986) *Cancer Res.* **46**, 2532-2540.
30. Moore, R.B., Chapman, J.D., Mokrzanowski, A.D., Arnfield, M.R., McPhee, M.S. and McEwan, A.J. (1992) *Brit. J. Cancer* **65**, 491-497.
31. Gonzalez, S., Arnfield, M.R., Meeker, B.E., Tulip, J., Lakey, W.H., Chapman, J.D. and McPhee, M.S. (1986) *Cancer Res.* **46**, 2858-2862.
32. Hirsh, B.D., Walz, N.C., Meeker, B.E., Arnfield, M.R., Tulip, J., McPhee, M.S. and Chapman, J.D. (1987) *Photochem. Photobiol.* **46**, 847-852.

RG2 RAT GLIOMA MODEL: INFLUENCE OF THE BLOOD-BRAIN BARRIER ON BIODISTRIBUTION OF BORONSULFHYDRYL

D. Schuepbach [1], P.R. Gavin [1,2], L.G. Salford [3], C. Ceberg [3], and H. Fankhauser[1]

[1] Department of Neurosurgery, CHUV, 1011-Lausanne, Switzerland
[2] On leave from Veterinary Clinical Medicine & Surgery, Washington State University, Pullmann, WA
[3] Department of Neurosurgery, University Hospital, Lund, Sweden

INTRODUCTION

The accumulation of borocaptate sodium (BSH) and other boronated molecules in brain tumors and normal brain is dependent, in part, upon the blood-brain barrier (BBB). Current investigations into the biodistribution of various boron carriers in animals and humans are done on intact tumors not subjected to previous surgery, whereas actual Boron Neutron Capture Therapy (BNCT) takes place shortly after surgical debulking (Hatanaka et al, 1992). The RG2 intracerebrally transplantable rat glioma (Krajewski et al, 1986) was used to develop an experimental model which would allow the simulation of the situation in patients undergoing BNCT 2 to 4 weeks after craniotomy for gross tumor resection. Surgical removal of the transplanted tumor was therefore attempted under general anesthesia approximately 1 week prior to the BSH administration, and the animals were allowed to recover. This model was established to study the tissue boron concentration in minute samples after intravenous administration of BSH, and its relationship to the degree of disrupture of the BBB by the intact tumor, by surgery alone, and by the combined effect of tumor and surgery. The BBB was assessed by the use of intravenous sodium fluorescein (SF). The model should permit the quantitative investigation of the interference of steroids and mannitol with SF and BSH biodistribution. These two drugs are commonly used in clinical practice and are known to influence the BBB. This work should result in the establishment of base line values concerning the behaviour of BSH in a biological system close to clinical reality which would allow a rapid and quantitative comparison of BSH with other boronated molecules.

MATERIALS AND METHODS

The RG2 tumor was stereotactically implanted into the brains of 8-10-week-old male or female Fisher rats CD 344. After desinfection with 70% ethyl alcohol, skin incision, and

Advances in Neutron Capture Therapy, Edited by
A.H Soloway *et al.*, Plenum Press, New York, 1993

exposure of the sutures of the vertex of the skull, a 1.5-mm drill hole was made 2 mm anterior and 2 mm to the right of the bregma. Suspended tumor cells were held in a precision syringe, 0-10 microliters, equipped with a 30 gauge needle. A volume of 5 microliter containing some 5000 tumor cells was injected into the subcortical white matter of the right frontal lobe. An average of 22 days after implantation, a right frontal craniectomy was performed under general anesthesia with 5% Ketamine and 2% Xylamine intraperitoneally, without endotracheal intubation, but with the animal's head held in the stereotactic frame for immobilization. The right frontal bone was drilled off with a woodcarver's tool, and the dura was incised under the operating microscope. The anterior part of the right frontal lobe, containing the tumor, was then removed to the falx and to the skull base, leaving intact the easily visible olfactory bulb. Hemostasis was achieved by tamponade with a hemostatic fabric. The resection cavity was filled with a plug of fresh autologous subcutaneous adipose tissue. No closure of the dura was attempted. The skin was simply sutured in 2 layers. One week later, animals were divided into several groups and injected through the right femoral vein with BSH, with or without SF. Standard doses were for BSH 60 mg ^{10}B/kg body weight in physiological saline (6 mg/ml), for SF 10 mg/kg in physiological saline (5 mg/ml). Groups of 5 animals each were used. Non-tumor-bearing surgical controls were treated in the same way.

Tumor-bearing rats without resection were used to investigate the influence of mannitol and steroids on the BBB and on the BSH tissue accumulation. Standard doses were for methylprednisolone 3 mg/kg in physiological saline (1 mg/ml), for mannitol 250 mg/kg in physiological saline (50 mg/ml), and for dexamethasone phosphate 1.5 mg/kg in physiological saline (0.5 mg/ml). Methylprednsiolone and mannitol were given concomitantly through the femoral vein with BSH and SF, dexamethasone was given intraperitoneally, at 48 h and 24 h prior to tissue harvesting. Tumor controls were used for histological examination.

In all animal groups, sacrifice and tissue harvesting took place 15 minutes after drug injection. Samples (tumor, adjacent brain, brain of the contralateral hemisphere, bone, muscle, skin, liver and blood plasma) for global boron and SF analysis were frozen at -70° C. Formaldehyde fixed samples were stained with hematoxyline-eosine for histological evaluation. Unstained, deparaffinated, formaldehyde fixed sections were examined for SF with the UV-microscope. Quantitative SF measurements in plasma and homogenized tissues, diluted with phosphate buffer (200 mM), at a pH ranging between 7.0 and 7.6, were done by spectrofluorometry. Boron analysis was carried out with Inductively Coupled Plasma-Atomic Emission Spectroscopy (ICP-AES).

RESULTS

Tumor bearing rats showed the first clinical signs about 20 days after implantation in the form of tonic-clonic convulsions, various degrees of starvation and diminished physical activity. By day 21 to 30, the unoperated tumor was fatal. By day 22 the diameter of the tumor was approximately 0.5 cm. Histologically, a solid, partially necrotic tumor with an infiltrating margin towards the adjacent brain and meninges could be observed (Fig. 1). There was no periventricular or meningeal spread.

Surgical removal of the tumor under the operating microscope proved to be feasible without major difficulties, after adaptation of the procedure to adjust for initially encountered difficulties. In order to reach the tip of the frontal lobe and to standardize the surgical procedure, the intracerebral implantation site was moved forward and closer to the surface than described for this model (Hemler et al, 1992). Operative mortality was less than 10% and there were no obvious neurological deficits in the survivors. After exposure

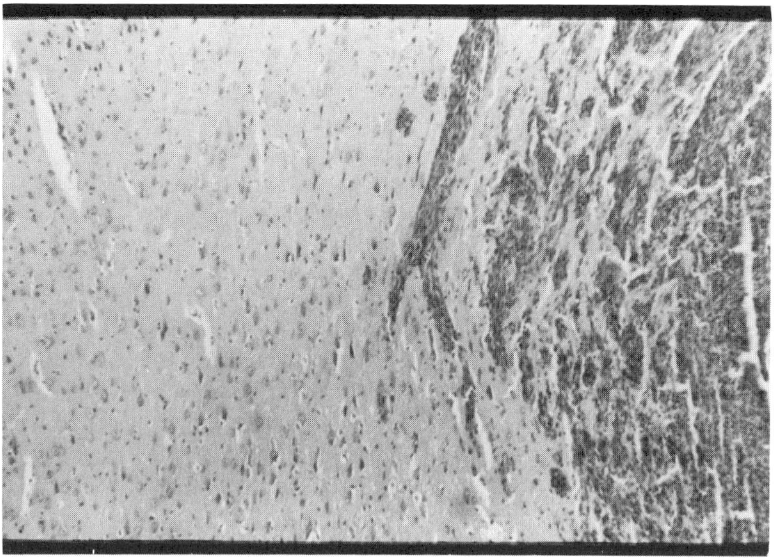

Fig 1. A transplanted rat glioma infiltrating adjacent to normal brain; infiltrating tumor cell nests can be clearly seen (22 days post implantation; hematoxylin-eosin staining, magnification 125x)

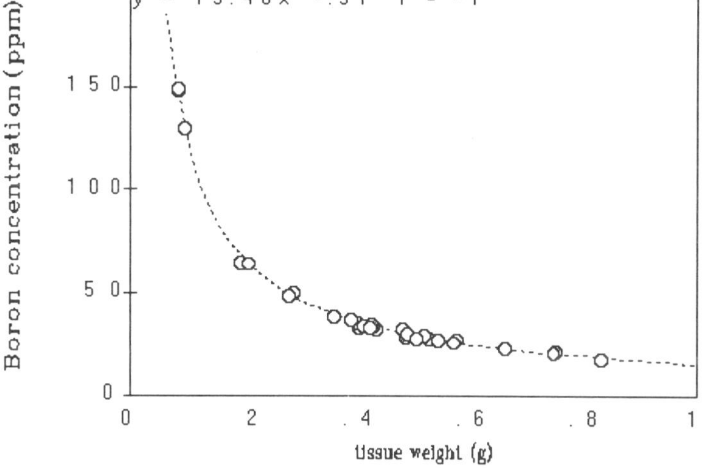

Fig 2. Correlation between tissue weight (x-axis) and measured boron concentration (y-axis) in specimen from normal rat brain, 15 minutes after intravenous administration of BSH.

of the frontal lobe, the tumor was visible as a slightly yellowish discoloration and disturbance of the normal gyral pattern. In SF injected animals the tumor was easily identified as a brightly fluorescent mass upon inspection under a Wood's lamp. Anterior frontal lobectomy under clean rather than sterile conditions, and closure of only the skin over the exposed brain was well tolerated. There were no infections.

Despite marked macroscopic fluorescence of the tumor in vivo and high SF concentrations measured by spectrofluorometry in tumors, but not in normal brain, microscopic examination of formaldehyde fixed sections only showed fluorescence confined to blood vessels in tumor and normal brain.

After surgical debulking there was a tumor recurrence within 5 to 10 days, which was usually fatal 1 to 3 days after its obvious clinical manifestation. Microscopic examination at this stage showed prominent necrotic areas in the tumor.

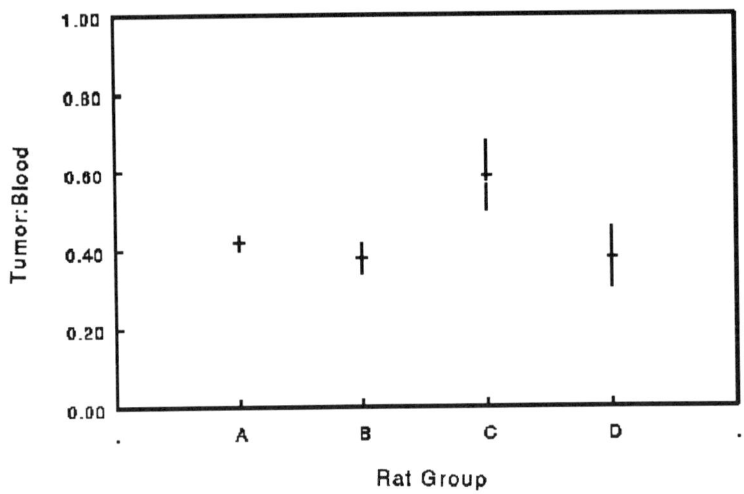

Fig 3. Tumor-to-blood ratio for SF in primary (A,B,C) and recurrent (D) tumor bearing rats. A: SF; B: SF and mannitol; C: steroids, SF and mannitol; D: SF. Bars indicate standard error of the mean.

Processing of tissue samples and analyses for boron by ICP-AES was carried out in two remote laboratories with two different ICP-AES machines. No cross controls were done at this stage. This has yielded unexpected results, demonstrating critical but yet unidentified problems with the analysis for boron of very small samples by ICP-AES. Boron concentrations for tumor, brain and muscle measured in one place were roughly 10 times higher than in comparable samples, prepared in a supposedly identical manner. They also showed a strong negative correlation between measured boron concentration and sample size, with a correlation coefficient of -1 (Fig. 2). For this reason some 200 samples could not be used for the intended evaluation. A search for an identical correlation between boron concentration and sample size in the second laboratory did not show any significance, with a correlation factor of -0.36. These problems will be discussed below.

Useable results were nevertheless obtained for quantitative assessment of SF in various situations. Four groups of 5 tumor bearing animals each were investigated for blood and tissue SF after intravenous administration of SF, either untreated or after additional administration of mannitol alone, or steroids and mannitol. One group was investigated 1 week after surgical tumor debulking. The tumor-to-blood ratio for SF in these 4 situations is depicted in figure 3. This ratio was approximately 0.4, except in the group treated concomitantly with steroids and mannitol, where it rose to 0.6 (Fig.3). The mean ratio for tumor-to-normal brain of the controlateral hemisphere in all groups ranged from 7.8 to 8.

Comparative boron and SF analysis was successful in 5 tumor-bearing rats treated with dexamethasone 48 and 24 hours before administration of BSH and SF. It showed that boron and SF concentrations in tumor, normal brain of the controlateral hemisphere, and blood were closely proportional, with SF concentrations (parts per million or ppm) constantly 5 to 6 times lower than for boron (ppm) which corresponds well with the 6 times lower administered dosage (mg/kg) of SF compared to boron.

DISCUSSION

This work in progress has so far demonstrated that it is possible to use the RG2 model to realistically simulate the situation in patients subjected to BNCT. The model has the advantage of using an intracerebral, infiltrating, glioma-like tumor, of being relatively simple and cheap, of presenting a high take rate after implantation of a small volume of cells, and of offering the possibility for microsurgical tumor removal with a high survival rate and low morbidity.

Nevertheless, some basic problems with assessment of boron in minute biological samples and with microscopic imaging of disruption of the BBB have to be solved.

Inability to demonstrate SF in tumor parenchyma outside vessels with UV-microscopy probably relates to redistribution and wash-out during formaldehyde fixation. SF, similar to BSH, must be considered as a highly diffusible, small molecule. Fixation with aldehydes destroys intracellular free element distributions (Zierold et Schäfer, 1988) and is supposed to do the same with small diffusible molecules. The problem may be solved by the use of frozen sections, but this will lead to undesired loss of histological resolution. It is our intention to study the human BSH biodistribution related to the BBB in a manner similar to the rat model. SF has therefore been chosen as BBB marker since this compound is currently used in humans and therefore does not need any further toxicity studies, but optimal doses for BBB assessment may be higher by a factor of 2 or 3 than those used for retinal angiography.

There may be several reasons for a negative correlation between size of minute samples and boron content in one of two laboratories. Some negative correlation may be expected to be real since larger tumors contain more necrotic areas; in human gliomas these are known to show low boron concentrations (Stragliotto and Fankhauser, 1992). But in our case, this was clearly not the explanation, since the absolute concentrations were much higher than expected, even for a non-necrotic tumor. An alternate explanation may be the use, during the digestion and measurement processes, of chemicals and dilutants which are not completely boron free. Totally boron free chemicals are not routinely available in laboratories and are expensive. Incorrect dilution or insufficient cleaning of the tissue digestion vessels may equally contribute to artificially increased boron concentrations. Finally, a mathematical error may have been committed. Current investigation into the problem seems so far to exclude all above mentioned sources of error. At this time it can only be said that periodic inter-laboratory cross checks of analytical methods are mandatory.

The tumor-to-blood ratios for boron and SF in the steroid treated rat group was below 1 for both drugs, possibly indicating that the time interval after drug injection has to be longer than 15 minutes in order to reach a ratio above 1. In a clinical investigation

(Stragliotto et Fankhauser, 1992) as well as in a quantitative study on pharmacokinetics and biodistribution of BSH in the RG2 model (Hemler et al, 1992), the tumor concentration approaches the blood concentration after 18 and 20 hours, respectively.

Our first valid results of ICP-AES measurement of boron and spectrophotometric measurements of SF in the same animals indicate a good correlation of absolute concentrations for both substances in tumor, normal brain, and blood. These small, hydrophilic molecules therefore appear to pass the BBB in a similar fashion.

To make full use of this tumor model it will be necessary to further refine the experimental variables in order to better meet the relevant clinical condition in patients. This means, for instance, that a series of measurements will have to be done at an earlier stage after tumor debulking, before obvious recurrence of the tumor. It also includes the investigation of clinically relevant drug administration schedules, such as longer intervals between administration of BSH, steroids and mannitol, and tissue sampling, as well as higher doses and repeat injections. In this context the model has limitations, since the very short tumor doubling time as compared to spontaneous human gliomas precludes experiments over more than a few days.

Finally, we expect this model to become extremely useful for comparative screening of new boron carriers when comparing them with BSH, and for future work on subcellular distribution of boron compounds using microanalytical methods.

ACKNOWLEDGEMENTS

This work was undertaken in collaboration with the institute of analytical chemistry of the University of Lausanne, Switzerland, and the ECN in Petten, The Netherlands.

This work has been supported by the Fonds National de la Recherche Scientifique, the Ligue Suisse contre le Cancer, the Ligue Vaudoise contre le Cancer, the Zyma Foundation for the Advancement of Medical and Biological Sciences, and the Commission of the European Communities.

REFERENCES

Hatanaka, H, Sano, K, Yasukochi, H, Clinical results of Boron Neutron Capture Therapy, in: "Progress in Neutron Capture Therapy for Cancer", B.J. Allen, ed., Plenum Press, New York, 1992

Hemler, R.J.B., Ceberg, C.P., Brun, A., Gabel, D., Larsson, B., Persson, B.R.R., and Salford, L.G., A quantitative study on pharmacokinetics and biodistribution of BSH in a rat glioma model, in:"Progress in Neutron Capture Therapy for Cancer", B.J. Allen, ed., Plenum Press, New York, 1992

Krajewski, S., Kiwit, J.C., Wechsler, W., 1986, RG2 glioma growth in rat cerebellum after subdural implantation, J. Neurosurg. 65(2):222-229

Stragliotto, G., and Fankhauser, H., Biodistribution of boron sulfhydryl (BSH) in patients with intracranial tumors, in : "Progress in Neutron Capture Therapy for Cancer", B.J. Allen, ed., Plenum Press, New York, 1992

Zierold, K., and Schäfer, D., 1988, Preparation of cultured and isolated cells for x-ray microanalysis, Scanning Microsc. 2:1775-1790

IN VIVO PHARMACOKINETICS OF A BORON NEUTRON CAPTURE AGENT IN A TUMOR-BEARING RAT VIA BORON-11 MRS AND MRI

G.W. Kabalka*, G.-Q. Cheng*, C. Anderson*, P. Bendel†,
P. Micca‡, and D. N. Slatkin‡

*Departments of Radiology and Chemistry,
University of Tennessee, Knoxville, TN
†Department of Chemical Physics,
Weizmann Institute of Science, Rehovot, Israel
‡Medical Department, Brookhaven National Laboratory, Upton, NY

INTRODUCTION

We have developed boron-11 MRI and MRS techniques of potential use in boron neutron capture therapy (BNCT)[1,2] for noninvasive quantification and localization of BNCT agents such as the sodium salt of the dimer of the closododecaboranyl mercaptide anion, $B_{24}H_{22}S_2^{4-}$ (BSSB)[3].

Infusion of a rat with BSSB for 7 to 9 days via an intraperitoneal osmotic pump results in boron concentrations of 30-100 μg per gram of tissue[4]. The verification and quantification of such low levels of boron *in vivo* remain difficult and challenging problems. An obstacle to imaging boron *in vivo* by MRI is the very short transverse relaxation time (T_2 < 1 ms), which is shorter than the minimal available time-to-echo (TE) in the standard spin-echo and gradient-echo imaging protocols.

To overcome these obstacles, we recently developed an optimal method for imaging elements with short T_2 relaxation times and utilized it to image boron-11 *in vivo* in a rat[5]. The method is more rapid than techniques involving conventional back projection methodology[1,6]. We now wish to report the results of a pharmacokinetic study in which boron-11 MRI and MRS were used to monitor the *in vivo* lifetime (in a rat) of the BSSB BNCT agent which is currently in clinical use.

MATERIALS AND METHODS

Tumor bearing, Fischer 344 male rats (Taconic Farms, Germantown, New York) were utilized in the study. The GS-9L cell line was derived from a *N*-nitrosomethylurea-induced rat gliosarcoma which is carried in cell culture[7]. Solid tumors were initiated on the rear legs by implanting 2.5×10^5 cells/0.1 ml growth medium subcutaneously. Administration of the BSSB was initiated when the tumors were ~10 cc in size.

Two milliliter capacity osmotic pumps (Alza Corp., Model 2ML1, Palo Alto, CA) were filled with a 48 mg/ml solution of $Na_4B_{24}H_{22}S_2$ with a natural abundance (~80 atom%) of boron-11. The pumps were implanted surgically into the peritoneal cavities of ~250 gram rats. Seven days after pump implantation, the pumps were removed immediately prior to MR analyses.

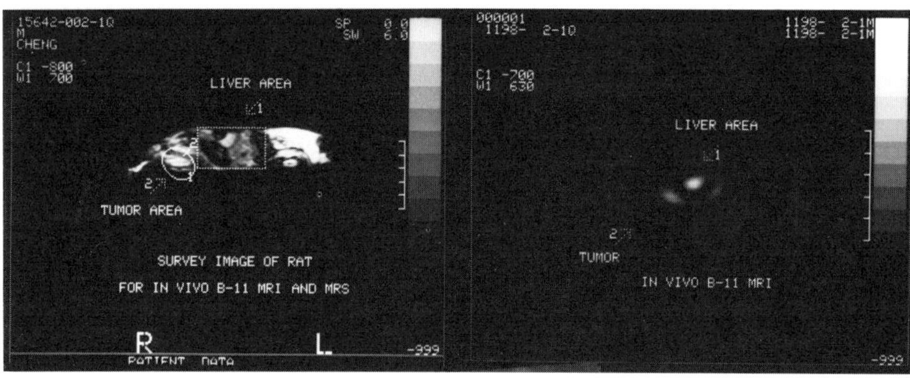

Figure 1. Sagittal hydrogen-1 MRI of the BSSB treated rat using a body coil with the boron-11 solenoid coil in place (left). [The regions of interest (ROI) are marked]. Sagittal boron-11 MRI of the same rat taken immediately after the hydrogen-1 MRI (right).

In vivo MR studies were conducted at room temperature utilizing a GYREX 2T, 90 cm-bore, whole-body MR unit operating at 26 MHz for boron-11 (81 MHz for hydrogen). A six turn solenoidal coil (10 cm diameter) was used for both imaging and spectroscopy experiments. The imaging protocol was described in our earlier studies; 256 averages were used at each of the 19×20 phase encoding steps which results in a total scan time of 1 hour[5]. Boron-11 MRS was carried out using standard pulse-acquire sequences. Data acquisition started 200 μs after the 90° pulse and 4000 averages were employed at an interval of 200 ms resulting in a scan time of 14 minutes.

RESULTS

The H-1 MRI and the boron-11 MRI survey images of the tumor-bearing rat are presented in Figure 1. In the H-1 MRI, liver and tumor areas are marked with a rectangle and an oval respectively. The corresponding boron-11, *in vivo* MRI is also displayed in Figure 1. The boron-11 MRI reveals that the BSSB agent localizes in both the tumor and the liver of the rat. Biopsy of a rat which had been treated under the same protocol, revealed that the initial concentration of boron-11 was approximately 100 μg/g in the liver and 30 μg/g in the tumor immediately after removal of the intraperitoneal pump. The MR results were also used to estimate the the boron-11 concentrations by normalizing the MR responses of each tissue (area/weight) versus a reference sample containing 34 μg/g of BSSB.

The B-11 spectra (over time) from the whole rat body are presented in Figure 2. The majority of the boron agent is located in the liver area as noted earlier. The corresponding boron spectra from the tumor area (marked with an oval on the survey image) is also displayed in Figure 2. The spectra were taken over a period of 16 hours following the removal of the osmotic pump from the rat body. Times are indicated in the figure.

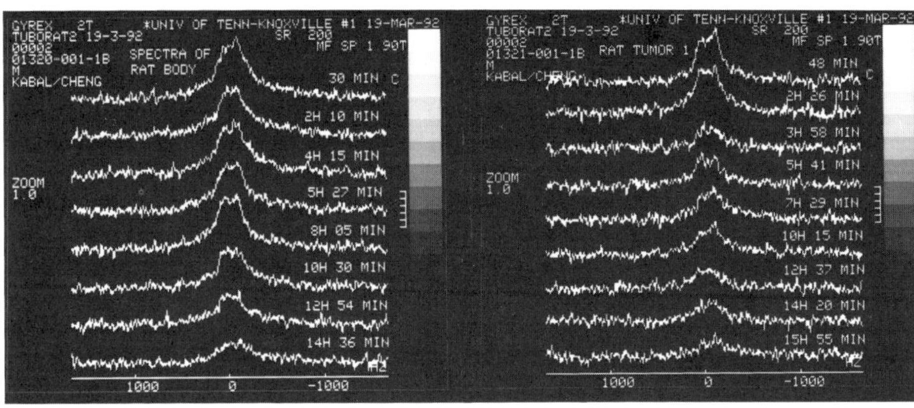

Figure 2. *In vivo* whole body boron-11 nuclear magnetic resonance spectra taken over time of the BSSB treated rat whose MRI are presented in Figure 1 (left). *In vivo* boron-11 nuclear magnetic resonance spectra of the tumor taken over time of the same BSSB treated rat (right).

DISCUSSION

The boron-11 MRI (Figure 1) reveals the distribution of boron in the rat body. Boron-11 MRS (Figure 2) proved to be more sensitive and permitted monitoring the boron concentration in both the liver and tumor.

BSSB exhibits a biexponential decay from both the liver and the tumor. In the liver, the intensity of the boron spectra decreases gradually over a period of 14 hours post removal of the osmotic pump. The boron-11 magnetic resonance signal exhibits a different behavior in the tumor; the intensity of the boron signal rapidly declines during the first two hours and then decreases more slowly. The minor fluctuation in spectra intensities were caused by repositioning the rat in the coil and by excretion of the BSSB through the kidneys. The differing behavior of boron-11 in tumor and liver implies a difference in boron diffusion, binding, and metabolism within these tissues.

CONCLUSION

Boron-11 MRI and MRS are valuable techniques for studing the biodistribution and pharmacokinetics of BNCT agents noninvasively. The MR techniques could prove superior and simpler to implement than BNC radiography. B-11 MRS is quite sensitive to boron-11 at low concentrations whereas B-11 MRI is capable of revealing spatial localization of BNCT agents. In this preliminary study, the *in vivo* disappearance rates of the boron agent in tumor and liver exhibited different behaviors. MR techniques could assist in establishing the optimum timing and duration of neutron irradiation following the infusion of a boron-containing drug for BNCT.

ACKNOWLEDGEMENTS

We thank the Department of Energy for support.

REFERENCES

1. G.W. Kabalka, M. Davis, and P. Bendel, Boron-11 MRI and MRS of Intact Animals Infused with a Boron-Neutron-Capture Agent, *Magn. Reson. Med.* 8:89 (1988).
2. P. Bendel, M. Davis, E. Berman, and G. W. Kabalka, A Method for Imaging Nuclei with Short T2 Relaxation: Application to Boron-11 MR Imaging of a BNCT Agent in an Intact Rat, *J. Magn. Res.* 88:369 (1990).
3. D. Slatkin, P. Micca, A. Forman, D. Gabel, L. Weilopolski, and R. Fairchild, Boron Uptake in Melanoma, Cerebrum, and Blood from $Na_2B_{12}H_{11}SH$ and $Na_4B_{24}H_{22}S_2$ Administered to Mice, *Biochem. Pharmacol.* 35:1771 (1986).
4. D. N. Slatkin, P. L. Micca, B. H. Laster, and R. G. Fairchild, Distribution of Sulfhydryl Boranes in Mice and Rats, in: *BNL-Report-51995*. Brookhaven National Laboratories, Long Island (1986).
5. G.W. Kabalka, G.-Q. Cheng, P. Bendel, P.L. Micca, and D.N. Slatkin, In-Vivo Boron-11 MRI and MRS Using $(B_{24}H_{22}S_2)^{4-}$ in the Rat, *Magn. Reson. Imaging.* 9:969 (1991).
6. G.H. Glover, J.M. Pauley, and K.M. Bradshaw, Boron-11 Imaging with a 3-dimensional Reconstruction Method, *J. Magn. Reson. Imaging.* 2:47 (1992).
7. H.H. Schmidek, S.L. Neilsen, A.L. Schiller, and J. Messer, Morphological Studies of Rat Brain Tumors Induced by N-Nitrosomethylurea, *J. Neurosurg.* 34:335 (1971).

PHARMACOKINETICS AND TISSUE DISTRIBUTION OF BORONOPHENYLALANINE FOLLOWING INTRAPERITONEAL INJECTION IN NUDE RATS WITH INTRACEREBRAL MELANOMA

Alfred E. Staubus[1], Khalid Matalka[2], Rolf F. Barth[2], and Michael Q. Bailey[3]

[1]Division of Pharmaceutics and Pharmaceutical Chemistry, College of Pharmacy, The Ohio State University
[2]Department of Pathology, College of Medicine, The Ohio State University
[3]Department of Veterinary Clinical Sciences and Department of Veterinary Anatomy and Cellular Biology, College of Veterinary Medicine, The Ohio State University, Columbus, OH 43210

INTRODUCTION

Para-boronophenylalanine (BPA) has been used as a boron delivery agent for boron neutron capture therapy (BNCT) of melanoma[1]. As an analogue of phenylalanine, a melanin precursor, it has been postulated that melanoma cells selectively take up BPA. The effectiveness of BNCT is dependent, in part, upon the concentration gradient of boron in the tumor in comparison with the surrounding normal tissues and blood. Consequently, it is important to determine the pharmacokinetics and tissue distribution of BPA and other boron compounds in order to evaluate the potential effectiveness of BNCT agents.

METHODS

Animals and Tumor Cell Line

The human melanoma cell line MRA 27 was derived from a 60 year old Norwegian male and has been propagated both in vitro and in vivo in nude mice and rats. MRA 27 cells were grown in McCoy's 5A media (Gibco, Grand Island, NY) supplemented with 10% fetal bovine serum (Hyclone, Logan, UT), 100 units/mL penicillin, 100 μg/mL streptomycin, and 2 mM L-glutamine were tested for mycoplasma contamination by DNA-fluorochrome stain and UV microscopy (Bionique Testing Laboratory, Saranac Lake, NY). Six to eight week-old athymic female nude rats of NIH-rnu strain were purchased from the Animal Production Branch, National Cancer Institute, Frederick, MD. The rats were maintained under specific pathogen-free conditions and fed sterilized food and water.

Advances in Neutron Capture Therapy, Edited by
A.H Soloway *et al.*, Plenum Press, New York, 1993

Implantation

A stereotactic implantation procedure, previously used for studies on BNCT of a rat glioma[2], was employed. MRA 27 cells were injected into the right caudate nucleus of nude rats at a concentration of 10^6 or 2 X 10^5/10 μL of serum-free McCoy's 5A medium containing 1% agarose with gelling temperature of ‹30°C.

Pharmacokinetics and Tissue Distribution Studies

For injection, either D,L-BPA or L-BPA (both Callery Chemical Co., Callery, Pa) was converted to a more water-soluble fructose complex. BPA and fructose were mixed in a 1:1 molar ratio to yield a final concentration of 120 mg of BPA (6.3 mg of boron)/2 mL of water. The pH was adjusted to 8.8 with 6N NaOH. Two mL of complex were administered intraperitoneally (i.p.) to 150 g rats 37 days following intracerebral implantation of the MRA 27 cells. These doses corresponded to 800 mg/kg of BPA. For the i.p. administration of D,L-BPA, samples of blood, brain, tumor, skin, liver, kidneys, muscles, eyes and skull were obtained at sacrifice from four animals at each post-injection time (1, 3, 6, 9, 12, and 18 h). The detailed pharmacokinetic and tissue distribution study utilized the less expensive racemic D,L mixture while, for comparative purposes, the one point distribution study utilized the purified L enantiomer. For the i.p. administration of L-BPA, samples were obtained at sacrifice from 4 animals at 6 h post-injection. Boron concentrations in the samples were determined by means of direct current plasma atomic emission spectroscopy (DCP-AES)[3]. Pharmacokinetic analysis was performed on the resultant geometric mean boron blood concentrations using a nonlinear regression program (NONLIN).

RESULTS AND DISCUSSION

Peak boron blood concentrations were obtained at the first sampling time (1 hour) indicating rapid absorption of the BPA from the peritoneal cavity. Boron blood concentrations exhibited a biexponential decay (Figure 1) and, consequently, were fitted

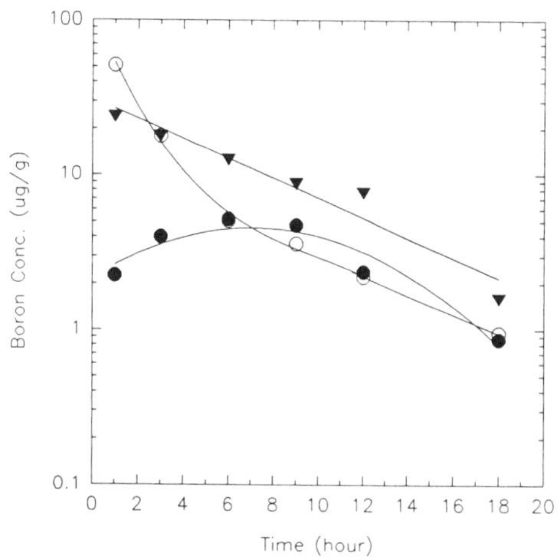

Figure 1. Boron concentrations in blood (○), brain (●), and tumor (▼) following a single i.p. injection of 120 mg/2 mL of D,L-BPA (6.3 mg of boron), as a fructose formulation, in nude rats carrying 37 day-old i.c. MRA 27 human melanoma.

to the classical two compartment open model system with first-order elimination from the central compartment and with an assumption of a rapid (0.1 min) zero-order input.

The resultant NONLIN fit (Table 1) revealed a rapid disposition half-life ($t_{1/2\alpha}$) of 0.91 h and a slow disposition half-life of ($t_{1/2\beta}$) of 5.29 h. Since the NONLIN fit assumed a rapid zero-order input function, which would be expected to overestimate the area under the curve during the absorption phase, the resulting apparent total body blood clearance (Cl_T/F) is probably underestimated at 521.2 mL/min if the extent of bioavailability is essentially complete (F=1) from the i.p. injection. Similarly, the apparent volume of distribution terms, V_1/F (60.6 mL), Vdss/F (118.3 mL) and Vd_β/F (238.6 mL), are probably underestimated. These apparent volumes of distribution of BPA in rats having a mean body weight of 150 g are relatively large and indicate rather extensive tissue uptake of BPA.

Table 1. Pharmacokinetic parameters of boron after i.p. administration of D,L-BPA in nude rats carrying i.c. melanoma

Parameter	Symbol	Unit	Value
Half-life of rapid disposition phase	$t_{1/2\alpha}$	h	0.91
Half-life of slow disposition phase	$t_{1/2\beta}$	h	5.
Apparent total body clearance	Cl_T/F	mL/min	521.2
Apparent volume of distribution of the central compartment	V_1/F	mL	60.6
Apparent volume of distribution at steady state	Vdss/F	mL	118.3
Apparent volume of distribution at distribution equilibrium	Vd_β/F	mL	238.6

Tumor levels of boron, however, exhibited monoexponential decay ($t_{1/2}$ = 5.64 h) with the terminal time points (6-18 h) in an apparent distribution equilibrium with boron levels observed in the blood. Tumor to blood boron concentration (T/Bl) ratios within the terminal phase ranged from 2.4 to 3.8. Normal brain tissue reached a peak of boron concentration 6-9 hrs after BPA administration and declined monoexponentially, essentially superimposed on the blood boron concentrations. Tumor to normal brain boron concentration (T/Br) ratios within the terminal log-linear phase averaged 2.4.

The best composite ratio for BNCT was observed 6 hrs post-administration of D,L-BPA. At that time the tumor boron concentration was 13.7 μg/g of tissue and the T/Bl and the T/Br ratios were 2.4 and 2.5 respectively. When the same amount of L-BPA was administered i.p. under similar condition, the tumor boron concentration increased to 24.9 μg/g (Table 2). The T/Bl and the T/Br ratios were 2.5 and 2.8 respectively. Furthermore, all other normal tissues exhibited boron concentrations 1.5-1.7 fold higher with the L-BPA than with the D,L-BPA.

Table 2. Distribution of p-boronophenylalanine (BPA) 6 hours following administration to nude rats carrying intracerebral human melanoma cell line MRA 27

Tissue	Boron concentration (μg/g $\pm$ SD)	
	D,L-BPA	L-BPA
Brain (Br)	5.4 $\pm$ 1.6	8.8 $\pm$ 0.5
Blood (Bl)	5.7 $\pm$ 3.1	9.9 $\pm$ 3.6
Tumor (T)	13.7 $\pm$ 5.8	24.9 $\pm$ 2.4
T/Bl ratio	2.4	2.5
T/Br ratio	2.5	2.8

CONCLUSIONS

The pharmacokinetic analysis obtained from our studies indicate that the optimum time for neutron irradiation in subsequent nude rat survival studies should be ~6 hours following BPA administration. At ~6 hours following intraperitoneal administration, our studies have shown that doses of 800 mg/kg of D,L-BPA in a fructose formulation exhibited peak boron tumor concentrations of 13.7 μg/g and had T/Bl and T/Br ratios of 2.4-2.5. Corresponding doses of L-BPA exhibited 1.8-fold greater peak tumor concentrations (24.9 μg/g) but also 1.5 to 1.7-fold greater boron concentrations in other normal tissues. Consequently, the pharmacokinetics of L-BPA appear to be different from the pharmacokinetics for either D-BPA or D,L-BPA. Optimally, tumor boron concentrations should be at least 20 μg/g[4]. In order to obtain these high tumor boron concentrations using BPA, it will be necessary to administer larger doses intravenously. It also may be necessary to administer the L-BPA compound rather than the less expensive racemic mixture. Additional pharmacokinetic studies are needed to evaluate the effect of dose, route of administration (intravenous vs. intraperitoneal), frequency of administration (single intravenous bolus vs. multiple intravenous dosing vs. prolonged intravenous infusion), and enantiomeric form on both boron tumor concentrations and on the T/Bl and T/Br ratios.

ACKNOWLEDGEMENTS

The authors wish to thank Dianne Adams, Joan Rotaro, Mary Ross, and David Carpenter for their technical assistance. Supported by Department of Energy grant DE-AC02-76CH000016 and by the Ohio State University Comprehensive Cancer Center/Cancer Research Hospital and Institute.

REFERENCES

1. Mishima, Y., Ichihashi, M., Honda, C., Shiono, M., Nakasawa, T., Obara, H. et al. (1992) Progress in neutron capture therapy for cancer, New York, Plenum Press, pp. 577-583.

2. Clendenon, N.R., Barth, R.F., Gordon, W.A., Goodman, J.H., Alam, F., Staubus, A.E., Boesel, C.P., Yates, A.J., Moeschberger, M.L., Fairchild, R.G., and Kalef-Ezra, J.A., Neurosurg., **1990**, 26, 47-55.

3. Barth, R.F., Adams, D.M., Soloway, A.H., Mechetner, E.B., Alam, F., and Anisuzzaman, A.K.M., Anal. Chem., **1991**, 63, 890-893.

4. Barth, R.F., Soloway, A.H., Fairchild, R.G., and Brugger, R.M., Cancer, **1992**, 70, 2995-3007.

UPTAKE OF ^{10}B IN GLIOSARCOMAS FOLLOWING THE INJECTION OF GLUTATHIONE MONOETHYL ESTER AND SULFHYDRYL BORANE

Darrel D. Joel, Daniel N. Slatkin and Jeffrey A. Coderre

Medical Department
Brookhaven National Laboratory
Upton, New York 11973

INTRODUCTION

The sulfhydryl borane $Na_2{}^{10}B_{12}H_{11}SH$ (BSH) was developed as a capture agent for BNCT about 20 years ago and is the compound currently used clinically in Japan for BNCT of malignant brain tumors, with some apparent success [1].

We found, however, that tumor ^{10}B concentrations following the infusion of the oxidized BSH, a disulfide dimer ($Na_4{}^{10}B_{24}H_{22}S_2$), were nearly twice those obtained following administration of equal amounts of boron as BSH[2]. Also, the rate of decrease of tumor ^{10}B concentration was slower after dimer infusion than after BSH infusion. Furthermore, when BNCT was administered to rats bearing intracerebral gliosarcomas, at a time when the average tumor boron concentration was the same, animals infused with dimer had a significant longer median survival time and a higher percentage of "cures" than rats infused with BSH[3].

Dimer, on the other hand, induces a moderately severe, but reversible, hepatotoxicity which may complicate its use in humans. It has been suggested that dimer-induced hepatotoxicity may be the consequence of free radical formation[4]. Intracellular glutathione plays an important role in defense against radical-mediated tissue injury. Although the administration of glutathione leads to only a small increase in intracellular glutathione, the administration of glutathione monoesters (GSH-ME) results in significant increases in cellular glutathione[5]. Glutathione monoesters have been reported to have a protective effective on cisplatin toxicity[6] and on radical-induced acute pancreatitis[7]. We, therefore, investigated the possibility of reducing dimer-induced hepatotoxicity by pre-administration of GSH-ME. The results indicate that not only does the pre-administration of GSH-ME markedly reduce dimer-induced hepatotoxicity, but also results in nearly a doubling of tumor boron concentration. Furthermore, GSH-ME markedly increases tumor boron uptake and retention following administration of BSH.

Advances in Neutron Capture Therapy, Edited by
A.H Soloway *et al.*, Plenum Press, New York, 1993

MATERIALS AND METHODS

The monoethyl ester of glutathione was prepared from glutathione and ethanol by the method of Anderson and Meister[8] using sulfuric acid as the catalyst. Crystalline GSH-ME preparations were assessed by thin layer chromotography, stored in a desiccator, and dissolved in water immediately prior to use.

The transplantable rat gliosarcoma (9L) and the preparation of BSH and dimer used in these BNCT experiments have been previously described[2,9]. Boron concentrations in blood and tissue samples (0.3 to 1.0 g) were analyzed by prompt-gamma spectroscopy.

Clonogenic cell survival assays were done on intracerebral tumors treated 16 days after initiation. Within 5 minutes after irradiation, rats were killed with an overdose of anesthesia and the intracerebral gliosarcomas were removed aseptically, minced and incubated with trypsin-EDTA (0.05% trypsin-0.53 mM EDTA in Hank's Balanced Salt Solution, without calcium or magnesium) for 30 min at 37°C. Fragments of tumor tissue were removed by centrifugation and aliquots of the single-cell suspension were diluted, plated and incubated for 10 days. The plating (colony-forming) efficiency of cells from similarly-harvested, unirradiated tumors, relative to the number of viable (Trypan Blue-excluding) cells plated, was in the 50-60% range.

EXPERIMENTS

Effect of GSH-ME on Dimer-Induced Hepatotoxicity

Sixteen rats bearing subcutaneous gliosarcomas were divided into 4 groups. One group was injected with GSH-ME 2 hours prior to the administration of dimer; the second group was injected with saline 2 hours prior to the administration of dimer; the third group was injected only with GSH-ME; and the fourth group was untreated. Serum enzyme levels of aspartate aminotransferase (AST) and alanine aminotransferase (ALT) were used as indicators of soft-tissue damage. ALT is a liver-specific enzyme in rats and an increase in activity is indicative of hepatocellular damage. As shown in Table 1, preadministration of GSH-ME had a significant

Table 1. Effect of Glutathione-Monoester (GSH-ME) Administration on Serum Alanine Aminotransferase (ALT), Serum Aspartate Aminotransferase (AST) and Tumor ^{10}B Concentrations of Rats Injected with Sulfhydrylborane Dimer.

	ALT (mu/ml)	AST (mu/ml)	Tumor ^{10}B (μg/g)[1]
Saline + Dimer[2]	>4000	>4000	12.6 ± 0.9
GSH-ME[3] + Dimer	415 ± 260	640 ± 300	30.7 ± 1.7
GSH-ME	58 ± 13	143 ± 78	-
Untreated	59 ± 3	146 ± 54	-

[1]Tumors were removed six hours after dimer injection.
[2]Dimer was administered as a single intraperitoneal injection of 35 μg ^{10}B/gram body weight.
[3]GSH-ME (5 μ moles/gram body weight) was injected intraperitoneally two hours before dimer injection.

protective effect against dimer-induced hepatotoxicity. This was also evident upon gross examination of the liver. GSH-ME treated rats had livers which were indistinguishable from control rats whereas livers from rats receiving saline and dimer had a mottled appearance with scattered foci of hemorrhage. In separate studies, GSH-ME administered 2 hours after dimer provided no protection.

Effect of GSH-ME on Tumor Boron Concentration Following Administration of BSH

The unexpected observation that the average boron concentration in tumors from rats injected with GSH-ME and dimer was more than double that found in tumors from rats injected with saline and dimer (Table 1) led us to test the effect of GSH-ME administration on tumor boron concentration following the injection of BSH. Ten rats, each bearing 4 subcutaneous gliosarcomas, were divided into 2 groups and given either GSH-ME or saline 2 hours prior to a single injection of 40 $\mu g\,^{10}B/gbw$ as BSH. The results are summarized in Figure 1.

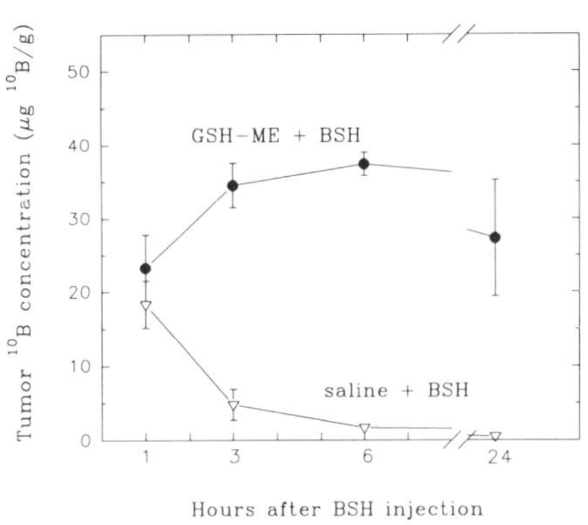

Figure 1. Effect of preadministration of glutathione monester (GSH-ME) on tumor ^{10}B concentration following injection of BSH at a dose of 40 $\mu g\ ^{10}B/gbw$.

Whereas the average boron concentration in tumors from rats pretreated with saline decreased rapidly, boron concentrations in tumors from rats pretreated with GSH-ME increased with time and remained high even 24 hours after the injection of BSH. Blood boron levels were correspondingly high.

Clonogenic Survival Following BNCT

The above findings suggest that preadministration of GSH-ME may alter the distribution of BSH leading to an increased intracellular localization and retention of boron. This hypothesis was tested indirectly by assaying clongenic cell survival following BNCT. Two groups of rats bearing 16-day intracerebral gliosarcomas were injected with either BSH alone or GSH-ME and BSH, using dosage schedules that yielded the same average tumor boron concentration in both groups. Three rats were exposed to 20 Gy x-rays for comparison. The results are shown in Table 2.

Although the average tumor boron concentration was the same in both groups of rats, a greater than 10-fold reduction in clonogenic cell survival was observed in tumors from GSH-ME treated rats. The marked reduction in clonogenic cell survival is undoubtedly due to an increase in intracellular boron leading to higher radiation doses to tumor cell nuclei from the $^{10}B\ (n,\alpha)^7Li$ reaction and further emphasizes the importance of microdistribution of boron in BNCT.

Table 2. The Average Number of Colony-Forming Units Derived in vitro from Intracerebral Gliosarcomas Treated with BNCT in vivo.[1]

Boron Compound Infused	Colony-Forming Units Per 10,000 Cells[2]	
	5.0 MW - min[3]	7.5 MW - min
BSH	370 ± 100 (n=6)	170 ± 80 (n=5)
GSH-ME + BSH	30 ± 20 (n= 7)	10 ± 7 (n=5)
X-Rays (20 Gy)	30 ± 10 (n=3)	-

[1]The average tumor ^{10}B concentration in each group of rats was ~ 30 $\mu g/gram$ and the average thermal neutron fluence was 2 x 10^{12} (5.0 MW-min) or 3 x 10^{12} (7.5 MW-min).
[2]Corrected for plating efficiency of untreated control tumors harvested at the same time.
[3]Reactor power in megawatts (MW) χ in irradiation time.

ACKNOWLEDGEMENTS

We thank the excellent assistance of M. M. Nawrocky, P. L. Micca, B. E. Pyatt, M. S. Makar, and Science and Engineering Research Students P. M. Johnson, A. S. Narang, and L. A. Acosta Garcia. Research supported under contract No. DE-AC02-76CH00016 with the United States Department of Energy.

REFERENCES

1. H. Hatanaka, S. Kamano, K. Amano, S. Hojo, K. Sano, S. Egawa, H. Yasukochi, Clinical experience of boron-neutron capture therapy for gliomas-a comparison with conventional chemoimmuno-radiotherapy, in: Boron-Neutron Capture Therapy for Tumors, H. Hatanaka, ed., Nishimura and Co., Niigata, Japan (1986).
2. D.D. Joel, D.N. Slatkin, R.G. Fairchild, P.L. Micca, and M.M. Nawrocky, Pharmacokinetics and tissue distribution of the sulfhydryl boranes (monomer and dimer) in glioma-bearing rats, Strahlentherap. Onkol. 165:167 (1989).
3. D.D. Joel, V. Benary, R.M. Brugger, and J.A. Coderre, Some aspects of BNCT research at the Brookhaven National Laboratory, in: "Proceedings of the International Workshop Towards Clinical Trials of Glioma with BNCT", D. Gabel and R. Moss, eds., Plenum Press, New York (In Press).
4. G.R. Wellum, E.I. Tolpin, A.H. Soloway, and A. Kaczmarczyk, Synthesis of μ-disulfido-bis (undecahydrocloso-dodecaborate) (4-) and of a derived free radical, Inorg. Chem. 16:2120 (1977).
5. M.E. Anderson, F. Powrie, R.N. Puri, and A. Meister, Glutathione monoethyl ester: preparation, uptake by tissue and conversion to glutathione, Arch. Biochem, Biophys. 239:538-548 (1985).
6. M.E. Anderson, A. Noganuma, and A. Meister, Protection against cisplatin toxicity by administration of glutathione ester, FASEB 4:3251 (1990).
7. B.A. Neuschwander-Tetri, L.D. Ferrell, R.J. Sukhabote, and J.H. Grendell, Glathione monoethyl ester ameliorates caerulein-induced pancreatitis in the mouse, J. Clin. Invest. 89:109 (1992).
8. M.E. Anderson, and A. Meister, Glutathione monoesters, Anal. Biochem. 183:16 (1989).
9. D.D. Joel, R.G. Fairchild, J.A. Laissue, S.K. Saraf, J.A. Kalef-Ezra, and D.N. Slatkin, Boron neutron capture therapy of intracerebral rat gliosarcomas, Proc. Natl. Acad. Sci. 87:9808 (1990).

BIODISTRIBUTION OF A BORONATED PORPHYRIN IN BC-1 MAMMARY CARCINOMA

Julia L. Mallesch[1,2], Douglas E. Moore[1], Stephen B. Kahl[3], Rick Thorne[4] and Barry J. Allen[2]

[1]Pharmacy Department, University of Sydney NSW 2006 Australia
[2]Biomedicine and Health Program, Ansto NSW 2234 Australia
[3]Department of Pharmaceutical Chemistry, UCSF, California, USA
[4]Oncology Department, Royal North Shore Hospital, NSW 2065 Australia

INTRODUCTION

Breast cancer is the most common invasive neoplasm and the second most common cause of death from cancer in women. In breast cancer, local control is a function not only of the size of the tumour but also of associated features, e.g. satellite nodules, and a great majority of patients die with disseminated disease. For many years the standard treatment for breast cancer has been surgery, coupled with post-operative radiotherapy and chemotherapy, where indicated. The use of radiotherapy has reduced the incidence of recurrence but there has been little effect on survival rates. Chemotherapy is also used to treat recurrent breast cancer but the use of these drugs has inherent problems.

Boron neutron capture therapy (BNCT), as a binary modality, has the potential to be more selective in targeting local disseminated cells and may improve survival rates and prolong recurrence free periods. In addition, due to the selective nature of the therapy, there may be a reduction in necrosis and fibrosis, avoiding long term effects from radiotherapy. In particular, BNCT may find application in the treatment of metastases to the nodes, bone, muscle, skin, lung and liver.

Porphyrins have long been known to locate in a wide variety of solid tumours although the precise biochemical basis remains unclear. The boronated protoporphyrin, BOPP, was used to evaluate the uptake into the BC-1 mammary carcinoma to assess the potential of using BNCT with porphyrins in the management of breast cancer.

MATERIALS AND METHODS

The BC-1 tumour originated as a spontaneous tumour in the breast of a Dark Agouti (DA) female rat. Histologically the tumour is an anaplastic adenocarcinoma forming some glandular patterns with little connective tissue stroma. The tumour tends to grow rapidly and

Advances in Neutron Capture Therapy, Edited by
A.H Soloway *et al.*, Plenum Press, New York, 1993

many areas of necrosis become evident. Large tumours invade the underlying tissues and ulcerate through the overlying skin[1].

Adult female DA rats of approximately 165 g were injected with BC-1 cells harvested from monolayer culture during the logarithmic stage of growth. Following a subcutaneous injection of approximately 3 to 5 x 10^6 BC-1 tumour cells in a 500 µL cell suspension into the area of the right inguinal fat pad, tumours appeared in 10-14 days and grew rapidly. Typically the tumour weight at the time of sacrifice was 1.7 ± 1.0 g (mean ± sd).

BOPP was synthesised as described previously[2] and suspended in isotonic saline at a concentration of 10 mg/ml prior to an 0.5 ml intravenous tail vein administration into the tumour bearing DA rats. BOPP is approximately 30% boron by weight.

To study the uptake kinetics, the tumour bearing rats were injected with 5 mg of the porphyrin i.v. and sacrificed at intervals up to 72 hours. The uptake of BOPP into the tissues was determined by inductively coupled plasma-atomic emission spectrometry (ICP-AES)[3].

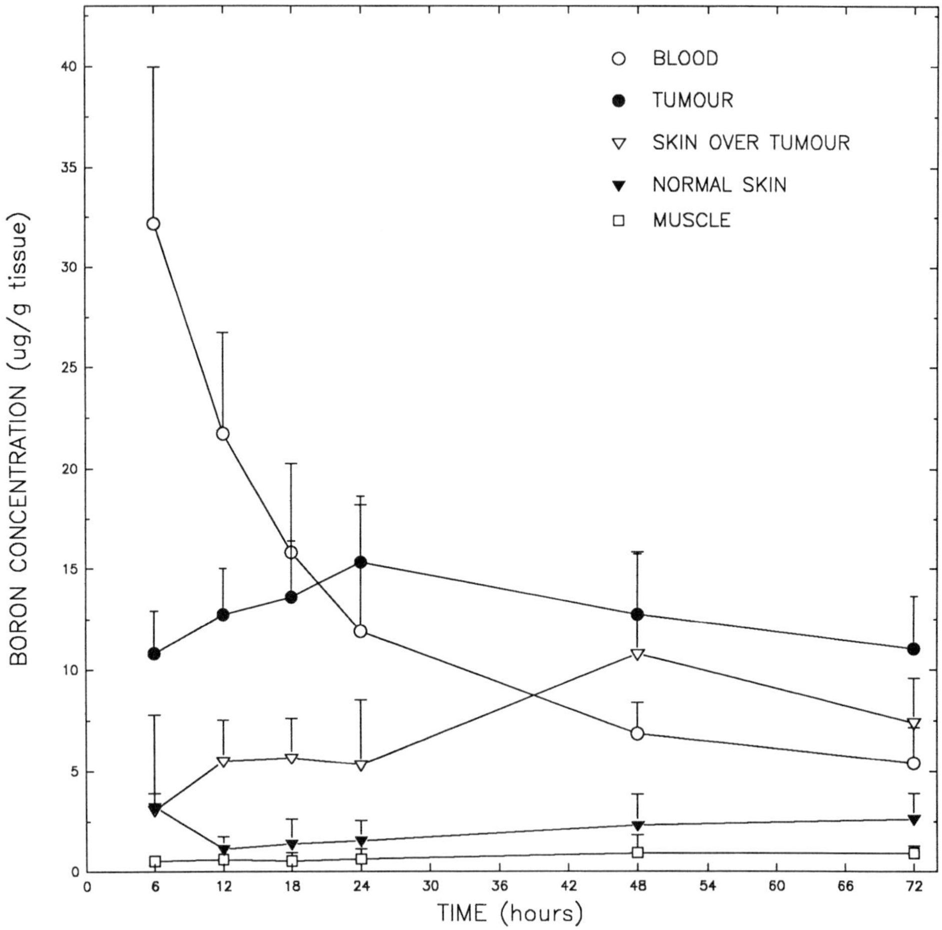

Figure 1. Biodistribution of BOPP in rats bearing BC-1 mammary carcinoma xenografts following 5 mg intravenous injection (mean ± sd: n=4).

RESULTS AND DISCUSSION

The boron concentration-time profiles for the various tissues following i.v. administration of 5 mg of BOPP are shown in Figure 1 and the pharmacokinetic parameters for the blood data are shown in Table 1. It is apparent that BOPP is a selective tumour localising agent. Maximal tumour boron concentrations were observed at 24 hours. Tumour levels remained relatively high throughout the experimental time course indicating the retention of BOPP in the tumour, even though blood levels fell substantially. Maximal tumour to blood ratio was found at 48 hours to be approximately 2:1.

The levels in the skin located over the tumour are substantially higher and exhibit different behaviour to the skin over the contralateral fat pad. This could be due to some leaching effect from the tumour into the surrounding tissues. The tumour possesses some fluid filled cysts which could explain this leaching process.

Table 1. Blood pharmacokinetic parameters in Da rat following 5 mg iv BOPP; assuming a one compartment intravenous model

Pharmacokinetic Parameters		
Model Used: Declining *e*-function		
BOPP dose	mg	5
Total B dose	mg	1.5
Theoretical initial concentration	μg/ml	39.74
Elimination constant	h^{-1}	0.046
Clearance half life	hrs	15.17
Apparent volume of distribution	l	0.038

The amount of boron found in the muscle and in the normal tissues is significantly below that of the blood. Therefore, BOPP does not readily diffuse through the blood vessel endothelium into these tissues. The tumour:muscle boron ratio is around 15:1 which indicates that BNCT could be of greater advantage in the treatment of metastatic cancer to the lymph nodes or to the skin. The low uptake of BOPP into the muscle indicates that the treatment of metastatic nodules would be more effective as a greater neutron dose could be directed to the tumour and the surrounding normal tissue would be spared. A capillary neutron dose reduction factor of three[4] would also apply because of the disparity of blood borne boron concentration with that in the surrounding tissues, but not in the tumour. Thus in principle a therapeutic ratio based on the effective boron concentrations could be as high as 45. This value could be reduced by the pressures of concomitant background radiation.

In addition, BOPP has been found previously to locate subcellularly into the mitochondria[5]. This localisation allows selective targeting of critical sub-cellular structures and should allow BNCT treatment of breast cancer to be highly dose effective, especially in the case of local disseminated disease. The uptake of BOPP into normal mammary tissue and the pharmacokinetics of BOPP in human mammary lines is still to be investigated.

ACKNOWLEDGMENT

Work supported in part by the Government Employees Assistance to Medical Research Fund.

REFERENCES

1. M.J. Mongelli, in thesis presented to the University of Sydney for the Degree of Bachelor of Science (Medicine) 1978
2. S.B. Kahl and M.S. Koo, *J. Chem. Soc. Chem. Commun.* 1768-1771, (1990)
3. S.R. Tamat, D.E. Moore, B.J. Allen, Determination of boron in biological tissues by inductively coupled plasma atomic emission spectrometry, *Anal Chem.* 59:1261-2164 (1987)
4. B.J Allen, Dose modification factors in boron neutron capture therapy. *Strahlenther. Onkol.* 1691:29-33, (1993).
5. J.S. Hill, S.B. Kahl, A.H. Kaye, S.S. Stylli, M.S. Koo, M.F. Gonzales, N.J. Vardaxis, C.I. Johnson. *Proc. Natl. Acad. Sci. USA,* 89:1785-1789 (1992)

BLOOD-BRAIN BARRIER PERMEABILITY IN PARTIALLY DEUTERATED RAT BRAIN

Yoshinobu Nakagawa,[1] Teruyoshi Kageji,[1]
Hiroshi Hatanaka,[2] Maki Moritani,[3] Katsuji Kitamura,[3]
Keizo Matsumoto,[3] and Mitsue Kobayashi[4]

1 Department of Neurosurgery, National Kagawa Children's Hospital, Kagawa 765 Japan
2 Teikyo University, Tokyo 173
3 Tokushima University, Tokushima 776
4 Research Reactor Institute, Kyoto University, Osaka 590-04

INTRODUCTION

It is well known that partial deuteration of body water can increase thermal neutron penetration in tissue (Hatanaka 1986). However, if deuteration of brain tissue alters the blood-brain barrier (BBB) permeability, the advantage of BNCT might be minimized. In our previous studies, we reported that there are two different conditions for the opening of the blood-brain barrier. Typical findings were observed in an experimental ischemic model (Nakagawa et al. 1987, 1990) and kaolin induced hydrocephalus in rat (Nakagawa et al. 1985). In the early stage of brain edema, very small substances, such as ionic lanthanum (MW 138.9) or sodium fluorescein (MW 376), pass through the tight junctions between the endothelial cells of the small vessels. On the other hand, as disruption of the blood-brain barrier advanced, leakage of large substances such as horse radish peroxidase (MW 43000) and /or Evans blue (combined with serum albumin) was demonstrated by electron microscope. If under deuteration large molecular substances, such as HRP or albumin, leak out from the vessels, we have to suppose the leakage of the boron compound. In order to investigate the change of the BBB function under partial deuteration, we measured brain water content and permeability of protein.

Advances in Neutron Capture Therapy, Edited by
A.H Soloway *et al.*, Plenum Press, New York, 1993

MATERIAL AND METHODS

Wistar rats of either sex weighing 150-180 g were used. All animals were divided into 6 groups (1 control and 5 experimental groups). Each animal was given light water as drinking water except for a few days before the experiment. The animals in group 2 were given heavy water using a gastric tube for one day before the experiments, for two days in group 3 , for three days in group 4, and for four days in group 5. The dosage of the heavy water was 0.1ml/ g body weight/day. On the day of the investigation, under general anesthesia of pentobarbital, a polyethylene catheter was inserted into the femoral vein and each animal received 125-I labeled albumin as a tracer of protein. Then light water (0.1ml/g body weight) in the control group or heavy water (99% in saline , 0.1ml/g body weight) in the experimental group was given by drip infusion through the femoral vein for 60 minutes. One hour after drip infusion of light or heavy water, all animals were killed and brain samples were obtained from three different areas of the brain (cortex, brain stem and cerebellum) and the kidney. Water content of the samples was measured by the freeze dry method. Radioactivity of 125-I in blood and brain samples was determined with a gamma scintillation counter. Heavy water was measured by infrared spectroscopic analysis in the same materials.

RESULTS

D_2O concentration in blood, brain cortex and kidney showed no remarkable differences in each group (Fig. 1). Mean values of D_2O concentration in blood were 17.9% in group 2, 22.7% in group 3, 30.2 % in group 4 , and 36.5 % in group 5 . Water content in the brain tissue

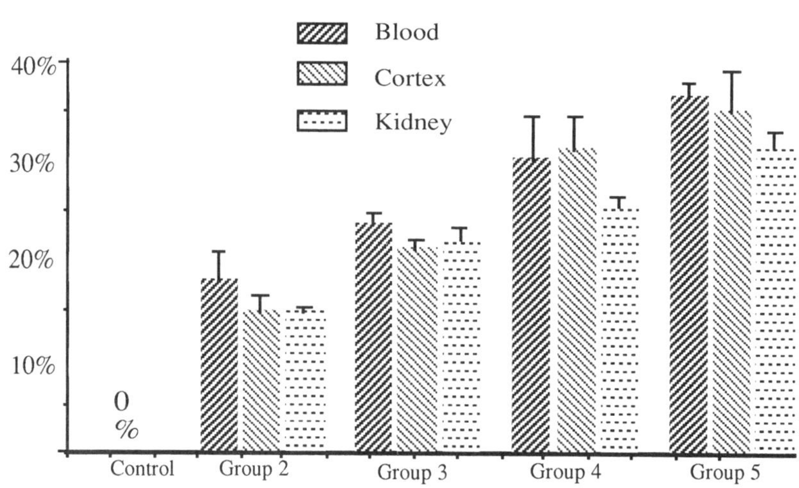

Figure 1. D_2O concentration

increased in the cortex in group 2, 3, 4 and 5; however, there were no statistical differences in each group (Fig. 2- 4). The ratio of 125-I in the brain tissue to blood showed significant differences between control and group 5 (Fig. 5). This indicated that the deuterated brain tissue over 10% showed mild leakage of water, which means the early stage of brain edema. Over 30% deuteration caused leakage of protein which raises suspicion of leakage of a boron compound from the vessels into the extracellular space around the normal brain tissue. As for BBB permeability, deuteration below 30% of body water may be still within the safety level in BNCT.

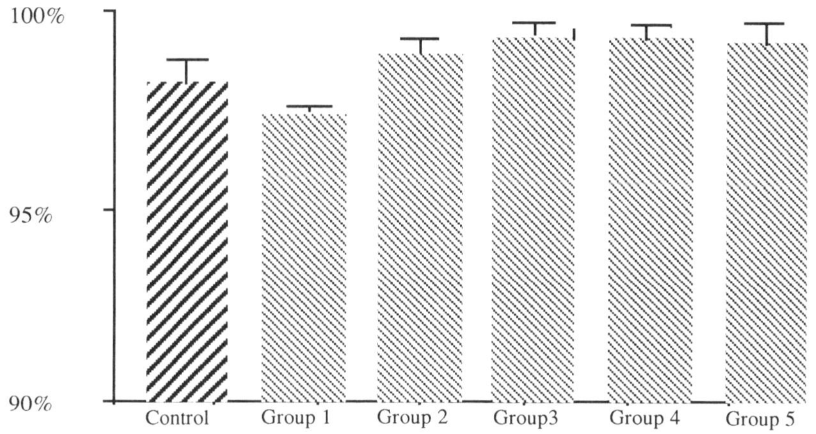

Figure 2. Water content in cortex

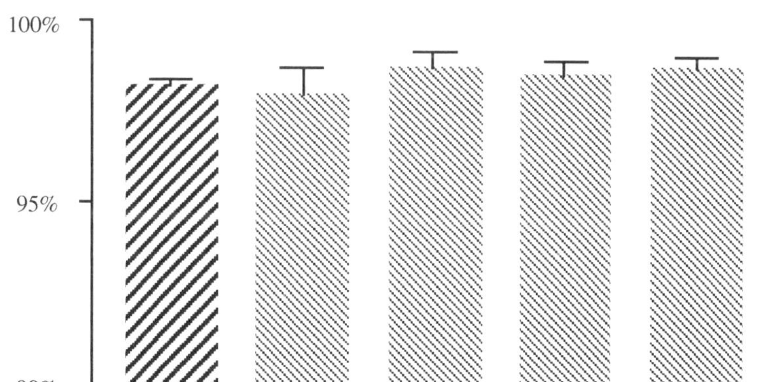

Figure 3. Water content in cerebellum

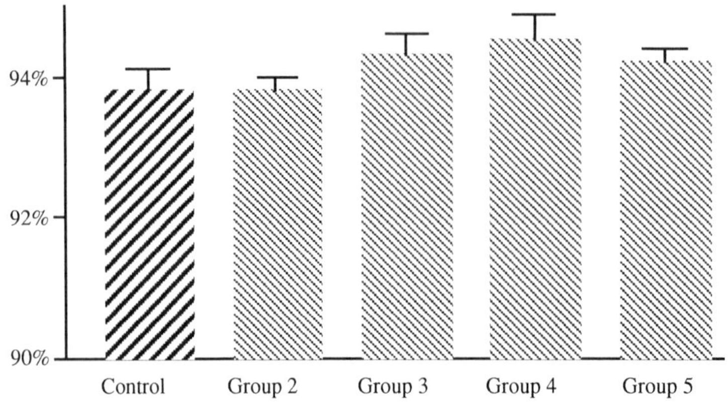

Figure 4. Water content in brain stem

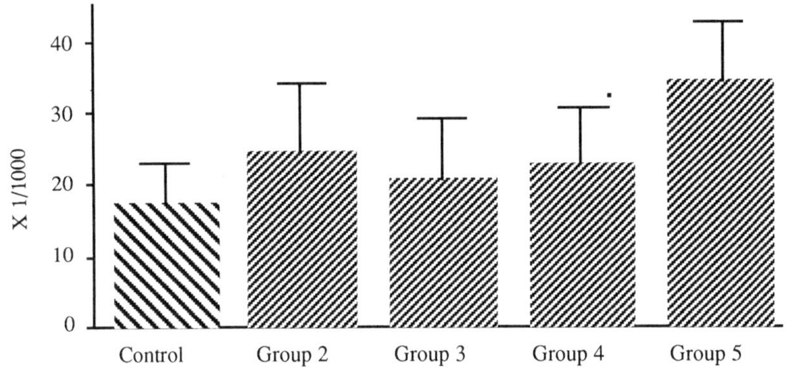

Figure 5. Ratio of 125-I in brain tissue and blood

REFERENCES

Hatanaka. H., Introduction, in Boron-Neutron Capture Therapy forTumors. Hatanaka. H., ed Nishimura, Niigata, p1-28, 1986

Nakagawa,Y., Cervos-Navarro,J., Artigas,J., Tracer Study on a Paracellular Route in Experimental Hydrocephalus , Acta Neuropatholo (Berl) 65;247-254, 1985

Nakagawa, Y., Fujimoto, N., Matsumoto.K., Cervos-Navarro,J., Blood-Brain Barrier Permeability in Acute Cerebral Ischemia after MCA Occlusion and Reperfusion in the Rat, in "Stroke and Microcirculation p 271-276," Cervos-Navarro & Ferszt, ed., Raven Press, New York. 1987

Nakagawa,Y., Fujimoto,N., K.Matsumoto, K., Cervos-Navarro J., Morphological Changes in Acute Cerebral Ischemia after Occlusion and Reperfusion in the Rat in: "Advances in Neurology Vol 52;Brain Edema p21-27 " ed by Long et al Raven Press, New York. 1990

Kobayashi, M., Decrease in Deuterium Content of Heavy Water in Contact with Air. Annual reports of the Research Reactor Institute, Kyoto University, Vol 23, p188-193. 1990

CARDIOVASCULAR TOXICITIES ASSOCIATED WITH INTRAVENOUS ADMINISTRATION OF P-BORONOPHENYLALANINE FORMULATIONS

T. R. LaHann, C. Sills, G. Hematillake, T. Dymock and G. Daniell

Center for Toxicology Research
Department of Pharmaceutical Sciences
College of Pharmacy
Idaho State University
Pocatello, Idaho, 83209

INTRODUCTION

^{10}Boronophenylalanine (BPA) has been shown to be an effective boron delivery drug for boron neutron capture therapy (BNCT) of experimental and spontaneously occurring cancers[1]. For successful BNCT, tumor boron concentrations of at least 20 ppm are thought to be necessary, but higher tumor boron levels are desirable. Calculations[2] indicate that for a given neutron exposure, each doubling of the tumor boron concentration should increase tumor cell kill by a factor of about 10,000. Thus, significant therapeutic benefit would be achieved by delivering the largest possible dose of BPA to tumor. Factors that limit the amount of BPA deliverable to tumor are 1) the aqueous solubility and 2) the intrinsic toxicity of the BPA. BPA is poorly soluble in aqueous solution (<4 g/L at pH 7.4), but formation of a BPA-fructose complex dramatically improves water solubility[3] and allows delivery of high levels of boron to tissues[3]. Little is known about the toxicity of BPA, in part because its poor water solubility has made it difficult to deliver large amounts of drug to animal tissues. Soloway et al.[4], Mishima et al.[5], Taniyama et al.[6] and Meek[7] have all studied the acute toxicity of BPA. Reported LD_{50} values for BPA ranged from 640 mg/kg[6] (i.p., pH 1 solution, rats) to 1,520 mg/kg[4] (i.v., pH 10 solution, mice). Chronic oral and s.c. dosing of BPA suspensions was well tolerated[7], although the apparent safety of BPA suspensions may more reflect BPA's relative insolubility at physiological pH than its intrinsic toxicity profile. Nevertheless, the prevailing view seems to be that BPA is a relatively safe drug. While assessing the biodistribution of novel BPA-cyclodextrin formulations, we noted that i.v. infusion of this formulation could impair cardiovascular and pulmonary function in dogs[8]. To better understand the significance of such an observation, we undertook a study of the effects of high dose BPA infusions on the function of the cardiovascular and pulmonary systems. In this paper, we report the approximate maximally tolerated dose for i.v. infusion of a BPA-fructose complex and the effects of high dose BPA infusions on cardiovascular function in the rat.

METHODS

Male Sprague-Dawley and Long-Evans rats were anesthetized (pentobarbital 50 mg/kg i.p or urethane 1.2 g/kg i.p.). Arterial blood pressure was monitored via an indwelling cannula (carotid or femoral artery). Cardiac left ventricular pressure was measured from a cannula threaded through the right carotid artery. Drug was infused via either the right jugular or femoral vein, and the left jugular vein was cannulated for measurement of central venous pressure. Heart rate was calculated from the EKG R-wave, 3-lead EKG being recorded from subdermal electrodes. The trachea was cannulated for artificial respiration (70 breaths/minute, tidal volume: 2.5 - 3.0 cc/stroke). Rectal temperature was clamped at 37.5 $\pm$ 1 degree C. Cardiac and stroke output were monitored by a 2 mm transit-time ultrasound flow probe (Transonic 6E) placed around the ascending aorta. D,L-p-boronophenylalanine (Callery Chemical Company, Pittsburgh PA) was prepared as a 6.0 $\pm$ 0.1 mg/ml (mean $\pm$ SEM, N=4) solution in pH 7.4, 0.1 M phosphate buffered saline by transiently increasing the pH of the suspension to about 10 and then readjusting pH to 7.4. BPA was prepared as a 173.2 $\pm$ 3.8 mg/ml (mean $\pm$ SEM, N=18, range: 148.6 - 196.9) solution in pH 7.4 fructose/distilled water by adding fructose to a suspension of 200 mg/ml BPA in a 1:1 molar ratio, transiently increasing the pH of the suspension to about 10 and then readjusting the pH to 7.4. A solution osmotically equivalent to 186 mg/ml BPA was prepared by dissolving mannitol and fructose (1:1 molar ratio) in pH 7.4 distilled water. Formulations were passed through a 0.22 micron filter and were prepared immediately prior to use. Reported concentrations of BPA are based on measurement of boron (by inductively coupled plasma atomic emission spectrometry) in the final, filtered product. Target rates for i.v. infusion of BPA-fructose were 8.5, 12.8 or 17 ml/kg/hr, and delivered volumes were within 7% of targeted volumes. The infused volumes, corrected for the actual concentration of the infused solutions corresponded to actual BPA doses of 1,694 mg/kg, 2,306 $\pm$ 144 mg/kg or 2,972 $\pm$ 124 mg/kg. The target rate for infusion of the BPA-buffer formulation was 17 ml/kg/hr for a target BPA dose of 102 mg/kg and an actual dose of 99.0 $\pm$ 1.7 mg/kg (N=5). When a continuous drug infusion protocol was not employed, rats dosed with BPA-fructose received a 35 minute post-BPA infusion of quarter-normal saline. Physiological data were recorded on an AstroMed MT-95000 thermal recorder and simultaneously digitized and stored on magnetic tape for later analysis on an 80486-type computer. Data were analyzed using algorithms developed with the software analysis program "DATAPAC II".

RESULTS AND DISCUSSION

I.V. Infusion of BPA - Lethal Dose

Anesthetized rats infused with an average of 2,972 mg/kg of BPA (N=5) over a one hour period, survived without grossly observable impairment of function, provided that surgical stress was minimized. BPA was delivered as a BPA-fructose complex, the observation period was 7 days and minimal stress was defined as no instrumentation of the animals except an i.v. cannula for infusion of drug. Animals dosed with 2,306 mg/kg BPA (N=7) survived until sacrifice on day 2. Anesthetized rats infused with 4,334 $\pm$ 56 mg/kg of BPA (target dose: 1,586 mg/kg/hr for 3 hours, BPA-fructose complex) died (N=4). Three rats died by the end of the infusion and one rat died approximately 24 hrs after the infusion. Rats infused with a mannitol/fructose/water solution that mimicked the volume and osmotic load of a 4,827 mg/kg dose of BPA (3 hour infusion, target rate: 8.5 ml/kg/hr, N=3) survived the 7 day observation period without grossly observable impairment of function. Rats infused with saline (8.5 ml/kg/hr for 3 hours, N=2) or with fructose-distilled water (172 mg fructose/ml; 8.5 ml/kg/hr for 3 hours, N=2) also survived the 7 day observation period. These experiments indicate that death associated with the BPA infusions was not a result of volume loading or of altered systemic osmolarity. Rats instrumented for recording of cardiovascular and pulmonary function and infused with the BPA-fructose complex died after receiving doses of BPA that were non-lethal in minimally stressed rats. Three of five rats instrumented for recording of cardiovascular parameters and infused with 2,181 $\pm$ 30 mg/kg of BPA died within 6 hours of infusion. Physiological stress apparently increases the toxicity of BPA, rendering smaller doses of BPA more likely to kill the rats. This suggests that patients afflicted with cancer may be more sensitive to the toxic effects of BPA than healthy subjects would be.

Cardiovascular Effects of BPA

The rapid death after infusion of high doses of BPA implicated cardiovascular or pulmonary dysfunction or a combination of both as the primary system failure responsible for death. Artificial support of respiration failed to protect against the lethal effect of BPA, implicating cardiovascular impairment as the causative factor. BPA formulations elevated arterial blood pressure (BP), partially as a result of the volume load infused, but BPA increased BP more than did equal volumes of control solution. After infusion of large amounts of BPA, the pressor response was followed by a gradual, sustained decline of BP (figure 1). Only minor changes in central venous pressure were observed. Like BP, cardiac output (CO) increased during BPA infusion, but then gradually and continuously declined until the animals died (figure 1). Peak stroke output (SO; maximum flow per cardiac contraction) initially increased, but then declined markedly. Total peripheral resistance (TPR, a measure of the extent of vascular constriction) increased over time (figure 1), possibly to compensate for the fall in CO. CO, SO and TPR were little affected by the mannitol infusion.

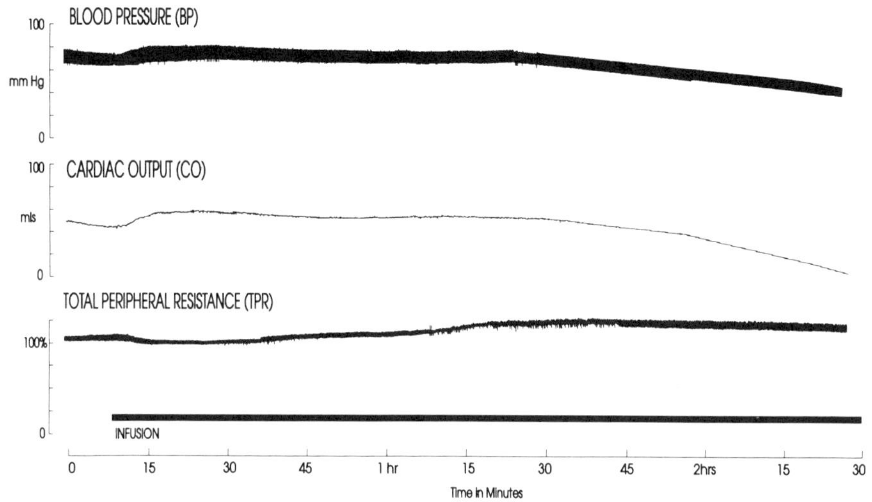

Figure 1. Blood pressure, cardiac output and total peripheral resistance before and during a 3 hour infusion (bar) of BPA (4,394 mg/kg) to a Long-Evans rat.

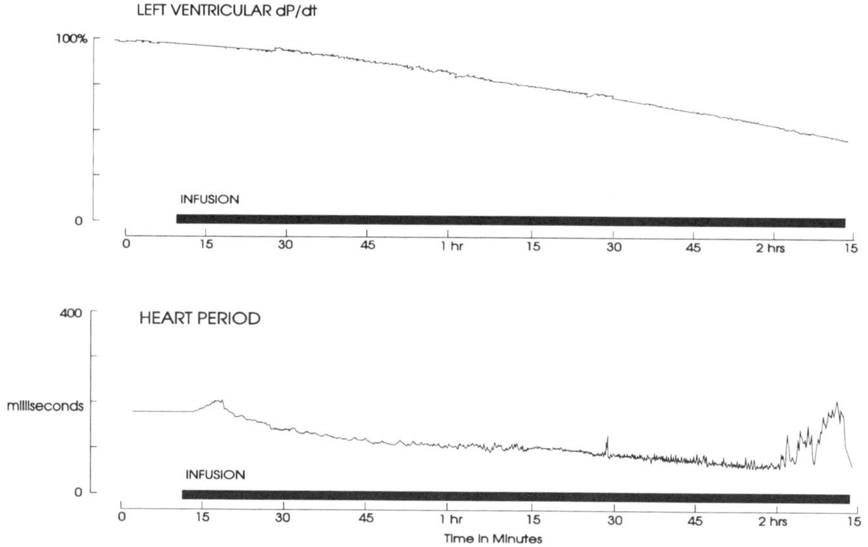

Figure 2. Left ventricular dP/dt and heart rate before and during a 3 hour infusion (bar) of BPA (4,394 mg/kg) to a Long-Evans rat.

BPA also altered left ventricular dP/dt and heart period (the time between heart beats) (figure 2). In some animals, BPA exerted a positive inotropic effect, but as the infusion proceeded, cardiac contractility always diminished. Heart period decreased gradually (i.e., heart rate increased) until just prior to death, which always appeared to be preceded by ventricular fibrillation.

Usually, cardiovascular parameters were little affected in rats receiving doses of BPA less than 100 mg/kg. However, there may be a small subpopulation of animals in which BPA doses as low as 100 mg/kg impair cardiovascular function. BPA-induced cardiovascular lability (figure 3) has been observed in only two rats, but an abnormal blood pressure response and a Biots-type breathing pattern were also observed in one of three dogs dosed with a BPA-Molecusol® formulation[8]. Vehicle-induced cardiovascular lability has not been observed in either rats or dogs, suggesting that the lability associated with BPA infusions was unlikely to be a measurement artifact. Although only a few animals have displayed cardiovascular lability after BPA administration, because lability has been observed in both dogs and rats, it may be that a small subpopulation of animals are particularly sensitive to BPA. This sensitivity might have a genetic basis, or may reflect a reduction in the animal's functional reserve. If a reduced functional reserve is the predisposing factor, then compared to healthy subjects, cancer patients may be at greater risk of BPA-induced toxic reactions.

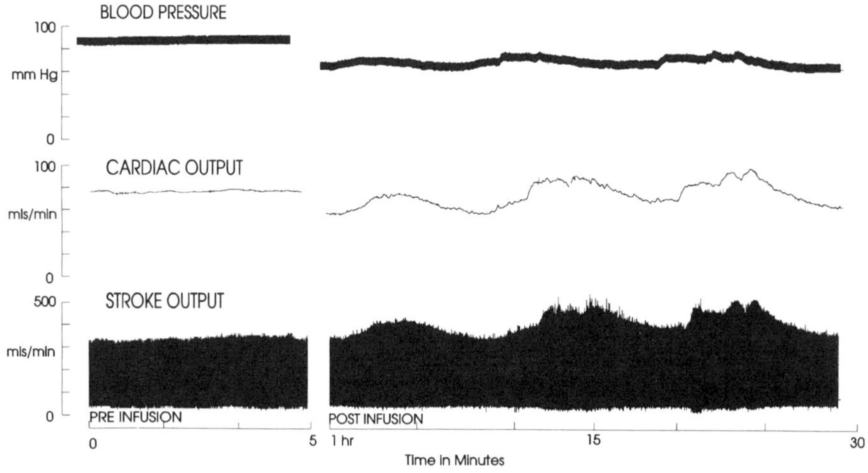

Figure 3. Blood pressure, cardiac output and stroke output before and after infusion of BPA (100 mg/kg) to a Sprague-Dawley rat.

CONCLUSIONS

Intravenous administration of BPA doses up to about 3,000 mg/kg appear to be well tolerated when delivered over a one hour infusion period to minimally stressed rats. Infusion of more than 4,000 mg/kg over a three hour time period resulted in death, probably as a result of BPA-induced impairment of cardiovascular function. Stress seemed to exacerbate BPA's toxicity so that doses of BPA that were not lethal in minimally stressed animals, did kill when administered in the presence of significant physiological stress. Finally, it is possible that a subpopulation of animals may exist that is abnormally sensitive to BPA and that administration of BPA doses as low as 100 mg/kg to these animals may elicit cardiovascular lability.

ACKNOWLEDGEMENTS

This study was performed under the auspices of the U.S. Department of Energy, Office of Energy Research, DOE Idaho Field Office, Contract No. DE-AC07-76ID01570

REFERENCES

1. Barth, R., Soloway, A. & Fairchild, R. (1990) Boron Neutron Capture Therapy for Cancer. *Scientific American* **263**, 100-107.
2. Wheeler, F., Griebenow, M., Wessol, D., Nigg, D. & Anderl, R. (1989) A Stoichastic Model for High LET Response for BNCT, in *Basic Life Sciences: Clinical Aspects of Neutron Capture Therapy*, eds. Fairchild, R., Bond, V. & Woodhead, A. (Plenum Press, New York) pp. 165-178.
3. LaHann, T., Lu, D., Sills, C., Daniell, G., Kraft, S., Gavin, P. & Bauer, W. (in press) Bioavailability of Intravenous Formulations of p-BPA in Dog and Rat, in *Proceedings of the Fifth International Symposium on Neutron Capture Therapy*, eds. Barth, R. & Soloway A. (Plenum Press, New York).
4. Soloway, A., Wright, R. & Messer, J. (1961) Evaluation of Boron Compounds for Use in Neutron Capture Therapy of Brain Tumors. *J. Pharmacol. Exp. Therap.* **134**, 117-122.
5. Mishima, Y., Kakihana, H., Okamoto, M., Yoshino, K. & Konno, K. (filed 1979) Auxiliary Agents for Neutron Capture in Skin, *Japanese Patent Application 30209-1979.*
6. Taniyama, K., Fujiwara, H., Kuno, T., Saito, N., Shuntoh, H., Sakaue, M. & Tanaka, C. (1989) Acute and Subacute Toxicity of ^{10}B-Paraboronophenylalanine, *Pig. Cell Res.* **2**, 291-296.
7. Meek, A. (1988) Investigational New Drug Application for p-Boronophenylalanine, *United States Food and Drug Administration Investigational New Drug Application.*
8. LaHann, T., Bauer, W., Gavin, P. and Lu, D. (in press) Pharmacokinetics and toxicity of a p-boronophenylalanine-cyclodextrin formulation delivered by i.v. infusion to dogs, in *Polymeric Drugs and Drug Delivery Systems*, ed. Ottenbrite, R. (American Chemical Society, Washington).

COPYRIGHT

TREATMENT OF INTRACEREBRAL MALIGNANT MELANOMA USING A RAT MODEL AND L-BORONOPHENYLALANINE AS THE CAPTURE AGENT

M.Q. Bailey,[1] K.Z. Matalka,[1] R.F. Barth,[1] J.A. Coderre,[2] A.H. Soloway,[1] J.H.Goodman,[1] A.E.Staubus,[1] E.K.Rofstad[3]

[1] The Ohio State University, Columbus, OH 43210, USA
[2] Brookhaven National Laboratory, Upton, NY 11973
[3] Norwegian Radium Hospital, Oslo, Norway

INTRODUCTION

Malignant melanoma is a highly metastatic tumor to the skin, liver, lung and especially the brain. Patients with brain metastases have a poor prognosis despite aggressive chemo- and radiotherapy. Many of these patients have disseminated melanoma but eventually succumb to their cerebral metastatic disease.

Boron neutron capture therapy (BNCT) is a potential therapeutic modality for brain tumors. Boronophenylalanine (BPA) is a boron containing aromatic amino acid that was synthesized to target neoplastic tissue. It has been shown that BPA accumulates in amelanotic tumors at a boron-10 (^{10}B) concentration (15-30 μg/g) that is believed to be adequate to sustain a lethal $^{10}B(n,\alpha)^7Li$ reaction. The purpose of the present study was to evaluate the theraputic efficacy of BNCT using BPA on a human melanoma cell line implanted intracerebrally in nude rats.

MATERIALS AND METHODS

Rat Brain Tumor Model

A brain tumor model utilizing nude rats has been developed employing the human melanoma cell line MRA 27. This cell line was derived from a 60 year old Caucasian male and has been propagated both *in vitro* and *in vivo* in nude mice and rats. MRA 27 cells were grown in McCoy's 5A media supplemented with 10% fetal bovine serum, 100 units/ml penicillin, 100 g/ml streptomycin, and 2 mM L-glutamine. The cells were periodically tested for mycoplasma contamination using DNA-fluorochrome staining and UV microscopy. Nude rats , 6 to 8 weeks old, were anesthetized with a 1.2/1 mixture of ketamine/xylazine, and were stereotactically injected with MRA 27 cells into the right

Advances in Neutron Capture Therapy, Edited by
A.H Soloway *et al.*, Plenum Press, New York, 1993

caudate nucleus at a concentration of $2x10^5/10\ \mu l$ of media containing 1% agarose at a gelling temperature of $<30°C$ as previously described. An injection time of 10 seconds was employed followed by rapid filling of the central screw hole with bone wax following withdrawal of the needle. The operative area was flushed with betadine solution and the incision was closed with a single sterile dermal clip. Median survival time (MeST) of nude rats implanted with MRA 27 was ~44 days.

Cesium 137 Irradiation Studies

In vivo gamma irradiations were carried out with a Picker Cesium 137 (Cs-137) teletherapy unit. The Cs-137 teletherapy used a 2 cm (OD) cone at 15 cm SSD with an output of 100 cGy/min. Gamma irradiations were performed 31 days following implantation of 2×10^5 MRA 27 cells. Three groups of rats with 5-6 rats per group were used. Group I received a single fraction of 1200 cGy, Group II received 9 fractions of 200 cGy, divided over eleven days, Group-III were untreated controls. The gamma photon beam was collimated to the skull. The body, excluding the head, was shielded with 5 mm thick lead.

Neutron Capture Therapy Irradiation

The Brookhaven Medical Research Reactor (BMRR) thermal neutron beam was used for BNCT experiments. The reactor power was maintained at 1.25-MW during the irradiation of all rats. BNCT was carried out 30 days following implantation of 2×10^5 of MRA 27 cells. Six groups of rats with 9-10 animals per group were used. Groups 1 and 2 received irradiation of 6 or 8 MW-min respectively. Groups 3, 4, and 5 received 4, 6 , or 8 MW-min of irradiation 6 hours post i.p. administration of 120 mg, 95% ^{10}B-enriched L-BPA (897 mg/kg, 47.2 mg of boron/Kg of body weight) in a fructose complex at pH 8.8. From previously reported work this was determined to be the best composite time. Tumor concentrations of ^{10}B (23.7 $\mu g/g$) were within the range considered effective for BNCT demonstrating a monoexponential decay in the tumor that would allow sufficient retention of ^{10}B[5]. Group 6 served as untreated controls. All rats were anesthetized with a 1.2/1 mixture of ketamine/xylazine and placed supine in a body shield-head stabilizer, designed by Dr. D. Slatkin, as described elsewhere. A 1.15 cm diameter collimated neutron beam was used to irradiate the rats head.

Dosimetry of the 1.2 MeV CS-137 gamma radiation was determined with a Capentec 192X rate exposure meter with a 0.6 ml Farmers replacement ionization chamber. The mixed radiation during BNCT includes thermal neutrons (~0.025 eV), fast neutron (>10,000 eV), photons and heavy particles that were generated in the presence of boron, $^{14}N(n,p)^{14}C$ and $^{10}B(n,\alpha)^7Li$ reactions. As described previously, the n_{th} fluences were measured in a killed rat using a 3 mm long gold wire that had been implanted into the cerebrum to a depth of 4-5 mm below the skull surface . The dose resulting from the $^{10}B(n,\alpha)^7Li$ and $^{14}N(n,p)^{14}C$ reactions were calculated using data from the measured n^{th} flux, assuming uniform boron distribution and a tissue concentration of 2.6%. The extrinsic photon and fast neutron doses were measured by using paired tissue equivalent plastic chambers with TE gas (Rossi gas) and graphite chambers filled with CO_2. The calculated doses in cGy are based on the assumption of uniform ^{10}B distribution the tumor and normal tissues.

Therapeutic response

Therapeutic response of the treatment regimens were evaluated by the percentage of increased life span (%ILS), determined from the Median Survival Time (MeST) in days following tumor implantation. The %ILS was determined by the formula:

$$\%ILS = \frac{[(\text{MeST of treated group - MeST of untreated group})}{(\text{MeST of untreated group})]} \times 100 \quad (1)$$

RESULTS

The MeSTs for rats treated with a single dose of gamma photons of 1200 cGy or 9 fractions of 200 cGy each (200 cGy x 9) were 86 d and 79 d, respectively compared to untreated controls with a MeST of 46 days (Fig 1). These differences in survival time were significant at $p<0.01$ for irradiated groups compared to 46 d for untreated animals. The %ILS of the rats that received 1200 cGy and 200 cGy x 9 were 83% and 67% respectively; these were not significantly different from each other.

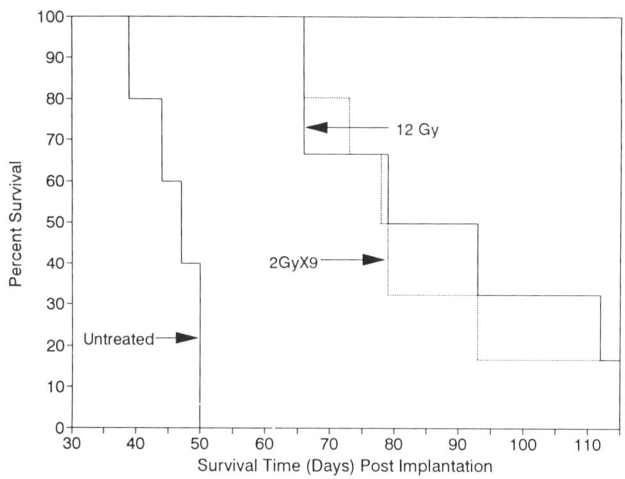

Figure 1. Survival Time Post Intracerebral Implantation of MRA 27 (Days). Gamma Photon Irradiation.

Animals treated with neutrons but without BPA 6 and 8 MW-min (273 and 364 cGy) showed an enhanced MeST (76d and 93d) that were significantly different than the non-irradiated controls but not the gamma irradiation groups. Animals treated with BPA and 4, 6, and 8 MW-min of irradiation (502, 754, and 1005 cGy) had higher MeST (170, 182 and 262 days) and %ILS (286, 314 and 495%) than gamma irradiated rats (Fig 2).

DISCUSSION

Extensive animal studies have been carried out on the treatment of extracerebral malignant melanoma, however there is little, if anything in the literature on the treatment of intracerebral melanoma. This model seems to be very useful for the evaluation of therapeutic modalities for the treatment of intracerebral melanoma.

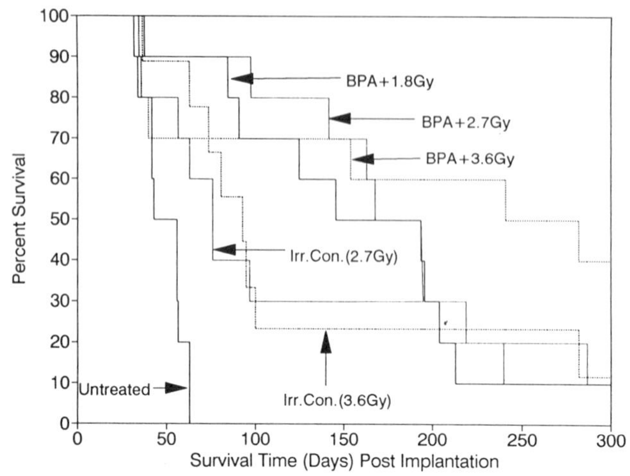

Figure 2. Survival Time Post Intracerebral Implantation of MRA 27 (Days) using BNCT Thermal Neutron Irradiation.

BNCT of the intracerebral malignant melanoma model shows clear evidence of therapeutic efficacy. The radiation dose in BNCT, groups 3, 4, and 5 were (296, 443, and 591 cGy) to the brain and (309, 464, and 618 cGy) to the blood in contrast to the tumor dose (502, 754, and 1005 cGy). In all cases examined, BPA administration 6 hours before BNCT showed a positive increase in life span that was greater than non-irradiated controls or, rats that had received gamma radiation or neutron radiation only. Rats that received BNCT at 8 MW-min (1005 cGy to the brain) demonstrated a 40% survival at greater than 300 days. These differences in radiation doses to the tumor and surrounding normal tissues resulted in sparing of normal tissues below tolerance, and illustrates the potential advantage of BNCT over other forms of radiation therapy. This data strongly suggests that large animal studies using BPA-BNCT for the treatment of intracerebral malignant melanoma are essential to further define both therapeutic efficacy and normal tissue effects of BNCT of intracerebral melanoma.

ACKNOWLEDGMENTS

This work has been supported by the U.S. Department of Energy grant D.E.-ACO2-76CH000016 and The Ohio State University Comprehensive Cancer Center/Cancer Research Institute and the Office of Research and Graduate Studies.

REFERENCES

1. H.R. Snyder, A.J. Reedy, and Lennarz, *W.J. Am. Chem Soc.* 80:835-838 (1958).
2. J.A. Coderre. J.D. Glass, R.G. Fairchild, P.L. Micca, I Fand, and D.D. Joel, *Cancer Res.* 50:138-141 (1990).
3. R.G. Fairchild, D.N. Slatkin, J.A. Coderre, P.L. Micca, B.H. Laster, S.B. Kahl, P. Som, I Fand, and F. Wheeler, *Pigment Cell Res.* 2:309-318 (1989).

4. N.R. Clendenon, R.F. Barth, W.A. Gordon, J.H. Goodman, F. Alam, A.E. Staubus, C.P. Voesel, A.J. Yates, M.L. Moeschberger, R.G, Fairchild, and J.A. Kalef-Ezra, *Neurosurgery* 26:47-55 (1990).
5. K. Matalka, M.Q. Bailey, R.F. Barth, A.E. Staubus, D.M. Adams, A.H. Soloway, S.M. James, J.H. Goodman, J.A. Coderre, R.G. Fairchild, and E.K. Rofstad, in "Advances in Neutron Capture Therapy," New York: Plenum Publishing Co., in press (1992).
6. D.N. Slatkin, J.A. Kalef-Ezra, S.K. Saraf, and D.D. Joel, in "Neutron Beam Design: Development and Performance for Neutron Capture Therapy," J.A. Harling, J.A. Bernard, and R.G. Zamenhof, eds. New York: Plenum Publishing Co., Vol. 54, 317-320 (1990).
7. J.A. Coderre, D.N. Slatkin, P.L. Micca, and J.R. Ciallella, *Radiat. Res.* 128:177-185 (1991).
8. D.D. Joel, R.G. Fairchild, J.A. Laissue, J.A. Saraf, J.A. Kalef-Ezra, and D.N. Saltkin, *Proc. Nat. Acad. Sci., U.S.A.* 87:9808-9812 (1990).

EFFECTS OF FRACTIONATION OF BORON NEUTRON CAPTURE THERAPY IN A MOUSE GLIOMA MODEL

Guido R. Solares,[1] Robert G. Zamenhof,[1] David E. Wazer,[1] Thomas A. DiPetrillo,[1] Stephen C. Saris,[2] Guillermo Cano,[2] Hywel Madoc-Jones,[1] and Otto K. Harling[3]

[1]Department of Radiation Oncology, New England Medical Center and Tufts University School of Medicine, Boston, MA, 02111 USA
[2]Department of Neurosurgery, New England Medical Center and Tufts University School of Medicine, Boston, MA, 02111 USA
[3]Nuclear Reactor Laboratory, Massachusetts Institute of Technology, Cambridge, MA, 02139 USA

INTRODUCTION

Mice with glioma 261 intracerebral tumors were fed D,L-3-(*p*-boronophenyl)alanine (BPA) and irradiated with total BNCT tumor doses of 500 to 5000 RBE-cGy using single and multiple fractions of thermal neutrons to investigate the effect of fraction size and interfraction interval on survival.

MATERIALS AND METHODS

Tumors Model

Cryopreserved tumor fragments of the glioma 261 (GL261) cell line were obtained from the Division of Cancer Treatment Repository (Frederick, Maryland). This is a murine brain tumor that was induced in the brains of C57BL/6 mice after intracerebral implantation of methylcholanthrene pellets.[1] Although these tumors do not express glial fibrillary acidic protein (GFAP) positivity, electron microscopy indicates that they contain the intermediate filaments seen in glial cells. They grow well after intracerebral inoculation in mouse brain, demonstrate a moderate degree of invasiveness, and result in death of the animal from mass effect in approximately three weeks. Tumors were grown in the flanks of C57BL/6 mice and maintained by serial subcutaneous transplantation.

Generation of Brain Tumors

Tumors were removed aseptically from the flanks of C57BL/6 mice and minced into 1 mm pieces. Approximately 12 week old female C57BL/6 mice received intracerebral inoculation of GL261 tumor fragments after induction of general anesthesia by inhalation of

Advances in Neutron Capture Therapy, Edited by
A.H Soloway *et al.*, Plenum Press, New York, 1993

methoxyfluorane. A 19 gauge spinal needle was used to inject a single tumor fragment approximately 2 mm deep to the dura just posterior to the coronal suture and lateral to the midline. Placement of bone wax in the skull defect and cleansing of the operative site with 70% ethanol prevented extracranial growth of tumor in most mice. Mice with 5 day old tumors were used in all experiments. Tumors were approximately 1 mm in diameter at this time and the mice were neurologically normal.

Boron Compound (BPA)

95% ^{10}B-enriched D,L-3-(*p*-boronophenyl)alanine (BPA) was obtained from Callery Chemical Company (Pittsburgh, PA). The drug was prepared as a slurry in distilled water immediately prior to oral feeding. After induction of general anesthesia by inhalation of methoxyfluorane, 2.5-6 mg BPA/g mouse (0.12-0.28 mg ^{10}B/g body weight) was delivered in 0.5 cc via an orogastric stent. For animals receiving more than one fraction, BPA was administered again before each fraction.

Neutron Irradiation and Dosimetry

Irradiations were carried out using a thermal neutron beam in the medical room of the 5 MW (Thermal) Massachusetts Institute of Technology Research Reactor (MITR-II). This facility provides a high thermal neutron flux with relatively small contamination from higher energy neutrons (fast and epithermal) and photons. Animals were fed BPA 3 h prior to irradiation and immediately before irradiation anesthetized by intraperitoneal (IP) injection of 0.4 ml of Avertin (0.2 mg/10 g). To prevent leakage of body fluids animals were confined in polyethylene bags and arranged in a circle on an acrylic plate. Their bodies were perpendicular to the axis of the vertical neutron beam and polyethylene bags filled with enriched ^{6}Li-carbonate shielded all but the heads.

All subsequently described BNCT doses are total doses to tumor and are expressed as equivalent absorbed dose, i.e., absorbed dose in cGy multiplied by RBE weighting factors. Table 1 presents the dose rates for each of the important BNCT components. Fast neutron, epithermal neutron, and gamma doses were measured by the paired ion chamber technique.[2] The ^{10}B concentration in tumor was determined in a previous biodistribution study[4] in which it was found that an average tumor ^{10}B concentration of 30 ± 5 μg/g is present 3 hours after administrating 2.5 mg BPA/g mouse.

Although the RBEs used in this work are smaller than those presently recognized, at the time that this work was done the RBE values recommended in reference 5 were adopted. Nonetheless, since most of the total dose is due to the ^{10}B reaction and most of the background dose is due to the gamma component, changes in the RBEs will not significantly affect the survival curves.

Table 1. Dosimetry of the Thermal Beam.[1]

DOSE COMPONENT	RBE	Tumor RBE Dose Rate (RBE-cGy/min.)	Tumor Physical Dose Rate (cGy/min.)	% of Total Dose (%)
Fast and Epithermal neutrons	1.6	2.1 ± 0.3	1.3 ± 0.2	3.3
$^{14}N(n,p)^{14}C$ reactions	1.6	1.9 ± 0.4	1.2 ± 0.3	2.9
$^{10}B(n,\alpha)^{7}Li$ reactions[2]	2.3	47.7 ± 14.4	20.7 ± 6.3	74.6
Gammas	1.0	12.2 ± 0.7	12.2 ± 0.7	19.1

[1] Thermal neutron flux of 1.32 ± 0.3 x 10^9 neutrons cm^{-2} sec^{-1} at reactor power of 4.5 MW.

[2] Assuming a tumor ^{10}B concentration of 30 ppm.

Experimental Design

To investigate the effect of fraction size on survival, non-treated controls and single fraction treatments of 1000, 2000, 3000, 4000 and 5000 RBE-cGy were compared to similar total doses delivered in two equivalent fractions within a two-day period and to doses of 3000 and 4000 RBE-cGy delivered in three daily fractions. The effect of interfraction interval on survival was studied by comparing single fractions of 1000 and 3000 RBE-cGy to 500 (x2) and 1500 (x2) RBE-cGy with interfraction intervals of 1 and 8 days. Five or seven animals were always used per group. The Mantel-Hanszel test [6] was used for statistical analysis.

RESULTS AND DISCUSSION

Fraction Size Effect on Survival

All single fraction groups survived significantly longer ($p < 0.04$) than untreated control animals at all dose levels. There was a trend of increasing survival with increasing dose to tumor (Fig. 1). One of six animals in the 2000 (x2) RBE-cGy group and one in the 5000 RBE-cGy group were survivors at 125 days. Only two groups, 2000 (x2) and 2500 (x2) RBE-cGy, of the animals that received two fractions were statistically significant ($p < 0.05$) compared to untreated animals. The three-fraction groups were not significantly different from controls.

There was not a significant benefit (or detriment) on survival for animals treated with two fractions as opposed to one fraction (Fig. 2). Animals treated with 3000 and 5000 RBE-cGy of single and fractionated dose showed no significant differences. The group exposed to a single fraction of 4000 RBE-cGy had a median survival of 32.6 days, which was less than the median survivals of the 3000 RBE-cGy group (48.3 days) and the 5000 RBE-cGy group (60.7 days). There is no explanation for this; however, in a similar experiment done by our group,[7] animals treated with a single fraction of 4000 RBE-cGy did have higher median survival than animals receiving a single fraction of 3000 RBE-cGy.

BNCT treatment with three fractions (Fig. 2), showed a detrimental effect on survival. Animals treated with 1000 (x3) RBE-cGy had a median survival of 20.2 days comparable to the median survival of 17.2 days observed in the control group. Animals treated with 1300 (x3) RBE-cGy had a median survival of 15.3 days, less than controls. These two groups received normal brain doses of about 1700 and 1900 RBE-cGy, respectively. Histopathological studies done by our group on normal mice treated by BNCT with comparable high doses, showed that early complications in these animals was due to substantial radiation damage to the gastro-intestinal tract with no apparent early radiation effects noted in the brain.

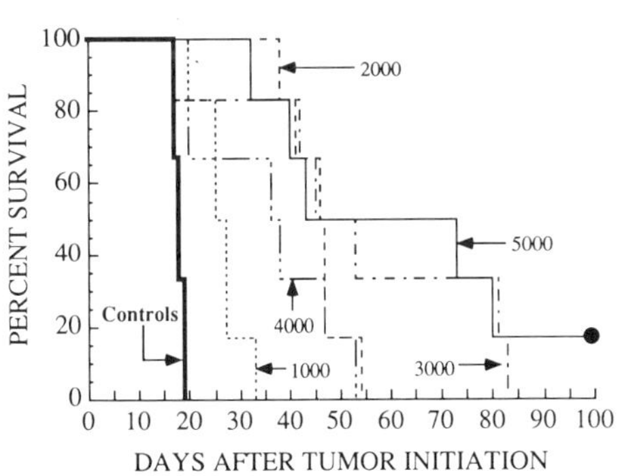

Figure 1. Single fraction survival.

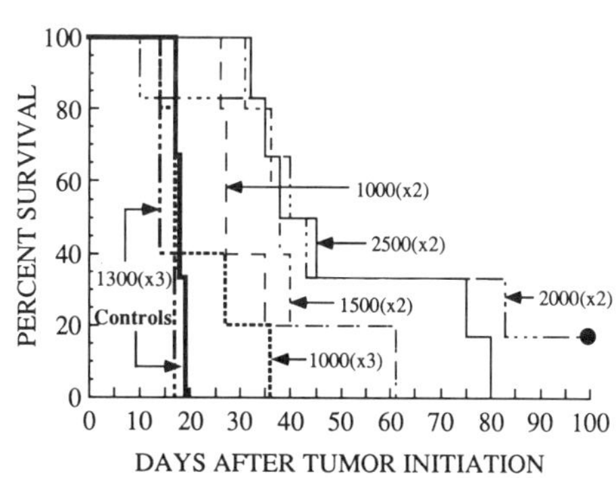

Figure 2. Two and three-fraction survival.

Interfraction Interval Effect on Survival

All groups treated with fractionated BNCT survived longer than the untreated control group (Fig. 3). There was a trend toward longer survival with increased radiation dose. The mean survival for the untreated control group was 21 days as compared to 27, 23, 49 and 48 days for the 500x2 (1 day), 500x2 (8 day), 1500x2 (1 day) and 1500x2 RBE-cGy (8 day) groups, respectively. Only group 500x2 RBE-cGy (8 day) was not significant ($p < 0.1$) relative to the control group. There was no significant benefit (or detriment) on survival for animals treated with interfraction intervals of one day as opposed to eight days nor for animals treated with a single fraction compared to two fractions.

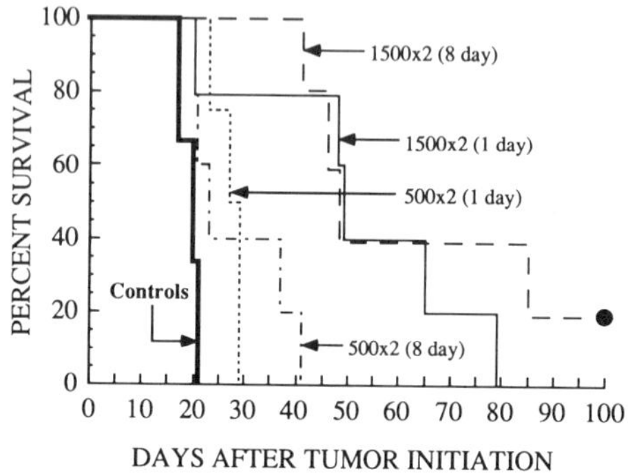

Figure 3. Two-fraction survival with interfraction intervals of one and eight days.

ACKNOWLEDGMENTS

The work reported here was supported by the U.S. Department of Energy under grants #DE-FG02-87ER-60600 and #DE-FG07-80ER-10770.A015.

REFERENCES

1. J.I. Ausman, W.R. Shapiro, and D.P. Rall, Studies on the chemotherapy of experimental brain tumors: development of an experimental model. *Cancer Res. 30:* 2394 (1970).
2. F.H. Attix, "Introduction to Radiological Physics and Radiation Dosimetry," John Wiley & sons, Inc., New York (1986).
3. G.R. Solares, "High resolution alpha track autoradiography and biological studies of boron neutron capture therapy," Ph.D. Thesis, Department of Nuclear Engineering, Massachusetts Institute of Technology, Cambridge, MA (1991).
4. G.R. Solares, R. Zamenhof, S. Saris, D. Wazer, S. Kerley, M. Joyce, H. Madoc-Jones, L. Adelman, O. Harling, Biodistribution and pharmacokinetics of *p*-boronophenylalanine in C57BL/6 mice with GL261 intracerebral tumors, and survival following neutron capture therapy, *in*: "Progress in Neutron Capture Therapy for Cancer," B. Allen, D. Moore and B. Harrington, eds., Plenum Press, New York (1992).
5. O. Harling, R. Zamenhof, J. Bernard, eds., "Neutron Beam Design, Development, and Performance for Neutron Capture Therapy," Plenum Press, New York (1990).
6. N. Mantel and W. Hanszel, Statistical aspects of the analysis of data from retrospective studies of disease, *J. Natl. Cancer Inst.*, 22:719 (1959).
7. S.C. Saris, G.R. Solares, D.E. Wazer, G. Cano, S. Kerley, M. Joyce, L. Adelman, O.K. Harling, H. Madoc-Jones, R. Zamenhof, Boron neutron capture therapy for murine brain tumors, *Cancer Res.* 52:4672 (1992).

CALIFORNIUM-252 BRACHYTHERAPY OF INTRACEREBRAL MELANOMA WITH AND WITHOUT ADMINISTRATION OF BORONOPHENYLALANINE UTILIZING A NUDE RAT MODEL

Khalid Z. Matalka[1], Rolf F. Barth[1], Michael Q. Bailey[1]
Yosh Maruyama[2], Jose M. Feola[2], and Jacek Wierzbicki[2]

[1]The Ohio State University, Columbus, OH 43210, U.S.A.
[2]University of Kentucky, Lexington, KY 40536, U.S.A.

INTRODUCTION

Californium-252 brachytherapy has been used clinically to treat bulky localized tumors such as cervical carcinomas and glioblastoma multiforme[1,2]. Results from the initial therapy trials have been promising, and have stimulated others to consider this therapeutic modality[1,2].

Beach et al.[3] have suggested that the combination of ^{252}Cf brachytherapy with boron-10 loading of the tumor could increase the tumor radiation dose due to a boost from the $^{10}B(n,\alpha)^{7}Li$ reaction. We have developed a nude rat brain model utilizing a human melanoma cell line MRA 27, which when implanted intracerebrally, grows progressively and kills the host[4]. This model simulates melanoma metastatic to the brain, which occurs in up to 50% of patients with disseminated melanoma[5]. Our previous experience with the implanted intracerebral (i.c.) melanoma in rats clearly indicated that boron neutron capture therapy (BNCT) was capable of significantly prolonging life or even eliminating implanted brain tumors in rats[4,6,7]. In the present study, we have evaluated the effects of ^{10}B loading of the MRA 27 melanoma by i.p. administration of boronophenylalanine (BPA) on ^{252}Cf brachytherapy.

MATERIALS AND METHODS

Animals and Cell Line. The human melanoma cell line MRA 27 was derived from a 60 year old Norwegian male and has been propagated both *in vitro* and *in vivo* in nude mice and rats. Six to eight week-old athymic female nude rats of NIH-rnu strain were purchased from the Animal Production Branch, National Cancer Institute, Frederick, MD. The rats were maintained under specific pathogen-free conditions and fed sterilized food and water.

Advances in Neutron Capture Therapy, Edited by
A.H Soloway *et al.*, Plenum Press, New York, 1993

Tumor Implantation. A stereotactic implantation procedure has been employed. Briefly, nude rats were sedated with a 1.2/1 mixture of ketamine/xylazine and a plastic screw was embedded in the skull. MRA 27 cells were injected through a central hole of the plastic screw into the right caudate nucleus at a concentration of 2 X $10^5/10$ μl of serum-free McCoy's 5A medium containing 1% agarose at a gelling temperature of <30 °C. The implantation procedure employed an injection time, of ~5 seconds, filling of the screw hole with bone wax following withdrawal of the needle, and flushing of the operative field with betadine before closing the scalp incision with a single sterilized clip.

Irradiation Studies. All irradiation experiments were carried out in the Radiation Therapy Oncology Center, at the University of Kentucky-Lexington, KY. Nude rats were irradiated 29 days (Exp. I) or 24 days (Exp. II) following i.c. implantation of 2 x 10^5 of MRA 27 cells. Animals were divided into four groups of 6-8 rats each in both experiments. In Exp. I, group 1a received a total radiation dose of 200 cGy of ^{252}Cf (134 cGy of neutrons and 66 cGy of γ); group 2a received the same dose as group 1a 5.5 hr following i.p. administration of 120 mg of 95% ^{10}B-enriched L-BPA; group 3a received a radiation dose of 1200 cGy of γ photons from a ^{137}Cs source; and group 4a served as untreated control. In Exp. II, group 1b received a total radiation dose of 341 cGy of ^{252}Cf (230 cGy of neutrons and 111 cGy of γ); group 2b received the same dose as group 1b 5.0 hr following i.p. administration of 120 mg of 95% ^{10}B-enriched L-BPA; group 3b received a radiation dose of 600 cGy of γ photons from a ^{137}Cs source; and group 4b served as untreated controls. The rats were anesthetized with 1.6/1 mixture of ketamine/xylazine, placed supine with their heads attached to a lucite plate and were then brought in contact with the radiation source.

RESULTS AND DISCUSSION

Uptake Studies. In this study, 120 mg of a fructose complex of ^{10}B-enriched L-BPA in 2 ml of H_2O were injected i.p. into rats 30 days following i.c. implantation of the MRA 27 melanoma. Six hours later the tumor ^{10}B concentration was 23.7 $\pm$ 2.3 μg/g compared to 9.4 $\pm$ 3.4 and 8.4 $\pm$ 0.5 μg/g for blood and brain respectively[6,7].

Irradiation Studies. In Exp. I, ^{252}Cf and ^{137}Cs irradiations were initiated 29 days following tumor implantation at which time the tumor weighed ~ 100 mg. Kaplan-Meier survival plots for the ^{252}Cf, ^{10}B + ^{252}Cf, ^{137}Cs treated rats, and the untreated animals are shown in Fig. 1. The median survival time (MeST) for the untreated rats (group 4a) was 33 d, 43 d for animals irradiated with 200 cGy of ^{252}Cf (group 1a), 54 d for rats that had received 120 mg of ^{10}B-BPA 5.5 hr prior to irradiation (group 2a), and 71 d for those that had received 1200 cGy of ^{137}Cs (group 3a). The percentages of increased life span (%ILS) for groups 1a, 2a, and 3a were 30, 64, and 115% respectively. The prolongation in survival times of all irradiated groups (1a, 2a, and 3a) were statistically significant compared to untreated animals ($p \leq 0.05$-0.01). Although the enhanced survival of rats treated with BPA prior to ^{252}Cf irradiation compared to rats irradiated with ^{252}Cf was not statistically significant, there was a statistically significant trend, as indicated by Jonckheere's test, towards increased survival times between group 1a, 2a, and 4a ($p < 0.01$).

In Exp. II, ^{252}Cf and ^{137}Cs irradiations were initiated 24 d following tumor implantation at which time the tumor weight was ~ 90 mg. Kaplan-Meier plots for all treated and untreated animals are shown in Fig. 2. The MeST for the untreated rats (group 4b) was 35 d, 47 d for those irradiated with 341 cGy of ^{252}Cf (group 1b) or 600 cGy of

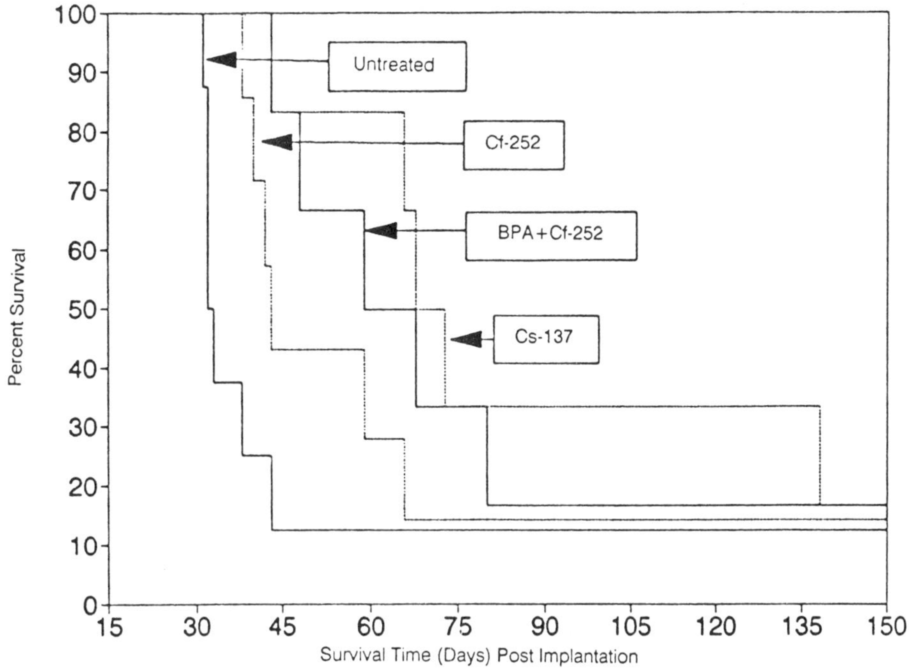

Figure 1. Exp. I. Kaplan Meier plots of rats carrying i.c. MRA 27 human melanoma tumor following ^{252}Cf or ^{137}Cs brachytherapy. Three groups were irradiated with 200 cGy of ^{252}Cf (gp 1a), 200 cGy following 5.5 hr of i.p. administration of 120 mg of ^{10}B-BPA (gp 2a), or 1200 cGy of ^{137}Cs (gp 3a). Group 4a served as untreated controls.

^{137}Cs (group 3b), and 56 d for those receiving 120 mg of BPA 5 hr prior to irradiation with 341 cGy of ^{252}Cf (group 2b). %ILS for groups 1b, 2b, and 3b were 34, 60, and 34% respectively. The prolongation in survival times of all irradiated groups were statistically significant compared to the untreated rats ($p \leq 0.05$-0.001). This trend towards increased survival times between groups 1b, 2b, and 4b was highly significant ($p < 0.005$). Furthermore, the difference in MeSTs between BPA treated ^{252}Cf group (2b) and ^{252}Cf group (1b) were also significant ($p \leq 0.05$). In addition, the MeST of group 2b was significantly different ($p \leq 0.01$) compared to the MeST of group 3b, which had received a γ photon dose of 600 cGy of ^{137}Cs. All rats that died were autopsied, and in every instance death was due to an expanding intracerebral melanoma originating at the implantation site.

In the present study, the tumor ^{10}B concentration was 23.7 μg/g and the distances from the ^{252}Cf sources were 1 cm, and 0.5 cm in Exp. I and Exp. II respectively. We have calculated that the doses from $^{10}B(n,\alpha)^{7}Li$ reaction in Exps. I and II were 1.3 and 3.8 cGy respectively, and the total enhancement dose would have been 0.65% and 1.1% respectively. These calculations are in agreement to MCNP calculations and thermal neutron flux measurements[3,8]. However, the survival data showed a 19%-25% increase in MeSTs when BPA was administered i.p. prior to irradiation with ^{252}Cf. This might have been due to a more favorable microdosimetric distribution of BPA in the MRA melanoma[9], and a higher relative biological effectiveness (RBE) than 2.3 for the $^{10}B(n,\alpha)^{7}Li$ reaction, which had been based on the assumption of uniform boron distribution within the tumor[10].

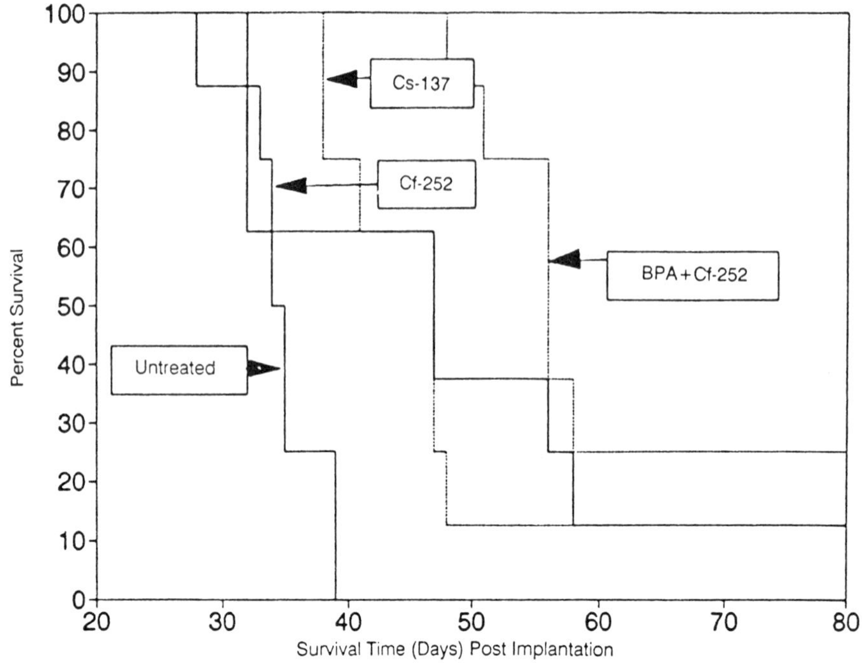

Figure 2. Exp. II. Kaplan Meier plots of rats carrying i.c. MRA 27 human melanoma tumor following ^{252}Cf or ^{137}Cs brachytherapy. Three groups were irradiated with 341 cGy of ^{252}Cf (gp 1b), 341 cGy following 5.0 hr of i.p. administration of 120 mg of ^{10}B-BPA (gp 2b), or 600 cGy of ^{137}Cs (gp 3b). Group 4b served as untreated controls.

Data from the present study suggest that dose enhancement of ^{252}Cf brachytherapy by means of the ^{10}B(n,α)^{7}Li reaction is feasible. Better dose enhancement could be obtained if the tumor ^{10}B concentrations were > 50μg/g[8]. This might be achievable by direct intratumoral injection of boronated compounds, and then allowing the boron containing compound to clear from the surrounding normal tissues. Improved positioning and external moderation of the ^{252}Cf sources might further augment the dose from the ^{10}B(n,α)^{7}Li reaction. Further studies are required to determine whether this is indeed the case.

ACKNOWLEDGMENTS

This work has been supported by U.S. Department of Energy grant DE-AC02-76CH000016 and The Ohio State University Comprehensive Cancer Center/Cancer Research Hospital and Institute. We thank Dr. M. Moeschberger for helpful suggestions relating to statistical analysis of the data.

REFERENCES

1. Maruyama, Y., Van Nagell, J.R., Kryscio, R., Yoneda, J., Beach, J.L., Feola, J.M., Donaldson, E.S., Hanson, M., & Gallion, H. (1986) Nuclear Science Applications **2**:467-482.

2. Chin, H.W., Maruyama, Y., Young, A.B., Beach, J.L., Tibbs, P., & Markesbery, W. (1986) Nuclear Science Applications **2**:585-598.
3. Beach, J.L., Schroy, C.B., Ashtari, M., Harris, M.R., & Maruyama, Y. (1990) Int. J. Radiat. Oncol. Biol. Phys. **18**:1421-1427.
4. Matalka, K.Z., Barth, R.F., Bailey, M.Q., Staubus, A.E., Adams, D.M., Soloway, A.H., James, S.M. Goodman, J.H., Coderre, J.A., Fairchild, R.G., and Rofstad, E.K. (1992) in Progress in Neutron Capture Therapy for Cancer, eds Allen B.J., Moore D.E. & Harrington B.V. (Plenum Press) pp. 429-434.
5. Akslen, L.A., Heuch, I., & Hartveit, F. (1988) Invasion Metastasis **8**: 193-204.
6. Matalka, K.Z., Bailey, M.Q., Barth, R.F., Staubus, A.E., Soloway, A.H., Moescberger, M.L., Coderre, J.A. & Rofstad, E.K. Cancer Res. (Submitted)
7. Bailey, M.Q., Matalka, K.Z., Barth, R.F., Coderre, J.A., Carpenter, D.E., Adams, D.E., Goodman,J.H., Staubus, A.E. & Rofstad, E.K. in These Proceedings
8. Yanch, J.C., Zamenhof, R.G., Wierzbicki, J. & Maruyama, Y. (1992) in Progress in Neutron Capture Therapy for Cancer, eds. Allen B.J., Moore D.E., Harrington B.V. (Plenum Press) pp. 191-194.
9. Gabel, D., Foster, S., & Fairchild, R.G. (1987) Radiat. Res. **111**, 14-25.
10. Gabel, D., Fairchild, R.G., Larsson, B. & Borner, H.G. (1984) Radiat. Res. **98**, 307-316.

IN VIVO STUDIES IN NCT WITH A BORONATED PORPHYRIN AND TUMOR GROWTH DELAY AS AN END POINT

B.H. Laster[1,2], S.B. Kahl[3], L. Warkentien[1] and V.P. Bond[1]

[1] Medical Department, Brookhaven National Laboratory, Upton, NY
[2] Department of Radiation Oncology, SUNY at Stony Brook Stony Brook, NY
[3] Department of Pharmaceutical Chemistry, University of California at San Francisco, San Francisco, CA

INTRODUCTION

The robust carrying capacity of the porphyrin molecule and its propensity for localizing in tumor justified the pursuit of synthesizing a porphyrin labeled with boron for use in BNCT. However, problems associated with poor solubility impeded the utility of the molecule. Until BOPP was synthesized (1), porphyrins were promising, but impractical.

After in vitro experiments had demonstrated the biological efficacy of BOPP and had confirmed its intracellular localizing ability (2), in vivo studies were carried out using BALB/c mice. The KHJJ murine mammary carcinoma was selected because, like glioblastomas, it is radioresistant (TCD_{50} of ~54 Gy)[1]. Irradiation of tumors to the TCD_{50} in a single fraction was precluded because the accompanying whole body dose is lethal. This problem was overcome by the use of fractionated doses of radiation.

MATERIALS AND METHODS

BOPP was administered either as three 0.5 ml injections per day over 2 days, or by continuous i.v. infusion, 2 ml per day over three days (3) for a total dose of ~42 μg ^{10}B/gbw. Boron-10 distribution in the tumor at the time of irradiation was

[1]The tumor was kindly provided by Dr. Sara Rockwell.

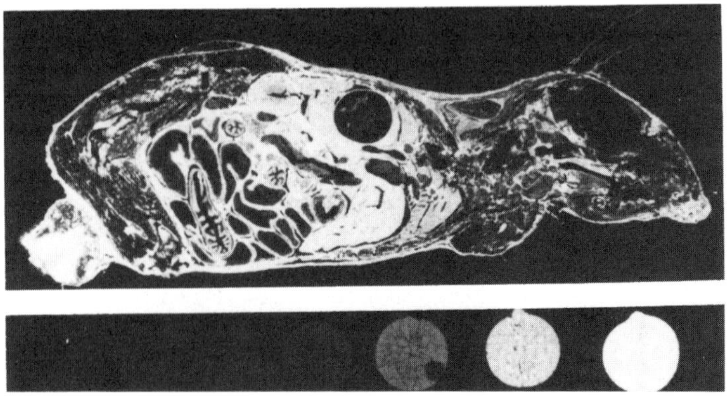

Figure 1

~20 μg/g. Fig. 1 is a neutron capture radiogram showing boron distribution with BOPP. Note that normal brain is free of boron, indicating that BOPP respects the blood-brain barrier.

Irradiations were carried out at the thermal beam port at the Brookhaven Medical Research Reactor (BMRR) at a reactor power of 1 MW. Mice were anesthetized with ketamine and rompun, then taped to a LiF-epoxy collimator with a centrally-truncated hole (1.5 cm aperture) to accommodate the tumor which was carried on the thigh of the mouse. 6.4 MW min was used for the 3 fraction experiment; 10.3 MW min. for the 5 fractions.

Because a tumor control dose could not be delivered, the end point selected for these experiments was tumor growth delay. This necessitated knowledge of the growth characteristics of the tumor. Tumor was implanted on the thigh of thirty mice and volume was determined by daily measurements of length (l), width (w), and thickness (h), corrections for the hemi-ellipsoidal shape of the tumor were made using the formula V=0.524 lwh, as recommended in (3). The tumor volume ,V, increased exponentially with time, t (days), according to the equation:

$$V = V_0 e^{(0.35-0.50V_0)\,t} \tag{1}$$

This relationship pertained during the course of a ten-fold increase from the initial volume, V_0, where $0.02\ cm^3 < V_0 < 0.28\ cm^3$. The analysis of tumor growth delay was based on the time span between the first irradiation and a ten-fold increase in tumor volume. The growth delay ratio (GDR) is the ratio of time to reach $10 \cdot V_0$ for the treated animal relative to the unirradiated control, normalized to the same initial V_0. After the data were analyzed in this manner, multiple linear regression analysis was applied and the GDR was found to fit the equation:

$$GDR = 0.996\, e^{(0.127\,t + 0.012\,[^{10}B])} \tag{2}$$

The resulting correlation coefficient was 0.982.

Physical dose rates and effective doses for the various mixed-field radiations are shown in Table 1, along with the total effective dose per MW min for each component radiation. The effective dose was determined using Dose·RBE/MW min. as described in (4). Dosimetric techniques are detailed in (5).

Table 1. Dose Rate at 2 mm Depth in a Murine Tumor Irradiated at the BMRR* Power Level = 1 MW

Component	Gy (per min)	Effetive dose rate RBE x Gy/min**	Percent of total effective dose rate
Fast Neutrons from the reactor	.22	.44	13.5
Extrinsic photons	.085	.085	2.7
$^{14}N(n,p)^{14}C$ reaction products	.11	.22	6.7
2.2 MeV photons from $^{1}H(n\gamma)^{2}H$	.03	.03	0.91
^{7}Li and α from $^{10}B(n,\alpha)^{7}Li$	1.01	2.5	76.0
TOTAL		3.3	100.0

*A uniform and identical extracellular and intracellular mass concentration of $2.4 \cdot 10^{-2}$ ^{14}N, $1.03 \cdot 10^{-1}$ ^{1}H and $2 \cdot 10^{-5}$ ^{10}B was assumed (fraction by weight). Electron equilibrium was assumed to exist at this depth.

**An RBE value of 2 was assumed for the fast neutrons and the $^{14}N(n,p)^{14}C$ reaction. A value of 2.5 was used for the $^{10}B(n,\alpha)^{7}Li$ reaction.

Table 2. Tumor growth delay as a function of irradiation time and dose (BMRR)

Irradiation Time (MW min) (n=fractions)	$\mu g\,^{10}B/gbw$	Initial tumor volume (cm^3)	Growth to $10 \cdot V_0$ (days)	Growth Delay Ratio	Dose (Gy•RBE)
0	----	0.074± .013	7.4	----	0
6.4[a] (n=3)	----	0.074± .013	17.6± 8	2.4± .12	5.1
0	----	0.104± .024	7.8	----	0
6.4[b] (n=3)	49	0.104± .024	28.2± 2.6	3.82± .40	21.1
0	----	0.050± .005	7.1	----	0
10.3[c] (n=5)	----	0.050± .005	22.5± 1.5	3.27± .21	8.2
0	----	0.043± .004	7.0	----	0
10.3[d] (n=5)	37	0.043± .004	44.5± 1.5	6.33± .22	34.0

Letters refer to the number of animals in the experiment: a=21; b=10; c=18; d=17.

Table 3. Tumor growth delay - γ-irradiation.

Initial Tumor Volume (cm_3)	Growth to $10 \cdot V_0$ (days)	Growth Delay Ratio	Dose (Gy)
0.090±.036	8.9±2.9	-----	0
0.129±.065	23.4±8.2	2.63	27.2

RESULTS AND DISCUSSION

Comparisons are made between a three and a five fraction experiment. In the former, BOPP increased the effectiveness of the neutron irradiations by ~60% at 6.4 MW minutes of irradiation, and in the latter, ~95% at 10.3 MW minutes. Relative to the untreated control mice, tumor growth was retarded by factors of 4 and 6, respectively. Results are shown in Table 2.

Although the biological efficacy of BOPP has thus been demonstrated in an in vivo system, it is uncertain that the conditions for exhibiting the advantage of the compound to its fullest extent were met. BOPP appears to remain systemic for long periods of time. The adrenal glands, spleen or liver might be the storage organs for BOPP, since high levels of the drug are present in these organs and are observed in tumor after a considerable period of time (unpublished data). It is likely that the drug recirculates systemically. In these experiments, BOPP was delivered prior to the first fraction. If BOPP has similar propensity for acting at the site of the endothelium as does hematoporphyrin derivative, vascular damage from the first fraction could have prevented the recirculation of the compound (6). Therefore, the advantage of boron retention in tumor would have been lost. To overcome this obstacle, we will have to consider alternative modes of delivering neutrons, e.g., hypo-fractionation. At any rate, the boronated porphyrin, BOPP, is a clear candidate for clinical application in NCT, and futher studies are warranted.

The above assessment of the efficacy of a boron-containing compound was accomplished using only megawatt minutes as a measure of the amount of exposure, i.e., it was not necessary resort to absorbed dose, RBE or the products of the two. In a companion paper, in which an in vitro system was used, we demonstrated the substantial difficulty of using the Dose·RBE approach for this purpose. We now show that the same difficulties exist when an in vivo system such as that employed here is used in an attempt to use the Dose·RBE approach.

In preliminary experiments, we first compared the tumor growth delay obtained in an x-ray study using the murine mammary tumor model. Two groups (six mice each), one x-irradiated and one control, were used. Tumors were irradiated to a total dose of 27.2 Gy using a 100 kVp x-ray source filtered with 1.5 mm Al. Data are shown in Table 3. The GDR in this experiment was 2.63, yielding a photon equivalent dose (PED) of 27.2 Gy. With reactor irradiations, a total dose of 5.1 Gy yielded a GDR of 2.4 (Table 2). Based upon the RBE listed in Table 1, the Dose·RBE was given as 5.1. The discrepancy between the PED of 27.2 Gy and the D_{EF} of 5.1 Gy, is a factor of 5. Clearly, using the Dose·RBE method in these preliminary experiments and RBE values that have frequently been used, did not predict the correctly the severity of biological effect. We intend to seek an explanation for the discrepancies in additional experiments. Fortunately, the effectiveness of boron compounds in vivo can be established using MW min. as a measure of exposure, without the need to resort to the Dose·RBE method.

ACKNOWLEDGEMENTS

We are deeply indebted to Dr. M.S. Koo for his assistance with compound synthesis, Dr. J. Kalef-Ezra for beam dosimetry and data analysis, to Dr. P. Som for providing the neutron capture radiogram, to Ms. A. Ruggiero for assitance with the manuscript, and particularly to our deceased friend and colleague, Dr. Ralph G. Fairchild, for valuable discussions during the course of this work.

REFERENCES

1. S.B. Kahl and M.S. Koo, Synthesis and properties of tetrakis-carborane-carboxylate esters of 2,4-bis-(α,ß dihydroxyethyl)deuteroporphyrin IX. Progess in Neutron Capture Therapy for Cancer. B.J. Allen, D.E. Moore, B.V. Harrington, (eds.) Plenum Press 223-226 New York (1992)
2. R.G. Fairchild, S.B. Kahl, B.H. Laster, J. Kalef-Ezra, and E.A. Popenoe, *In vitro* determination of uptake, retention, distribution, biological efficacy, and toxicity of boronated compounds for neutron capture therapy: A comaprison of porphyrins with sulfhydryl boron hydrides. Cancer Res. 50:4860-4865 (1990)
3. S.C. Rockwell, R.F. Kallman, and L.F. Fajardo, Characteristics of a serially transplanted mouse mammary tumor and its tissue-culture-adapted derivative. J. Natl Cancer Inst 49:735-749 (1972)
4. B.H. Laster, H.B. Liu, C.R. Gordon, and V.P.Bond, A comparison of the dose·RBE and the biological dosimetry approach for treatment planning in BNCT. Proceedings of the Fifth International Symposium on Neutron Capture Therapy, September (1992)
5. B.H. Laster, S.B. Kahl, J.A. Kalef-Ezra, V. Benary, M.S. Koo, P. Som, and R.G. Fairchild, Biological efficacy of a boronated porphyrin for neutron capture therapy in a murine tumor, Cancer Res. (1993)
6. M.C. Berenbaum, G.W. Hall and A.D. Hoyes, Cerebral photosensitization by haematoporphyrin derivative. Evidence for an endothelial site of action. Br. J. Cancer 53:81-89

BORON NEUTRON CAPTURE THERAPY:
SKIN AND SPINAL CORD TOLERANCE

G.M. Morris and J.W. Hopewell

CRC Normal Tissue Radiobiological Research Group
Research Institute, (University of Oxford)
Churchill Hospital, Oxford. OX3 7LJ, U.K.

INTRODUCTION

As part of the effort to realise the full therapeutic potential of boron neutron capture therapy, more information is needed on the radiation tolerance of normal tissues to proposed BNCT treatment schedules.

MATERIALS AND METHODS

AERE Harwell, U.K.

The neutron capture agent sodium mercaptoundecahydro–closo–dodecaborate (BSH) was administered by intravenous infusion over a 10 minute period at a dose of 100mg/kg body weight. Tissue boron content was determined using inductively coupled plasma atomic emission spectrometry (ICPAES). The cold thermal beam (0.025eV) was produced using a liquid argon, sulphur and aluminium filter on the H6 port of the DIDO reactor. The 20mm x 15mm irradiation field encompassed vertebrae T_{12} to L_2.

Thirty six Sprague–Dawley rats aged twelve weeks were used in the BSH pharmacokinetic study. Blood samples were taken at time intervals of 1, 2, 3, 4, 6 and 8h after infusion.

The irradiation jig consisted of an air–tight box providing shielding from thermal neutrons. Long term anaesthesia was achieved by pumping a fluothane (2%) and oxygen gas mixture throught the box. A total of 32 Sprague Dawley rats aged 12 weeks, infused with 100mg BSH/kg body weight, were irradiated for 3, 4, 4.5 and 5.0h. Estimated total physical absorbed doses to the skin were 31, 42, 47 and 52Gy; and to the spinal cord (blood) were 19, 25, 28 and 31Gy. At 21 months after irradiation, surviving rats were perfusion fixed and the spinal cord removed for histological analysis.

Advances in Neutron Capture Therapy, Edited by
A.H Soloway *et al.*, Plenum Press, New York, 1993

Brookhaven National Laboratory, U.S.A.

BSH was administered to twelve week old Fisher 344 rats by intravenous infusion (10min) at a dose of 100mg/kg body weight. The capture agent p-boranophenylanaline (BPA) was given intragastrically at a dose of 1500mg/kg body weight. The boron content of tissue samples was determined using prompt gamma spectrometry.

In the BSH pharmacokinetic study a total of 36 rats were used. Blood samples were taken at time intervals of 0.5, 1.5, 2, 3, 4, 6 and 8h after infusion. Samples of skin, oesophagus and trachea were taken at intervals of 0.5, 1.5 and 2h.

The thermal beam on the Brookhaven Medical Research (BMR) reactor has a considerably higher neutron flux (~x30) than the thermal beam on the AERE DIDO reactor. In contrast to the DIDO beam the BNL beam contains fast neutron and gamma components. The spinal cord field was moved to encompass vertebrae T_2 to C_4, to avoid potential damage to the gut, liver and lungs.

A jig and associated body-irradiation shielding has been constructed enabling rats to be irradiated with minimal dose to the thorax and abdomen. Exposure times in the range 5–30 MWmins have been used. Total absorbed physical dose rates to the skin using the beam-alone, beam plus BSH and beam plus BPA were 0.6, 8.7 and 1.4Gy/MWmin respectively. Eight rats were irradiated per dose group.

RESULTS AND DISCUSSION

Pharmacokinetic Studies

In the Sprague Dawley rats there was a relatively rapid initial decline in blood boron content from 136 ± 7μg/g at the 1h sampling point to 98 ± 4μg at the 4h sampling point. Thereafter there was a plateau in blood boron levels until 8h after BSH infusion (Fig. 1). The biological half life ($t_{½}$) was 18.5h. This value represents the time taken for 50% of the boron (^{10}B) from BSH to be cleared from the blood. T½ values were calculated directly from the boron clearance curves and were not computer generated. In contrast to the Sprague Dawley rats the boron (^{10}B) from the BSH was cleared rapidly from the blood of Fisher 344 rats and the $t_{½}$ was ~2h.

This would appear to be a potential strain difference, as identical BSH compounds and drug delivery methodologies were used. The Sprague Dawley rats were ~100gm

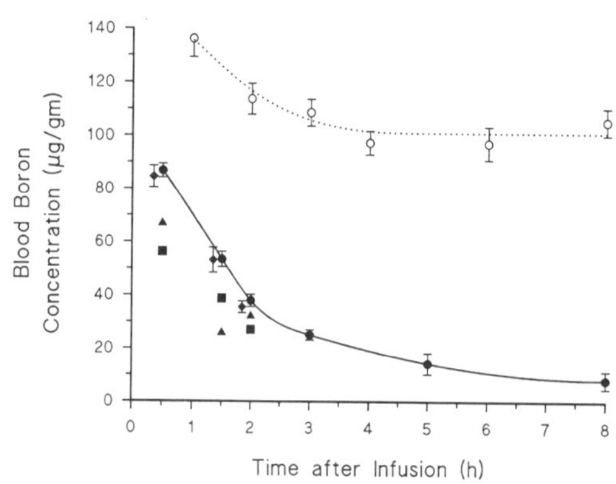

Figure 1. Time related changes in blood boron concentration (μg/gm) in Fisher 344 rats after the infusion of 100μg/kg of BSH ●Blood, ♦ Skin, ▲Oesophagus, ■ Trachea, ----- Results obtained in Sprague Dawley rats. (Error bars ±SE).

heavier than the Fisher 344 rats at the time of BSH administration. As a consequence, the Sprague Dawley rats were given a higher dose of BSH (relative to blood volume) than the Fisher 344 rats. This, together with the higher levels of subcutaneous fat present in the Sprague Dawley rats, probably contributed to the slower rate of blood boron clearance seen in this strain. Blood boron clearance rates have also been found to be variable in other strains of rat. Clendenon et al (1990) using inbred CD Fisher 344 rats obtained a $t_{½}$ value of 37 mins after the administration of 50mg/kg of BSH; while Abe et al (1986) indicated a $t_{½}$ value of 6h after the infusion of 100mg/kg of BSH in the Sprague Dawley rats. In the dog (Gavin et al, 1989) a $t_{½}$ value of 5h was estimated after the infusion of 55mg/kg of BSH. In common with the blood, the boron content in the skin of Sprague Dawley rats remained constant over the observation period. It was a factor of 0.6 lower than the blood. However in Fisher 344 rats comparable levels of boron were found in the skin relative to the blood (Fig. 1). In a previous study on the rat (Abe et al, 1986) boron levels in the skin were found to be a factor of ~3 lower than in the blood at 12h after the intravenous injection of 50mg/kg of BSH. Boron levels in dog skin, after the infusion of 55mg/kg of BSH, were 0.3–1.0 times those for the blood at 6h and 12h (Gavin et al, 1989).

The time course of ^{10}B accumulation in and removal from the blood, after BPA administration, has been established by Coderre et al (1992). An intragastric dose of 1500mg BPA results in a blood and CNS ^{10}B concentration of ~9μg/g from 6 to 14h after administration.

Radiobiological Studies

Thermal neutron exposure times (DIDO beam) of ≥4h in the presence of BSH resulted in vigorous skin reactions (Fig. 2).

Skin breakdown (moist desquamation) occurred over a similar time course to that seen after x-irradiation (Hamlet and Hopewell, 1982). However, in complete contrast to x-irradiation, there was a clear biphasic skin reaction in the Sprague Dawley rat. The second wave reaction was more severe than the first, with dermal necrosis developing at 18 to 38 weeks after irradiation (Fig. 2). The severity of the response was dose dependent with persistent dermal necrosis after exposure times of 4.5 and 5h. These findings suggest long term damage to the dermal vasculature.

The thermal beam on the BMR reactor has a relatively high fast neutron component amounting to 45% of the total absorbed physical dose at the skin surface (in the absence of neutron capture agent). At present data is only available for the first wave skin reactions in Fisher 344 rats. The time of onset of moist desquamation was between 1.5

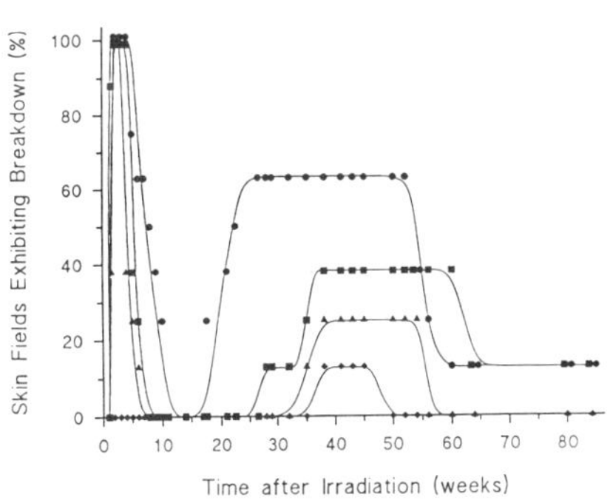

Figure 2. Percentage of skin fields showing breakdown with time after thermal neutron expsoure times (DIDO beam) of 3h (♦), 4h (▲), 4.5h (■) and 5.0h (●) in the presence of BSH.

and 2 weeks after irradiation with the beam alone or in the presence of BSH or BPA (Fig. 3). The occurrence of moist desquamation was over an identical time scale in the Sprague Dawley rats. The time taken for the skin to heal was dependent on the exposure time (Fig. 3).

It is possible to make some preliminary inferences about relative biological effectiveness (RBE)/compound factor values for the skin from the available data. The dose at which 100% of X–irradiated fields on the dorsal skin of the Sprague Dawley rat exhibit moist desquamation (ED_{100}) is 30Gy (Ross, unpublished). An analogous skin response in terms of the severity of first wave skin reaction and the time taken for the skin to heal, was achieved after a thermal neutron (DIDO beam) exposure time of 4h in the presence of BSH (42Gy). This gives a compound factor for BSH of ~0.7. Assuming a similar ED_{100} value for moist desquamation in Fisher 344 rats, some tentative RBE/compound

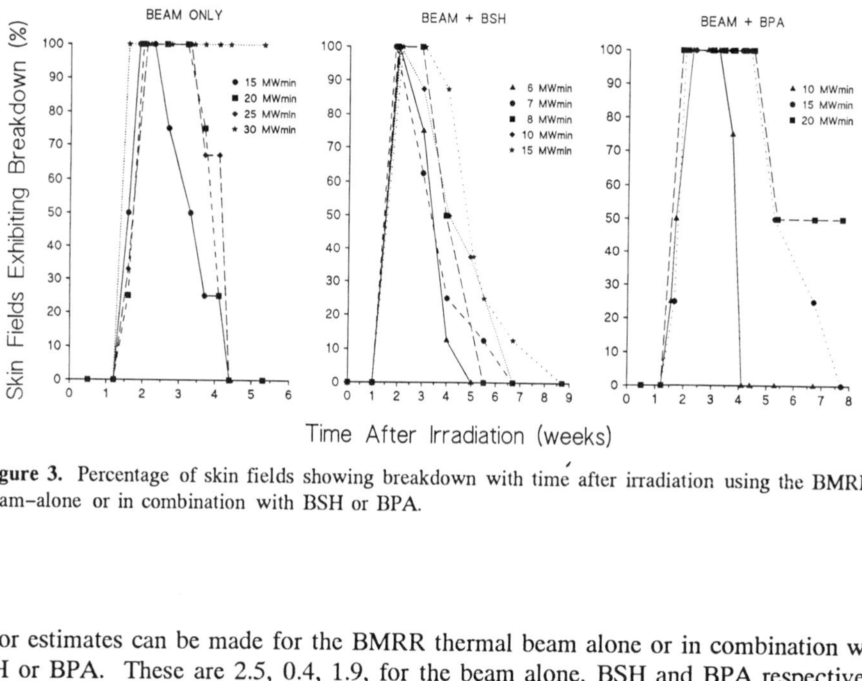

Figure 3. Percentage of skin fields showing breakdown with time after irradiation using the BMRR beam–alone or in combination with BSH or BPA.

factor estimates can be made for the BMRR thermal beam alone or in combination with BSH or BPA. These are 2.5, 0.4, 1.9, for the beam alone, BSH and BPA respectively. The compound factor estimates for BSH in rat skin are in close accord with the estimate of 0.5 for dog skin (Gavin et al., 1991).

In a previous study (Hopewell et al., 1987) in which a comparable length of spinal cord was irradiated to that of the present study, a single dose of 21.5Gy of X–rays was sufficient to produce paralysis in 50% of rats. The estimated physical absorbed doses to the blood at the centre of the spinal cord of rats irradiated using the DIDO thermal beam in the presence of BSH were in the range 18–31Gy. There was no evidence of neurological dysfunction over the duration of the 84 week observation period and the spinal cords appeared normal in histological sections. It can be concluded from these findings that the compound factor for BSH in the spinal cord CNS has a value of <1.

The spinal cords of the Fisher 344 rats have been irradiated on the BMRR thermal beam using considerably higher doses. The neurological end point is limb paralysis. At present animals have been monitored for 4 months after irradiation. The onset of limb paralysis is not anticipated until ≥5 months after irradiation.

REFERENCES

Abe, M., Amano, K., Kitamura, K., Tateishi, J. and Hatanaka, H., 1986, Boron distribution analysis by alpha–radiography, J. Nuclear Med. 27:677.

Coderre, J.A., Joel, D.D., Micca, P.L., Nawrocky, M.M. and Slatkin, D.N., 1992, Control of intracerebral gliosarcomas in rats by boron neutron capture therapy with p–boronaphenylalanine, Radiat. Res. 129:290.

Clendenon, N.R., Barth, R.F., Gordon, W.A., Goodman, J.H., Alam, F. et al., 1990, Boron neutron capture therapy of a rat glioma, Neurosurgery, 26:47.

Gavin, P.R., Kraft, S.L., Wendling, L.R. and Miller, D.L., 1989, Canine spontaneous brain tumours – a large animal model for BNCT, Strahlenther. Onkol. 165:225.

Gavin, P.R., Wheeler, F.J., Huiskamp, R., Siefert, A., Kraft, S. and Dehaan, C., 1991, Large animal model studies of normal tissue tolerance using an epithermal neutron beam and borocaptate sodium. Proceedings of the European Collaboration on Boron Neutron Capture Therapy. Plenum Press, New York.

Hamlet, R. and Hopewell, J.W., 1982, The radiation response of skin in young and old rats, Int. J. Radiat. Biol. 42:573.

Hopewell, J.W., Morris, A.D. and Dixon–Brown, A., 1987, The influence of field size on the late tolerance of the rat spinal cord to single doses of X–rays, Brit. J. Radiol. 60:1099.

AN INTRA-PANCREATIC AND HEPATIC NUDE MOUSE CANCER XENOGRAFT MODEL FOR BORON NEUTRON CAPTURE THERAPY

Julia L. Mallesch[1,2], Dino Chiaraviglio[2,3], Barry J. Allen[2] and Douglas E. Moore[1]

[1]Pharmacy Department, University of Sydney NSW 2006 Australia
[2]Biomedicine and Health Program, Ansto NSW 2234 Australia
[3]University of Pavia, Italy

INTRODUCTION

The cure of many cancers depends on their early detection and surgical excision. This is not the case for primary pancreatic cancer and hepatic metastases which have very poor prognoses. Cancer cells infiltrate the liver and current treatment protocols are unable to selectively kill these cells. Even though the affected organs are deep seated in the patient, NCT with an epithermal beam of the appropriate characteristics, used externally or intra-operatively, might represent an approach to the treatment of the cancers. In the development of an animal model system, the subcutaneous inoculation of pancreatic and hepatic tumour cell lines into nude mice has many limitations. An alternative approach is to inoculate internal organs with cancer cells to produce a different, but potentially more realistic, behaviour of that tumour type. This may then provide a superior indication of the potential for targeting these cancers with boron compounds.

Our aim has been to develop a new experimental model in order to assess the potential of Neutron Capture Therapy for the management of pancreatic and liver carcinomas and hepatic melanoma metastases. Initial work has involved the establishment of the model and the subsequent biodistribution of D,L-p-boronophenylalanine administered as the fructose derivative (BPA-Fructose). BPA is a phenylalanine derivative which is selectively taken up by melanoma cells[1].

MATERIALS AND METHODS

BPA-fructose pH 7.8 was prepared by complex formation with fructose. 12 mg of D,L-BPA was dissolved in 0.5 ml of 0.3 M D-fructose solution.

For reproducibility we have been using the established Harding Passey melanoma cell line to set up the model and to do the first experiments. This cell line has been used by our laboratory in previous research and our results will be analysed in the light of those results. Once this model has been established other pancreatic and liver carcinoma human cell lines will also be tested and biodistribution studies with BPA-Fructose will be performed.

Harding-Passey murine melanoma cells were harvested from monolayer culture during the logarithmic stage of growth. Each mouse was injected with a 0.1 ml suspension of 1.5 x 10^6 cells in DMEM media containing 10% foetal calf serum.

Athymic nude mice (Balb/c Nu/Nu; 6-8 weeks old) were anaesthetised with ketamine and xylazine. A midline incision was made and the pancreas or the liver isolated. The tumour cells were injected slowly between the peritoneal membranes containing the pancreas, or directly into the parenchyma of the right lobe of the liver. Subcutaneous tumours were grown by inoculation of 1.5 x 10^6 tumour cells into the outer thigh. One to two weeks after inoculation, 12 mg of BPA-fructose was injected intraperitoneally, then the mice were sacrificed at intervals up to 24 hours for dissection and tissue digestion followed by boron analysis by inductively coupled plasma atomic emission spectrometry (ICP-AES)[2].

RESULTS

The primary tumour obtained from the inoculation site is a discrete nodule which can be easily separated from the surrounding healthy tissue, providing enough material for boron analysis by ICP-AES. Other normal tissue samples were also obtained and assayed for comparison.

The BPA uptake into the tissues following ip injection of 12 mg BPA is shown in Figure 1. Selective uptake into the tumour with respect to blood is seen for all inoculation sites. BPA pharmacokinetics in the intrapancreatic and subcutaneous tumours are essentially the same, showing a rapid decline in tumour boron levels over time. However the concentrations of the BPA in the normal pancreas (Figure 2) are similar to and follow the same kinetics as for the intrapancreatic tumours. The tumour:normal pancreas ratio is less than one.

The boron concentration in the intrahepatic tumours follows a different BPA pharmacokinetic pattern. The level in the tumour rises to a peak at 4 hours and then declines. The boron concentrations in the normal liver are however significantly lower than in the intrahepatic tumour. The boron concentration ratio of melanoma to normal liver for these tumours is of the order of 3 to 5.

CONCLUSION

The pancreatic inoculations proved to be very successful. The tumours grow at a faster rate than the subcutaneously injected control tumours. The normal pancreas has a greater uptake of BPA-fructose than either the subcutaneous melanoma controls or the intrapancreatic melanoma models. The hepatic inoculations were less successful and produced smaller tumours although the tumour to liver boron ratio is in the range of 3-5. Because the tumour to pancreas boron ratio was less than 1 for intrapancreatic melanoma, hormonal stimulation and/or inhibition would be required to achieve selective tumour uptake of BPA.

ACKNOWLEDGMENT

Work supported in part by the Government Employees Assistance to Medical Research Fund.

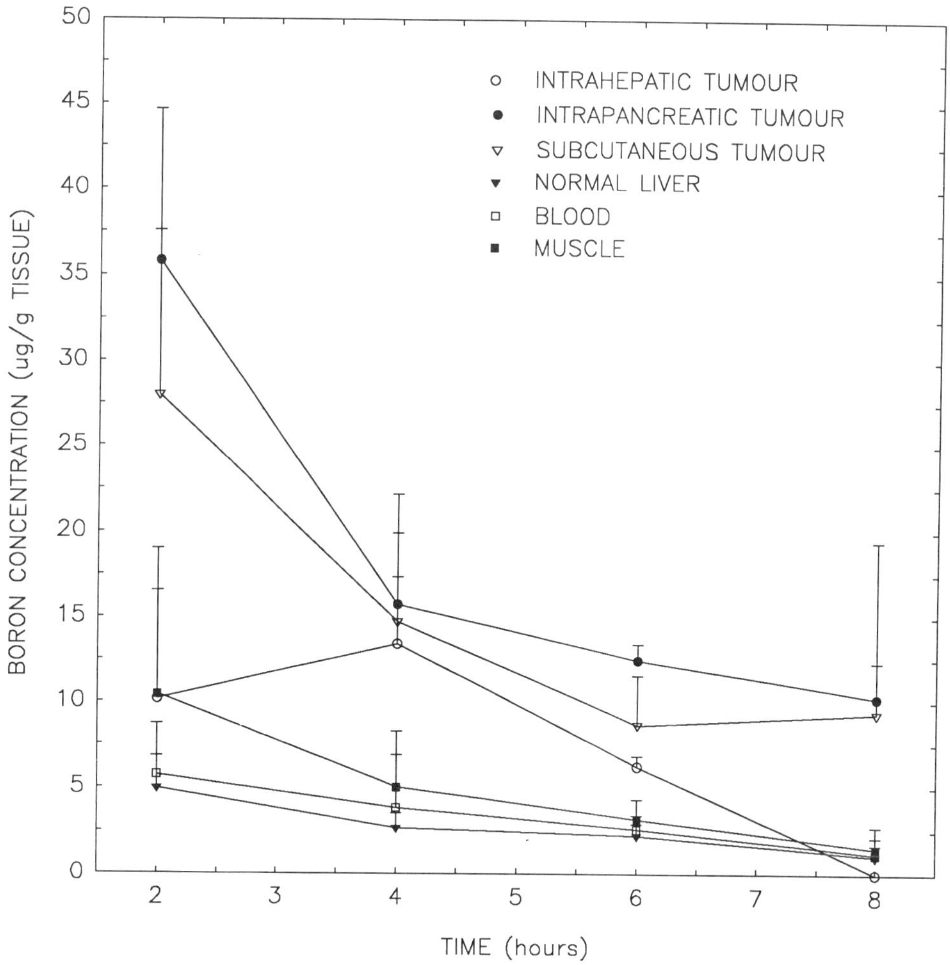

Figure 1. The effect of tumour inoculation site on the biodistribution of BPA-fructose in nude mice with Harding-Passey murine melanoma xenografts
(intrahepatic, n=3 (except for t=4hrs, n=7); intrapancreatic n=5; subcutaneous n=4;).

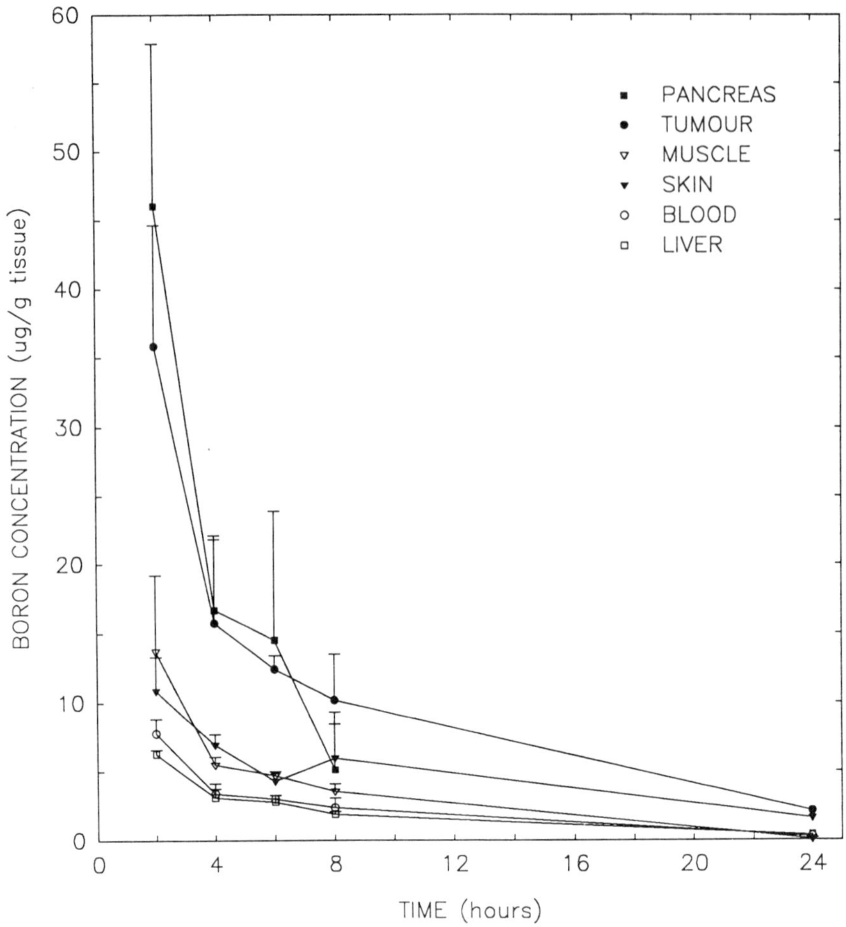

Figure 2. The biodistribution of BPA-fructose in nude mice with intrapancreatic Harding-Passey murine melanoma xenograft (n=5).

REFERENCES

1. Allen, B.J.; Corderoy-Buck, S.; Mallesch, J.L.; Crotty, K.; and Moore, D.E. Local control of subcutaneous melanoma xenografts in nude mice by neutron capture therapy, Melanoma Research 2:253-262; 1992
2. S. R. Tamat, D. E. Moore and B. J. Allen, Determination of Boron in Biological Tissues by Inductively Coupled Plasma Atomic Emission Spectroscopy, Anal. Chem., 59:1261-2164 (1987).

BORON NEUTRON CAPTURE THERAPY (BNCT) OF F98 GLIOMA BEARING RATS USING BORONOPHENYLALANINE (BPA) AS A CAPTURE AGENT

Khalid Z. Matalka[1], Rolf F. Barth[1], Alfred E. Staubus[1], Melvin Moeschberger[1], and Jeffrey A. Coderre[2]

[1]The Ohio State University, Columbus, OH 43210, U.S.A.
[2]Medical Department, Brookhaven National Laboratory
Upton, Long Island, NY 11973, U.S.A.

INTRODUCTION

Fifty percent of primary malignancies of the central nervous system are anaplastic astrocytomas and glioblastoma multiforme[1]. The prognosis of patients with these tumors is dismal despite aggressive surgery followed by radio- and chemotherapy, and characteristically the mean survival time of these patients still is approximately 12 months[2,3].

We previously have studied the effects of BNCT on the F98 rat glioma using sodium borocaptate (BSH or $Na_2B_{12}H_{11}SH$) as the capture agent[4]. Coderre et al.[5] have shown an apparent cure rate of 44% of 9L-gliosarcoma bearing rats following BNCT with BPA as a capture agent. Since the F98 glioma closely simulates human glioblastoma in its biologic behavior, and has been refractory to all therapeutic agents[4,6,7], we thought it would serve as a useful model to further define the efficacy of BPA as a capture agent for the F98 rat glioma.

MATERIALS AND METHODS

Rat Brain Tumor Model. Six-week-old male Fischer rats weighing 160-180 g were purchased from Charles River (Wilmington, MA). A previously described stereotactic implantation procedure was employed[4]. Briefly, rats were sedated with a 1.2/1 mixture of ketamine/xylazine and a plastic screw was embedded into the skull. F98 cells were injected into the right caudate nucleus at a concentration of 10^5 cells/10 μl of serum-free Dulbecco's medium containing 1% agarose at a gelling temperature of < 30°C. The implantation procedure employed a 5 second injection time, rapid filling of the screw hole with bone wax following withdrawal of the needle, and flushing of the operative site with betadine before closing the scalp incision with a single sterilized clip.

Pharmacokinetics and Tissue Distribution Studies. A solution of either D,L or L forms of BPA (Callery Chemical Co., Callery, PA), was converted to a fructose complex by mixing 1:1 molar ratio of BPA and fructose to enhance the solubility of the compound at pH 8.8. F98 tumor-bearing rats were injected i.p. with D,L-BPA at a dose of 897 mg/kg of body weight, which was equivalent to 47.2 mg of B/kg of body weight, 15 days following intracerebral implantation of 10^5 tumor cells. Animals were killed 3, 6, and 9 hours later and samples of blood, brain, tumor, skin, liver, kidneys, and muscle were taken, weighed, and then frozen until boron determinations were carried out. Boron concentrations were determined by means of direct current plasma atomic emission spectroscopy, as described in detail elsewhere[8].

***In Vivo* Irradiations.** BNCT was initiated 7 days following i.c. implantation of 10^5 F98 cells. Rats were divided into seven groups of 8-10 animals each. All irradiations were carried out at the Brookhaven Medical Research Reactor. The reactor power was maintained at 2 MW during the irradiation of all rats. Groups 1, 2, and 3 received 4, 6, or 8 Megawatt-minutes (MW-M) of irradiation respectively, which are equivalent to physical doses of 182, 273, 364 cGy. Groups 4, 5, and 6 received 4, 6, or 8 MW-M of irradiation 6 hrs following i.p. administration of 803 mg/kg of body weight of ^{10}B-enriched L-BPA (36.7 mg ^{10}B/kg). Animals in group 7 were untreated controls.

RESULTS AND DISCUSSION

Pharmacokinetics and Tissue Distribution Studies. Between 3 and 9 hr post administration of 897 mg/kg of D,L-BPA, blood boron concentrations exhibited monoexponential decay with a half-life ($t_{1/2}$) of 4.3 hr and a terminal rate constant of 0.16 hr^{-1} (Fig.1).

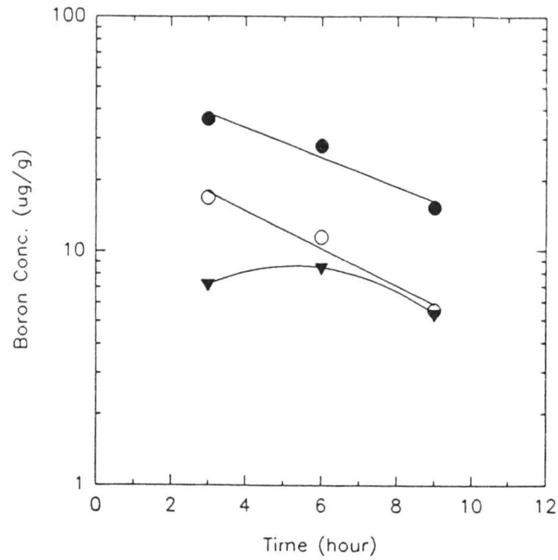

Fig. 1. Blood (○), brain (▼), and tumor (●) concentration-time profiles of BPA as measured by boron concentrations (± 1SD), after a single i.p. injection of 897 mg/kg body weight of D,L-BPA into F98 glioma bearing rats (3-4 rats at each time point).

Tumor boron concentrations also exhibited monoexponential decay with a $t_{1/2}$ of 5.28 hr. The best composite ratios of tumor to brain (T/Br) and tumor to blood (T/Bl) were 3.6 and 2.6 respectively, at 6 hours post administration of BPA at which time the average tumor boron concentration was 30.6 μg/g. Similar results were obtained when 803 mg/kg of ^{10}B-enriched L-BPA (36.7 mg ^{10}B/kg) was administered to tumor bearing rats, although T/Bl and T/Br ratios of ^{10}B were somewhat higher (3.5 and 3.9 respectively).

Tumor Therapy Experiments. Reactor irradiations were carried out 7 days following i.c. tumor implantation at which time the tumor weighed ~ 20-30 mg. All of the irradiated animals showed a transient weight loss of 10 g at four days following irradiation, but regained this by the 7-10th day. The physical doses in Table 1 represent the contributions of fast neutrons, γ photons, $^{14}N(n,p)^{14}C$, and $^{10}B(n,\alpha)^{7}Li$ reactions.

Table 1. Survival times of F98 glioma bearing rats following irradiation

Group	Treatment BPA	Treatment MW-M	Physical dose (cGy)	MST[a]	MeST[b]	%ILS (M[c])	%ILS (Me[d])
1	-	4	182	28	27	17	13
2	-	6	273	32	33	33	38
3	-	8	364	38	38	58	58
4	+	4	579	34	32	42	33
5	+	6	889	45	37	88	54
6	+	8	1159	64	59	167	146
7	-	-	-	24	24	-	-

[a,b]Mean and Median Survival Time were determined from groups of 8-10 rats each.
[c,d]%ILS were defined relative to MST and MeST.

Kaplan-Meier plots for BNCT experiments are shown in Fig. 2. The median survival time (MeST) for the irradiated control groups 1, 2, and 3 (182, 273, 364 cGy) were 27, 33, and 38 d respectively compared to 24 d for the untreated rats (group 7). The prolongation in survival times of group 1, 2, and 3 compared to untreated rats was significant ($p \leq 0.025$-0.0001). For BNCT treated groups 4, 5, and 6, the total estimated tumor doses were 579, 889, 1159 cGy respectively and the MeST were 32, 37, and 59 d. These values, when compared to 24 d for untreated rats (group 7), were highly significant ($p \leq 0.0093$-0.0001). The enhanced MeSTs for BNCT treated groups 5 and 6 compared to their matched irradiated controls (groups 2 and 3) were significant with p values of 0.0175 and 0.0277 respectively. However, there was no significant difference between BNCT treated group 4 (BPA + 4 MW-M) and its matched irradiated control group 1 (4 MW-M).

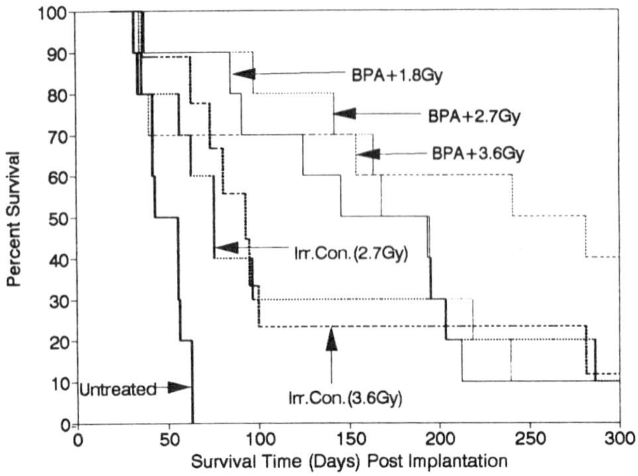

Fig. 2 Kaplan-Meir plots of BNCT and irradiated and unirradiated rats

The percentage of increased life span (%ILS) determined from MeST or mean survival time (MST) of each BNCT group was higher than its matched irradiated counterpart (Table 1). For all BNCT treated groups, the %ILS determined from MST was higher than %ILS determined from the MeST. However, for irradiated controls %ILS were the same whether calculated from either MST or MeST, suggesting that there were variations in the therapeutic response between individual rats treated with BNCT. Necropsies were performed on all rats. and in all instances death was due to an expanding tumor at the site of implantation.

In the present study, the data obtained indicate that although BPA may be useful as a capture agent for intracerebral melanoma[9-11] or 9L-gliosarcoma[5], it is less certain whether this is the case for high grade malignant gliomas. Furthermore, since BPA may be taken up by critical neuroanatomic structures in the brain such as the substantia nigra, in depth normal tissue radiobiologic studies are absolutely essential before one could contemplate using BPA as a capture agent to treat any kind of brain tumor, whether primary or metastatic.

Acknowledgments

This work has been supported by U.S. Department of Energy grant DE-AC02-76CH000016 and The Ohio State University Comprehensive Cancer Center/Cancer Research Hospital and Institute.

REFERENCES

1. Burger, P.C, Vogel, F.S. & Green, S.B. (1985) Cancer **56**, 1106-1111.
2. Leibel, S.A. & Sheline, G.E. (1987) J. Neurosurg. **66**, 1-
3. Kornblith, P.L. & Walker, M. (1988) J. Neurosurg. **68**, 1-17.
4. Clendenon, N.R., Barth, R.F., Gordon, W.A., Goodman, J.H., Alam, F., Staubus A.E., Boesel, C.P., Yates, A.J., Moeschberger, M.L., Fairchild, R.G. & Kalef-Ezra, J.A. (1990) Neurosurg. **26**, 47-55.

5. Coderre, J.A., Joel, D.D., Micca, P.L., Nawrocky, M.M., & Slatkin, D.D. (1992) Radiat. Res. **129**, 290-296.
6. Kaneko, S., Clendenon, N.R., & Kartha, M. (1983) Med. Chir. (Tokyo) **23**, 910-923.
7. Tzeng, J-J, Barth, R.F., Clendenon, N.R., & Gordon, W.A. (1990) Cancer Res. **50**, 4338-4343 (1990).
8. Barth, R.F., Adams, D.M., Soloway, A.H., Mechetner, E.B., Alam,F.& Anisuzzamam A.K.M. (1991) Analytical Chem. **6**, 890-893.
9. Matalka, K.Z., Bailey, M.Q., Barth, R.F., Staubus, A.E., Adams, D.M., Soloway, A.H., James, S.M., Goodman, J.H., Coderre, J.A., Fairchild, R.G. & Rofstad, E.K. (1992) in Progress of Neutron Capture Therapy for Cancer, eds Allen B.J., Moore D.E. & Harrington (Plenum Press) pp.429-434.
10. Bailey, M.Q., Matalka, K.Z., Barth, R.F., Coderre, J.A., Soloway, A.H., Carpenter, D.E., Adams, D.M., Goodman, J.H., Staubus, A.E. & Rofstad, E.K. in These Proceedings
11. Matalka, K.Z., Bailey, M.Q., Barth, R.F., Staubus, A.E., Soloway, A.E., Moescberger, M.L., Coderre, J.A. & Rofstad, E.K. Cancer Research (Submitted)

HISTOLOGICAL EXAMINATION OF MURINE MELANOMA XENOGRAFTS IN THE NUDE MOUSE MODEL: PRE AND POST NEUTRON CAPTURE THERAPY

Julia L. Mallesch[1,2], Kerry Crotty[3], Douglas E. Moore[1] and Barry J. Allen[2]

[1]Department of Pharmacy, University of Sydney NSW 2006 Australia
[2]Biomedicine and Health Program, Ansto NSW 2234 Australia
[3]Department of Medicine, University of Sydney NSW 2006 Australia

INTRODUCTION

Boron neutron capture therapy combines elements of radiotherapy and chemotherapy in order to achieve the selective irradiation of tumour tissue. A non-toxic boron compound with affinity for the tumour is combined with thermal neutrons to achieve selective tumour kill amongst normal tissue cells.

The biological effects of γ irradiation on tumour damage is well documented[1,2]. However limited histological data has been published for BNCT treated tumours. The aim of the present study was to investigate the biological effect of BNCT on a subcutaneous murine melanoma xenograft implanted in the nude mouse. Histopathological methods were used to investigate the extent of damage, and to see if any viable melanoma cells remained 300 days after BNCT treatment.

MATERIALS AND METHODS

Following the inoculation of 1.5 million Harding-Passey melanoma cells on the outer surface of the thigh, subcutaneous (sc) murine melanoma xenografts were grown in the nude mouse. The mice were divided up into three groups with eight animals in each group. In the control group of untreated mice, the tumours were allowed to grow to about 0.5 ml volume, and the mice subsequently sacrificed. The second group of mice were irradiated with 1.1×10^{13} n cm^{-2}, while the third group received 12 mg D,L-p-boronphenylalanine.HCl (BPA.HCl) intraperitoneally, 4 hours before the same neutron exposure. Irradiations occurred at T_o = 5,7,11,13 days post inoculation, for tumour sizes from 2 to 10 mm in diameter[3].

Advances in Neutron Capture Therapy, Edited by
A.H Soloway *et al.*, Plenum Press, New York, 1993

Melanomas selectively accumulate aromatic amino acids from extracellular fluid for use as precursors in the synthesis of melanin. The boronated analogue of phenylalanine, BPA, has been shown to accumulate selectively in the Harding-Passey melanoma. BPA has been demonstrated to be effective in BNCT of animal models[3,4,5] and has recently been used clinically in Japan for BNCT of human malignant melanoma.

For the control mice and for the mice which received neutron irradiation only, no local controls were seen. Six of the eight BNCT treated mice exhibited long term tumour regression. Three of the BNCT mice lived in excess of three hundred days before being sacrificed for histological analysis. Tumour growth was not observed even though the irradiated tumour beds showed a dense black mass still visible under the skin.

Upon sacrifice the tumour beds were excised and fixed in 10% phosphate buffered formalin. 5 μm thick serial paraffin sections were cut and stained with haematoxylin and eosin (H&E). Immunohistochemical staining with S-100 monoclonal antibody stain using the alkaline phosphatase method[6] was also undertaken to determine the presence of viable melanoma.

RESULTS AND DISCUSSION

Viable malignant Harding-Passey melanoma is shown in Figure 1. As for other cancer cell lines, the Harding-Passey melanoma is more variable in size and shape than normal cells and exhibits a high cell density. The individual cells are relatively uniform in appearance with little cytoplasm and prominent nuclei showing moderate pleomorphism. These cells have a high mitotic index. Harding-Passey tumours of this size (200-300 mg) do not possess significant necrotic regions.

Histological examination of BNCT treated tumour bed sections stained with H&E indicated the absence of mitoses (Figure 2). However the presence of an occasional pleomorphic cell in close proximity to capillaries is found amongst the tumour debris. These abnormal cells are surrounded by what appear to be histiocytes containing melanin and possibly connective tissue or fibroblasts.

Sections were stained with S-100 antibody to determine if the pleomorphic cells were viable melanoma. S-100 immunochemical stain targets a particular amino acid sequence found in a number of cell types, especially melanoma. The Harding-Passey melanoma showed moderate to strong staining of the pleomorphic cells (Figure 3) whereas the surrounding histiocytes were not stained. The cytological characteristics and the positive reaction to the S-100 stain indicate that the pleomorphic cells were intact melanoma cells.

Pleomorphic cells were observed in tumour sections of two out of the three mice. One of the mice gave no response with the S-100 stain. There was evidence of calcium deposits indicating damaged tissue or a long standing inflammatory process.

Further work is required to elucidate the reproductive nature of the pleomorphic cells given suitable conditions and therefore to assess their potential for tumour recurrence. They have remained dormant for over three hundred days and are surrounded by residual tumour debris. The pleomorphic cells have atypical nuclei and lack mitoses, and are more indicative of cellular obsolescence rather than decreased differentiation. These cells may have been chemically traumatised by toxins released from the degradation of the surrounding tumour cells. Alternatively, radiation induced DNA damage may have occurred resulting in reproductive death. Even though there are capillaries present near the cells in some of the sections they may be truncated, and extravascular transport of nutrients to the cells could be inhibited.

The next step would be to try an *in-vivo/in-vitro* experiment where the treated tumours are excised, suspended and grown in culture medium. This experiment would determine whether these abnormal cells have the potential to divide given suitable conditions. As well this experiment will also enable us to elucidate the optimum conditions for the therapy by varying the administered boron concentration and the neutron fluence.

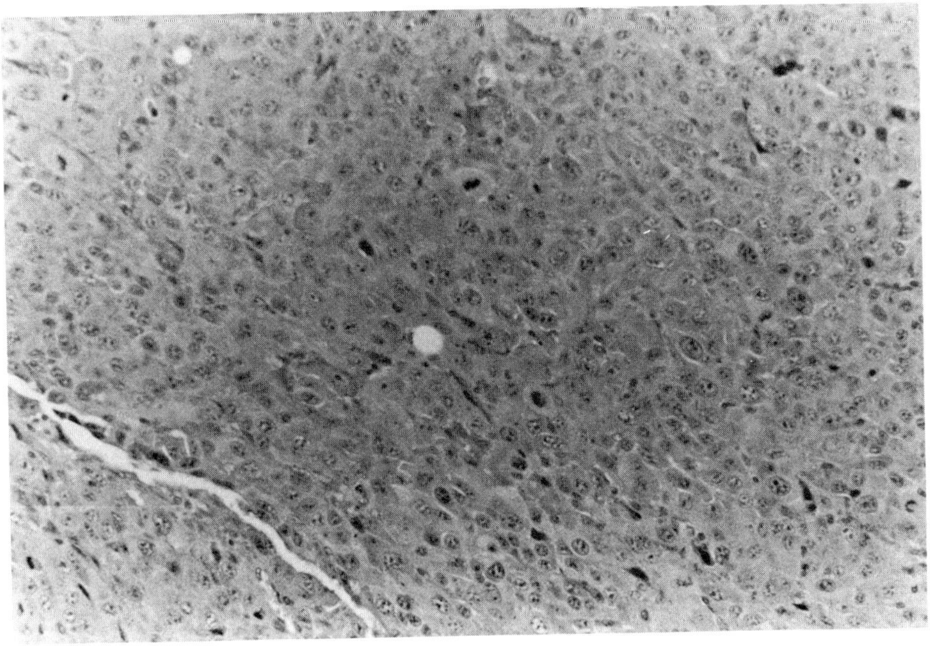

Figure 1. Viable malignant Harding-Passey murine melanoma (H&E stain; magnification x20)

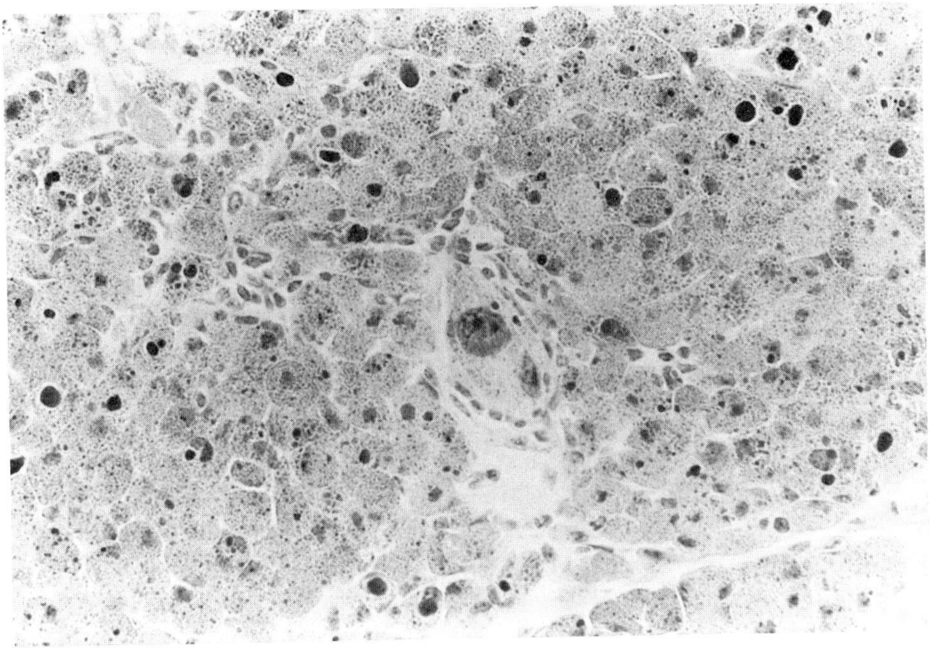

Figure 2. Pleomorphic cells in BNCT treated Harding-Passey murine melanoma visualised by H&E stain (magnification x20).

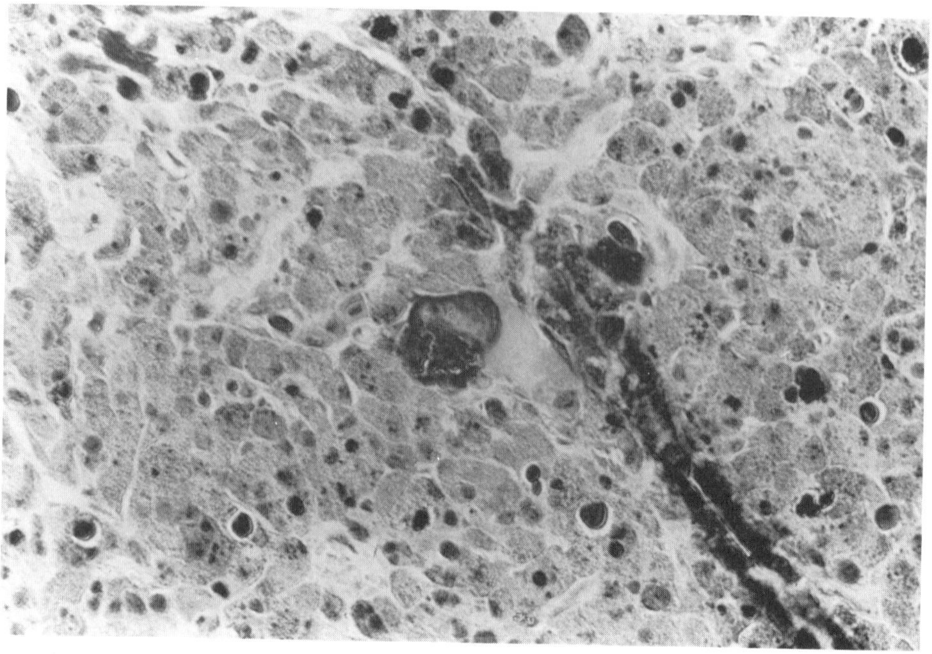

Figure 3. Pleomorphic cells stained positive with S-100 antibody. (S-100 stain; magnification x20)

CONCLUSION

Successful regression of malignant melanoma xenografts in nude mice has been achieved by BNCT after ip administration of enriched ^{10}B-D,L-BPA. Immunohistological examination shows that local control of the tumour can be achieved although some intact melanoma cells survive. It is not known if these cells have proliferative capacity.

ACKNOWLEDGMENTS

Work supported in part by the Government Employees Assistance to Medical Research Fund. The assistance of Sue Corderoy-Buck is acknowledged.

REFERENCES

1. K.H. Falkvoll, Histological study of the regrowth of a human melanoma xenograft exposed to single dose irradiation. *APMIS*. 98:758-764 (1990)
2. O.K. Solesvik, E.K. Rofstad and T. Brustad, Vascular changes in a human malignant melanoma xenograft following single dose irradiation. *Radiat. Res.* 98:115-128 (1984)
3. B. J. Allen, S. Corderoy-Buck, D. E. Moore, Y. Mishima and M. Ichihashi, Local Control of Murine Melanoma Xenografts in Nude Mice, Ed. B. J. Allen et al. in: Progress in Neutron Capture Therapy for Cancer, Plenum, N.Y. (1992), 425-428.
4. J.A. Coderre, D.N. Slatkin, P.L. Micca and J.R. Ciallella, Boron neutron capture therapy of a murine melanoma with p-boronophenylalanine: Dose response analysis using a morbidity index. *Radiat. Res.* 128:1-9 (1991)
5. J.A. Coderre, J.D. Glass, R.G. Fairchild, U. Roy, S. Cohen and I. Fand, Selective targeting of boronophenylalanine to melanoma in Balb/c mice for neutron capture therapy. *Cancer Res.* 47:6377-6383 (1987)
6. A.J. Cochran, D. Wen, H. Herschman and R. Gaynor, Detection of S-100 protein as an aid to the identification of melanocytic tumours. *Int. J. Cancer.* 30:295-297 (1982)

EVALUATION OF DOSE COMPONENTS FOR HEALTHY TISSUE TOLERANCE STUDIES ON DOGS AT THE HFR PETTEN

P. Watkins[1], R.L. Moss[1], A. Siefert[1], R. Huiskamp[2], P. Gavin[3] and M. Konijnenberg[4]

[1]Institute for Advanced Materials, JRC Petten, The Netherlands
[2]The Netherlands Energy Research Foundation, Petten, The Netherlands
[3]Veterinary Dept., Washington State University, USA
[4]Netherlands Cancer Institute, Amsterdam, The Netherlands

INTRODUCTION

Before the start of clinical trials of BNCT on glioma patients at the Petten reactor, certain preconditions must be determined. In particular the tolerance of healthy brain tissue exposed to the epithermal neutron beam requires investigation. In these studies, beagle dogs have been subjected to different levels of irradiation and ^{10}B, the latter in the form of BSH. To support this work a treatment planning tool is being developed to predict the various dose components within the treatment volume. A Monte Carlo code, MCNP[1], has been used to simulate the particle transport and to predict the different dose rate distributions. The dose rates generated by MCNP are manipulated with a processing code, TREAT, to give irradiation times, peak dose positions and to display the required data in a graphical format.

This paper explains the basic methodology used in the system and a practical case is presented for one of the healthy tissue tolerance dogs. Doses, both physical and RBE weighted, have been produced for pre-treatment planning studies.

OUTLINE OF THE CALCULATIONAL METHOD

The principal objective of this work has been to develop a computational model which is capable of evaluating the various dose components at all positions within a target volume from a predetermined neutron and photon source. The model must be able to account for the complex geometries of the beagle dogs and other targets being irradiated and also has to model the particle transport accurately. Such a model is effectively a treatment planning system which are traditionally composed of the following three main parts which are used iteratively to produce acceptable dose distributions for medical treatment;

- **Problem Visualisation.** MRI or CT images are used with an interactive viewing facility to add collimator specifications, wedge placement, etc. The beam parameters are taken as fixed input and multiportal irradiations can be modelled by most systems.

- **Dose Calculation.** From the predefined beam particles are propagated through the arrangement of wedges, collimator and target and dose distributions are generated.

- **Display of results.** The doses thus calculated are displayed, usually superimposed on the initial MRI or CT images to identify regions with excessively high or low doses.

Existing treatment planning systems often emphasise the visualization and display stages with rather crude models of the particle transport. One dimensional, ray tracing methods with corrections for scattering are still widely used. For photon treatments these may well be acceptable because of the relatively long mean free path of such particles. However for neutrons the mean free paths are much shorter and scattering is very much more important. Indeed, BNCT relies completely upon the epithermal neutron scattering to generate thermal flux at depth in a target. The large amount of neutron scattering also means that neutrons rapidly diffuse out of the incident beam, transforming an initial one-dimensional geometry into three dimensions. Furthermore, BNCT is inherently a mixed field technique so both neutron and photon transport must be modelled, including the generation of photons at neutron interactions. Therefore the present work attempts to model the particle transport very accurately so that reliable results can be obtained. The "user-friendliness" and speed of the system may suffer but these matters will be addressed in future developments.

Given the need for a detailed particle transport model in potentially very complex geometries, the choice of a continuous Monte Carlo technique is natural and the MCNP code has been selected. MCNP provides a very accurate model of the particle transport and allows neutron and photon fluxes to be recorded, or "tallied", over selected surfaces or volumes within the target. A useful feature of the code is the ability to fold a wide variety of energy dependent functions into the flux tallies, using the DE, DF and FM cards, which can be used to convert fluxes into dose rates as shown in Table 1. Many different tallies can be recorded allowing various neutron and photon dose rates to be extracted from a single MCNP simulation.

Table 1. Derivation of the neutron and photon dose rates in the MCNP simulations.

Fast Neutron Dose	Convolute ICRU dose function with neutron flux using DE and DF cards.
$^{10}B(n,\alpha)^{7}Li$ Dose	FM card gives energy deposition rate and converts to Gy/hr.
$^{14}N(n,p)^{14}O$ Dose	FM card gives energy deposition rate and converts to Gy/hr.
Photons	Convolute ICRU dose function with neutron flux using DE and DF cards.

The MCNP calculations generate the various dose rates over a mesh that is superimposed on the target volume, typically 1 x 1 x 0.5 cm voxels. Two calculations are required; one for the neutron component, which also generates photon dose distributions from the capture gammas, and the second for the photons originating from the reactor core. The neutron and photon dose rates relate to a single portal irradiation and to a specific boron concentration in the target volume. Additional portals may be evaluated by further MCNP simulations using the same scoring mesh for the target. The various neutron and photon dose rates from a number of different MCNP simulations are then combined using the processing code TREAT, which is under development at Petten. Apart from the MCNP dose rates for the different components, TREAT also corrects for the "real" boron concentration, and will apply RBE values to the dose rates if requested. After making these corrections, TREAT locates the peak dose rate within the target, RBE weighted or otherwise. Based on this value, an irradiation time is evaluated in order to reach some specified peak target dose.

USE OF MCNP/TREAT FOR BEAGLE IRRADIATIONS

The treatment planning system of MCNP plus TREAT has been used extensively at Petten to evaluate the dose distributions in the beagles that have been used for the healthy tissue tolerance study. The MCNP model is shown in figure 1. A simplified shape for the dog head is used, being the size of an "average" dog. A scoring mesh of 9 x 11 x 14 cells is imposed over the head giving a voxel size of 1 x 1 x X cm^3 where the component X varies with the depth into the target.

At present a simplified scheme is used with a single portal and the incident core gamma field is assumed to be uniform at 1 Gy/hr over the whole target volume. From measured values of the free beam this is known to be a slight overestimate of the real situation.

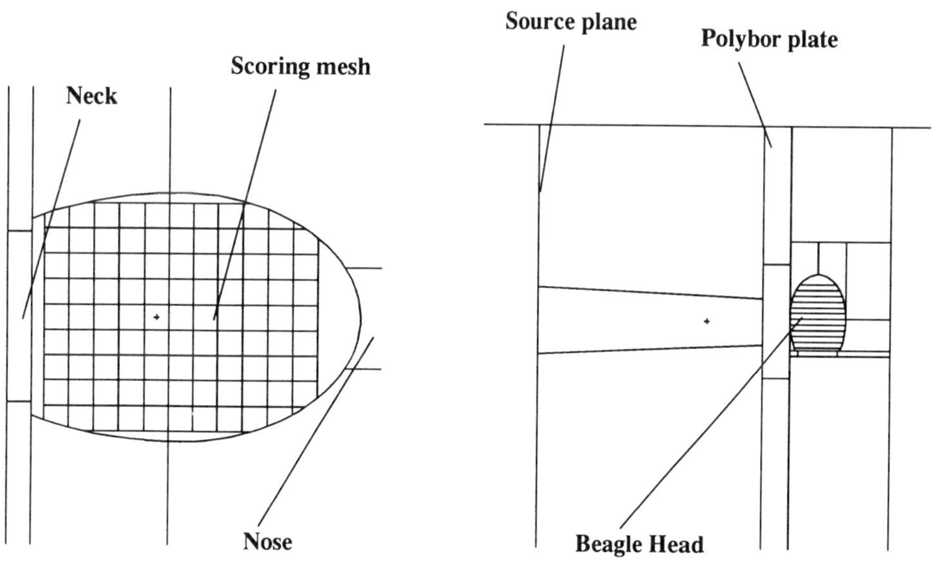

Figure 1. MCNP model of the dog head as used for the healthy tissue tolerance study.

For the beagle irradiations physical dose rates and doses have been used. The results from TREAT can be presented in one of a number of forms; tables of the different dose components at specified positions, two-dimensional intensity plots at various depths and one-dimensional line graphs, as shown in figure 2, where the different dose components can be seen as a function of depth into the beagle head.

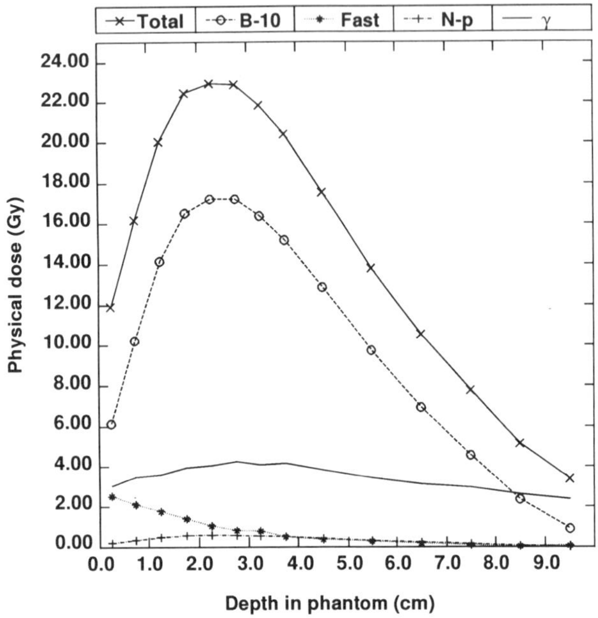

Figure 2. Depth-dose distributions for the dog model from MCNP/TREAT.

CONCLUSIONS

A computational model has been developed at Petten using MCNP and a processing code TREAT to evaluate the various dose components for BNCT irradiations. The model is still in the early stages of development but the emphasis has been to simulate the particle transport accurately. Future work will extend the graphical capabilities of TREAT to provide a more user-friendly method of displaying the final results. Other work will concentrate on simplifying the problem specification to MCNP and speed up particle transport calculations.

The model has been used at Petten for the healthy tissue tolerance study with beagle dogs. Present results are very promising and in good agreement with measured activation foils on the dogs or within phantoms.

REFERENCE

1. Briesmeister, J.F, (Ed), "MCNP - A General Purpose Monte Carlo Code for Neutron and Photon Transport", LA-7396-M, Revision 2, September 1986.

DOSE EFFECT COMPARISONS BETWEEN HFR AND BMRR IRRADIATED DOGS WITH RESPECT TO HEALTHY TISSUE TOLERANCE

René Huiskamp[1], Patrick R. Gavin[2], Floyd J. Wheeler[3], Axel Siefert[4], and Katharina H.I. Philipp[1]

[1]Netherlands Energy Research Foundation ECN, Petten, The Netherlands
[2]Washington State University, Pullman WA, USA
[3]EG&G Idaho Inc., Idaho Falls ID, USA
[4]Commission of the European Communities, JRC, Petten, The Netherlands

INTRODUCTION

Epithermal neutrons beams are being developed for the application of boron neutron capture therapy (BNCT) of deep seated tumors, like glioblastoma and astrocytomas, through the intact skin. Epithermal neutrons will be moderated by the tissue mass between skin and tumour to produce the thermal neutrons necessary for the $^{10}B(n,\alpha)^7Li$ reaction in the target tissue. Although the neutron capture cross-sections of elements in normal tissue are several orders of magnitude lower that for boron, the high abundance of hydrogen and nitrogen will cause a significant contribution to the total absorbed radiation dose through the $^1H(n,\gamma)^2H$ and the $^{14}N(n,p)^{14}C$ reaction, respectively. Due to inevitable incomplete filtration, an epithermal beam will also contain a fast neutron component, i.e. neutrons with energies $\geq$ 10 keV, and a γ-photon component originating from the reactor and produced in structural and filter materials. Therefore, the resultant radiation consists of a complex of low and high LET radiation of which the constituents vary rapidly with depth in tissue. Together with the dependence of the biological effects on the extra- and intracellular localization of boron, the assessment of the tolerance of normal tissues in large animal models necessary prior BNCT with borocaptate sodium (BSH) of brain tumors is very complicated.

Based on the ongoing canine healthy tissue tolerance study at the Brookhaven Medical Research Reactor (BMRR) using the epithermal beam without BSH, the relative biological effectiveness (RBE) of the fast neutron beam component has been determined for skin reactions. In addition, a "compound factor", i.e geometry x RBE, for the $^{10}B(n,\alpha)^7Li$ reaction was derived for dogs irradiated at the BMRR with the epithermal beam and BSH (Gavin et al., 1993a, 1993b, in press). Currently, a healthy tissue tolerance study with BSH is being carried out at the HB11 epithermal beam of the High Flux Reactor at Petten. The present papers describes prelimenary dose effect comparisons between High Flux Reactor

Advances in Neutron Capture Therapy, Edited by
A.H Soloway *et al.*, Plenum Press, New York, 1993

(HFR) and BMRR irradiated dogs with respect to healthy tissue tolerance in order to refine the BSH compound factors and the fast neutron RBE for skin and brain.

MATERIALS AND METHODS

A total of thirty dogs have been irradiated at the Brookhaven Medical Research Reactor (BMRR) reactor. Fifteen of these dogs were irradiated with the epithermal neutron beam alone (EPI dogs). Fifteen dogs were irradiated with the epithermal beam while the blood boron concentration was 30-62 ppm ^{10}B following an intravenous administration of BSH (BORON dogs). A description of the dosage groups as well as the observed biological effects have been previously reported (Gavin et al.,1992). To date, fifteen dogs have been irradiated at the High Flux Reactor in Petten, The Netherlands. Three dogs have been irradiated with the epithermal beam alone and twelve dogs have been irradiated at blood boron level varying from 25-62 ppm ^{10}B (see Siefert et al., these proceedings).

The physical and dosimetric characteristics of both beams and field sizes used are reported elsewhere (Nigg et al., 1992; Harker et al., these proceedings; Watkins et al, these proceedings).

The relative biological effectiveness (RBE) for fast neutrons was derived using the EPI dog tolerance studies by subtracting the physical doses multiplied by their RBE of the various beam components and compared with historical values for conventional photon irradiation. The RBE's used in the calculations were equal to 2.7 and 1.0 for the $^{14}N(n,p)^{14}C$

Table 1. Calculation of fast neutron RBE for skin reactions using epithermal beam irradiated only (EPI) dose groups.

Response type	Biological endpoint	Ref. X-ray dose (Gy)[1]	Radiation component	Phys. dose (Gy)	Eq. dose (RBE*Gy)[3]	Fast neutron RBE[4]
Early skin reactions	Moist desquamation (11.0 Gy EPI dogs)	18.3	Other[2]	0.52	0.52	
			N(n,p)	1.35	3.65	
			Gamma	4.49	2.99	
			Fast n	3.18		3.5
	Moist desquamation + ulceration (13.7 Gy EPI dogs)	24.4	Other[2]	0.65	0.65	
			N(n,p)	1.69	4.55	
			Gamma	5.61	3.76	
			Fast n	3.97		3.9
Late skin reactions	50% skin necrosis (16.5 Gy EPI dogs)	36.6	Other[2]	0.78	0.78	
			N(n,p)	2.03	5.48	
			Gamma	6.76	4.53	
			Fast n	4.78		5.4
	>50% skin necrosis (16.5 Gy EPI dogs)	> 36.6	Other[2]	0.78	0.78	
			N(n,p)	2.03	5.48	
			Gamma	6.76	4.53	
			Fast n	4.78		>5.4

[1]Park et al.(1974).

[2]Other: dose coming from neutron capture reactions by other elements, primarily Na and Cl.

[3]RBE other = 1.0; RBE N(n,p) = 2.7; DRF gamma = 0.67.

[4]$$RBE_{fast\ neutrons} = \frac{\{\text{Ref. X-ray dose} - [(\text{Eq. dose gamma}) + (\text{Eq. dose N(n,p)}) + (\text{Eq. dose other})]\}}{\text{Phys. dose fast neutron}}$$

reaction and other capture reactions respectively. The "compound factor", i.e. geometry* RBE, for the $^{10}B(n,\alpha)^{7}Li$ reaction was derived from the BORON dog tolerance study using a similar procedure as described for the calculation of the fast neutron RBE. The fast neutron RBE derived from the EPI dog tolerance study was used as additional input leaving the equivalent boron capture dose as variable (Gavin et al., 1993a,1993b, in press).

For dogs irradiated with the HFR epithermal neutron beam, equivalent doses were calculated using BMRR derived dose-modifying factors and the observed biological effects obtained with both beams were compared.

RESULTS AND DISCUSSION

The fast neutron RBE for early and late skin reactions were derived from the epithermal neutron only irradiated dogs and using published data on the skin tolerance doses determined for 200 kVp X-rays (Park et al.,1974). Since the gamma dose rate in the BMRR epithermal neutron beam is about 0.02 Gy/min, a gamma dose rate factor (DRF) for skin of 1.5 was introduced in the present calculations (Fu, 1991). RBE values equal to 3.4-3.9 and ≥5.4 were calculated for early and late skin effects respectively (Table 1).

To date no detectable neurological signs have been observed in the 11 Gy and 13.7 Gy EPI dogs. The 16.4 Gy EPI dogs had to be euthanitized due to dermal necrosis. Since both skin and brain tolerance are determined by the vasculature response after irradiation, the RBE for fast neutron component in brain is assumed to be identical to that calculated for the skin. In addition, a gamma DRF of 1.5 for brain was used (Fu,1991). The resultant peak equivalent dose to brain for the 11 Gy and 13.7 Gy EPI dogs was equal to 16.4 and 20.4 RBE*Gy respectively which is far above the reported tolerance of the brain for X-irradiation (ED50=14.9 Gy, Fike and Gobbel, 1991). Since the peak equivalent dose in these animals is located at the surface of the brain and mainly determined by the fast neutron dose which rapidly declines with depth and volume effects may play a role, the majority of the brain will be exposed to tolerable equivalent doses. An equivalent dose estimation 1 cm beyond the peak equivalent dose, i.e. 1 cm within the brain of a Labrador dog, was calculated to get a more representative equivalent dose estimation for brain tissue. The equivalent dose calculated at a position 1 cm in the brain was equal to 13.5 and 16.7 RBE*Gy for the 11 and 13.7 Gy EPI dogs respectively.

The compound factor for the $^{10}B(n,\alpha)^{7}Li$ dose was also determined 1 cm in the brain, using the published and presently calculated dose- modifying factors for the other beam components and compared with published X-ray on skin and brain tolerance. The obtained compound factors for BSH for skin and brain damage were equal to 0.51 and 0.47 respectively (Table 2).

Table 2. Calculated BSH compound factor for CNS necrosis defined at 1 cm depth in the brain and skin reactions.

Biological endpoint (dose group)	BSH Compound factor[1]
Skin moist desquamation w & w/o ulceration (38 Gy BORON dog group)	0.51
Dermal necrosis (64 Gy BORON dog group)	0.52
CNS necrosis (27 Gy BORON dog group)	0.47

[1]Dose-modifying factors used for compound factor calculation: DRF Gamma = 0.67; RBE other = 1.0; RBE fast n = 5.0; RBE N(n,p) = 2.7

Table 3. Summary of the evaluation of HFR dog irradiations to date (October 1992).

irradiation date	exposure time	blood [^{10}B] (ppm)	Peak phys. dose (Gy)	Equivalent dose at 2.25 cm depth (RBE*Gy)	Observed effect
91/7	1.77 hrs	38.5	17.5	14.7	brain necrosis
91/7	1.77 hrs	62.3	24.9	18.2	brain necrosis
92/4	1.60 hrs	50.7	19.2	14.9	non-lethal damage
92/4	1.60 hrs	47.6	18.3	14.4	non-lethal damage
92/4	1.60 hrs	52.0	19.6	15.1	none
92/4	1.43 hrs	58.7	19.2	14.2	non-lethal damage
92/4	1.43 hrs	53.2	17.9	13.7	none
92/4	1.93 hrs	0	6.2	9.9	none
92/4	1.93 hrs	0	6.2	9.9	none
92/4	1.93 hrs	0	6.2	9.9	none

To date 15 Beagle dogs have been irradiated using the HFR epithermal neutron beam. Two dogs irradiated last year, developed lethal brain necrosis. Of the 13 dogs irradiated in April and May this year, 2 out of 3 dogs that were irradiated at 19 Gy physical dose with 50 ppm ^{10}B showed lesions in their brain comparable with those observed in the 27 Gy BORON dogs irradiated at the BMRR. Their calculated equivalent doses, calculated also 1 cm in the brain, ranged from 14.4-14.9 RBE*Gy and are in good agreement with the calculated equivalent dose for the BMRR irradiated dogs. One out of 2 dogs that were irradiated at 17 Gy physical dose with 50 ppm ^{10}B. The calculated equivalent doses are listed in Table 3. Three dogs irradiated with the epithermal beam alone at 6.2 Gy physical dose did not show any lesions. The other dogs irradiated in May are currently under evaluation. Although the dose description used in this study is arbitrary, the resulting dose-modifying factors allow prediction of biological effects.

The observed skin effects of the HFR irradiated dogs are reported elsewhere in these proceedings (Philipp et al, these proceedings). The calculated equivalent doses for the skin and the observed effects are in agreement with historical X-ray data.

In conclusion, the use of a dose rate factor for the gamma component and the derived fast neutron RBE and compound factors for skin and brain, and a dose description beyond the thermal peak allow a good equivalent dose description in relation to the observed effects. The used dose-modification factors allow inter-comparisons of healthy tissue tolerance studies at different beams. Future results from both healthy tissue tolerance studies will refine the dose-modifying factors used for the equivalent dose description in BNCT.

REFERENCES

Fike, J.R. and Gobbel, G.T., 1991, Central nervous system radiation injury in large animals models, in: Radiation Injury to the Nervous System,Gutin,P.H., S.A. Leibel, G.E. Sheline, ed., pp.69-87, Raven Press, New York.

Fu, K.K., 1991, Influence of dose rate on normal tissue tolerance, in: Radiation Injury to the Nervous System, Gutin,P.H., S.A. Leibel, G.E. Sheline, ed., pp.69-87, Raven Press, New York.

Gavin, P.R., Wheeler, F.J., Huiskamp, R., Siefert, A., Kraft, S.L., and DeHaan, C.E., 1992, Large animal

model studies of normal tissue tolerance using an epithermal beam and borocaptate sodium. In: Proc. European Collaboration on Boron Neutron Capture Therapy, Petten, Plenum Press, New York, in press (1993a).
Gavin, P.R., Huiskamp, R., Wheeler, F.J., Kraft, S.L. and DeHaan, C.E., Large animal normal tissue tolerance using an epithermal neutron beam and borocaptate sodium. Strahlentherapie und Onkologie 169, in press (1993b).
Gavin, P.R., Kraft, S.L., DeHaan, C.E., Griebenow, M.L., and Moore, M.P., 1992, A large animal model for boron neutron capture therapy. In: Proc. of the 4th Intern. Symposium on Neutron Capture Therapy, Sydney, B.J. Allen, D.E. Moore, and B.V. Harrington, B.V., ed., pp.479-484, Plenum Press, New York.
Harker, Y.D., Anderl, R.A., Amaro, C.R., Watkins, P.R.D., and Voorbraak, W.P., Neutron spectrum measurements at the High Flux Reactor (Petten, The Netherlands) and the Brookhaven Medical Research Reactor, Ibid.
Harker, Y.D., Amaro, C.R., Randolph, P.R., and Wheeler, F.J., Neutron measurements in canine and cylindrical phantoms at the Brookahven Medical Research Reactor and the High Flux Reactor (Petten), Ibid.
Nigg, D.W., Wheeler, F.J, and Randolph, D.J. Characterization of the BMRR and PPBF epithermal neutron beams in phantom using three-dimensional deterministic radiation transport theory, In: Proc. of the 4th Intern. Symposium on Neutron Capture Therapy, Sydney, B.J. Allen, D.E. Moore, and B.V. Harrington, B.V., ed., pp.479-484, Plenum Press, New York.
Park, R.D., T.R. O'Brien, Baker,B.B., and Morgan,J.C., 1974, Single dose irradiation of canine skin. Vet. Radiol. 15: 108.
Philipp, K.H.I., Gavin, P.R., Casado, J., Siefert, A., Moss, R.L. and Huiskamp, R., Developments in the healthy tissue tolerance studies as a precondition for clinical application of boron neutron capture therapy, Ibid.
Siefert, A., Casado, J., Philipp, K., Huiskamp, R., Gavin, P.R., Dühmke, E., and Moss, R.L., Brain effects observed in the healthy tissue tolerance studies in BNCT with borocaptate sodium at the epithermal neutron beam of the HFR, Petten, Ibid.
Watkins, P., Moss, R.L., Siefert, A.S., Huiskamp, R., Gavin, P.R. and Konijnenberg, M., Evaluation of dose components for the healthy tissue tolerance studies on dogs at the HFR Petten, Ibid.

DEVELOPMENTS IN HEALTHY TISSUE TOLERANCE STUDIES AS A PRECONDITION FOR CLINICAL APPLICATION OF BORON NEUTRON CAPTURE THERAPY

Katharina H.I. Philipp,[1,2] Patrick R. Gavin,[3] Juan Casado,[4] Axel Siefert,[4] Raymond L. Moss,[4] and René Huiskamp[1]

[1]Netherlands Energy Research Foundation (ECN), Petten, The Netherlands
[2]Ludwig Maximilian University, Munich, Germany
[3]Washington State University, Pullman, WA, USA
[4]Commission of the European Communities,JRC, Petten, The Netherlands

INTRODUCTION

During the course of the joint project of the European Community towards realisation of a clinical facility for treating patients with BNCT, an epithermal neutron beam has been developed at the High Flux Reactor in Petten. This is designed for use in treating deep-seated brain tumours such as glioblastomas and astrocytomas through the intact skin (and skull). This paper covers the initial studies of epithermal irradiation and boron neutron capture using a canine model and borocaptate sodium (BSH) as boron carrier. The goal of this investigation is the estimation of the tolerance of normal tissues, such as brain, skin and bone marrow.

MATERIALS AND METHODS

As an animal model, normal beagle dogs were infused over one hour with 95% B-10 enriched BSH supplied by Centronics. Neutron beam irradiations were performed when the mean blood boron concentrations were 0, 25 and 50 mg ^{10}B/kg bw. At each boron level two to three dose groups were investigated where the groups are defined by peak physical dose. Additional experiments were performed at the 25 ppm level in the course of which the whole dose was given in four fractions separated by 24 hrs. For fractionated irradiation infusion was repeated prior to each fraction.

The boron content of the blood was analyzed by two different methods: 1.) prompt-gamma ray spectroscopy, and 2.) Inductively Coupled Plasma-Atomic Emission Spectroscopy. Prior to irradiation, a high dose of Dexamethasone was applied i.v. Anaesthesia was introduced with Acepromazine and Methadon and maintained with 1.2%

Advances in Neutron Capture Therapy, Edited by
A.H Soloway *et al.*, Plenum Press, New York, 1993

Isoflurane in an oxygen/N_2O mixture. The dogs were positioned on the irradiation table and for dosimetry, activation foils and TLDs were attached at various positions. An ECG was connected to monitor the animal during irradiation. Beam shutters were opened as soon as the blood boron-10 concentration has reached the projected level.

Table 1. Irradiation programme

Mean blood B-10 conc. (ppm)	Peak phys. dose (Gy)	Peak equiv. dose (Gy)	Location equiv. peak (cm depth)	Equiv. dose at 1.75 cm (Gy)	Peak phys. dose rate (Gy/h)	Irrad. time (h)	# of dogs
	17	13.08		12.92		1.43	5
50	19	14.62	1.25	14.44	11.9	1.6	5
	21	16.16		15.96		1.76	3
25	13.5	14.85	0.25	13.77	7.5	1.8	5
	15	16.5		15.30		2	5
25	13.5	14.85	0.25	13.77	7.5	1.8	5
(4 fract.)	15	16.5		15.3		2	5
0	6.2	14.95	0.25	12.1	3.2	1.93	5
	6.8	16.4		13.27		2.12	5

Assumed RBE values and compound factors are:
$^{10}B(n,\alpha)$= 0.33, Gamma and others = 1.0, N(n,p)= 2.7, fast neutrons = 4.5

RESULTS

The irradiation and post-treatment magnetic resonance imaging (MRI) procedures were generally well tolerated by all animals. During infusion of BSH some dogs, mainly of the 50 ppm group, showed a slight indisposition which in one case led to vomiting. Heartbeat was generally very hard and strong but frequency was not obviously altered.

Two dogs were killed six months after irradiation when extensive neurological changes were observed on the MRI scans. Clinical symptoms were less obvious at this time. These two dogs had been irradiated in summer '91. Both dogs were irradiated for 1.77 hours with peak physical doses of 17.5 and 24.9 Gy. First neurological symptoms occurred six months after irradiation. Main findings were loss of proprioception and other reflexes, reduced sensitivity and disorientation. Temporarily hyperreflexy was also observed. Since the neurological symptoms did not improve even after application of high doses of corticosteroids and the MRI scan showed extensive changes mainly of the white matter, both dogs were euthanized. The findings in MRI and histology are reported in these proceedings (Siefert et al.).

Thirteen additional dogs were irradiated in April and May of this year. All animals were irradiated approximately at the projected boron levels resulting in peak physical doses varying from 6.2 Gy (0 boron group) to 19.4 Gy (50 ppm boron group). Two dogs in the 25 ppm group received the dose in four equal fractions separated by 24 hours each. Three animals were irradiated with the epithermal beam alone.

Six months post irradiation two dogs of the 50ppm/19Gy group and the one dog of the 50ppm/17Gy group which actually received (due to the higher mean blood boron-10 concentration) 19.4 Gy peak physical dose showed changes in the MRI scans. There was no correlation with the neurological examination since all dogs behaved normally (see also Siefert et al., 1992). None of the other dogs have shown any neurological changes to date, i.e. five to six months post irradiation.

Table 2. Doses and onset of epilation

Projected dose group: mean blood [B-10] (ppm)	Projected dose group: peak phys. dose (Gy)	Irradiation time (hrs)	# dogs	Actual mean blood [B-10] (ppm)	Calculated* peak phys. dose (Gy)	Equiv. dose at 0.25 cm (Gy)**	Onset of epilation (days p. irr.)
				54	18	13.7	27
	17	1.43	2	59	19.4	14	32
				51	19.4	15.2	32
	19	1.6	3	48	18.4	15.15	38
50				50	19.3	15.2	32
				62	25	17.4	19
	21	1.76	2	41	18.5	16.3	22
				26	13.8	15.7	25
25	13.5	1.8	3	26	14	15.7	32
				25	13.5	15.6	32
				27	3.5	3.9	
				25	3.1	3.9	20
				26	3.2	3.9	
				27	3.5	3.9	
25	4 x 3.38	4 x 0.45	2				
(4 fractions)				26	3.5	3.9	
				22	3.2	3.7	20
				24	3.2	3.9	
				27	3.5	3.9	
							25
0	6.2	1.93	3	0	6.2	15.7	26
							26

* dose calculated with 3D Monte Carlo computer model (MCNP)

** RBEs, dose rate and compound factors used are: B-10(n,α)= 0.51, gammas= 0.67, N(n,p)= 2.7, fast neutrons= 5

Table 3. Skin scoring

Numerical grading of acute skin reaction:		X-ray control (Park et al. 1974)
0	no observable changes	
1	erythema	6.1 Gy
2	dry desquamation	12.2 Gy
3	moist desquamation	18.3 Gy
4	ulcers / skin loss	>24.4 Gy
Chronic radiodermatitis (stable after 120 days):		
0	no observable changes	
1	greying of hair but good hair growth	6.1 Gy
2	greying of hair but thinning hair growth	12.2 Gy
3	elpilation and atrophy	18.3 Gy
4	recurrence of chronic ulcers	>24.4 Gy

For calculation of skin equivalent doses RBE's, dose rate and compound factors were used as reported in these proceedings by R. Huiskamp et al. Skin dose did not vary greatly and there was no obvious dose dependence for onset of epilation.

To be able to judge skin damage qualitatively we developed a numerical grading system for skin reactions (see table 3) based on a scoring system described by Park et al. (1974) for beagles. Observed reactions in our dogs and calculated skin doses between 14 and 17 equivalent Gy agree rather well with the historical X-ray results reported by Park et al. (1974). The most severe acute reaction was total hair loss in the irradiated area and dry desquamation. For chronic damage the highest scoring was 2, which means that after 120 days most of the hair had regrown but instead of brown was more or less white.

There were no signs of bleeding or haemorrhage at the time of maximum platelet depression and no improvement in blood parameters could be observed from fractionated irradiation. Due to the high corticosteroid dose given just before each irradiation, the white blood cell counts peak on day one. The maximum depression in the lymphocyte concentration occurred within the first two days after irradiation, when it dropped to 25% of the average initial value. Both, the total leucocyte count and the lymphocyte count were slightly subnormal at day 8.

Only moderate changes could be observed in the number of platelets. A first nadir in the thrombocyte count was seen around day 20-25 with 40% of normal values. Final regeneration started at the end of the third week post-irradiation and reached normal values at day 30. None of the observed reductions in blood cell counts exceeded those reported by Nothdurft et al. (1992) for X-rays.

No changes in the pharmacokinetics of BSH in the blood were observed after multiple application. There are no obvious differences in pharmocokinetics between 25 and 50 ppm groups.

CONCLUSIONS

The results obtained to date from the ongoing healthy tissue tolerance study at Petten indicate that:

- -skin is not a limiting factor for BNCT with the Petten epithermal neutron beam,
- -the time related pattern of blood cell changes as well as quantitative differences seem to be independent of the applied doses,
- -the whole body dose is well tolerated,
- -although only two dogs of the fractionated group have been irradiated, no signs of tissue-sparing effect of fractionation on the depression of blood cell counts could be observed,
- -the pharmacokinetics of BSH in blood do not change after multiple application.

REFERENCES

Siefert, A., Casado, J., Philipp, K.H.I., Huiskamp, R., Gavin, P.R., Dühmke, E. and Moss, R.L., 1992, Brain effects observed in the canine healthy tissue tolerance studies for BNCT with borocaptate sodium at the epithermal neutron beam of the HFR, Petten, *These proceedings*.

Huiskamp, R., Gavin, P.R., Siefert, A. and Philipp, K.H.I., Dose effect comparison between HFR and BMRR irradiated dogs with respect to healthy tissue tolerance,1992, *These proceedings*.

Park, R.D., O'Brien, T.R., Baker, B.B.,and Morgan,J.P.,1974, Single dose X-irradiation of canine skin, *Vet. Radiol.* 15:108

Nothdurft, W., Selig, C., Fliedner, T.M., Hintz-Obertreis, P., Kreja, L., Krumwieh, D., Kurrle, R., Seiler, F.R. and Weinsheimer, W.,1992, Haematological effects of rhGM-CSF in dogs exposed to total body irradiation with a dose of 2.4 Gy, *Int.J.Radiat.Biol.* 61:519.

BRAIN EFFECTS OBSERVED IN THE CANINE HEALTHY TISSUE TOLERANCE STUDIES FOR BNCT WITH BOROCAPTATE SODIUM AT THE EPITHERMAL NEUTRON BEAM OF THE HFR, PETTEN

Axel Siefert,[1,3] Juan Casado,[1] Katharina Philipp,[2]
René Huiskamp,[2] Patrick R. Gavin,[4] Eckhart Dühmke,[3]
Raymond L. Moss[1]

[1]Commission of the European Communities, Joint Research Centre (JRC), Petten, The Netherlands.
[2]Radiobiology and Radioecology, Netherlands Energy Research Foundation (ECN), Petten, The Netherlands.
[3]Dept. of Radiotherapy, University of Göttingen, Federal Republic of Germany.
[4]Dept. of Veterinary Medicine, Washington State University (WSU), Pullman WA, USA.

INTRODUCTION

Before the first clinical trials of BNCT with borocaptate sodium (BSH) on patients suffering from malignant brain tumours can be performed at the epithermal neutron beam of the HFR Petten, the tolerance of healthy tissue exposed to this beam has to be determined. Therefore irradiation experiments are currently being performed on beagle dogs at different mean blood ^{10}B concentrations. To monitor radiation induced brain damage after irradiation, a method has to be employed which provides sufficient sensitivity for detecting radiation induced damage to tissue. Magnetic Resonance Imaging (MRI) is a non-invasive method which yields excellent discrimination between grey and white matter.[1] Information on the integrity of the blood-brain-barrier can be obtained by using paramagnetic contrast agents like Gadolinium-DTPA (Gd-DTPA).

This paper summarizes the results of the MRI's performed on the irradiated beagle dogs up to August 1992 and compares these results with the findings of histopathology.

MATERIALS AND METHODS

Irradiation experiments are being performed at 0, 25 and 50 ppm mean blood ^{10}B concentrations. In most of the experiments the irradiation is applied as a single fraction. In some experiments, at the 25 ppm ^{10}B level, the total dose is delivered in four fractions, separated by 24 h. At each ^{10}B concentration, two to three dose groups are being investigated (Table 1). Details of the experimental setup of the "Healthy Tissue Tolerance Studies" are reported elsewhere[2] and need not be covered here. Descriptions of the skin reactions and the total body response can be found in these proceedings.[3]

Magnetic Resonance Imaging (MRI) is being performed at fixed time intervals after irradiation to detect radiation induced damage of the brain. A 4.7 T system of the "National *in vivo* NMR Facility" at the Bijvoet Center for Biomolecular Research in Utrecht, The Netherlands, is used. The magnet (Oxford Instruments) has a 33 cm horizontal bore (clear working space) and an active shielded gradient (18 mT/m) connected to a SISCO 200/400 spectrometer. Data analyses are performed on SUN workstations.

Axial images of the dog's head are obtained using a spin-echo sequence in a multislice technique (slice thickness: 5 mm, gap 0 mm; slice-selection pulse: SINC 5 ms; slice-selection gradient: 4.5 mT/m). The predominantly T2 weighted images are acquired with an echo-time (TE) of 60 ms and a repetition-time (TR) of 1000 ms. For T1 weighted images a TE of 30 ms and TR of 800 ms are used. The T1 weighted images are repeated 2 minutes after intravenous administration of 0.1 mmol/kg bodyweight Gd-DTPA.

MRI is performed on each beagle before irradiation, which serves as a control. MRI's are obtained 24 hours after the end of the final irradiation of the fractionated dogs and within the first 4 weeks for all other dogs in order to evaluate acute radiation damage of the brain. For monitoring the late radiation effects, MRI's are taken 4, 5, and 6 months after irradiation. In cases where neurological symptoms occur additional MRI's are carried out.

Table 1. Scheduled irradiation programme and irradiations performed for the Healthy Tissue Tolerance Studies.

mean blood ^{10}B concentration [ppm]	peak physical dose [Gy]	number of dogs	no. of irradiations performed (date)
50	17	5	2 (4/92)
	19	5	3 (4/92)
	21	3	0
	23	2	2 (7/91)
25	13.5	5	3 (4/92;5/92)
	15	5	0
25/ 4 fractions	13.5	5	2 (5/92)
	15	5	0
0	6.2	5	3 (4/92)
	6.8	5	0

At the end of the observation period (1-2 years) or the occurrence of severe neurological symptoms, the animals are euthanized. After necropsy histological slides are prepared by standard techniques. Slices of 8 µm thickness are stained with Haematoxylin-Eosin and Nissl. Histopathological findings are compared with the results of MRI.

RESULTS

A total of 15 beagles were irradiated from July 1991 to May 1992 (Table 1). MRI's taken 24 hours after and within the first 4 weeks after irradiation did not show any abnormalities. No additional signal enhancement was present in the MRI's obtained 24 hours post irradiation after the administration of Gd-DTPA, indicating an intact blood-brain barrier. All other MRI's that were taken in the period 4 weeks to 4 months after irradiation were normal.

MRI's obtained from the first two dogs after the onset of neurological symptoms 6 months post irradiation revealed extensive signal enhancing regions throughout the white matter in the predominantly T2 weighted images, indicating increased amounts of free water in the tissue (Figure 1). No severe changes were present in the T1 weighted images without application of Gd-DTPA. After the injection of Gd-DTPA, multiple signal enhancing areas in the white matter, primarily above the lateral ventricles, were present (Figure 1).

Neurological symptoms progressed in the following days, leading to the euthanization of both animals.

Histology of the brains disclosed areas of extensive necrosis and hemorrhage around the lateral ventricles, surrounded by edematous tissue. Glia proliferation, swelling of endothelial cells and thickening of the basement membrane were present around capillaries. The location of these histological findings corresponded well with the location of the signal intense regions in the T1 weighted images after administration of Gd-DTPA.

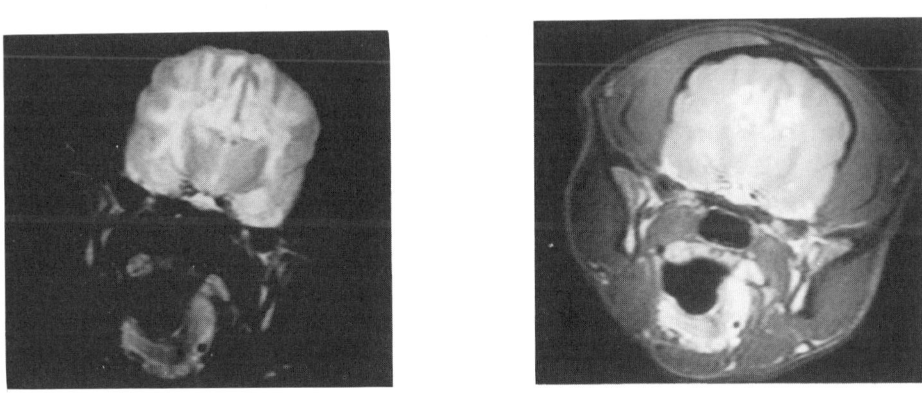

Figure 1. Axial MRI's obtained from a beagle six months post irradiation after the onset of neurological symptoms (**left:** predominantly T2 weighted image with signal intense areas throughout the white matter and **right:** T1 weighted image after injection of Gd-DTPA revealing signal enhancing regions in the white matter above the lateral ventricles).

DISCUSSION

Magnetic Resonance Imaging is an excellent method for revealing radiation induced changes in the brain after BNCT. Edema of the brain is visualized in T2 weighted images by the additional signal that originates from free water, whereas areas of hemorrhage and necrosis can be seen in the T1 weighted images after the administration of Gd-DTPA due to the accumulation of this contrast agent in regions where the blood-brain-barrier is damaged.

No changes were present in the MRI's obtained shortly after irradiation. This is an important observation since it proves that BNCT with BSH does not cause an acute leakage of the blood-brain-barrier at the applied dose levels. A leakage of the blood-brain-barrier would be prohibitive for repeated BSH administration which is essential for fractionated irradiations.

It has been demonstrated[4] that severe late neurological damage occurs in beagle dogs between 5 and 6 months after X-irradiation of the brain. Using BNCT we have observed a similar time period for the development of severe neurological damage. Therefore it is too early to draw final conclusions from the 13 beagles which were irradiated in April and May 1992. The MRI's due to be taken 5 and 6 months post irradiation will give more information whether or not the applied dose levels exceed the tolerance of the healthy brain.

ACKNOWLEDGEMENT

This study has been financially supported by the Commission of the European Communities. Part of this research was carried out at the "National *in vivo* NMR facility" at the Bijvoet Center for Biomolecular Research of the University of Utrecht, The Netherlands. The Gd-DTPA (Magnevist) was provided free of charge from Schering AG, Berlin, Germany.

REFERENCES

1. S.W. Atlas. "Magnetic Resonance of the Brain and Spine," Raven Press, New York, (1991).
2. A. Siefert, J. Casado, R.L. Moss, P.R. Gavin, K. Philipp, R. Huiskamp, and E. Dühmke, Healthy tissue tolerance studies for BNCT at the High Flux Reactor in Petten - first results, in: "Boron Neutron Capture Therapy: Toward Clinical Trials of Glioma Treatment," D. Gabel and R.L. Moss, eds., Plenum Press, New York (in press)
3. K. Philipp, P.R. Gavin, J. Casado, A.Siefert, R.L. Moss, R. Huiskamp, Developments in the healthy tissue tolerance studies as a precondition for clinical application of boron neutron capture therapy, Ibid.
4. J.R. Fike, C.E. Cann, R.L. Davis, J.K. Borcich, T.L. Phillips, and L.B. Russel, Computed tomography analysis of the canine brain: effects of hemibrain X-irradiation, Rad. Res. 99:294-310 (1984).

PHARMACOKINETICS OF BOROCAPTATE SODIUM IN CANINE HEAD DETERMINED BY ^{11}B MAGNETIC RESONANCE IMAGING AND SPECTROSCOPY

K. M. Bradshaw,[1] M. P. Schweizer,[2] G. H. Glover,[3] and J. R. Hadley[2]

[1]Neutron Technology, Boise, ID, 83702
[2]Medical Imaging Research Laboratory, University of Utah
Salt Lake City, UT
[3]Dept. of Radiology, Stanford University, Stanford, CA

INTRODUCTION

As new boron compounds are introduced for BNCT, requirements to conduct pharmacokinetic studies more quickly and noninvasively will intensify. The purpose of this paper is to present a method for determining the *in vivo* distribution and kinetics of $Na_2B_{12}H_{11}SH$ (Borocaptate Sodium or BSH) and to monitor it's deployment in real time. To obtain pharmacokinetic data in "real time" the method must be non invasive and fast enough to sample the changing environment. Magnetic Resonance (MR) had been suggested from the beginning as an ideal non invasive method, but, because of the low signal-to-noise (SNR) for boron imaging (about 10^6 times less than proton imaging) and fast relaxation of the nuclei (less than 1 msec), problems seemed insurmountable. With the advances in MRI technology, a solution has been found that makes the premise of this research feasible. Instead of using two-dimensional chemical shift imaging[1,2] (2D-CSI) which takes too long to acquire image data, we are proposing the use of a three-dimensional (3D) projection reconstruction method. The details of this method have been reported elsewhere[3] and will not be repeated here. The method meets the necessary requirements of being non invasive and fast, with data acquisition on the order of minutes instead of hours.

To demonstrate the viability of using boron 3D projection reconstruction imaging to conduct pharmacokinetic research, *in vivo* studies were conducted involving dogs with and without tumors. The BSH uptake and elimination of different tissue types were compared qualitatively by comparing the relative boron MRI signals.

METHODS

MR Imaging Techniques

Using an idea from computed tomography, projection reconstruction has been adapted to an MRI system allowing faster acquisition of the data within the constraints of the fast decay time of the signal. The boron data are acquired within 140 μsec of the start of the pulse sequence.

Advances in Neutron Capture Therapy, Edited by
R.F. Barth and A.H. Soloway, Plenum Press, New York, 1993

^{11}B is used in all imaging and spectroscopy techniques reported here since ^{10}B is eight times less sensitive for equal number of nuclei[4]. Still ^{11}B imaging is about 10^6 times lower in SNR than achieved with proton imaging requiring special attention to SNR. Using a 0.5 dB low noise preamplifier and quadrature techniques, sufficient improvements in the SNR were made that allowed imaging of ^{11}B concentrations below 25 ppm using projection reconstruction. The most difficult challenge we had was the acquisition of the fast decaying MRI signal. Because the decay of the MRI signal from ^{11}B in BSH bound to protein (T2 relaxation time) is on the order of 600 µsec, the signal dies out before it can be acquired using normal imaging sequences such as "spin echo". Techniques that required less pre data acquisition time to establish the imaging environment such as whole head spectroscopy and chemical shift imaging were tried but did not meet our imaging requirements.

With our projection reconstruction technique we can achieve an anisotropic volume of 7.5 mm x 7.5 mm x 15.0 mm. A typical image series (scan) takes approximately eight minutes, although faster scanning is possible. During an eight-minute scan, the 3D data are collected with a GE Signa™ 1.5 Tesla MRI system and then reconstructed on a Sun™ workstation. The resulting boron images can be superimposed over corresponding proton images of higher resolution or displayed separately. Using region of interest (ROI) measurements, the intensity data of the boron image is correlated with a known reference within the imaging volume, and the relative boron concentration is then determined. With this technique, pharmacokinetic studies are performed by taking a series of scans over time and comparing the same locations from scan to scan. Whole head spectroscopy is used to determine large volume kinetics.

In Vivo Study Design

Dogs with and without gliosarcoma tumors have been tested for BSH uptake and distribution using the following protocol:

1. The dog is initially immobilized with Telazol and then maintained on 1.5-2% isoflurane anesthesia for the remainder of the experiment.
2. Proton sagittal and axial T1 and T2 weighted images are obtained for anatomical referencing and tumor enhancement.
3. BSH is infused by IV over 1 hour. Solution is 100 mg ^{11}B/kg in 11 ml saline/kg (kg-dog weight).
4. Blood is sampled every 10 minutes during infusion and every 20 minutes during elimination phase of BSH.
5. Whole head spectroscopy is performed before each blood sampling using 100 averages.
6. ^{11}B images are collected every 20 minutes after infusion.
7. Tissue taken at necropsy and blood sampling taken during the experiment are analyzed by Inductive Coupled Plasma-Atomic Emission Spectroscopy.

RESULTS

Differences between the blood and tissue kinetics are seen by comparing the boron in blood with the whole head spectroscopy (Figure 1). Both ^{11}B concentration in blood and ^{11}B MR signal intensity are normalized to the maximum level observed. There is strong correlation during uptake, indicative of a concentration driven gradient , and a longer retention of boron in tissue for the first hour of the elimination phase. This last observation infers that the MR signal comes primarily from the tissue and that there is some mechanism slowing the release of BSH from tissue. This general kinetic behavior is corroborated by the invasive studies performed by Kraft et. al[5].

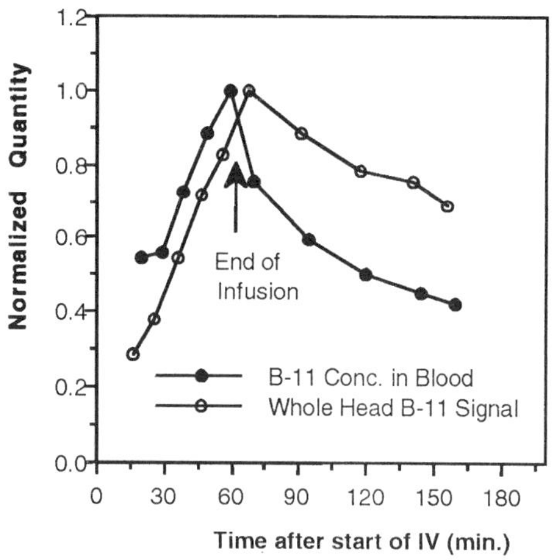

Figure 1. Comparison of boron spectroscopy signal from whole head with boron in whole blood.

ROI measurements of a 48 cm field of view (FOV) boron image with an isotropic volume (voxel) of 1.5 cm x 1.5 cm x 1.5 cm are shown in Figure 2. Here several different tissues are compared, with the more vascularized tissue showing faster changes over time, as expected. The intensity of each measurement was normalized against the average intensity from a vial filled with 50 ppm boric acid which was placed on the dog's head over the eyes. Since these data re from a dog whose blood brain barrier was intact there was negligible concentration of BSH in the brain. For comparison, the end point of the graph for muscle (lines 3 and 4) corresponds to about 25 ppm as measured by ICP-AES.

Examination of the images containing tumors offers additional understanding of the distribution of boron in the dog's head when the blood brain barrier has been destroyed by a

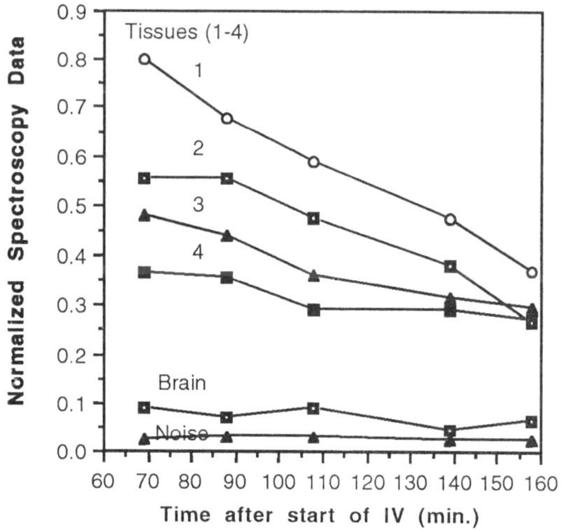

Figure 2. Comparison of boron signal intensity of different tissues of equal volumes after end of infusion of BSH. 1 and 2 are from tongue and nasal mucosa respectively, and 3 and 4 are from masticator muscle. Boron concentration in brain is negligible.

tumor. Figure 3 shows a dog with a gliosarcoma located in the left hemisphere. A T2 weighted proton image on the right is provided for referencing anatomy. Comparing the intensity data from this image with data taken about one hour later shows that the boron in the surrounding tissue decreased by 30% whereas the boron decreased only by 14% in the tumor. This indicates a preferential retention of BSH in tumor. The normal brain still showed negligible amounts of boron.

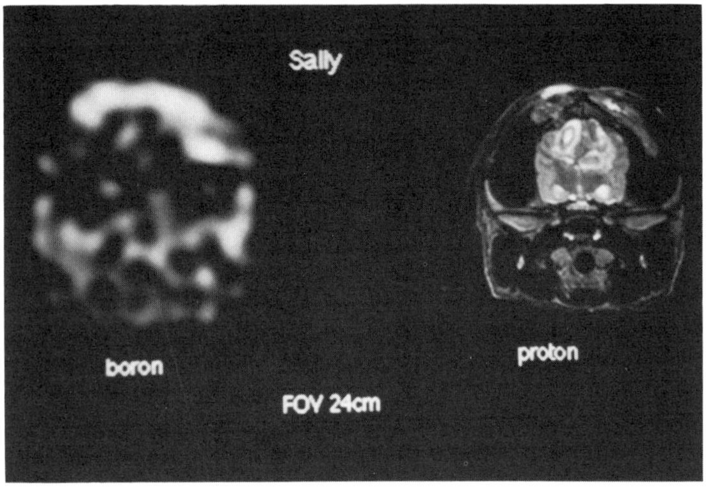

Figure 3. A 24 cm FOV boron image showing a gliosarcoma in the left hemisphere of the brain. Resolution is 7.5 mm x 7.5 mm x 15 mm.

CONCLUSIONS

Using the projection reconstruction MR imaging technique, qualitative comparisons of boron in tissue (concentrations less than 25 ppm) can be obtained which also compares favorably with known invasive studies. With a scan time of 8 min. 28 sec., the image data provide a fair "snap-shot" of the kinetic changes during the elimination phase. Since the biological half-life of BSH is about four hours, significant changes in boron concentration in tissue will not have occurred during each scan time. Future studies will focus on developing analytical methods to determine the absolute concentration of boron *in vivo* using the intensity of the MR signal. This will require using a standard boric acid reference that will be in the FOV of the boron image as well as investigation of potential intensity changes with different tissues.

The smallest anisotropic volume that has been achieved is 7.5mm x 7.5 mm x 15.0 mm. Further improvements in the quadrature splitter and the elimination of extra coax cable, as well as faster repetition rates will improve the SNR making a smaller isotropic volume possible.

ACKNOWLEDGMENTS

This research was supported in part by the U.S. Department of Energy under contract DE-AC07-76ID01570 (Idaho National Engineering Laboratory). We thank Dr. John Hilton, and the Johns Hopkins University for the gliosarcoma cells. We appreciate the assistance of the following individuals from the University of Utah, Dr. Richard Tippets,

Department of Neurosurgery, Suzanne Johnson and Tamer Ghanem (animal assistance), Dr. Cheryl Harris, Department of Pathology and Dr. Peng-Peng Tang, Department of Medicinal Chemistry.

REFERENCES

1. Kabalka, GW, Davis M, Bendel P, "Boron-11 MRI and MRS of Intact Animals Infused with a Boron Neutron Capture Agent," *Magn Reson Med* 1988;8:231-237.

2. Bradshaw, KM, Richards TL, "In vivo Chemical Shift Imaging of Canine Brain Tumors After Injection of Na2B12H11SH (abstr)," In: Book of abstracts; Society of Magnetic Resonance in Medicine 1990, Berkeley, Calif: Society of Magnetic Resonance in Medicine, 1990; 1166.

3. Glover, GH, Pauli, JM, Bradshaw, KM, "Boron-11 Imaging with a Three-dimensional Reconstruction Method," *JMRI,* Jan./Feb., pp. 47-52, 1992.

4. Bovey, FA; *in* "Nuclear Magnetic Resonance Spectroscopy," 2nd Ed. p3, Acad. Press, 1988.

5. Kraft SL, Gavin PR, DeHaan CE, Leathers CW, Bauer WF, Dorn RV. The biodistribution of boron in canine spontaneous intracranial tumors following borocaptate sodium infusion. In: Progress in NCT for Cancer. Allen BJ et al, Eds. Plenum Press, New York, 1992. pp. 479-484.

BIOAVAILABILITY OF INTRAVENOUS FORMULATIONS OF P-BORONOPHENYLALANINE IN DOG AND RAT

T.R. LaHann, D.R. Lu, G. Daniell, C. Sills,[1] S.L. Kraft,[2] P.R. Gavin,[3] and W.F. Bauer[4]

[1]Center for Toxicology Research
Department of Pharmaceutical Sciences
Idaho State University
Pocatello, Idaho, 83209

[2]College of Veterinary Medicine
Kansas State University
Manhattan, Kansas, 66506

[3]College of Veterinary Medicine
Washington State University
Pullman, Washington, 99164

[4]Idaho National Engineering Laboratory
Idaho Falls, Idaho, 83415

INTRODUCTION

Boron neutron capture therapy (BNCT) is being developed as a treatment for malignant melanoma. P-boronophenyialanine (BPA) was initially proposed as a boron (B) delivery drug for BNCT of malignant melanoma because it was postulated that this B-containing amino acid, by mimicking a melanin precursor, would selectively accumulate in melanoma cells. BPA does seem to selectively accumulate in melanocytes, apparently as a result of uptake by an amino acid transport system[1]. For successful BNCT, tumor B concentrations of at least 20 ppm are thought to be necessary, but higher tumor B levels are desirable. Calculations[2] indicate that for a given neutron exposure, each doubling of the tumor B concentration should increase tumor cell kill by a factor of about 10,000. Thus, even modest increases in the amount of B in tumor cells can dramatically improve the effectiveness of BNCT as a cancer treatment.

If amino acid transport systems accumulate BPA, then high extracellular concentrations of BPA should increase the amount of BPA entering melanocytes. I.V. infusion of BPA is the simplest route for delivering BPA to the tumor cells, but is not widely used because at physiological pH, BPA exhibits poor water solubility (<4 ppm). Possible options for increasing BPA solubility include host-guest formation[3] and use of organic complexes[4]. The purpose of the work reported here was to 1) determine the extent to which the aqueous solubility of BPA is increased by host-guest complex formation with the cyclodextrin 2-hydroxypropyl-β-cyclodextrin and by organic complex formation with fructose and 2) test the hypothesis that i.v. formulations which enhance BPA solubility will increase the delivery of B to tissue sites.

METHODS

D,L-p-boronophenylalanine was purchased from Callery Chemical Company (Pittsburgh PA) and 2-hydroxypropyl-β-cyclodextrin (Molecusol®) was purchased from Pharmatec, Inc. (Alachua, FL). For solubility determinations, an excess of BPA was always used. All formulations were sonicated, adjusted to a final pH of 7.4 and filtered through a 0.22 micron filter. The BPA concentration of filtered solutions was calculated from B levels which were measured by inductively coupled plasma atomic emission spectrometry. Formulations used in rat studies were prepared immediately before use; formulations for dog experiments were prepared within 3 days of use. High performance liquid chromatography studies demonstrated that aqueous BPA solutions were stable at physiological pH for at least 3 days. Male Sprague-Dawley rats (300-400 g) and female Beagle dogs (8.3-11.0 Kg) were anesthetized, a vein cannulated for infusion of drug, and either the cephalic vein (dog) or the femoral artery (rat) cannulated for blood sampling. Drug was infused to dogs at a target rate of 17 ml/kg/hr for a one hour period, while rats were infused at target rates of 8.5, 12.8 or 17 ml/kg/hr for 1-3 hours. Urine was collected by urinary catheter (dogs) or bladder cannulation (rats). Plasma and serum samples were collected from dogs prior to, during (every 20 minutes) and after the infusion (8 to 10 samples over 72 hrs). Plasma and serum samples were collected from rats at hourly intervals for 3-7 hrs. Each dog received three treatments: 1) BPA in buffer, 2) BPA in Molecusol® and 3) Molecusol® alone, with at least a three week wash-out period between infusions. Each rat received only a single drug formulation. Pharmacokinetic analysis of the data was conducted using the computer program PharmK (v.1.2.). Unless otherwise specified, BPA and B levels are reported as mean $\pm$ SEM of three experiments.

RESULTS AND DISCUSSION

Solubility Studies

B levels in distilled water and in 0.1 M, pH 7.4 phosphate buffer were below the limits of detection (<0.008 mg/ml BPA). The solubility of BPA in the buffer was 2.2 $\pm$ 0.4 mg/ml (N=7), but varied as a function of the lot of BPA used. Different BPA lots displayed aqueous solubilities ranging from 1.15 to 3.7 mg/ml. By briefly increasing the pH of the BPA-buffer suspension to about 10 before adjusting it to 7.4, more BPA could be solubilized. A pH 7.4 BPA solution of 6.0 $\pm$ 0.1 mg/ml (N=4) prepared by manipulation of pH did not precipitate for several hours. BPA solubility in Molecusol®/neutral water solutions increased with increasing concentrations of Molecusol® (10-45% w/v) and since 45% was approximately iso-osmotic, this concentration was selected for *in vivo* experimentation. The BPA concentration in 45% Molecusol®/water was 6.4 $\pm$ 0.1 mg/ml. Briefly elevating the pH of the BPA-Molecusol® suspension to about 10 and then readjusting pH to 7.4 solubilized more BPA (11.5 $\pm$ 1.0 mg/ml). However, the greatest increase in solubility was achieved by complexing BPA with fructose. A 1:1 molar ratio of BPA and fructose in pH 7.4 distilled water solubilized less than 10 mg/ml BPA. Briefly increasing the pH of a 200 mg/ml BPA suspension to about 10 and then readjusting pH to 7.4 solubilized 173.2 $\pm$ 3.8 mg/ml BPA (N=18, range: 148.6 - 196.9). Neither BPA-fructose nor BPA-Molecusol® formulations showed signs of precipitation over a 7 day observation period and when added to rat plasma, showed no macroscopic evidence of precipitation.

Dog Studies

The plasma B concentrations, plotted as a function of time during and after i.v. infusion of BPA-buffer (1.15 mg/ml) and BPA-Molecusol® (5.7 mg/ml) are shown in Figure 1.

In previous dog experiments, either serum or plasma B levels were measured. Significant differences between serum and plasma B concentrations were not observed in the current experiments, indicating that either could be used. The peak serum concentration achieved with the BPA-Molecusol® formulation (16.5 $\pm$ 1.9 ppm B) was higher than that achieved with the BPA-

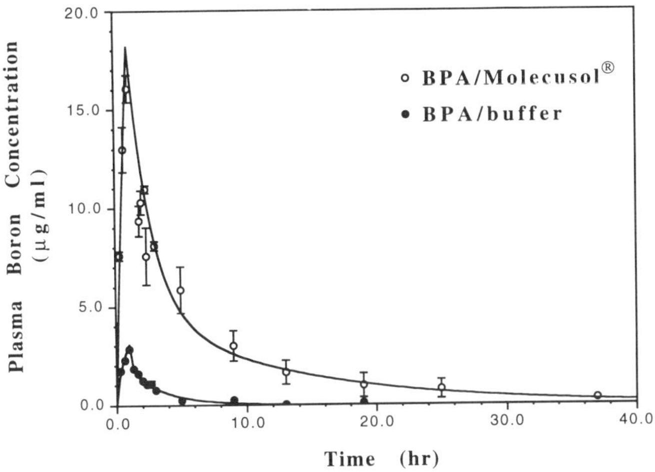

Figure 1. Plasma B profiles after infusion of BPA-water-Molecusol® (●) and BPA-buffer (○) formulations into dogs.

buffer formulation (3.8 ± 0.3 ppm B) and higher than the peak values achieved with subcutaneous administration of a 200 mg/kg suspension (peak serum level: 3.4 ± 0.5 ppm B, N=5) or oral administration of a 50 mg/kg suspension (peak serum level: 2.0 ± 0.6 ppm B). A two compartment model for i.v. infusion was used to fit the data. This model predicts that the plasma concentration versus time curve will be biexponential: $C_p = A(e^{aT}-1)e^{-at} + B(e^{bT}-1)e^{-bt}$. During the infusion period, T = t, and during the post-infusion period T becomes a constant (i. e. the duration of the infusion). For the BPA-Molecusol® formulation, the parameters were calculated to be: A = 30.83; B = 59.27; a = 0.58 and b = 0.08, while for the BPA-buffer formulation, the parameters were calculated to be: A = 1.28; B = 5.67; a = 4.43 and b = 0.38. The areas under the plasma concentration-time curve (AUCs), the half lives (b phase), apparent volumes of distribution and clearances were 90.10 and 6.95 μg hr/mL, 8.43 and 1.83 hr, 0.20 and 0.12 L/kg, and 0.05 and 0.14 L/hr/kg for the BPA-Molecusol® formulation and for the BPA-buffer formulation, respectively. Measurement of urine and blood B levels suggested that over the first 4 hours post-infusion, most of the B remained in the body, but resided in tissues other than blood. Assuming that total tissue load 4 hours after termination of infusion (tissue B_{4hrs}) can be calculated as: (total B infused)-[(B in blood $_{4hrs}$) + (urinary excretion of B_{4hrs})], and that the drug distributes equally throughout the dog tissues (i.e., the dog mimics a single, well-stirred compartment), the average tissue B_{4hrs} achieved with the BPA-Molecusol® formulation can be estimated as 4.5 ± 0.9 ppm. Blood B levels at this time were 3.4 ± 1.1 ppm. For the BPA-buffer infusion, average tissue B_{4hrs} was 0.9 ± 0.1 ppm, with blood B_{4hrs} of 0.4 ± 0.2 ppm.

Rat Studies

Plasma, serum and tissue samples were collected from rats infused with buffer only, with BPA in buffer (mean BPA concentration: 1.2 mg/ml to mimic dog experiments), with BPA in pH adjusted buffer (mean BPA concentration: 6.0 mg/ml), and with BPA in fructose/water (mean BPA concentration: 173.2 mg/ml). BPA-buffer was infused at a target rate of 17 ml/kg/hr for one hour and actual rates were within 5% of the target value. The BPA-fructose complex was infused at target delivery rates of 12.8 and 17 ml/kg/hr, and delivered BPA doses were 2,217 ± 39 mg/kg (N=15) and 2,917 ± 61 mg/kg (N=8). Actual infusion rates were within 7% of target values. Rats infused at target doses of 12.8 and 17 ml/kg/hr of BPA survived if subject to minimal surgical stress, but rats infused with 4,334 ± 56 mg/kg (target dose: 1,586 mg/kg/hr for 3 hours) all died. Toxicities associated with this formulation are discussed in an accompanying paper[5]. Rats infused

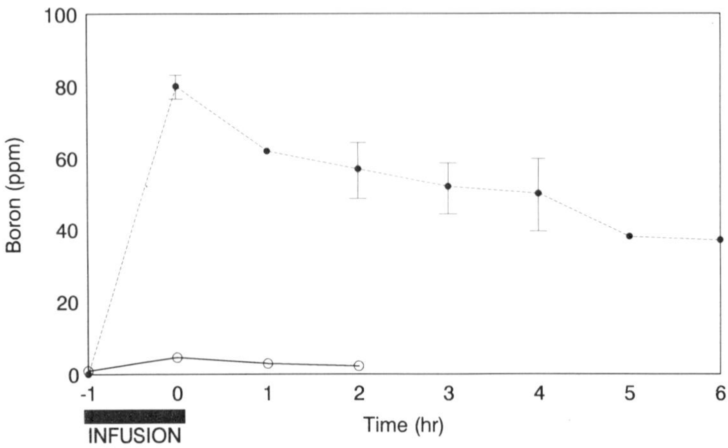

Figure 2. Time course of serum B after infusion of 2,063 ± 55 mg/kg (N=3) of BPA-fructose-water (●) and 99 ± 1.7 mg/kg (N=5) of BPA-buffer (○) formulations into rats. Equal volumes of formulation were infused over a one-hour period (bar).

with BPA doses of 2,148 ± 45 mg/kg (N=6) were sacrificed 6 hours after termination of the BPA infusion, and tissue samples collected for boron analysis. Figure 2 illustrates the difference in the time course of serum B when equal volumes of the BPA-fructose complex and the BPA-buffer solution (6 mg/ml) were infused.

Serum B values were not significantly different from plasma B levels. If the assumption is made that BPA distributes equally throughout the tissues of the rat (i.e., the rat is viewed as a single, well-stirred compartment) the average tissue distribution of BPA at the termination of the BPA-fructose infusion can be estimated to be 1.58 mg/g, for an average tissue B concentration of 75.7 ppm. Using the same assumption, the average tissue concentration of BPA after infusion of BPA-buffer, can be estimated to be 0.075 mg/g, for an average tissue B concentration of 3.6 ppm. Twenty ppm B is considered the lower limit for effective BNCT. Analysis of tissue samples demonstrated that even 6 hours after i.v. infusion of target doses of 2,148 ± 45 mg/kg BPA (N=6), tissue B levels greatly exceeded the requirement for effective BNCT (e.g., heart: 42 ± 7 ppm, muscle: 56 ± 10 ppm, lung: 64 ± 19 ppm, liver: 54 ± 14 ppm). Tissue B levels achieved 2 hours after infusion of the 6 mg/ml BPA-buffer solution were much lower (e.g., heart: 2.3 ± 0.2 ppm, muscle: 3.3 ± 1.6 ppm, lung: 2.7 ± 0.4 ppm, liver: 2.9 ± 0.3 ppm) and tissue levels of B achieved 2 hours after infusion of 1.2 mg/ml BPA in buffer were lower yet (heart: 1.0 ppm, muscle: 0.8 ppm, lung: 1.1 ppm, liver: 1.0 ppm, N=2). Rats receiving control infusions (no BPA) had tissue B levels near or below the detection limit for B (0.4 ppm).

CONCLUSIONS

Our studies suggest that novel formulations of BPA should increase the delivery of B (BPA) to malignant melanoma cells, while eliciting little or no systemic toxicity. Of the options investigated, BPA complex formation with fructose appears most promising. Using the fructose formulation, BPA doses of about 3,000 mg/kg could be administered to rats without inducing obvious toxicities. Tissue B levels measured 6 hours after termination of the infusion were still 2 to 4 fold greater than the minimum levels (20 ppm) deemed necessary for effective BNCT. Given these results, we postulate that i.v. infusion of high doses of BPA delivered as a BPA-fructose complex will increase the B loading of melanoma cells and facilitate BNCT of malignant melanóma. Studies to test this hypothesis are underway.

ACKNOWLEDGEMENTS

The authors wish to thank Dr. E. Jarvi (ISU) for performing the HPLC analysis, and Ms. C. Johnson and Mr. V. Sweet (WSU) for assisting with the animal experiments. This study was performed under the auspices of the U.S. Department of Energy, Office of Energy Research, DOE Idaho Field Office, Contract No. DE-AC07-76ID01570.

REFERENCES

1. Coderre J., Glass, J., Fairchild, R., Roy, U., Cohen, S. & Faud, I. (1987) Selective Targeting of BPA to Melanoma for Neutron Capture Therapy, *Cancer Res.* **47**, 6377-6383.
2. Wheeler, F., Griebenow, M., Wessol, D., Nigg, D. & Anderl, R. (1989) A stoichastic model for high LET response for BNCT, in *Basic Life Sciences: Clinical Aspects of Neutron Capture Therapy*, eds. Fairchild, R., Bond, V. & Woodhead, A. (Plenum Press, New York), pp 165-178.
3. Brewster, M., Simplins, J., Hora, M., Stern, W. & Bodor, N. (1989) The Potential Use of Cyclodextrins in Parenteral Formulations, *J. Parent. Sci. Technol.* **43**, 231-240.
4. Yoshino, K., Suzuki, A., Mori, Y., Kakihana, H., Honda, C., Mishima, Y., Kobayashi, T. & Kanda, K. (1989) Improvement of Solubility of p-Boronophenylalanine by Complex Formation with Monosaccharides, *Strahlenter. Onkol.* **165**, 127-129.
5. LaHann, T., Sills, C., Hematillake, G., Dymock, T. & Daniell, G. (in press) Cardiovascular toxicities associated with i.v. administration of p-BPA formulations, in *Proceedings of the Fifth International Symposium on Neutron Capture Therapy*, eds. Barth, R. & Soloway A. (Plenum Press, New York).

COPYRIGHT

COMPUTER MODELING THE BORON COMPOUND FACTOR IN NORMAL BRAIN TISSUE

P.R. Gavin,[1] R. Huiskamp,[2] F.J. Wheeler,[3] and M.L. Griebenow[3]

[1]Washington State University, Department of Veterinary Clinical Medicine & Surgery, Pullman, WA 99164-6610 USA
[2]Netherlands Energy Research Foundation, Petten, The Netherlands
[3]Idaho National Engineering Laboratory, EG&G Idaho, Inc., INEL BNCT Program, P.O. Box 1625, Idaho Falls, ID 83415-1575 USA

INTRODUCTION

The macroscopic distribution of borocaptate sodium ($Na_2B_{12}H_{11}SH$ or BSH) in normal tissues has been determined and can be accurately predicted from the blood concentration.[1,2] The compound para-borono-phenylalanine (p-BPA) has also been studied in dogs and normal tissue distribution has been determined.[3] The total physical dose required to reach a biological isoeffect appears to increase directly as the proportion of boron capture dose increases.[4] This effect, together with knowledge of the macrodistribution, led to estimates of the influence of the microdistribution of the BSH compound. This paper reports a computer model that was used to predict the compound factor for BSH and p-BPA and, hence, the equivalent radiation in normal tissues. The compound factor would need to be calculated for other compounds with different distributions. This information is needed to design appropriate normal tissue tolerance studies for different organ systems and/or different boron compounds.

MATERIALS AND METHODS

A Monte Carlo computer model of the capillaries was used to determine the physical and equivalent dose to the endothelial cell nucleus as a function of the boron level and distribution pattern (Figure 1). Charged-particle transport methods used have been described.[5]

Known macroscopic and inferred microscopic information from the canine studies were used in the models to predict the effect of the epithermal beam and the $^{10}B(n,\alpha)$ reaction on the endothelium. The equivalent dose to the capillary nucleus was held constant at 12 equivalent Gy. The RBE values used for nitrogen capture (n,proton), $^{10}B(n,\alpha)^7Li$, and fast neutrons were 2.7, 2.3, and 4.5, respectively. The first two values are from the literature and the fast-neutron RBE was derived from skin reactions in dogs.[6,7]

Advances in Neutron Capture Therapy, Edited by
A.H Soloway *et al.*, Plenum Press, New York, 1993

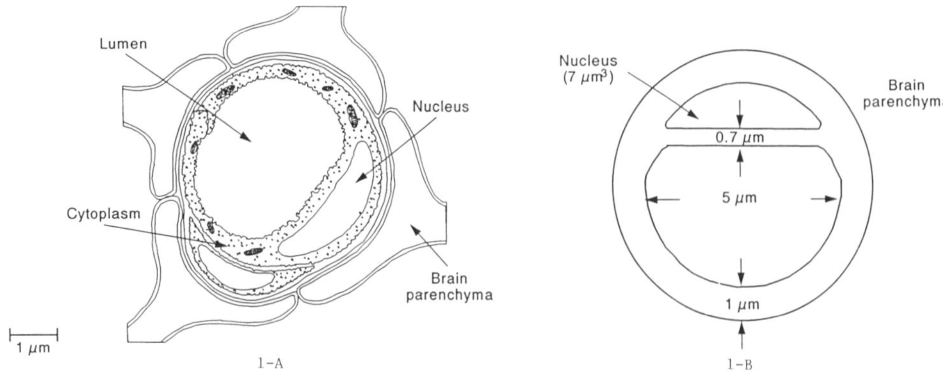

Figure 1. Brain capillary model. A = artist's drawing of histologic structure; B = computer model.

RESULTS

Using the BMRR epithermal beam parameters, the capillary model, with a distribution similar to BSH of 25 ppm in the lumen and 1 ppm in the endothelial cell cytoplasm and surrounding parenchyma, resulted in a capillary endothelial-nuclear physical dose of 10.5 Gy with a peak physical dose of 20 Gy. The calculated equivalent dose to the endothelial nuclei was 16.54 equivalent Gy (Table 1). The percentages of boron capture physical dose to the capillary endothelial nucleus from interregional transport of ^{10}B capture fragments is illustrated in Figure 2-A.

The model resulted in a capillary endothelial-nuclear physical dose of 11.6 Gy when the boron was 8 ppm and uniformly distributed in the blood, endothelial cell, and surrounding parenchyma (i.e., a distribution pattern observed with p-BPA). The calculated equivalent dose to the endothelial nuclei was 19.51 equivalent Gy (Table 1). The percentage physical dose contributions to the endothelial cell nucleus from the transport of ^{10}B capture fragments is depicted in Figure 2-B.

The fraction of the total boron dose that struck the endothelial nucleus, multiplied by the RBE of 2.3, resulted in a compound factor of 0.40 for the first case with asymmetric microdistribution. The second case, with uniform microdistribution, resulted in a compound factor of 1.96.

DISCUSSION

Some inferences of the microscopic distribution can be drawn from the macroscopic distribution of BSH and p-BPA in the blood and brain. The normal blood-brain-barrier excluded the boron from the normal brain parenchyma following an intravenous administration of BSH. The primary portion of the barrier is the tight junctions between the endothelial cells.[8] This fact, together with the hydrophilic nature of the compound, would dictate that the compound is also excluded from the endothelial cytoplasm. The boron associated with p-BPA readily crossed the blood-brain-barrier and was present in the normal brain parenchyma in the same concentration as the blood, indicating a large extravascular component. Since it is the boron element that is analyzed rather than the parent phenylalanine compound, the mode of boron passage into the brain is unknown. Phenylalanine is a precursor to the neuroamines and there are active transport mechanisms for these compounds in the brain endothelial cells. In any case, since the boron was

Table 1. Brain capillary endothelial dose

Type of compound distribution	Type of calculation	Dose					
		10 B(n,alpha)	14 N(n,p)	Fast neutron	Gamma	Other	Total
Non-uniform[a]	Physical (Gy)	2.20	0.88	0.90	5.04	0.60	9.62
(BSH in brain)	Equivalent dose (Eq*Gy)[c]	5.07	2.38	4.05	5.04	0.60	17.14
Uniform[b]	Physical (Gy)	3.43	0.88	0.90	5.04	0.60	10.85
(p-BPA in Brain)	Equivalent dose (Eq*Gy)[c]	7.89	2.38	4.05	5.04	0.60	19.96

[a]25 ppm boron in the endothelial lumen, 1 ppm in the endothelial cytoplasm and brain parenchyma.
[b] 8 ppm boron (l-BPA) in the endothelial lumen & cytoplasm, & 8 ppm in the brain parenchyma.
[c]RBE's used for equivalent dose calculation: 2.3 for 10 B(n,alpha); 2.7 for 14 N (n,p); & 4.5 for fast neutrons.

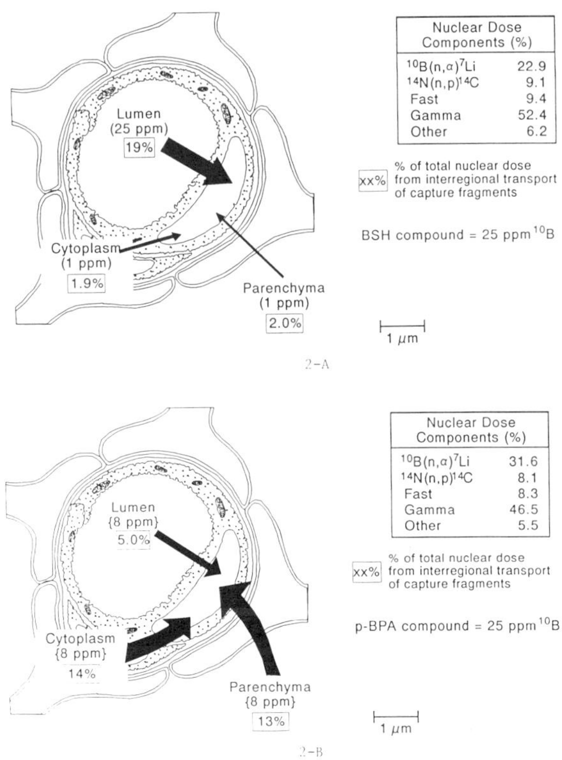

Figure 2. A = Percentage of endothelial nucleus doses from interregional transport of ^{10}B capture fragments. Nonuniform boron distribution that occurs in the brain following BSH infusion. With a tumor-to-blood ration of 1:1, the tumor concentration would be 25 ppm; B = Percentage of endothelial nuclear dose from interregional transport of ^{10}B capture fragments. Uniform boron distributions that occur in the skin following BSH infusion, or in skin and brain following p-BPA administration. Calculations also used 25 ppm in tumor (tumor-to-blood ratio of 3:1.

present in approximately equal concentrations in the blood and brain, there must be similar concentrations in the endothelial cytoplasm.

The capillary endothelial model resulted in the endothelial nuclear radiation doses, as depicted in Figures 2-A and 2-B. The lumenal concentrations used for Figure 2-A would be similar to those that occur with a tumor concentration of 25 ppm and a tumor-to-blood ratio of 1, as observed with the BSH compound. The lumenal concentration used in Figure 2-B would be similar to those that occur with a tumor concentration of 25 ppm and a tumor-to-blood ratio of 3, as observed with the p-BPA compound.

The vascular computer model for the endothelial response resulted in compound factors for the two scenarios that were markedly similar to those calculated from the dog-head irradiation studies using brain necrosis endpoints.[6] The nonuniform compound distribution (BSH in brain) resulted in a computed compound factor of 0.33-0.40 that would model endothelial dose response observed in the canine brain necrosis.[6] The compound factor is the product of RBE and microscopic physical dose divided by total or macroscopic physical dose. This compound factor would indicate that approximately 83% of the peak physical boron capture dose (roughly 5 of 6 fission events) do not contribute to the BSH central nervous system endothelial nuclear dose. This sparing is greater than that previously reported, but less than that observed in the canine normal tissue brain studies.[9,10]

The uniform boron distribution (p-BPA in skin and brain) resulted in a compound factor of 1.96 that would model the endothelial dose seen in the canine skin necrosis (BSH in skin).[6] Skin necrosis, that can occur approximately 3 months following irradiation, has been attributed to vascular damage to the dermis.[11] The canine skin response indicated an increased compound factor of BSH in skin compared to brain, but a value of only 0.5 was obtained.[6] The skin boron concentration is the total boron per gram of a complex tissue. The tissue analyzed contained hair, epidermis, dermis, and subcutaneous fat. The discrepancy between the biologic and computer compound factors for BSH in skin indicates that the boron was not uniformly distributed in relation to the dermal vessels. Microscopic quantification of the boron in skin is needed before this discrepancy can be fully explained.

Brain necrosis with BSH is logically due to the capillary BNCT endothelial dose and this has been supported by histologic analysis of the affected brains. Skin necrosis that can occur approximately three months following irradiation has also been attributed to vascular damage to the dermis.[11] The uniform boron distribution model simulates the p-BPA distribution in the brain. Additional critical targets, including the myelin-producing oligodendrogliocytes, may decrease the tolerance of the brain with this compound.[12] The calculated endothelial nuclear physical and equivalent doses shown in Table 1 were substantially higher in the uniform distribution model, even with one-third of the lumenal concentration of boron.

The vascular sparing of the capillary endothelial nuclei has been predicted.[8,9] The results from the dog irradiations with BSH support this sparing effect; however, the different distribution of the compounds indicate a need to determine this effect, or compound factor, for each tissue and compound.[13]

ACKNOWLEDGEMENT: Work supported in part by the U.S. Department of Energy, Office of Energy Research, under DOE Field Office, Idaho, Contract No. DE-AC07-76IDO1570.

REFERENCES

1. S.L. Kraft, P.R. Gavin, C.E. DeHaan, C.W. Leathers, W.F. Bauer, D.L. Miller, and R.V. Dorn III, Borocaptate sodium: A potential boron delivery compound for boron

neutron capture therapy evaluated in dogs with spontaneous intracranial tumors, Proceedings, National Academy of Sciences, USA. 89:11973-11977 (1992).

2. A. Siefer, J. Casado, R.L. Moss, P.R. Gavin, K. Philipp, R. Huiskamp, and E. Duhmke, Healthy tissue tolerance studies for BNCT at the High Flux Reactor Petten: First results, Proceedings of the European Collaboration on Boron Neutron Capture Therapy Workshop, Petten, The Netherlands, September 18-20, 1991, Plenum Press, New York (1991).
3. P.R. Gavin, S.L. Kraft, C.E. DeHaan, R.D. Sande, M. Papageorges, and W.F. Bauer, Spontaneous canine oral melanoma: A large animal model for BNCT, Proceedings of the Fourth International Symposium on Neutron Capture Therapy, Sydney, New South Wales, Australia, December 4-7, 1991, Plenum Press, New York (1990).
4. P.R. Gavin, S.L. Kraft, C.E. DeHaan, M.L. Griebenow, and M.P. Moore, A large animal model for boron neutron capture therapy, Proceedings of the Fourth International Symposium on Neutron Capture Therapy, Sydney, New South Wales, Australia, December 4-7, 1991, Plenum Press, New York (1990).
5. D.W. Nigg, F.J. Wheeler, and P.D. Randolph, Characterization of the BMRR and PBF epithermal-neutron beams in phantom using three-dimensional derministic radiation transport theory, in: Progress in Neutron Capture Therapy for Cancer, B.J. Allen, D.E. Moore, B.V. Harrington, eds., Plenum Press, New York, pp 145-149 (1992).
6. P.R. Gavin, F.J. Wheeler, R. Huiskamp, A. Siefert, S.L. Kraft, and C.E. DeHaan, Large animal model studies of normal tissue tolerance using an epithermal beam and borocaptate sodium, Proceedings of the European Collaboration on Boron Neutron Capture Therapy Workshop, Petten, The Netherlands, September 18-20, 1991, Plenum Press, New York (1991).
7. J.O. Archambeau, Swine skin: A model to evaluate dose recovery from different radiations, in: Clinical Aspects of Neutron Capture Therapy, R.G. Fairchild, V.P. Bond, A.D. Woodhead, eds., Plenum Press, New York, Basic Life Sciences 50:9-20 (1989).
8. P.A. Stewart, C.L. Farrell, R.F. DelMaestro, and C.G. Ellis, The effect of cellular microenvironment on vessels in the brain, Part I: Vessel structure in tumor, peritumor, and brain from humans with malignant blioma, Int J Radiat Biol. 60:125-130 (1991).
9. R.A. Rydin, O.L. Deutsch, and B.W. Muray, The effect of geometry on capillary wall dose for boron neutron capture therapy, Phys Med Biol, 21:134-138 (1976).
10. K. Kitao, Vascular wall dose from boron neutron capture reaction, in: Boron Neutron Capture Therapy for Tumors, H. Hatanaka, ed., Nishimura Co., Ltd., Nigata, Japan, pp 191-197 (1986).
11. J.W. Hopewell, The importance of vascular damage in development of late effects in normal tissues, Radiation Biology in Cancer Research, R.E. Meyer and H.R. Withers, eds., Raven Press, New York, pp 449-459 (1980).
12. A.J. van der Kogel, Effects of high-LET radiations on the CNS, Proceedings of the European Collaboration on Boron Neutron Capture Therapy Workshop, Petten, The Netherlands, September 18-20, 1991, Plenum Press, New York (1991).
13. R. Gahbauer, R., RBE in normal tissue, Proceedings of the European Collaboration on Boron Neutron Capture Therapy Workshop, Petten, The Netherlands, September 18-20, 1991, Plenum Press, New York (1991).

COMPUTER MODELING THE BORON COMPOUND FACTOR IN BRAIN TUMORS

P.R. Gavin,[1] R. Huiskamp,[2] F.J. Wheeler,[3] and M.L. Griebenow[3]

[1]Washington State University, Department of Veterinary Clinical Medicine Surgery, Pullman, WA 99164-6610 USA
[2]Netherlands Energy Research Foundation, Petten, The Netherlands
[3]Idaho National Engineering Laboratory, EG&G Idaho, Inc., INEL BNCT Program, P.O. Box 1625, Idaho Falls, ID 83415-1575 USA

INTRODUCTION

Dogs with spontaneous brain tumors have been used to study the potential of boron neutron capture therapy (BNCT) using an epithermal neutron beam with borocaptate sodium ($Na_2B_{12}H_{11}SH$ or BSH).[1,2] The compound para-borono-phenylalanine (p-BPA) has also been studied in dogs with oral melanomas and normal tissue distribution has been determined.[3] It has been suggested that the term "compound factor" be used to describe the product of the microscopic distribution and relative biologic effectiveness (RBE) of the $^{10}B(n,\alpha)$ reaction to predict the normal and tumor tissue responses.[4] This paper reports the computer model that was used to predict the compound factor for BSH and p-BPA and, hence, the equivalent radiation in tumors and the relation to normal tissue response. The compound factor would need to be calculated for other compounds with different distributions.

MATERIALS AND METHODS

The charged-particle transport methods used for dose calculations are described in these proceedings.[5] A Monte Carlo computer model was developed to represent a tumor. This model was one of a matrix of 10 μm-diameter spheres representing the tumor cells. The cell nucleus was a 5 μm-diameter sphere and the interstitial space had a thickness of 0.5 μm (Figure 1). Therefore, the model is represented as concentric spheres with radii of the nucleus, cytoplasm, extracellular space, and cytoplasm of adjacent cells of 2.5, 5.0, 5.5, and 11.0 μm, respectively. The tumor model was used for two different sets of parameters that placed 25 ppm boron into the tumor. The first set (BSH) used a

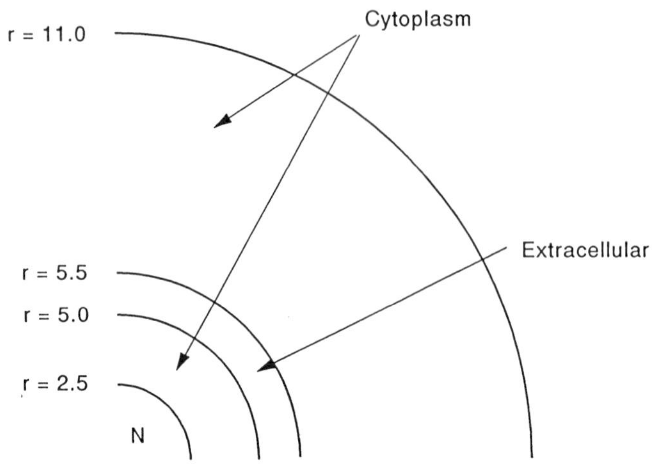

Figure 1. Computer model of tumor. Relative volumes of nucleus, cytoplasm, and extracellular space are 10.7%, 75.1%, and 14.2% respectively.

tumor-to-blood ratio of 1 and limited the tumor distribution to 90% to the intercellular space and 10% to the tumor cell cytoplasm. The second set (p-BPA) used a tumor-to-blood ratio of 3 and the tumor concentration was 80% in the extracellular space, 17.5% in the tumor cell cytoplasm, and 2.5% in the nucleus.

Known macroscopic and inferred microscopic information from the canine studies were used in the models to predict the effect of the epithermal beam and the $^{10}B(n,\alpha)$ reaction on the tumor. The tumor depth for calculational purposes was 2 cm beyond the thermal neutron peak. The equivalent dose to the capillary nucleus was held constant at 12 equivalent Gy to calculate the relative effect on the tumor as the boron concentration was varied. The RBE values used for nitrogen capture (n,proton), $^{10}B(n,\alpha)^7Li$, and fast neutrons were 2.7, 2.3, and 4.5, respectively. The first two values are from the literature and the fast-neutron RBE was derived from skin reactions in dogs.[6,7]

RESULTS

The tumor model with the first set of parameters matched the nonuniform boron normal tissue distribution, as found with BSH in the blood and brain.[1,2] The tumor dose was calculated on the midline 2 cm distal from the peak physical dose. The tumor-cell nuclear equivalent dose was calculated using a constant brain capillary endothelial equivalent dose of 12 equivalent Gy with varying tumor-boron concentrations ranging from 10-100 ppm (Table 1). The equivalent dose to the tumor nuclei increased dramatically, compared to the capillary endothelial nuclei, as the boron concentration in the blood was increased (Figure 2). The calculated compound factor for the boron in the tumor was 0.91.

The tumor model with the second set of parameters used a uniform boron distribution similar to that observed with BSH in the skin and blood and with p-BPA in the skin, brain, and blood.[1-3,8,9] The tumor-cell nuclear equivalent dose was calculated as above with a constant equivalent brain endothelial cell nuclear dose of 12 equivalent Gy and for tumor-boron concentrations ranging from 10-100 ppm (Table 1). A less pronounced effect of tumor nuclei-equivalent dose to endothelial nuclei-equivalent dose was observed as the boron concentration was increased. The calculated compound factor for p-BPA in the tumor was 1.16.

Table 1. Brain tumor dose as a function of boron distribution normalized to endothelial Iso-dose

Type of compound distribution	Tumor:blood ratio	% compound extracellular	% compound in cytoplasm	% compound in nucleus	ppm 10B tumor	Macro peak equiv. dose (Eq*Gy)	Macro tumor equiv. dose (Eq*Gy)	Micro endothelial nuclear equiv. dose (Eq*Gy)	Micro tumor nuclear equiv. dose (Eq*Gy)	Tumor nuclear equiv. dose/endothelial nuclear equiv. dose
Non-uniform BSH	1.1	90	10	0	10	20.7	15.1	12.0	10.1	0.84
					25	31.0	22.3	12.0	12.5	1.04
					50	42.1	31.3	12.0	15.0	1.25
					100	54.2	40.5	12.0	18.6	1.55
Uniform (p-BPA)	3:1	80	17.5	2.5	10	19.5	14.2	12.0	10.3	0.89
					25	25.7	19.0	12.0	11.9	0.99
					50	30.8	22.9	12.0	13.1	1.09
					100	35.3	26.3	12.0	14.3	1.19

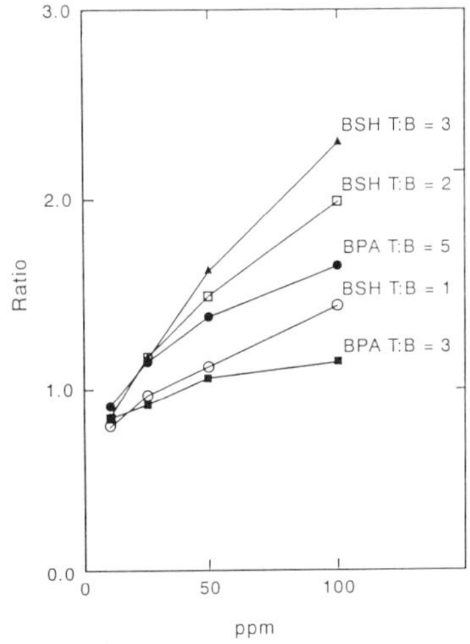

Figure 2. Plot of therapeutic advantages of brain tumor nuclear equivalent dose to brain vascular endothelial nuclear dose with various compounds and tumor-to-blood ratios vs boron concentrations in the tumor.

DISCUSSION

The microscopic tumor-boron distribution of the compound was purposely favorably set for the model that would simulate the uniform normal brain-boron distribution and 3:1 tumor-to-brain ratios that occur with p-BPA.[3,8,9] The calculated compound factor for this scenario was 1.16, substantially better than the nonuniform microscopic boron distribution compound factor of 0.91 calculated for BSH distribution. There is no conclusive evidence that BSH in tumors is intracellular. The hydrophilic nature and pharmacokinetic data would suggest that most of the compound may be extracellular in the interstitial space.[1] Cell culture data has indicated some boron within the cells incubated in medium containing BSH.[10] Therefore, the model parameters of 90% extracellular and 10% intracellular for the BSH compound were chosen. Pharmacokinetic and cell culture data have indicated that the boron associated with p-BPA is also probably not permanently incorporated into the cell cytoplasm;[11] however, the intracellular localization of the boron was artificially doubled for the p-BPA compound to compare the compound's effects on tumors and brain microvasculature.

The tumor-to-blood ratios reported by others[8,9,12] are in excess of the 1:1 used for BSH and the 3:1 used for p-BPA. Therefore, additional data was obtained from the tumor model with the same nonuniform microscopic boron tumor distribution as before, but with tumor-to-blood ratios of 2:1 and 3:1 for BSH and 5:1 for p-BPA with the uniform distribution (Figure 2). Concentrations of boron in spontaneous canine brain tumors can exceed 50 ppm with conventional dosages and administration schedules.[1] Concentrations of boron following p-BPA administration have generally not been associated with tumor concentrations greater than 30-40 ppm.[8,9] These results suggest that BSH may be the better compound for treating brain tumors.

The results shown in Figure 2 demonstrate that the increased equivalent dose to the endothelial cell nucleus with p-BPA overwhelms the increased tumor equivalent dose from improved tumor-to-blood ratios and improved microscopic tumor cell boron localization. The increase of boron concentrations in tumor (and blood) boron concentrations has a large effect on the equivalent tumor dose compared to the equivalent brain endothelial vascular dose. This increase in therapeutic gain was more pronounced with the nonuniform boron distribution observed with BSH in the brain.

ACKNOWLEDGMENT. Work supported in part by the U.S. Department of Energy, Office of Energy Research, under DOE Field Office, Idaho, Contract No. DE-AC07-76IDO1570.

REFERENCES

1. S.L. Kraft, P.R. Gavin, C.E. DeHaan, C.W. Leathers, W.F. Bauer, D.L. Miller, and R.V. Dorn III, Borocaptate sodium: A potential boron delivery compound for boron neutron capture therapy evaluated in dogs with spontaneous intracranial tumors, Proceedings, National Academy of Sciences, USA. 89:11973-11977 (1992).
2. P.R. Gavin, S.L. Kraft, C.E. DeHaan, M.L. Griebenow, and M.P. Moore, A large animal model for boron neutron capture therapy, Proceedings of the Fourth International Symposium on Neutron Capture Therapy, Sydney, New South Wales, Australia, December 4-7, 1991, Plenum Press, New York (1990).
3. P.R. Gavin, S.L. Kraft, C.E. DeHaan, R.D. Sande, M. Papageorges, and W.F. Bauer, Spontaneous canine oral melanoma: A large animal model for BNCT, Proceedings of the Fourth International Symposium on Neutron Capture Therapy, Sydney, New South Wales, Australia, December 4-7, 1991, Plenum Press, New York (1990).
4. R. Gahbauer, R., RBE in normal tissue, Proceedings of the European Collaboration on Boron Neutron Capture Therapy Workshop, Petten, The Netherlands, September 18-20, 1991, Plenum Press, New York (1991).
5. D.E. Wessol, F.J. Wheeler, R. Babcock, and J.D. Starkey, Medical image reconstruction and geometry generation software, Proceedings of the 5th International Symposium on Neutron Capture Therapy, Columbus, Ohio, September 14-17, 1992, Plenum Press, New York (1992).
6. P.R. Gavin, F.J. Wheeler, R. Huiskamp, A. Siefert, S.L. Kraft, and C.E. DeHaan, Large animal model studies of normal tissue tolerance using an epithermal beam and borocaptate sodium, Proceedings of the European Collaboration on Boron Neutron Capture Therapy Workshop, Petten, The Netherlands, September 18-20, 1991, Plenum Press, New York (1991).
7. J.O. Archambeau, Swine skin: A model to evaluate dose recovery from different radiations, in: Clinical Aspects of Neutron Capture Therapy, R.G. Fairchild, V.P. Bond, A.D. Woodhead, eds., Plenum Press, New York, Basic Life Sciences 50:9-20 (1989).
8. J.A. Coderre, J.D. Glass, R.G. Fairchild, U. Roy, S. Cohen, and I. Fand, Selective Targeting of boronophenylalanine to melanoma in BALB/c mice for neutron capture therapy, Cancer Res 47:6377-6388 (1987).
9. J.A. Coderre, J.D. Glass, R.G. Fairchild, P. Micca, I. Fand, and D.D. Joel, Selective delivery of boron by the melanin precursor analogue p-boronophenylalanine to tumors other than melanoma, Cancer Res, 50:138-141 (1990).
10. R. Verrijk, I.J.H. Smolders, R.H. Huiskamp, W.C. deBruijn, and A.C. Begg, Pharmacokinetics of boron in animals and cell cultures using ICP-AES and boron localization with electron spectroscopic imaging, Proceedings of the 5th International Symposium on Neutron Capture Therapy, Columbus, Ohio, September 14-17, 1992, Plenum Press, New York (1992).
11. M. Papageorges, C. Elstad, G.G. Meadows, and W.F. Bauer, Effect of tyrosine and phenylalanine deprivation on the uptake of para-boronophenylalanine in melanoma cells, Proceedings of the Fourth International Symposium on Neutron Capture Therapy, Sydney, New South Wales, Australia, December 4-7, 1991, Plenum Press, New York (1990).
12. H. Fankhauser and G. Stragliotto, Biodistribution of known sulfhydryl (BSH) on patients with intracranial tumors, Proceedings of the Fourth International Symposium on Neutron Capture Therapy, Sydney, New South Wales, Australia, December 4-7, 1991, Plenum Press, New York (1990).

DETERMINATION OF FAST-NEUTRON DOSE DISTRIBUTIONS IN THE CANINE CENTRAL NERVOUS SYSTEM

David W. Nigg, Jean M. Moran, and Floyd J. Wheeler

National Center for BNCT Measurement and Development
Idaho National Engineering Laboratory
EG&G, Idaho, Inc.
P.O. Box 1625
Idaho Falls, Idaho 83415

The Idaho National Engineering Laboratory (INEL) Center for Boron Neutron Capture Therapy (BNCT) Measurement and Development conducts, sponsors, and coordinates research and development efforts in a variety of BNCT-related areas. Current activities include an ongoing series of large-animal (canine) model irradiation experiments conducted in collaboration with the Washington State University School of Veterinary Medicine[1]. These experiments employ the epithermal-neutron beam at the Brookhaven Medical Research Reactor (BMRR)[2] and have the purpose of determining normal tissue tolerance and spontaneous brain tumor response to the unique mix of radiation components that characterize BNCT.

The normal tissue tolerance experiments employ borocaptate sodium (BSH) as the boron carrying pharmaceutical. The experimental animals for the tolerance experiments are typically Labrador Retrievers without brain tumors. Tolerance levels in skin and central nervous system (CNS) tissue with various concentrations of boron have been established and initial estimates of the relative biological effectiveness (RBE) of the various radiation dose components have been made[1]. Irradiations designed to bring the CNS of the dogs to expected tolerance levels without boron have, however, been hampered by skin damage caused by the moderately high fast neutron component of the BMRR beam. Surface tissue is dose-limiting for the zero-boron irradiations. In addition, the dogs employed for this experiment series have significantly more cover tissue over the brain (primarily because of the temporalis muscle) than is the case for humans. This cover tissue attenuates the fast neutron component in the treatment beam to levels in the brain itself that make it difficult to accurately estimate the RBE of this important beam component from the observed biological response of the brain to the total mixed irradiation field that is present.

During a recent National Academy of Sciences Institute of Medicine peer review of the canine program it was suggested by the referees that some of the tolerance experiments could be performed using dogs with the scalp and temporal muscle surgically reflected prior to irradiation in order to avoid the problems noted above. This is being seriously considered and the results of some exploratory calculations to determine the peak CNS fast-neutron dose for canine subjects with and without scalp and temporal muscle reflection have been completed and are summarized here.

The calculations were performed using the recently-developed bnct_edit/rtt_MC BNCT analytic dosimetry software package[3]. This software package employs the Monte Carlo technique to compute three-dimensional neutron and photon flux and absorbed dose distributions within a BNCT irradiation volume that is geometrically described by non-uniform rational B-spline surface equation fits to the various regions of interest, as outlined (for this application) on a series of axial medical images of a typical Labrador Retriever head. Figure 1 shows the spline surface fits to

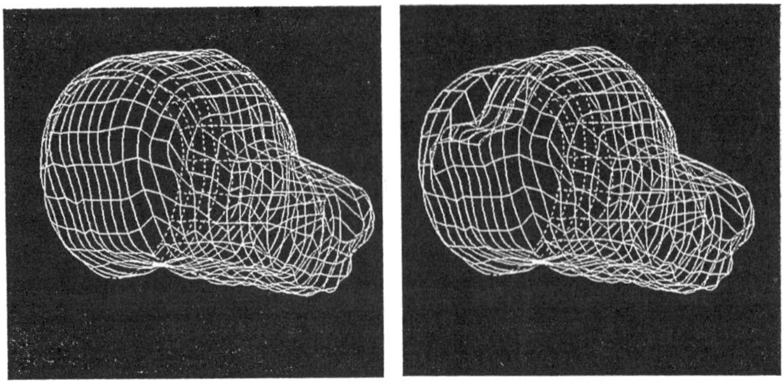

Figure 1. Reconstructed outer surfaces for the intact and surgically-modified canine head models.

the outside surface of the canine head for both the intact dog and for a head model that has been modified to simulate the removal of the skin and temporal muscle overlying the right hemisphere of the brain. Both models also include surface reconstructions of various interior structures (brain, eyes, sinus cavities, etc.). These have been deleted from Figure 1 for clarity.

Three calculations were performed. All three assumed the presence of a 5 cm by 10 cm BMRR beam delimiter aperture over the right hemisphere of the canine brain with the long dimension of the aperture lined up parallel to the longitudinal axis of the head. In the first case the intact dog model is used and the beam impinges at an angle of incidence that is perpendicular to the top of the head. The second case employs the reflected-scalp model, again with the beam perpendicularly-incident on the top of the head. In the third case scalp reflection is assumed and the beam is incident on the head at an angle that brings the exit plane of the delimiter as close as possible to the exposed skull surface overlying the brain. At first glance, this would be expected to increase the fast neutron dose to the brain relative to the case with perpendicular beam incidence.

Calculated results for the fast neutron dose in the three cases are illustrated in Figures 2-4. In all cases the fast neutron dose is normalized to a peak total physical dose in the brain (with no boron) of 10 Gy. As expected, scalp reflection increases the fast neutron component of the total dose significantly (about a factor of 2), assuming perpendicular beam incidence for both the intact (Figure 2) and reflected-scalp (Figure 3) cases. The relative difference in fast neutron components between these two cases has been independently confirmed using three-dimensional deterministic calculations based on the canine head model described in Reference 4. It is interesting to note, however, that with scalp reflection, a decrease in the relative fast neutron component is observed when the beam is obliquely incident (Figure 4) rather than perpendicularly-incident (Figure 3). This may be explained by the fact that the delivery rates of all dose components increase with oblique beam incidence, but the delivery rates of those components that result from thermal neutron interactions increase more than is the case for the fast-neutron component. Hence when one normalizes to a specified total peak dose, the relative fast neutron component decreases for the case with oblique incidence. That this would be the case was not immediately obvious prior to the completion of these calculations. It is a consequence of the complex nature of the neutron transport phenomena associated with BNCT. This illustrates the importance of performing detailed three-dimensional planning calculations in connection with all in-vivo BNCT irradiations.

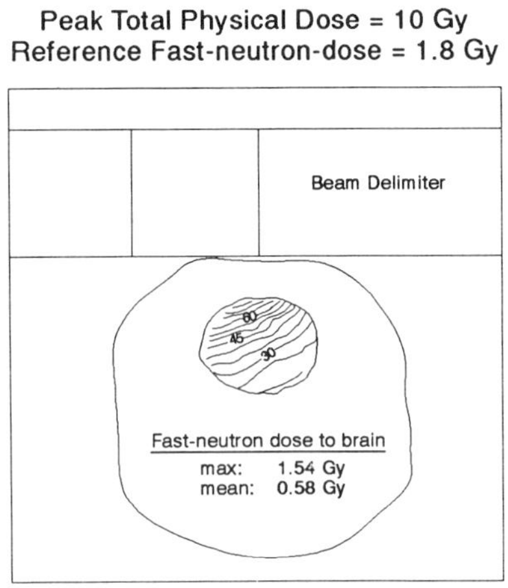

Figure 2. Canine brain fast-neutron-dose percentage contours for the intact canine head model.

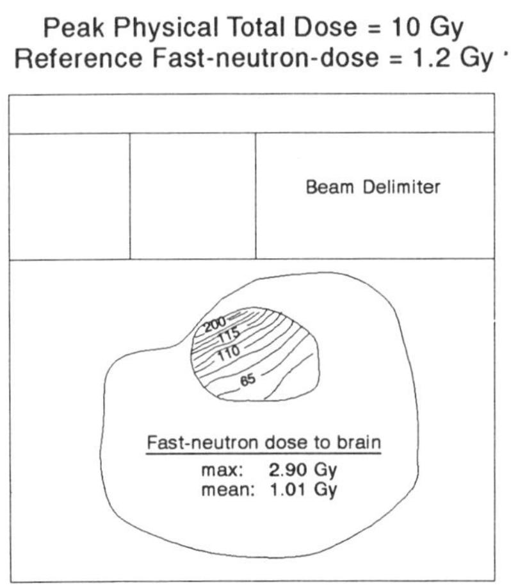

Figure 3. Canine brain fast-neutron-dose percentage contours for the surgically-modified canine head model.

Peak Total Physical Dose = 10 Gy
Reference Fast-neutron-dose = 1.5 Gy

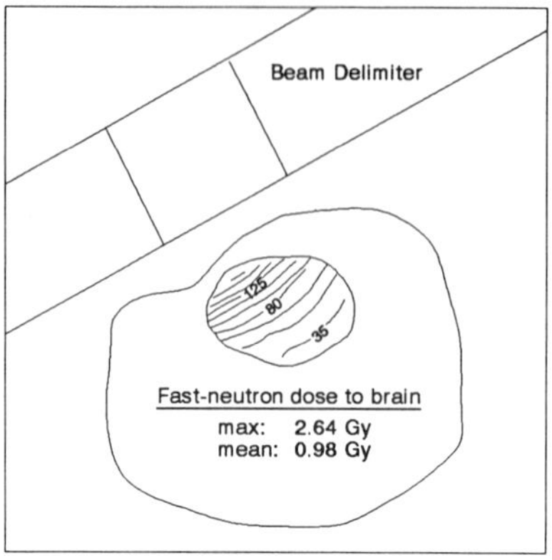

Figure 4. Canine brain fast-neutron-dose percentage contours for the surgically-modified canine head model with a rotated beam angle of incidence.

ACKNOWLEDGEMENTS

Ken Bradshaw, working in collaboration with the University of Utah, provided the sequence of canine magnetic resonance images used to construct the calculational models employed in this study.

This study was performed under the auspices of the U.S. Department of Energy, Office of Energy Research, DOE Field Office, Idaho, under Contract Number DE-AC07-76ID01570.

REFERENCES

1. P.R. Gavin, et. al., Large animal model studies on the use of BNCT for the treatment of brain tumors, *Proceedings of the Fifth International Symposium on Neutron Capture Therapy*, OSU, September 1992, (To be published).

2. D.W. Nigg, et. al., Demonstration of three-dimensional deterministic radiation transport theory dose distribution analysis for boron neutron capture therapy, *Medical Physics,* 18:1, 43-53, January 1991.

3. D.W. Nigg, Analytical dosimetry and treatment planning for BNCT, *Proceedings of the Fifth International Symposium on Neutron Capture Therapy*, OSU, September 1992, (To be published).

4. J.M. Moran, et. al., Macroscopic geometric heterogeneity effects in radiation dose distribution analysis for BNCT, *Medical Physics*, 19:3, 723-732, June 1992.

DOSE OPTIMIZATION FOR BORON NEUTRON CAPTURE THERAPY OF SPONTANEOUS CANINE BRAIN TUMORS

Carol A. Atkinson, Floyd J. Wheeler
Idaho National Engineering Laboratory
Idaho Falls, Idaho 83415-1575

INTRODUCTION

The ultimate success of Boron Neutron Capture Therapy (BNCT) will depend, among other things, upon the ability to calculate, *a priori*, accurate and optimal radiation dose distributions in the target volume. Twenty-one dogs with spontaneously-occurring brain tumors have been treated, at the Brookhaven Medical Research Reactor (BMRR) in a collaborative effort between the Idaho National Engineering Laboratory (INEL), Washington State University (WSU), and Brookhaven National Laboratory (BNL). Because these 21 treatments were designed to demonstrate safety rather than efficacy, the dogs involved did not necessarily receive irradiations corresponding to a treatment plan designed to maximize the estimated cell kill at the target location. Four of these dogs (with intra-axial tumors above the tentorium) were chosen as test cases to determine the effects of optimizing treatment. Although the tumors present in all four dogs were of the same type, they varied significantly in size and location. Optimum treatment was found by varying boron concentration and incident neutron beam position and size. The effect of using a beam having a "cleaner," more forward-peaked angular distribution [Georgia Tech Research Reactor (GTRR)] was also examined.

DISCUSSION

A brief description of the four dogs chosen for this study and their "as irradiated" parameters is given in Table 1. A standard calculational model was developed using regular-geometry primitives[1] and a calculated spectrum for BMRR.[2] The following variations, representing several variables on which the irradiation would be physically dependent, were applied to this calculational model to determine an optimal treatment plan for each dog: the beam port opening was varied between a 10 x 10 cm square opening and circular openings of 7.62, 6.35, 5.08, and 3.81 cm; the head was positioned as close to the beam port opening as possible while simultaneously trying to minimize the distance between the tumor and the beam port; the ^{10}B concentration was increased to 100 ppm; the boron tumor-to-blood ratio was varied between 1 and 5; and selected cases were repeated using a calculated spectrum for the GTRR.[3] Monte Carlo methods for neutron and gamma transport[1] were used to determine flux distributions throughout each model.

Generally-accepted relative biological effectiveness (RBE) values for boron, nitrogen, gamma, and fast neutron were applied to the results of these calculations to determine an effective dose. The following RBE values were assumed: 2.3 for ^{10}B at the target location[4], 2.7 for nitrogen[5], 4.5 for fast neutron[6], and 0.6 for gamma.[7] Due to geometric considerations, a compound

Advances in Neutron Capture Therapy, Edited by
A.H Soloway *et al.*, Plenum Press, New York, 1993

Table 1. Results of selected calculations used in determining an optimal treatment plan for four previously irradiated dogs having spontaneously-occurring brain tumors.

Dog[a]	Case and *"head position"*	Average ^{10}B Concentration (ppm)	RBE Dose[b] (cGy)	Distance from Beam Port	Beam Port Size (cm)	Therapeutic Ratio[c] (Log Cell Kill)		Cell Kill Advantage[d]	
						BMRR	GTRR	T:B=1	T:B=5
"Pugee" Krebs Portland Boxer 60 lbs Tumor Oligodendroglioma	1 *"as irradiated"*	18.5	physical dose goal 1900 actual 1911	1 cm	10 x 10	1.88 (7.89)	2.35 (8.22)	2.16	2.26
Volume (ml) 7.71 Depth (cm) 1.9 - 4.6	2 *"as irradiated"*	18.5	1200	1 cm	10 x 10	1.88 (6.98)	2.35 (8.74)	57.04	138.80
	3 *"as irradiated"*	100.0	1200	1 cm	10 x 10	3.65 (13.59)	4.12 (15.32)	53.30	85.14
	4 *"as irradiated"*	100.0	1200	1 mm	10 x 10	3.87 (14.42)			
	5 *"moved"*	100.0	1200	1 mm	10 x 10	3.94 (14.67)	4.19 (15.60)	8.55	20.26
	6 *"moved"*	100.0	1200	1 mm	7.62 diam.	3.77 (14.03)	4.05 (15.06)	10.84	58.87
"Mugsy" Marshall Miniature Boxer 25 lbs Tumor	7 *"as irradiated"*	21.6	physical dose goal 1900 actual 1933	1 cm	10 x 10	2.15 (8.81)	2.54 (8.63)	0.66	0.57
Astrocytoma Volume (ml) 1.45 Depth (cm) 2.7 - 4.3	8 *"as irradiated"*	21.6	1200	1 cm	10 x 10	2.15 (8.00)	2.54 (9.46)	28.90	61.97

Dog[a]	Case and *"head position"*	Average ^{10}B Concentration (ppm)	RBE Dose[b] (cGy)	Distance from Beam Port	Beam Port Size (cm)	Therapeutic Ratio[c] (Log Cell Kill)		Cell Kill Advantage[d]	
						BMRR	GTRR	T:B=1	T:B=5
	9 *"as irradiated"*	100.0	1200	1 cm	10 x 10	3.83 (14.25)	4.20 (15.62)	23.79	36.48
	10 *"as irradiated"*	100.0	1200	1 mm	10 x 10	3.80 (14.16)			
	11 *"moved and rotated"*	100.0	1200	1 mm	10 x 10	4.09 (15.23)	4.10 (15.26)	1.09	3.31
	12 *"moved and rotated"*	100.0	1200	1 mm	7.62 diam.	3.76 (13.98)	4.13 (15.38)	24.89	219.14
"Brandy" Grudzinski Boxer 50 lbs Tumor Oligodendroglioma Volume (ml) 3.37 Depth (cm) 4.3 - 6.4	13 *"as irradiated"*	24.9	physical dose goal 1900 actual 2169	1 cm	10 x 10	1.27 (5.47)	1.52 (5.58)	1.30	1.85
	14 *"as irradiated"*	24.9	1200	1 cm	10 x 10	1.27 (4.72)	1.52 (5.67)	8.93	96.82
	15 *"as irradiated"*	100.0	1200	1 cm	10 x 10	2.14 (7.98)	2.34 (8.72)	5.44	45.74
	16 *"as irradiated"*	100.0	1200	1 mm	10 x 10	2.13 (7.94)			
	17 *"moved and rotated"*	100.0	1200	1 mm	10 x 10	3.18 (11.83)	3.46 (12.86)	10.70	26.06
	18 *"moved and rotated"*	100.0	1200	1 mm	7.62 diam.	3.01 (11.21)	3.19 (11.87)	4.57	53.14

continued

Table 1. Continued.

Dog[a]	Case and *"head position"*	Average ^{10}B Concentration (ppm)	RBE Dose[b] (cGy)	Distance from Beam Port	Beam Port Size (cm)	Therapeutic Ratio[c] (Log Cell Kill)		Cell Kill Advantage[d]	
						BMRR	GTRR	T:B=1	T:B=5
"Scuzzy" Jenkins Sheltie 40 lbs Tumor	19 *"as irradiated"*	85.3	physical dose goal 2300 actual 2357	1 mm	10 x 10	3.51 (11.63)	3.88 (11.68)	1.14	0.94
Astrocytoma Volume (ml) 5.04	20 *"as irradiated"*	85.3	1200	1 mm	10 x 10	3.51 (13.06)	3.88 (14.44)	23.78	50.24
Depth (cm) 2.0 - 4.5	21 *"as irradiated"*	100.0	1200	1 mm	10 x 10	3.67 (13.66)	4.02 (14.97)	20.20	39.60
	22 *"moved and rotated"*	100.0	1200	1 mm	10 x 10	3.88 (14.44)	4.38 (16.32)	75.39	86.09
	23 *"moved and rotated"*	100.0	1200	1 mm	7.62 diam.	3.75 (13.95)	4.06 (15.12)	14.67	84.72

a. Tumor volume and depth calculated from computational model.
b. The following RBE values were assumed: 2.3 for ^{10}B at the target location, 2.7 for nitrogen, 4.5 for fast neutron, and 0.6 for gamma. A compound factor of 0.33 was used for ^{10}B in healthy tissue. (References for the use of these values are given in the text of the article.)
c. Therapeutic ratio is defined to be the RBE dose to the tumor divided by the RBE dose at the thermal peak. Log cell kill has been defined to be the negative natural log of the fractional survival at the tumor location. Values given assume a boron tumor-to-blood ratio of one (T:B=1).
d. Cell kill advantage is defined to be the number of remaining cells at the tumor location using the BMRR beam divided by the number of remaining cells at the tumor location using the GTRR beam. A value of 1.0E+09 cells per cm^3 was used to determine the number of remaining cells at the tumor location.

factor of 0.33 was used for ^{10}B in healthy tissue.[6] The doses were normalized to 12 Gy RBE at the thermal peak. Therapeutic ratio and log cell kill were computed for comparison between treatment plans developed for the same neutron source; cell kill advantage was computed for comparison between the two different neutron sources.

Results of Monte Carlo calculations for these four dogs are given in Table 1. For the less forward-peaked beam, BMRR, it appears that a generalized treatment plan could be determined to produce optimal tumor cell kill: high ^{10}B concentrations; large beam port opening; and the head positioned to minimize the distance between the tumor and the beam port. However, for irradiations planned in a more forward-peaked beam, GTRR, a generalized plan is not so obvious. This can be seen in the results for "Mugsy" (Cases 9, 11, and 12) where tumor cell kill could be improved by increasing the distance between the head and the beam port and using a smaller opening.

Under almost all circumstances, the more forward-peaked beam will produce a greater cell kill at the tumor location as illustrated by the values given for the cell kill advantage in Table 1. To illustrate another important feature of using a "cleaner", more forward-peaked beam, the boron tumor-to-blood ratio (illustrative of the ability to get higher concentrations of boron to the tumor) was increased from 1 to 5 and cell kill advantage was again computed. In almost all cases, the cell kill advantage increases, sometimes dramatically, as the tumor-to-blood ratio increases.

CONCLUSIONS

To see significant increases in estimated tumor cell kill, optimal treatment plans specific to each patient are recommended. The determination of these plans depends on several parameters: depth and location of the tumor within the brain, concentration of ^{10}B that can be delivered to the tumor mass, and the angular distribution and energy spectrum of the neutron source. It is also seen that the importance of beam quality is magnified as ^{10}B delivery systems improve.

REFERENCES

1. F.J. Wheeler and D.W. Nigg, "Three-Dimensional Radiation Dose Distribution Analysis for Boron Neutron Capture Therapy," *Nuclear Science and Engineering, 110*, January, 1992, pp. 16-31.
2. F.J. Wheeler, D.K. Parsons, B.L. Rushton, and D.W. Nigg, "Epithermal Neutron Beam Design for Neutron Capture Therapy at the Power Burst Facility and the Brookhaven Medical Research Reactor," *Nuclear Technology, 92*, October, 1990, pp. 106-117.
3. D.W. Nigg, G.J. Storr, and F.J. Wheeler, "Physics Parameters for an Epithermal-Neutron Beam at the Georgia Institute of Technology", *Proceedings of the Fifth International Symposium on Neutron Capture Therapy, Columbus, Ohio, September 14-17, 1992*, these proceedings.
4. D. Gabel, R.G. Fairchild, B. Larsson, K. Drescher, and W.R. Rowe, "The Biological Effect of the $^{10}B(n,\alpha)^7Li$ Reaction and its Simulation by Monte Carlo Calculation," *Proceedings of the First International Symposium on Neutron Capture Therapy, Cambridge, Massachusetts, October 12-14, 1983*, BNL 51730, Brookhaven National Laboratory, R.G. Fairchild and G.L. Brownell, editors, p.128.
5. J.O. Archambeau, Swine skin: A model to evaluate dose recovery from different radiations, *in:* "Clinical Aspects of Neutron Capture Therapy," Plenum Press, New York, 1989, pp.9-20.
6. P.R. Gavin, F.J. Wheeler, R. Huiskamp, A. Siefert, S.L. Kraft, and C.E. DeHaan, "Large Animal Model Studies of Normal Tissue Tolerance Using an Epithermal-Neutron Beam and Borocaptate Sodium," *Proceedings of the International Workshop on Macro and Microdosimetry and Treatment Planning for Neutron Capture Therapy, Massachusetts Institute of Technology, Cambridge, Massachusetts, October 31 - November 1, 1991*, to be published.
7. K.F. Fu, Influence of dose rate on normal tissue tolerance, *in:* "Radiation Injury to the Nervous System," P.H. Gutin, S.A. Leibel & G.E. Sheline, eds., Raven Press, New York, 1991, pp.69-87.

COPYRIGHT

CELLULAR UPTAKE, SUBCELLULAR DISTRIBUTION AND *in vitro* TOXICITY OF CARBORANYL URIDINES

Liang Liu, Rolf F. Barth, Werner Tjarks, Albert H. Soloway, and David H. Ives

The Ohio State University, Columbus, Ohio 43210,U.S.A.

INTRODUCTION

It has been known that the biological effects of BNCT can be maximized if ^{10}B can be delivered to the nuclei of tumor cells so that there is a greater probability that alpha particles and ^{7}Li nuclei will discharge their energy within the cell nucleus[1,2]. Bromo- and iododeoxyuridine are preferentially taken up by rapidly proliferating neoplastic cells compared to slowly or non-proliferating normal cells[3]. These observations have provided the rationale for the synthesis of a group of carborane containing derivatives of uridine[4,5]. 2'-O-(o-carboran-1-ylmethyl)uridine (CBU-2') was synthesized by inserting a carboranyl moiety into the 2'-position on the carbohydrate portion of uridine to form a uridine analogue. This theoretically might be incorporated into DNA by the thymidylate synthetase pathway, and into RNA by normal RNA anabolism[3]. *In vitro* studies on the cellular uptake and persistence of CBU-2' with F98 glioma cells showed that this compound attained high uptake and persisted at significant levels for up to 48 hours[6]. The purpose of the present study was to further evaluate CBU-2' using a series of biological and biochemical methods to determine the selectivity, mechanism of uptake, subcellular distribution and *in vitro* toxicity of this compound.

MATERIALS AND METHODS

Cellular Uptake Assays. F98 rat glioma and U-87 MG human glioma, B16 melanoma, SP2/0 myeloma, 3T3 and MDCK fibroblasts were grown in DMEM supplemented with 10% fetal bovine serum to semi-confluency for tumor cells and confluency for non-tumor cells. Following this, CBU-2' (Fig 3) was added to the flasks at a final concentration of 6.75 μg boron/ml and the cells were cultured for an additional 16 hours. They were then washed 2x with 0.01 M PBS, at pH 7.4, disaggregated with 0.25% trypsin and again washed 2x with PBS. Cellular boron concentrations were measured by means of direct current plasma-atomic emission spectroscopy (DCP-AES)[7].

Advances in Neutron Capture Therapy, Edited by
A.H Soloway *et al.*, Plenum Press, New York, 1993

Effects of Metabolic Inhibitors on Uptake of [^{3}H]-uridine and CBU-2' by F98 Glioma Cells. F98 glioma cells were cultured in 96 well microplates at 37°C to semi-confluency. Following this, rotenone, which blocks electron transport[8], and NBMPR (6-[(4-nitrobenzyl)thio]-9-*β*-*D*-ribofuranosylpurine) and dipyridamole, both of which interfere with nucleoside membrane transport[9], were added and the cultures were pulsed with 0.2 μCi of [^{3}H]-UdR and incubated for an additional 18 hours at 37°C. The cells then were harvested with a Skatron cell harvester and [^{3}H]-UdR uptake was determined with a Beckman LS 7000 beta scintillation counter. The effects of metabolic inhibitors on the uptake of CBU-2' were determined by culturing F98 cells in 150 mm^2 flasks with 10μM of each inhibitor and CBU-2' (6.75 μg B/ml) of CBU-2' for 16 hours at 37°C. Following this, the cells were harvested, washed, and boron concentrations were measured by DCP-AES.

Subcellular Distribution of CBU-2' in F98 Glioma Cells. F98 glioma cells, which had been exposed to CBU-2', were resuspended in 1% Nonidet P-40, and then centrifuged. Boron concentrations in both supernatants (membrane and cytoplasm) and sediments (nuclei) were determined by means of DCP-AES. The cytosolic fraction was separated from the cell membrane and nuclear fractions by repetitively freezing and thawing the cells, which had been resuspended in distilled water. Disruption of the cell membranes was determined visually by microscopic examination. After centrifugation, boron concentrations in the supernatant (cytoplasm) and pellets (membrane and nuclei) were measured by DCP-AES. Cellular RNA and DNA were isolated and precipitated by means of phenol and ethanol extraction[10].

Cellular Toxicity of CBU-2'. F98 glioma cells were cultured in 96 well microplates at 37°C to semi-confluency, and varying concentrations of CBU-2' or methotrexate were then added to each well and then pulsed with 0.2 μCi of [^{3}H]-TdR. Cells were harvested 18 hours later with a Skatron cell harvester and [^{3}H]-TdR uptake was determined.

RESULTS

Cellular Uptake of CBU-2'. In this study neoplastic and non-neoplastic cells were incubated with CBU-2' for 16 hours at 37°C and cellular uptake of CBU-2' ranged from 46.2 to 75.1 μg B/g cells for tumor cells and approximated 63.6 μg B/g cells for MDCK fibroblasts (Table 1). The results indicate that there was no apparent selectivity of CBU-2' for neoplastic cells.

Table 1. *In vitro* **Cellular Uptake of CBU-2' by Neoplastic Cells and Fibroblasts**

Cell line	Cellular boron uptake (μg B/g cells)[a]
F98 glioma	75.1±11.8[b]
U-87 MG glioma	72.1± 4.7
B16 melanoma	46.2± 5.2
SP2/0 myeloma	73.6± 4.5
MDCK fibroblasts	63.6±22.4

[a]One gram cells=10^9. [b]Mean ± SD.

Effects of Metabolic Inhibitors on Uptake of [³H]-UdR and CBU-2' by F98 Glioma Cells. Rotenone, and NBMPR and dipyridamole were used to determine the metabolic basis for the cellular uptake of CBU-2'. In order to establish a baseline, the effects of these inhibitors on the uptake of [³H]-UdR was determined using F98 glioma cells. Dipyridamole and rotenone produced a 96% and 69% inhibition respectively of the uptake of [³H]-UdR (Fig 1), compared to a 1% and 9.5% inhibition of the uptake of CBU-2'. These results indicate that uptake of CBU-2' was neither nucleoside transport nor energy dependent.

Cellular Toxicity of CBU-2'. The *in vitro* toxicity of CBU-2' for F98 glioma cells is summarized in Fig 2. As measured by % reduction of [³H]-TdR uptake, the IC_{50} was $3x10^{-5}M$ for CBU-2' compared with $>1x10^{-3}M$ for methotrexate. The result indicates that CBU-2' had direct cytostatic and/or cytotoxic effects on F98 glioma cells.

Subcellular distribution of CBU-2' in F98 Glioma Cells. The subcellular distribution of CBU-2' in F98 glioma cells was determined by separating the cell membrane fraction from cytosolic and nuclear fractions by the physical and chemical methods described above.

Table 2. Subcellular Distribution if CBU-2' in F98 Glioma Cells

Cellular component	Cellular boron (μg B/g cells)[a]	Distribution of boron in cellular component (%)
Whole cells	75.1±11.8	
Cell membrane	56.8±10.8	75.6
Cytoplasm	11.7±0.8	15.6
Nucleus	6.6±0.2	8.8
RNA/DNA	NM[b]	0.0

[a] One gram cells=10^9. [b] NM means non-measurable.

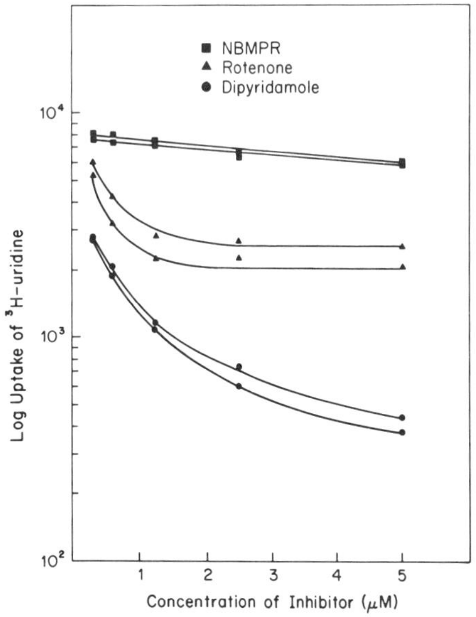

Fig. 1

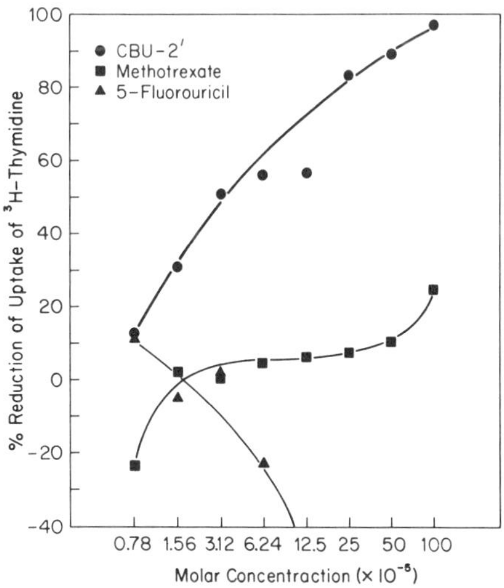

Fig. 2

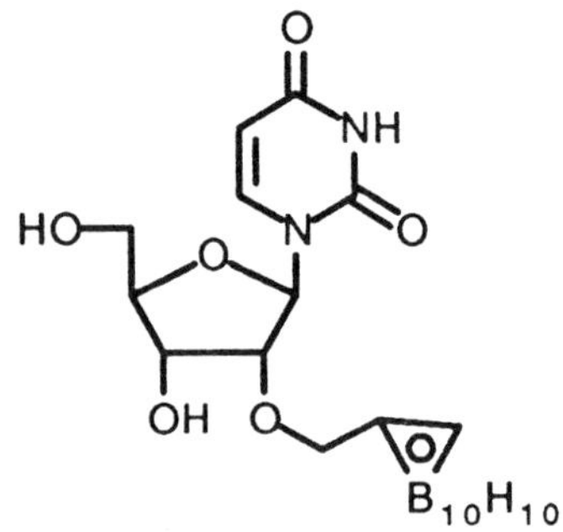

Fig 3. 2'-O-(o-carboran-l-ylmethyl)uridine (CBU-2')

75.6% of the recoverable boron was localized in the cell membrane fraction, 15.6% in cytosolic fraction and 8.8% in the nuclear fraction. CBU-2' was not measurable in the RNA and DNA fractions (Table 2).

DISCUSSION

We previously have reported that CBU-2' was taken up by F98 glioma cells in large amounts (84.9-94.2 μg B/gm cells) and persisted for up to 48 hours at a significant level (11.8 μg B/gm cells)[6]. In order to determine whether the uptake of this compound was proliferation dependent, canine MDCK and murine 3T3 fibroblasts were used as control cells. When MDCK cells attain confluency, they become contact inhibited and enter into and remain in G_0[11]. Consequently, the rates of DNA and RNA synthesis of confluent MDCK cells in G_o are much lower than those of proliferating cells. Since CBU-2' was synthesized as a derivative of uridine, it was expected that proliferating tumor cells would show much higher levels of uptake compared to non-proliferating MDCK cells[12]. However, MDCK cells showed the same uptake as a variety of tumor cells, and it therefore can be concluded that uptake of CBU-2' was neither proliferation dependent nor selective for malignant cells.

Nucleosides are taken up by cells by an active membrane transport process in which transport proteins and energy are required. This process is saturable and can be blocked by metabolic inhibitors such as rotenone and dipyridamole[13,14]. Our data indicate that the uptake of [^{3}H]-UdR, but not CBU-2', by F98 glioma cells was inhibited by dipyridamole, suggesting that CBU-2' was not biochemically recognized as a uridine derivative. Since the carboranes are highly lipophilic, a passive transport process may have been involved in the uptake of this compound. At this point in time the *in vitro* data that we have accumulated suggest that CBU-2' may not be a selective enough compound to target brain tumors. Nevertheless, before any final conclusions can be made, it is essential to carry out *in vivo* pharmacokinetic studies following systemic injection of the compound, and these studies will be carried out by us in the new future.

o-carborane has a cage structure with a $C_2H_{12}B_{10}$ molecule. The insertion of a carboranyl methyl group at the 2' position of uridine changed its biochemical properties compared with deoxyuridine so that it did not enter into the normal metabolic pathways of uridine. This would explain why the cellular uptake of CBU-2' was not dependent upon the nucleoside membrane transport system, and why it was not detectable in the DNA or

RNA fractions of F98 glioma cells. It has been reported that the insertion of a large group into the 5 position of uridine did not affect the cellular uptake of the compound, nor its incorporation into nucleic acids[15]. Yamamoto[16] and Schinazi[17] have synthesized nucleosides with the carborane moiety at this location, and the latter has observed that the nucleoside can be converted to a monophosphate. There is, however, no evidence that such compounds are incorporated into nucleic acids and structures are needed that will function as true analogues of thymidine and can be incorporated into DNA. Efforts are being made in a number of laboratories to synthesize the next generation of these compounds, which still appear to have great potential for targeting proliferating tumors.

ACKNOWLEDGMENTS

This work has been supported by U.S. Department of Energy grants DE-FG02-90ER60972 and DE-AC02-76CH000016 and National Institutes of Health grant R01-CA53896.

REFERENCES

1. Barth, R.F., Soloway, A.H. & Fairchild, R.G. (1990) Cancer Res. 50:1061-1070.
2. Kobayashi, T. & Kanda, K. (1982) Radiat. Research. 91:77-94.
3. Richard, P. (1988) Annu. Rev. Biochem. 57:349-374.
4. Anisuzzaman, A.K.M., Alam, F. & Soloway, A.H. (1990) Polyhedron. 9:891-892.
5. Tjarks, W., Anisuzzaman, A.K., Liu, L., Soloway, A.H., Barth, R.F., Perkins, D.J.& Adams,D.M. (1992) J. Medicinal. Chem. 35:1628-1633.
6. Liu,L., Barth, R.F., Soloway, A.H., Anisuzzaman, A.K.M., Alam,F., Tjarks, W., Zha, X.H. & Morrison,G.H. (1991) Proc. Amer. Assoc. Cancer. Res. 34:2418.
7. Barth, R.F., Adams, D.M., Soloway, A.H., Mechetner, E.B., Alam, F. & Anisuzzaman, A.K.M. (1991) Anal.Chem. 63:890-892.
8. Krueger, M.J., Singer, T.P., Casuda, J.E. & Ramsay, R.R. (1990) Biochem. Biophys. Res. Commmun. 169:123-128.
9. Phillis, J.W., O'Regam, M.H. & Walter, G.A. (1989) Brain Res. 481:309-316.
10. Sambrook, J., Fritsch, E.F. & Maniatis, T. (1989) in Molecular Cloning: Rapid isolation of total RNA from mammalian cells. Cold Spring Harbor Laboratory. pp7.3-7.10.
11. Madin, S.H. & Darby, D.B. (1958) P.soc.Exp.M. 98:574-578.
12. Soloway, A.H., Anisuzzaman, A.K.M., Liu, L., Barth, R.F., Alam, F. & Tjarks, W. (1992) in Progress in Neutron Capture Therapy for Cancer. Plenum Press, New York. pp207-210.
13. Oriffith, D.A., Conant, A.R. & Jarvis, S.M. (1990) Biochem Pharmacol. 40:2297-2302.
14. Phillis, J.W., O'Regam, M.H. & Walter, G.A. (1989) Brain Res. 481:309-316.
15. Grienl, H., Hayden, W., Schwarz, W., Bachmayer, H. & Rosenwirth, B. (1985) Eur.J. Med. Chem. 20:105-110.
16. Yamamoto, Y., Nemoto, H., Seko, T., Takamatsu, S. and Rong, F.G. (1992) in Progress in Neutron Capture Therapy for Cancer. Plenum Press, New York and London. pp219-222.
17. Schinazi, R.F., Goudgaon,N., Soria, J. & Liotta, D.C. (1992) in Fifth International Symposium on Neutron Capture Therapy for Cancer. pp11.

IN VIVO AND *IN VITRO* UPTAKE OF BORONATED COMPOUNDS BY B16-BL6 MURINE MELANOMA

C. A. Elstad[1], K. R. Meinkoth[1], B. A. Mathison[1], G. G. Meadows[1], D. H. Kinder[1], M. F. Hawthorne[2], K. Shelly[2], D. A. Feakes[2], and W. F. Bauer[3]

[1]Department of Pharmaceutical Sciences, College of Pharmacy, WSU, Pullman, WA 99164-6510
[2]Department of Chemistry and Biochemistry, UCLA, Los Angeles, CA 90024
[3]INEL BNCT Research Programs, EG&G Idaho, Idaho Falls, ID 83415-2208

INTRODUCTION

A crucial requirement for effective boron neutron capture therapy (BNCT) of cancer is the selective localization of high concentrations of boron in tumor tissue relative to adjacent normal tissue, thus minimizing damage to normal tissues caused by the neutron beam. Several promising boron compounds are presently available as boron-delivery agents for neutron capture therapy of malignant melanoma. More are expected to be developed in the near future. Therefore, there is a need to experimentally describe these boron-delivery compounds *in vitro* and *in vivo*, to screen these agents by critically comparing their efficacies with known boronated agents, and to effectively optimize their BNCT potential.

We have developed standardized *in vivo* and *in vitro* experimental assays to evaluate boron pharmacokinetics and distribution in tumor and control (normal) cell lines after exposure to selected boron-delivery compounds. We have shown that boronated compounds accumulate both *in vitro* and *in vivo* in the highly invasive and metastatic B16-BL6 (BL6) murine melanoma. This project has emphasized the potential of various boron compounds with affinities for melanin, such as para-boronophenylalanine (BPA) which is an analogue of natural melanin precursors[1,2], or of boron-delivery carriers such as liposomes which preferentially accumulate in neoplastic tissue[3-5].

MATERIALS AND METHODS

Tumor Cells and Culture Conditions

The well characterized BL6 variant of the B16 murine melanoma was chosen for these studies[6,7]. BL6 is highly invasive and highly metastatic. For routine culture, cells were grown *in vitro* in Dulbecco's modified Eagle's complete minimal essential medium supplemented with 10% heat-inactivated fetal bovine serum, sodium pyruvate, nonessential amino acids, 2-fold vitamin solution, L-glutamine, 100 U/ml penicillin, and 100 µg/ml streptomycin. MEM Selectamine Kits (GIBCO Laboratories, Grand Island, NY) were used to formulate all amino acid-limited media[8]. These media were prepared using dialysed FBS to eliminate free amino acids in the serum. To determine uptake of BPA, tumor cells were exposed in the culture medium to BPA at concentrations ranging from 25-100 mg/l. Cells were incubated in a humidified atmosphere of 5% CO_2 in air at 37°C and harvested using 4

Advances in Neutron Capture Therapy, Edited by
A.H Soloway *et al.*, Plenum Press, New York, 1993

mM EGTA to avoid the membrane-altering effects of trypsin. Cell viability was determined by trypan blue exclusion.

Animals and *In Vivo* Screening of Boron Compounds

Specific-pathogen-free, female B6D2F1 mice were purchased from Jackson Laboratories, Bar Harbor, ME, at 6-8 weeks of age and housed singly in the WSU Wegner Hall Vivarium, which is AAALAC accredited. Mice were inoculated subcutaneously into the dorsal hip with 1 x 10^6 viable BL6 tumor cells in 200 µl of Ca^{2+}- and Mg^{2+}-free saline solution, as described previously[9]. Approximately 1-2 weeks later, unanesthetized tumor-bearing mice were inoculated with 200 µl of boron compound into the dilated lateral tail vein. In other studies, BPA was also injected intraperitoneally as a slurry at a concentration of 750 mg/kg body weight. Following injection of boron compound, mice were sacrificed through a 72-hour time period. At each time point, tissues (tumor, blood, liver, spleen, brain, muscle, kidney, skin) from six mice were collected and analyzed by ICP-AES for boron content.

RESULTS AND DISCUSSION

We have shown that boronated compounds accumulate both *in vitro* and *in vivo* in BL6 murine melanoma. *In vitro* uptake of boron by BL6 increases with increasing levels of BPA (25 to 100 mg/l) (Figure 1). Levels of boron in cells exposed to 100 mg BPA/ml are significantly greater ($p<0.05$) than levels in cells cultured in 25 mg BPA/ml.
Figure 2 shows that this uptake is enhanced 7-fold when levels of tyrosine and phenylalanine are restricted to 4 µg/ml (22 µM tyrosine, 24 µM phenylalanine) which corresponds to an 88% reduction compared to levels found in normal medium[8].

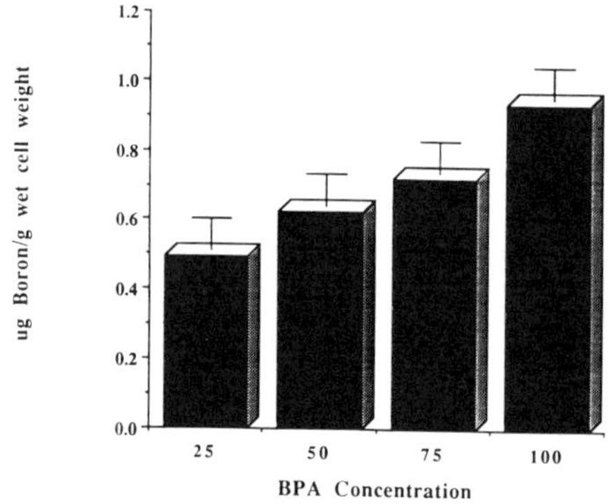

Figure 1. Cellular uptake of boron by B16-BL6 exposed *in vitro* for 6 hours to increasing concentrations (25-100 mg/ml medium) of BPA in the culture medium.

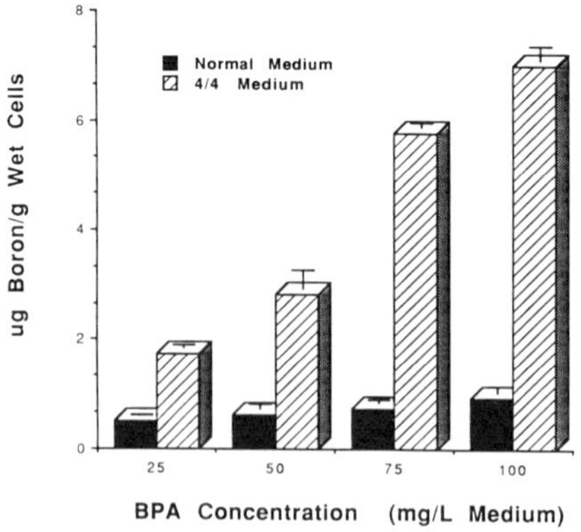

Figure 2. Cellular uptake of boron by B16-BL6 exposed *in vitro* for 6 hours to increasing concentrations of BPA in normal culture medium or medium restricted in tyrosine and phenylalanine to 4 µg/ml each.

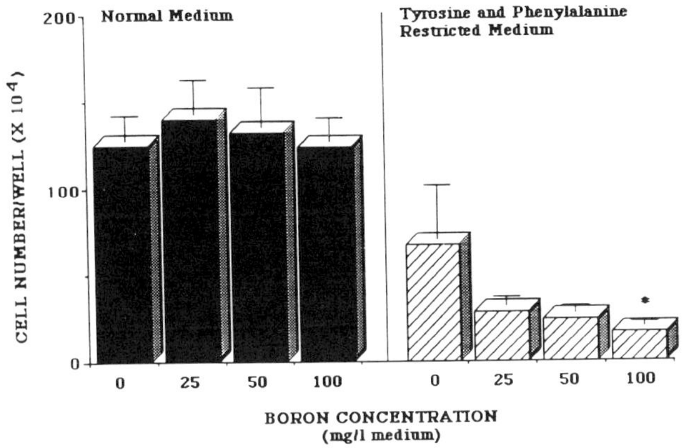

Figure 3. Growth of B16-BL6 on day 5 of culture in normal or tyrosine/phenylalanine-restricted medium and increasing concentrations of BPA.

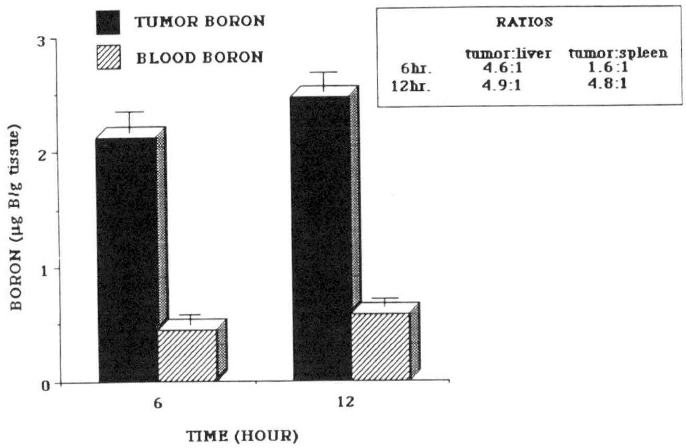

Figure 4. Distribution of boron in mouse tissues 6 and 12 hours after intraperitoneal injection of BPA as a slurry at a concentration of 750 mg/kg body weight. Tissues were collected from six mice at each time point.

This increased uptake is not due to any growth modulatory effects of BPA since *in vitro* growth of BL6 is not affected by exposure to exogenous BPA (Figure 3).

We have also investigated the *in vivo* biodistribution of BPA. After intraperitoneal injection into mice of 750 mg BPA/kg body weight as a slurry, respective tumor:liver and tumor:spleen ratios were 5:1 and 2:1 at 6 hours after injection and 5:1 and 5:1 at 12 hours after injection (Figure 4). Tumor boron levels at both times points were 2-2.5 µg B/g tissue.

The effectiveness of liposomal delivery of boronated compounds to BL6 murine melanoma tumors has also been investigated. Following intravenous injection of liposomal-encapsulated boron compounds into tumor-bearing mice, boron accumulation in subcutaneous BL6 tumors typically reaches maximum levels within 24 hours postinjection with concentrations approaching subtherapeutic levels for BNCT. A representative *in vivo* screening of a liposomal-encapsulated compound is shown in Figure 5. Tumor:normal tissue (muscle or brain) ratios generally range from 5:1 to 11:1. Liposomal delivery of low doses of boron has resulted in therapeutic quantities (≥15-20 ppm) of boron in other tumor systems, specifically the EMT6 murine mammary tumor[10]. These low doses minimize toxicities to the mice and may facilitate future experiments investigating alternative delivery protocols.

We are developing *in vitro* and *in vivo* assays which utilize tissue culture and small animals (mice) to determine the effectiveness of boron compounds for BNCT before more costly and time-consuming large animal studies are conducted. We are using these *in vitro* and *in vivo* techniques to enhance selective uptake of boron into tumor cells to optimize BNCT. Identification of alterable factors influencing boron uptake could have a significant impact on therapy of malignant melanoma as well as other cancers.

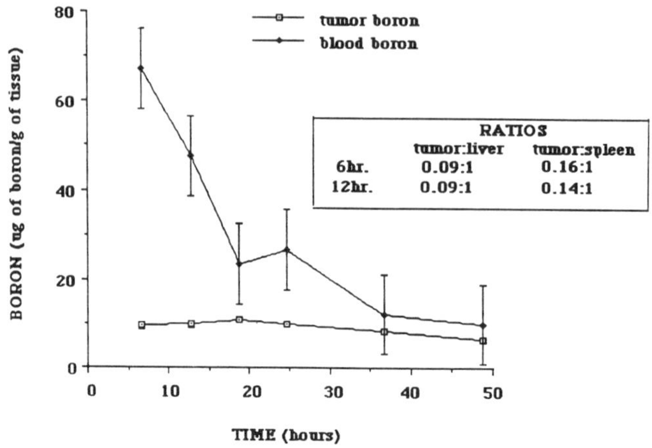

Figure 5. Accumulation of boron in tissues after intravenous injection of 200 μl of liposomal-encapsulated Na_2n-$B_{20}H_{18}$. Injected dose of boron = 212 μg B. Tissues were collected from six mice at each time point..

ACKNOWLEDGMENTS

These investigations have been funded by the U. S. Department of Energy, Office of Energy Research, DOE Idaho Field Office, Contract number DE-AC07-DO1570.

REFERENCES

1. F. Alam, B. V. Bapat, A. H. Soloway, R. F. Barth, N. Mafune, and D. M. Adams, Boronated compounds for neutron capture therapy, *Strahlenther. Onkol.* 165:121 (1989).
2. J. A. Coderre, J. D. Glass, R. G. Fairchild, P. Micca, and J. Kalef-Ezra, Boronophenylalanine for neutron capture therapy of melanoma, *Strahlenther. Onkol.* 165:215 (1989).
3. A. Gabizon and D. Papahadjopoulos, Liposome formulations with prolonged circulation time in blood and enhanced uptake by tumors, *Proc. Natl. Acad. Sci. USA* 85:6949 (1988).
4. D. A. Potter, C. M. Gorschboth, and P. d. Schneider, Liposome uptake by melanoma: In vitro comparison with hepatocytes, *J. Surgical Res.* 39:157 (1985).
5. A. Rahman and P. S. Schein, Use of liposomes in cancer chemotherapy, in: "Liposomes as Drug Carriers", G. Gregoriadis, eds, John Wiley & Sons, New York (1988).
6. I. R. Hart, The selection and characterization of an invasive variant of the B16 melanoma, *Am. J. Pathol.* 97:587 (1979).
7. G. Poste, J. Doll, I. R. Hart, and I. J. Fidler, *In vitro* selection of murine B16 melanoma variants with enhanced tissue-invasive properties, *Cancer Res.* 40: 1636 (1980).
8. C. A. Elstad and G. G. Meadows, Modulation of B16-BL6 murine melanoma metastatic phenotype by tyrosine and phenylalanine restriction in the absence of host selection pressures, *Anticancer Res.*, submitted (1992).
9. C. A. Elstad and G. G. Meadows, Phenotypic stability of B16-BL6 melanoma exposed to low levels of tyrosine and phenylalanine, *Anticancer Res.*, 10: 1313 (1990).
10. K. Shelly, M. F. Hawthorne, and P. G. Schmidt, Liposomal delivery of boron for BNCT, Proceedings of the 4th International Symposium on Neutron Capture Therapy for Cancer, 1991.

THE USE OF BORON NEUTRON CAPTURE

FOR THE TREATMENT OF PITUITARY TUMORS

Barry D. Albertson[1], Monica A. Millan[1], Stephen E. Binney[2], Gail B. Willeke[1], Steven W. Martsolf[2], Jennifer E. Johnson[2], and D. Lynn Loriaux[1]

[1]Division of Endocrinology, Diabetes, and Clinical Nutrition, Department of Medicine, Oregon Health Sciences University, Portland, OR
[2]Department of Nuclear Engineering, Oregon State University, Corvallis, OR

INTRODUCTION

Pituitary tumors comprise approximately 15% of all intracranial tumors[1]. The clinical presentation of these tumors is diverse; morbidity and mortality can be high. Current treatment modalities include conventional x-irradiation and/or neurosurgery, but are inadequate for the most part. Neither are uniformly successful and both are relatively non-specific. However, several features of these tumors make them ideal candidates for Boron Neutron Capture Therapy (BNCT). First, the tumors are localized in the pituitary gland, rarely metastacising to other areas of the body. Second, they are rarely larger than 2 cm in diameter. Third, the tumors are outside the blood brain barrier. And fourth, the pituitary tumor cells contain cellular membrane receptors that bind and internalize specific hypothalamic polypeptide releasing hormones. The ablity to target boron-10 (^{10}B) containing compounds to these pituitary tumor cells with releasing hormones and subsequently kill these cells with BNCT is currently being evaluated in pituitary tumor cells in vitro.

METHODS AND MATERIALS

CHEMICALS AND REAGENTS

Rodent pituitary tumor cells (AtT-20) were obtained from the American Type Culture Center, Rockville, MD. Cells are cultured in Dulbecco's Modified Eagle Medium (D-MEM) (Gibco Labs, Grnad Island, NY) with antibiotics added (penicillin/streptomycin) and 10% fetal bovine serum (FBS) (Gibco). All cell washings are performed with sterile

Advances in Neutron Capture Therapy, Edited by
A.H Soloway *et al.*, Plenum Press, New York, 1993

Dulbecco's phosphate buffered saline (D-PBF) (Gibco) in 15 ml conicle centrifuge tubes (Corning Labs, Corning, NY). Cells are harvested from T-150 culture flasks (Corning) with a rubber policeman, or from six-well cell culture dishes (Corning) with trypsin/EDTA (0.05%/0.53 mM) (Versene, 1:5000) (Gibco). Cells are irradiated in sterile, 20 ml polypropylene scintillation vials. After serveral washings cells are counted on a hemocytometer after staining with the supravital stain Trypan blue (Sigma Chemical Co, St. Louis, MO) using an Olympus CK2 inverted microscope.

CELL IRRADIATION FACILITY

Cell irradiations are conducted in the thermal column of the 1 MW Mark II Oregon State University TRIGA Reactor. Three stringers of the graphite thermal column have been removed resulting in a 10 cm x 30 cm cross sectional beam. The innermost end of this region is filled with a bismuth shielding assembly which holds the cell samples, followed by a 50 cm long graphite plug. The thermal neutron and gamma ray fields in the thermal column fall off exponentially with relaxation lengths of 30 and 80 cm, respectively. For these measurements, the rear bismuth shield and graphite plug were absent.

Bare and Cd-covered gold foils were used to measure the thermal and epithermal neutron fluxes at the sample location. These values are 5.2×10^{10} n-cm^{-2}-s^{-1} and 9.2×10^{8} n-cm^{-2}-s^{-1}-Δu, respectively, at 1 MW. The fast neutron flux, measured in a single unshielded graphite stringer using indium threshold foils[2], is 1.5×10^{8} n-cm^{-2}-s^{-1}. Thus, a reasonably pure thermal neutron beam has been achieved: $\phi_{th}/\phi_{epi} = 56$; $\phi_{th}/\phi_{fast} = 350$.

Gamma ray dose was measured with TLD-400 (CaF_2:Mn) thermoluminescent dosimeters. The gamma dose rate at 1 MW in the cell sample was 1.1 cGy/s. The ratio of gamma dose to the thermal neutron fluence was 2.0×10^{-11} cGy-cm^2.

CARBORANE CAGE SYNTHESIS AND RELEASING HORMONE CONJUGATION

$^{10}B_{10}$ carborane cages have been synthesized, purified, and supplied by Professor M. Frederick Hawthorne, Department of Chemistry and Biochemistry, UCLA. The uniqueness of the carborane cage lies in the incorporation of two carbon atoms combined as an integral part of the cage with an attached acetate reactive group and positioned so that the "cage" can be conjugated via the acetate to protein or polypeptide molecules like the releasing hormone ovine corticotrophin releasing hormone (oCRH). The conjugation of the cage is carried out with a classic peptide linkage, thereby joining the cage to the amino group on the terminal serine residue of oCRH. These synthetic steps and the final high pressure liquid chromatography purification of the conjugate are performed by Dr. Harry Chen, ERRB, NICHD, NIH, Bethesda, MD (Figure 1).

EXPERIMENTAL CELL LINE (AtT-20)

The AtT-20 rat pituitary corticotroph tumor cell line, established in 1953 by Furth et al.[3], responds to CRH in a dose dependent fashion with the production of the intracellular second messenger, cyclic adenosine monophosphate (cAMP) (Figure 2).

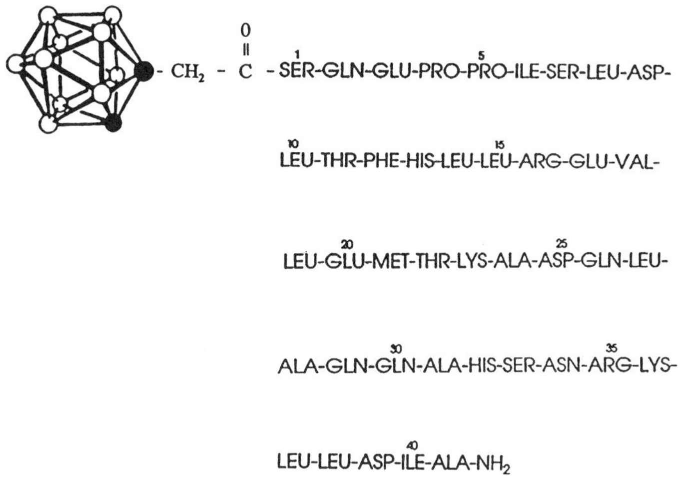

Figure 1. The amino acid sequence of oCRH is shown with the conjugated carborane cage ($^{10}B_{10}$). Carborane cage boron atoms are shown as open circles, carbon atoms are shown as black circles. The acetate reactive group is shown conjugated to the number 1 serine residue of oCRH.

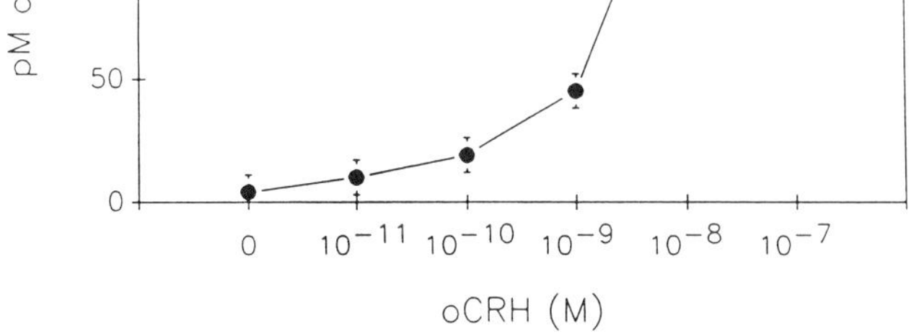

Figure 2. c-AMP dose response after a 10 minute exposure of AtT-20 cells to varying concentrations of oCRH. 500,000 AtT-20 cells were plated in 12-well cell culture dishes in D-MEM-10% FBS. Cyclic AMP was measured using a radioimmunoassay with antisera provided by the National Hormone and Pituitary Agency, NIAMD, Baltimore, MD. Each point represents the mean ± 1 SD of triplicate wells.

DISCUSSION AND RESULTS

RETENTION OF BIOLOGICAL ACTIVITY OF CORBORANE-CRH

Carborane conjugated ovine CRH (oCRH) has been shown to retain biological activity, evaluated by the ability to stimulate the production of intracellular levels of cAMP when compared to equimolar concentrations of pure oCRH in in vitro experiments using the AtT-20 cell line (Figure 3).

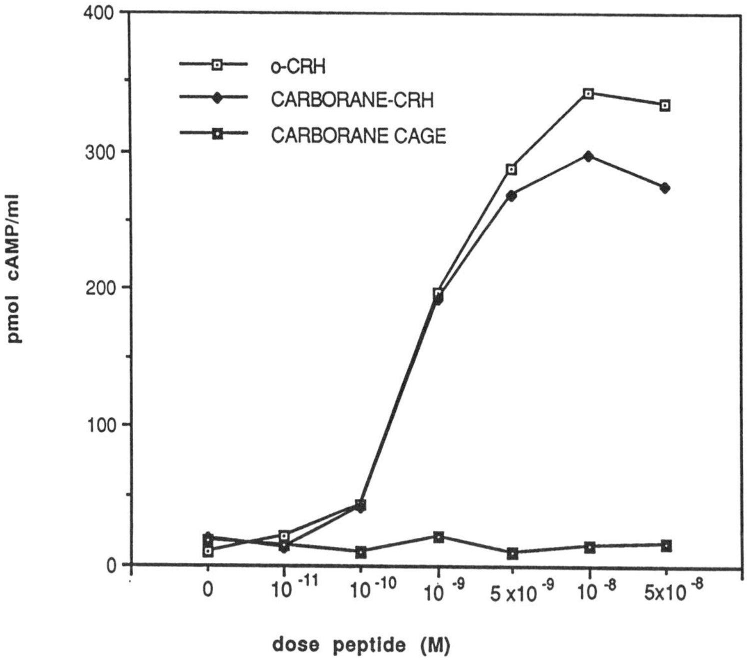

Figure 3. Cyclic AMP response of AtT-20 cells to oCRH, carborane-CRH, or carborane cage alone at various concentrations. Dose of peptide, carborane-CRH, or cage alone ranges from 10^{-11} to $5x10^{-8}$ M. Picomoles of cAMP produced by 500,000 AtT-20 cells cultured in 12-well culture dishes per ml media is shown on the Y-axis.

EXPERIMENTAL PARADIGM

AtT-20 cells are incubated in a water-jacketed CO_2 incubator in the presence of oCRH, carborane conjugated oCRH or unconjugated carborane cage and oCRH. After a specified length of time the cells are washed in D-PBS, removed from the T-150 culture flasks using a rubber policeman, and placed into sterile, plastic cell irradiation vials. The vials are positioned in the thermal column as described above. The cells are exposed for up to 10 minutes at different reactor powers. The cell vials are subsequently removed, cell media aspirated after centrifugation (800 rpm for 10 minutes), the cells washed in D-PBS, and replated in T-50 culture flasks in D-MEM with 10% FBS. Cell viability is assessed over time using trypan blue.

REACTOR DOSIMETRY

AtT-20 cells in D-MEM were placed in sterile 20 ml scintillation vials and irradiated under different radiation dose protocols to determine the maximal gamma and/or neutron dose that AtT-20 cells in our in vitro paradigm can withstand. The cells were washed and subcultured in D-MEM for 5 days. Live AtT-20 cells (those that are attached to the plastic culture dish and exclude trypan blue) were removed from the dish using Versene and counted in a hemocytometer. The results of this pilot experiment are shown in Figure 4.

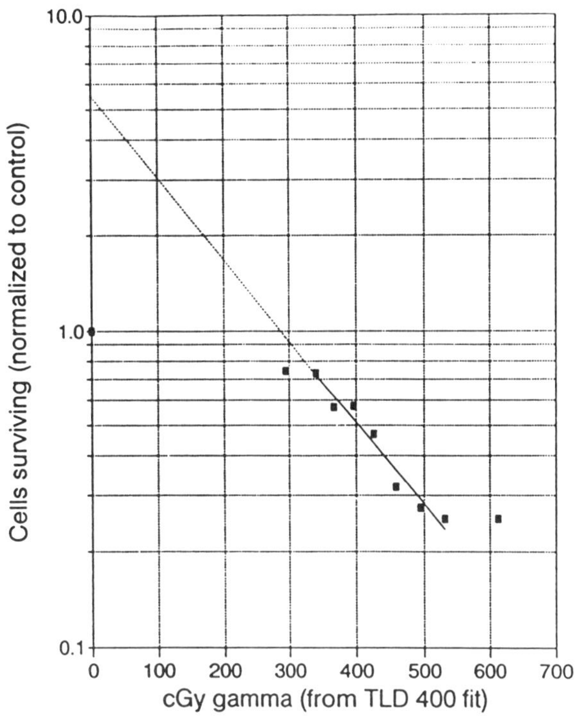

Figure 4. AtT-20 cell survival after 5 days, following exposure to varying gamma doses in the reactor thermal column. The data suggest that the 1/e dose for AtT-20 cells was 170 cGy with an extrapolation number of 5.5. The initial AtT-20 cell number at day 0 was 5 x 10^6.

BNCT EFFECT ON AtT-20 CELLS

AtT-20 cells were preincubated with oCRH ($5x10^{-7}$ M) or carborane conjugated oCRH analog ($5x10^{-7}$ M) for 18 hours at 37 C, washed with D-PBS, and irradiated for 4 minutes at 500 kW (representing a gamma and neutron dose that does not appreciably harm the cells) as derived from figure 4. The cells were removed from the thermal column, washed twice with D-PBS and subcultured in T-50 flasks with D-MEM-10% FBS for 8 days. Cell survivability was assessed at that time using trypan blue. The results are shown in Figure 5.

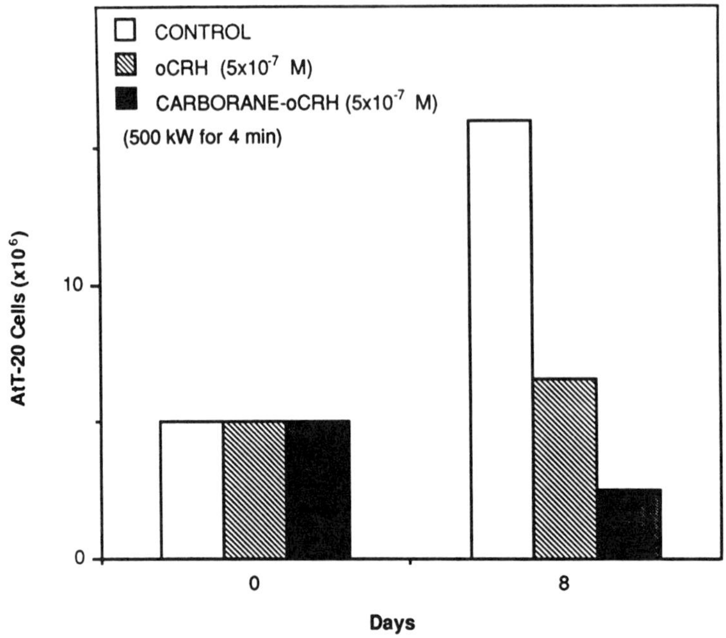

Figure 5. Control AtT-20 cells receiving no irradiation grew over an 8 day period from 5 x 10^6 cells/flask to approximately 18 x 10^6 cells/flask. Cells preincubated in the presence of 5 x 10^{-7} M oCRH grew over the same period of time from 5 x 10^6 to 6.5 x 10^6 cells. However, cells preincubated with carborane-oCRH decreased in number from day 0 to 2.5 x 10^6 cells/flask (day 8), suggesting a BNCT effect.

CONCLUSIONS

These preliminary data suggest that pituitary tumor cells can be killed using BNCT by linking enriched boron to hypothalamic releasing hormones. Strategies to decrease even further the contaminating gamma dose delivered to the AtT-20 cells are being tested. Cell loading experiments are currently under way (in conjunction with cellular boron quantitation using inductively coupled plasma-mass spectrometry and radio-iodine labelling of the carborane-CRH conjugate) to maximize the amount of carborane cage retained by the pituitary tumor cells so that the tumor cell to non-tumor cell ^{10}B ratios are as high as possible. These studies will provide important insight into the design of in vivo experiments using rodents and sub-human primates to demonstrate the efficacy of BNCT for the treatment of pituitary tumors in humans.

REFERENCES

1. K. Thapar, K. Kovacs, E.R. Laws, and P.J. Muller, Pituitary Adenomas: Current Concepts in Classification, Histopathology, and Molecular Biology. The Endocrinologist 3(1): 39-57, 1993.

2. Y.D. Harker, R.A. Anderl, G.K. Becker, and L.G. Miller, Spectral Characterization of The Epithermal Neutron Beam At The Brookhaven Medical Research Reactor. Nuc Sci Eng 110: 355, 1992.

3. J. Furth, E.L. Gadsen, and A.C. Upton, ACTH Secreting Transplantable Pituitary Tumors. Proc Soc Exp Biol Med 84: 253, 1953.

CHARACTERISTICS OF MELANOMA CELLS REGARDING BINDING OF CARBORANYLALANINE AND EFFECTS OF BORON NEUTRON CAPTURE

Orn-Anong Pettersson[1], Peter Lindström[1], Pär Olsson[1], Jörgen Carlsson[1], Stefan Sjöberg[1] and Bengt S. Larsson[2]

[1] Division of Physical Biology, Department of Radiation Sciences; Box 535, S-751 21 Uppsala, Sweden
[2] Department of Toxicology, Biomedical Centre, Uppsala University, S-751 21 Uppsala, Sweden

INTRODUCTION

Carboranylalanine is a boron containing amino acid which is similar in structure to phenylalanine and p-boronophenylalanine, BPA. The phenyl ring is exchanged for a carborane[1] which is a "cage-like" arrangement of ten boron atoms. BPA is an interesting substance presently considered for therapy of malignant melanomas[2,3] and it is possible that carboranylalanine might be of even higher interest for this purpose since it contains ten boron atoms. The binding of L- and D-carboranylalanine to cultured mouse B16 and human IGR1 melanoma cells was analysed. A glioma cell line was used for comparison. BPA was used as reference substance. The comparative tests were made by using the same concentration of ^{10}B in the culture medium independent of the type of substance. Possible binding to melanin of carboranylalanines was also studied.

MATERIALS AND METHODS

Mouse B16- and human IGR1-melanoma and human glioma U-343MGa cells were analysed. They were cultured as monolayers (5% CO_2, 37°C) in F10 medium with 10% foetal calf serum, 2 mM glutamine, 100 units/ml penicillin and 100 µg/ml streptomycin. The medium was changed 3 times a week.

L- and D-carboranylalanine ($C_5H_{17}B_{10}O_2N$) were synthesized, using natural boron, according to methods published separately[1,4]. BPA (L-p-boronophenylalanine (95%^{10}B)) was a gift from J.A. Coderre, Medical Department at Brookhaven National Laboratory, N.Y., USA.

The boron substances were dissolved, the pH adjusted to 7.4, diluted with F-10 medium to 10 µg ^{10}B/ml and sterile filtered (0.2 µm). The cells were plated in 35 mm dishes. The medium was changed to medium containing the boron compound and the cells were incubated 20 hours. The cells were then washed six times with cold F-10 medium.

To investigate whether the boron substances were internalized, the cells were treated with 0.7 ml acetic acid buffer pH 2.5 at 4°C for 6 minutes, then

Advances in Neutron Capture Therapy, Edited by
A.H Soloway *et al.*, Plenum Press, New York, 1993

washed twice with 0.4 ml of the same solution. This procedure removed substances that had not been internalized[5]. The cells were then digested by incubation with 0.7 ml 1M NaOH at 37°C for 1h and washed twice with 0.4 ml. The samples were pooled and ^{10}B analysis was made by neutron capture radiography.

A small volume, 0.2 µl, of each sample was dropped on a polycarbonate sheet (CR-39) which was pretreated with a 70% 6M NaOH, 30% alcohol solution at 70°C for 45 minutes. After air drying, the CR-39 sheets were vacuum packed and sent for irradiation at the R2-0 reactor in Studsvik, Sweden. The sheets were irradiated with a thermal neutron fluence of $5x10^{12}$ n/cm^2 at the radiography position[6] and then etched with 6.25 M NaOH 70°C for 1.5 h. The track distributions were evaluated by image analysis (Microscale TM/TC, Digithurst GmbH Ltd, Royston, England). The numbers of tracks were estimated for the total area of the drops[7].

The toxicity of L- and D-carboranylalanine was analysed by cell counting in an electronic cell counter (Coulter, model ZM) at different time intervals after a 20-hour exposure to the boron compounds. Growth curves were constructed.

The melanoma cells B16 and IGR1 were irradiated with ^{60}Co gamma (0-10 Gy, 1 Gy/min) or with 200 MeV helium ions (with range modulated Bragg peak) from the Gustaf Werner cyclotron, Uppsala university (0-6 Gy, 5 Gy/min). The average LET for the helium ions was about 30 keV/µm. Three dishes were plated for each dose and five for the control. The plating efficiency varied in the range of 0.10-0.25 for IGR1 and 0.15-0.35 for B16 cells. The survival curves were analysed with the single hit-multi target and linear-quadratic theoretical models according to principles previously described by Albright[8].

The B16 cells were incubated with boron substances, 10 µg ^{10}B/ml, for 20 hours and washed twice with normal medium. Thereafter the culture bottles were filled with normal medium without boron. Control cells were treated in the same way. The cells were irradiated at 37°C with neutrons at the thermal neutron facility at the R2-0 reactor in Studsvik. The cell flasks were placed at the "living samples position"[6]. One set of cells were placed in the reactor hall, as controls. After irradiation the cells were processed for clonogenic growth as described above. The plating efficiency of the controls was not significantly changed by the boron substances. Six dishes were plated for each dose.

RESULTS

Several experiments were made to determine the intracellular binding of the boron compounds. The results given in figure 1 are pooled data from all experiments. Both L- and D-carboranylalanine bound to all tested cell lines. BPA showed only small binding to the B16 cells. The cultured human glioma cells bound all the tested substances and the highest binding value was for BPA.

Various concentrations of L- and D-carboranylalanine were given to the B16 cells in the toxicity tests. The growth curves showed that the doubling times of the B16 cells were not changed as a result of the L- and D-carboranylalanine treatments. The doubling times were in the range of 1.0-1.4 days.

Survival curves after gamma or helium ion irradiations showed that the relative biological effectiveness (RBE) values were in the ranges 2.1-2.4 and 1.5-1.7 for the B16 and IGR1 cells respectively. Survival curves after neutron irradiations are shown in figure 2. Administration of L-carboranylalanine or BPA gave decreased survival in relation to the controls. D-carboranylalanine and BTU-1 did not give significant effects in relation to the controls.

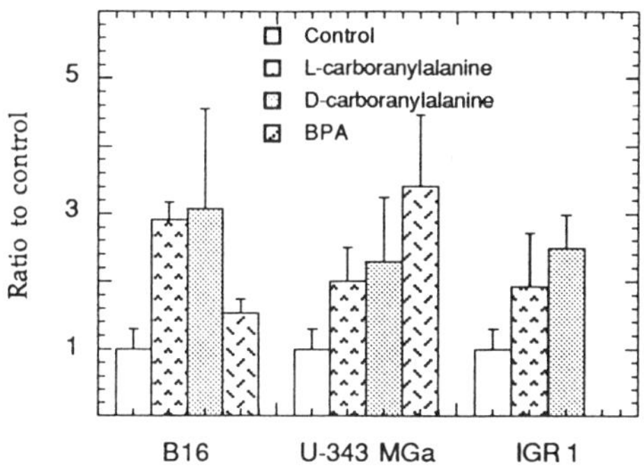

Figure 1. Intracellular binding of different boron compounds. The figure shows pooled data from five independent experiments after application of L- or D-carboranyl-alanine or BPA to B16, IGR1 or U-343MGa cells. Mean values and standard deviations are shown.

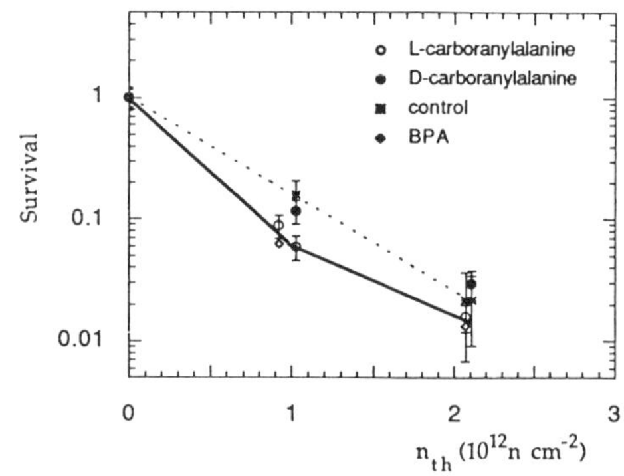

Figure 2. Survival curves for B16 cells after irradiation with thermal neutrons. The cells were incubated for 20 hours with 10 μg ^{10}B/ml of L- or D-carboranylalanine or BPA. Each point corresponds to the analyses of six cloning dishes. Mean values and estimated standard deviations are shown.

DISCUSSION

L- and D-carboranylalanine accumulated in all tested cells and are of interest for further studies regarding BNCT. This seems to be the case not only for treatment of melanomas but possibly for other tumors as well.

The mechanism for intracellular binding of L- and D-carboranylalanine is not clear. Cell free tests showed that both forms were not incorporated into nascent melanin but that they had some binding to preformed melanin. The binding was relatively low compared to some non-boronated substances[9] but might be sufficient for BNCT of pigmented tumour cells *in vivo.*

The accumulation of L- and D-carboranylalanine could be due to an active uptake over the cell membranes similar to the mechanism proposed for BPA[3]. Other mechanisms might also be involved and this must analysed in more detail. Further investigations will also be made to analyse how much boron was removed through the acid washing procedure.

Carboranylalanine did not seem toxic after a 20-hour exposure to 10 μg ^{10}B/ml, corresponding to 0.47 mM carboranylalanine, since the growth rate and plating efficiency of the exposed cells were not significantly changed in relation to the control cultures. The concentrations of these substances could be reduced by using enriched ^{10}B instead of natural boron. The ^{10}B concentration of 10 μg/ml will, in this case, correspond to only about 0.1 mM of carboranylalanine.

The irradiations of B16 cells with thermal neutrons showed that L-carboranylalanine and BPA gave significant decreases in survival compared to the controls, while no significant changes were seen for D-carboranylalanine. This is intriguing since BPA gave a therapeutical effect even though only small amounts were bound to the cells while D-carboranylalanine gave no therapeutical effect in spite of better binding.

One explanation is that the substances had different intracellular localisations. If, for example, BPA is mainly localized near the cell nucleus while D-carboranylalanine is mainly localized near the plasma membrane such effects could be obtained. Furthermore, the intracellular distributions of boron might have changed during the neutron irradiation. This has to be analysed in further studies.

ACKNOWLEDGEMENTS

The authors thank Per Svensson and the staff at Studsvik Nuclear AB, Sweden, for help with the neutron irradiations, Kellie Russell, Department of Radiation Sciences, Uppsala University for help with the helium ion irradiations and Dr. René Huiskamp, Petten, Holland, for help with ICP-AES analyses. The work was financially supported by grants from the Swedish Cancer Society; 1176-B90-03XA, 3009-B91-01XAB, 980-B90-02XB and 2366-B91-06XBB.

REFERENCES

1. Andersson, A., Andersson, J., Burgman, J. O., Capala, J., Carlsson, J., Condé, H., Crawford, J., Graffman, S., Grusell, E., Holmberg, A., Johansson, E., Larsson, B.S., Larsson, B., Liljefors, T., Lindström, P., Malmqvist, J., Pellettieri, L., Pettersson, O., Pontén, J., Roberti, A., Russel, K., Reist, H., Salford, L., Sjöberg, S., Stenerlöw, B., Strömberg, P., and Westermark, B. Program for BNCT with accelerator-produced keV neutrons and related chemical and biological studies. *In*: Progress in neutron capture therapy for cancer, eds.: Allen, B. J., Moore, D. E., and Harrington, B. V., pp 41-52. New York and London: Plenum Press., 1992

2. Mishima, Y., Ichihashi, M., Tsui, M., Sasumu, H., Ueda, M., Honda, C., and Suzuki, T. Treatment of malignant melanoma by selective thermal neutron capture therapy using melanoma-seeking compound. J. Invest. Dermatol., *92* (Suppl.): 3215-3255, 1989.

3. Coderre, J. A., Glass, J. D., Packer, S., Micca, P., and Greenberg, D. Experimental boron neutron capture therapy for melanoma: Systemic delivery of boron to melanotic and amelanotic melanom. Pigm. cell Res., *3*: 310-318, 1990.

4. Lindström, P., and Sjöberg, S. Asymmetric routes to (R)- and (S)-o-carboranylalanine. Manuscript 1992

5. Andersson, A., Holmberg, A., Carlsson, J., Pontén, J., and Westermark, B. Binding of epidermal growth factor-dextran conjugates to cultured glioma cells. Int. J. Cancer, *47*: 439-444, 1991

6. Pettersson, O., Svensson, P., Larsson, B., and Grusell, E. Studsvik thermal neutron.facility. *In*: Progress in neutron capture therapy for cancer., eds.: Allen, B. J., Moore, D. E. and Harrington, B. V., pp 91-93. New York and London: Plenum Press., 1992.

7. Laurent-Pettersson, M, Corel, M-N, Girard, N, Abraham, R, Gabel, G, Thellier, M., and Delpech, B. Immunoreactivity of boronated antibodies. J. Immuno. Met. *126* : 95-102, 1990

8. Albright, N. Computer programs for the analysis of cellular survival data. Radiation Res., *112*: 331 - 340, 1987.

9. Larsson, B. Mechanisms of accumulation of foreign substances in melanin. Binding to preformed melanin and incorporation into growing melanin. Acta Universitatis Upsaliensis, *43*: 1-52, 1979.

IRRADIATION OF SINGLE CELLS WITH INDIVIDUAL HIGH-LET PARTICLES

John M. Nelson and Leslie A. Braby

Radiation and Chemical Physics
Biology and Chemistry Department
Pacific Northwest Laboratory
Richland, WA 99352

INTRODUCTION

The dose-limiting normal tissue of concern when irradiating head and neck lesions is often the vascular endothelium within the treatment field. Consequently, the response of capillary endothelial cells exposed to moderate doses of high LET particles is essential for establishing exposure limits for neutron-capture therapy. In an effort to characterize the high-LET radiation biology of cultured endothelial cells, we are attempting to measure cellular response to single particles. The single-particle irradiation apparatus, described below, allows us to expose individual cells to known numbers of high-LET particles and follow these cells for extended periods, in order to assess the impact of individual particles on cell growth kinetics. Preliminary cell irradiation experiments have revealed complications related to the smooth and efficient operation of the equipment; these are being resolved. Therefore, the following paragraphs deal primarily with the manner by which high LET particles deposit energy, the requirements for single-cell irradiation, construction and assembly of such apparatus, and testing of experimental procedures, rather than with the radiation biology of endothelial cells.

BACKGROUND AND RATIONALE

Low-LET radiations, such as X and γ rays, deposit energy uniformly throughout a large tissue volume because each interaction deposits only a small amount of energy, and large numbers of interactions occur in each cell within the irradiated volume. Even at very low doses of low LET radiations, we find many charged particle tracks per cell. In fact, multiple tracks in cells occur at doses as low as about 0.01 Gy. In contrast, high-LET particles deposit large amounts of energy, but only along their tracks. At moderate doses of α particles, *e.g.* 0.4 Gy, there may be an average of only a single charged-particle track per cell. Track-segment irradiations, in which large populations

Advances in Neutron Capture Therapy, Edited by
A.H Soloway *et al.*, Plenum Press, New York, 1993

of cells are exposed to a flux of particles, have demonstrated that the cell survival curve for cells exposed to high-LET particles appears is a simple exponential (Barendsen, 1968). Similar studies have shown that an average of 3 to 7 tracks through the nucleus is required to inactivate a cell. The fact that it takes several tracks to produce one lethal event suggests that a large portion of the nucleus may be relatively insensitive to radiation; however, this is not consistent with our current view that lethal events evolve from DNA breaks, and that the DNA in interphase cells is distributed more or less uniformly throughout the entire nuclear volume. The fact that the survival curve for cells exposed to high LET radiations is essentially exponential, is also not consistent with the idea that damage from two or more tracks may interact. As a single track produces many double-strand breaks, the critical damage leading to cell inactivation is surely more complex than a simple break or pair of breaks.

Very low doses of high-LET radiations can lead to considerable differences in the fraction of the irradiated population which receives no dose at all, as illustrated in Figures 1a and 1b. In these figures, each cell nucleus is labeled with the calculated dose (in Gy), calculated by microdosimetric calculations, received when the whole population has been exposed to a dose of α particles. Figure 1a shows the cellular dose distribution after a dose of 0.4 Gy; Figure 1b shows the same for a dose of 0.04 Gy. For these examples, the mean specific energy, *i.e.*, the mean dose produced in the cell nucleus with exactly one track, was 0.4 Gy; the dose in Figure 1b is precisely one tenth of that specific dose. In Figure 1a, approximately one third of the cells received no dose, a third of them received doses ranging between zero and about 0.6 Gy, the maximum dose produced by a single track, and about a third of them were traversed

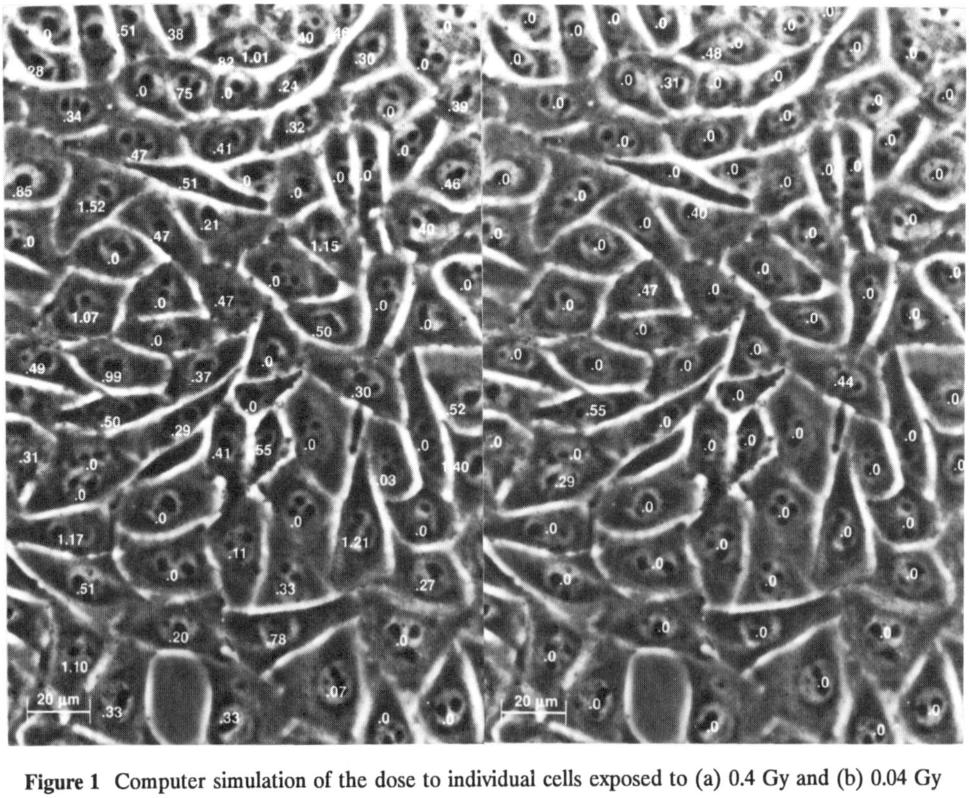

Figure 1 Computer simulation of the dose to individual cells exposed to (a) 0.4 Gy and (b) 0.04 Gy of α particles.

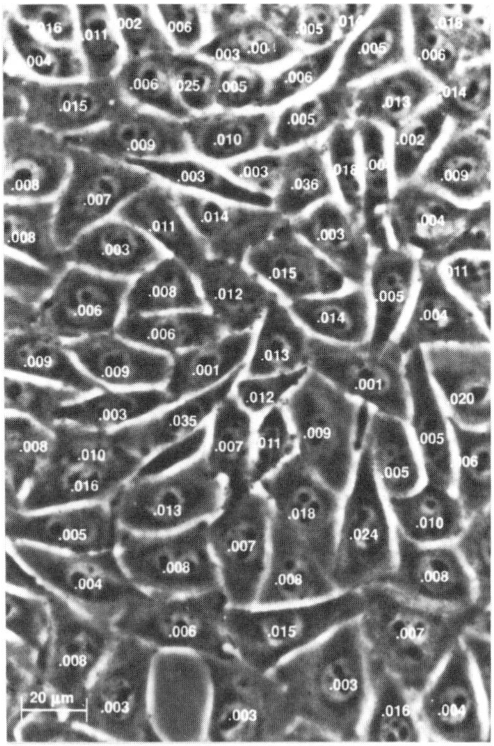

Figure 2 Computer simulation of the dose to individual cell nuclei exposed to only 0.01 Gy of low-LET γ rays.

by two or more tracks, receiving a dose which is usually more than 0.4 Gy. At the lower dose (Figure 1b) about 90% of the cells are not hit at all and the remaining 10% receive doses between zero and 0.6 Gy. In this case, few cells are found to have received double tracks. Because of this large unirradiated fraction, spontaneous cellular events become a major contributor to the overall biological effect, masking radiation effects.

Given equivalent doses of low-LET radiation, dose distribution is significantly different and the fraction of cells receiving some dose is greatly increased. Figure 2 shows the effect of only 0.01 Gy, of γ rays. In this case we find little variation between individual cells (0.003 to 0.018 Gy to each). Doses shown here were calculated from microdosimetric assumptions (7μm diameter spherical nucleus, 96 KeV/μm α particles or ^{60}Co spectrum), chosen to illustrate the dosimetric differences noted; a mean of one track per cell nucleus would be attained at a different dose for different cells and different particles.

For doses up to about 0.4 Gy of α particles, an increase in the dose relates primarily to an increase in the number of cells being traversed by an α particle. In the much higher range of therapy doses, an increase in dose relates to more tracks through the nucleus, rather than a further increase in the number of cells being hit. Beam-segment irradiation experiments with larger doses of high-LET particles yield radiobiological data with reasonable small variances. Linear extrapolation from such experiments cannot reliably predict the effects of small numbers of particles, as we have no assurance that this dose-effect relationship is linear at low doses; we cannot simply

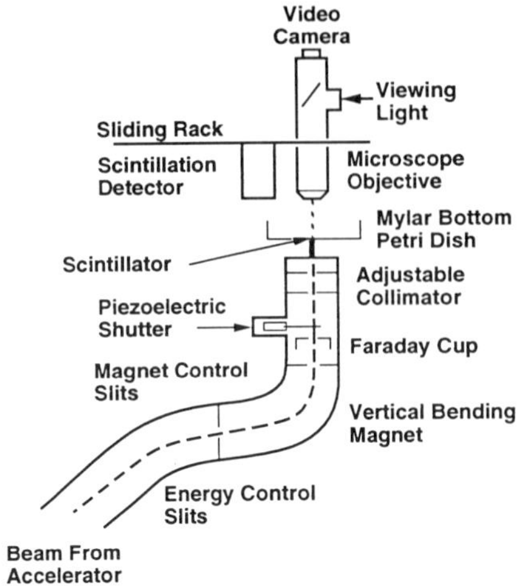

Figure 3 Schematic diagram of the particle irradiator. The beam is bent vertically upward; single particles pass through the nuclei of cells growing on the Mylar® film.

assume that because it appears linear at high doses, it will continue to be linear all the way to zero. It is also not possible to define biological effects from low-dose track-segment irradiations at very low doses, because of the large variances in energy deposited at those low doses. Consequently, we have chosen to determine radiation sensitivities and other radiobiological parameters for very small numbers of high-LET particles, which pass through or stop within individual cell nuclei.

THE MICROBEAM IRRADIATION APPARATUS

The single-cell irradiation apparatus is illustrated schematically in Figure 3. The proton or α-particle beam is produced by a 2 MV National Electrostatics, Corp. (NEC) tandem accelerator, and is bent vertically. In its present configuration, the beam is collimated to a square beam 5μm on a side by two sets of knife edges set 6 and 22.5 cm below the exit window. It exits vertically upward through the center of a 500 μm aperture covered by a 1.5 μm Mylar® film. Immediately above this window is a small piece of scintillation film which produces light signal as each particle passes. This scintillator is used to count the individual particles. Cells to be irradiated are allowed to attach to a Mylar® film, which sits immediately above and in contact with the scintillator. The proximity of the Mylar® film to the scintillator is designed to reduce scatter. The whole assembly is surrounded by a Zeiss Axiomat® microscope equipped with phase optics for epi-illumination microscopy. The output from a video camera mounted on the microscope headpiece is digitized by an image processor and displayed on a monitor for cell recognition. An image intensifier is used to amplify the very small fraction of light reflected by the cells with this optical system. This greatly improves the quality of the reflected-light video image available.

During each cell irradiation, the objective lens is replaced by a photomultiplier tube (PMT). Once the PMT is in place, a very fast-acting piezoelectric shutter opens and allows particles to pass through the aperture, beginning the irradiation. While cells are being positioned, the PMT is covered and away from any light source, but even then, it picks up occasional stray photons which raise the PMT background level.

Although 3 MeV protons are presently being used, several other particles are also available with this apparatus (Table 1). Note that a significant amount of energy is lost as the beam passes through the exit window, the plastic scintillator, and the film to which the cells are attached; this energy loss is sufficient to cause lithium ions to stop within the nucleus of attached and even relatively flattened cells. Figure 4 shows an image of the beam, made by exposing a piece of the radiachromic film which turns blue with exposure to ionizing radiations. The film was positioned at the location where cells would be growing. The scintillations picked up by the PMT are counted by a pulse counter connected to the piezoelectric shutter. After a predetermined number of particles have passed through the scintillating film, the shutter is again closed. This piezoelectric shutter closes in about 200 μseconds, allowing irradiations consisting of a single particles to be made with beams of around 500 particles per second.

Table 1 Particle Beams Available for Single-Cell Irradiation Studies.

Particle	*Initial Energy (MeV)*	*Energy in Cell (MeV)*	*Stopping Power (KeV/μm)*
Proton	2-4	1.8-3.9	9.8-17.6
Deuteron	2-4	1.7-3.8	17-32
Alpha	3-6	1.7-5.2	83-230
Lithium	3-6	0-3	Stop in nucleus

Cells to be irradiated are plated into specially fabricated Petri dishes constructed with a 1.5 μm (0.06 mil) Mylar® bottom. After 60-90 minutes to allow attachment, the culture dish is installed in the apparatus. Each cell can be positioned directly over the 5 μm exit window, with a resolution of about 0.2 μm. The culture medium remains over the cells throughout the entire procedure.
Cells are viewed through the microscope, and the resolution of our video system is set by the image processor at about 1 μm, *i.e.*, each pixel represents 1 μm on the Mylar® film.

Typical Irradiation Procedure: After a plate of cells is mounted into the X-Y positioning arm and the box containing the microscope is sealed, identification, positioning, and irradiating cells is done remotely from a computer console. When all cells in the microscope viewing frame have been identified, the nucleus of the first cell is positioned over the beam exit window, the microscope objective lens is replaced with the PMT, and the beam shutter opened. When the desired number of particles has passed, the shutter is closed and the next cell nucleus is moved into position. When all marked cells have been irradiated, the dish is moved to the next video frame and the procedure repeated until all cells in the specified region of the dish have been exposed. Following exposure, the culture dish is removed from the irradiation apparatus and placed into a miniature incubator built onto the stage of a Zeiss® inverted microscope, and maintained at 37°C under a water-vapor saturated atmosphere consisting of 5% CO_2 in air. Movement of this microscope stage is controlled to within 1 μm accuracy allowing return to each of several hundred selected positions. Digitized video images are recorded at specific intervals over several days and assembled to form time-lapse sequences in which we can observe the growth of irradiated cells.

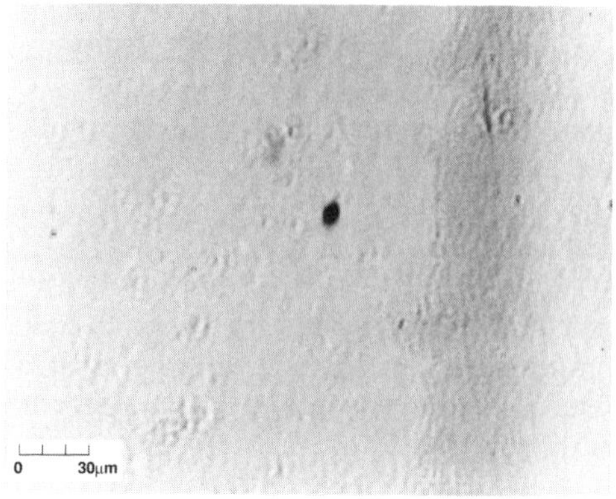

Figure 4 Image of a 7-8 μm beam made by placing a piece of radiachromic film on the Mylar® surface of a culture dish, opening the shutter, and exposing it to a large number of particles.

ACKNOWLEDGEMENT

This work was supported by the Office of Health and Environmental Research (OHER), US Department of Energy, under Contract DE-AC06-76RLO 1830.

REFERENCES

Barendsen, G.W., 1968, Responses of cultured cells, tumors, and normal tissues to radiations of different linear energy transfer. pp. 293-356 *in*: "Current Topics of Radiation Research; vol. IV," M. Ebert and A. Howard, eds., North-Holland Publishing Company, Amsterdam.

Raju, M.R., Eisen, Y., Carpenter, S., Inkret, W.C., 1991, Radiobiology of α particles. III. Cell inactivation by α-particle traversals of the cell nucleus. Radiat. Res. 128(2): 204-209.

THERMAL NEUTRON CAPTURE THERAPY USING $^{10}B_1$-BPA: INDUCEMENT OF MARKED LETHAL EFFECT ON MELANOMA CELLS IRRESPECTIVE OF X-RAY SENSITIVITY

M.Ichihashi[1], H.Utsumi[2], H.Tsukamoto[1], Y.Mishima[1], T.Kobayashi[3]

[1]Special Institute of Cancer Neutron Therapy, Dept. Dermatology, Kobe Univ. School of Medicine, Kobe, 650, Japan
[2]Radiation Biol. Center, Kyoto Univ. Kyoto; Japan
[3]Kyoto Univ. Res. Reactor, Osaka; Japan

ABSTRACT

We studied the effect of neutron capture therapy using ^{10}B-compound on X-ray sensitive and X-ray resistant cell lines, and found a high lethal effect of boron neutron capture therapy in comparison with conventional ionizing radiation. The P-39 line was sensitive to X-ray radiation, thermal neutron radiation, and extremely sensitive to bleomycin treatment, quite similar to ataxia telangiectasia cells, whereas the G-361 line was resistant to X-ray radiation, thermal neutron radiation and bleomycin treatment.

The G-361 cell line exposed to thermal neutron radiation after treatment with $^{10}B_1$-BPA (100μg/ml medium), however, showed similar sensitivity to the P-39 cell line.

These results strongly indicate the ^{10}B thermal neutron radiation induces irrepairable cellular damage (resistant to repair ability) in tumor cells; therefore, ^{10}B thermal neutron radiation more efficiently inactivates melanoma cells than conventional X-ray radiation does.

INTRODUCTION

Thermal neutron capture therapy (NCT) using ^{10}B-compounds has theoretically been shown to be extremely effective in inactivating malignant tumour cells[1]. We have already treated more than ten human melanoma patients by NCT using ^{10}B-paraborono-phenylalanine (^{10}B-BPA)[2]. Several stage-I patients have had no recurrence after a few years[3]. There are a few basic scientific reasons to explain the high lethal and selective effect of NCT using ^{10}B-BPA:
(1) the selective accumulation of ^{10}B atoms and the selective bombardment by $^{10}B(n, \alpha)^7Li$ reaction within a tumor cell, (2)extremely low energy of thermal neutron radiation (0.024eV). Individual tumor cells, however, usually exhibit varying degrees of sensitivity to both X-ray and fast neutron radiation, possibly due to individual differences in the ability of the cells to repair the damage caused by radiation.

We have already demonstrated that B-16 melanoma cells pretreated with $^{10}B_1$-BPA and exposed to thermal neutron radiation lose the ability to repair sublethal and potentially lethal damage[4]. To further demonstrate a highly lethal effect of ^{10}B plus thermal neutron radiation on X-ray sensitive (P-39) and resistant (G-361) human melanoma cell lines, we compared the lethal effect of thermal neutron radiation with or without $^{10}B_1$-BPA pretreatment.

METHODS

The lethal effect of thermal neutron radiation and of boron neutron capture therapy (BNCT) on X-ray sensitive and resistant human melanoma cells was evaluated based on D_o value from survival curves.

1.Cells used in this study

- P-39: X-ray sensitive human melanoma cell line
- G-361: X-ray resistant human melanoma cell line
- N2KYSV: normal human fibroblast line transformed by Sendai virus
- ATKYSV: ataxia telangietasia cell line transformed by Sendai virus

These cells were maintained in alpha-modified MEM supplemented with penicillin (100 IU/ml), HEPES (20mM), and 10% fetal calf serum (Irvine Scientific, Santa Ana, CA).
Plating efficiencies of P-39 and G-361 were 25% and 32%, respectively.

2.Boron chemical
L-$^{10}B_1$-paraboronophenylalanine was used to target melanoma cells (Fig.1).

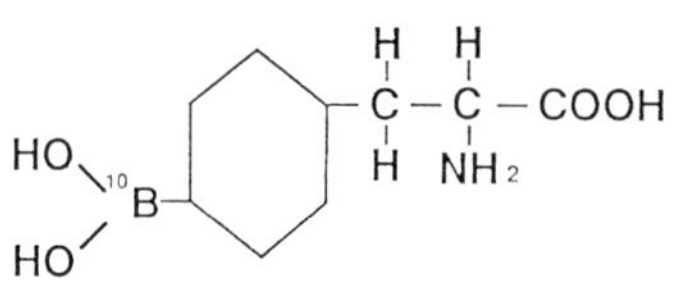

Figure 1. Chemical structure of L-$^{10}B_1$-paraboronophenylalanine

3.Radiation
a.X-ray radiation

A suitable number of cells harvested from an actively growing phase were inoculated into 60-mm plastic dishes. After growing overnight, the cells were irradiated with 50kV X-ray (10mA, 0.2mmAl filtration, 1100cGy/min) after the removal of the medium, and cultured for approximately two weeks to obtain colonies with more than 20 cells.

b.Thermal neutron radiation

The actively growing cells in 100-mm plastic dishes were incubated with the medium containing 0 or 20 μg L-$^{10}B_1$-BPA/ml for 20 hours until harvesting by trypsin treatment.
The cells were suspended at the density of 6×10^5 cells/ml of culture medium. One ml of cell suspension was transfered into a screw-capped teflon tube. The teflon tubes were located at different positions to obtain various doses of neutron radiation[5]. For thermal neutron irradiation, the thermal column of KUR with the flux of $7.9 \times 10^5 n/cm^2$ S.MW was used. After irradiation, an appropriate number of cells were seeded in triplicate dishes and cultured for 14 days at 37°C.

RESULTS

1. To select human melanoma cell lines sensitive and resistant to X-ray radiation, we tested 5 human melanoma cell lines(Fig.2). P-39 with D_0 value of 76 cGy and G-361 with a D_0 value of 160 cGy were selected and used for the X-ray sensitive and X-ray resistant cell lines, respectively. Both fast- and slow-type potentially lethal damage repairs of G-361 were 2 - 3 times greater than those of P-39 (data are not shown).

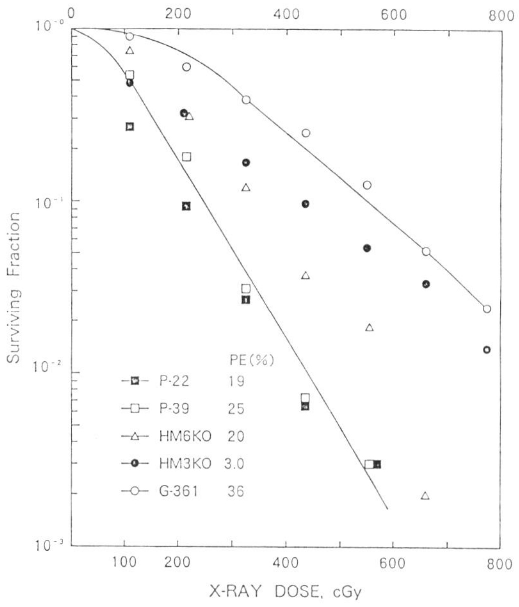

Figure 2. Survival curves of five different human melanoma cell line exposed to X-ray radiation (0-780 cGy).

2. P-39 cells sensitive to X-ray radiation were also sensitive to bleomycin treatment, and G-361 cells resistant to X-rays were also resistant to bleomycin treatment. These cell lines, however, exhibited identical sensitivity to ultraviolet (UV) radiation (Fig.3).
3. P-39 cells showed a higher sensitivity to thermal neutron radiation than did G-361 cells.
Both cell lines, however, had identical sensitivity to thermal neutron radiation when the cells were pre-treated with 200 μgL-$^{10}B_1$-BPA/ml before irradiation (Fig.4).

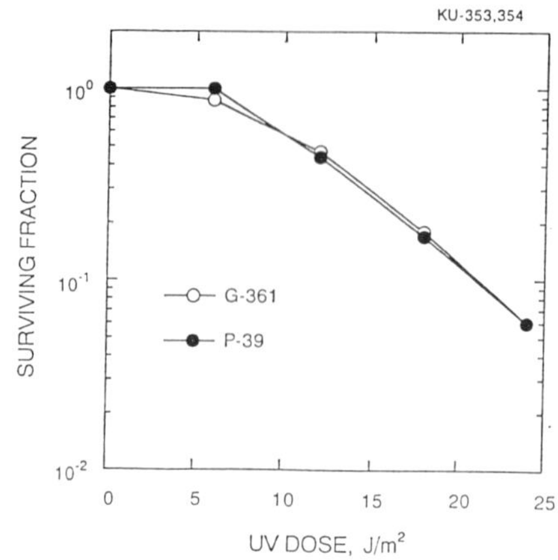

Figure 3. Survival curves of G-361 and P-39 human melanoma cells following single dose of far-UV light (predominantly 254 nm).

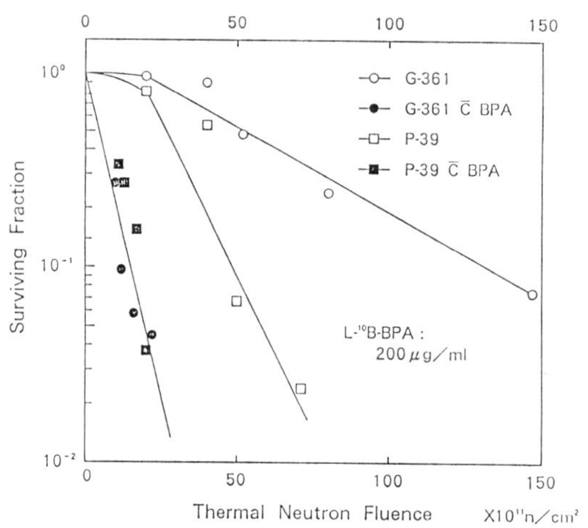

Figure 4. Identical D values of X-ray resistant and X-ray sensitive melanoma cell lines exposed to NCT using ^{10}B-BPA.

DISCUSSION

The X-ray resistant cell line was also resistant to the X-ray mimetic chemical bleomycin and to thermal neutron radiation; however, both X-ray resistant and sensitive cell lines showed identical sensitivity to $^{10}B(n, \alpha)^7Li$ reaction when the cells were pretreated with $^{10}B_1$-BPA and subsequently exposed to thermal neutron radiaiton.

Based on the assumption that recovery from sublethal damage (SLD) and potentially lethal damage (PLD) plays an important role in tumor cell survival, thermal neutron capture therapy using $^{10}B(n, \alpha)^7Li$ reaction appears to offer appreciable therapeutic advantage over X-ray and gamma ray therapy, since boron neutron capture therapy produces a high LET radiation without SLD and PLD repairs.

REFERENCES

1. Locher GL: Biological effects and therapeutic possibilities of neutrons. Am J Roentgenol 36:1-13, 1936
2. Mishima Y, Ichihashi M, Hatta S, Honda C, Yamamura K, Nakagawa T: New thermal neutron capture therapy for malignant melanoma: Melanogenesis-seeking ^{10}B molecule-melanoma cell interaction from in vitro to first clinical trial. Pigment Cell Res 2:226-24, 1989
3. Mishima Y, Honda C, Ichihashi M, Obara H, Hiratsuka J, Fukuda H, Karashima M, Kobayashi T, Yoshino K: Treatment of malignant melanoma by single thermal neutron capture therapy with melanoma-seeking ^{10}B-Compound. Lancet 2:388-389, 1989
4. Utsumi H, Ichihashi M, Funasaka Y, Mishima Y: Potentially lethal damage repair of B-16 melanoma cells pre-treated with $^{10}B_1$-paraboronophenylalanine after exposure to thermal neutron radiation. Neutron Capture Therapy. Ed. by Hatanaka H, Nishimura, 1986, pp247-257
5. Ichihashi M, Ueda M, Hayashibe K, Hatta S, Tsuji M, Mishima Y: In vitro radiobiological evaluation of selective killing effects of $^{10}B_1$-paraboronophenylalanine·Hcl in the thermal neutron capture therapy of malignant melanoma cells. KURRI-TR-260:1-13, 1985

BIBENZIMIDAZOLES AS RADIOPROTECTORS AGAINST BNC-INDUCED DNA DAMAGE

Anne Corder and Roger F. Martin

Molecular Sciences Group
Peter MacCallum Cancer Institute
Melbourne, Australia

INTRODUCTION

For several years we have been exploring the potential of DNA ligands, (that is DNA binding compounds), as radiomodifiers for use in cancer radiotherapy. There are three main aspects to this program, one of these is the development of ^{10}B- and ^{157}Gd-labelled DNA ligands for neutron capture therapy[1,2]. Both the other aspects of the program concern more conventional radiotherapy and radiomodifiers of ionising photons. We are studying some ligands which are radioprotectors[3] and others that are radiosensitisers[4]. The radioprotectors are of relevance to neutron capture therapy as well as to conventional radiotherapy, and this paper will focus on such radioprotectors.

All the DNA ligands are bibenzimidazoles that bind in the minor groove of DNA, exemplified by Hoechst 33258 and Hoechst 33342, which only differ by one substitutent; the latter has an ethoxy group instead of a hydroxyl on phenyl ring. Exposure of cultured cells to concentrations as low as 20μM Hoechst 33342, results in substantial protection against ionising radiation[3,5,6]

We have also investigated the radioprotection with purified DNA[3], using pBR322 plasmid DNA which provides a convenient assay for DNA strand breakage. Such experiments show that both Hoechst 33342 and Hoechst 33258 protect against radiation-induced DNA single-strand breaks. Many experiments of this sort, and experiments in which DNA damage is analysed on DNA sequencing gels, have provided us with some information about a mechanism of radioprotection[3]. This in turn has lead to the development of new synthetic ligands, some of which are more potent than the original Hoechst 33342.

So in summary, in the context of gamma rays, quite low concentrations of these ligands yield a degree of protection which would require much higher concentrations of classical radioprotectors such as, aminothiols. Another important feature of the ligands is that since they bind tightly to DNA they have a limited penetration through cell layers[7], raising the possibility of topical application to critical normal tissues. This is important in the context of neutron capture therapy, where skin can be a dose-limiting tissue.

However, all the data we have on the radioprotectors is with low LET radiation, so the obvious question to address is whether the compounds are also effective as radioprotectors of high LET radiation. This question is the focus of this paper; to date we only have results for experiments involving radiation damage to purified DNA.

Advances in Neutron Capture Therapy, Edited by
A.H Soloway *et al.*, Plenum Press, New York, 1993

EXPERIMENTAL

The experiments involve irradiation of plasmid DNA in a ^{10}B-borate buffer, so the boron neutron capture reaction is essentially being used as a source of high LET radiation in solution. Further details of the use of plasmid DNA to assay for DNA strand breaks are described in an accompanying paper[8].

The final concentrations of the components in the solutions irradiated in the thermal neutron beam at the MOATA reactor were as follows:

- pBR322 DNA; 50μg/ml,
- ^{10}B-borate buffer (60mM), and
- Hoechst 33342; 25μM (omitted from some control samples).

After irradiation, 0.5μg samples were assayed by agarose gel electrophoresis to determine DNA single-strand breaks (ssbs) and DNA double-strand breaks (dsbs). The quantitative results obtained from densitometry of negative photographs of ethidium bromide-stained gels were analyzed by programme 1R, Version 1990, BMDP Statistical Software Inc.,(Los Angeles, California). The statistical analysis included fitting the curves shown in Fig. 1.

RESULTS AND DISCUSSION

The results are shown in Fig 1. The top panel shows the disappearance of intact plasmid DNA with increasing radiation dose; reflecting induction of DNA single-strand breaks. There is a marginally lower level of DNAssbs in the presence of the protector, but the difference is not significant. Regression analysis of the slopes yielded a P value (F ratio) of 0.79. A further analysis, of the dose-dependent yield of DNA ssbs (after correction for The component of 'lost' intact plasmid due to DNA dsbs) also failed to yield a significant difference.

The lower panel shows the assay for DNA-double strand breaks, detected as an increase in the appearance of linear plasmid molecules with increasing radiation dose. There is a clear indication of protection of DNA double-strand breaks, and although the difference is not large, it is consistent at all dose points. The gradients for the control and protected samples are 0.0321±0.0015 (standard error) and 0.0251±0.00094 Gy^{-1}, respectively. regression analysis yielded a P value (F ratio) of 0.0032. It is also interesting to compare the yield of single- and double-strand breaks in the unprotected samples. At the radiation dose that yields one DNA ssb per molecule (ie. when 37% of the plasmid molecules are intact), the yield of linear molecules is 12.7%. Thus there are about 3 DNA double-strand breaks per 100 single-strand breaks. For similar experiments with low LET radiation[9], this ratio is about 1 in 100. This difference is consistent with the high LET nature of the irradiation conditions used in the present experiments.

CONCLUSIONS

Hoechst 33342 protects against BNC-induced DNA double-strand breakage at least in purified DNA. We are now proceeding to cell culture experiments, and to the investigation of the new radioprotectors which appear to be more effective than Hoechst 33342. It seems possible that such radioprotectors may have some potential in protecting against high LET radiation, and more particularly the possibility of protection of skin in BNCT.

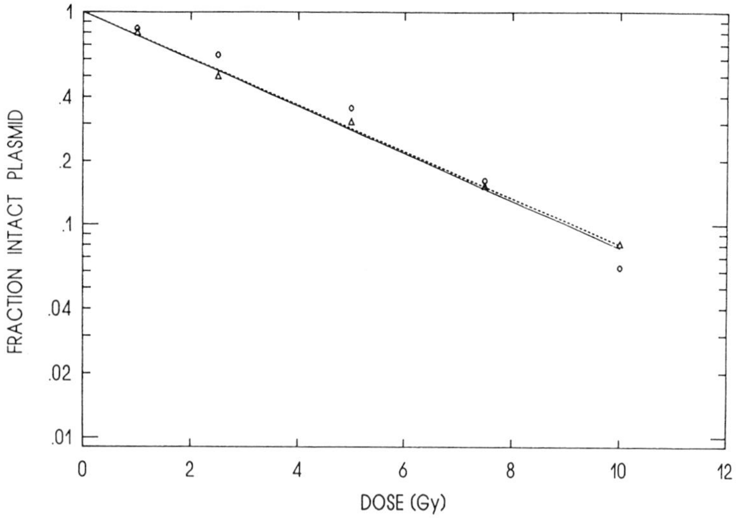

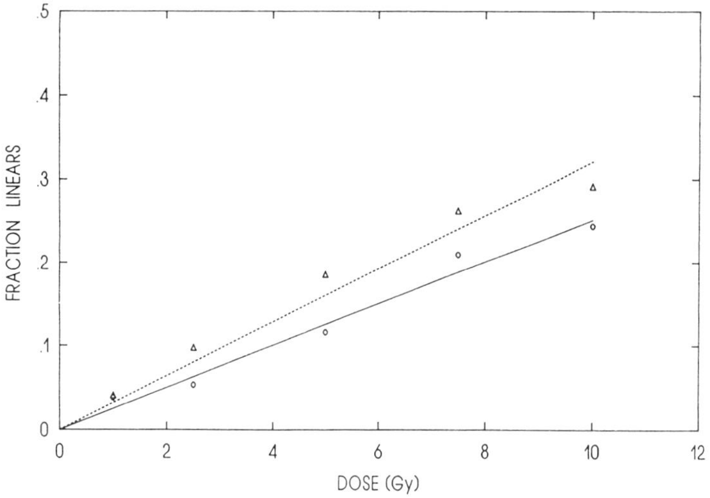

Fig. 1 The upper panel shows the disappearance of intact pBR322 DNA, reflecting the induction of DNA ssbs; radiation-only controls (---Δ---), and test samples with Hoechst 33342 (—o—). The lower panel shows the appearance of linear pBR322, reflecting induction of DNA dsbs, with the same symbol key as for the upper panel.

ACKNOWLEDGEMENTS

This project was supported by the Anti Cancer Council of Victoria, and by a grant from the Australian Institute of Nuclear Science and Engineering, which enabled access to the MOATA reactor at Ansto. The authors thank Dr. Jane Matthews and Jenny Smith from the PMCI Statistical Centre for the statistical analysis of the results.

REFERENCES

1. T. Kobayashi and K. Kanda. Analytical calculation of boron-10 dosage in cell nucleus for neutron capture therapy, *Radiat. Res.* 91: 77 (1982).
2. D.E. Charlton. Energy deposition in small ellipsoidal volumes by high-LET particles: application to thermal neutron dosimetry, *Int. J. Radiat. Biol.* 59: 827 (1991).
3. L. Denison, A. Haigh, G. D'Cunha and R.F. Martin. DNA ligands as radioprotectors: molecular studies with Hoechst 33342 and Hoechst 33258, *Int. J. Radiat. Biol.* 61: 69 (1992).
4. R.F. Martin, V. Murray, G. D'Cunha, M. Pardee, E. Kampouris, A. Haigh, D.P. Kelly and G.S. Hodgson. Radiation sensitisation by an iodine-labelled DNA ligand, *Int. J. Radiat. Biol.* 57: 939 (1990).
5. P.J. Smith and C.O. Anderson. Modification of the radiation sensitivity of human tumour cells by a bis-benzimidazole derivative, *Int. J. Radiat. Biol.* 46: 331 (1984).
6. S.D. Young and R.P. Hill. Radiation sensitivity of tumour cell strand in vitro or in vitro with the benzimidazole fluorescence Hoechst 33342. *Brit. J. Cancer* 60: 715 (1989).
7. R.E. Durand, Use of Hoechst 33342 for cell selection from multicellular systems, *J. Histochem. Cytochem.* 30: 117 (1982).
8. A. Whittaker, D.P. Kelly, M. Pardee, A. Corder, H. Meriaty, B.J. Allen, R.F. Martin. Synthesis and and evaluation of ^{157}Gd-labelled DNA ligands. **These proceedings.**
9. R.E. Krisch, M.B. Flick and C.N. Trumbore. Radiation Chemical mechanisms of Single-and Double-strand break formation in irradiated SV40 DNA, *Radiat. Res.* 126: 251 (1991).

RBE: A USEFUL CONCEPT IN NEUTRON THERAPY AND BNCT?

André Wambersie[1], Hans G. Menzel [2], Reinhard Gahbauer[3]

1 Université Catholique de Louvain, Cliniques Universitaires St-Luc, 1200-Brussels, Belgium

2 Commission of the European Communities, Radiation Protection Programme, 1049-Brussels, Belgium

3 Ohio State University, A. James Cancer Hospital, Columbus, Ohio, 43210, USA

INTRODUCTION: THE CONCEPT OF RELATIVE BIOLOGICAL EFFECTIVENESS (RBE)

An RBE value is the result of a radiobiological experiment and is thus associated with an experimental uncertainty. The definition of RBE is clear and unambiguous. However, the biological system, type and level of effect, the dose and the experimental conditions in which a given RBE value has been obtained have to be specified. In the present discussion, and except when specifically mentioned, RBE values are always expressed relative to ^{60}Co γ-rays (or high-energy X-rays).

RBE VALUES FOR FAST NEUTRON THERAPY

1. Survey of Experimental Data

Since the beginning of fast neutron therapy and even before, a large number of determinations of RBE values were made. Some general conclusions can be derived from the available experimental data :

- large differences in RBE values are observed (ranging from about 2 to 5);
- the neutron RBE increases with decreasing dose;
- the neutron RBE values are larger for late responding tissues than for early responding tissues. In other words, the neutron RBE's are higher for late complications than for early tolerance.

The RBE also varies to a large extent with neutron energy in the energy range used in therapy (about 40% for some systems), which raises difficulties when exchanging

Advances in Neutron Capture Therapy, Edited by
A.H Soloway *et al.*, Plenum Press, New York, 1993

information between different neutron therapy facilities [2]. The neutron energy spectrum has then to be accurately specified [4].

2. The Reference RBE

Due to the wide variation of RBE with biological system and criterion, dose and experimental conditions, it is necessary to select reference conditions for RBE specification when transferring or exchanging information [2] [5].

These reference conditions should be as close as possible to clinical situations and therefore the following conditions appear to be the most relevant :

- dose level : 2 Gy (photon equivalent) per fraction;
- biological system : a system and endpoint "representative" of the RBE for "average" or "overall" late tolerance of normal tissues.

The RBE defined by these reference conditions has been called: **reference RBE.**

3. The Clinical RBE

In contrast to the reference RBE, which is based on a pure radiobiological concept, the **clinical RBE** is a clinical and operational concept. The term "clinical RBE" has been used and understood, especially in the US, as a ratio of the absorbed dose which would be given in a photon treatment to the neutron dose which is actually prescribed at a given neutron therapy facility (and for a given tumor localisation).

A first approximation of the neutron dose to be delivered clinically is obtained by dividing the photon dose by the reference RBE. However, dose prescription and treatment planning depend also on other factors such as beam penetration or, in general, geometrical factors. If the physical selectivity of the neutron beams used is significantly inferior to that achieved with photons, a dose reduction factor taking into account the inferior physical selectivity of the neutrons has to be applied. Finally, when clinical experience becomes available, it should be taken into account when selecting the neutron dose to be delivered to the patient.

The "clinical RBE" is then a quantity that the radiotherapist has to select when prescribing the irradiation (e.g. in terms of monitor units). Although it is a dose ratio for two radiation qualities, it is not in a strict sense a RBE: it is the reference RBE (as defined above) weighted by physical selectivity (when needed) and clinical experience (when available) [5]. It implies a value based on the judgment and the clinical experience of the radiotherapist.

RBE IN BORON NEUTRON CAPTURE THERAPY (BNCT)

The neutron beam for BNCT is complex and consists of different components: fast, epithermal, and thermal neutrons and γ rays [1]. Thermal neutrons induce protons through the ^{14}N (n, p)^{14}C reaction and γ rays through the ^{1}H (n, γ) ^{2}H reaction, but mainly they induce an irradiation by α-particles (and ^{7}Li-particles) selectively in some cells that are in the vicinity of the incorporated boron compounds.

Part of the discussion about RBE presented above for fast neutrons is relevant for BNCT applications. However, there are two additional problems which are specific to BNCT : the first one is related to the poor penetration of the thermal and epithermal neutrons, the second one is related to the heterogeneous distribution and the short range of the α-particles. We will restrict the discussion to the applications of BNCT in the treatment of brain gliomas.

1. Absorbed Dose and RBE Variations in Depth

In BNCT, the radiation dose will decrease rapidly with depth, mainly due to the poor penetration of the epithermal and thermal neutrons (let us exclude for the moment the α-particles from boron disintegration). As a consequence, in the human brain the dose distribution (but also the radiation quality) will significantly change with depth, even if several irradiation fields are applied.

RBE values for late tolerance of the brain in different species of mammals are of course most relevant for BNCT of gliomas, especially when they are obtained in conditions similar to that planned in the clinical applications (e.g. number of fractions). However, there may be serious limitations for extrapolation to the clinical situation. Spatial variations of absorbed dose, radiation quality and RBE depend critically on shape and composition of the irradiated volumes and thus on brain size, location of brain nuclei and irradiation geometry. On the other hand, the brain tolerance to radiation varies from one brain region to another one and it is specially low for the grey central nuclei (poorer vascularisation). The depth of these critical regions are of course different for different brain sizes and animal species.

2. The Alpha-Particles

The radiation quality problem of the disintegration products from the neutron capture in boron has two aspects :

a) the range of the α-particles is 9 μm (and the range of ^{7}Li particles 6 μm) so that each particle traverses only one or two cells (diam. about 10 μm). The cells traversed experience a very high energy deposition whereas neighbouring cells will have no energy deposition. The dose distribution, on a cellular and subcellular level, is therefore highly unhomogeneous and any macroscopic dose quantity or concept must necessarily have shortcomings [**3**] . In addition, the LET of the particles varies along the particle tracks and therefore corresponding variations of radiation quality have to be expected.

b) the spatial origin of the disintegration products is also highly unhomogeneous and depends on the concentration and distribution of ^{10}B nuclides and the spatial fluence distribution of thermal and epithermal neutrons. This unhomogeneity, paired with the short range of the released charged particles, must put any dose concept based on macroscopically defined quantities into question.

It should be noted that RBE is the ratio of two absorbed doses, i.e. a macroscopically defined quantity which implies the existence of an appropriate averaging procedure. The large local unhomogeneities of absorbed dose in BNCT pose major problems if the conventional quantities and concepts are to be used [**3**].

Systematic radiobiological investigations are now being performed in the different centres preparing BNCT applications in order to assess brain tolerance, first to the neutron beam (without boron), and then after administration of different concentrations of boronated compounds.

CONCLUSION

In fast neutron therapy, introduction of the concept of reference RBE is justified because it facilitates a clearer exchange of information between neutron therapy centres.

However, the neutron dose prescribed to the patient is the result of the clinical judgment and experience of the radiotherapist and takes into account not only the reference RBE but also the (often inferior) physical selectivity of the neutron beams as well as the clinical experience when available (especially late sequellae).

The term "clinical RBE" has been used, especially in the United States, to define the ratio of the photon doses and neutron doses actually prescribed to patients at a given neutron therapy centre and for a given tumor location (or recommended in protocols). The clinical RBE defined in that way is then a operational quantity, implicitely including a clinical judgment, and distinct from the reference RBE which is a pure radiobiological concept.

In BNCT, the highly unhomogeneous spatial distribution and the short range of the α-particle tracks raises difficult problems in dosimetry and radiobiology. In particular, the concept of absorbed dose, which is a macroscopically defined quantity, necessarily has shortcomings. The same is true for the RBE, which is defined as a a ratio of two absorbed doses.

However, the available radiobiological data on α-particles [RBE, OER (Oxygen Enhancement Ratio), repair capacity, etc.] are definitely useful in selecting the optimum conditions of application of BNCT. In addition, the RBE of the BNCT beam (without boron) for late tolerance of the brain is useful information, although it may change with depth in terms of the radiation quality due to the poor penetration of the epithermal and thermal neutrons.

A final decision on the prescribed irradiation will have to be taken at a certain point in the programme. From a very practical point of view, the prescribed irradiation could be expressed in e.g. terms of "monitor units" or exposure time to the neutron flux and amount of administered boron (as well as fraction number and eventually dose rate). In BNCT, as in fast neutron therapy, the final decision will always include clinical judgment. However, all the available information (physical, pharmacological, radiobiological, etc.) will help to select the prescription of the irradiation conditions in an optimal way. The RBE data are needed, but will play an as yet not fully understood and probably limited role in the final decision.

REFERENCES

1. Fairchild R.G., Bond V.P., Woodhead A.D. (Eds). Clinical Aspects of Neutron Capture Therapy. Plenum Press, New York, 1989.
2. Hall E.J., Astor M., Brenner D.J. Biological intercomparisons of neutron beams used for radiotherapy generated by p →Be in hospital-based cyclotrons. The British Journal of Radiology, **65**, 1992, 66-71.
3. International Commission on Radiation Units and Measurements (ICRU). Radiation quantities and units. ICRU Report 33, 1980. ICRU Publications 7910 Woodmont avenue, Suite 800, Bethesda, Maryland 20814-3095, USA.
4. Pihet P., Norman C., Gueulette J., Menzel H.G., Wambersie A. Microdosimétrie des faisceaux de neutronthérapie d(50)+Be et p(65)+Be à Cyclone, Louvain-la-Neuve. In Radiophysique, XXIV Congrès de la Société Française des Physiciens d'Hôpital, Tours, 1985, 269-277.
5. Wambersie A., Battermann J.J. Practical problems related to RBE in neutron therapy. In: K.H. Kärcher, H.D. Kogelnik and T. Szepesi (Eds), Progress in Radio-Oncology III, ICRO (International Club for Radio-Oncology), Vienna, Austria, 1987, pp. 155-162.

A COMPARISON OF THE DOSE·RBE AND THE BIOLOGICAL DOSIMETRY APPROACHES FOR TREATMENT PLANNING IN BNCT

B.H. Laster[1,2], H.B. Liu[1], C.R. Gordon[1], and V.P. Bond[1]

[1]Medical Department, Brookhaven National Laboratory
Upton, NY 11973
[2] Department of Radiation Oncology, University Hospital
Stony Brook, NY 11794

Treatment planning for clinical trials with boron neutron capture therapy (BNCT) is complicated substantially by the fact that the radiation field generated by the activating external neutron beam is composed of several different types of radiation, i.e., fast neutrons, recoil protons from elastic collisions with hydrogen, gamma rays from the reactor and from neutron capture by body hydrogen, protons from nitrogen capture, and the products of the NCT interaction. Furthermore, the relative contribution of each type of radiation varies with depth in tissue. Because each of these radiations has its own RBE, and the RBE of the fast neutron component will not be constant as the neutron spectrum changes with depth, the problem of predicting the severity of the biological effect, in depth, becomes complex indeed.

In order to attack this problem, Monte Carlo calculations of dose, checked against benchmark measurements, are employed. Two approaches are then used to assess the severity of the effect. In the first, the effective dose (D_{EF}) is determined by summing the products of (D·RBE) for each radiation. The other approach, "biological dosimetry", involves placing cells at the location for which the D_{EF} was calculated. Using a dose-response curve from a low-LET radiation, e.g. ^{137}Cs gamma rays ($D_{\gamma,Cs}$), the "photon equivalent dose" (PED, or D_P) can be determined. If the RBE values used are correct, the D_{EF} and the D_P should be essentially identical.

In our preliminary experiments, the agreement was not satisfactory. The object of this paper is to describe these initial efforts in some detail, and then outline our plans for attempting to reconcile the differences. Because we regard Monte Carlo as an essential part of treatment planning, it is imperative that the afore-mentioned discrepancies be addressed.

The first step was to establish an apparent RBE (RBE_a) for fast neutrons in the epithermal beam of the Brookhaven Medical Research Reactor (BMRR). Cells were irradiated, in air, at the epithermal beam of the BMRR in a configuration and geometry that yielded ~4% thermal neutrons. Thus, the major dose components were

Table 1. Physical doses for BMRR phantom experiment (all doses given in Gy)

Position	D_{nit}	$D_{f,epi}$	D_γ	D_{boron}	D_{tot}
1	0.522	0.720	2.349	-----	3.591
% D_{tot}	14.5	20.0	65.4	-----	~100
2	0.621	0.414	3.114	-----	4.149
% D_{tot}	14.9	9.9	75.0	-----	~100
3	0.450	0.243	2.880	-----	3.573
% D_{tot}	12.6	6.8	80.6	-----	~100
4	0.270	0.135	2.115	-----	2.520
% D_{tot}	10.7	5.4	83.9	-----	~100
1*	0.174	0.264	0.783	2.178	3.399
% D_{tot}	5.1	7.8	23.0	64.1	~100
2*	0.207	0.138	1.038	2.496	3.879
% D_{tot}	5.3	3.6	26.8	64.3	~100
3*	0.150	0.081	0.960	1.809	3.000
% D_{tot}	5.0	2.7	32.0	60.3	~100
4*	0.090	0.045	0.705	1.074	1.914
% D_{tot}	4.7	2.4	36.8	56.1	~100

* with 30 μg/g enriched boric acid

fast neutrons,, gamma rays and protons from the n,p reaction with nitrogen. An RBE of 1 was used for gamma, and an RBE_a of 2 for the protons, a value that has been estimated (1), and measured (2)[1]. For the BMRR experiments, survival was first plotted as a function of MW minutes. This was subsequently converted to dose (D) by means of Monte Carlo calculations, which yielded a conversion factor of Gy/MW min (3). Cell survival curves for the BMRR and ^{137}Cs gamma rays are shown in Figs. 1 and 2. Note the absence of effect from boric acid with the ^{137}Cs gamma rays. Irradiation techniques and cell culture assay methods are detailed elsewhere (4,5). The 10% level of survival was selected for the determination of both D_{EF} and D_P.

Assuming that the RBE_a of 2 for the 0.6 MeV protons from nitrogen capture is correct, the RBE_a for the fast neutrons was determined by means of the following equation:

$$D_{EF} = D_{prot} \cdot RBE_a + D_\gamma \cdot RBE + D_{neut} \cdot RBE_a \qquad (1)$$

Substituting the D_P for the D_{EF}, and using the RBE's above, the RBE_a for fast neutrons was calculated to be 3.6.

[1]The term RBE_a is applied here since the measurements in (2) were carried out in a mixed-field.

We then placed cells at depth (1.4, 3.6, 5.8, and 8 cm) in a lucite phantom, and the RBE for neutrons was again determined. The arrows in Fig. 3 show the sequential positions of the cell holders for these irradiations (position 1 is nearest the beam). Physical doses are given in Table I. The RBE_a for fast neutrons at position 1, calculated by substituting D_P for the D_{EF} in equation 1, was found to be 4.8. Thus, in order to make D_{EF} equal to D_P, it was necessary to adjust the RBE for fast neutrons upward, from 3.6 to 4.8.

To determine an RBE_a for the boron reaction products, the following equation was used:

$$RBE_{a,boron} = D_P / D_{boron} \tag{2}$$

When 30 $\mu g\,^{10}B/g$ boric acid was used, the RBE_a was calculated to be 3.6, closely approximating that reported earlier (6). As shown in Table 2, the D_{EF} is equal to the D_B in the first two cell positions only (1,1*,2,2*). For positions 3 and 4, major adjustments in the apparent RBE's were required in order to make D_{EF} equal to D_P.

Several factors may be influencing the apparent effectiveness of the various radiations at depth in the phantom. The gamma dose, relative to the fast neutron dose, increases with depth (Table 1). Also, the energy spectrum of the fast neutrons changes with depth, which should increase the RBE. The high and low LET radiations may act synergistically (7), and the RBE of 2, assumed for recoil protons, may not be correct. Because the absolute value for the dose of neutrons decreases rapidly with depth, the calculated neutron dose may be imprecise. The neutron dose in depth is low, so it may result in using a large number to adjust the dose.

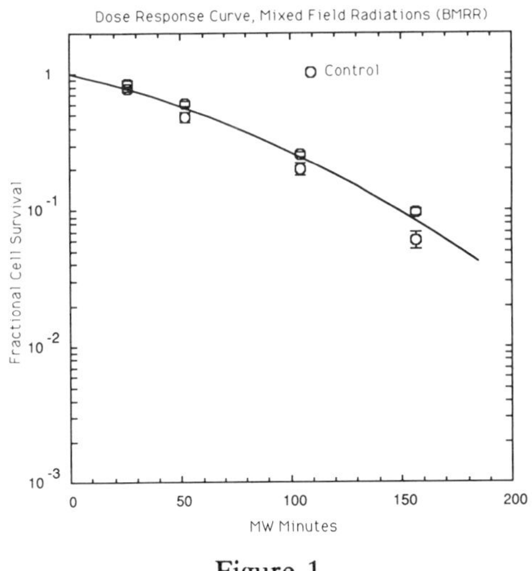

Figure 1

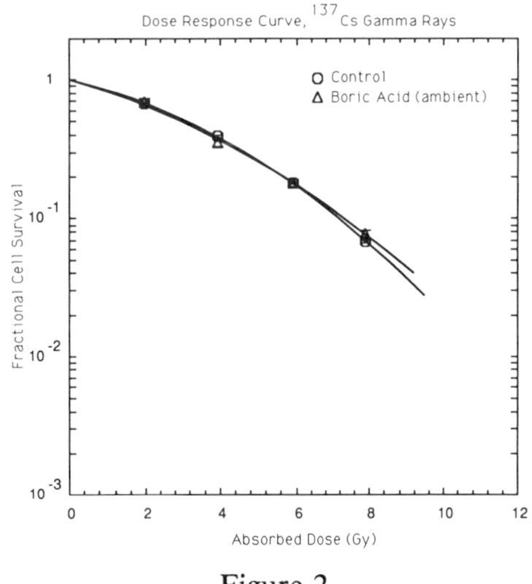

Figure 2

Table 2. RBE_a for neutron and boron doses

Position	Physical dose (Gy) BMRR	Per cent survival	Absorbed dose (Gy) ^{137}Cs	RBE_a neutrons (fast)	RBE_a Boron-10 (30μg/g)
1	3.6	9.6	7.2	4.8	-----
2	4.1	17.6	6.2	4.8	-----
3	3.6	12.3	6.7	11.9 (?)	-----
4	2.5	29.8	4.2	11.4 (?)	-----
1*	3.4	2.4	10.3	4.8	3.6
2*	3.9	2.2	10.2	4.8	3.6
3*	3.6	2.6	10.3	31.0 (?)	3.6
4*	1.9	13.5	6.5	38.9 (?)	3.6

RBE nitrogen, assumed 2
RBE gamma, assumed 1
*with 30 $\mu g^{10}B/g$ boric acid

It is evident that the use of Monte Carlo calculations and assumed RBE's did not provide an accurate assessment of the effective dose in our experiments designed to simulate patient treatment planning for BNCT. The difficulties attendant upon the approach include, a) the accuracy of the RBE values for all components are in question because they can not be determined separately, and b) additional mechanisms involved in the interactions may change the severity of effect (e.g. synergism).

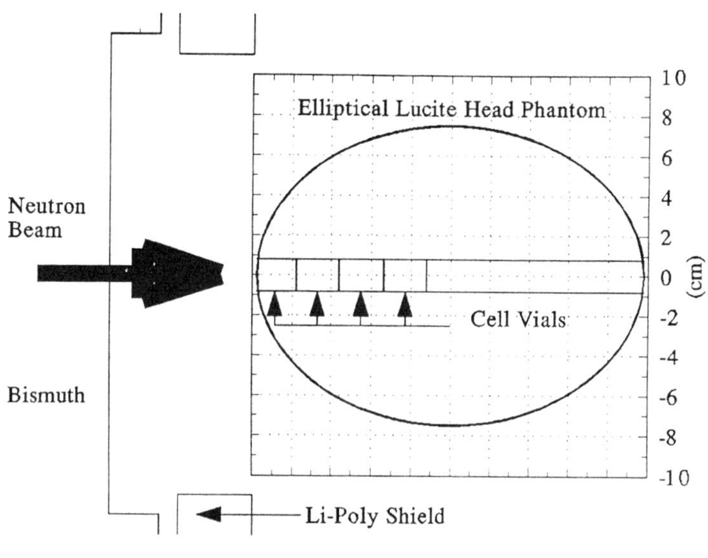

Figure 3

It is our contention that a cell system which is highly reproducible with respect to its response to photon irradiation should be used as a bench mark measurement of the PED, and thus of the true D_{EF} for BNCT radiations. We intend to pursue these studies further by positioning cells at given regions of interest in a phantom that is tissue equivalent for neutrons. Their response will be evaluated with and without boron, under conditions of homogeneous and heterogeneous intracellular boron distribution. It is anticipated that the difficulties outlined above can be resolved so that the approach using calculations, measurements, RBE and D_{EF} can provide a realistic and precise prediction of the severity of biological effect at any pre-determined site.

ACKNOWLEDGEMENTS

This research was supported by U.S. DOE contract DE-AC02-76CH00016.

REFERENCES

1. R.G. Zamenhof, B.W. Murray, G.L. Brownell, G.R. Wellum and E.I. Tolpin, Boron neutron capture therapy for the treatment of cerebral gliomas.I:Theoretical evaluation of the efficacy of various neutron beams. *Med. Phys.* 2:47-60 (1975).
2. D. Gabel, R.G. Fairchild, B. Larsson and H.G. Borner, The relative biological effectiveness in V79 Chinese hamster cells of the neutron capture reactions in boron and nitrogen. *Radiat. Res.* 98:307-316 (1984).
3. H.B. Liu and R.M. Brugger, Dose calculation and treatment planning for the Brookhaven NCT facility, *in*: Proceedings of the Fifth International Symposium on Neutron Capture Therapy. Plenum Press (1993).
4. B.H. Laster, S.B. Kahl, E.A. Popenoe, D.W. Pates and R.G. Fairchild, Biological efficacy of low-density lipoprotein for boron neutron capture therapy as measured in cell culture. *Cancer Res*. 51:4588-4593 (1991).
5. R.G. Fairchild, S.B. Kahl, B.H. Laster, J.Kalef-Ezra and E.A. Popenoe, *In vitro* determination of uptake, retention, distribution, biological efficacy, and toxicity of boronated compounds for neutron capture therapy:A comparison of porphyrins with sulfhydryl boron hydrides. *Cancer Res.* 50:4860-4865 (1990).
6. M.A. Davis, J.B. Little, K.M.M.S. Ayyangar and A.R. Reddy, Relative biological effectiveness of the $^{10}B(n,\alpha)^{7}Li$ reaction in HeLa cells. *Radiat. Res.* 43:534-553 (1970).
7. M. Zaider and H.H. Rossi, The synergistic effects of different radiations, *Radiat. Res.* 83:732-739 (1980).

A PC-BASED COMPUTER PROGRAM FOR THE ESTIMATION OF THE RADIATION DOSE IN *IN VITRO* AND *IN VIVO* BORON NEUTRON CAPTURE IRRADIATION EXPERIMENTS

Ruud Verrijk[1], Rene Huiskamp[2] and Adrian C. Begg[1]

[1] Division of Experimental Therapy
The Netherlands Cancer Institute, Plesmanlaan 121
1066 CX, Amsterdam, The Netherlands

[2] Netherlands Energy Research Foundation, ECN
P.O. Box 1, 1755 ZG Petten, The Netherlands

INTRODUCTION

In Boron Neutron Capture Therapy (BNCT) microdosimetry of charged particle radiation depends on total boron concentration and intracellular boron distribution. Due to the inhomogeneity of boron distribution in cells, radiation doses to both tumor and normal tissue are influenced by boron and nitrogen concentrations and intracellular distributions, cell volume and shape, nuclear size and geometrical structure of the tissue. For correct calculation of the radiation dose in BNCT, these factors should be taken into account. Several computer models have been developed previously in order to estimate the absorbed dose from charged particles in BNCT (Gabel et al. 1987; Kobayashi and Kanda, 1982). In these models, however, single values for mean Relative Biological Effectiveness (RBE) are used to convert high LET radiation doses to isoeffective photon equivalent doses. The RBE depends on both LET and endpoint, such as surviving fraction of tumor cells or normal tissue tolerance (Barendsen et al. 1966). Since LET is not constant along the track of a charged particle, the RBE cannot be considered constant for particles generated by boron and nitrogen neutron capture. Experimental RBE data to be used in BNCT have been gathered, but without consensus (Gabel et al. 1984; Fukuda et al. 1987).

A computer program designed to run on a microcomputer has been written in Turbo Pascal to determine energy deposition in cell nuclei resulting from charged particle emission after boron or nitrogen neutron capture in nuclear, cytoplasmic and extracellular compartments. This computer model goes beyond former models in estimating a microscopic

Advances in Neutron Capture Therapy, Edited by
A.H Soloway *et al.*, Plenum Press, New York, 1993

RBE for each individual charged particle track segment that traverses a cell nucleus. Another refinement is the implementation of dynamic modelling, which offers a more realistic simulation of cell and tissue geometry. This was approached by varying cell geometry and arrangement parameters within a simulation.

METHODS

Stopping power tables of α, ^{7}Li particles and protons were generated with PC-TRIM (TRansport of Ions in Matter), a microcomputer version of TRIM-88 (Ziegler et al. 1970). Bar-shaped cells were represented in three-dimensional space as an array of 3 x 3 x 3 cells with the central cell as the target cell. This allows calculation of cross fire contributions from charged particles originating in neighbouring cells. Two similar models were additionally developed for the simulation of capture reactions in a cell monolayer or in a single suspension cell in culture (Figure 1).

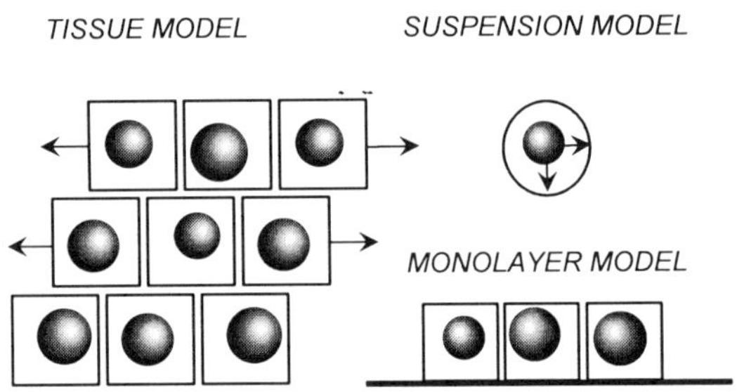

Figure 1. Two dimensional representation of modelling of cells in tissue, monolayer and suspension structure. Arrows illustrate positioning of nuclei and cells in dynamic modelling.

Cellular geometry can be entered in the program as constants or as dynamic variables within a simulation. The relative shift of the three cell layers in X and Y direction, and the nuclear diameter and position of the nucleus in cells, can be entered in the same manner. Dynamic variation of variables was based on random distribution within user-defined limits. A series of capture reactions was generated in the cell array at random locations. The assumption was made that charged particles are emitted back-to-back in a straight line in random directions. Typically, 20,000-100,000 capture events are generated for 2% statistical uncertainty. When an ion track has traversed the target cell nucleus partly or completely, energy deposition was scored using the stopping power tables. Subsequently, the location of the capture reaction was determined to be from nuclear, cytoplasmic or extracellular origin and whether the capture reaction took place in the central target cell or in one of the adjacent cells. Using this procedure, fractional contributions from nuclear, cytoplasmic and extracellular regions were calculated for both boron and nitrogen related capture reactions. Since hydrogen in cells or tissue can be assumed to be distributed homogeneously, hydrogen neutron capture was not modelled this way. Due to the homogeneity of the gamma radiation at a cellular scale, and the possibility to measure gamma radiation doses, there is no need to model this on a microdosimetric level.

For every ion track, the LET-RBE relationship as described by Barendsen et al.(1966) was used to calculate micro-RBEs. Based on the LET in the stopping power tables and the endpoint (in most cases cell survival level) the RBE was determined at every location in the ion track. Since most tracks do not traverse the nucleus completely, a micro-RBE was calculated for that part of the ion track that deposits energy in the nucleus. This track-individual micro-RBE was then used to convert the radiation dose deposited in the nucleus to a photon equivalent dose. With track-individual micro-RBEs mean RBEs can be calculated for boron capture reactions in the nucleus, the cytoplasm and the extracellular space. Subsequently, local boron and nitrogen concentrations, neutron cross sections and neutron fluence are used to calculate a boron and nitrogen capture related photon equivalent radiation dose. Gamma radiation and fast neutrons (including RBE) are added to the neutron capture radiation dose to give the total dose.

RESULTS

As an example for micro-RBEs, tumor cells were modelled with 12 x 12 x 12 μm as base dimensions with a nuclear volume of 30% of the cell volume. The intercellular distance was chosen in such a way that the interstitial volume accounted for 14% of the total tissue volume. No boron was assumed in the cell nuclei. Table 1 shows micro-RBEs for three cellular compartments for charged particles from boron and nitrogen neutron capture.

To validate the model, photon equivalent dose rates were calculated and compared with data from two publications on *in vitro* irradiation experiments at the Brookhaven Medical Research Reactor with V79 cells (Gabel et al. 1984; Fairchild et al. 1990). The neutron fluence can be calculated from the thermal neutron flux (5.45×10^9 $n_{th}.cm^{-2}.sec^{-1}$) and the irradiation time. For more detailed information, such as fast neutron and gamma radiation dose rate, we refer to both papers. Figure 2 shows comparisons for boric acid,

Table 1. Micro-RBEs of charged particles in Neutron Capture in relation to the origin of capture.

ALPHA PARTICLES	Nuclear	Cytoplasmic	Extracellular
Direct	2.21	2.21	2.28
Cross Fire	2.48	2.41	2.35

LITHIUM IONS	Nuclear	Cytoplasmic	Extracellular
Direct	1.90	2.27	2.62
Cross Fire	2.81	2.77	2.68

PROTONS	Nuclear	Cytoplasmic	Extracellular
Direct	1.58	1.83	2.00
Cross Fire	2.26	2.27	2.21

assuming a homogeneous boron distribution, and comparisons for BOPP (a boronated porphyrin) and BSSB (the dimer of borocaptate sodium) assuming an exclusively cytoplasmic location for boron in washed cells. Cytoplasmic boron concentrations were derived from boron measurements as reported in the paper by Fairchild et al. Simulation data points were constructed with the linear quadratic model for cell survival. Due to our choice of α and β values for V79 cells with X-rays, being 0.128 Gy^{-1} and 0.046 Gy^{-2} respectively, simulation data tend to be somewhat more curvy.

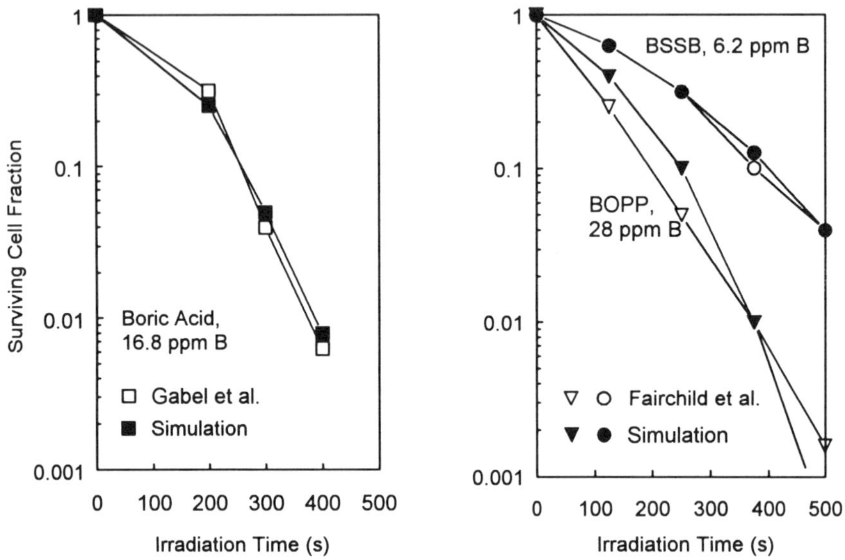

Figure 2. Comparison of V79 cell survival data of two published *in vitro* irradiation experiments with simulation data

DISCUSSION

The microdosimetry program presented here appears to predict radiation doses and photon equivalent doses reasonably well, as shown by the comparison with two *in vitro* irradiation data sets. Planned irradiation experiments will further determine the program's abilities to estimate radiation doses in irradiated animals. In order to give an accurate prediction, however, data on boron microdistribution and cell histology should be available for a given boron-containing drug in tumor and normal tissues at risk.

It is recognized that modelling cells as bar-shaped bodies with a spherical nucleus is an oversimplification. Taken over a considerable number of cells, however, this may still be a good average approach, since average internuclear distances, for example, can be made similar to that with spherical or ellipsoidal cells. The computer program models cells in a tissue-like structure (or as monolayer or single cells in culture) and is not able to model blood capillaries. Other mathematical models have been developed to calculate the radiation dose to capillary endothelial cells (Wheeler et al. 1989). Especially in glioma BNCT treatment, brain capillary endothelial cells seem to be the dose limiting factor that eventually causes severe side effects. Due to the small diameter of most capillaries, charged particle equilibrium is not reached for boron compounds that are excluded from the normal brain (like borocaptate,

BSH), resulting in a sparing effect on the endothelial tissue. When the capillary wall is not the dose limiting tissue, however, the microdosimetry model presented here is valuable in calculating photon equivalent radiation doses in both *in vitro* and *in vivo* systems.

In the future, a three-dimensional photon equivalent radiation dose profile could be calculated by combining simulation results with three-dimensional neutron energy and flux distributions in a tumor or normal tissue volume.

REFERENCES

Barendsen, G.W., Koot, C.J., van Kersen, G.R., Bewley, D.K., Field, S.B. and Parnell, C.J., 1966, The effect of oxygen on impairment of the proliferative capacity of human cells in culture by ionizing radiations of different LET, *Int. J. Radiat. Biol.* 10:317.

Fairchild, R.G., Kahl, S.B., Laster, B.H., Kalef Ezra, J. and Popenoe, E.A., 1990, In vitro determination of uptake, retention, distribution, biological efficacy, and toxicity of boronated compounds for neutron capture therapy: a comparison of porphyrins with sulfhydryl boron hydrides, *Cancer Res.* 50:4860.

Fukuda, H., Kobayashi, T., Matsuzawa, D., Kanda, K., Ichihashi, M. and Mishima, Y., 1987, RBE of a thermal neutron beam and the $^{10}B(n,\alpha)^{7}Li$ reaction on cultured B-16 melanoma cells, *Int. J. Radiat. Biol.* 51:167.

Gabel, D., Fairchild, R.G., Larsson, B.S. and Börner, H.G., 1984, The relative biological effectiveness in V79 Chinese hamster cells of the neutron capture reactions in boron and nitrogen, *Radiat. Res.* 98:307.

Gabel, D., Foster, S. and Fairchild, R.G., 1987, The Monte Carlo simulation of the biological effect of the $^{10}B(n,\alpha)^{7}Li$ reaction in cells and tissue and its implication for boron neutron capture therapy, *Radiat. Res.* 111:14.

Kobayashi, T. and Kanda, K., 1982, Analytical calculation of boron-10 dosage in cell nucleus for neutron capture therapy, *Radiat. Res.* 91:77.

Wheeler, F.J., Griebenow, M.L., Wessol, D.E. and Nigg, D.W., 1989, Analytical modeling for neutron capture therapy, *Strahlenther. Onkol.* 165:186.

Ziegler, J.F., Biersack, J.P. & Littmark, U., 1970, The stopping and range of ions in solids (Pergamon Press, New York).

NEW DIMENSIONS OF BORON THERMAL NEUTRON CAPTURE THERAPY IN NEUROSURGERY

Hiroshi Hatanaka

Dept. of Neurosurgery
Teikyo University
2-11-1 Kaga, Itabashi-ku
Tokyo 173, Japan

Boron-neutron capture therapy (BNCT) for glioblastoma and other gliomas has been practiced by the author for almost a quarter of a century and more than 100 patients have been treated. (Table 1) As presented elsewhere in this Symposium, 9 patients with various types of brain tumors out of 53 patients treated by BNCT before May, 1987 have survived more than 10 years. Among the nine, two with glioblastoma have been in excellent condition for 20 and 15 years respectively; a child with a glioma, grade III in the frontal lobe has grown up to become a draftswoman after 11 years; three others with recurrent gliomas (eventually glioblastomas) have survived longer than 10 years although they have developed delayed radiation necrosis in areas surrounding the tumors due to overlapping radiotherapies.

Table 1. All cases treated by BNCT in Japan (1968 - October 14, 1992)

Hatanaka	118	brain tumor	115
		AVM	2
		extracranial	1
Oda	7	brain tumor	7
Nakagawa	6	brain tumor	6
Mishima	12	melanoma	12
TOTAL	=	143 cases	

As reported in detail in Chapter XXVI (pathological study) of Boron Neutron Capture Therapy for Tumors[1] (a publication subsidized by the Ministry of Education, Science and Culture), viable tumor residues were not found at autopsy in a sphere where more than 2.5 x 10^{12} neutrons (in n.v.t.) per cubic centimeter had been delivered. In the 18 cases examined in that report no brain damage attributable to neutron capture was recognized.

If the neutron beam aperture is large enough such a sphere had a depth of 6 cm. This is why the author has been calling 6 cm the Maximum Therapeutic Depth[2] by the current standard regimen of boron-thermal neutron capture therapy with mercaptoundecahydrododecaborate.

As described on pages 25 and 387 of the same volume, use of deuterium water to replace water content in the brain, either by drinking or by intravenous or intraventricular injection, will prevent attenuation of neutron flux.

By heavy water replacement of brain water content and multiportal irradiation, capture therapy with time-tested thermal neutrons will be as valuable as epithermal capture therapy, the biological safety of which has not been well established and which is not easily available in every reactor.

In 1986 an interim clinical result was published by the author on 77 patients treated by BNCT in Chapter XXV of the same volume. Until that time, all the malignant glioma patients who came to the author were treated by BNCT except when either boron-10 or a reactor were unavailable. The patients who could not avail themselves of boron or a reactor had to be treated by conventional radiotherapy, thus forming a control group. The important feature of the results was that for grade III-IV gliomas (excluding previously treated ones) the 5-year survival of the BNCT group was 19% (n = 38, average age = 50.2 ± 1.9). Of the thirty-eight, the 12 patients whose tumors were within the Maximum Therapeutic Depth (6 cm from surface) showed a 5-year survival of 58% (n = 12, average age = 49.8 ± 4.1). The control group died out before 7 years after treatment and their 5-year survival was 5% (n = 46, average age = 43.5 ± 2.1).

After April 1985, the number of patients exceeded the capacity of the author, making it difficult to treat every incoming patient with grade III-IV gliomas because of limited access to reactors and limited boron-10 supply. After that period, clinical analysis has not been attempted because the policy of patient selection has not been as non-selective as before, and because the use of heavy water was started.

In spite of the lack of more recent BNCT statistics, the author's general impression is as follows: If treated within 2 weeks after diagnosis, and if the tumors are at the surface of the brain, then a glioblastoma patient can expect a 50% 5-year survival. If appropriate treatment of a glioblastoma is delayed, these rapidly growing neoplasms will quickly spread in the entire nervous system making it difficult to treat with any type of radiotherapy or chemotherapy.

This statement assumes a satisfactory surgical debulking as an inherent part of BNCT. The procedure has to be performed avoiding iatrogenic tumor seedings. Brain stem gliomas, which do not allow even a debulking surgery, have been treated with some success even without any debulking. (Figure 1) To further improve the therapy, more potent boron compounds or new means to transport boron to tumor cells should be developed.

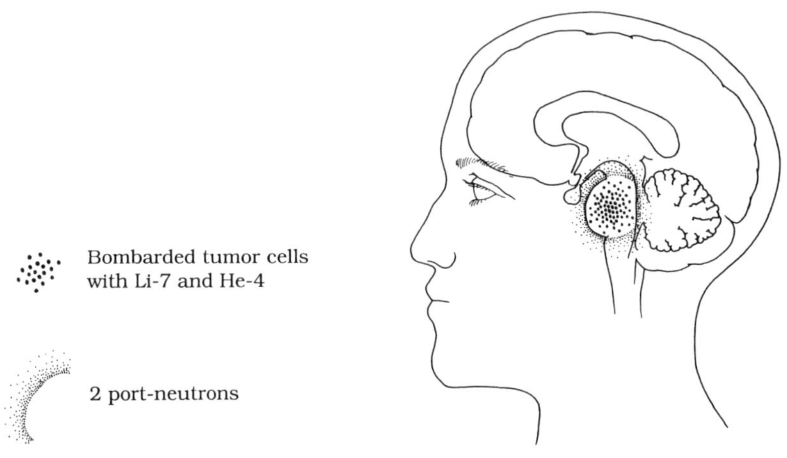

Figure 1. Pons tumor (astrocytoma infiltrating in the normal tissue)

Apart from glioblastoma treatment which seems to have reached the state of the art with the current boron compound and thermal beam facility, the following diseases are newly in sight as new dimensions of BNCT.

Multiple cerebral metastases from cancers can be easily controlled by a single radiation with BNCT. Two patients with cerebral metastases from lung cancer were treated by BNCT. In both, all the cerebral tumors became invisible on CT by a single irradiation of neutrons. None of them recurred, but the primary cancer in the lungs killed the patients. BNCT was so effective that their terminal-stage agony due to pulmonary exacerbation made the author doubt the value of such an effective treatment of cerebral metastases which could have given the patients a chance for substantial "euthanasia".

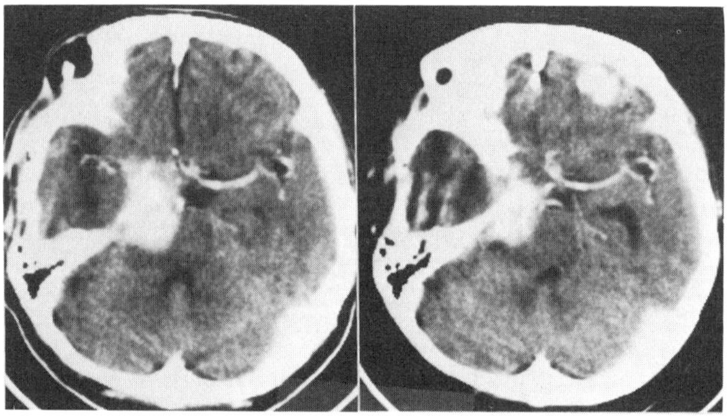

Figure 2. A case of cavernous sinus meningioma.
Two surgical attempts to remove the tumor had failed. Before neutron capture therapy the patient suffered hypesthesia in the left half of the lips, double vision, and bitemporal hemianopsia. The last was a postoperative sequela.
Left: before BNCT; Right: two years after BNCT. Double vision disappeared within 2 months after BNCT.

Although pathologically benign, some meningiomas are considered "clinically malignant" because meningiomas are radioresistant and some of them defy total extirpation, slowly incapacitating the patients over several years before death liberates them. A total of six such patients were treated. Four patients were treated more than 10 years ago. Each of them responded well to the treatment. Two are still alive, and in one of them the original tumor site appears hollow on CT. The other has not since visited the author. The last two patients who were treated in the past 2 years by BNCT had so-called skull-base meningiomas [3,4,5] which had been operated on twice unsuccessfully. "Skull-base surgery" is a fashionable topic in neurosurgery, but the reality is that quite a few of the operated patients suffer from postoperative neurological deficits while there is no guarantee of total extirpation. By BNCT, however, the pre-existing neurological symptoms due to the tumor have been slowly alleviated and the size of the tumors appear to be slowly shrinking on CT follow-ups. (Figure 2) Now that a case treated more than 10 years ago demonstrates a hollow cavity, it can be postulated that skull-base meningiomas, too, will be totally eliminated without any neurological sequelae if a technical problem of delivery of thermal neutrons to the lesion is solved. (Figure 3)

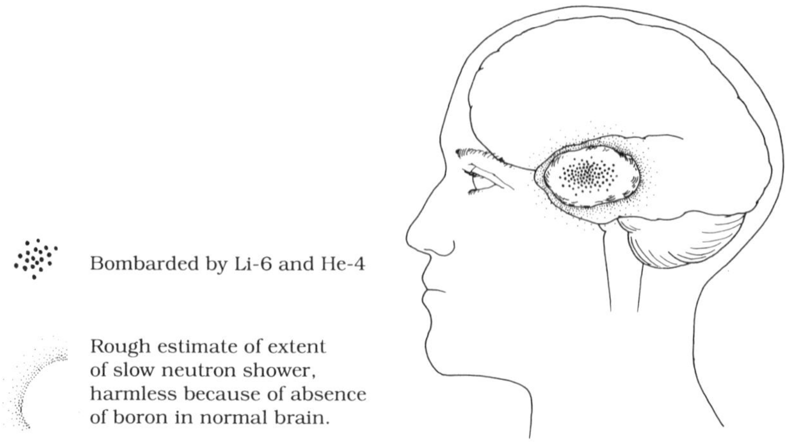

Figure 3. Cavernous sinus meningioma

Other benign intracranial tumors will be treated in the future, particularly <u>bilateral neurinoma (or neurofibroma)</u> where total extirpation will most probably make the patient totally deaf, or <u>craniopharyngioma</u> (Figure 4) which according to available statistics very few infant patients could survive until adulthood.[6,7] Even though some neurosurgeons such as Sweet, Symon, and Yasargil are achieving good surgical results, good operative results are still exceptional. (The author's result of 82% 10-year survival is also a good one.) Post-operative radiotherapy is advocated by some, but it has been controversial in view of the radiation hazards for physical and mental growth and due to persistence of neurological deficits. The author postulates that a highly specific BNCT of craniopharyngioma will be superior to other radiotherapies if boron-10 is instilled into the tumor cyst because the short range of heavy particles will not cause injury.

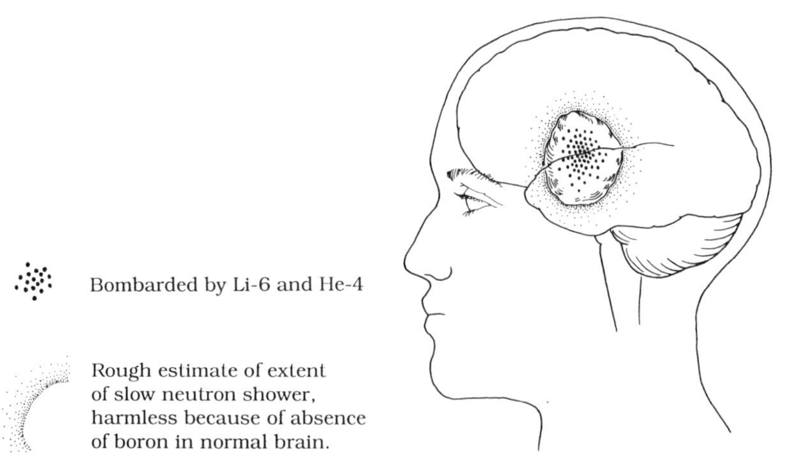

Figure 4. Craniopharyngioma (BNCT by instilling boron in the cyst)

Not only the above-mentioned intracranial tumors, but all other benign yet inexcisable tumors will be treated by BNCT in future — quite often by a direct infusion of boron into the tumor.

Cerebral arteriovenous malformation (AVM) is not a rare disease. According to autopsy studies, an incidence of 0.35% of the general population is reported (Jellinger).[8] It causes cerebral and subarachnoid hemorrhage, epilepsy, paresis, mental growth retardation, visual impairment, and in the worst case, cardiac insufficiency. As the separate table[9] indicates, it appears that the ratio of the operated to the diagnosed ranges from 50% to 85%. Even the author, who, according to the table, was apparently the most aggressive in surgical excision of AVM, has been unable to excise 10% of all the diagnosed AVMs he saw. It is the size but not the depth of an AVM that determines operability (Chou).[10] A direct operation on a giant AVM is not only hazardous, but also painfully exhausting to a neurosurgeon. The first BNCT on a giant AVM was attempted in 1991, and the interim result was quite encouraging. The second case was treated in 1992.

In conclusion, it seems that almost all the intracranial tumors, benign and malignant, will be the target of BNCT. Not only neoplasms but also vascular problems like AVM may respond to BNCT. All these recent findings make us quite excited.

REFERENCES

1. H. Hatanaka. "Boron-Neutron Capture Therapy for Tumors," H. Hatanaka ed., Nishimura, Niigata (1985).
2. H. Hatanaka, K. Sano, and H. Yasukouchi. Clinical results of boron-neutron capture therapy. in: "Progress in Neutron Capture Therapy for Cancer," B. Allen, D.E. Moore, and B.V. Harrington, ed., Plenum Press, New York (p.562) (1992).
3. M. Samii. Space occupying lesions of the middle skull base. in: "Surgery of the Skull Base," M. Samii and W. Dref ed., Springer, Berlin-Heidelberg (1989).
4. V.V. Dolenc. Cavernous sinus tumors. in: "Anatomy and Surgery of the Cavernous Sinus," Springer, Wien-New York (1989).
5. K.V. Wied and N. Eskinja. Meningioma and parasellar pituitary adenoma affecting the cavernous sinus: radical tumor extirpation? in: "Surgery of the Sellar and Paranasal Sinuses," M. Samii ed., Springer, Berlin-Heidelberg-New York-London-Paris-Tokyo (1991).
6. W.H. Sweet. The radical operation for removal of cranipharyngioma. ibid.
7. H. Hatanaka. Present and future of clinical surgery. Neurosurgery — in Japan and the world. Progress in neurosurgery in the past 50 years. Journal of the Japanese Society for Clinical Surgery, 50 (suppl.):61 (1989).
8. K. Jellinger. The morphology of centrally-situated angiomas. in: "Cerebral Angiomas," Pia, Gleave, Grote and Jierski ed., Springer, Berlin-Heidelberg-New York (1975).
9. H. Hatanaka, F. Tovarys, and S. Kamano. A case of arteriovenous malformation, grade IV, with two giant aneurysms in the dominant side speech and motor-sensory area — evaluation of grades, operability with microsurgical technique and pre- and post-operative performance status in 37 cases of cerebral vascular malformations. Neurological Surgery, 8:701 (1980).
10. S. Chou. Personal communication (1979).

THERMAL NEUTRON CAPTURE THERAPY FOR BRAIN STEM AND SKULL BASE TUMORS: HEAVY WATER AND MULTIPORTAL NEUTRON DELIVERY

Yoshinobu Nakagawa,[1] Teruyoshi Kageji,[1]
Hiroshi Hatanaka,[2] Kanji Mukai,[3] Katsuji Kitamura,[3]
Fumio Tsuji,[4] Keiji Kanda,[5] Tooru Kobayashi[5]

1 Department of Neurosurgery
National Kagawa Children's Hospital, Kagawa 765
2 Department of Neurosurgery, Teikyo University,
3 Department of Neurosurgery, Tokushima University,
4 Department of Anesthsiology, National Okayama Hospital
5 Research Reactor Institute, Kyoto University

INTRODUCTION

It is widely believed that thermal neutrons cannot ideally treat a patient with deep seated brain tumors because of the attenuation of neutron flux in the brain tissue (Barth et al., 1990 Hatanaka 1992). However, thermal neutron sourses are far more easily available than epithermal facilities. We have been trying to improve the efficacy of thermal neutron therapy by several technicalities. We report two cases treated by two portal irradiation and partial deuteration of body water (Hatanaka 1986).

CASE STUDIES

Case 1. The patient is a 44 year-old Caucasian male suffering left hemiparesis and dysarthria. A thalamic tumor was demonstrated on CT. Histological diagnosis after stereotactic biopsy was astrocytoma. Subsequently, he had a conventional radiotherapy. However, 11 months after the irradiation, he again developed left hemiparesis, and the recurrence of the tumor was evidenced by CT. Histologically glioblastoma was proved. He and his family

Advances in Neutron Capture Therapy, Edited by
A.H Soloway *et al.*, Plenum Press, New York, 1993

wanted BNCT and he was referred to us. On admission to the hospital, MRI showed a large mass lesion in the left temporo-parietal area (Fig. 1-a: MRI T-1 plain). The lesion was clearly enhanced by contrast medium, and infiltration of the tumor into the midbrain was also demonstrated (Fig. 1-b: MRI T-1 Gd+). MRI T-2 image showed remarkable peri-tumoral edema not only in the left hemisphere but also in the right hemisphere. The tumor was very large and infiltrated not only into the midbrain but also into the contralateral hemisphere. Then we decided to combine bilateral radiation and partial deuteration of body water. On the basis of the phantom experiment reported by one of the authors (T.K.) (1992), we calculated the dose of neutron fluence. The dose on the surface of the brain of the left side was 1.2×10^{13} and 1.8×10^{13} on the skull of the right side. The flux in the midline was estimated to be more than 9.3×10^{12}. The left parietal bone flap was removed and the neutron fluence of the left side was less than that of the right side. We measured boron content using gamma spectrometry at the reactor. Fig. 2 demonstrates the time course of boron content in blood and D_2O concentration. Just before irradiation, boron content in blood was 9.7 ppm and D_2O concentration was 12.3 %. During the irradiation heavy water in the form of physiological saline was continuously infused, and then D_2O concentration increased to 14.1% by the end of radiation. On the other hand, boron content in blood decreased 7.4 ppm. Before irradiation a gold wire was inserted into the cortex vertically near the brain tumor to measure neutron fluence. We calculated the half value layer by using the gold wire. It was about 2.5 cm. Ordinarily, the range of the half value layer which we adopt for BNCT is 1.7 to 2.0 cm. This significantly improved neutron penetration should be the result of partial deuteration and two portal irradiation. Postoperatively this patient had a good clinical course, and postoperative CT scan showed no remarkable change. However, in the 5th week, his condition suddenly deteriorated. CT scan indicated gross edema of the tumor causing the shift of the midline structure. It is important to note that the brain edema occurred only around the tumor and not in the opposite hemisphere (Fig. 3). His condition improved after treatment with prednisolone and he could go home. Unfortunately, debulking of the necrotic mass was not done, and he died 2 months after the BNCT.

Case 2 is a 48-year-old female. When the patient was referred to the authors, she was tetraparetic with total opthalmoplegia and auditory disturbance. Six years before, a meningioma at the base of the skull was found on CT and she underwent two craniotomies in an attempt to excise the tumor. She became unable to walk after then. MRI demonstrated a large tumor lying on the clivus which compressed the brain stem remarkably. This patient was treated by the same method as Case 1. The patient underwent a preliminary debulking of the tumor and a gold wire was placed in the tumor. We confirmed neutron fluence using this gold wire after BNCT. There was a significant difference between the attenuation rate in the cortex and that of experimental data using a phantom(Fig. 4). The neutron flux at the deepest point (on the midline) was only about 40% of the surface flux. Such an improvement of neutron penetration must be attributed to the partial deuteration and multiportal irradiation. The MRI taken 1 year after BNCT shows a lower profile of the tumor. Clinically, this patient has demonstrated a striking improvement of neurological findings and she can now walk with a stick.

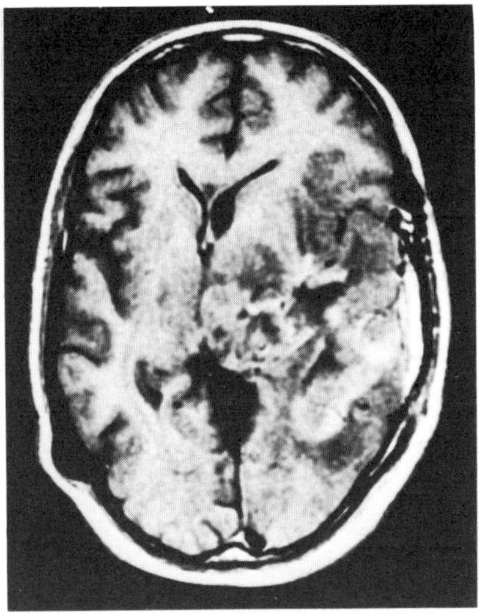

Figure 1 a. MRI T-1 plain.
A large mass lesion in the left temporo-parietal area was demonstrated.

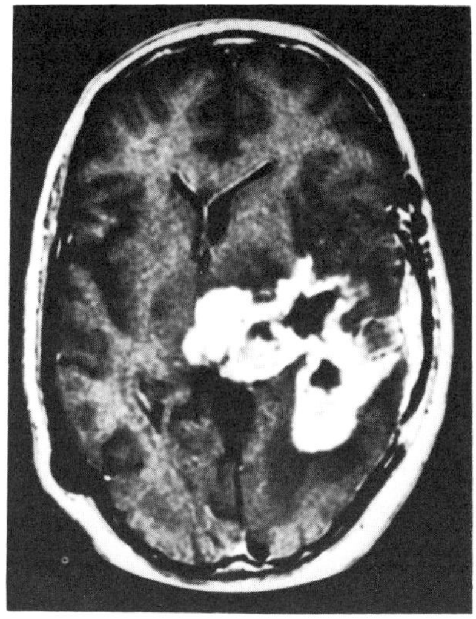

Figure 1 b. MRI T-1 Contrast-enhanced.
Note the clearly enhanced lesion and infiltration into the midbrain.

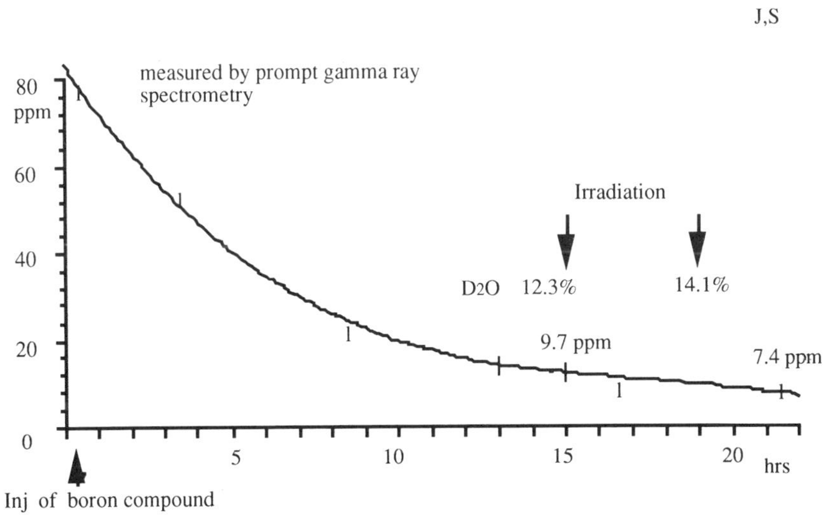

Figure 2. Content of boron in the blood

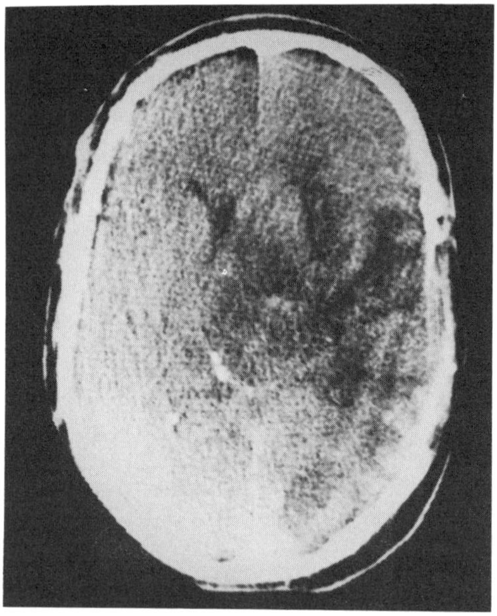

Figure 3. Follow up CT scan at 5 weeks after BNCT. Although a remarkable shift of the midline structure can be seen edema of the brain tissue is only around the tumor.

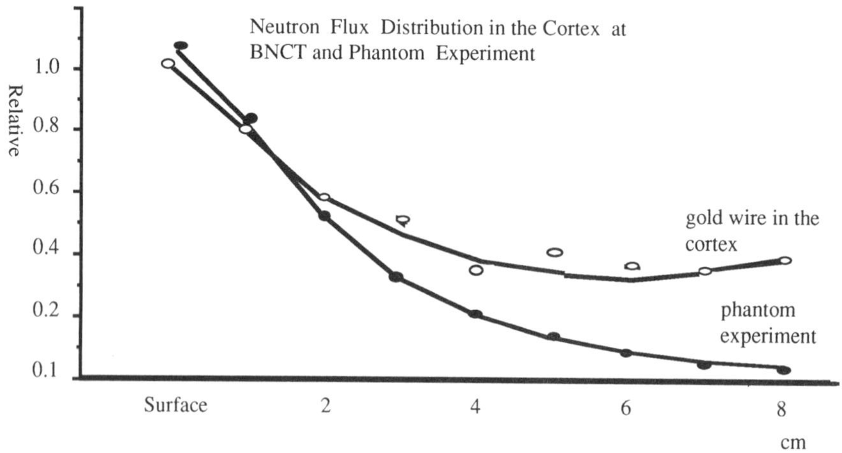

∗∗ Phantom experiment was done using single radiation in saline

Figure 4. Effect of deuteration and bilateral irradiation.

CONCLUSION

The combination of partial deuteration of body water and multiportal neutron irradiation will improve the efficacy of thermal neutron capture therapy for brain stem and skull base tumors

REFERENCES

Hatanaka. H., Boron- Neutron Capture Therapy for Tumors, in "Glioma" , Karim, A., & Laws, ER. ed. Springer-Verlag , Berlin, New York, p233-270 (1991).

Barth, R.F., Soloway, A.H., Fairchild, R.G., Boron neutron capture therapy for cancer, Scientific American Oct: 68-73 (1990).

Hatanaka. H., Introduction, in Boron-Neutron Capture Therapy forTumors. Hatanaka. H., ed. Nishimura, Niigata, P1-28 (1986).

A TREATMENT PROTOCOL FOR BNCT OF MALIGNANT GLIOMAS WITH BOROCAPTATE SODIUM (BSH) AND EPITHERMAL NEUTRONS

Heinz Fankhauser,[1] Patrick R. Gavin,[1,3] Axel Siefert,[2] and Giuseppe Stragliotto [1]

[1]Department of Neurosurgery, CHUV, Lausanne, Switzerland
[2]Commission of the European Communities, Joint Research Center, Petten, The Netherlands and Department of Radiotherapy, University of Göttingen, Germany
[3]On sabbatical leave from Washington State University, Pullman, WA

INTRODUCTION

The research program of the European Concerted Action on Boron Neutron Capture Therapy (BNCT) for Tumors is now approaching a clinical trial with epithermal neutrons and borocaptate sodium (BSH) for ^{10}B delivery. The therapeutic beam prepared with a liquid argon filter has become available for normal tissue tolerance studies in healthy dogs injected with BSH early in 1992. It is therefore timely to draft a protocol for the initial patient trial.

The protocol can be divided into two parts. The first part includes the description of all parameters excluding the BNCT itself, such as patient selection criteria, pre-radiation work-up, surgery, follow-up, evaluation of results, treatment modalities in case of complications and tumor recurrence. The second part comprises the description of the parameters of BNCT itself, which means dose and administration schedule for BSH, description of the setup at the neutron beam port, radiation time, and fractionation scheme.

Both aspects have been discussed at various meetings of the European Concerted Action. Aspects of patient selection, surgical treatment and follow-up have been the subject of a publication which is in press at the time of the this symposium.[1] It will not be included in the present report, but a short summary will be given thereafter. This paper will focus on aspects of radiation parameters.

The initial treatment protocol will be designed as a phase I normal tissue tolerance study for epithermal neutrons and BSH. The study design nevertheless includes the possibility to induce a tumor response, since eligible patients will actually be those which are likely to be included into future phase II and phase III studies. Follow-up will therefore have to include screening of normal tissues as well as tumor parameters. At this stage an evaluation of the tumor response is nevertheless not the goal, and a control group treated with conventional radiation is not included. Every effort will have to be made to differentiate between radiation damage and tumor recurrence or progression.

The study will be designed to determine the tolerance of normal human cranial tissues, especially brain and skin, to the epithermal neutron beam of the High Flux reactor at Petten. In order to be considered for the study, patients have to present radiologically with a likely malignant supratentorial glioma. They have to be scheduled for craniotomy for gross total resection. The tumor should be in a favorable location for reoperation in

Advances in Neutron Capture Therapy, Edited by
A.H Soloway *et al.*, Plenum Press, New York, 1993

case of recurrence. During the initial debulking procedure, a biodistribution study for BSH in tumor and normal tissues will be performed, but the results will have no influence on the decision to proceed with BNCT, nor on radiation parameters. Due consideration will nevertheless be given to the observed individual blood pharmacokinetics, since the blood boron level during BNCT is likely to be critical for normal tissue tolerance. Patients will be definitely included into the protocol after confirmation of the histological diagnosis of glioblastoma, anaplastic astrocytoma, anaplastic oligodendroglioma, or anaplastic ependymoma. The extent of resection will be confirmed by an early post-operative contrast MRI study. The patient has to be in a neurological condition sufficiently good to allow travel to Petten by public transportation. BNCT will take place within 3 weeks after surgery. Follow-up will take place every 6 weeks. Main follow-up parameters will be clinical examination and MRI, but due consideration will be given to other critical systems, especially pituitary and visual functions. In case of radiological evidence of tumor regrowth and/or radionecrosis, reoperation will be performed, unless strongly contraindicated. Further treatment after regrowth and/or radiation necrosis will be left with the decision of the neurosurgeon in charge, but any form of radiation is strongly advised against.

Provision will be made to obtain early consent for reoperation under the above conditions. Special attention will be given to the necessity of autopsy in the case of fatal outcome, since as a normal tissue tolerance study, histological demonstration of, and differentiation between recurrent tumor and radiation effects is mandatory.

PROTOCOL

Patients will be radiated in the epithermal neutron beam of the HF reactor at Petten from two lateral, opposite fields with equal exposure time. The treatment will be administered in 4 fractions at 1 day intervals. An equal amount of BSH will be administered intravenously 24 hours before each fraction. The duration of radiation in initial patients will be determined from a normal tissue tolerance study in healthy beagle dogs. This study has been started in early 1992. Its goal is to define the dose-response curve for asymptomatic radiation damage to the dog brain, namely alterations in the white matter, consisting in high signal abnormalities on T2 weighted MR images, and followed by abnormal contrast enhancement on T1 weighted images. This curve will be established for two target boron levels in blood, namely 25 and 50 ppm, obtained by intravenous infusion of BSH 24 hours before radiation in the definite, therapeutic epithermal neutron beam.

Table 1 lists the groups of dogs and number of animals in each group which are considered a minimum to obtain the desired information. By October 1992, the first 15 dogs had been radiated and 4 of them have so far developed clinical and MRI manifestations of cerebral radiation damage. All surviving dogs will be observed at least during 6 months before sacrifice and histological examination.

Table 1. Healthy tissue tolerance study in dogs.

Number of dogs	Blood boron concentration	Peak physical dose
10	no boron	6-7 Gy
11	25 ppm	13-15 Gy
10	25 ppm	13-15 Gy, 4 fractions
14	50 ppm	13-21 Gy

The dose-response curve will be established for a single fraction, but one group of dogs will get 4 BSH administrations and 4 fractions of radiation, in an identical manner as planned for humans.

Care will be taken to use an identical setup at the beam port for both, dogs and humans, including reactor power, filter, collimator, shielding and distance from the beam port.

As soon as sufficient data is available to define the dose-response curve, the radiation time corresponding to the approximate 5% toxicity level for the asymptomatic radiation changes will be estimated. This time will be diminished by 20% in order to obtain the total radiation time for the initial patients.

Dogs are radiated through a single portal from the vertex. The corresponding radiation time will be divided by 8 in order to obtain the time for a single fraction of radiation from one lateral field.

The precise exposure time of each fraction of radiation will be decided on-line after instant determination of the blood boron level at the start of each exposure by the prompt gamma method. The target blood boron level in humans is 25 to 50 ppm. The necessary dose of BSH to an individual patient likely to achieve this will be determined from the pharmacokinetics study currently under way in a large number of patients,[2] and from the result of the patient's own BSH study performed during the initial operation. Further adjustment of the dose may be necessary according to the blood boron level observed on the preceding day of radiation.

Dose escalation for radiation time in consecutive groups of patients will be done in steps of 10%. The dose will be increased at earliest 3 months after completion of the previous dose level. This means that at the time of radiation of each patient the observation interval for the 10% lower dose level is at least 3 months and for the 20% lower dose level at least 6 months.

Each exposure level will require 4 patients, in which the required blood boron concentrations have been reached.

The protocol will be terminated normally when a 100% response for asymptomatic MRI changes is achieved after 6 months follow-up or less. If 2 out of 10 patients die from radiation damage within 6 months after the treatment, the protocol will be interrupted prematurely and modified.

DISCUSSION

This paper is a complement to a previous report containing a proposal of principles on which the first clinical trial with epithermal neutrons for BNCT of brain tumors in the Petten reactor may be based.[1] It deals with the elaboration of the BSH administration schedule and radiation parameters.

BNCT always includes various dose components with unknown or controversial RBE values for each component. The concept of RBE itself has questionable validity in this context.[2] The microdosimetry is most relevant but unknown, mainly as a consequence of lack of knowledge of the subcellular microdistribution of boron in tumors and normal tissues. It becomes obvious that a classical, inductive approach to the determination of radiation parameters in initial patients is not feasible with BNCT, or at least not reliable. A consensus has therefore been reached inside the European Concerted Action that the safest way is to rely upon normal tissue tolerance determined in large animals with defined blood boron levels, exposed to the therapeutic epithermal neutron beam.

The dog is certainly not the optimal animal, mainly because of the small size of its brain and head, because of different tissue composition of the latter compared to humans, and because of the difference in radiation sensitivity of the skin of the two species. Beagle dogs nevertheless represent a reasonable compromise when considering feasibility and cost-effectiveness. Furthermore, the integration of valuable data already available from radiation of normal and tumor bearing dogs with epithermal neutrons at Brookhaven National Laboratory represents a significant safety factor.[3,4] For instance, the identification of the first detectable asymptomatic radiation damage in the normal brain, namely T2 MRI changes in the white matter, stem from these experiments.[3] The latter will be used in the proposed protocol. In the dog these changes can be expected within 6 months after radiation. Although asymptomatic, potentially transient, and not linked to definite delayed radiation damage, they are probably very close to symptomatic and severe radiation necrosis. This very steep dose-response curve is typical for particle radiation and has led us to propose a radiation dose escalation in steps of only 10% increase of radiation time.

Various safety factors have to be introduced when passing from dog to human BNCT experimentation, since the animal may be more tolerant. These safety factors in our proposal include 1) selection of a non-symptomatic endpoint for normal tissue tolerance, 2) selection of a 5% dose response, further diminished by 20% in radiation time for initial patients, 3) use of the final therapeutic beam setup in both experiments, including reactor power, filter, collimator, distance from portal, and room shielding, and 4) use, in humans, of a bilateral radiation portal and 4 fractions of radiation, by simply dividing the total radiation time without any further adjustment to take advantage of (yet unproven) increased normal tissue tolerance.

Bilateral portal certainly brings improved coverage of deeper parts of the brain with neutrons, with a lesser dose to superficial structures, but its influence on normal tissue tolerance is uncertain, since the radiated volume is larger. For our normal tissue tolerance study it is certainly advantageous, since it will allow observation of the normal hemisphere which will get the same exposure as the neoplastic brain.

Equally, the effect of fractionation is disputable. although it certainly brings an improved tolerance to the gamma, and possibly to the fast neutron component of the treatment, the clinical relevance of this improvement may be negligible due to the relatively small contribution of photons to the total physical dose in the Petten beam. Contaminating gamma rays and fast neutrons in BNCT may even play a role in tumor control, since they may reach infiltrating tumor cells which are excluded from boron uptake. Finally, there is concern, although probably unjustified,[4] about the influence of radiation fractions on the BBB and on the biodistribution of BSH during consecutive administrations. Furthermore, fractionation needs repeat BSH administration and this may lead to progressive accumulation of boron in the edematous peritumoral brain due to bulk flow and diffusion, with consecutively decreased normal tissue tolerance. This question has not yet been fully investigated, but this is now under way in patients.[5] Again, it has been decided that the most efficient and safest way to get information on normal tissue tolerance with fractionation is to include the intended fractionation scheme in the dog study. Unless a negative effect is demonstrated in this experiment, fractionation will be maintained, since this is considered an important factor for the acceptance of the protocol by standards of conventional radiotherapy. In the opposite case, we will have to work out whether a single high dose or a continuous low dose rate radiation will be the next choice.

Neither the ongoing normal tissue tolerance study in dogs, nor the ^{10}B-biodistribution study in humans[5] reproduces faithfully the situation in those patients which are eligible for the planned BNCT trial. As compared to intact brain in normal dogs, normal brain tolerance in humans may be altered by the preceding surgical trauma, which is known to induce contrast enhancement related to neovascularity, and probably to increase boron accumulation in this area. This should affect only brain in the immediate surroundings of the resection cavity. It nevertheless stresses the importance to study not only the normal hemisphere, but to obtain histological verification of the previous tumor site in all our patients, in order to differentiate between tumor recurrence or/and radionecrosis.

The aim of the proposed study is not only to get the best and most detailed information on normal tissue tolerance with BNCT, but also to get this information as quickly and with as few animals and patients as possible.

ACKNOWLEDGMENT

This work has been supported by the Fonds National de la Recherche Scientifique, the Ligue Suisse contre le Cancer, the Ligue Vaudoise contre la Cancer, the Zyma Foundation for the Advancement of Medical and Biological Sciences, and the Commission of the European Communities.

REFERENCES

1. H. Fankhauser, P.R. Gavin, A. Siefert, and G. Stragliotto, A treatment protocol for BNCT of malignant glioma with borocaptate sodium (BSH) and epithermal neutrons. *Strahlenther. Onkol.* (in press)

2. A. Wambersie, Relative biological effectiveness : a useful concept in BNCT?, *in:* "Neutron Capture Therapy," R.F. Barth, ed., Plenum Press, New York (this volume)

3. P.R. Gavin, A large animal model for BNCT, *in:* "Boron Neutron Capture Therapy : Towards clinical trials of glioma," R.F. Barth, ed., Plenum Press, New York (in press)

4. R. Huiskamp, P. Gavin, F. Wheeler, A. Siefert, and K. Philipp, Dose effect comparison between HFR and BMRR irradiated dogs with respect to healthy tissue tolerance, *in:* "Neutron Capture Therapy," R.F. Barth, ed., Plenum Press, New York (this volume)

5. G. Stragliotto, D. Schüpbach, P.R. Gavin, and H. Fankhauser, Update on biodistribution of borocaptate sodium (BSH) in patients with intracranial tumors, *in:* "Neutron Capture Therapy," R.F. Barth, ed., Plenum Press, New York (this volume)

EXPERIENCES OF BNCT USING THE KUR

Keiji Kanda, Tooru Kobayashi, Keizo Akuta, and Koji Ono

Research Reactor Institute, Kyoto University
Kumatori-cho, Sennan-gun, Osaka 590-04, Japan

INTRODUCTION

The Kyoto University Reactor (KUR, 5MW, Tank-Type) was constructed and attained the first criticality in June 1964. The Heavy Water Thermal Neutron Facility attached to the KUR has been remodeled for boron neutron capture therapy (BNCT)[(1)] and the first treatment was performed by Dr. Hatanaka in May 1974[(2)].

The KUR is the only research reactor of MW class owned by universities in Japan and opened to joint use programs for all university researchers. The facility is only the standard field of thermal neutrons approved by an IAEA Consultant Meeting[(3)], and is always very crowded with researchers.

All technology-related BNCT physics developed by KURRI was transferred to Musashi Institute of Technology and the reactor (MuITR, 100kW, TRIGA-II) has been remodeled for BNCT[(4)] and had been utilized for some hundred patients till the end of 1989, when there was trouble with water leakage.

Since 1990, the KUR has been utilized for BNCT but only six days per year because of conflict with other research programs.

FACILITY

The Heavy Water Thermal Neutron Facility is shown in Figure 1. The main characteristics are[(5)] :

Thermal neutron flux :	3×10^9n/cm^{2}sec
Gamma ray dose rate :	0.8Sv/hr
Cadmium ratio of gold :	~5000

Advances in Neutron Capture Therapy, Edited by
A.H Soloway *et al.*, Plenum Press, New York, 1993

BREAKDOWN OF TREATMENT

Since 1990, the following twenty four patients were treated by seven medical doctors.

Breakdown by tumor :	brain tumor 19, melanoma 5
Breakdown by nationality :	Japanese 17, German 3, Australian 3, American 1
Breakdown by Sex :	Male 13, Female 11
Average by age :	44.25 years old
Breakdown by chief medical doctor :	H. Hatanaka 7, Y. Nakagawa 5, Y. Oda 5, Miyatake 1, M. Takagaki 1, Y. Mishima 4, M. Ichihashi 1

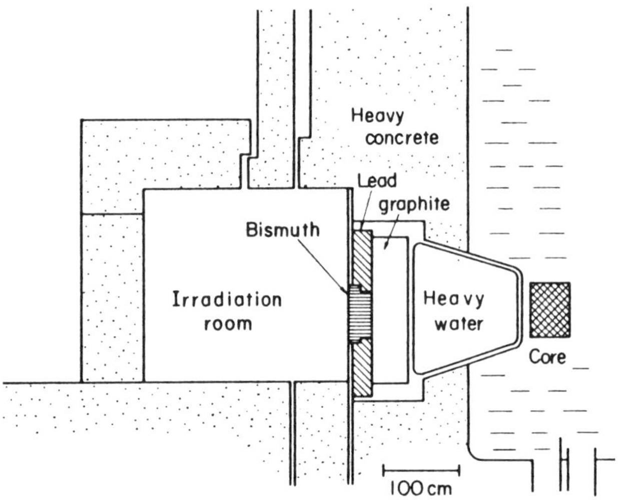

Figure 1. Heavy water facility of the KUR.

Table 1 shows a record of these treatments. As mentioned above, the KUR can be used for BNCT only six days per year. The irradiation days are announced to chief doctors several months ahead. Sometimes we have applications of five patients for one day, and then the Medical Committee selects patients to be treated. On July 13, 1992, for example, three patients were treated in a day. The time schedule is shown in Table 2.

Table 1. Record of Cancer Treatments Using KUR since February 1990
(As of July 13, 1992)

No.	Date	Age	Sex	Nationality	Cancer*	Neutron Fluence	MD**
1	Feb. 10, 1990	64	M	Japanese	G	1.67×10^{13}	H.H.
2	"	45	M	Japanese	G	1.97×10^{13}	H.H.
3	Feb. 13, 1990	61	M	Japanese	M	1.00×10^{13}	Y.M.
4	June 9, 1990	9	M	German	G	1.57×10^{13}	H.H.
5	Aug. 31, 1990	61	M	German	G	1.80×10^{13}	H.H.
6	Oct. 15, 1990	32	M	Japanese	G	1.80×10^{13}	Y.O.
7	Nov. 9, 1990	57	M	Japanese	M	1.35×10^{13}	Y.M.
8	Jan. 28, 1991	59	M	American	G	2.06×10^{13}	Y.N.
9	"	30	M	Japanese	G	1.85×10^{13}	Y.O.
10	June 10, 1991	65	F	Japanese	M	0.83×10^{13}	Y.M.
11	"	64	F	Japanese	G	2.17×10^{13}	S.M.
12	July 15, 1991	44	M	Australian	G	1.76×10^{13}(R) 0.92×10^{13}(L)	Y.N.
13	Sept.17, 1991	8	F	Japanese	G	1.48×10^{13}	Y.N.
14	"	61	M	German	G	1.66×10^{13}	H.H.
15	Sept.18, 1991	58	F	Japanese	G	1.98×10^{13}	Y.O.
16	Oct. 21, 1991	43	F	Japanese	G	1.90×10^{13}	M.T.
17	"	48	F	Japanese	G	1.90×10^{13}(R) 1.00×10^{13}(L)	H.H.
18	Nov. 11, 1991	75	M	Japanese	M	8.40×10^{12}(R) 5.80×10^{12}(L)	Y.M.
19	"	6	F	Australian	G	1.60×10^{13}	H.H.
20	June 9, 1992	36	M	Australian	G	1.52×10^{13}	Y.N.
21	"	42	F	Japanese	G	1.00×10^{13}	Y.O.
22	July 13, 1992	42	F	Japanese	G	1.88×10^{13}	Y.O.
23	"	1	F	Japanese	G	8.0×10^{12}	Y.N.
24	"	51	F	Japanese	M	8.8×10^{12}	M.I.

* G : glioblastoma or astrocytoma, M : melanoma

**

H.H. : H. Hatanaka (Teikyo University Hospital)	G	7 patients
Y.N. : Y. Nakagawa (National Kagawa Children's Hospital)	G	5
Y.O. : Y. Oda (Kyoto University)	G	5
S.M. : S. Miyatake (Kyoto University)	G	1
M.T. : M. Takagaki (Kyoto University)	G	1
Y.M. : Y. Mishima (Mishima Institute)	M	4
M.I. : M. Ichihashi (Kobe University Hospital)	M	1

Table 2. Schedule of BNCT on July 13, 1992

Hour	Patient 1 Glioblastoma	Patient 2 Glioblastoma	Patient 3 Melanoma
8	Arrival		
9	Meeting		
10	Operation		
11		Arrival	
12	Irradiation (11:14~13:35)	Meeting	
13		Operation	
14	Treatment		
15			
16	Leaving	Irradiation (15:19~18:31)	Arrival
17			Meeting
18			
19			
20		Treatment	
21			Irradiation (20:35~21:51)
22		Leaving	
23			Treatment
24			Leaving

REFERENCES

1. K. Kanda, T. Kobayashi, K. Ono, T. Sato, T. Shibata, Y. Ujeno, Y. Mishima, H. Hatanaka, and Y. Nishiwaki, "Elimination of Gamma Rays from a Thermal Neutron Field for Medical and Biological Irradiation Purposes," IAEA-SM-193/68, IAEA, Vienna, Austria, March 10-14, 1975, *Biomedical Dosimetry,* p.205.

2. H. Hatanaka, W.H. Sweet, "Slow Neutron Capture Therapy for Malignant Tumors. Its History and Recent Development," *ibid.,* p.147.

3. K. Kanda, K. Kobayashi, S. Okamoto, and T. Shibata, "Thermal Neutron Standard Field with a Maxwellian Distribution Using the KUR Heavy Water Facility," *Nucl. Instr. Meth.*, 148:535 (1978).

4. O. Aizawa, K. Kanda, T. Nozaki, and T. Matsumoto, "Remodeling and Dosimetry on the Neutron Irradiation Facility of the Musashi Institute of Technology Research Reactor for Boron Neutron Capture Therapy," *Nucl. Technol.*, 48:150 (1980).

5. T. Kobayashi, K. Kanda, Y. Ujeno, and M.R. Ishida, "Biomedical Irradiation System for Boron Neutron Capture Therapy at the Kyoto University," *in : Neutron Beam Design, Development, and Performance for Neutron Capture Therapy,* O.K. Harling J.A. Bernard and R.G. Zamenhof ed., Plenum Press, New York (1990).

CLINICAL EXPERIENCE WITH BNCT FOR MALIGNANT BRAIN TUMORS

Yoshifumi Oda,[1] Masao Takagaki,[1] Zhen Hua Zhang,[1]
Haruhiko Kikuchi,[1] Tooru Kobayashi,[2] Keiji Kanda[2]

[1]Department of Neurosurgery
Faculty of Medicine
Kyoto University Hospital
[2]Research Reactor Institute
Kyoto University
Sakyo-ku, Kyoto 606-01, Japan

INTRODUCTION

Twenty-four patients, 19 with brain tumors and 5 with malignant melanomas, have been treated at the Kyoto University Reactor by boron neutron capture therapy (BNCT) in the past two years. The purpose of this paper is to summarize the seven cases we treated and to examine the most suitable conditions for patients, boron compounds, and irradiation modification for successful BNCT.

MATERIALS AND METHODS

All our cases are clinically classified as malignant astrocytomas, grade 3 or 4. The treatment schedule is shown in Figure 1. Sodium borocaptate (BSH) for the boron compound was purchased from Boron Biologicals, Inc. and kindly supplied by Shionogi Pharmaceutical Company.

6:00	PREDONISOLONE 100mg
18:00	PREDONISOLONE 100mg
19:00 – 21:00	5mg/500ml OF BSH ($Na_2B_{12}H_{11}SH$) 100mg BSH/kg b.w. DRIP
NEXT DAY 13:00	START OF IRRADIATION (16HRS AFTER ADM. OF BSH)
	SURFACE NEUTRON FLUX TO NORMAL BRAIN: 18Gy

Figure 1. Schedule for administration of BSH.

Advances in Neutron Capture Therapy, Edited by
A.H Soloway *et al.*, Plenum Press, New York, 1993

A 100 mg/kg body weight of BSH containing 45 mg of purified boron-10 was administered about 17 hours before BNCT. Irradiation doses were regulated to be under 18 to 19 Gy on the surface of the normal brain. In this situation, tumor doses were estimated to be 20 Gy at a depth of 5 cm when the boron concentration in the tumor tissue was about 25 ppm.

The patient's skin was covered by ^6LiF to protect against the effect of the boron neutron capture reaction. Because of this procedure, none of our cases had skin trouble, even though they had multiple surgical procedures.

To improve the penetration of neutrons into the tissue, a Ping-Pong ball was placed into the dead space after tumor removal. A phantom study shows an 80% increase in neutron flux at 4 cm in depth and a 30% increase in gamma rays (Figure 2).
In most of the cases, gold wires were placed in a crosswise fashion around the dead space so as to just fit on the brain surface. We confirmed the actual flux in each case. The thermal neutron flux at the deepest position was always on the order of $1 \times 10^{12} n/cm^2/sec$.

RESULTS

The clinical results are shown in Table 1. Three patients died of the brain tumor and two of other causes. The last two cases have remained in good condition, though the observation period has been short. Tumors that extended to a depth greater than 5 cm or that disseminated intraventricularly showed a tendency toward early regrowth, and effective factors that improved the prognosis of the patients could not be found.

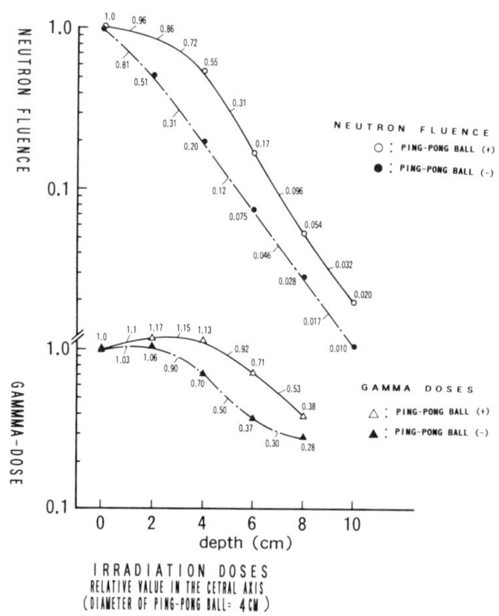

Figure 2. Irradiation Doses: relative value in the central axis (Ping-Pong ball dia. = 4 cm).

Table 1. The seven cases.

NAME	PRIMARY/ RECURR.	ASTRO GRADE	PERIOD UNTIL BNCT	REST OF TUMOR	LATERA-LITY	VENTR DISS	SITE	PROGN	DATE	CAUSE OF DEATH	RADIA-TION	CHEMO-TX	OTHER TX
1. C. H. 32Y MALE (BNCT90/10/15)	REC	3	55DAYS	8*8	BIL	(+)	FP	DEAD	67DY	TUMOR	60Gy	ACNU	TNF
2. Y. H. 30Y MALE (BNCT91/1/28)	REC	4	80	4*7	BIL	(+)	F	DEAD	179	TUMOR	(+)	?	
3. C. O. 64Y FEM. (BNCT91/6/10)	PR	4	55	5*5	BIL	(+)	FTP	DEAD	108	SEPSIS	(-) (+)*	CDDP, MTX VP-16	
4. N. K. 58Y FEM. (BNCT91/9/18)	PR	4	53	4*2	UNI	(-)	T(R)	DEAD	266	PNEUMONIA	(-)	CDDP, MTX VP-16	
5. Y. K. 43Y FEM. (BNCT91/10/21)	PR	4	54	0*0	UNI	(-)	F(L)	DEAD	208	TUMOR	(-)	ACNU	
6. N. S. 41Y MALE (BNCT92/6/9)	REC	4	252	0*0	UNI	(-)	F(L)	ALIVE	98+	PS=2	54Gy	ACNU, PVB	TNF
7. M. T. 42Y FEM. (BNCT92/7/13)	REC	4	42	3*3	UNI	(-)	TO	ALIVE	64+	PS=2	57Gy		IFN

* ADD. RADIATION AFTER BNCT

CASE PRESENTATIONS

Case 2 was a thirty-year-old male. He was operated on one month after the onset of focal seizure on the left side of his face and had radiochemotherapy. However, eight months later the left hemiparesis gradually increased (Figure 3). He was reoperated on, but the disease progressed.

This patient might not have been an optimal candidate for BNCT. The boron-10 concentration in the tissue was measured by prompt gamma spectroscopy. About 10 ppm of boron was confirmed in the serum, but only 6.8 pm in the tumor sample. The RBE doses in the tumor were over 27.8 Gy at 2 cm in depth, compared with 16.2 in normal brain tissue, though the doses at a depth of 5 cm were not sufficient to kill tumor cells. On the other hand, the tumor enhancement ratio (TER) to normal brain was 1.7 and 1.4 at 2 and 5 cm in depth, respectively. For four months after BNCT, the patient's condition was stable, but it then deteriorated. He expired 179 days after the treatment.

Case 7 was a forty-two-year-old female with right hemiparesis and motor dysphasia. A huge tumor located in the left temporo-occipital area was partially removed, and radiochemotherapy was performed. However, even with the radiotherapy, the symptoms progressed during the three months after the primary surgery, and our hospital was consulted (Figure 4).

After decompression by partial removal of the tumor and lobectomy, the mass effect slightly improved before BNCT. With the use of a slow-injection method, the boron concentration in the serum was higher than in the previous case, though the value in the tumor was 10 ppm.

The RBE doses at a depth of 2 cm reached 27.4 Gy, but only 8.1 Gy at 5 cm (Figure 5). TER is also shown, which is 1.5 to 2.0 (Figure 6). As in case 2, we could get good TER, but the absolute tumor dose was not sufficient.

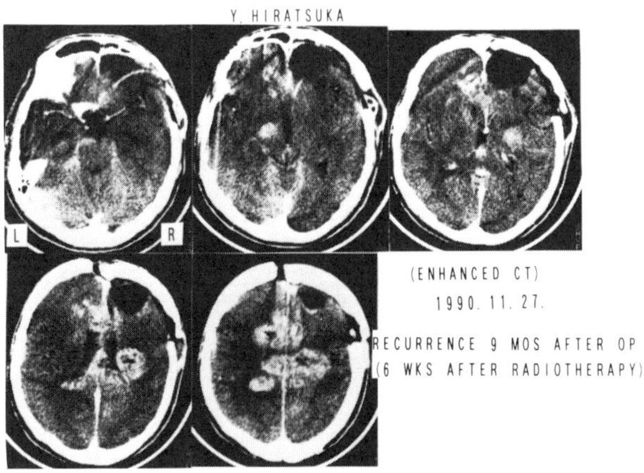

Figure 3. CT scan of case 2 on recurrence.

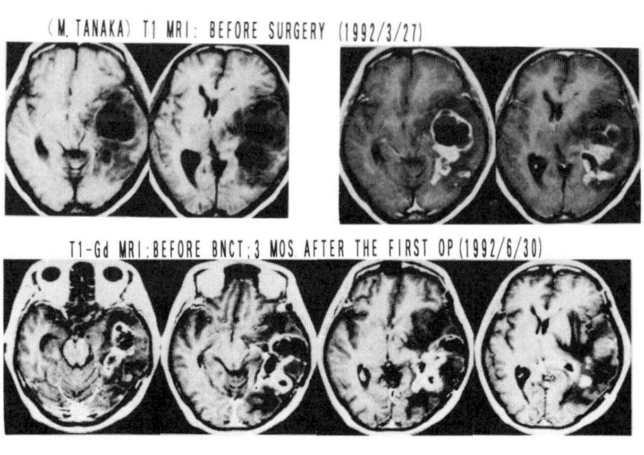

Figure 4. Pre BNCT MRI of case 7.

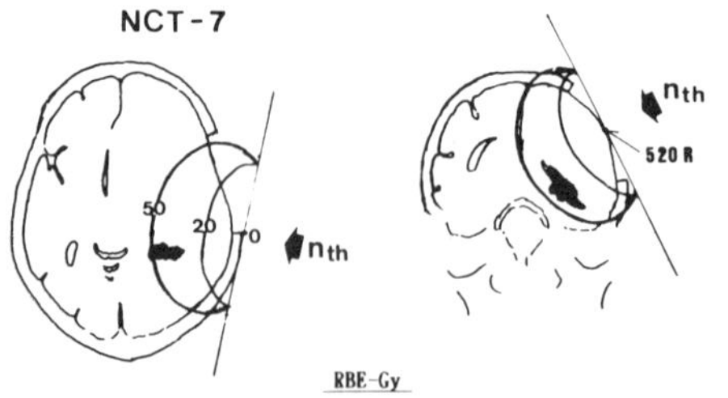

RBE-Gy

DEPTH(mm)	0	20	50
Brain	21.1	13.2	5.2
Tumor	—	27.4	8.1

Figure 5. Schematic drawing of BNCT.

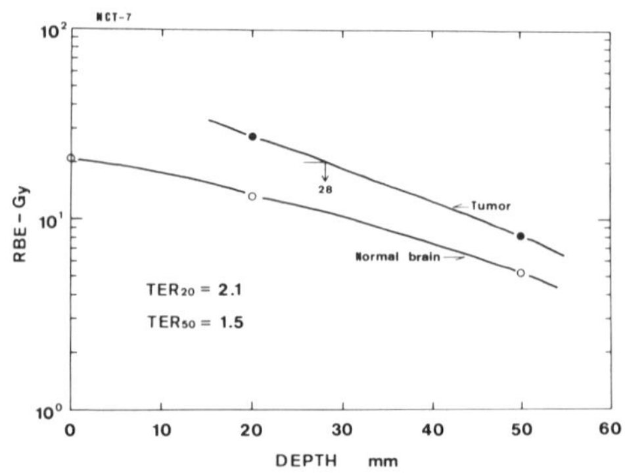

Figure 6. RBE doses of case 7.

After BNCT the patient's neurological signs improved, and the right hemiparesis disappeared. She has had only alexia without agraphia for three months after BNCT. The MRI also showed marked improvement (Figure 7).

Case 5 caused much discussion. The patient was forty-three years old and initially had a convulsive seizure. A CT scan showed a right frontal tumor, which was almost totally removed in a second operation (Figure 8). Boron-10 concentration in the tumor was 5 ppm but we thought

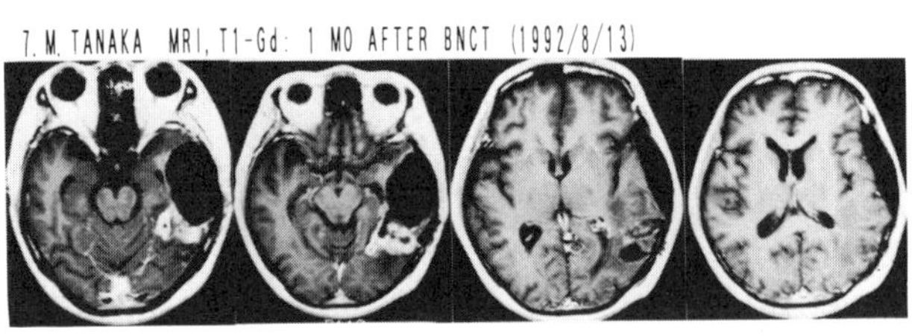

Figure 7. Post BNCT MRI of case 7.

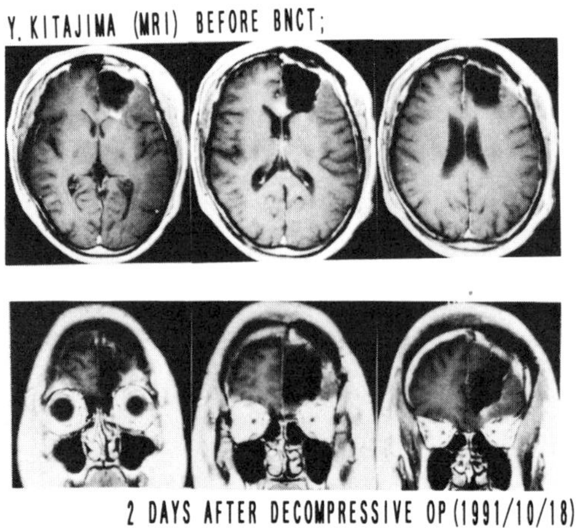

Figure 8. Pre-BNCT MRI of case 5.

that the rest of the tumor was very small, and a part of the resected tumor contained normal brain tissue. The radiation field was 9 x 13 cm^2, and thermal neutron fluence was adequate. The simulation curve showed that tumor doses were 10.3 to 13.3 Gy at 5 cm in depth. TER was 2.1 at 5 cm, although the 20 Gy ("tumor therapeutic dose") zone was at 4.1 cm in depth.

Although the ventricle was opened in the second operation, it was sealed with oxycellulose and fibrin glue. The patient's pretreatment condition was not very bad, and the waiting period until BNCT (54 days) was not as long as in some of the other cases (see Table 1). We thus thought it reasonable to expect a complete cure by BNCT in case 5; but four months later, CT and MRI studies (Figure 9) revealed that the tumor had disseminated not only into the ventricle but also into the contralateral frontal lobe through the corpus callosum. Having refused extensive chemotherapy, the patient died 208 days after BNCT. At the autopsy it was found that most of her brain had been invaded by active tumor cells. The radionecrotic area was barely discernable.

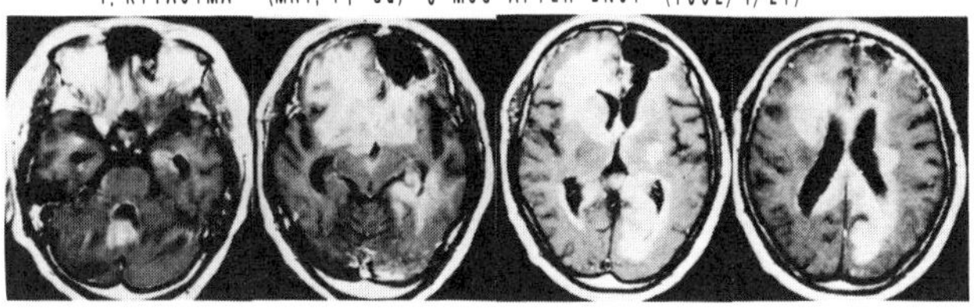

Figure 9. Case 5, recurrent stage.

CONCLUSION

The penetration of thermal neutrons into the tissue is enhanced by placing a Ping-Pong ball into the dead space. And a moderate surrounding effect of the thermal neutrons can be expected.

The neutron flux is sufficient at 5 cm in depth from the surface. The boron-10 concentrations in tumor tissues of our seven cases showed relatively low values, lower than 10 ppm. This might be one of the reasons for our poor results in comparison with Hatanaka's encouraging reports.[1]

We could not measure the concentration of boron-10 in virgin tumor tissues because we could not get a sufficient a supply of BSH. It will be necessary to measure the tissue concentration of BSH in virgin tumors, and in nontumor brain tissue if possible.

In our experience the uptake of BSH into brain tumors was not as high as shown by other authors.[1] This point indicates the need to study more-effective boron compounds.[2] In addition, the efficacy of gadolinium compounds for neutron capture therapy[3] and the ability of phenobarbital to protect normal brain from radiation damage are now being tested in other projects.

ACKNOWLEDGMENTS

This study was supported by grants-in-aid from the Ministry of Education and Culture, the Ministry of Health and Welfare of Japan, and the Kyoto University Foundation.

REFERENCES

1. H. Hatanaka, "Neutron Capture Therapy," Nishimura Publishing Company, Niigata (1986).
2. Y. Yamamoto, T. Seko, H. Nakamura, H. Nemoto, et al., Synthesis of carboranes containing nucleoside bases: Unexpectedly high cytostatic and cytocidal toxicity toward cancer cells, *J Chem Soc Chem Comm.* 157-58 (1991).
3. M. Takagaki, Y. Oda, H.Kikuchi, T. Kobayashi, and K. Kanda, Gadolinium neutron capture therapy for brain tumors: Biological aspects, *in:* "Biological Aspects of Brain Tumors," K. Tabuchi, ed., Springer Verlag (1991).

CURRENT STANDARD BORON NEUTRON CAPTURE THERAPY WITH SHIONOGI'S MERCAPTOUNDECAHYDRODODECABORATE FOR MALIGNANT GLIOMA

H. Hatanaka,[1] J.H. Goodman,[2] and M.E. Miner[2]

[1] Department of Neurosurgery, Teikyo University
Kaga, Itabashi-ku, Tokyo 173, Japan

[2] Department of Neurosurgery, Ohio State University
Columbus, Ohio 43210

INTRODUCTION

The authors reviewed malignant glioma patients who have lived longer than 10-20 years after current standard boron-neutron capture therapy(BNCT).

METHOD

A total of 37 patients with cerebral gliomas (mostly grade III-IV) were treated by one of the authors (H.H.) between August 1968 and April 1982, 10 years before the date of the present study. Of the thirty-seven, sixteen had been treated by conventional radiotherapy. The method of BNCT was consistent throughout this period and until March 1985.[1] Before March 1985, all the arriving glioma patients were treated uniformly by boron-neutron capture therapy by H.H. unless some patients with gliomas had to be treated by conventional therapies either because a reactor was unavailable or because the boron compound supply was short. (Since March 1985 the history of past radiotherapy and the depth of the tumor have been taken into consideration when selecting patients for BNCT, and the treatment method also has changed. Boron-10 carrier used for this clinical series was sodium monomercaptoundecahydro-closo-dodecaborate made from 95 - 97% enriched boron-10 supplied by the Shionogi Research Laboratories.

RESULTS

Survival here means the elapsed time after the patient's first surgery. Six glioma patients lived or have lived longer than 10 years, out of the 37 cerebral glioma patients operated on before May 1982 (16.2%). Out of these 37 patients, sixteen had been treated by conventional radiotherapy.

Among the six survivors, two died at 17 and 12 years, but without evidence of recurrence of their tumors. The other four are still alive. Out of these six 10-year-plus survivors, three had been previously treated by conventional external radiotherapy and they developed radiation damage which

Advances in Neutron Capture Therapy, Edited by
A.H Soloway *et al.*, Plenum Press, New York, 1993

incapacitated all three and led ultimately to the death of two of them. The remaining three, who had no previous radiation history, have been active and have no evidence of tumors.

1) Case M.T.

A 50-year-old male. Glioblastoma (grade IV glioma). On June 20, 1972 a left frontal craniotomy was performed, and 20 grams of tumor was excised. The residual tumor was irradiated by BNCT on June 29. Preoperative symptoms (aphasia and right hemiparesis) disappeared. After discharge from the hospital, the patient has been active as a farmer for 20 years. In the meantime, he has retained his driving license. His ailments in the 20 years since treatment were a left-side cataract probably due to the scattered radiation, and back pain due to an old ruptured disc. Because the bone flap had been discarded, the skin flap covering the defect developed a circulation disturbance in the center which required repair by a plastic surgeon 15 years after BNCT. Otherwise, for the past 20 years the patient has remained free from any neurological problem or the radiation dementia which is common among patients who undergo conventional radiotherapy. There is no sign of tumor residue on follow-up CTs.
(Max. surface fluence: 1.0×10^{13}, boron-10 dose: 40 mg ^{10}B/kg, T/B ratio: 0.56, boron-10 in tumor: 15.3 μg ^{10}B/g)

2) Case R.N.

A 60-year-old female. Glioblastoma grade IV, or astrocytoma grade III according to different pathologists. On July 5, 1977 she underwent a right parietal craniotomy and biopsy of the tumor. BNCT was conducted on July 19. Ever since she has been in excellent condition. She has held one or two part-time jobs, one as a cleaning woman, the other as a dishwasher in a laboratory, or an animal farm assistant. More recently since she became 70 years old, she is working as a messenger between a photo-shop and the author H.H.'s office (3/4 hour by bus) 3 times a week.
(Surface fluence: 1.3×10^{13}, boron-10 dose: 41, T/B: 1.05, boron-10 in tumor: 14.0)

3) Case K.N.

A 25-year-old male. Initially glioblastoma (later identified as cellular ependymoma by other pathologists). The patient underwent a right frontal craniotomy at a university hospital. The diagnosis there was glioblastoma. (The specimen was later examined by Prof. K. Jellinger of Vienna and Prof. Y. Urano of Tokyo, who independently gave opinions that it was cellular ependymoma, grade II glioma; one of the authors [H.H.] also prefers cellular ependymoma to glioblastoma.) Soon after a conventional radiation of 750 cGy the patient was transferred and underwent BNCT on February 16, 1979. He worked as a post office employee for at least 3 years. Then he began to suffer frequent epileptic seizures and became left-hemiparetic. He had to resign from his job after 5 years. Seven years after BNCT, his vision deteriorated abruptly. A craniotomy revealed a large necrotic mass which turned out to be necrotic tissue of the tumor and the brain matter. After debridement the frequency of the seizures diminished. After 10 years he remained mostly in bed. He passed away quietly 12 years after BNCT.
(Surface fluence: 1.6×10^{13}, boron-10 dose: 44, T/B: 1.15, boron-10 in tumor: 35.2)(The first craniotomy : January 8, 1979)

4) Case T.M.

An 11-year-old female. Oligoastrocytoma, grade III. On September 18, a right frontal craniotomy and a partial removal of the tumor were performed. BNCT was conducted on October 19. She grew up mentally and physically. She graduated from a technical school after high school and got a job as a draftswoman at a civil engineering company. By now she has a huge cavity of 7 cm x 6 cm in the right frontal area on CT without any contrast en-

hanced tissue, in comparison to the initial tumor of 9 cm in diameter with a multicystic tumor content.
(Surface fluence: 1.3 x 10^{13}, boron-10 dose: 46, T/B: 1.43, boron-10 in tumor: 11.2)

5) Case Y.T.

A 35-year-old female. Glioblastoma (recurrence of grade II astrocytoma). This patient had been treated with a full-course of whole-brain radiotherapy for astrocytoma, grade II at another hospital. First craniotomy: September 24, 1981. This patient was treated by BNCT on March 20, 1986 for recurrence. By this time the histology was glioblastoma. The patient was demented due to the past radiation and was unable to leave the bed. Although only temporarily, after BNCT she could enjoy some improvement — to the extent that she could walk without assistance or cook food at home; soon, however, she became bedridden again. She then underwent repeated operations (necrotomy) at a local hospital and has been almost constantly hospitalized. She is alive 7 years after BNCT and 10 years after the initial treatment.
(Surface fluence: 1.3 x 10^{13}, boron-10 dose: 50, T/B: 1.42, boron-10 in tumor: 13.9)

6) Case R.K.

A 46-year-old male. Glioblastoma (recurrence of oligodendroglioma). Because of an oligodendroglioma in the left frontal lobe, he was treated by the first craniotomy (December 25, 1972) and with 3000 cGy to the whole brain together with a radiosensitizing drug. He enjoyed an almost full recovery except for a mild speech problem. Then 10 years later he came back with a recurrent tumor and with dementia. This time the histology was glioblastoma. BNCT was conducted (April 25, 1986). He improved again but not fully. A year later he had frequent seizures and became bedridden by the 3rd post-BNCT year. He died 17 years after the initial treatment and 7 years after BNCT.
(Surface fluence: 1.6 x 10^{13}, boron-10 dose: 40, T/B: 1.89, boron-10 in tumor: 21.6)

DISCUSSION

It is interesting that in one case the tumor to blood ratio (T/B) was not higher than one. But the patient (Case M.T.) apparently did not suffer any radiation damage. The neurological function (speech and motor function) improved after BNCT rather than deteriorated, although the tumor was in the left posterior frontal motor cortex (motor speech area).

Boron concentrations obtained from biopsy specimens do not necessarily represent the tumor uptake of boron because it is impossible to measure the uptake in all parts of the tumor. The following analytical result depends on such incomplete samples.

The author (H.H.) had reported an average of 26.5 $\pm$ 3.4 μg ^{10}B/g of tumor for the therapy hours (12-24 hours after boron-10 infusion, average 17.5 $\pm$ 0.3 hrs).[2] This value obtained after intraarterial infusion may be significantly higher than the values obtained by intravenous infusion; also, H.H.'s boron analysis data were obtained at the time of actual BNCT, a week or two after the first tumor excision, which should have improved the blood supply to the tumor because of the decompressive effect of the surgery.

Steroid administration should not have caused any spread in boron uptake among patients, because in Hatanaka's series all patients are unanimously given a large dose of steroid (prednisolone) before and after surgery.

Boron concentration in blood can be influenced by age, as children excrete boron promptly. The average in Hatanaka's series was 12.7 $\pm$ 0.8 μg ^{10}B/g of blood.

Beside the 6 patients with gliomas, 3 patients with other radioresistant tumors have been alive and well for more than 10 years. (A chondrosarcoma patient 15 years, a meningioma patient 14 years, and another meningioma patient also 14 years.)

It is clear from the present review of these long-term survivors that neutron fluence of less than 1.5×10^{13} at the cortical surface is well-tolerated by human brains under the current standard regimen of BNCT. However, the presently available patients' data do not yet allow us to pin-point the maximum tolerable neutron fluence level. There is no doubt that the previously radiated patients' brains could not tolerate BNCT, even though some of these patients certainly lived longer than 10 years.

CONCLUSION

BNCT is well-tolerated by patients' brains and did not affect a child's mental or physical growth. The current standard technique of BNCT is capable of curing glioblastoma if properly applied.

REFERENCES

1. H. Hatanaka, S. Kamano, K. Amano, K. Sano, S. Egawa, and H. Yasukouchi. Chapter XXV. Clinical experience of boron-neutron capture therapy for gliomas — A comparison with conventional chemo-immuno-radiotherapy. in: "Boron-Neutron Capture Therapy for Tumors," H. Hatanaka ed., Nishimura, Niigata (1986).
2. H. Hatanaka, K. Amano, H. Kanemitsu, I. Ikeuchi, and T. Yoshizaki. Chapter V. Boron uptake by human brain tumors and quality control of boron compounds. ibid.

ENDOVASCULAR BORON NEUTRON CAPTURE RADIOTHERAPY FOR GIANT ARTERIOVENOUS MALFORMATIONS

Hiroshi Hatanaka[1], Yoshinobu Nakagawa[2], and Seigo Sakaki[3]

[1]Dept. of Neurosurgery, Teikyo University, Tokyo 173
[2]Dept. of Neurosurgery, National Kagawa Children's Hospital, Zentsuji 765
[3]Dept. of Neurosurgery, Kagoshima University, Kagoshima 890

Cerebral arteriovenous malformation (AVM) is not a rare disease. According to autopsy studies, even as much as 0.35% of the general population is reported to have this disease (Jellinger).[1] It causes cerebral and subarachnoid hemorrhage, epilepsy, paresis, mental growth retardation, visual impairment, and in the worst case, cardiac insufficiency.

As Table 1 indicates, it appears that the ratio of those operated on to those diagnosed ranges between 50% and 85%.[2] Even the author (H.H.), who was apparently the most aggressive in surgical excision of AVM, has been unable to excise 10% out of all referred AVMs. It is the size but not the depth of an AVM that determines operability (Chou).[3] A direct operation on a giant AVM is not only hazardous, but also painfully exhausting to a neurosurgeon.

Radiosurgery with Lexell-Larsson's[4] gamma-knife is limited to small AVMs of 2-3 cm in diameter, to avoid brain damage.[5] Radiosurgery with a cyclotron cannot give radical dose of protons to a large AVM also for the same reason. Only 20% of the proton-beam-treated AVMs are reported to have become invisible on angiograms (according to Kjellberg).[6,7]

We treated a case with a large AVM (7 x 5 x 3 cm) with endovascular boron-neutron capture therapy (BNCT) for the first time in June 1991. (Figure 1) Lest the AVM should bleed with a powerful single radiation dose to the vascular wall of the AVM, it was planned to deliver a relatively small dose (less than 10 Gy, 1/3 of the dose desired by the authors) three times at 6-12 month intervals. The interval is considered necessary in order to wait for the thrombotic obliterating process in the AVM vessels to be completed. (Rupture of excessively BNCT-radiated tumor blood vessels has been encountered by one of the authors [H.H.] with a meningioma, a glioblastoma, and a primitive neuroectodermal tumor which were treated with an intensive dose of neutron capture therapy. Two of these cases resulted in the death of the patients after massive bleeding within the tumors.)

After an intravenous drip of sodium ^{10}B-mercaptoundecahydrododecaborate (1.7 gram) on the night before neutron capture therapy, the first course of treatment was conducted. Within 5 days the contrasted CT evidenced a slight fading of the contrast-visualization of the AVM nidus. (Figure 2)

Another CT scan obtained a month later showed even more reduction of contrast-enhancement of the AVM nidus. The feeding arteries or draining veins appeared to be slightly reduced in diameter. Then the scan made a

Table 1. Operability, Excisability and Results of Arteriovenous Malformation

Main operator	Operated cases in diagnosed cases	Total extirpation in operations	Operative mortality	Post-op morbidity
Olivecrona	125/ ?	81/125 (64.8%)	7.2%	40.0%
Tönnis	?	56/134 (41.8%)	8.9%	33.0%
McKissock	36/78 (46.2%)	?	2.8%	22.2%
Krenchel	?	79/ ?	8.9%	44.3%
Mattson	20/34 (58.8%)	12/20 (60.0%)	20.0%	30.0%
French	42/ ?	?	7.0%	16.0%
Kunc	68/117 (58.1%)	57/68 (83.8%)	11.8%	32.4%
Yasargil	45/ ?	?	4.4%	17.8%
Morello	102/154 (66.2%)	88/102 (86.3%)	7.8%	18.7%
Sano	>165/205 (>80.5%)	?	4.7%	21.8%
Handa	78/129 (60.5%)	34/78 (43.6%)	2.6%	20.5%
Kitamura	87/121 (71.9%)	78/121 (64.5%)	1.1%	24.1%
Suzuki	133/155 (85.8%)	111/133 (83.5%)	3.8%	26.3%
Drake	50/ ?	35/50 (70.0%)	2.0%	12.8%
Stein	21/28 (75.0%)	?/21	4.8%	33.3%
Trumpy	42/89 (47.2%)	25/42 (59.5%)	9.5%	23.8%
Kawabuchi	51/90 (56.7%)	42/51 (82.4%)	0.0%	9.8%
Sundt	19/37 (51.4%)	?/19	0.0%	26.3%
Wilson	83/ ?	65/83 (78.3%)	9.8%	43.3%
Hatanaka	33/38 (86.8%)	29/33 (87.9%)	0.0%	9.1%

(1992)

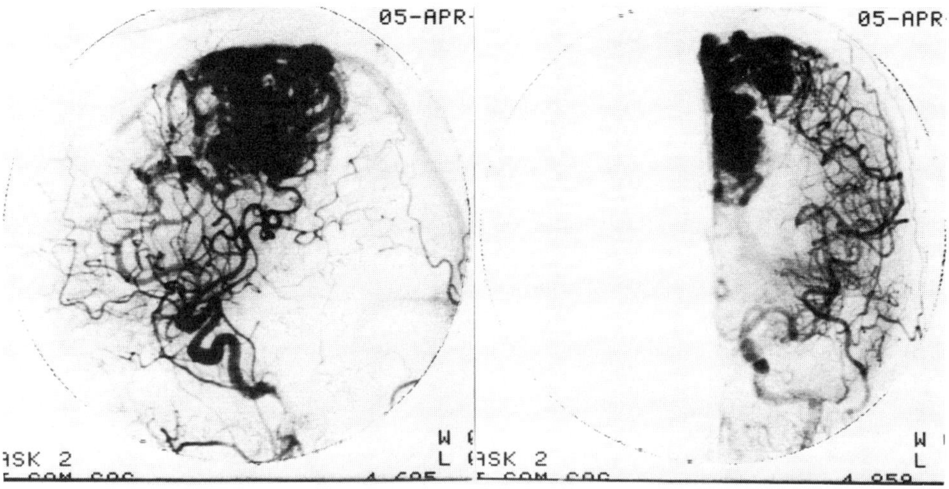

Figure 1. Angiograms before BNCT. A large AVM is visualized in the parietal area of the brain.

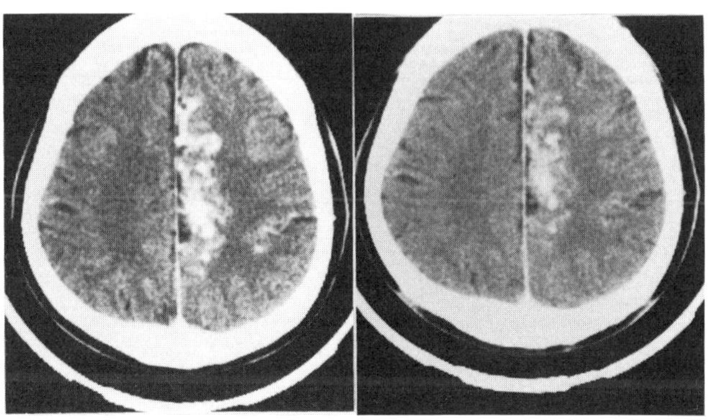

Figure 2. Contrast-enhanced CT before and after the first course of planned BNCT. Although three courses had been planned, the patient's headache was ameliorated within days after the first course.

few months later confirmed the partially occluded nidus. Several CTs obtained at various intervals after the first course of BNCT gave an impression that the thrombotic occlusive process had reached a plateau, although the 2nd course had been planned for September 1992. Within days after the first course of BNCT the patient's headache was undoubtedly ameliorated, and he started working within 2 months as a garbage truck worker.

The rationale for BNCT as a selective and effective treatment of AVM is not yet known. One explanation is geometry.[8,9] Only a fraction of radiations of the $^{10}B(n,\alpha)^{7}Li$ reactions that take place in the circulating blood inside the normal cerebral vascular lumen during brain exposure to thermal neutrons is normally absorbed by the small vascular wall, for geometrical reasons. According to Kitao's or Rydin's calculations, only one-third is absorbed by the small vascular wall because such secondary particles travel in all directions and thus affect the surrounding vascular wall with a probability of approximately one-third. On the other hand, all the secondary radiations produced inside the AVM vessel lumen will be absorbed, because the AVM nidus, particularly its A-V fistula area, is a conglomerate of fine vessels which are conjoined together and which will absorb most of the heavy particles without allowing them to be lost into the normal brain substance. There is a second possibility. An AVM may be made of premature tissues. The endothelium of AVM vessels may take up boron-10 more readily than the more mature cerebral blood vessels. (Figure 3)

As to the rationale of the treatment, nobody knows the answer. As it is impossible to reproduce this disease in animal experiments, and as we do not know of any strain of animal which develops this disease, it is difficult to elucidate the mechanism of the apparently selective effect of BNCT on AVMs. If we present an example based on the geometry theory, an AVM receives 1187 cGy at 2 cm below the surface, while the surrounding brain receives only 684 cGy because boron-10 content in the brain matter is practically zero, and the normal cerebral vessel endothelium will receive only 852 cGy. (Usually the arteriovenous fistulae are 1-2 cm below the brain sur-

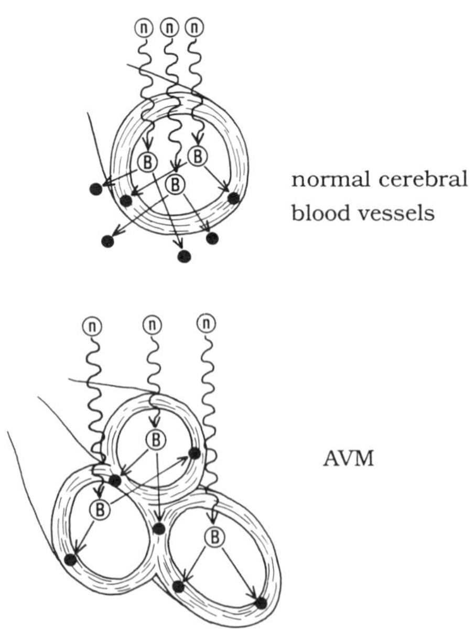

Figure 3. Schematic drawing of geometrical difference in vascular structure.

face, according to the author H.H.'s operative experience on 49 malformations.) If an RBE of 3.7 is considered, these values will be even more favorable for more serious damage to AVM and for preservation of normal vessels (2549, 1306, 684 RBE dose for AVM, normal vessel, and brain substance). Tolerance of the normal brain vessel is easily understood by these simple calculations. The mechanism of the effect of BNCT on AVMs will be difficult to clarify, as it involves vasospasms and thrombosis of blood vessels, which themselves are full of enigmas.

The present problem in developing a most effective BNCT for AVM is how to decide optimal doses.

The application of BNCT to AVM is a very unusual use of the therapy, since in BNCT for tumors it is important to protect cerebral blood vessels.[10,11]

REFERENCES

1. K. Jellinger. The morphology of centrally-situated angiomas. in:"Cerebral Angiomas," Pia, Gleave, Grote and Jierski ed., Springer, Berlin-Heidelberg-New York (1975).
2. H. Hatanaka, F. Tovarys, and S. Kamano. A case of arteriovenous malformation, grade IV, with two giant aneurysms in the dominant side speech and motor-sensory area — evaluation of grades, operability with microsurgical technique and pre- and post-operative performance status in 37 cases of cerebral vascular malformations. Neurological Surgery, 8:701 (1980)
3. S. Chou. Personal communication (1979).
4. B. Larsson. Radiological fundamentals in radiosurgery. in:"Radiosurgery: Baseline and Trends," L. Steiner et al. ed., Raven Press, New York (1992).
5. L. Steiner and C. Lindquist. Radiosurgery in cerebral arteriovenous malformation. Neurosurg. State. Art. Rev. 2:329 (1987).
6. R.N. Kjellberg. Proton beam therapy for arteriovenous malformations of the brain. in:"Operative Neurosurgical Techniques, Indication, Methods and Results," ed. 2, H.H. Schmidek and W.H. Sweet ed., Grune and Stratton, New York (1988).
7. R.N. Kjellberg. Personal communication (1989.
8. K. Kitao. A method for calculating the absorbed dose near interface from $^{10}B(n,\alpha)^{7}Li$ reaction. Rad. Res., 61:304 (1975).
9. R.A. Rydin, O.L. Deutsch, and B.W. Murray. The effect of geometry on capillary wall dose for boron-neutron capture therapy. Phys. Med. Biol., 21:134 (1976).
10. H. Hatanaka, T. Masuzawa, K. Amano, and K. Sano. Basic and clinical studies on boron-neutron capture therapy. Nippon Acta Neuroradiolica, 9(1):37 (1968).
11. H. Hatanaka. "Boron-Neutron Capture Therapy for Tumors," H. Hatanaka ed., Nishimura, Niigata (1986).

SELECTIVE MELANOMA THERMAL NEUTRON CAPTURE THERAPY FOR LYMPH NODE METASTASES

Y. Mishima[1,2], C. Honda[2], M. Ichihashi[2], M. Shiono[2], N. Wadabayashi[2], H. Obara[3], J. Hiratsuka[4], H. Fukuda[5], H. Karashima[6], K. Kanda[7], T. Kobayashi[7] and K. Yoshino[8]

[1]Dept. Dermatology, Kobe Kaisei Hospital, Nada-ku, Kobe 657, Japan
[2]Dept. Dermatology, Kobe U. School of Medicine, Japan
[3]Dept. Anesth., Kobe U. School of Medicine, Japan
[4]Dept. Radiology, Kawasaki U.; Japan
[5]Dept. Radiology, Tohoku U.; Sendai Japan
[6]Hyogo Medical Center, Akashi; Japan
[7]Kyoto U. Res. Reactor, Osaka; Japan
[8]Dept. Chem., Shinshu U., Matsumoto, Japan

INTRODUCTION

The success of treatment of tumors with NCT lies primarily with the development of suitable vectors to optimally carry ^{10}B to the tumor site, efficient methods of their administration, and effective neutron beam delivery.

A significant advance in the treatment of malignant melanomas was initiated with our development (Mishima[1], 1976) of the dopa analogue $^{10}B_1$-para-borono-phenylalanine ($^{10}B_1$-BPA) as the carrier of ^{10}B.

Melanin is formed from dopa as a result of the accentuated tyrosinase activity occurring within the premelanosomes of malignant melanoma cells. To exploit this bio-chemical phenomenon in terms of therapeutic potential we initially synthesized three molecular hybrid compounds of the precursor dopa analogue and ^{10}B. In light of our extensive studies of *in vitro* radio-biological analysis of cultured B16 melanoma cells and *in vivo* studies using melanoma-bearing hamsters as well as human patients we have found $^{10}B_1$-BPA to be the most effective medium for accumulating ^{10}B into the melanoma[2,3]. Because of the high dopa accumulating property of melanosomes within melanoma cells, $^{10}B_1$-BPA is readily absorbed into the lesion rendering it vulnerable to destruction by the reaction between thermal neutrons and the ^{10}B.

Since beginning clinical trials in 1987[4], we have successfully treated 12 cases of malignant melanoma (Mm) using our selective melanoma thermal neutron capture therapy (NCT). During this time we have accumulated much theoretical and practical information concerning the dynamics of ^{10}B concentration and absorbed energy dosimetry both within the melanoma and the covering skin, the tolerance to NCT of normal skin covering the lesion[5], and comparison of radio-therapeutic data among cases. These findings have lead us to develop optimal increase in concentration of $^{10}B_1$-BPA in the target lesion and also an

Advances in Neutron Capture Therapy, Edited by
A.H Soloway *et al.*, Plenum Press, New York, 1993

improved means of thermal neutron beam delivery. Enhanced penetrating capacity of the thermal neutron beam has also been accomplished by the oral administration of D_2O to the patient prior to irradiation.

NCT FOR HUMAN METASTATIC MALIGNANT MELANOMA IN LYMPH NODES

We would like to discuss the recent advances in the application of NCT in the light of one of our most recent cases.

The eleventh of our twelve successfully treated patients was a 75 year old man with bilateral inguinal lymph node metastasis. He noticed an ulcerative eruption on his right heel about February, 1991 which was diagnosed as malignant melanoma. The heel lesion was surgically removed at a local hospital on July 18, 1991, however, about 2 months after the operation, a rapidly growing subcutaneous tumor developed in his right inguinal area. An operation of lymph node dissection was scheduled at another hospital, but the patient sought a non-operative treatment with NCT. Just before NCT, two tumors were palpated in the right and one tumor in the left inguinal areas, respectively.

Computer tomography of a cross-section taken at the center of the lesions showed that the right lesion was 27mm and the left lesion was 15mm from the surface to the deepest portion (figs. 1, 2).

Our previous NCTs had involved ordinary melanomas which were in the dermis as well as the epidermis and were relatively shallow. However, deep seated tumors, such as this in the lymph node, are difficult to treat with a therapy such as NCT due to the limited penetrating power of our currently available thermal neutrons. To overcome this problem we needed to:

i) establish an optimal collimation method so that there is greater direction of thermal neutron beams and minimal damage to surrounding healthy tissue,

ii) determine the most effective method of administering $^{10}B_1$-BPA to ensure a greater concentration of ^{10}B in the tumor, and

iii) find out how to increase the penetration of thermal neutrons.

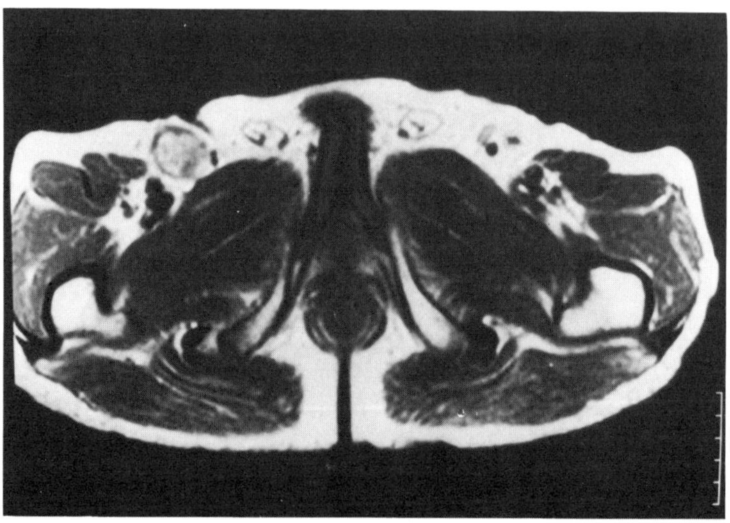

Figure 1 Computer tomograph of cross section through affected lymph nodes.

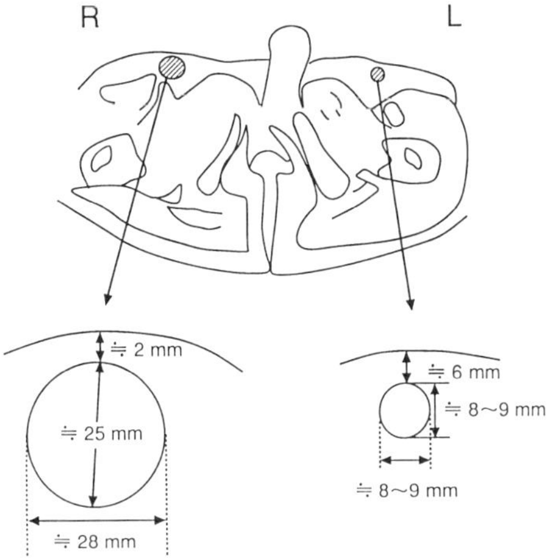

Figure 2 Schematic diagram of the affected area showing difficult location of tumors.

Improved Method of Collimation

To most effectively guide the neutron beam to the target site a modified method of collimation was required. Previous methods of collimation using materials such as cadmium metal or boron composites were not ideal because of the production of gamma rays in the neutron reaction with these materials. ^{6}LiF sheet was found to be most satisfactory[6]. As shown in figure 3, the modified arrangement of the collimator allows closer positioning of the patient to collimated and more parallel thermal neutron source.

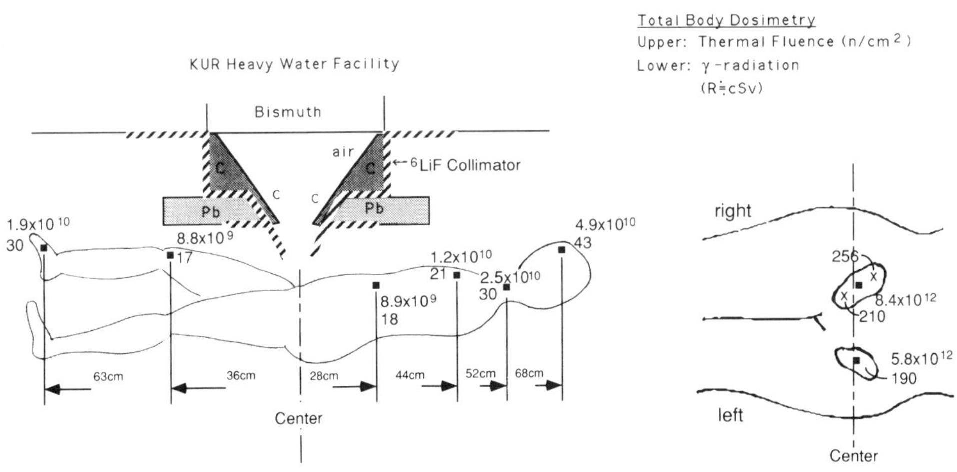

Fig. 3 A collimator of flexible ^{6}Li F sheet was used to direct the thermal neutron beam without significant energy loss from γ-ray production. Figures show total body dosimetry .

Increasing Penetrating Ability of Thermal Neutrons by Deuteriation

To determine factors involved in increasing penetrating power of thermal neutrons we have also performed animal experiments[7].

Melanoma bearing hamsters were allowed to freely drink tap water containing 30% heavy water for five days. The blood deuteriation of the hamsters ranged from 6.3% to 14.9%. We expected this deuteration (the replacement of ordinary water, H_2O, with heavy water, D_2O) to contribute to the therapeutic effect of irradiation on the tumor by increasing the penetrating ability of thermal neutrons by reducing the H(n,γ)D reaction. The enhanced penetration would also allow us to reduce the total radiation dose, thus minimizing damage to surrounding skin.

After irradiation there was a greater thermal neutron flux distribution in the vertical direction of the tumor area in the deuterated animals than in non-deuterated animals. The deuterated animals consequently showed more rapid tumor regression than did the non-deuterated animals (fig. 4).

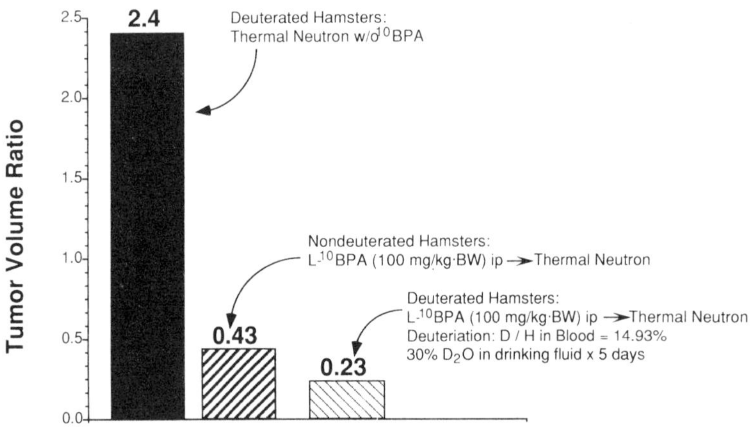

Fig. 4 Comparitive ratios of tumor volume to original volume in deuteriated and non-deuteriated hamsters nine days after NCT

Encouraged by the success of administering D_2O to hamsters, and after confirming its safety, we instructed our patient to orally ingest 30% D_2O in tap water for five days prior to treatment. We also administered a 30% D_2O·0.9% NaCl water solution by drip infusion, immediately prior to and during irradiation for a total D_2O concentration of 4.4% in blood during NCT (fig. 5).

Clinical Efficacy of our $^{10}B_1$-BPA Melanoma NCT with D_2O Administration

Along with administration of D_2O, a $^{10}B_1$-BPA·fructose complex was given by drip infusion five hours prior to irradiation and concluded 2 hours before irradiating the right nodule (fig. 5). We treated the left nodule in essentially the same manner.

The solubility of $^{10}B_1$-BPA in water is too low for medical use and the hydrochloric acid salt used in the past to improve solubility causes the patient pain and must be administered with anesthetic. To overcome this problem the use of $^{10}B_1$-BPA·fructose complex was developed by us[8].

This method was effective in treating the lymph node metastasis as observed by the regression of tumors (fig. 6). At present, no re-growth of either tumor has been detected almost one year after irradiation.

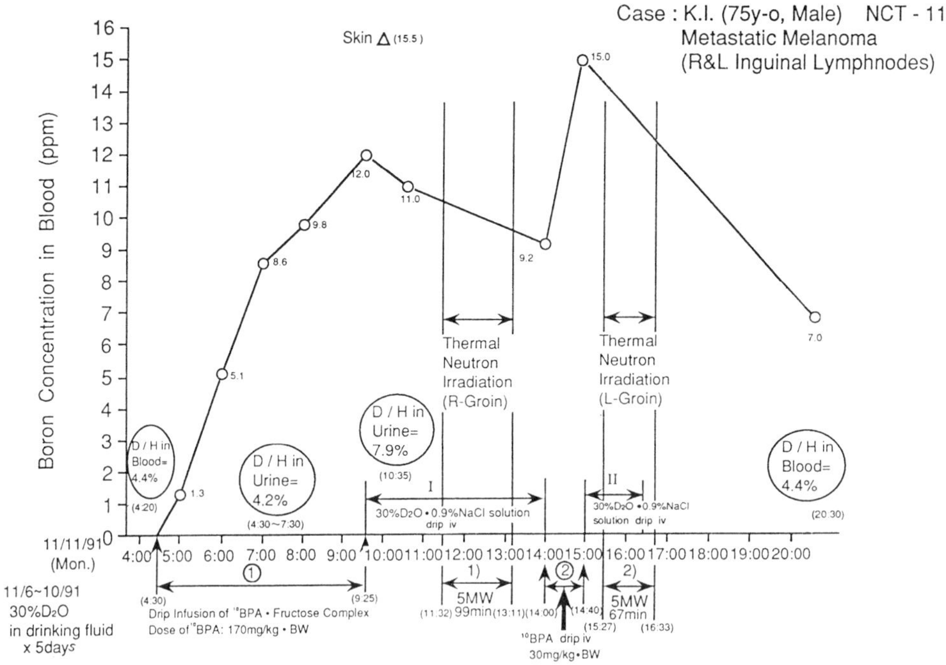

Figure 5 Diagram showing administration of ^{10}B-BPA and the time course of ^{10}B concentration in blood.

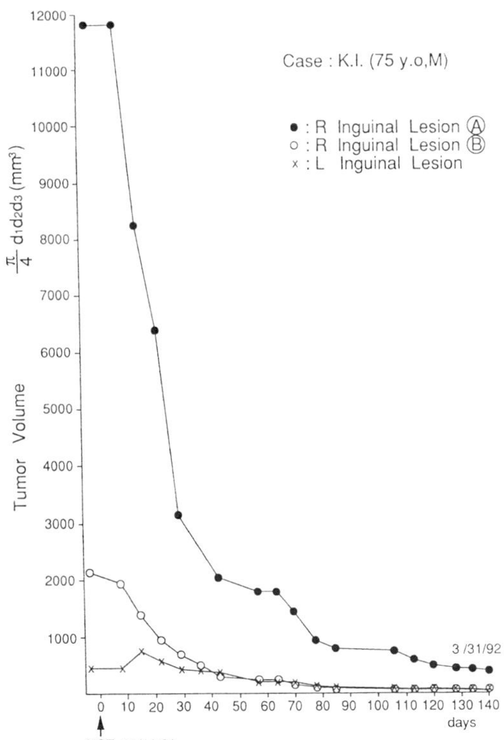

Figure 6 Regression curves of metastatic melanomas in inguinal lymph nodes after deuteriation and NCT.

REFERENCES

1. Y. Mishima and T. Shimakage, Thermal neutron capture treatment of malignant melanoma using ^{10}B-dopa and $^{n}B_{12}$-chlorpromazine compounds, *in:* "Pigment Cell," 2:394, V. Riley, ed., S. Karger, Basel (1976).
2. M. Ichihashi, H. Fukuda, J.K. Brown, M.H. Mountford, B.J. Allen, J.G. Wilson, and Y. Mishima, In vitro evaluation of for mealanoma at Moata - joint work between Japan and Australia BNCT reasearch teams *in:* "Progress in Neutron Capture Therapy for Cancer," B.J. Allen *et al.* ed., Plenum Press, New York (1992).
3. Y. Mishima, M. Ichihashi, T. Nakanishi, M. Tsuji, M. Ueda, T. Nakagawa, and T. Suzuki, Cure of malignant melanoma by single thermal neutron capture treatment using melanoma seeking compounds: ^{10}B / melanogenesis interaction to *in vitro / in vivo* radiobiological analysis to preclinical studies, *in:* "Proc. 1st International Symposium on NCT," R.G. Fairchild and G.L. Brownell, eds., U.S. Government Printing Office (1984).
4. Y. Mishima, C. Honda, M. Ichihashi, H. Obara, J. Hiratsuka, H. Fukuda, H. Karashima, T. Kobayashi, K. Kanda andY. Yoshino, Treatment of malignant melanoma by single neutron capture treatmentwith melanoma-seeking ^{10}B-Compound, *Lancet* 11:388-389 (1989).
5. H. Fukuda, K. Ando, C. Honda, M. Ichihashi, Y. Mishima, J. Hiratsuka, T. Kobayashi, and K. Kanda, Tolerance limits of normal skin treated by a single thermal neutron capture therapy, *in:* "Progress in Neutron Capture Therapy for Cancer," B.J. Allen *et al.* ed., Plenum Press, New York (1992).
6 T. Kobayashi, S. Hatta, C. Honda, K. Yamamura, T. Akiyoshi, K. Kanda, and Y. Mishima, Estimation of absorbed doses in human malignant melanoma treated by neutron capture therapy with special references to vertical direction, *Pigment Cell Res.* 2:361-364(1989).
7. C. Honda, M. Shiono, M. Ichihashi, N. Wadabayashi, Y. Mishima, K. Yoshino, T. Kobayashi, K. Kanda, and H. Karashima, Deuteriation trials for neutron capture therapy of deep-seated melanoma lesions, *in:* "Proc. of Sixth Int. Symp. on NCT 9/14~17/92, Columbus," R.F. Barth ed., Plenum Press, New York (in press).
8. K. Yoshino, A. Suzuki, Y. Mori, H. Kakihana, C. Honda, Y. Mishima, T. Kobayashi, and K. Kanda, Improvement of solubility of p-boronophenylalanineby complex formation with monosaccharides, *Strahlenther. Onkol.*165:127 (1989).

DEUTERIATION TRIALS FOR NEUTRON CAPTURE THERAPY OF DEEP-SEATED MELANOMA LESIONS

Chihiro Honda[1], Masahiro Shiono[1], Nobutoshi Wadabayashi[1], Masamitsu Ichihashi[1], Yutaka Mishima[2], Kazuo Yoshino[3], Tooru Kobayashi[4], Keiji Kanda[4], and Hiroshi Karashima[5]

[1]Dept. of Derm., Kobe University School of Medicine, Kobe 650, Japan
[2]Mishima Inst. for Derm. Research, Kobe, Japan;
Dept. Dermatology, Kobe Kaisei Hospital, Nada-ku, Kobe 657, Japan
[3]Dept. Chem., Shinshu Univ., Matsumoto, Japan
[4]Res Reactor Inst., Kyoto Univ., Osaka, Japan
[5]Dept. of Radiology, Hyogo Med. Center for Adults, Akashi, Japan

INTRODUCTION

We have succeeded in curing primary cutaneous malignant melanomas of human patients by neutron capture therapy (NCT) using $^{10}B_1$-para-boronophenylalanine (^{10}BPA)[1,2]. Since primary cutaneous melanomas are usually within 10mm in thickness, we can irradiate the whole primary lesions with thermal neutrons in a sufficient dose for eradication. In further applying our NCT to deep-seated melanoma lesions, especially those metastasizing into the lymph nodes, partial replacement of hydrogen (H) with deuterium (D) (deuteriation or deuteration) is expected to improve the penetration of thermal neutrons into the depth of the tumors due to the decreased H(n,g)D reaction and very low cross section of D for neutron capture.

In this study, we tried deuteriation *in vivo* using hamsters and investigated the influence of deuteriation on the therapeutic effect of NCT and thermal neutron penetration.

MATERIALS AND METHODS

(1) Melanoma-Bearing Hamsters

Tumor tissue fragment of Greene's melanoma was implanted subcutaneously into the thigh of each female Syrian golden hamster.

(2) Deuteriation

Since it is reported[3,4] that animals could drink up to 30% deuterated water continuously with no serious side effects, while higher levels elicit toxic effects, we decided to test 30% D_2O (deuterium oxide, heavy water) in drinking fluid.

The deuterated drinking water, 30% D_2O, was prepared by mixing 300ml D_2O (99.8%D) with 700ml fresh tap water.

The hamsters were given *ad libitum* either ordinary tap water or 30% deuterated tap water.

(3) ^{10}BPA

L-^{10}BPA·fructose complex solution (pH 7.4), whose ^{10}BPA and fructose concentrations were 30 mg/ml and 66 mg/ml respectively, was prepared. L-^{10}BPA was administered intraperitoneally in a dose of 100 or 200 mg/kg·BW.

(4) Deuterium Analysis

D_2O analyses of tissue samples were performed by combined use of gas chromatography and platinum catalyst described by elsewhere[5] after purification of water from the samples (by vacuum sublimation).

(5) Melanoma Treatment by NCT

Greene's melanoma-bearing hamsters were divided into groups of deuteriated and non-deuteriated hamsters. We deuteriated the hamsters by giving 30% D_2O for five days prior to irradiation, and then ordinary tap water was supplied after irradiation. Non-deuteriated hamsters drank ordinary tap water throughout the experiment.

With the body shielded by LiF tiles, the melanoma bearing region of the thigh was exposed to thermal neutron irradiation at Kyoto University Reactor, 4 hours after i.p. injection of L-^{10}BPA (100 mg/kg·BW) or without ^{10}BPA administration.

On the day of irradiation the distance from the skin surface to the deepest portion of the tumors was approximately 10~15mm. Tumor size was measured following irradiation.

(6) Boron Assay

We analyzed tissue boron concentrations by inductively coupled plasma atomic emission spectrometry (ICP-AES).

(7) Neutron Dosimetric Experiment

With a gold wire inserted into the melanoma region from the skin surface in the vertical direction of the irradiation, the melanoma region was irradiated with thermal neutrons under the same conditions as the melanoma treatment experiment described above.

RESULTS AND COMMENTS

Blood deuteriation of hamsters by drinking 30% D_2O *ad libitum* for 5 consecutive days ranged from 6.3% to 14.9%, which proved to be rather lower than the measurements in mice reported elsewhere[3]. Melanoma deuteriation, which was almost consistent with blood deuteriation, ranged from 11.3% to 16.4% after 13days ingestion of 30% D_2O.

In the treatment experiment by NCT (the average neutron fluence on the skin surface: 1.2×10^{13} n/cm^2) using ^{10}BPA, melanoma growth was more remarkably inhibited in deutriated hamsters than in non-deuteriated hamsters, in view of growth delay time, which is defined as the days required for the tumor to re-grow to its initial volume (Fig. 1).

Selective ^{10}BPA uptake by melanoma cells was hardly affected by ingestion of 30% D_2O for 5~15 days (Fig. 2).

The neutron dosimetric experiment demonstrated that deuteriation of hamsters increased the neutron flux by approximately 10 to 15% in the deepest portion of the subcutaneous melanomas (Fig. 3).

Thus, we conclude that short-term pre-irradiation deuteriation can enhance the anti-melanoma effect of NCT using ^{10}BPA by increasing the neutron dose to depth of a tumor, without affecting ^{10}BPA accumulation in melanoma.

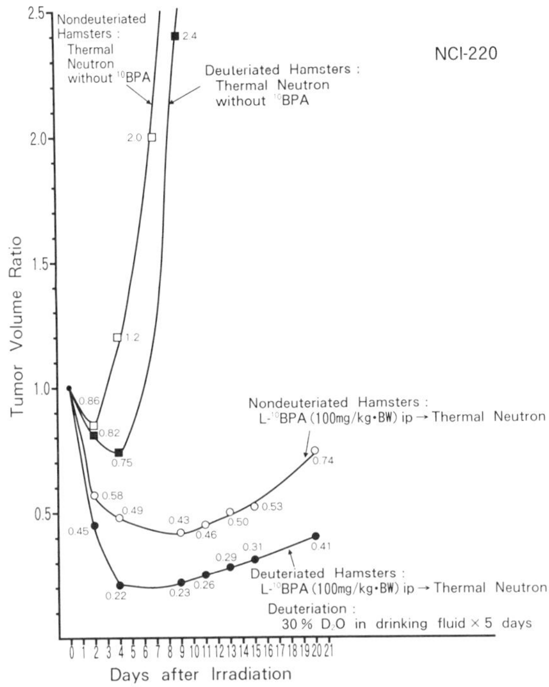

Fig. 1 Effect of deuteriation on melanoma growth inhibition by thermal neutron capture therapy using ^{10}BPA.

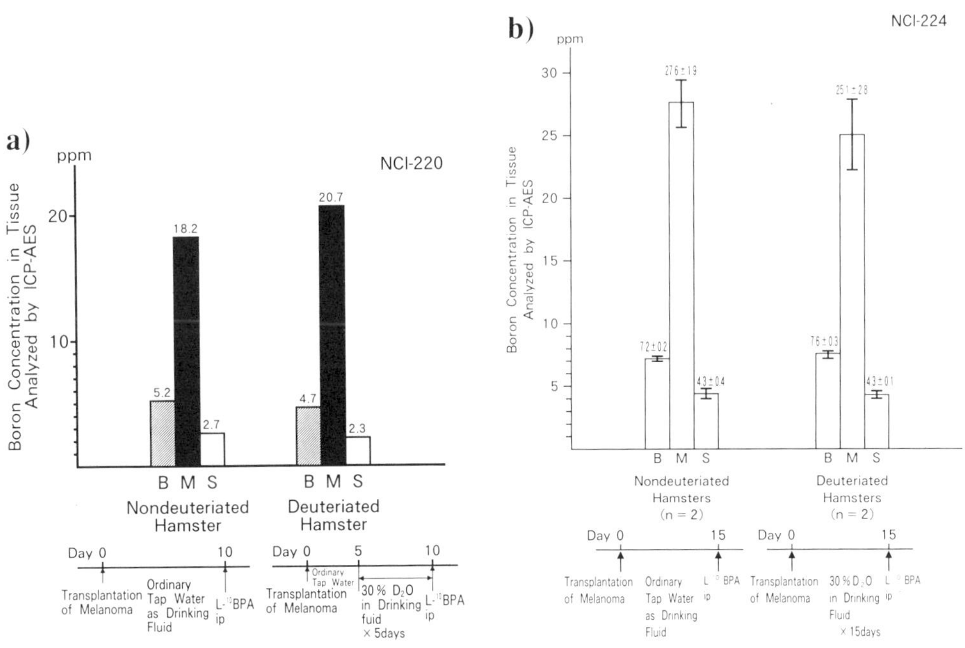

Fig. 2 Boron concentrations in blood (B), melanoma (M), and skin (S) in deuteriated and non-deuteriated melanoma-bearing hamsters:

a) 4 hours after i.p. injection of L-^{10}BPA·fructose complex (dose of ^{10}BPA: 100 mg/kg·BW)

b) 3 hours after i.p. injection of L-^{10}BPA·fructose complex (dose of ^{10}BPA: 200 mg/kg·BW)

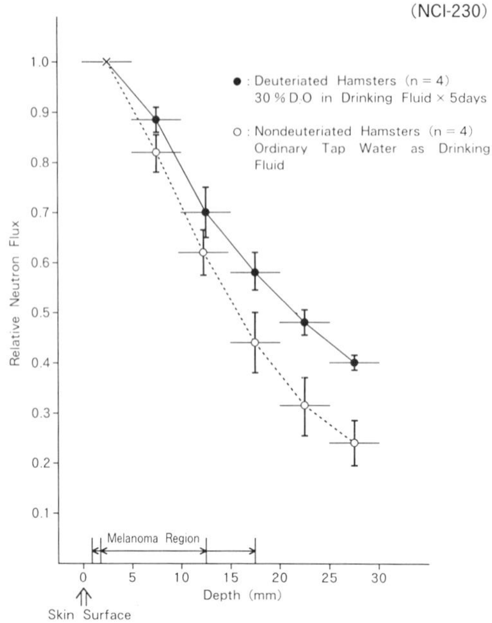

Fig. 3 Neutron flux distributions in the vertical direction of the melanoma regions of deuteriated and non-deuteriated melanoma-bearing hamsters by thermal neutron irradiation.

REFERENCES

1. Y. Mishima, C. Honda, M. Ichihashi, H. Obara, J. Hiratsuka, H. Fukuda, H. Karashima, T. Kobayashi, K. Kanda, and K. Yoshino, First cure of primary malignant melanoma in man by single thermal neutron capture therapy using melanoma-seeking ^{10}B-compound, *Lancet II* 388-389 (1989).
2. Y. Mishima, A post melanosomal era: Control of melanogenesis and melanoma growth, *Pigment Cell Res. (Suppl.)* 2:3-16 (1992).
3. J.J. Katz, H.L. Crespi, D.M. Czajka, A.J. Finkel, Course of deuteriation and some physiological effects of deuterium in mice, *Am. J. Physiol.* 203:907-913 (1962).
4. A. Hodel, J.O. Gebbers, H. Cottier, J.A. Laissue, Effects of prolonged moderate body deuteration on proliferative activity in major cell renewal systems in mice, *Life Sci.* 30:1987-1996 (1982).
5. M. Mochizuki, S. Noda, T. Morishita, A simple method for determination of heavy water by combined use of gas chromatography and platinum catalyst, *Radioisotopes* 36:163-168 (1987).

CORRELATION OF BSH UPTAKE WITH CT-SCAN CONTRAST ENHANCEMENT IN PATIENTS WITH INTRACRANIAL TUMORS

Giuseppe Stragliotto,[1] Heinz Fankhauser,[1] Patrick R. Gavin,[1,2] and Reto A. Meuli [3]

[1]Dept of Neurosurgery, and [3]Radiology, CHUV, Lausanne, Switzerland
[2]Sabbatical leave from Washington State University, Pullman, WA, USA

INTRODUCTION

CT-scan is a widely used primary imaging technique for intracranial pathology. A iodinized contrast medium can be injected, and is retained into the physiologic vascular system of the brain. Only structures with a defective blood-brain barrier (malignant intracerebral tumors) or lacking this barrier (as meningiomas) show an uptake of contrast medium. Intravenous contrast enhancement of tumors on CT-scan images was correlated with subsequently measured boron levels, with the aim to define a simple predictive test for BSH accumulation in intracranial neoplasms.

METHODS

30 CT-scan pre- and post-contrast were gathered, including all types of tumors, from patients who received BSH, 95% ^{10}B enriched at a dose of 5 mg B/kg. 400mg/ml of Iopamidol, 100ml per patient was injected about 15-30 minutes prior to imaging. CT images were reviewed by 2 of us (GS and PRG) and a consensual code was used. Densities were plotted on a relative scale: 1= low contrast enhancement 2= medium enhancement 3= high contrast enhancement, and compared to tumor-to-blood ratios for boron. Tumors were grouped in low grade intracerebral tumors, high grade intracerebral tumors, metastases, meningiomas.

RESULTS

CT-scan contrast enhancement was the highest and most homogeneous in meningiomas and metastases. Glioblastomas reached an intermediate intensity with much heterogeneity. No contrast enhancement was measurable in low grade astrocytomas.

Advances in Neutron Capture Therapy, Edited by
A.H Soloway *et al.*, Plenum Press, New York, 1993

Relative contrast mean index for astrocytoma was 1.5 (±0), mean tumor to blood ratio was 1.1 (±0.1). Mean contrast index for glioblastoma was 1.83 (±0.83), and tumor to blood ratio was 0.97 (±0.76). Mean contrast index for meningioma was 3 (±0), and tumor to blood ratio was 1.94 (±1.31). In the calculation of tumor to blood ratio, delay after boron administration varies. By plotting these values together, we found a straight regression line with a 0.98 correlation coefficient (Figure 1).

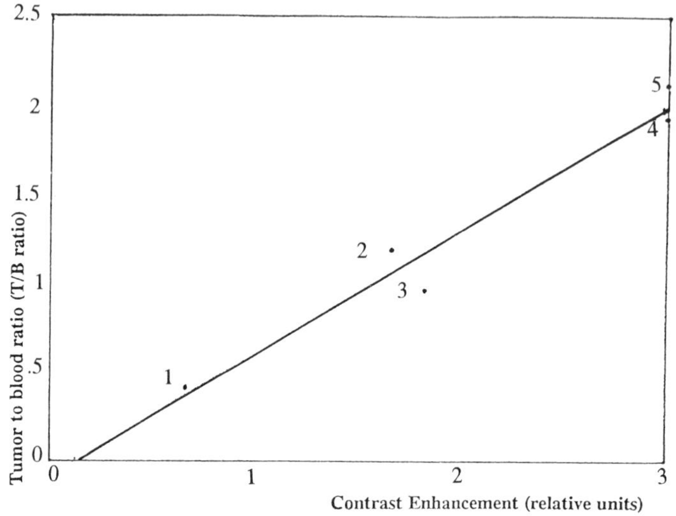

Fig. 1: Tumor to blood ratio of human intracranial tumors vs. CT-scan contrast enhancement. Legends: 1) low grade astrocytoma. 2) Anaplastic astrocytoma. 3) Glioblastoma. 4) Meningioma. 5) Brain metastasis.

DISCUSSION

A good correlation between contrast enhancement and tumor to blood ratio was found when using mean values of relative contrast index and mean tumor to blood values for a group of tumor. Meningiomas, as well as low grade intracerebral tumors, were quite homogeneous (in this latter case, because there was no contrast enhancement and no boron uptake). Glioblastomas showed a high heterogeneity, both in contrast uptake and in boron uptake. Therefore, taken individually, contrast enhancement could not predict boron content in tumors, especially for the glioblastoma group.

Explanations for this include individual variability in tumor blood supply and stroma factors. For instance, several cases of meningioma had cerebral angiography prior to operation and boron administration. Indeed, a correlation existed between the degree of vascularity of the tumor (including preoperative embolization) and the actual boron uptake. In addition, several days may have occurred between CT scan imaging and boron administration to the patients; in the same way, the total amount of contrast medium may have varied in respect to mg of contrast/patient weight. Different CT machines were used, and some patients had a second operation at time of imaging and boron administration.

This simple study confirms that tumor uptake of boron is variable. CT scan contrast enhancement may be used as a screening for patients, whose tumor could potentially accumulate boron. Absence of contrast enhancement indicates that the patient is not appropriate for BNCT with BSH.

ACKNOWLEDGEMENTS

This work has been supported by the Fonds National Suisse de la Recherche Scientifique, the Ligue Suisse contre le Cancer, the Ligue Vaudoise contre le Cancer, the Zyma Foundation for the Advancement of Medical and Biological Sciences, and the Commission of the European Communities.

UPDATE ON BIODISTRIBUTION OF BOROCAPTATE SODIUM (BSH) IN PATIENTS WITH INTRACRANIAL TUMORS

Giuseppe Stragliotto,[1] Daniel Schüpbach,[1] Patrick R. Gavin,[1,2] and Heinz Fankhauser,[1]

[1]Dept of Neurosurgery, CHUV, Lausanne, Switzerland
[2]Sabbatical leave from Washington State University, Pullman, WA, USA

INTRODUCTION

The use of borocaptate sodium (BSH) for Boron Neutron Capture Therapy (BNCT) has been studied since 1968. BSH has shown differential accumulation in rodent tumor models both subcutaneously and intracerebrally implanted. BSH administration to humans for BNCT has also been reported [1,2,3,4], including at this symposium [see report by D. Haritz]. In 1990, we presented the biodistribution of BSH in 28 patients with various intracranial tumors [5]. This compound showed accumulation in most high grade intracerebral tumors and low grade extracerbral tumors with a variable kinetic; a delay of at least 12 hours after administration was usually necessary to obtain a positive tumor to blood ratio. In order to obtain more substantial information on the distribution pattern, and to determine an adequate dose to be given before neutron radiation, we have extended this investigation to a larger number of patients.

MATERIALS AND METHODS

Seventy patients with intracranial tumors were studied. About half of the patients had malignant intracerebral tumors including 28 malignant gliomas, 3 medulloblasomas and 7 metastases. Other intracerebral tumors included 9 low grade astrocytomas. Low grade extracerebral tumors included 16 meningiomas and 3 accoustic schwannomas. Two other patients turned out to have an intracerebral hematoma instead of the expected tumors.

BSH, 95% ^{10}B enriched (Callery, Pittsburg, USA and, lately, Centronics, New Addington, U.K.), diluted in 50 ml saline, was administered intravenously in 30 minutes. The dose of ^{10}B was 5 mg/kg weight in the first 45 patients, and then increased up to 50 mg ^{10}B/kg weight.

Delays between injection of BSH and tissue sampling at operation ranged from 3 to 72 hours; since our previous study had showed a clear tendency to an improved tumor to blood ratio after a delay of 18 hours and more, we choosed an interval of 24 hours for a dose escalation study.

Tissues obtained for analysis included tumors, preferably with different locations, peritumoral normal brain, dura mater, muscle, bone, skin, cerebrospinal fluid, cyst fluid blood and urine. The samples were quick frozen and stored at -80°C.

The analysis was performed by Inductively Coupled Plasma-Atomic Emission Spectroscopy for all samples. Most solid samples, particularly small ones, were also analyzed by alpha-track neutron capture radiography.

Toxicity of BSH was evaluated clinically only, since laboratory values were inevitably altered by the surgical procedure, and concurrent medical treatment.

RESULTS

Table 1 lists the detailed results of boron determination in tumors and other tisssues in all patients, including dose and interval from administration. This interval between BSH administration and tissue sampling refers to tumor and blood. For technical reasons related to intracranial surgery, time to collection for other tissues may have deviated up to 3 hours from the indicated value.

Heterogeneity of boron tumor levels were related to heterogeneity of histological pattern, particularly in malignant gliomas. This was taken into account by refraining to use average values of all tumor samples. An obvious necrotic part was not averaged with a highly proliferating cellular specimen of the same tumor. For our calculations, we therefore have used the highest value for tumors and the value for blood at the time of tumor sampling.

The concentration of boron in blood decreased with time from the end of BSH administration. For a same dose of BSH, blood boron content was 3 to 5 times higher at 3 hours than at 24 hours after administration. Patients receiving a larger dose of boron had higher boron concentration in blood. Boron was still detectable in blood 7 days after a single administration.

Maximum tumor concentrations of boron were obtained in high grade intracerbral tumors. The highest value was 19.5 ppm, 24 hours after a single boron dose of 50 mg/kg. For all administered amounts of BSH, the maximal concentration in tumor was measured at 3 hours and decreased steadily thereafter. With a boron dose of 5 mg/kg, the high grade tumors contained 10 ppm at 3 hours, and 5.5 ppm at 24 hours after administration. The maximal concentration measured in an extracerebral tumor was 7.7 ppm at 13 hours and less than 2 ppm at 24 hours. The low-grade intracerebral tumor contained 5.4 ppm at 4 hours and only 0.2 ppm at 27 hours.

Clearance of boron from the high grade gliomas led to a decrease of the concentration by a factor of 2 in 24 hours; during the same interval, the blood boron decreases by a factor of 3 or more. This difference gave rise to a tumor to blood ratio which was below 1 at 3 hours and increased to 1.5:1 or 2:1 at 24 hours. Some extracerebral tumors follow the same pattern. Low grade intracerbral tumors always presented with a tumor to blood ratio below 1.

By increasing the administered BSH dose, the tumor concentration increased, but the tumor to blood ratio did not improve. This is illustrated in patients studied 24 hours or more after injection:

- 4 patients received 5 mg boron/kg; the maximal tumor concentration after 24 hours was 5.5 ppm, and the maximal tumor to blood ratio was 2.5 (range from 0.3 to 2.5).

- 2 patients received 15-20 mg boron/kg; the tumor concentration was 5 ppm after 48 hours in one, and 3.4 ppm at 72 hours in the other. The tumor to blood ratio was 1.6 and 1.9 respectively.
- 4 patients received 25-30 mg boron/kg; the tumor concentration was 12.7 ppm after 24 hours, and 7.1 ppm at 48 hours. Tumor to blood ratio ranged from 0.9 to 2.
- 2 patients received 40-50 mg boron/kg; the tumor concentration was 19.5 ppm at 24 hours, respectively 6 ppm at 48 hours. Tumor to blood ratio was 1.2 and 0.9 respectively.

Boron in peritumoral normal brain was found in lower concentration than in tumor and blood at all times. The highest value was 5.8 ppm at 24 hours, after an administered dose of 50 mg boron/kg. The tumor to brain ratio was always 3 or higher. Normal brain at a distance of more than 2-3 centimeters from the tumor margins was not available, for obvious reasons. However, cerebrospinal fluid (CSF), which is protected by a tight and selective meningo-vascular barrier similar to the blood brain barrier, was available in half of the patients, including those with malignant glioma. CSF contained very low amounts of boron. Tumor to CSF ratio ranged from 10 to 30 for high grade tumors; it was only 2:1 in low grade astrocytomas.

No major side effects were encoutered even at the highest BSH dose. Solubility became critical with dilution of more than 4 or 5 g of BSH in 50 ml saline. Furthermore, highly concentrated solutions sometimes caused irritation at the injection site. Two patients suffered from nausea and chest pain during infusion; these symptoms were not accompanied by any change in vital functions and disappeared spontaneously after temporary interruption of infusion or slowing down of the infusion speed.

DISCUSSION

The need to precisely assess the boron concentration in tumors in vivo appears to be a prerequisite for the progress of BNCT. However, a number of questions and remarks coming out of our study have led us to a somewhat different conclusion. 1) We have been analyzing small samples of heterogeneous tumors; this means that the measured boron content is at best a close approximation of the real concentration in a particular region of the tumor. This conclusion has also been reached when looking at much larger tumor samples in dogs [4,6]. 2) The analytical methods are not sensitive enough to determine the boron concentration in a single cell, and even if this were possible, we would find a marked heterogeneity throughout the tumor. Biologically relevant microdosimetry therefore cannot be applied realistically. 3) In our study we were actually measuring samples which would belong to the surgically debulked part of the tumor in case of BNCT. 4) Tissue sampling in our patients was performed during profound alterations of homeostatic conditions induced by surgical trauma, blood loss, and the administration of intravenous fluids, blood transfusions, steroids, mannitol and others. This may have altered the kinetics of BSH, as it does for other plasma bound compounds [7]. One of our patients developed perioperative disseminated intravascular coagulopathy; the large volume of blood and fluid replacement led to a twofold decrease of the blood boron concentration as compared to the expected value. Furthermore, if we could measure the boron kinetics in blood, the effect of transfusion on the tumor boron content could not be investigated before and after transfusion. 5) The analyses of our specimens only indicated isolated time points in a single patient, but they may not reflect boron kinetics over time for that individual. Furthermore, due to numerous variables between individuals, extrapolation from one patient to another may not be safe enough to

Table 1 **Boron concentration (ppm) in tumor and normal tissues from 67 patients**

Case	Histology	Dose	Interval	Tumor	T/B	Ratio	Brain	Skin	Dura	Muscle	Bone	CSF
		$mg^{10}B/kg$	hours		Blood							
1	Glioma	5	2	6.4	17	0.4	-	5.8	-	3	0.6	-
2	"	5	3	7.2	16	0.5	-	-	-	-	-	-
3	"	14	4	8.8	12.7	0.7	1.5	-	7	-	-	-
4*	"	5	12	4	8.7	0.5	-	-	7	-	0.4	0.2
5	"	5	13	3.1	3	1	-	4.5	-	1	0.1	0.2
6	"	5	13	1.7	5.2	0.3	0.5	-	-	2	-	0.2
7	"	2.5	17	1	0.6	1.7	0.2	-	-	2	-	
8	"	5	17	1	3.1	0.3	-	-	-	7	-	-
9*	"	5	22	1.7	2.9	0.6	-	1.1	-	0.8	1.7	-
10	"	24	22	12.7	11.8	1.1	-	9.2	-	7	1.7	-
11*	"	5	23	2.7	1.3	2	1	3	-	-	-	0.1
12*	"	5	23	7.5	3	2.5	0.5	-	-	1.5	-	-
13	"	25	23	6.5	8.1	0.9	-	10.4	-	9.3	2.3	-
14	"	5	24	5.5	9	0.6	-	3.4	-	-	-	-
15	"	50	24	19.5	15	1.2	5.8	9.2	-	6.6	-	0.6
16	"	23	32	4.3	5.5	0.9	-	-	-	-	-	-
17	"	15	48	5	3.1	1.6	-	-	-	1.8	-	-
18	"	25	48	7.1	3.4	2	-	4.4	4	2.6	-	-
19	"	43	48	6	7	0.9	-	-	-	-	-	
20	"	29 (2x)	48	2	5.1	0.5	-	3	-	-	-	0
21	"	18	72	3.4	1.8	1.9	-	2.9	-	1.4	-	-
22	anaplastic astrocytoma	5	3	10	11	0.9	-	7	-	-	-	0
23	"	5	18	2.6	2.6	1	-	1	7	3	-	0
24	"	5	19	1.5	1.3	1.2	1	2.5	3	-	-	0.2
25*	"	5	19	2.2	1.6	1.4	0.5	0.6	-	2	0.2	0.2
26	"	35	20	17	5.6	3	-	11	-	5.4	4	-

Case	Histology	Dose mg^{10}B/kg	Interval hours	Tumor	T/B Blood	Ratio	Brain	Skin	Dura	Muscle	Bone	CSF
27	medulloblastoma	5	13	2.6	4	0.7	-	3.6	-	3.4	-	-
28	"	5	18	1.1	1.3	0.8	0.3	-	-	1	-	-
29	"	5	18	2	1.5	1.3	-	6	-	2	-	-
30	metastasis	5	3	5	16.2	0.3	-	-	-	-	-	-
31	"	26	7	15	35	0.4	-	23	36	10	4.1	-
32	"	5	12	4.6	4	1.1	-	-	-	3.6	2.6	-
33	"	5	13	3	4.5	0.7	-	3.1	-	3.2	-	-
34	metastasis	5	16	2.7	7.2	0.4	-	2.4	4	2.6	-	-
35	"	5	17	3	1.5	2	0.3	-	-	-	0	0.2
36	"	5	19	4	1	4	0.1	1.7	6	3.1	-	-
37	astrocytoma II	5	4	5.4	15	0.4	-	-	-	-	-	0
38	"	5	12	5.5	9	0.6	0.3	-	-	7	-	-
39	"	5	16	1.5	1.5	1	-	0.7	-	2	-	-
40	"	14	18	2.5	4.8	0.5	-	1	-	-	-	-
41	"	5	19	0.3	0.5	0.6	0.2	-	-	2	-	0
42	"	30	48	1	2.9	0.35	0.3	2.5	14	1.4	0	-
43	"	5	27	0.2	0.9	0.2	0.2	-	-	2.5	-	0.1
44	hemangioblastoma	5	4	6.2	8.2	0.8	-	3	-	6	-	-
45	"	5	17	5	2.8	1.8	-	-	3	3.6	-	0
46	meningioma	5	13	7.7	6	1.3	-	6.1	5.8	5.1	-	-
47	"	5	13	4.8	7	0.7	-	1.4	-	-	2.1	-
48	"	5	16	4.4	7.5	0.6	-	-	7.2	6.4	4.1	-
49	"	5	17	4.2	2.1	2	-	-	3.5	-	-	0.1
50	"	5	17	3.7	1.2	3	-	4.4	-	2.8	-	0
51	"	5	18	1	3.3	0.3	0.3	4	1.2	1	-	0.1
52	"	5	20	4	1	4	0.3	2.6	-	2.4	-	0.1
53	"	5	20	1.9	2.5	0.8	-	1.9	1.7	0.9	0.3	0.3
54	"	5	22	2	2.4	0.8	-	-	-	1.5	-	0.2

(continued)

Table 1 **Boron concentration (ppm) in tumor and normal tissues from 67 patients** *(continued)*

Case	Histology	Dose $mg^{10}B/kg$	Interval hours	Tumor	T/B Blood	Ratio	Brain	Skin	Dura	Muscle	Bone	CSF
55	"	5	23	1.6	0.5	3.2	0.3	-	3	-	-	0.1
56	"	10	26	3.1	3.2	1	-	0.5	-	-	0	-
57	"	10	26	2.5	5.4	0.5	-	1.2	5.2	-	0	-
58	"	14	27	3.2	3.1	1	-	-	2	-	0.3	-
59	"	24	46	6.8	6	1.1	-	-	-	-	-	-
60	schwannoma	5	12	7	7.7	0.9	-	4	-	3.5	-	0.3
61	"	5	18	1.2	1	1.2	-	-	3	-	-	0.1
62	"	5	19	2	1.6	1.3	-	-	4	-	-	0.2
63	infarction	5	26	(1)	3.6	0.3	-	-	3	-	-	0.1
64	radiation necrosis	37	20	3.8	15.7	0.2	-	-	-	-	-	-
65	not analyzed	24 (2x)	48									
66	"	24	-									
67	"	17	26									

*recurrent tumors, (2x)= double injection.

serve as a basis for actual treatment planning. A biodistribution study seems mandatory for each patient scheduled for BNCT.

These arguments indicate to us that the investigation of the macroscopic boron distribution in brain tumors is not the ultimate goal of such studies. Instead, we consider the collection of information concerning boron concentration in normal tissues exposed to neutrons during BNCT, including skin, muscle, bone, blood, and especially normal brain surrounding the tumor, whether edematous or not, as dose limiting. Skin and muscle boron content appear to be related directly to blood boron concentrations. Therefore, a long delay after BSH administration would be advantageous to reduce radiation damage to skin and endothelium. Normal peritumoral brain, histologically tumor free, is usually edematous. Fortunately, the measured boron concentration in this area was no more than one third of the concentration in the adjacent tumor, most often less. The highest brain value was 5.8 ppm, compared to the highest tumor value of 19.5 ppm. However, it is planned to deliver BNCT in several fractions after repeated administrations of BSH, likely in 24 hours intervals. Therefore, we should know boron concentration in peritumoral normal brain after several BSH administrations. The study of boron kinetics after 2 administrations is now under way in our patients. A preliminary indication that bulk flow through edematous brain does not carry significants amounts of BSH comes from our observation that boron content in cerebrospinal fluid was always low or unmeasurable, even after an administration interval of 48 hours.

The main lessons of this study concern: 1) the differential kinetics of BSH in the tumor as compared to blood and other normal tissues, and 2) the low boron concentrations found in peritumoral normal brain and cerebrospinal fluid at all times after BSH administration. In our previous study [5], we found that tumor boron concentration were highest shortly after administration of BSH, favoring a passive distribution from blood to tumor across a permeable blood-brain barrier (tumors that do not enhance on CT-scan do not accumulate boron). The present study indicate a slower release of boron out of the tumor, compared to blood and normal tissue clearance. Despite the fact that maximum tumor concentrations stay far below maximum blood concentrations, a positive tumor to blood ratio is reached in many cases of malignant intracerebral tumors after an interval of 18 hours or more after administration of the boron compound. However, this was not achieved in all cases, nor was it in all measured areas of an individual tumor. Furthermore, the optimum time point for BNCT after BSH administration will have to be decided upon, taking into consideration not only the best tumor to blood ratio but also the rapidly declining absolute boron concentration.

As a rule, we conclude that we can expect a macroscopic boron concentration of 20 ppm, in a highly proliferating and well vascularized part of malignant intracerebral tumor, 24 hours after intravenous administration of 50 mg B/kg as BSH. The tumor to blood ratio is then approximately 1.5, and a tumor to normal brain ration of 3 or more is likely. In order to reach higher tumor concentrations, repeated administrations may be necessary, rather than larger single doses, in order to avoid drug intolerance. Considering the numerous variations, some of which have been demonstrated in this study, it is not safe to extrapolate tumor concentrations from blood measurements in individual patients. However, boron content in normal tissues as skin and muscle follow closely boron kinetics in blood.

ACKNOWLEDGMENTS

This work was undertaken in collaboration with the Institute of Analytical Chemistry of the University of Lausanne, Switzerland, the Division of Physical Biology in Uppsala, Sweden, the Department of Chemistry of the University of Bremen, Germany, and the ECN in Petten, The Netherlands.

This work has been supported by the Fonds National de la Recherche Scientifique, the Ligue Suisse contre le Cancer, the Ligue Vaudoise contre le Cancer, and the Commission of the European Communities.

REFERENCES

1. Finkel GC et al.: Distribution of ^{10}B after infusion of $Na_2{}^{10}B_{12}H_{11}SH$ into a patient with malignant astrocytoma: Implication for Boron Neutron Capture Therapy. Neurosurgery 24: 6-11, 1989

2. Hatanaka H et al.: Boron uptake by human brain tumors and quality control of boron compounds. In Hatanaka H (ed): Boron Neutron Capture Therapy for Tumors, Niigata, Japan, Nishimura, 1986, pp 77-106

3. Hatanaka H et al.: Clinical results of Boron Neutron Capture Therapy. In Allen BJ (ed): Progress in Neutron Capture Therapy for Cancer. Plenum Press, New-York, 1992, pp 561-568

4. Haritz D et al.: Distribution of BSH in patients with malignant glioma. In Allen BJ (ed): Progress in Neutron Capture Therapy for Cancer. Plenum Press, New-York, 1992, pp 557-560

5. Stragliotto G, Fankhauser H: Biodistribution of Boron Sulfhydryl (BSH) in patients with intracranial tumors. In Allen BJ (ed): Progress in Neutron Capture Therapy for Cancer. Plenum Press, New-York, 1992, pp 551-556

6. Kraft SL et al.: Biodistribution of boron in canine spontaneous intracranial tumors following borocaptate sodium infusion. In Allen BJ (ed): Progress in Neutron Capture Therapy for Cancer. Plenum Press, New-York, 1992, pp 537-540

7. Kraft SL et al.: Biodistribution of boron in normal canine tissues following borocaptate sodium administration and the effect of plasma exchange. In Allen BJ (ed): Progress in Neutron Capture Therapy for Cancer. Plenum Press, New-York, 1992, pp 489-492

RESULTS OF CONTINUED CLINICAL INVESTIGATIONS OF BSH IN PATIENTS WITH MALIGNANT GLIOMA

Dietrich Haritz[1,2], Detlef Gabel[2], Harold Klein[3], and René Huiskamp[4]

[1]Dept. of Neurosurgery, ZKH St. Jürgen-Straße, Bremen, Germany (Director: Prof. Dr. med. K. Piscol)
[2]Dept. of Chemistry, University of Bremen, Bremen, Germany
[3]Inst. of Clinical Neuropathology, ZKH Ost, Bremen, Germany
[4]ECN Petten, Petten, The Netherlands

INTRODUCTION

The development of Boron Neutron Capture Therapy (BNCT) as an alternative, more successful treatment modality for malignant glioma (WHO grade III-IV) is the aim of the European Collaboration on BNCT. As a part of it the group in Bremen started a clinical phase-I-study to investigate the pharmacokinetics, biodistribution and possible toxicity of the boron compound $Na_2B_{12}H_{11}SH$ (BSH). Data about the distribution in tissues eventually exposed to the neutron beam and preliminary results of the pharmacokinetics of BSH have been reported previously (1).

We wish to report here on the continued clinical investigations of BSH in patients with malignant glioma. Especially, results concerning the spatial determination of BSH in relation to the histopathological status of the tissue will be reported. Of major interest are the patient-specific variations of blood pharmacokinetics and urinary excretion, and their implications for the time point of therapy.

PROTOCOL

BSH is infused at a maximal concentration of 30 mg boron per kg body weight. Permission was obtained to administer BSH twice at a maximal concentration of 15 mg boron per kg each. Blood samples are taken at predetermined intervals up to 168 hours (post start of infusion). For determining alternative excretion pathways feces are collected in addition to urine samples.

BORON ANALYSIS AND HISTOLOGICAL INVESTIGATIONS

The concentration of boron in blood and tissue is measured by quantitative neutron capture radiography (QNCR) (3), in urine and feces by induced plasma-atomic emission spectroscopy (ICP-AES) at ECN Petten. In addition to the described method in order to correlate the concentration of boron with the type of tissue (1) we developed a method for areal correlation between the boron concentration and the histopathological diagnosis. After obtaining a tissue slice for QNCR the remaining tissue is cut with a thickness of 8 μm, mounted on a glass slide, fixed in methanol, stained and evaluated neuropathologically in a 1-mm grid. The boron concentration determined by QNCR can thus be compared directly, over the area of the embedded tissue, with the differential histopathological diagnosis.

Advances in Neutron Capture Therapy, Edited by
A.H Soloway *et al.*, Plenum Press, New York, 1993

RESULTS OF THE TOXICITY INVESTIGATION

In none of the 24 patients that entered the study to date, any measurable acute toxic side effect of BSH was found. The different blood tests (1) revealed no significant differences compared to the control group. This was found for the patients with an administered dose of 30 mg boron per kg as well as for the patients with a double administration of BSH. Moreover the patients did not complain of any subjective disorders like nausea or vomiting.

BORON PHARMACOKINETICS

The characteristics of the course of the boron concentration in blood showed no differences, in principle, for the patients who received an increased dose of BSH or received the compound twice. Fig. 1 shows the boron concentration in blood as a function of time for a patient receiving two infusions of BSH (15.3 mg boron/kg) with a 48 hours interval(1).

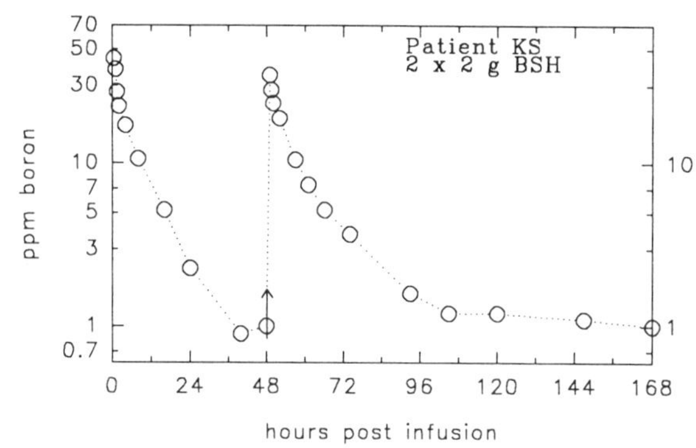

Figure 1. *Pharmacokinetics of BSH in blood. Data for one of the patients infused with 2 g BSH twice (= 2 · 15.3 mg ^{10}B/kg). Decline of concentration in blood follows a biphasic kinetic.*

Previously, it was shown that between 38 % and 96 % of the boron administered is excreted through the urine, the remainder being excreted through an unknown pathway or retained in the body. There is no correlation between the boron concentration in blood and the amount of boron excreted in urine (r = 0.1474)as shown in Fig. 2. For the patients HN and HS was found around 15 ppm boron in blood. Patient HS had excreted only 2% of the given dose, whilst patient HN had excreted more that 80% during the same interval. It is therefore probable that the boron compound is retained in a deeper compartment with variations from patient to patient.

Some boron is excreted in feces (see Fig. 3). Over the observation time around 1500 mg were excreted in urine and feces. Boron is thus excreted also through an entero-hepatic pathway. Moreover, boron not excreted over the observation time is retained in a deeper compartment.

BORON DISTRIBUTION IN TISSUE

Using the biopsy needle guided technique (1), compound accumulation within the tissue can be correlated to its histological status. For correlating larger areas, we used cryosections obtained adjacent to QNCR sections, for histological analysis. Fig. 4 shows that even within tumorous tissue, large variations in boron concentration must be expected, with some relation to the histological status.

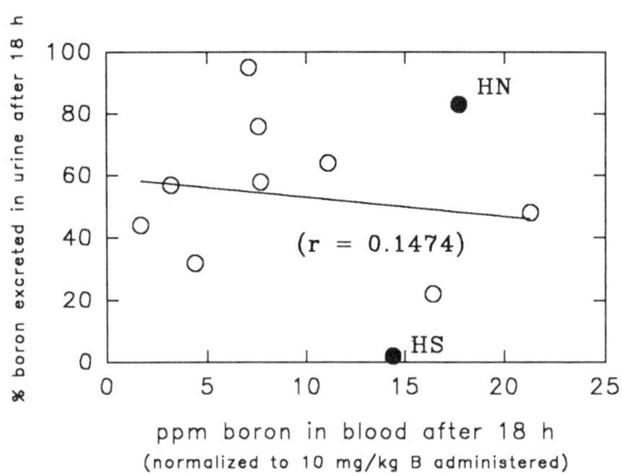

Figure 2. *Relation between boron concentration in blood and amount excreted 18 hours after infusion. Data for 11 patients (out of 25). Dotted circles are representing the patients HN and HS. Pay attention to large variations of boron concentration in blood at the same time.*

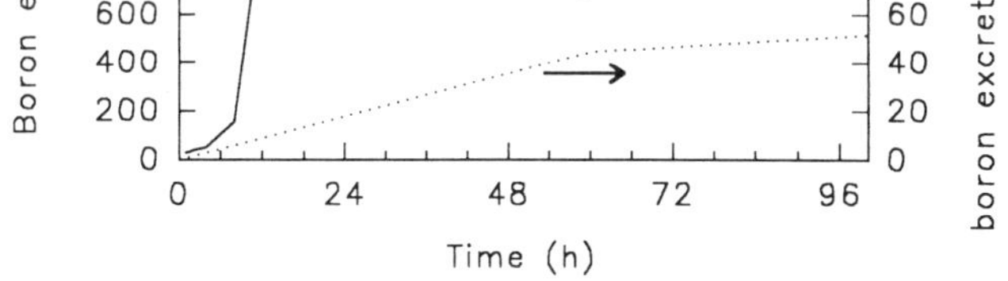

Figure 3. *Excretion of BSH in urine and feces. Data for one of the patients. Normal line is representing excretion in urine, dotted line excretion in feces. Around 150 mg ^{10}B of the total dose is retained in the body.*

				G >100	T >100	Td >100	T 85	Ts 10	T 5	Ts 4		
		Tn 7	T 9	Ts 6	T 13	Td 22	Td 16	GT 17	T 20	T 25	T 15	
	T 13	Ts 10	T 7	T 7	T 9	T 11	T 13	T 15	T 19	Ts 19	Ts 54	T 9
T 5	Ts 5	Ts 8	Ts 13	T 20	T 19	T 13	Ts 63	T 18	Tn 64	Ts 61	Ts 52	T 3
	Ts 8	T 9	Ts 11	T 8	T 11	T 14	Tn 14	T 2	Td 45	T 16	Td 35	
		Ts 11	Ts 8	T 12	Td 16	Ts 13	Td 13	Td 14	T 35	T 9	Tn 11	
		T 10	T 16	T 16	GT 12	T 11	Td 49	Td 49	GT 12	T 4		
			T 10	T 22	Td 16	T 10	Td 29	Td 19	Td 15			
				GT 10	Td 15	T 20	Td 41	Td 26	GT 13	T 16		
				T 17	Ts 22	T 22	Td 23	Td 20	Td 21			
							Td 18					

Figure 4. *Histological diagnosis and boron concentrations in ppm evaluated in a 1-mm-grid. G = blood vessel with erythrocyts, T = tumor tissue, Ts = spongeous tumor tissue, Td = dense tumor tissue, Tn = necrotic tumor tissue, GT = tumor tissue and blood vessel.*

DISCUSSION

The continued clinical investigations of boron distribution on patients with malignant glioma showed that large variations of the blood concentrations over the observation period must be expected. Fig. 2 illustrates vastly the different concentrations of boron in blood 18 hours after cessation of the infusion. This is of importance for the timing of the treatment. Using Prompt-Gamma Spectroscopy the actual boron concentration in blood must be determined right before the start of irradiation.

It was shown (see Fig. 2 and 3) that the excretion of boron in urine is independent from the concentration in blood and the retained amount of boron in the body differs from patient to patient.

REFERENCES

1. D. Haritz, D. Gabel, H. Klein, R. Huiskamp, and O. Petterson. BSH in patients with malig nant glioma. Distribution in tissues, comparison between BSH concentration and histology, *in:* Boron Neutron Capture Therapy: Towards Clinical Trials of Glioma. D.Gabel, R.L. Moss, ed., Plenum Press, New York, 163-174 (1992) (in press).
2. D. Haritz, D. Gabel, and K. Piscol. The distribution of BSH in patients with malignant glioma, *in:* Progress in Neutron Capture Therapy for Cancer. B.J. Allen, D.E. Moore, B.V. Harrington, eds., Plenum Press, New York, 557-560 (1992).
3. D. Gabel, H. Holstein, B. Larsson, L. Gille, G. Ericson, D. Sacker, P. Som, and R.G. Fairchild. Quantitative neutron capture radiography for studying the biodistribution of tumor-seeking boron-containing compounds, *Cancer Res.* 47:5451-5456.

^{111}In- AND ^{99m}Tc-LABELED LOW DENSITY LIPOPROTEIN UPTAKE BY HIGH-GRADE GLIOMAS

Merja Kallio,[1] Jaana Leppälä,[2] Tuomo Nikula,[5] Päivi Nikkinen,[3] Helena Gylling,[4] Markus Färkkilä,[1] Jukka Hiltunen,[5] Juha Jääskeläinen,[2] and Kristian Liewendahl[3]

[1]Departments of Neurology, [2]Neurosurgery, [3]Clinical Chemistry, Division of Nuclear Medicine, [4]II Department of Medicine, Helsinki University, 00290 Helsinki, Finland and [5]MAP Medical Technologies Inc., Otakaari 3A, 02150 Espoo, Finland.

INTRODUCTION

The prerequisite for the boron neutron capture therapy (BNCT) of brain tumors is the specific and high accumulation of boron into the tumor. One of the suggested possible carriers of boron into the tumor is low density lipoprotein (LDL). The rationale for using LDL is that all growing tissues need cholesterol, likewise brain tumors, and LDL is the main carrier of cholesterol in the body.

LDL is a spherical particle with a diameter of 22 nm. It contains an oily core of about 1500 molecules of cholesterol esters. Its hydrophilic coat contains about 800 molecules of phospholipid, 500 molecules of unesterified cholesterol, and a 400 kDa protein called apoprotein-B. LDL can be boronated by substituting the cholesterol esters in the core of LDL by boronated fatty acids (1).

LDL is taken into cells by receptor mediated endocytosis (2). In the central nervous system LDL receptors are found for example in vascular endothelial cells (3) and also brain tumors. The number of LDL receptors in brain tumors has varied over a wide range, but seems to be increased when compared with normal brain (4). Glioma cell lines have also shown increased uptake of cholesterol via LDL receptors (5).

It is not known how much LDL is taken up by brain tumors in vivo. In order to find out we performed gamma camera imaging of the brain after intravenous injection of autologous LDL labeled with radioisotopes.

Advances in Neutron Capture Therapy, Edited by
A.H Soloway *et al.*, Plenum Press, New York, 1993

PATIENTS AND METHODS

Four patients with recurrent high-grade glioma have so far entered the study. LDL was separated from blood samples by sequential ultracentrifugation. The autologous LDL was labeled with ^{111}In for two patients and with ^{99m}Tc for the other two. ^{111}In was coupled to LDL with DTPA and ^{99m}Tc by reducing the cysteine based S-S groups to sulfhydryls using the ascorbic acid method (6, 7). 1.3 and 3.0 mCi of ^{111}In and 18.6 and 35 mCi ^{99m}Tc were injected, respectively. Gamma camera imaging was started 2 hours and 20 hours after the intravenous injection of the radiolabeled LDL. In order to assess the magnitude of nonspecific accumulation of protein due to blood brain barrier disruption present in high grade gliomas a control study with ^{99m}Tc-labeled human serum albumin (HSA), (15 mCi), was obtained from the other patient studied with ^{99m}Tc-LDL. Radioactivities in blood were measured during the first 24 hours after the injection of the labeled compounds.

Two patients were studied with ^{111}In-labeled LDL (Table 1). They had been operated for high-grade glioma one and two years earlier, respectively, and after operation both had received whole brain irradiation of 40 Grays with a tumor boost of 20 Grays. Symptoms due to tumor recurrence were noted and recurrent tumor was diagnosed neuroradiologically; in the other patient the recurring tumor mass was small.

Two other patients were studied with ^{99m}Tc-labeled LDL (Table 1). They also had recurrent glioma. The other patient had been operated twice and had received radiotherapy, whereas the other patient had been operated only once prior to radiotherapy.

Table 1. Patients, amounts of isotope injected, and tumor vs non-tumor activity ratios.

Sex	Age	Time from last operation	Label	Injected dose	Tumor vs non-tumor activity ratios (hours from injection) ≈2 hours	≈20 hours	≈24 hours
male	23 years	1 year	^{111}In	1.3 mCi	1.5		1.2
male	57 years	2 years	^{111}In	3.0 mCi	1.2		1.1
male	35 years	1 year	^{99m}Tc	18.6 mCi	1.4	2.7	
female	42 years	1 year	^{99m}Tc	35 mCi	1.5	2.5	

RESULTS

The patients studied with ^{111}In-labeled LDL showed slight uptake of isotope in the tumor area. The tumor area to opposite non-tumor area activity ratios were 1.5 and 1.2 after 2 hours, 1.2 and 1.1 after 24 hours (Table 1). These low activity ratios apparently represent nonspecific protein accumulation due to blood brain barrier disruption. Marked accumulation of radioactivity was observed in kidneys and bladder in accordance with previous observations (6). As LDL is not excreted via the kidneys this finding suggested that there was in vivo release of label. These results lead us to change over to ^{99m}Tc-label.

The two patients studied with ^{99m}Tc-labeled LDL showed uptake of isotope in the tumor area so that the tumor vs. non-tumor activity ratios were 1.4 and 1.5 after 2 hours, and 2.7 and 2.5 after 20 hours for the respective patient (Table 1). The activity ratios in the ^{99m}Tc-HSA control study in the other patient were 1.4 at 2 hours and 1.7 at 20 hours, respectively.

The blood activities after injection of ^{99m}Tc-labeled LDL were 86% at 1, 71% at 4, 65% at 8, 55% at 12, and 44% at 21 hours from the injection. Approximately 99% of the radioactivity detected in the blood was coupled to the protein moiety at 21 hours.

DISCUSSION

BNCT should not be started until a reasonably specific carrier for boron has been found. The results of the ^{99m}Tc-LDL gamma camera imaging data indicate that glioma tissue does take up LDL in vivo. The uptake is significantly higher than that of the ^{99m}Tc-HSA control study. The uptake of LDL increases at least up to 20 hours and the radioactivity persists in the tumor area implying that the LDL is taken up intracellularly and not only accumulating in the tumor area due to diffusion across the disrupted blood brain barrier. The blood clearance is identical to that of ^{125}I-labeled LDL, the standard reference label, indicating good binding of the radioisotope to LDL (8). As efficient BNCT requires sufficient difference in boron concentrations between blood and tumor, LDL seems promising. Our results warrant further studies for the development of LDL as a boron carrier into brain tumors for BNCT.

REFERENCES

1. Kahl, S. B. & Callaway J. C. (1989) *Strahlenter. Onkol.* **165,** 137-139.
2. Brown M. S. & Goldstein J. L. (1986) *Science* **232,** 34-47.
3. Méresse S., Delbart C., Fruchart J-C. & Cecchelli R. (1989) *J. Neurochem.* **53,** 340-345.
4. Rudling M. J., Angelin B., Peterson C. O. & Collins V. P. (1990) *Cancer Res.* **50,** 483-487.
5. Murakami M., Ushio Y., Mihara Y., Kuratsu J., Horiuchi S. & Morino Y. (1990) *J. Neurosurg.* **73,** 760-767.
6. Rosen J. M., Butler S. P., Meinken G. E., Wang T. S. T., Ramakrishnan R., Srivastava S. C., Alderson P. O. & Ginsberg H. N. (1990) *J. Nucl. Med.* **31,** 343-350.
7. Thakur M. L., Eshbach J., Wilder S., John E. & McDewitt M. R. (1992) "IXth International Symposium on Radiopharmaceutical Chemistry" (Conservatoire National des Arts et Mediérs, Paris), pp. 365-367.
8. Vallabhajosula S., Paidi M., Badimon J. J., Le N-A., Goldsmith S. J., Fuster V. & Ginsberg H. N. (1988) *J. Nucl. Med.* **29,** 1237-1245.

THE BIODISTRIBUTION OF p-BORONOPHENYLALANINE-FRUCTOSE IN HUMAN PATIENTS WITH GLIOMA AND METASTATIC MELANOMA

Julia L. Mallesch[1,2], Douglas E. Moore[1], Barry J. Allen[2], William H. McCarthy[3], Robert Jones[4] and Warwick A. Stening[5]

[1]University of Sydney, NSW 2006, Australia
[2]Ansto, PMB 1, Menai, NSW 2234, Australia
[3]Royal Prince Alfred Hospital Camperdown, NSW 2050, Australia
[4]Prince of Wales Hospital, High St. Randwick, NSW 2031, Australia
[5]Prince Henry Hospital, Little Bay, NSW 2036, Australia

INTRODUCTION

Modern diagnostic techniques and the introduction of radiotherapy and chemotherapy have not improved the prognosis for certain cancers where life expectancy after diagnosis is less than 12 months. High grade brain tumours and metastatic melanoma to the brain fall into this category. Their common feature of infiltrating malignant cells contraindicates curative surgery and radiotherapy.

BNCT offers the advantage of localised treatment of the tumour with minimal damage to surrounding normal tissue. In addition it has potential to control subclinical metastases. While animal models are useful in selecting the most effective boron compounds, the human patient is the ultimate test of efficacy. Our current work has focused on racemic p-boronophenylalanine fructose (BPA.f) biodistribution in patients with metastatic melanoma or high grade glioma patients. We report here on the distribution and pharmacokinetics of the first ten patients.

PROTOCOL

For patients to be eligible to participate in the study they must have a pre-operative diagnosis of a high grade glioma or metastatic melanoma. They were informed about the objective of the study and asked to participate on a voluntary basis. The protocol was approved by the participating hospital ethics committees.

1) Metastatic melanoma: 960 mg D,L-BPA preparation complexed to fructose (BPA.f) was intravenously (iv) injected as a 30 ml bolus 1-4 hours before surgery.

2) High grade glioma: (i) Patients 1-3: 960 mg D,L-BPA.f was injected as a 30 ml iv bolus 1-4 hours before surgery. (ii) Patients 4-6: 20-30 mg/kg D,L-BPA.fructose is injected as an iv bolus 3-7 hours before surgery.

BPA.f was prepared aseptically. Each multidose vial held 960 mg unenriched boron in the form of D,L BPA in 30 ml of a 0.3 M D-fructose solution at pH=8. Each batch was tested for pathogens and pyrogens before use. The drug was dispensed on an "individual patient use" basis.

Blood samples were collected for 24 hours after injection of the BPA. During the operation, tumour and other appropriate normal tissues, eg skin, dura, periosteum and surrounding normal brain samples were collected. All tissue and blood samples were digested, diluted, then analysed for boron content using inductively coupled plasma-Atomic emission spectrometry[1] (ICP-AES).

Due to the low administered dose and the rapid clearance of boron from the blood, the scheduling of the tissue sampling after the injection of the BPA was at the earliest practical time. The preferred sample weight was 400 mg for the blood and tissue, but in some cases this could not be achieved. The minimum sample weight of solid tissues was designated as 100 mg. In the analysis, the tissue sample was digested and made to a volume of 5 ml. The sensitivity limit of ICP-AES technique for boron is 0.015 ppm boron. For 5 ml dilution, this limit corresponds to 0.085 μg B in the sample. For sample weights of 100 mg, the sensitivity limit was 0.75 μg/g, but for 50 mg samples it increases to 1.5 μg/g. Thus a result of "no boron detected" (ND) corresponds to a limit determined by the size of the available sample.

RESULTS

Tables 1 and 2 list the tissue boron concentrations (expressed as μg B/g tissue) as well as tumour:blood ratios, for the melanoma patients and glioma patients respectively. The results for the cerebral melanoma patient are given in both tables as this patient has data pertinent to both groups.

DISCUSSION

Metastatic Melanoma

Considerable tumour uptake was seen in all melanotic or amelanotic melanomas. The variability between the tumour nodules in different sites of a patient's body reflect a

Table 1. Metastatic melanoma patient data

TUMOUR TYPE	DOSE BPA mg/kg	TIME (h)~	TUMOUR SITE [B]		BLOOD [B]	SKIN [B]	MUSC [B]	FAT [B]	TUMOR/ BLOOD
1:CUTANEOUS		2.8	ABD^	2.24	0.45	0.56	-	ND	5.1 ± 1.4
AMELANOTIC		2.8	GROIN	2.12	0.45		-	0.45	4.8 ± 1.1
	16.6	3.2	CHEST-TP*	5.66	0.36		ND	-	15.7 ± 4.6
		3.2	CHEST-TC+	1.48	0.36		-	-	4.1 ± 0.4
		4.7	BACK	1.02	0.15		0.08	-	6.9 ± 0.3
2:CUTANEOUS		2.5	MAND	1.56	0.47	-	ND	1.23	3.3 ± 0.4
MELANOTIC		2.5	CHEEK	2.94	0.47	0.0	-	-	6.3 ± 0.6
	9.9	2.8	GROIN	0.73	0.47	-	-	ND	1.6 ± 0.4
		2.9	BACK	0.50	0.47	-	-	ND	1.1 ± 0.2
		2.8	SCALP	ND	0.47	0.22	-	-	ND
3:CUTANEOUS AMELANOTIC	11.1	3.1	GROIN	0.70	0.68	0.35	ND	ND	1.0 ± 0.3
4:CEREBRAL	10	2.8	TC	1.91	0.39	0.10	-	-	4.9 ± 0.3
MELANOTIC		2.8	TP	3.87	0.39				10.0 ± 0.8
MEAN ± SD	12.6 ± 3.2			1.9 ± 1.5	0.4±0.1				5.0 ± 4.1

~ = Time interval between BPA injection and sampling; ^ABD = Abdomen; *TP = Tumour periphery; +TC = Tumour centre; #MAND = Mandibular; - = No sample; ND = No detected boron. All tissue sample [B] expressed as μg B/g tissue.

heterogeneous nature of the melanoma, and variations in the degree of melanisation and calcification were observed. The boron levels in the normal tissues such as skin, muscle and fat were also variable and in many cases boron levels were undetectable at the dose given.

For one of the cutaneous melanoma patients as well as the cerebral melanoma patient, significant differences were seen in the boron levels of the tumour centre compared to the growing tumour margin. This may reflect a more necrotic tumour centre. The tumour:blood ratios seen for the melanoma patients are encouraging but the variable nature of the uptake highlights the importance of knowing the tumour boron concentration before neutron irradiation. The cerebral metastatic tumour demonstrated good boron uptake and tumour: blood and tumour:brain ratios were of the order of 10 to 1. This result is extremely encouraging and, although it may not be typical of all cerebral melanomas, BNCT would be indicated for this patient.

High Grade Brain Tumours

The administered dose for the first two patients was the same as for the melanoma patients. However no detectable tumour boron levels were seen even though the blood values were significant. Tumour samples obtained for ICP-AES analysis from the glioma patients were smaller than for the melanoma patients and the boron content was below the limits of detection.

Table 2. High grade brain tumour patient data.

TUMOUR TYPE	DOSE BPA mg/kg	TIME (h)*	TU [B]	BL [B]	SK [B]	MU [B]	BR [B]	DU [B]	BO [B]	PE [B]	TU/BL RATIO
1:GLIOMA GRADE II	12.6	5-6.5	ND	0.66- 0.48	-	ND	ND	-	ND	-	ND
2: GLIOMA GRADE III	13.7	1.0	ND	0.5	ND	ND	0.12	-	0.57	-	ND
3: GLIOBLASTOMA MULTIFORME	20.6	2.2	ND	0.74	ND	0.70	ND	ND	0.35	ND	ND
4: ANAPLASTIC ASTROCYTOMA	20	7.6 7.7	TC 1.38 TP 2.01	0.41 0.54	ND	-	ND	ND	ND	ND	2.6±0.7 4.2±1.7
5: ASTROCYTOMA GRADE II	30	4.4 4.9	TC 1.66 TP 1.44	1.03 0.92	ND	1.07	0.92	ND	0.81	-	1.6±0.2 1.6±0.3
6: GLIOBLASTOMA MULTIFORME	30	2.4	0.58	0.71	ND	-	-	-	1.32	2.12	0.8±0.1
MEAN ± SD	21.2±6.9		1.4±0.5	0.7±0.2							2.2±1.2
7: METASTATIC MELANOMA	10	2.8	TC 1.91 TP 3.87	0.39	0.1	-	0.24	ND	ND	0.66	4.9±0.3 10.0±0.8

* = Time interval between BPA injection and tumour sampling; TU = Tumour; BL = Blood; SK = Skin; MU = Muscle; BR = Brain; DU = Dura; BO = Bone; PE = Periosteum; - = No sample; ND = No detected boron; TC = Tumour centre; TP = Tumour periphery. All tissue sample boron concentrations [B] expressed as µg B/g tissue.

A modified regimen was used for patients 3-6. The BPA.f dose was increased to 20-30 mg/kg. To investigate whether there is a delayed absorption of the BPA.f into glioma, BPA.f was administered 7 hours before surgery in patient 4 (anaplastic astrocytoma). Although tumour boron levels were now detectable it was considered that the improvement was due to the dosage increase, not the change in the time of tissue sampling. The dose of BPA was further increased to 30 mg/kg to aid detection of boron levels in other normal tissues.

At the increased dose level, enhanced boron uptake into the tumour was seen in two patients but not to the same magnitude as for the melanoma patients. Generally the glioblastoma multiforme tumours exhibited very poor uptake. These tumours are often very necrotic and this may contribute to the poor boron levels. The uptake into the viable cells or in other subclinical metastases is not known. With the astrocytomas tumour:blood ratios of greater than one were observed. In one of these patients the increased boron uptake in the growing tumour periphery was higher than in the tumour centre.

The blood boron levels for the final three glioma patients were below or equivalent to the levels seen in the tumour. With a few exceptions the levels in the other normal tissues were below that of the blood and the tumour. The levels of boron found in normal brain and muscle were generally equivalent to the blood levels. No detectable levels were found in the dura or the scalp but only small samples could be obtained for these tissues (≈50 mg) and boron detection was at the limits of sensitivity. Significant boron levels were found in the periosteum for patients 5 and 6.

Comparison of our data with that of Coderre[2] and Mishima[3] shows comparable normalised blood levels. While our patient numbers are still small, the data suggests that a superior tumour:blood ratio with the use of L-BPA rather than with racemic BPA is found. It will be valuable to increase the number of human patients in our own data set so that we may achieve some unequivocal conclusions as to the usefulness of BPA.

CONCLUSION

Selective tumour uptake of D,L BPA.f was observed for melanoma patients. Our results indicate that BNCT with intravenous BPA.fructose could be effective for metastatic melanoma, especially for the single case of cerebral metastasis reported in this paper. Uptake into glioma is found to be less selective than for melanoma.

REFERENCES

1. S.R. Tamat, D.E. Moore, B.J. Allen, Determination of boron in biological tissues by inductively coupled plasma atomic emission spectrometry, *Anal Chem.* 59:1261-2164 (1987)
2. J.A. Coderre, A phase 1 biodistribution study of p-boronophenylalanine. In: Proceedings of an International Workshop and Plenary Meeting, "Towards Clinical Trials of Glioma with BNCT", 18-20 September, 1992, Petten, The Netherlands, Plenum Press, (in press).
3. Y. Mishima, M. Ichihashi, C. Honda, M. Shiono, T. Nakagawa, H. Obara, J. Shirakawa, J. Hiratsuka, K. Kanda, T. Kobayashi, T. Nozaki, O. Aizawa, T. Sata, H. Karashima, K. Yoshino and H. Fukuda, Advances in the control of human cutaneous primary and metastatic melanoma by thermal neutron capture therapy, in: "Progress in Neutron Capture Therapy," B.J. Allen *et al*, ed., Plenum Press, New York, 1992; 577-583.

ON THE PROGRESS TOWARDS CLINICAL TRIALS AT PETTEN (POST-SYDNEY)

R.L. Moss

High Flux Reactor Division
Joint Research Centre
Petten, 1755 ZG, The Netherlands

INTRODUCTION

The paper summarises the progress over the last 2 years towards the implementation of clinical trials at Petten. At the last International Symposium on NCT at Sydney in December 1990, it was reported that an epithermal neutron beam had been installed at the High Flux Reactor in August of that year and despite being in an operational state, the planned research and experimental programme had still to begin[1].

This paper presents the current status of the Petten project, with comments on the development of the clinical set-up for the start of the clinical trials, now planned to start towards the end of 1993. Most of the activities and results given here are presented in more detail by colleagues in the appropriate sessions in these proceedings.

THE HIGH FLUX REACTOR AT PETTEN AND BEAM TUBE HB11

The High Flux Materials Testing Reactor (HFR) at Petten is owned and managed by the Commission of the European Communities. For the last 25 years, the HFR has been utilised primarily for the irradiation testing of materials and nuclear fuels for the European civil nuclear power programme. The reactor operates at 45MW and is cooled and moderated by light water.

For BNCT applications, an epithermal neutron beam was designed and installed in 1990, at the beam tube HB11 at the north side of the HFR reactor vessel[2]. The beam tube configuration and surrounds were chosen not only because of their nuclear

Advances in Neutron Capture Therapy, Edited by
A.H Soloway *et al.*, Plenum Press, New York, 1993

characteristics but also because of the large working area at the exit side of the beam where an irradiation room for performing experiments and later for the eventual treatment of patients could be readily accommodated. Furthermore, the facility faces the reactor building's emergency exit, which in effect gives unhindered access from the outside.

The design of the filter configuration was performed[3] jointly between the Petten group and groups at AEA Harwell (UK) and JRC Ispra (Italy). The following combination of filter materials to produce the optimal beam characteristics was chosen: 15cm Al; 5cm S; 1cm Ti; 0.1cm Cd; and 150cm liquid Ar. The design calculations predicted a beam with a neutron flux intensity of 1.07×10^9 $cm^{-2}s^{-1}$, a mean neutron energy of 7.75 keV, and a gamma dose rate of 68.9 cGy hr^{-1}.

STATUS OF THE PROJECT IN EARLY 1991

Due to stringent safety criteria, as laid down by the local safety committees, permission to use the beam at full reactor power could not be granted until extra safety provisions, such as back-up systems, emergency procedures etc. had all been documented and demonstrated. In addition, the uncertainties attached to using liquid argon meant that initially permission to open the beam could only be granted at low reactor power. Apart from some initial dosimetry measurements at low power, improvements were made to the shielding of the irradiation room, and to the operation of the beam shutters. Some 12 months following installation, permission was given and measurements could be performed at full reactor power. The improved irradiation room, also in its current configuration, is shown in Fig.1.

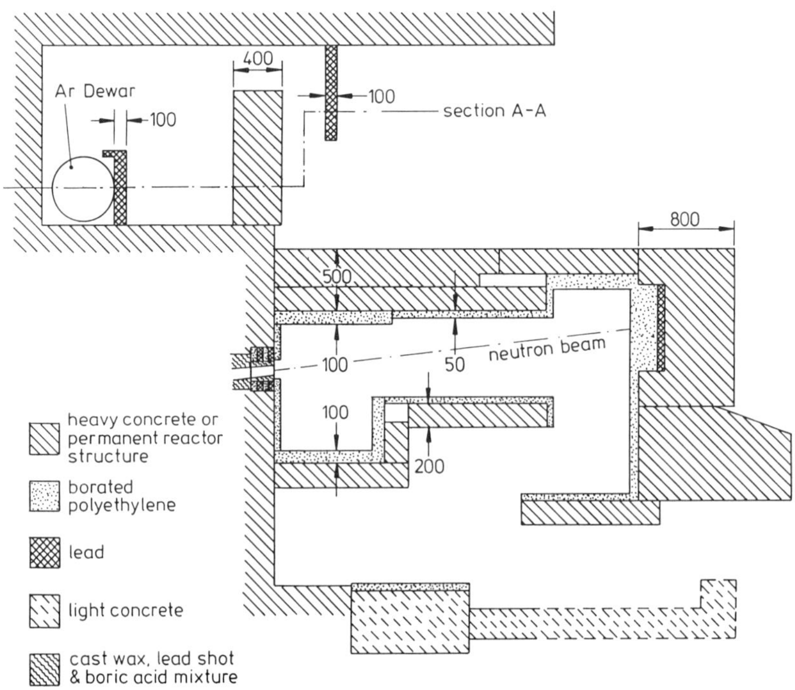

Figure 1. Cross-section through present irradiation room at the Petten HB11/BNCT facility. (Dimensions in mm)

FIRST MEASUREMENTS AT FULL REACTOR POWER, SUMMER 1991

In July (1991), a collaborative effort between INEL USA, AEA Harwell UK, ECN and JRC Petten carried out various measurements using stacked resonance-energy activation foils, proton recoil proportional counters and thermoluminescence detectors (TLD). The resultant exercise[4] gave a total neutron flux intensity of 4.9 x 10^8 cm^{-2} s^{-1}; an epithermal neutron flux intensity of 3.8 x 10^8 cm^{-2} s^{-1} ; a fast neutron dose of 1.5 Gy/hr and a gamma dose of 1.2 Gy/hr. The mean neutron energy was 10.4 keV. The neutron flux intensity was a factor of 2 less than predicted. The major discrepancy was due to the fact that natural argon contains just 0.34% of ^{36}Ar, the rest being ^{40}Ar. The contribution of ^{36}Ar to the overall capture cross-section of natural argon had been overlooked. The neutron capture cross-section for ^{36}Ar is approximately 100x greater than that for ^{40}Ar.

The measurements confirmed that the beam is highly collimated with an angular spread of less than 10 degrees.

Experiments with two plastic phantoms were also performed[5]. The first was a realistic model of a standard beagle head and the second, a solid cylindrical model. Drilled holes distributed in the phantoms allowed activation wires and TLD's to be placed. The results confirmed the lower than expected flux levels. In comparison with similar measurements at Brookhaven, the Petten flux levels were lower by about 30%, but at depth, the thermal neutron flux levels were similar, strongly indicating the advantage of a well-collimated beam in comparison with an isotropic beam, as at Brookhaven.

The complete experimental exercise during this summer period included the first attempts with cell culture irradiations and the first 2 irradiations in the healthy tissue tolerance study using beagle dogs (see later).

Despite having performed our first irradiations, it was proving difficult to operate and control the liquid argon system in a stable condition. Also, it was only possible to maintain the argon completely liquid for periods of 3 hours. It was therefore decided that the argon system had to be redesigned and modified. The resulting technical exercise took until April 1992 to complete.

SCOPE OF WORK, SINCE APRIL 1992

Since April 1992, the facility has been in excellent working order. There has been an intensive work schedule, whereby numerous irradiations have been performed. These are summarised as follows:

Beam Characterisation

Free Beam Measurements. Neutron and gamma dosimetry measurements have been performed on a monthly basis. Foil stacks, using 10 different activation materials, some covered with Cd, and along with TLD's, are placed in the centre of the beam at the therapy position. Of the 3 measurements performed to date, there is only a 1-2% variation in results, thus inferring beam stability with respect to intensity and spectrum changes.

Phantom Measurements. The beagle-head phantom exercise has been repeated using activation wires only. The results confirm the lower flux intensities measured the previous year. Measurements using fission chambers inserted into a polyethylene block, belonging to the Dutch Cancer Institute, and other measurements using a hollow, cylindrical phantom filled with tissue equivalent liquid, containing activation foils and TLD's distributed at depth, have also been performed (results pending).

Healthy Tissue Tolerance Studies

The healthy tissue tolerance study continued and, following the revised beam characteristics, a new matrix of experiments has been approved, and includes both single and multi-fractionated experiments. Following 15 irradiations so far, current results indicate, amongst others that the skin can be well tolerated by the projected doses of the epithermal beam (see Phillip, these proceedings), and that the healthy tissue tolerance dose can be detected by Magnetic Resonance Imaging (see Siefert, these proceedings).

Cell Culture Experiments

Cell culture experiments were restarted with the irradiation of V79 Chinese hamster cells in suspension placed in a hollow, cylindrical, polyethylene phantom filled with tissue equivalent liquid (see Huiskamp, these proceedings). This collaborative effort has continued with the irradiation of U251 mouse glioma, U87MG human glioma and B16 melanoma cells, using different concentrations of boric acid, BSH, BPTU-1 and BPA, again results pending.

On-line Monitoring of the Epithermal Beam

Efforts have started towards the development of an on-line monitoring system for the beam. The aim is to verify, through consistency, that the beam remains stable during therapy. Currently, twinned GM tubes and fission chambers have been inserted into the collimator wall behind the outer (secondary) beam shutter. Progress towards using a transmission chamber that covers the whole beam port, and a clock interlock system are also under development.

Treatment Planning

A treatment planning scheme is currently under development (see Watkins, these proceedings). For BNCT, conventional treatment planning schemes which usually employ a semi-empirical approach are of limited use only, as the neutrons in the incident beam, unlikely photons, will spread quite extensively in the target volume. Hence, the present code will be extended considerably by incorporating 3D treatment planning routines, with full 3D graphics capabilities.

Design of the Treatment Room for Patient Trials

The current set-up of the irradiation room, see Fig.1, is only suitable for the experiments described above. It will be necessary to enlarge the facility for patient irradiations. The design of the irradiation room is currently in progress (see Constantine, these proceedings). The maximum space available is approximately 4 x $5m^2$ in floor space, with a maximum height of 2.1m. The working area or treatment area around the beam exit will be of sufficient size to allow the head of the patient to be irradiated from two laterally opposed directions.

Apart from creating space, emphasis on shielding requirements cannot be understated. Through shielding calculations, the quantity and size of shielding material will need to be determined. Within the room, provisions will be made available to monitor the beam, position the patient, observe the patient, and place the patient for treatment.

For medical observation purposes, an instrumentation panel to display the essential beam parameters is planned. Access from outside to the reactor building, and hence the facility, will be via the reactor building's emergency exit. Currently, for clinical trials, no permanent hospitalisation facilities are foreseen. It is considered that temporary accommodation, ie. in the form of portacabins, will suffice.

CONCLUDING REMARKS

With the epithermal neutron beam at Petten now in an excellent working order, the goal to determine all the pre-conditions necessary for implementing clinical trials should be completed sometime during summer 1993. During this period the treatment room, currently in design, should be almost complete.

The work at Petten is central to the future of BNCT in Europe. With a successful demonstration of BNCT at Petten, it is anticipated that at other reactor and accelerator centres in Europe, other facilities will be developed for the treatment of tumours by BNCT.

REFERENCES

1. R.L.Moss, F.Stecher-Rasmussen, R.Huiskamp, L.Dewit and B.Mijnheer, "The Petten BNCT project", in : "Progress in Neutron Capture Therapy for Cancer", Ed. B.J.Allen et al, Plenum Press, Proc. Fourth Int. Symposium Neutron Capture Therapy for Cancer, Sydney (1990).

2. R.L.Moss, F.Stecher-Rasmussen, K.Ravensberg, G.Constantine, and P.R.D.Watkins, "Design, construction and installation of an epithermal neutron beam for BNCT at the high flux reactor Petten", idem.

3. P.R.D.Watkins, G.Constantine, F.Stecher-Rasmussen, W.Freundenreich, R.L.Moss, and R.Ricchena, "MCNP calculations for the design and characterisation of the Petten BNCT epithermal neutron beam", idem.

4. P.Watkins, M.Konijnenberg, G.Constantine, H.Rief, R.Ricchena, J.B.M.de Haas and W.Freundenreich, "Review of the physics calculations performed for the BNCT facility at the HFR Petten", Proceedings of an Int. Workshop on "BNCT: Toward Clinical Trials of Glioma Treatment", Petten, September 1991, Plenum Press.

5. R.L.Moss, A.Siefert, P.R.D.Watkins, G.Constantine and K.Philipp, "The Petten BNCT Facility : I - Phantom Dosimetry Techniques used in the Nuclear and Biological Characterisation of the Epithermal Neutron Beam and in the Healthy Tissue Tolerance Studies on the Canine Brain, and II - the Development towards a Treatment Planning System for the Treatment of Glioma Patients by BNCT", Proc. of an Int. Workshop on "Macro and Microdosimetry and Treatment Planning for Neutron Capture Therapy", November 1991, MIT, Cambridge, USA, Plenum Press.

DESIGNING A TREATMENT ROOM FOR BNCT CLINICAL TRIALS AT THE HFR, PETTEN

G. Constantine,[1] L. Dewit[2], R.L. Moss[1], B.J. Mijnheer[2], K. Ravensberg[3], F. Stecher-Rasmussen[3]

[1]Commission of the European Communities, Joint Research Centre, JRC, Petten, The Netherlands
[2]Netherlands Cancer Institute, Amsterdam, The Netherlands
[3]Netherlands Energy Research Foundation, ECN, Petten, The Netherlands

INTRODUCTION

The present irradiation facility adjacent to the outlet port of the HB11 epithermal, filtered beam of the High Flux Reactor (HFR) at Petten was envisaged as a low-cost, evolutionary stage on the way to realisation of BNCT. Locally known as the "dog tunnel", it is a labyrinth corridor leading to the beam exit port, constructed from blocks of heavy concrete, with a roof of lead bricks supported by 50 x 100 mm^2 channel section steel beams. The innermost leg of the labyrinth is lined with "polybor", sheets of polyethylene to which boric acid has been added during manufacture, to moderate and absorb scattered neutrons. The components closer to the reactor, including the filter elements and three independent beam shutter mechanisms, were built in their final form[1]. Following modifications to the system supplying liquid argon to the filter vessel, the complete installation has since given excellent service.

THE CLINICAL FACILITY

At the completion of the healthy tissue tolerance studies early in 1993 there will be a time window for building the irradiation facility described in this paper. Emphasis has been placed on achieving as large a treatment room as possible, and minimum cost, embracing re-use of existing materials where practicable. Features will include on-line beam dosimetry linked to automated timing of the irradiation period, a sliding shielded door for patient access, TV camera and sound communication between physician and patient, and a subsidiary labyrinth for rapid access by medical staff. These latter measures improve patient re-assurance, the last also renewing the possibility of a light path, with the help of mirrors, for direct viewing of the patient, which is proving invaluable in the present facility.

Advances in Neutron Capture Therapy, Edited by
A.H Soloway *et al.*, Plenum Press, New York, 1993

Design Constraints

The main engineering constraint imposed by the site of the HB11 facility is a severe restriction of available headroom. Several steel pillars in the vicinity support three floors, housing computers and ancillary equipment belonging to other experimental facilities, above the treatment room area. The first floor is carried on horizontal girders, below which there is a free space of only 2.51m. These floors also deny direct crane access, so all shielding components must be moved by forklift truck, airpads, cantilever arms using the reactor hall crane, or old-fashioned muscle power.

The shielding specification is based on MCNP calculations, and extrapolation of measurements on the present facility, taking into account a possible beam enlargement from 80 to 150mm diameter. Because of the low proportion of fast neutrons in the beam, the optimum shielding arrangement has neutron moderation and preferential capture by H and ^{10}B respectively in polybor, preceding the γ-ray absorption function. In practice 100mm polybor is needed where the shield is accessible from outside, though 50mm is enough for instance on the floor. A γ-ray shielding thickness of 156g cm^{-2} is adequate (450mm heavy concrete of density 3.5 g cm^{-3}, 200mm of steel, or 150mm of lead). Directly above the patient, who represents an intense source of 2.2 MeV capture γ-rays, the proximity of personnel on the first floor is minimal and a combination of 100mm steel and 100mm lead is called for.

Solutions

While the dog tunnel roof could be supported by quite small steel beams, the therapy facility will be much wider, as shown in Figure 1, where cross sections of both are compared. Beams adequate to carry the shielding weight would occupy much of the available space. The preferred solution is to combine support and shielding in the form of 100mm square cross section steel bars in a double layer with the vertical gaps staggered. The upper layer is replaced by lead bricks over the therapy position. The roof shielding must be placed progressively, starting from the reactor face, as subsequent access is very limited.

Placing 100mm square steel bars 5m long, each weighing nearly 400 kg, with their ends resting on the sidewalls as shown in Figure 1, presents special problems. A solution is to employ a counterweighted cantilever beam temporarily attached to each bar, to move the overall centre of gravity to a point accessible to the crane. A forklift truck lifts the counterweight to release the cantilever from the bar once it is in position. Many of the available heavy concrete blocks are 2.25m in height, giving sufficient room for the two layers of steel bars beneath the 2.51m to the underside of the first floor support girders. The headroom within the treatment room, taking into account linings of 50 and 100mm of polybor on floor and ceiling respectively, is thus a reasonable 2.1m. Figure 2 shows an over-all view of the therapy room.

ON-LINE MONITORING

This important requirement has been addressed by placing detectors in the beam-line. At present these include two fission chambers and two GM tubes located in the beam-shaping collimator assembly after the rotating shutter. This comprises a conical hole cast in the shielding material (lead shot, wax and boric acid), and further holes to house the detectors, as in Figure 3. It is intended to use these detectors to monitor

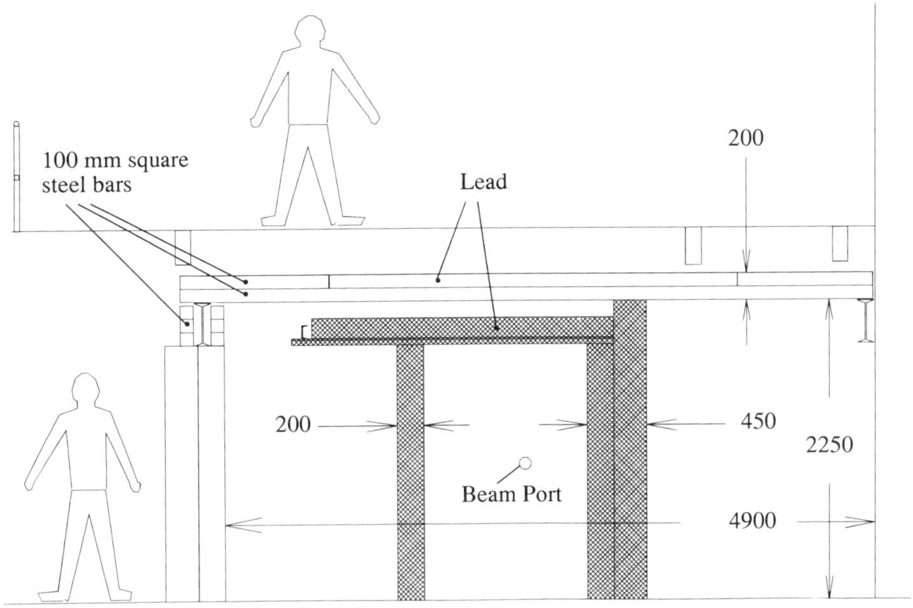

Figure 1. Section through the therapy facility immediately outside the beam port. Part of the "dog tunnel" is drawn hatched, for comparison. Polybor linings have been omitted. Dimensions are in mm.

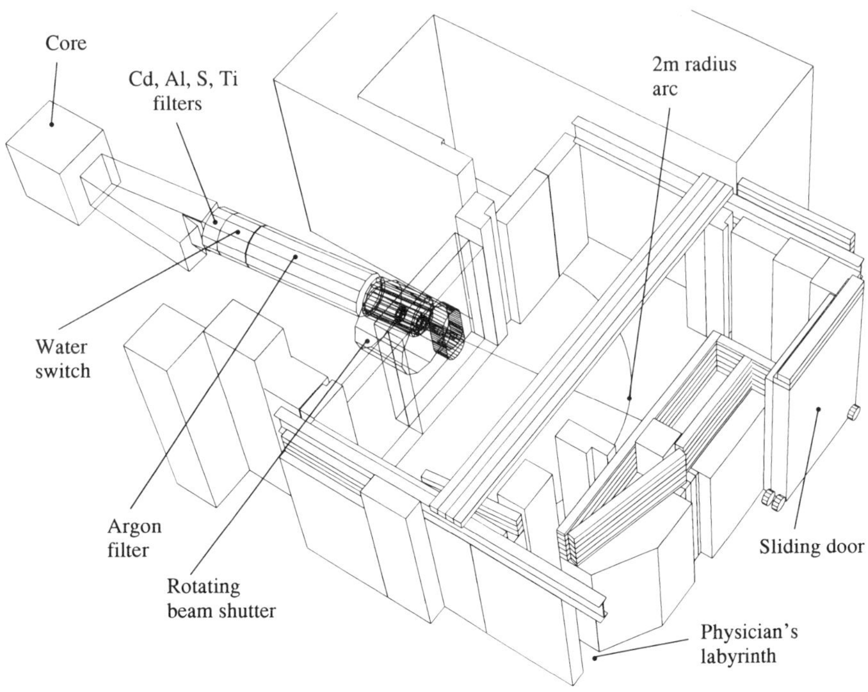

Figure 2. Over-all view of the therapy facility, showing beam-line, sliding door, physician's labyrinth and roof construction/support system. As an aid to clarity, all but four of the bottom layer of 100mm square steel, roof shielding bars have been omitted. Polybor linings are not shown.

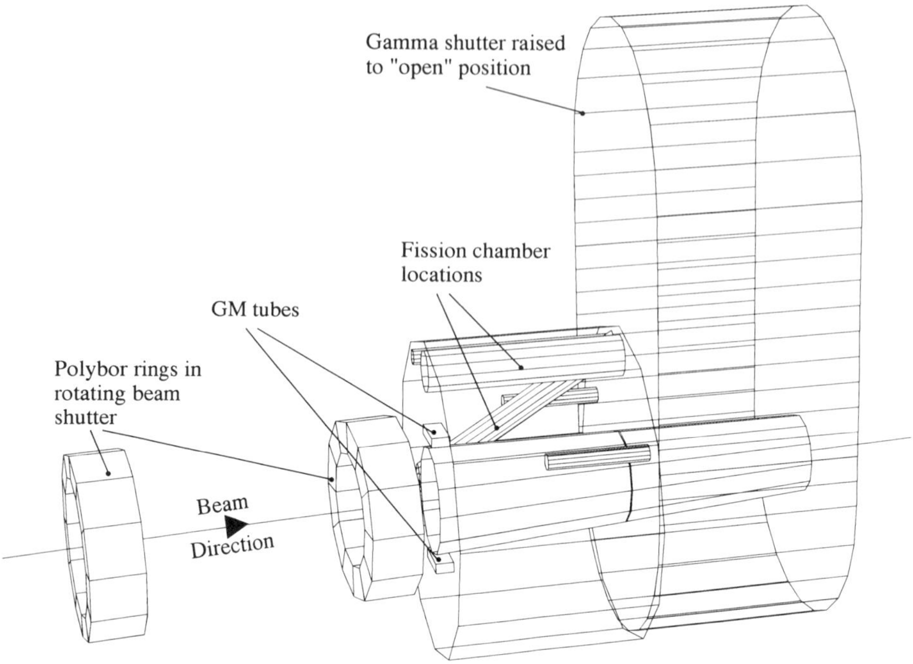

Figure 3. Location of on-line monitors.

the beam during the rest of the healthy tissue tolerance studies. Development work is also underway on a transmission chamber to monitor the full beam, which will be located immediately after the rotating shutter.

A drop in liquid level in the argon filter clearly has very serious consequences. It is hoped to rebuild the collimation system so that duplicated GM tubes are aligned with the potential gas space that would develop if the argon level falls, to give early warning of this and trigger emergency shutter closure.

CONCLUSIONS

The European Concerted Action BNCT project at Petten is fast approaching the clinical trials stage, and preparations are well in hand to build an effective treatment room with up-to-date facilities in readiness.

REFERENCES

1. R.L. Moss et al, Proc. Fourth Int. Symp. on Neutron Capture Therapy, Sydney, Australia (1990).

FEDERAL AND INSTITUTIONAL APPROVALS NECESSARY PRIOR TO A U.S. CLINICAL TRIAL OF NEUTRON CAPTURE THERAPY

R.G. Zamenhof[1], D.E. Wazer[1], J. Bernard Jr.[2], O.K. Harling[2], T.A. DiPetrillo[1], H. Madoc-Jones[1], J.A. Coderre[3], A.G. Meek[4]

[1]New England Medical Center, Boston, MA, U.S.A.; [2]Massachusetts Institute of Technology, Cambridge, MA, U.S.A.; [3]Brookhaven National Laboratory, New York, U.S.A.; [4]Stonybrook Medical Center, New York, U.S.A.

INTRODUCTION

The New England Medical Center (NEMC) - Massachusetts Institute of Technology (MIT) program in neutron capture therapy (NCT) in collaboration with the Brookhaven National Laboratory - Stonybrook program in NCT will shortly initiate a phase-I clinical trial of NCT for subcutaneous melanoma on the extremities. This paper briefly reviews the accepted methodology for evaluation of new therapeutic modalities in the U.S., describes the clinical protocol proposed, and discusses the various Federal and institutional approvals that we have had to obtain.

METHODOLOGY FOR CLINICAL TRIALS IN THE U.S.

The five stages of investigation from the conception of an innovative therapeutic modality to the routine implementation of the treatment may be summarized as follows:

Preclinical Toxicity Study

Carried out in partial requirement of obtaining an investigational new drug approval (IND) from the U.S. Food and Drug Administration (FDA), required in at least one rodent and one non-rodent animal model, up to drug doses at least an order of magnitude above the dose level anticipated as necessary for clinical use, or to dose levels which demonstrate the onset of significant toxicity, usually assessed by organ pathology. Terminates when an IND number for the drug or drug related therapy has been successfully obtained from FDA.

Phase-I Clinical Study

Involves dose-escalation, evaluation of safety, maximum acceptable toxicity, and pharmacokinetics in human subjects. Involves no requirement or criteria for assessing anti-tumor activity. Accountability to FDA.

Phase-II Clinical Study

Limited efficacy study in human subjects using dose levels established as safe in phase-I and specifically evaluating anti-tumor activity of the therapy. Accountability to FDA.

Advances in Neutron Capture Therapy, Edited by
A.H Soloway *et al.*, Plenum Press, New York, 1993

Phase-III Clinical Study

More extensive efficacy study in human subjects including comparison against most effective current therapy using randomized "blind" controls (where feasible). Accountability to FDA.

Phase-IV

Routine implementation of therapy. Technology transfer to other treatment sites where applicable.

An IND number has been obtained for the boron drug l,p-boronophenylalanine (l,BPA) and we are presently at the stage of obtaining final approvals for initiating a phase-I clinical study. Descriptions of the preclinical technical and biological accomplishments of the NEMC-MIT program can be found in (1-6).

CLINICAL PROTOCOL

Although our eventual intent is to apply NCT to primary and metastatic tumors of the central nervous system, the initial phase-I clinical study which we have proposed seeks to evaluate the toxicity of NCT in subcutaneous melanoma nodules on the extremities and thereby to lessen the potential injury that might accrue to subjects resulting from radiotoxicity. We propose using the same epithermal neutron beam at the MITR-II research reactor that eventually will be used for CNS irradiations. Subjects admitted to the study will be selected according to various inclusion criteria which will not be discussed here. They will first be given an oral "test" dose of l,BPA (400 mg/kg), after which blood, tumor, and subcutaneous normal tissue sampling will be carried out at periodic intervals. Bulk and microscopic boron-10 concentration distributions will be measured by prompt-gamma spectroscopy and high resolution alpha track autoradiography. To continue in the protocol the subject will need to have a boron-10 tumor uptake at 9 hours of >20 ppm (by weight) and a tumor-to-normal tissue boron-10 ratio of >3. The subject will then undergo CT scanning for treatment planning purposes and a Monte Carlo based treatment plan will be developed. This treatment plan will be made absolute through an in-phantom calibration carried out on the epithermal neutron beam at MITR-II. Nine hours prior to neutron irradiation the subject will be given a second oral l,BPA dose of 400 mg/kg, and approximately 1-2 hours prior to irradiation further blood, pericutaneous tissue, and tumor biopsy samples will be obtained and analyzed for boron-10. From these measured boron-10 concentrations, the previously derived boron-10 time/concentration curves, and the Monte Carlo treatment plan, the number of "neutron monitor units" required to reach the planned dose will be determined. The irradiations will be delivered in four daily fractions, with a further administration of l,BPA 9 hours prior to each irradiation and repeated tissue sampling and analysis. Maximum normal tissue dose will be escalated as follows: 10, 12.5, 15, 17.5, and 20 RBE-Gy, with three subjects per dose level. Radiotoxicity evaluations will be made according to the CHART system and defined stopping rules will be used. 4-6 months of observation at each dose level will precede progression to the next dose level.

REQUIRED FEDERAL AND INSTITUTIONAL APPROVALS

The committees and agencies that are required to review the clinical protocol often differ somewhat in their perspective. For example, as a broad generalization the institutional review boards (IRB) trust that what is proposed in terms of drug and radiation dose delivery can be accomplished competently and safely. They focus, however, on the potential injury to subjects that may result from the protocol, the ethical aspects of the protocol, the adequacy and clarity of informed consent forms, and the appropriateness of information that is conveyed to the research subjects by the investigators. The Nuclear Regulatory Commission is more oriented towards the safety of the protocol from the perspective of nuclear (reactor) safety, radiation safety (of personnel), and conformance as closely as possible to existing NRC regulations governing use of byproduct material. The FDA pharmacology division (which grants INDs) is primarily concerned about the chemical safety of l,BPA. However, since NCT is a binary therapy many discussions occurred as to whether l,BPA should be separately considered as a "drug," or whether the binary effect of l,BPA and neutrons should be considered in granting the final IND amendment for phase-I studies (the latter approach was adopted). The submission of appropriately oriented documentation to all the agencies was greatly facilitated by contacting the respective agencies by telephone or in writing to request guidance in the preparation of the documentation.

The following Federal and institutional approvals have been applied for, and are either approved or pending; also, a peer review panel examined our protocol.

1. NEMC Human Investigation Review Committee
2. NEMC Radiation Safety Committee
3. NEMC Clinical Study Unit Research Committee
4. NEMC amendment of NEMC's Broad Scope NRC license to include MITR-II
5. MIT Reactor Safeguards Committee
6. MIT Committee on Use of Humans as Experimental Subjects
7. MIT Committee on Radiation Exposure to Human Subjects
8. MIT NRC "Tech-Spec" change for MITR-II and new clinical protocols
9. FDA original IND approval for use of BPA in humans (sponsor: Dr. Allan G. Meek, SUNY-Stonybrook/BNL)
10. FDA IND amendment for raising doses of BPA to therapeutic levels and adding neutron irradiation, jointly through NEMC/MIT and SUNY-Stonybrook/BNL
11. FDA IDE (Investigational Device Exemption) compliance for MITR-II as a "significant risk" therapeutic radiation device
12. Peer review by 18 invited scientists, voluntarily organized by us in October 1991.

It should be noted that the MITR-II research reactor is regulated by NRC while the BNL medical reactor is regulated by the Department of Energy (DOE). Which additional approvals might be required under such circumstances and whether the NRC would claim any jurisdiction over the use of the BNL medical reactor for the irradiation of human subjects is presently undetermined.

SUMMARY

Of the formal approvals that are required all those listed above have been obtained with the exception of #4, which is currently pending. The peer review (#12 above) resulted in substantial modifications to our clinical protocol and we now believe that the legitimate concerns of the scientists on the review panel have been largely satisfied. With the completion of the three outstanding approvals we look forward to initiating a phase-I clinical trial of boron neutron capture therapy for subcutaneous melanoma on the extremities in early 1993.

ACKNOWLEDGEMENT

Research Supported by DOE Grant# FGO2-87ER-6060

REFERENCES

1. R.G. Zamenhof, H. Madoc-Jones, O.K. Harling, D. Wazer, S. Saris, J.C. Yanch, G. Solares, G. Rogers, "The neutron capture therapy research program at New England Medical Center and the Massachusetts Institute of Technology," *in*: "Progress in Neutron Capture Therapy for Cancer," B. Allen, D. Moore, B. Harrington, eds., Plenum Press, New York (1992).
2. S.C. Saris, G.R. Solares, D.E. Wazer, G. Cano, S.E. Kerley, M. Joyce, L.S. Adelman, O.K. Harling, H. Madoc-Jones, R.G. Zamenhof, Boron neutron capture therapy of murine brain tumors, *Cancer Research*, 52:4672, (1992).
3. J.R. Choi, R.G. Zamenhof, J.C. Yanch, R. Rogus, O.K. Harling, Performance of the currently available epithermal neutron beam at the MIT research reactor, *in*:: "Progress in Neutron Capture Therapy for Cancer," B. Allen, D. Moore, B. Harrington, eds., Plenum Press, New York (1992).

4. G. Solares, R.G. Zamenhof, S. Saris, D. Wazer, S. Kerley, M. Joyce, H. Madoc-Jones, L. Adelman, O.K. Harling, Biodistribution and pharmacokinetics of p-boronophenylalanine in C57BL/6 mice with GL261 intracerebral tumors, and survival following neutron capture therapy, *in*:: "Progress in Neutron Capture Therapy for Cancer," B. Allen, D. Moore, B. Harrington, eds., New York, Plenum Press (1992).
5. R. Rogus, O.K. Harling, I. Olmez, S. Wirdzek, Boron-10 prompt gamma analysis using a diffracted neutron beam *in* : "Progress in Neutron Capture Therapy for Cancer," B. Allen, D. Moore, B. Harrington, eds., Plenum Press, New York (1992).
6. R.G. Zamenhof, J. Brenner, J.C. Yanch, D. Wazer, H. Madoc-Jones, S. Saris, O.K. Harling, Treatment planning for neutron capture therapy of glioblastoma multiforme using an epithermal neutron beam from the MITR-II research reactor and Monte Carlo simulation, *in*: "Progress in Neutron Capture Therapy for Cancer," B. Allen, D. Moore, B. Harrington, eds., Plenum Press, New York (1992).

PREPARATIONS FOR CLINICAL TRIALS AT THE MIT REACTOR AND THE NEW ENGLAND MEDICAL CENTER

Otto K. Harling,[1] Ronald D. Rogus,[1] Jean-Michel Chabeuf,[1] Sam C. Yam,[1] Frédérique Lambert,[1] Jacquelyn C. Yanch,[1] Robert G. Zamenhof,[2] Guido Solares,[2] David E. Wazer,[2] and Hywel Madoc-Jones[2]

[1]Nuclear Reactor Laboratory, Massachusetts Institute of Technology (MIT)
Cambridge, MA 02139, USA
[2]Department of Radiation Oncology, New England Medical Center (NEMC)
Boston, MA 02111, USA

INTRODUCTION

The NEMC/MIT group is planning to initiate Phase I clinical trials of BNCT for deep-seated melanoma on the extremities. These trials will be carried out using the epithermal neutron beam at the 5 MW MIT Research Reactor (MITR-II). This paper outlines some of the preparations for these trials.

PREPARATIONS FOR PHASE I CLINICAL TRIALS

The Epithermal Beam

The epithermal medical beam at MITR-II has been improved through changes in the filtration. The current beam, designated M62, is shown in Fig. 1.

Beam filtration is accomplished by using aluminum, sulfur, and cadmium. Shielding of gamma rays is achieved with bismuth located close to the patient position.

The performance of the M62 beam was measured in the usual way using gold foils and dual ion chambers in a mid-thigh phantom. This phantom was made from a mold taken from a human leg which had a diameter of 17.5 cm. From these data a complete depth/dose distribution was obtained. Figure 2 presents the summary of results for the tumor and healthy tissue dose as a function of depth for a hypothetical unilateral irradiation where the healthy tissue dose is limited to a maximum of 2000 RBE cGy. We have assumed the use of L-BPA with a B-10 concentration of 30 ppm and 7.5 ppm, respectively, in tumor and normal tissue. Two sets of curves are presented in Fig. 2 to represent the range of RBE values which we feel brackets the actual range. The curve labeled "likely RBEs" uses 4.0, 2.3, 2.3, and 0.5 for the B-10, fast neutron, N-14 (n,p) reaction, and gamma RBEs, respectively. The curve labeled "pessimistic RBEs" uses 4.0, 4.0, 4.0, and

Advances in Neutron Capture Therapy, Edited by
A.H Soloway *et al.*, Plenum Press, New York, 1993

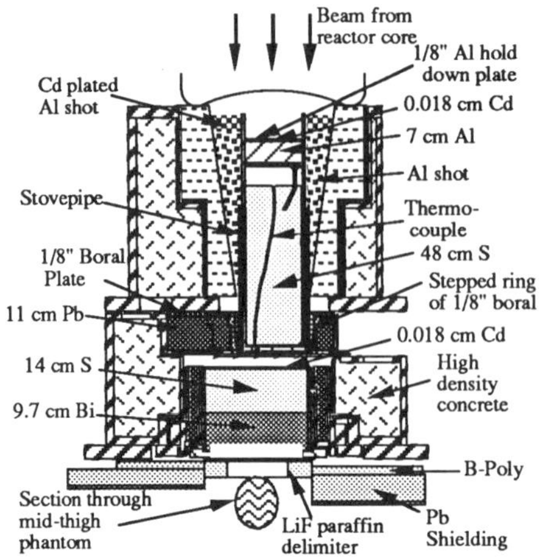

Figure 1. The M62 epithermal medical treatment beam at the MITR-II.

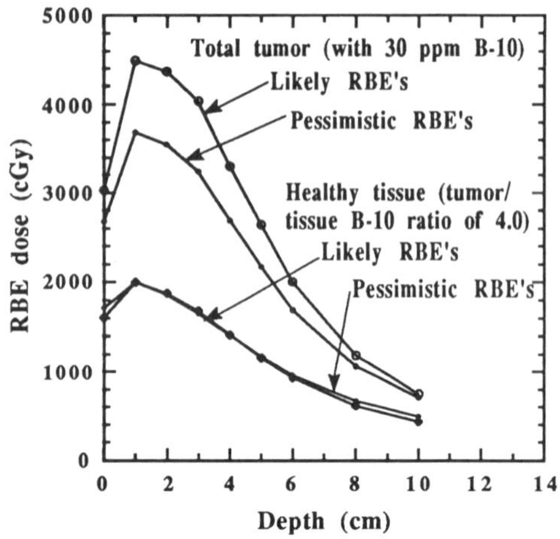

Figure 2. The dose vs. depth distribution of the M62 beam in a mid-thigh phantom.

1.0, respectively, for the same RBEs. The RBEs are based on our assessment of the relevant literature.

Treatment Planning

A Monte Carlo based three-dimensional clinical treatment planning computer program called NCTPLAN has been developed. The key features of this program are: full 3D computation of dose in mathematical phantoms custom designed for each individual patient; ability to superimpose isodose contours on actual anatomical CT images; and ability to compute and display specialized BNCT dosimetric figures of merit such as "advantage depth," and "advantage ratio," in one, two, and sometimes three dimensions; ability to provide fiducial patient information to assist in patient setup at MITR-II.

Dosimetry and Beam control

We are completing an on-line beam-monitoring system which combines real-time measurement of the intensity of the treatment beam with integrated patient dose. Two thermal neutron detectors, two epithermal neutron detectors, and a gamma-sensitive ion chamber are distributed around the periphery of the beam aperture as shown in Fig. 3a. The intensities measured by these detectors and their integrated intensities are simultaneously displayed on a PC-based color monitor, Fig. 3b. Analog and digital displays are used for each detector. The integrated intensities correspond to patient dose through intercalibration with phantom measurements and are stored on a magnetic disk and are also recorded on an independent set of scalers.

The different sensitivities and locations of the beam monitor detectors help assure that significant variations in the beam spectrum, geometric distribution, and intensity can be detected immediately. Since each dose fraction will require irradiation times in the range of ten minutes to one hour, we have not included automatic shutter closure when a targeted dose is reached as part of the current system. This is consistent with accepted procedures currently routinely used for brachytherapy.

While the beam monitoring system described above will be used as the primary system for assuring the correct dose during an irradiation, the reactor power control instrumentation provides us with a completely separate system for monitoring patient dose.

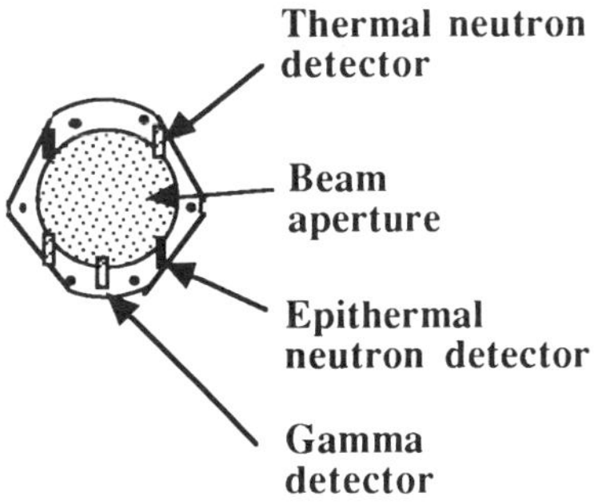

Figure 3a. Detectors around the beam aperture.

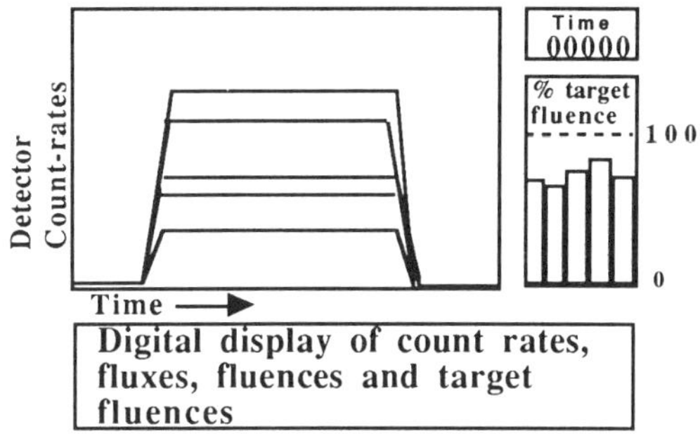

Figure 3b. Screen display for the on-line monitors.

Patient Positioning and Shielding

The patient positioning system is shown in Fig. 4. For brain irradiations, four lasers will be used to orient the patient's head, but for subcutaneous melanoma of the extremities, a plumb bob is judged adequate for neutron beam alignment.

To reduce dose to the body of the patient we have constructed beam delimiters with different size apertures. These are constructed of lithium-loaded paraffin. Further reduction of patient whole body dose would be achieved with flexible blankets of lithium fluoride or carbonate.

To reduce medical room activation, which is significant for medical and reactor personnel but not for the patient, we have installed boral plates (0.6 cm thick plates of aluminum plus B_4C) on the floor of the room.

Boron Quantification

An improved prompt gamma activation analysis facility has been constructed at MITR-II. A diffracted beam with mean energy of 0.015 eV and intensity $6x10^6$ n/cm^2-sec is available at the sample position. The beam is relatively free of fast neutrons and gamma rays and is further purified with a sapphire single-crystal filter. The n type Ge gamma-ray detector is shielded with Li-6 and lead. Due to the low background the Ge detector can be placed close to the sample, resulting in a relatively high efficiency for the detection of prompt gammas from the sample.

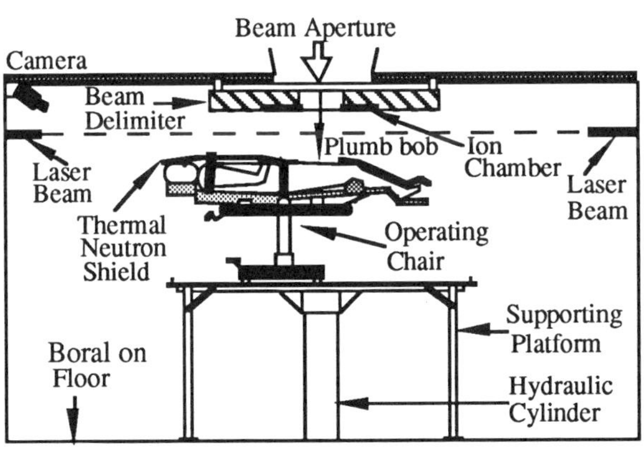

Figure 4. The MITR-II medical room, showing patient positioning system for irradiation of the upper leg.

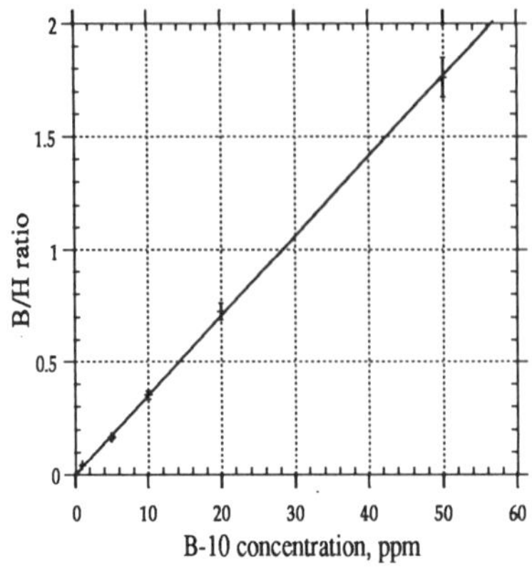

Figure 5a. Calibration curve for 0.1 ml samples.

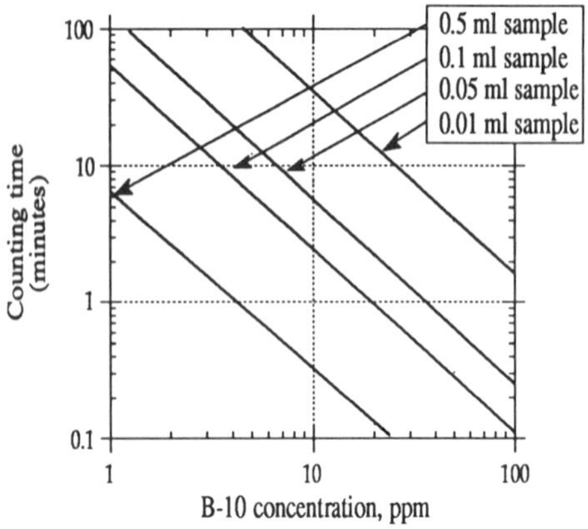

Figure 5b. Counting times for 10% uncertainty.

The facility is well suited for sample sizes ranging from a few milliliters to ten microliters, thus allowing the determination of relevant boron concentrations from typical blood samples as well as from punch and needle biopsies. The sensitivity for B-10 is 7.5 cps per μg of boron. Figure 5a shows the ratio of B-10 to H capture gamma rays as a function of the B-10 concentration in standard boric acid samples of 0.1 ml. Fig. 5b gives the counting times versus the B-10 concentration for sample sizes 0.01-0.5 ml assuming the measurement is made with statistical error of 10 percent (one standard deviation). A 50 microliter sample, typical of a punch biopsy, can be analyzed in five minutes for a concentration of 10 ppm.

The prompt gamma facility is expected to provide accurate and rapid analyses of B-10 for melanoma patients shortly before they are irradiated at MITR-II.

High resolution quantitative autoradiography (HR-QAR), which was developed by our group, allows subcellular imaging of B-10 at concentrations relevant to BNCT. We have combined HR–QAR with computer-based image processing and Monte Carlo methods to obtain realistic microdosimetric information, allowing the computation of compound factors which are expected to relate to the effect of BNCT *in-vivo*.

Approvals Needed for Clinical Trials

The following approvals have been obtained: 1) NEMC/Human Investigation Review Committee; 2) BUMC/Human Investigation Review Committee; 3) NEMC/Radiation Safety Committee; 4) MIT Reactor Safeguards Committee; 5) MIT Committee on the Use of Humans as Experimental Subjects; 6) MIT Committee on Radiation Exposure to Human Subjects; and 7) United States Food and Drug Administration approval for administration of BPA in human trials.

The following approvals are pending: 1) Nuclear Regulatory Commission approval of amended technical specifications for MITR-II; and 2) Nuclear Regulatory Commission approval of amendments to the NEMC medical license to permit use of MITR-II for patient irradiation.

ACKNOWLEDGEMENT

The work reported here was supported by the US Department of Energy under Contract No. DE-FG02-87ER-60600.

PROCEDURES FOR THE MEDICAL APPLICATION OF RESEARCH REACTORS

Keiji Kanda

Research Reactor Institute, Kyoto University
Kumatori-cho, Sennan-gun, Osaka 590-04

INTRODUCTION

In Japan five research reactors in four institutions - Research Reactor Institute, Kyoto University (KURRI); Atomic Energy Research Laboratory, Musashi Institute of Technology; Nuclear Engineering Research Laboratory, University of Tokyo; and Tokai Establishment, Japan Atomic Energy Research Institute - have been licensed to be utilized for boron neutron capture therapy, BNCT. These reactors have been modified for BNCT purposes. In this paper, the legal and other procedures for using existing reactors for BNCT are described. They are based on "The Law for the Regulation of Nuclear Source Material, Nuclear Fuel Material and Reactors".

LEGAL PROCEDURES

The following applications for BNCT are to be submitted to the Office of the Prime Minister or the Science and Technology Agency: (1) Permission for establishment, (2) Sanction of design and method of construction, (3) Sanction of safety regulations, (4) Authorization of a chief medical doctor, and (5) Notification of each treatment.

(1) Permission for Establishment

The application submitted to the Prime Minister contains the following items:

(1) The name and the address and, in case of a juridical person, the name of its representative,
(2) The purpose for which reactors are to be used,
(3) The type, the thermal output, and the number of reactors,
(4) The name and the address of the factory or the place of business where reactors are to be established,
(5) The location, structure, and equipment of reactors and their related facilities,
(6) The construction plan of reactor facilities,
(7) The type of nuclear fuel material to be used in reactors and the estimated annual consumption,
(8) The method of disposal of spent fuel.

The Prime Minister shall not give the permission unless he recognizes that the application adheres to each of the following quidelines:

(1) That reactors will not be used for non-peaceful purposes,
(2) That the permission will cause no hindrance to the planned development and utilization of atomic energy,
(3) That the application demonstrates sufficient technical ability and financial resources to establish reactors, and the technical ability to operate them competently,
(4) That the location, structure, and equipment of the reactor facilities are such that they will cause no hazard from nuclear fuel material and reactors.

(2) Sanction of Safety Regulations

The safety regulations of each reactor are to be rigorous enough to prevent accidents from nuclear reactors. The safety regulations of each reactor must have a section devoted to BNCT, where the priority of reactor safety to BNCT must be described.

(3) Authorization of a Chief Medical Doctor

Prof. H. Hatanaka was the first to be approved by Prime Minister Eisaku Sato in 1968. At that time, BNCT itself was reviewed. Prof. Y. Mishima and others were carefully reviewed later by the Science and Technology Agency.

Items to be reviewed are :

(1) Name,
(2) Affiliation,
(3) Personal history, especially as related to BNCT,
(4) Clinical experiences.

This review system was requested by Prime Minister Kakuei Tanaka in 1974. The above review takes about three months for each medical doctor.

(4) Notification of Each Treatment

After the decision of the Director General of KURRI mentioned below, the President of Kyoto University must give notification of each treatment to the Science and Technology Agency five days before irradiation. The format is as follows:

1. Date,
2. Place,
3. Reactor personnel on site the day of irradiation,
4. Names of the chief medical doctor and the clinical staff,
5. Patient (nationality, sex, age),
6. Kind of tumor,
7. Tumor location,
8. Total neutron fluence,
9. Schedule for the day.

Attachments :

1. Condition of the reactor facility,
2. Personal radiation control for clinical staff,
3. Patient's record,
4. Chief MD's personal record,

5. Detailed information on clinical staff,
6. Past record of BNCT at the site.

INDIVIDUAL INSTITUTIONAL PROCEDURES

The individual institutional procedures are different among the above four institutes. In this paper, the case at KURRI is described.

Medical Ethics Committee

Prior to the start of BNCT, the Medical Ethics Committee, a steering committee under the President of Kyoto University consisting of professors of medicine and the humanities, approved the method itself, after various discussions.

Medical Use Committee

The Medical Use Committee (chaired by the Director of Radiation Oncology, Research Laboratory, KURRI), consisting of professors from the medical schools of Kyoto University, Osaka University, and Tsukuba, and two Division Directors of Reactor and Radiation Control at KURRI, is held a week before each irradiation, and the chief doctors are invited. The Committee approves the treatments and decides the time schedule after hearing information from the chief doctors including confirmation of informed consent.

Reactor Safety Committee and Directors' Meeting

Immediately after the decision of the Medical Use Committee, the Reactor Safety Committee is held to examine reactor safety and radiation control related to the treatments. A Directors' Meeting, consisting of division directors, senior reactor supervisors, and others, is also held to check personal arrangements, the reactor operation schedule, etc. Finally, the Director General of KURRI issues the approval for each treatment to the chief medical doctor.

Report of Treatment

A week after the treatment, all chief doctors must report to the Director General of KURRI about the status of the patient. Then the President of Kyoto University reports to the Science and Technology Agency about the treatment for each patient, including (1) Date and time, (2) Patient, (3) Neutron and gamma ray doses, (4) Dose of doctors, (5) Status of patient and others.

At the end of year, all doctors are invited to KURRI in order to report on and discuss the treatments performed at KURRI.

OTHER PROCEDURES

After the Director General approves the treatment, he must notify the local governments before each irradiation.

Back-up hospitals are prepared for an emergency by the chief doctors or local governments.

A PROPOSAL FOR LOW DOSE RATE BEAM THERAPY FOR GLIOBLASTOMA

F. Ellis

Department of Radiotherapy, United Oxford Hospitals and
Lecturer, Oxford University (retired), Oxford,
United Kingdom[1]

ABSTRACT

In view of the poor results from conventional methods of treatment of glioblastoma and the necessity of a credible alternative in a randomized trial of BNCT, a new approach is suggested depending on known rates of proliferation of glioblastoma cells, taking account of cell loss and using low dose rate continuous radiation so as to maintain a continually destructive process for malignant cells. The dose should be as high as possible to destroy all malignant cells but within the tolerance dose of normal brain tissue.

INTRODUCTION

The BNCT campaign against glioblastoma multiforme is gathering momentum, and therefore preparations should be made for clinical trials. Unless the method has been shown to be successful in virtually all cases, and success means killing the primary tumor, there must be one or more reasonable alternatives. In dealing with glioblastomas, no methods so far reported have been sufficiently successful. Stemming from considerations of the natural history of glioblastomas and of the possible delivery of low dose rate radiation (LDR) in sufficient dose to destroy the tumour without exceeding normal brain tolerance, the principles of growth of glioblastoma and of the possible maximum tolerance dose of radiation to the brain are discussed. It is suggested that these doses be tested experimentally on animals and on patients. Two methods of LDR are suggested, one by beam radiation and the other intracavitary. In addition, the possibility of local toxic effects of necrotic tissue and the extent and rate of autolysis may be important. Treatment of glioblastoma by conventional beam therapy is always by dose rates of the order of 1-3 Gy/minute, while the usual low dose rate treatment in other parts of the body with relatively uniform dose distribution is of the order of 0.35 - 0.60 Gy/hour. Much experience with these low dose rates - necessitating long, usually

[1]Present address: 2 Bladon Close, Woodstock Road, Oxford OX2 8AD, United Kingdom

Advances in Neutron Capture Therapy, Edited by
A.H Soloway *et al.*, Plenum Press, New York, 1993

continuous treatment schedules-has bred a confidence in the results, if the dose is of the order of 60-80 Gy, the dose rate of ~.3 to 0.6 Gy/hour, and the dose distribution uniform throughout the volume treated.

Data[1] indicate that for HeLa cells the dose-rate effect begins to be different as the dose rate is reduced to ~ .13 Gy/min., which is equivalent to 7.8 Gy/hour. It changes rapidly, then until it reaches about 1.2 rad/min (.012 Gy/min) or .72 Gy/hr., after which the dose rate effect becomes less. The change between $d_o = 260$ and $d_o = 240$ is approximately from 1.38 Gy/hr to .18 Gy/hr. At exactly what level the low dose rate difference begins, and if there is one, is not certain, but that there is such a difference is suggested by recent work of Bedford[2]. It appears that at 0.4 Gy/hour cells progressed through the mitotic cycle but did not proceed to mitosis. Instead, the chromosomes fractionated. Because of this possibly different mechanism, and because of the large gaps between fractions, which are unavoidable in beam therapy with electromagnetic gamma radiation, it seems reasonable to try low-dose rate treatment with such beams.

LOW DOSE RATE TREATMENT

The usual methods of low dose rate treatment by removable or permanent implants must have greater risk in the brain than in other organs, and the application of low dose rate beam radiation obviously has difficulties. Nevertheless, it would seem that because it is possible to use such a technique and because immobilizing patients for long periods is used in Intensive Care Units, it would be reasonable to try low dose rate treatment as an alternative treatment for comparison with BNCT. BNCT is bound to be available for relatively few cases in the initial clinical trials. Low Dose Rate Beam (LDRB) therapy, however, even though difficult, would be possible in many centers. Questions arise, particularly the necessity immobilizating the patient for long periods, which is a problem for anesthetic departments to solve. The questions to ask are: 1. What dose-rate should be used? 2. What dose should be used? 3. How much should be treated?

1. The dose rate should be of the order of those used successfully for other types of tumours in clinical work for many years(~ about 0.35 - 0.60 Gy/hour).

2. To determine the dosage pattern one might consider two models the LQ[3] and the NSD-TDF[4,5] using a large number of fractions in the calculation to simulate low dose rate.

   (a) The LQ model is biologically based, depending on α/β ratio. The formula relating two schedules of equal effect is[6]

$$\frac{N_1 d_1}{N_2 d_2} = \frac{\alpha/\beta + d_2}{\alpha/\beta + d_1} \cdot \text{ If } \frac{25x2}{200xd_2} = \frac{1 + d_2}{1 + 2}$$

   then $d_2 = 0.5$ if $\alpha = 1$
   and $N_2 d_2 = 100$ Gy

   The value of α/β to be used is uncertain. It seems that a tolerance dose for LDR continuous radiation may be between 80 and 120 Gy. If 400 fractions with the same α/β ND = 116 Gy

   (b) The NSD concept is clinically based, and several models have been suggested for the CNS. Using these but (on the safe side) neglecting the time factor as with the LQ concept, they give the following values.

		N-200 Nd.(Gy)	N-400 Nd.(Gy)
Wara and Phillips[7]	$11.9N^{\cdot 377}T^{\cdot 068}$	88.1	114
Van der Kogel[8]	$12.5N^{\cdot 37}T^{\cdot 05}$	89	115
Shelline[9]	$10.16N^{\cdot 44}T^{\cdot 06}$	105	142
Wegg[10]	$8.60N^{\cdot 46}T^{\cdot 09}$	98	135

It has recently been suggested that a modified NSD type formula with an exponent of 0.41 for the fraction number appeared to provide an adequate fit over the range of fraction sizes from 25.0 Gy to 0.55 Gy for rat spinal cord and that the LQ formula underestimates the possible tolerable dose.[11] Not knowing the true values of α/β, but knowing them approximately and comparing the values given by both models, it seems that for multiple fractions approaching equivalence to continuous radiation, doses of 80 - 120 Gy might be safe, but that there is room for experiment starting at 60 Gy and increasing by steps of 20Gy to find the tolerance of the normal brain in animals, and then to determine the tolerance of patients with the experimental values as a guidance. To spread 80 Gy over 1 week requires a dose rate of 80/168 which is .476 Gy/hour. The radiation technique for this purpose could be based on a cobalt machine suitably screened with lead or on a cesium-137 appliance, similarly screened to achieve a suitable dose rate and using appropriate compensation for tissue contours and heterogeneity to ensure optimal accuracy of dosage. It is clear that the necessary dose may require treatment duration approaching 1 week. Probably the principle aim of the treatment would not be affected by suitable small intervals for changing fields and for toilet activities.

It would be reasonable, therefore, to carry out experimental work on animals, if it is practical, giving doses of the order of 60, 70, 80, 90, 100 Gy to the whole brain to determine the tolerance level. There is difficulty in determining an endpoint. Perhaps, the clinical difficulties are insurmountable, but with modern aids to exact localization, they should not be. Using compensation to ensure known dosage, it should be possible to transfer the information to treatment of patients with suitable precautions. All this seems a good deal of trouble, but the curing of glioblastoma is a worthwhile aim, and the above suggestions are less trouble than acquiring a nuclear reactor. Moreover, it is the type of treatment that can be carried out at any referring centre as an alternative aim of a randomized trial carried out by multiple referring centres.

Another possibility is the use of a local application into a cavity in the brain, such as is left after surgical excision. The author used this method many years ago in cooperation with J. Pennybacker, Chief of Neurosurgery, Oxford. The insert was a plastic sphere containing a radium tube, and aiming at 4,000 rad at the surface of the sphere. In accordance with the above discussion, the dose should be increased experimentally so as to give in animals doses of 50, 60, . . 110, Gy at the surface. Logically, if the tumor were completely removed surgically, the remaining cells should be in the tissue immediately impinging on the application. Clinically, it was convenient to arrange the dosage so that the routine theater-day in 1 week was the time scheduled for removal of the plastic container and the radium together. The presence of a necrotic mass in the brain seems to be something to avoid. One wonders if it has locally toxic properties? In any case, by these methods, it might be possible to determine the relative effect of the presence or absence of such a mass.

NATURAL HISTORY OF GLIOBLASTOMA

The natural history of glioblastoma is of considerable interest, relative to its cell kinetics.

(i) How long is the tumour present before it causes symptoms and signs?
(ii) How quickly does it progress after it causes clinical signs?
(iii) With cell loss, what kind of mass tends to accumulate?
(iv) What is the rate of autolysis and disappearance of dead cell and tissue?

The author formed the opinion several years ago that the gaps in a fractionated radiotherapy schedule could allow for significant proliferation of cells not damaged by the radiation. In the case of glioblastoma, the rate of progression suggested that this could have considerable effect, if it is assumed that the malignant cells close to the supplying vessels used most of the oxygen available so that cells beyond them, being deprived of oxygen are unable to synthesize DNA and therefore are resting. Hypoxic cells are well-established as relatively insensitive to radiation. After a dose of radiation inactivates the dividing cells, as the oxygen becomes available to the hypoxic cells, they could immediately start to divide. Hence, the gaps in treatment could be significant, allowing for unchecked proliferation of active cells. In the usual period of 6 weeks of a course of radiation, the gaps in 40 days are approximately as follows:

(i) In each interval between daily treatments, 23.5 hours
(ii) In each day of 3 treatments daily, about 23 hours
(iii) At each weekend, 3 days

RATE OF PROLIFERATION OF GIOLBLASTOMA CELLS

Obviously, the total gap is nearly the total time, and much will depend on the effects of a dose fraction on resting (hypoxic) cells and on the rate of proliferation of the cells. Hoshino et al.[12] studied four surgically removed glioblastoma specimens by double labelling using preoperative tritiated thymidine for the labelling index, and then after an exactly known interval carbon-14 was administered. The synthetic time (Ts) and the "birth rate" of the cells were determined. The most rapid birth rate was 1.7 cells/hundred cells/hour (T_{pot} = 42 hours), and the slowest was .5 cells per 100 cells per hour (T_{pot} = 139 hours); where T_{pot} is defined as the potential doubling time of the tumour assuming that there is no cell loss. The rapidity of cell division represented by these figures supported the general thesis outlined above. Gordon Steel (personal communication) has pointed out the need to consider cell loss and Jack Fowler (personal communication) stated that the general thesis influenced the thinking leading to CHART, although not having a neurosurgical department, CHART was not used for glioblastomas. Recently Hoshino and colleagues[13] have described the results of sequential labelling by intravenous infusion with iododeoxyuridine (IUdR) and bromodeoxyuridine (BUdR) of 57 human brain tumors. Of these, 13 were glioblastomas of which the shortest T_{pot} was 2 days, corresponding to the fastest birthrate 1.7 of a reported earlier 12. The longest T_{pot} was 18 days (432 hours) - a birthrate of .016 cells per hundred cells per hour. These values (1.7 and 0.16) have been used in subsequent calculations of the likely fastest and slowest times for proliferation to occur in clinical glioblastomas. The proliferation of cells allowing for 80% cell loss and assuming equilibrium at 2cm diameter (1 cm radius), we have volume of tumors = $4/3\ \pi r^3$ where r is the radius of each cell.

Table 1 shows the number of cells related to the number of divisions and using a cell loss factor of 80%, as if it is inherent in the cells and applies to the new cells formed in a division. If 80% is applied to all the cells all the time, the tumor would not grow. The clinical facts are that glioblastoma grows very rapidly after it first causes clinical symptoms, and that after most treatments by surgery and/or other methods rapid recurrence

Table 1. Cell Division and Clinical Growth of Glioblastoma (Assuming 80% cell loss)

No. of divisions n	Viable dividing cells(v) $(1.2)^{n-1}$	Products of division (2v) $2x(1.2)^{n-1}$	Viable cells for next division (v + 2v) $(1.2)(1.2)^{n-1}$	Cells lost (.8v) $.8(1.2)^{n-1}$	Accumulated cell loss $\frac{.8(1.2^n-1)}{1.2-1} = 4(1.2)^n$	Total cells in clinical volume [4] & [6] = $5(1.2)^n = N$
[1]	[2]	[3]	[4]	[5]	[6]	[7]
50	$7.583x10^{3}$	$1.616x10^{4}$	$9.100x10^{3}$	$6.067x10^{3}$	$36.40x10^{3}$	$45.5x10^{3}$
90	$1.115x10^{7}$	$2.229x10^{7}$	$1.338x10^{7}$	$8.917x10^{6}$	$53.50x10^{6}$	$66.9x10^{6}$
95	$2.774x10^{7}$	$5.547x10^{7}$	$3.328x10^{7}$	$13.313x10^{7}$	$13.313x10^{7}$	$16.614x10^{7}$
100	$6.902x10^{7}$	$13.803x10^{7}$	$8.282x10^{7}$	$5.52x10^{7}$	$33.128x10^{7}$	41.410x10
105	$1.717x10^{8}$	$3.435x10^{8}$	$2.061x10^{8}$	$1.374x10^{8}$	$8.243x10^{8}$	$10.304x10^{8}$
110	$4.273x10^{8}$	$8.546x10^{8}$	$5.128x10^{8}$	$3.419x10^{8}$	$20.511x10^{8}$	$25.639x10^{8}$
112	$6.153x10^{8}$	$12.307x10^{8}$	$7.384x10^{8}$	$4.922x10^{8}$	$29.537x10^{8}$	$36.921x10^{8}$
114	$8.861x10^{8}$	$17.722x10^{8}$	$10.633x10^{8}$	$7.089x10^{8}$	$42.533x10^{8}$	$53.166x10^{8}$
115	$1.276x10^{9}$	$2.552x10^{9}$	$1.531x10^{9}$	$1.021x10^{9}$	$6.125x10^{9}$	$76.56x10^{8}$

The formulae involving (1.2 for 80% loss) and the number of cell divisions n clearly cannot apply when the number of cells, deriving from 1 is small and the number of cell divisions is small. At what stage cell loss becomes 80% or 60% (1.4) or 20% (1.8) is impossible to define, but the formulae at the head of the columns can only be applied realistically when whole cells are involved, e.g. for 10 divisions $(1.2)^9 = 5.16$ cells.

Table 2. Cell Loss and Clinical Growth of Glioblastoma

The numbers of cell divisions related to diameter of a sphere, assuming 80% cell loss also the time in hours to reach those values using the shortest and slowest values of T_{pot}, as determined by Hoshino[13].

Number of cell divisions	N from [7], Table 1. Total no. of cells in the volume (autolysis excluded)	Diameter (D) of spherical volume of N in (mm) $\sqrt[3]{N} \div 100$	Time to T_{pot}42 hrs. 1.7 cells/100/hour $\frac{\text{Log N}}{\text{Log 1.017}}$	Time to T_{pot}434 0-16 cells/100/hr $\frac{\text{Log N}}{\text{Log 1.0016}}$
50	$45.5x10^{3}$	.357	636	6709
90	$66.9x10^{6}$	4.06	1069	11270
100	$41.41x10^{7}$	7.45	1177	12411
110	$25.64x10^{8}$	13.69	1285	13551
112	$36.92x10^{8}$	15.40	1307	13780
113	$44.30x10^{8}$	16.42	1318	13893
114	$53.17x10^{8}$	17.45	1328	14008
115	$6.40x10^{9}$	18.57	1339	14124
116	$7.66x10^{9}$	19.71	1350	14234
117	$9.19x10^{9}$	20.94	1361	14350

Radius of tumour = R(mm) Volume = $4/3\ \pi R^3$
N from [7], Table 1. e.g. from 3-4 cm diameter. Shortest time 54 hrs., longest = 24 days.

Volume of tumor= N $4/3\ \pi r^3$ where r = 5μ
$\therefore 4/3\ \pi R^3 \therefore R^3 = r^3N$
Diameter = 2R = $2rN^{1/3}$ If r = 5μ; then D = $(N/100)^{1/3}$

is the rule. The rate of removal of dead cells by phagocytosis is unknown. If any small portion of a tumor behaves as indicated in the Table 1, it would seem justifiable to adopt low dose rate continuous method of radiation up to a high dose, which has previouslybeen found by experimental confirmation to be within the tolerance limit of normal brain.

For Table 1 to be understood, it should be used by considering each category horizontally, as if the tumour is spherical. The columns follow logically from left to right to determine the number of cells and the volume. The volumes are shown in Table 2 together with the shortest and longest time (hours) of cell proliferation from the beginning and from one tumour diameter to another. For example, according to this reasoning, between 105 and 112 divisions, the diameter could increase from 2 to 3 cm diameter (spherical) in at least 76 hours (1307 - 1231 hours) and at most 800 hours (13780 - 12981) according to the most and least rate of division[13] i.e. 2.5 times increase in volume in 76 hours (3 days) to 800 hours (33 days) with 80% cell loss per division, but taking no account of loss of cells by apoptosis or removal of any cells mechanically - e.g., in the circulation.

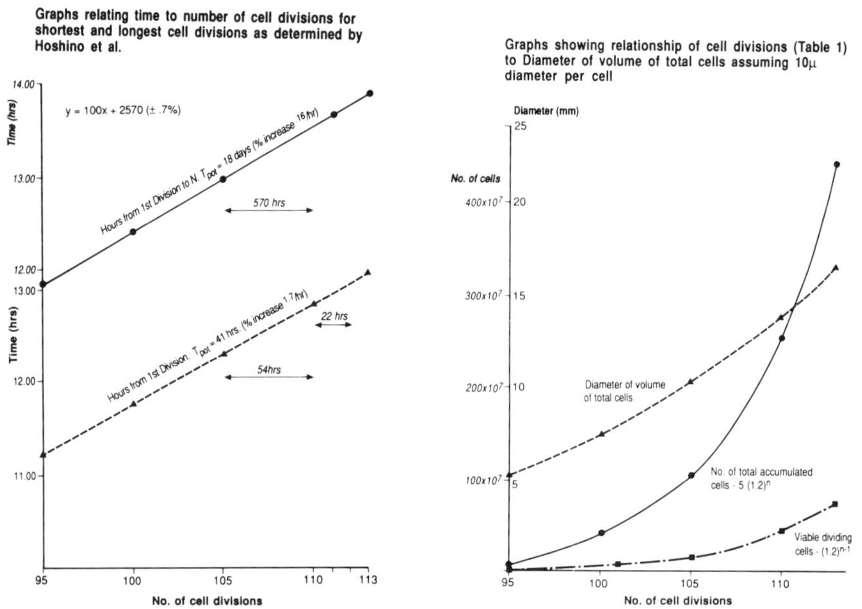

Figure 1

DISCUSSION

It is unfortunate that there are so many uncertainties in pathology and clinical features of glioblastoma. From the point of view of management of treatment there are several questions which should be investigated - if the investigation offers a prospect of answering the question. Important and relevant questions are as follows.

What is the smallest tumour which can be visualized by MRI or CT? Even a tumour 1 mm in diameter may contain 1 million cells, which would take approximately 500 - 5000 hours to develop during which, during the 3 - 30 weeks this would take, might give rise to cells separating off from the main tumour and only likely to be detectable if they proliferate sufficiently to be visible within the life-time of the glioblastoma patient. The same applies with less force to metastases outside the brain than to those within the CNS.

As a clinician with long experience of many cases of glioblastomas the author has seen one such metastasis in a rib, proven histologically by drill biopsy. Another uncertainty related to the visible image is the time taken for autolysis of dead cells. Is this relatively constant? The presence of a mass of necrotic cells in the brain may have a clinically damaging effect, the extent of which is not known, and so in any programs serious thought should be given to the removal of any such mass even if it involves a second craniotomy, provided that the quality of life can be sufficiently improved for a long enough time to justify such interference. The tolerance level of dosage by low dose rate continuous radiation should be within the province of radiobiologists to determine in animals, but variation of dosage by an isoeffect model related simply to dose-rate may be very difficult. For the NSD-TDF model the tolerance level was equated to that for fractionated radiation on connective tissue by clinical observation of the reaction to radium implant on the tonsillar region, in many cases, as compared with that from fractionated treatment by a cobalt 60 beam. It was from this that I considered that 6000 rad in 1 week was equivalent to 31 x 200 rad in 6 weeks, thus giving an NSD = 18, if doses are in Gy. The equation $2.4 \times 10^4 \times r^{-1.35}$ was determined by Orton[5] from this, and from data by Paterson and Mitchell, but what applies to connective tissue cannot be assumed to apply to brain. What endpoint will be convenient for clinical isoeffect determination related to the brain for low dose rate radiation may be possible in animals but impossible in patients, unless clinical accidents occur. The technique of ensuring radiation dosage precision with our knowledge of the physics and techniques using compensation in conjunction with radiological imaging should present no difficulty, but suitable procedures for prolonged immobilization and observation will only be determined as a result of considerable thought and practice in collaboration with those already expert in such work.

ACKNOWLEDGEMENTS

I thank Ms. Stephanie Walters for secretarial and Dr. R. Barth and Dr. R. Gahbauer for editorial assistance in the preparation of this manuscript.

REFERENCES

1. Hall, E.J. (1972). Radiation dose-rate: a factor of importance in radiotherapy. Brit J. Radiol 45 81-97.
2. Bedford, J. S. and Mitchell, J.B. (1977). Mitotic accumulation of HeLa cells during continuous irradiation. Radiation Res.
3. Hendry, J.H. and Thames, M.D. (1986). The tissue rescuing unit. Brit. J. Radiol. 59:628-630.
4. Ellis, F. (1969). Dose time and fractionation: a clinical hypothesis. Clinical Radiology 20: 1-7.
5. Orton, C.G. and Ellis, F. Brit. J. Radiol 47: 603. Amplification in the use of the NSD concept. Brit. J. Radiol. 46:529-537.
6. Fowler, J.F. (1984). Fractionated radiation therapy. Acta Radiologica 23: 209-216.
7. Wara, W.M., Phillips, T.L., Shelline, G.E. and Schoad, J.G. (1975). Radiation tolerance of the spinal cord. Cancer 35. 1558-1562.
8. Van der Kogel, A.J. (1977). Radiation tolerance of the rat spinal cord: time dose relationships. Radiology 122: 505-509.
9. Shelline, G.E. (1980). Irradiation injury of the human brain in Radiation Damage to the Nervous System Silbert, H.A. and Kagan, A.R. Raven Press New York. pp 39-58.
10. Wigg, D.R., Koschel, K. and Hodgson, G.S. (1981). Tolerance of the mature human nervous system to photon radiation. Brit. J. Radiol. 54. 787-798.

11. Shun Wong, C., Minken, S. and Hill, R.P. (1992). Linear quadratic model underestimates sparing effect of small doses per fraction in rat spinal cord. Radiotherapy and Oncology 23: 170-184.
12. Hoshino, T., Barker, M., Wilson, C.B., Boldney, E.B. and Fewer, D. (1972). Cell kinetics of human gliomas. J. Neurosurg. 37: 15-26.
13. Hoshino, T., Siatoyuki, I., Asai, A., Shuboya, M., and Prados, M.D. (1992). Cell kinetic analysis of human brain tumours by in situ double labelling with bromodeoxyuridine and iododeoxyuridine. Int. J. Cancer 50: 1-5.

APPENDIX TO EXPLAIN TABLE I

Assume cell loss of 80%

If .8 cell loss of all cells produced by division, then at first division 2 cells produced 1.6 cells lost .4 cells left ∴ growth impossible. It is not known when cell loss starts or if it is constant but *assume* it is .8 of half of new cells produced.

Then at 1st division 2 cells are produced .8 are lost and 1.2 remain to divide.

At 2nd division 1.2 cells → 2.4 of .8(1.2) are lost and 1.2 + .2(1.2) remain = $(1.2)^2$ capable of division

At 3rd division $(1.2)^2 \rightarrow (1.2)^2 + .2(1.2)^2 = (1.2)^3$ remain to divide $.8(1.2)^2$ are lost.

At nth division cells to divide = $(1.2)^{n-1}$ $(1.2)^{n-1} \rightarrow (1.2)^{n-1} + .2(1.2)^{n-1} = (1.2)^n$

Cells lost = $.8(1.2)^{n-1}$ (Col 5)

Cumulative cells lost = $.8[1+(1.2)+(1.2)^2+\ldots\ldots+(1.2)^{n-1}] = \dfrac{.8[(1.2)^n-1]}{[1.2-1]} = 4(1.2)^n - 4$ (Col 6)

Total cells = Cumulated cells lost + viable cells = Col 4 + Col 6 = Col 7 + $5(1.2)^n - 4$ i.e. $5(1.2)^n$

Col 7 makes no allowance for autolysis (rate unknown) so that increase as a proportion of existing tumour is likely to be more rapid than suggested in Table 2.

ROUNDTABLE DISCUSSION ON NEUTRON CAPTURE THERAPY

Moderators:
R. Gahbauer[1] and F. Ellis[2]

[1]Division of Radiation Oncology, The Ohio State University, Columbus, Ohio, 43210, U.S.A. and [2]Department of Radiotherapy, United Oxford Hospitals and Lecturer (retired), Oxford University, Oxford, United Kingdom

Panelists:
R. Withers[3], G. Laramore[4], E. Glatstein[5], and P. Rubin[6]

[3]Department of Radiation Oncology, U.C.L.A. Center for Health Sciences, Los Angeles, CA, [4]Department of Radiation Oncology, University of Washington Medical Center, Seattle, WA, [5]Department of Radiation Oncology, Simmons Cancer Center, University of Texas Southwestern Medical School, Dallas, TX, and [6]Department of Radiation Oncology, University of Rochester, NY

Dr. Gahbauer. Good afternoon. It is a pleasure for me to introduce the outstanding panel that the Organizing Committee was able to recruit to give us a global perspective and critique and encouragement to proceed with boron neutron capture therapy. If the aim was to get the global perspective, how better to achieve it than to get the previous Chief of the Radiation Oncology Branch, National Cancer Institute, Eli Glatstein, or Phil Rubin who has been the editor of the International Journal of Radiation Oncology, Biology and Physics for twenty years or more, and pioneers in the field of radiation therapy such as Rod Withers and Frank Ellis, who are experts in fields that relate to BNCT such as normal tissue tolerance and fractionation and, of course, George Laramore, who has very extensive experience in high LET radiation therapy. Before we start, however, I would like to briefly introduce the format. First, Frank Ellis would like to make a few comments from his very extensive experience in radiation therapy. Then, every member of the panel will present for approximately 15 minutes, after which we will allow one or two technical questions, if any clarifications relating to the talk are required. Then, at the end, the panel will be up in front again, and we will allow the discussion to continue with audience participation. It is now my pleasure to introduce Dr. Frank Ellis to give the introductory remarks.

Dr. Ellis. The reasons for this meeting derive from the work carried out in Boston and Tokyo. The attractive original idea for BNCT was formed and acted upon in Boston, but without the success needed to treat successfully the most deadly of cancers, glioblastoma. Professor Hatanaka followed up his training at the Massachusetts General Hospital with Dr. Sweet by trying to improve on certain technical aspects such as timing, type of boron compound and depth distribution of slow neutrons from a reactor, which were all considered by him. His success in some cases attracted attention, and to me, these, with confirmation of the diagnoses by Professor Jellinger in Vienna, seemed to open up possibilities of cure of glioblastoma. Were it not for the attention resulting from Hatanaka's work, carried out with considerable difficulties, this and other conferences on BNCT would not have taken place. The accumulation of statistically acceptable records in the beginning of such an activity is difficult, and I have had similar experiences in starting three different Radiation Oncology departments from nothing. The development of interest by so many scientists and clinicians and the amount of work being carried out in so many disciplines is either an indication of real scientific interest and direction, or of ineptitude on a scale worthy of politicians. Primary glioblastoma and brain metastases from other types of cancer remain the challenges, and much remains to be done. It is undisputed that with sufficient boron in the cell, and a sufficient thermal neutron flux, the tumor cell will be destroyed.

The problems are how:

(1) To get the boron selectively into all malignant cells and to persist there long enough.

(2) To provide enough thermal neutrons to obtain the necessary high LET ^{4}He and ^{7}Li nuclei.

(3) To avoid damage to normal brain structures by photons and fast neutrons.

I think an insight into the natural history of glioblastoma cells and tumors is essential. The time from the first symptom to death is usually only a few weeks. Why is its destruction by conventional methods so difficult? My own picture of the development of glioblastoma is first of progression from less malignant gliomata to a stage when the glioblastoma cells use up so much oxygen that cells a little further from the blood supply do not have enough oxygen for metabolism followed by division, and therefore, are "resting". Because of this, conventional radiation, which is sensitive to oxygen tension, and high LET radiation delivered to many cells at relatively insensitive phases of the cell cycle have been unsuccessful in killing all glioblastoma cells, even while causing severe damage to normal brain. This seems to me to be because conventional type fractionation with time gaps in the schedule permits oxygen availability resulting in division of these "resting" cells. This is consequent on the diminished metabolism of the "active" cells following a dose of radiation.

Thus, the selective presence of boron in malignant cells in sufficient quantity and subject to the necessary flux of thermal neutrons offers hope of more success. But any clinical trial requires a comparison, unless virtually 100% success is assured. Therefore, I think that the controls for randomization should be based on some picture, such as the above or a better one based on more expert knowledge. I think that the case for fractionation with the kinds of beams developed suggests that accelerated fractionation by beam therapy using gamma radiation or electrons or low dose rate continuous radiation to a sufficient dose by gamma or fast neutrons (in all cases within the limitations of morphologic and functional tolerance of the normal brain) should be the types of control series used. Certainly, repetition of "conventional" schedules would be virtually useless. I also think that as much autopsy material as possible should be made available, even if long term survivals occur indicating cure. It might be that with a proportion of local "cures", we may begin to see metastases outside the CNS.

(Papers were presented by each of the invited participants, after which the following discussion ensued.)

Dr. Coderre. Could I just get back that question of whether you need to get boron into every single cell. We've done experiments under conditions that were the same as those shown by Dr. Rubin, (the same boron distribution, and the same irradiation time), where we've produced 50% long term survival in the 9L gliosarcoma, except in this experiment, we sacrificed the animals five minutes after the irradiation, disaggregated the tumor, made single cell suspensions, and looked at the clonogenic survival of the tumor cells. So, it's a difficult question to answer whether there is boron in every single cell. I doubt very much that there is, but we have certainly not killed every cell in those tumors. We have reduced the cell survival by four or five logs, but there are certainly viable cells in those tumors immediately after the irradiation. The endpoint is prompt reproductive cell death. However, at one year there was no evidence of viable cells. Fifty percent of treated animals survived for one year, so, that's an indirect answer to the question: "Is there boron in every single cell?" I doubt it, but we are curing tumors, and there were viable cells there early on. The involvement of immune response is also possible.

Dr. Glatstein. But in point of fact, Dr. Coderre, you cured the animal of the tumor for one year. The rodent will live, perhaps, for two years. Hopefully, human beings live longer than that, and those are the kinds of issues that we must address. We all see patients whose tumor goes away and comes back after a year or two, and we must deal with that problem as well. Ten to twenty years is the real goal. I'm just simply saying that rodent tumors are merely a tool to help us compare to the human being, and I'm sure you don't argue with that, and what we really want and need is that kind of followup. With only a one year followup, you cannot make any conclusion about "cure". I mean by definition clonogenic cells, and certainly in tumors, not 100% of all cells are clonogenic, but that's another problem in trying to identify the distribution of your compound. What would you see in recurrence rates if the animals could survive up to 5 years? I respectfully suggest that no one knows.

Dr. Mishima. This Roundtable Discussion is interesting for me in a sense, since I am hosting the next International Symposium for Neutron Capture Therapy in Kobe. I'm participating in this symposium as well as collecting key information and logistics for the planning. Further to the fact that the members of the ISNCT are so multi-disciplined, we also obviously need well organized discussions to promote more profound understanding and data comparison and to diminish the current communication gap between BNCT specialists and general radiotherapists and other oncologists. Concerning the questions raised by non-specialists about physics, ^{10}B compounds, and radiobiological and radiotherapeutic data, I should say that most investigative data is already reported. Thus, I will briefly summarize the current state of information.

In 1970, I initiated my research to develop selective NCT of melanoma using its specific melanogenic activity following our synthesis of ^{10}B-boronophenylalanine as a boronated melanin precursor analogue. The pharmacodynamics of ^{10}B-BPA was studied together with $^{10}B(n,\alpha)^7Li$ induced absorbed energy analyses (Radiation Research, 91:77-94, 1982) first in cultured melanoma cell lines, then in melanomas transplanted into hamsters, and later in naturally occurring swine melanoma, which we eventually cured with combined use of single thermal neutron irradiation (J. Investigative Dermatology, 92:321S-25S,1989). Before starting the use of melanoma NCT on human patients we carried out various studies including acute- and sub-acute toxicity studies. We subsequently carried out pharmacodynamic studies on many of our operable and consenting melanoma patients, administering ^{10}B-BPA in half and full therapeutic doses. We have also done uptake and excretion studies of ^{10}B in melanoma, skin and surrounding tissues, both at cellular and subcellular levels (Pigment Cell Research 2:223-384, 1989). On average, two hours following administration of ^{10}B-BPA, ^{10}B will first be recovered in large amounts in the melanogenic subcompartments called coated vesicles where the melanin monomer is already synthesized. Then these coated vesicles transfer the ^{10}B into the premelanosomes within which the melanin polymer is finally synthesized. Twelve hours after intravenous injection of ^{10}B-BPA into human beings, the melanoma-to-blood ratio of ^{10}B is more than 11 to one.

I was fortunate, from 1970, to have received continuous support from our government. We have also modified our reactors with Dr. Kanda and many other physicists, all of whom are well known among participants, and we have determined the RBE of the $^{10}B(n,\alpha)^7Li$ reaction (International J. Radiation Biology, 51:167-175, 1987). In this session, someone has expressed concern about heterogeneity of cell population in melanomas. However, it should be noted that this reaction occurs within melanoma cells and that the alpha particles and lithium atoms travel a distance of 14μ from the point of ^{10}B activation, attacking all directions randomly, including neighboring cells, resulting in a rather homogeneous distribution of absorbed energy. Since treating our first human patient in 1987 (The Lancet, 2:388-389, 1989), we have successfully treated 12 cases of human malignant melanoma. Many of the patients with primary lesions showed complete regression after a single application of our selective NCT. Thank you. I hope this makes the following discussion go more smoothly.

Dr. Ellis. There are several people who would like to speak. I'd like to ask them to confine their remarks to one minute please. Dr. Larsson has been wanting to speak for some time.

Dr. Larsson. I think it should be said that the motif for BNCT research is not only the immediate requirement of better glioma therapy. Maybe 90% of the society members are more interested in the fact that boron is a good model in any study of molecular targeting, for example, receptor/mediated uptake of toxic substances. Because we try to mix these aspects, there is a failing appreciation of our field or research. I think there is no difficulty to defend the basic approach - with reference to the beauty of boron being a substance that could be used for analysis as well as for interventions at the cellular level. But let us concentrate on the clinical issue, today, because we have a fine team of interested clinicians here. The important question is: "Is it a good time now to start clinical trials, as we prepare for them in Europe within the European collaboration?"

Dr. Ellis. All right. Thank you. Does any member of the panel want to answer that.

Dr. Withers. The worst thing you could do would be to start a clinical trial prematurely and have it go badly. That's happened before. I suspect that you need some more development time, and should move into clinical trials in Phase I studies very slowly, indeed. Otherwise you'll make mistakes, and you won't have patients referred to you, and the chances for a definitive phase III randomized trial will be dead. So you have to move very slowly once you do start working with patients. That's the only generalization that I would like to make.

From a basic radiobiological point of view, the advantage of BNCT is also its Achilles heel. The precise localization of the high LET dose within cells containing boron requires not only that the compound be selectively, or at least preferentially incorporated into tumor cells, but that 100% of the tumor cells be labeled. If, for example, only 90%, or even 99% of tumor cells incorporate an adequate amount of boron, the maximum effect of BNCT would be equivalent to only a 90% or 99% debulking by surgery. If the tumor contained 10^{10} clonogenic cells, this amount of cytoreduction would be equivalent to only one-tenth or one-fifth of the dose needed for tumor control. In general, the approaches needed to overcome these problems would be to use BNCT as an adjuvant to less localized irradiation (e.g. external beam X-ray therapy) and/or to use a multitude of methods for incorporating boron into cells, exploiting a variety of mechanisms or pathways.

Dr. Rubin. Again, having struggled with this for more than two decades and being one of the co-founders of the RTOG, I feel the only thing that doesn't work, and the sure road to failure is if you don't do something. I don't know if you do something, if you're going to succeed. So, I feel the evidence is compelling, and what we should do is not therapy, but begin the Phase I and II pharmacokinetic studies. I think we ought to have these compounds, they have already begun, and that's the beauty of this methodology. You don't have to treat someone unless you show that there is a concentration of the boron compound, and that we will develop, I think, the ability to analyze and to model which patients should be treated in the future. So, I think we are at the edge - I didn't show all the slides that I had, but I think we should continue our animal work. We just got support for that. So, I think this has to be really pushed, and at the same time, do phase I and II studies for pharmacokinetics and concentrations in a variety of tumors.

Dr. Sweet. I want to congratulate Drs. Glatstein and Withers for emphasizing points that I think are crucial. We have absolutely no critical evidence, whatsoever, that the fingerlings or rootlets of invasive tumor in the normal brain take up an increased amount of any boron compound, or for that matter any other compound. We do have tactics available to us to find that out. Moreover, the likelihood is that a new complicated technique called secondary ion mass spectrometry (SIMS) will increase the accuracy and applicability of such studies. At present, the alpha particle track studies, including the perfected method developed at MIT, are important on that score. What we need is a critical examination of lobectomized or large tissue specimens containing not only the main mass of the glioblastoma, but also the streamers invading the otherwise normal brain. These are the cells we must kill. We have data on the main mass of tumor cells, but the surgeon takes that out and it is the remaining tiny invaders that we must study and learn to destroy.

Dr. Hatanaka. Thank you very much. I am very happy to have this opportunity. I'd like to ask the panelists if all of you have read this book, Boron Neutron Capture Therapy for Tumors. This book has an epilogue written by Professor Ellis. This volume is not the proceedings. This is a book written by 46 people who have been interested in this study, and this book has most of the answers to those criticisms raised by some radiotherapists. As is mentioned, I developed this technique more than 25 years ago.

Tomorrow, I am going to talk about the clinical results. I have nine patients who have lived between 10 and 20 years and 10-year survival rate was 17%.

Dr. Ellis I think at this meeting there are more than 200 registrants. There are chemists, engineers, physicists, radiobiologists, dermatologists, neurosurgeons, etc. The Roundtable Discussion at this meeting consisted entirely of four radiation oncologists, so it is obvious that the meeting has been slanted to the clinical aspects, because these are the important ones, and as far as I am concerned, the really important thing is to try to cure glioblastoma multiforme. In all of my life, I have never cured a single case, and it isn't for want of trying. There are various points I think we must know about, and apart from providing the means, we must know about the distribution of the boron, the ways of getting the most boron atoms into tumors to allow the slow neutrons to have their effect by producing the alpha particles, which have a range of approximately the diameter of the glioblastoma cells. All this started with Dr. Sweet and Dr. Brownell and then, Dr. Hatanaka, and I think it is true to say without Dr. Sweet's influence on Dr. Hatanaka first, and then Dr. Hatanaka's initiative in carrying out this treatment in Japan, this meeting and the previous international meetings would not have occurred. Some people have criticized Dr. Hatanaka's work because of inadequate and incomplete records. To me, if somebody had a glioblastoma, proved histologically, and the patient is alive, that's evidence, and I feel that is the really important thing if he can do it, and he did it without a lot of preparation. The fact that we are going to such a lot of trouble now to make sure that the chemistry, and the dosimetry, and the protection, and the beam are all right is an indication that Dr. Hatanaka's work has impressed the scientific community, and its influence has been tremendous. Another point I want to be sure about is whether single doses are better that fractionated schedules. I think, for instance, we must take in to account the natural history of the glioblastoma cells. Well, I am sure that there are many other questions, but I will have to draw these proceedings to a close, and so I thank all the speakers, the contributors to the discussion and the panel. Thank you very much for the very interesting discussion.

"SUMMING UP": THE PHYSICS OF NCT

Robert M. Brugger

Medical Dept., Brookhaven National Laboratory, Upton, NY 11973

INTRODUCTION

Neutron Capture Therapy (NCT) is a binary system for treating tumors in which both an agent, usually B-10, and thermal neutrons are directed to a tumor where they combine and release lethal radiation. Directing the agent to the tumor is mainly a chemistry and biology problem, while the production, targeting and effects of the neutrons is mainly a physics problem.

Drs Soloway and Barth have asked me to "sum up" the physics side of NCT as presented at this symposium, the neutron beams, the dosimetry and treatment planning, and the element identification. The following are my observations based on my experience and my interpretation of information presented at this symposium.

NEUTRON BEAMS

First, let me address neutron beams. While the Japanese continue to treat patients with beams of thermal neutrons, attention in most of the rest of the world is on epithermal neutron beams that will penetrate to deeper tumors. Dr. Ray Moss gave an excellent plenary presentation on beams at reactors. Starting with his table, the table has been expanded and modified somewhat to represent suggested modifications and to include epithermal neutron beams using accelerators as neutron sources. Table 1 shows the three epithermal beams of neutrons that now exist and several that are proposed. The values for the table that Moss chose to present are fluxes and doses in air at a patient position. This is a good choice since a beam can be completely defined and beams can be compared without reference to the geometry of a phantom.

The first column of Table 1 lists the reactors while the second column gives the maximum operating power. The third column, which is the epithermal neutron flux at the patient irradiation position, represents the neutrons that will penetrate into a patient and be moderated into thermal neutrons near the tumor. The general

Table 1. Epithermal Neutron Beams for NCT (Fluxes and doses in air patient position), September 1992

Neutron Source	Power* MW	Epithermal Neutron Flux $n/cm^2sec \times 10^{-9}$	Dose from Neutrons per Neutron $cGycm^2/n \times 10^{11}$	Dose from Gammas per Neutron $cGycm^2/n \times 10^{11}$	J/Ø
Available Beams					
BMRR	3	1.2	4.8	1.4	0.67
Petten	45	0.33	10.4	8.4	0.8
MITR-II	5	0.20	13	14	
Proposed Beams					
GTRR	5	4	1.5		
MURR	10	9.5	2.9	0.4	0.82
BMRR (1)	3	1.7	4.8		0.67
BMRR (2)	3	2.2	3.2		0.72
FIR - 1(TRIGA)	0.25	3.5	2.6		
JRC (TRIGA)	0.25				
Converter plate	0.005	1			
Accelerator Sources					
OSU	0.025	0.75	3.2		
MIT	0.010	1.0	12		
UM(^{7}Li)	0.025	0.9	4.3	2.7	0.66
UM(^{3}H)	0.025	1.25	2.5		

*For reactors, the core or converter plate power; for accelerators, the target power

consensus is that the target for this value should be equal to or greater than 1×10^9 n/cm^2 sec to allow complete treatments to be performed in an hour's time or less. Column four shows dose delivered by fast neutrons which is most critical for skin. The general consensus is that this value should be less than 1×10^{-10} $cGycm^2/n$ so that skin dose will not be the limiting dose. The NCT group at the Idaho National Engineering Lab argues that this value should be even lower, even as low as 2×10^{-11} $cGycm^2/n$, for significant therapy.

The fourth column of table 1 represents the dose from gamma rays in the beam. Usually this value can be made small, a few X 1×10^{-11} $cGycm^2/n$, which is small compared to the dose from capture gamma rays produced by the H(n,gamma) reaction in the patient. The value of the fifth column ranges from 0.5 for a completely isotropic flux to 1 for a well defined, forward directed beam. It is an advantage to have this value closer to 1 than to 0.5 because, when the beam is forward directed, a small increase in beam penetration is realized along with a gain of up to two in the flux of thermal neutrons in the patient compared to the epithermal neutrons in the beam.

Looking at table 1, one notes that the Brookhaven Medical Research Reactor (BMRR), as it existed at the time of the meeting, delivered the highest flux of epithermal neutrons and at the lowest fast neutron dose. The values are well within the limits discussed above. The other two beams are within, or near, the limits and all three beams are being prepared for introductory therapy studies. The Petten beam has better directionality than the BMRR and the Massachusetts Institute of Technology Reactor (MITR) beams. Since these three beams exist and are satisfactory for therapy, NCT with patients should be demonstrated using these beams.

Designs for epithermal neutron beams at a number of other existing reactors have been made. In addition, a design has been presented to enhance the effectiveness of the existing BMRR beam. These new higher flux beams could shorten the time for treatment thus increasing the convenience and tolerance for patients and reducing the chances for adverse effects from prolonged anesthesia. In addition, these beams target lower doses from fast neutrons to give more leeway between the tumor doses and the skin doses. Also, several of these beams are forward directed.

Several designs for epithermal neutron beams at lower power reactors , less than 1 MW, were presented at the symposium. The values presented for the designs appear very encouraging indicating relatively large epithermal neutron fluxes and low fast neutron doses. More epithermal neutrons and less fast neutron dose is projected per MW of reactor power than is achieved from the three presently operating epithermal beams. The geometrical arrangement and the material of the moderators may be better than for the present beams. However, little experimental data is available to confirm these predictions.

Several designs for epithermal neutron beams at accelerator sources of neutrons were also presented and here again the fluxes and doses appear encouraging. A few low power tests have been tried to confirm the predictions but tests at full power have not yet been realized. While the designs indicate that accelerators delivering greater than 10 mA of 2.5 MeV protons are possible, such accelerators are not a standard off-the-shelf item available for purchase. In addition, Li-7 targets, that can withstand greater than 25 kW of heat, have not been demonstrated. Other low energy charged particle reactions, that would make sturdier, more tolerant, targets, such as the Be(p,n) reaction, are being considered, but these reactions have other disadvantages.

A world of caution should be introduced at this point related to these beams of epithermal neutrons from low power reactors and from accelerators. Neither an epithermal neutron beam at a low power reactor nor a beam at an accelerator source has been demonstrated with a complete system at full power. The past experience with the three existing beams is that one gets less than one anticipates; fewer epithermal neutrons and a larger neutron dose. Part of the cause for this discrepancy may be that reality introduces geometrical and engineering restrictions not found in ideal designs. For beams at low power reactors and at accelerator sources, full size, complete, full power systems should be demonstrated before it is assumed that these systems can replace higher power reactors. The birds-in-hand, the existing three epithermal beams, should be used to demonstrate successful NCT before a rush is made to other sources.

Now a few comments about costs. One participant at a session suggested that the Georgia Tech Research Reactor (GTRR) could be modified for $0.5M while another participant estimates that $2.2M will be required. The latter value is probably closer to what will be experienced. Brookhaven National Lab (BNL) staff are estimating that the BMRR can be upgraded for $0.5M. Recent experience with building a new lower power reactor and building indicates that the cost of such a new reactor in the USA will be \$6M and will take 4 years to complete. For an accelerator-based system, a cost as low as \$1.0M has been suggested while one poster stated that a high current Radio Frequency Quadrapole (RFQ) accelerator alone may cost \$3-6M. The costs of proposed systems are still uncertain. It should not be assumed that an adequate accelerator system will be less costly than modifying an existing reactor or building a new reactor.

Several speakers at this meeting stated that a reactor could not be built at a hospital. I believe that this conclusion is wrong. If NCT is demonstrated to cure cancers or significantly lengthen life, and epithermal neutron beams from reactors are required for this success, reactors can and will be built at hospitals.

In Japan, thermal neutron beams are used for BNCT and throughout the world, thermal neutron beams are used for cell and animal studies. As an example, the Musashi reactor at which Hatanaka has treated a number of patients delivers 5.7×10^9 n/cm^2 sec with only a neutron dose of 0.8×10^{-11} cGy/n and a gamma dose of 2.3×10^{-11} cGy/n at a maximum core power of 100 KW. Thermal neutron beams are easier to achieve than epithermal beams and a number of thermal neutron beams exist. For the cell and animal studies, fluxes of thermal neutrons above 1×10^9 n/cm^2 sec are required and near 1×10^{10} n/cm^2 sec are desired. Experience at BNL indicates that the dose from fast neutrons, even though this appears low, is still a problem.

DOSIMETRY

Turning to the mixed beam dosimetry that is encountered in NCT, examples of both measurements and calculations were presented. The measurements and calculations are being compared and interrelated and good agreement is being achieved. The physical dosimetry required for NCT that is presently available is quite adequate to start therapy studies. A much larger uncertainty than any in the physical doses is introduced by the difficulty of predicting boron concentration and by the uncertainties of the biological effects of the various types of radiation. A general conclusion is that the physical dosimetry will be ready for NCT when patient treatments start.

The code Monte Carlo Neutron Photon (MCNP) has become the work horse and the standard for calculations of fluxes and doses for NCT. In some cases, faster deterministic codes are being used as support. It is important that the Department of Energy continue to support and upgrade the MCNP code.

For treatment planning, which provide the methods to show oncologists and radiologists the dose distributions in patients, the development of systems is progressing. At this symposium presenters from at least two groups showed the extensive progress they have made in developing their systems. Two other groups have presented their treatment planing systems at previous meetings. In most cases,

CT and MRI scans are used to determine the anatomy and the distribution of neutron scattering material in the patient. The displays are in color, three dimensional or slices, and can be rotated and adjusted in other ways. A general conclusion is that the treatment planning will be ready for NCT when patient treatments start.

ELEMENT IDENTIFICATION

In the development of NCT, there is a continuing need to identify and locate boron and other elements in biological samples, animals and patients at the PPM level. This identification needs to be made on both a macroscopic scale, for organs, blood, tumors, etc. and on a microscopic scale, cells, cytoplasm, nuclei, etc. On the macro scale, Prompt Gamma Neutron Activation Analysis (PGNAA), Inductively Coupled Plasma (ICP) spectroscopy, Direct Current Plasma (DCP) spectroscopy and other methods have been developed but these are not widely available to all investigators. On the micro scale for identifying and locating B in cells and elsewhere, a few methods are being developed such as autoradiography, track etch detection and electron microprobe scanning. A need exists for element identification for in-vivo measurements. Several are being developed such as the MRI identification of B-11. Of course any agents containing Gd-157 will be easily identified in-vivo with MRI. General conclutions are that better methods of element identification are needed and these need to be more conveniently accessible to user.

CONCLUSIONS

To "sum up" then, the development of the physical tools and methods are not now limiting the successful demonstration of NCT. Several epithermal neutron beams are available, dosimetry methods are in hand, and treatment planning methods can be ready when required. Methods of identifying elements have been demonstrated, but are not yet widely accessible. New methods could supply valuable additional information about the quantity and location of NCT elements, particularly on the microscopy scale and *in vivo*.

ACKNOWLEDGEMENTS

This research was supported by U.S. Department of Energy contract DE-AC02-76CH00016.

COMPOUND DEVELOPMENT AND ANALYTICAL STUDIES

Albert H. Soloway

College of Pharmacy
The Ohio State University
Columbus, Ohio 43210

INTRODUCTION

For Neutron Capture Therapy (NCT) to become a useful clinical modality for treating solid tumors for which current therapies are of marginal utility, it is of crucial importance that effective compounds be designed, synthesized, and evaluated. The two compounds that are being used clinically or are undergoing clinical pharmacokinetic studies are $Na_2B_{12}H_{11}SH$ (BSH) and p-boronophenylalanine (BPA). Both of these compounds are more than 30 years old; and, while that *per se* is not a negative factor in the clinical use of a compound, it is pertinent to determine whether either possesses the requisite minimal requirements and to elucidate the biochemical factors by which tissue localization and persistence may be achieved. The two critical requirements for useful compounds in BNCT are: (1) the attainment of selectivity for and persistence within malignant cells compared with that in blood and contiguous normal tissues, and (2) the achieving of tissue concentrations, in the case of boron-10 as the capture agent, of 10^9 boron atoms per cancer cell or approximately 30 micrograms of boron per gram of tumor. Since the mean free path of these high LET particles is very short, it would be optimal if the capture reaction were to occur near the tumor cell nucleus, but for B-10, this is not a *sine qua non.* However, such a location may reduce by at least twofold the necessary concentration in the malignant cell. A key question is what degree of tumor specificity must be obtained? In essence, what is the minimal requirement for a useful compound? This is not a simple question with an obvious answer. There can be a marked heterogeneity among tumor cells from: (1) a histological and biochemical standpoint; (2) their cell cycle status and; (3) the degree of hypoxia influenced by the variability of the blood supply to different parts of the tumor. Also, many compounds that have been synthesized were not specifically designed for BNCT but were determined pragmatically to possess potential, but only now are their subcellular concentrations being examined. The lack of this information has greatly limited the ability to make meaningful radiobiological calculations at the cellular level. Obviously, the higher the differential between tumor and normal tissue the better, but what is the minimal ratio and the necessary concentration in tumor below which BNCT will be of questionable utility? Also, the capture agent must localize in all viable tumor cells, both proliferating and non-proliferating, since the latter may be in a quiescent stage but still retain the potential for replication under appropriate biological conditions. Compound localization in necrotic debris is not useful since delivering radiation to such areas would not appear to contribute to tumor control. Finally, even if subcellular distribution of compounds were known, the RBE value of the capture reaction remains a point of contention.

Due to the current failures in existing therapies for primary malignant brain tumors and CNS metastases derived from other types of cancer, a great deal of effort in BNCT has concentrated in attempting to treat such malignancies. Initial approaches concentrated on the development of compounds that would be excluded from normal brain by the action of the blood-brain barrier (BBB) but not the tumor whose barrier mechanism may have been compromised. Now, it is clear that for successful tumor control, those tumor cells that may indeed be protected by the BBB must be effectively targeted while the normal vasculature is spared. The concentration on treating CNS tumors is derived from the fact that improving both the quality of life and life expectancy of such patients would demonstrate the potential utility of BNCT for treating other cancers.

In addition recently, efforts have been made primarily in Japan to use BNCT to treat melanoma, which for a variety reasons cannot be surgically resected. The compound used has been BPA, under the rationale that such a boron-containing aromatic amino acid has significant potential for localizing in this type of malignancy.

The increase in the number laboratories engaged in compound development throughout the world is a very important step in providing the number of chemists needed to ensure that more appropriate compounds with a strong biochemical rationale will be synthesized. In the past, compound screening has been carried out on existing boron structures whose syntheses were frequently unrelated to the needs of BNCT. Now, compounds are increasingly being designed with a specific biochemical objective in mind. It is essential that, as further information unfolds as to the biochemical differences between tumor and normal and hypoxic and normal oxic cells, this information is used in compound development. In essence, the biochemical, pharmacological, and analytical studies should provide important data that will guide the preparative chemist in new compound synthesis.

DISCUSSION

Compound Development

For the purpose of presenting an overview of the meeting from the standpoint of compound development and analytical studies, it is useful to examine the different types of compounds and the development of analytical methodologies, especially at the subcellular level, to determine the precise cellular location of the various boron compounds. With respect to compounds, it is pertinent for discussion to divide the group into two broad classifications: (1) small molecular weight molecules and (2) macromolecular weight and vesicle species. With respect to small molecular weight compounds, the symposium showed that an increasing number of investigators are interested in developing boron-containing amino acids. This stems in part from the more widespread use of p-boronophenylalanine (p-BPA), especially as its fructose complex, not only for melanoma but for other tumors. An important requirement is attaining high boron concentrations in tumor, thereby permitting somewhat increased levels in normal tissues in the beam path without compromising the radiobiologic dose differentials. In addition to the para isomer, the synthesis of the ortho and meta analogues of boronophenylalanine have been described. Since p-BPA accumulation in tumor is transient, there is increasing interest as to the metabolic fate of BPA and the basis for its uptake and transitory concentration differentials in tumor cells. In addition to examining amino acids with a single boron moiety, a number of investigators have been developing the synthetic methodology for preparing various carborane-containing analogues both as racemates and as enantiomers. Not only do such compounds possess ten times the boron content, but they offer the potential for becoming incorporated chemically into peptides. Carboranylalanine, whose synthesis was described more than 17 years ago, has now been prepared in several laboratories in useful amounts, as have certain carborane-containing aromatic amino acids that were described at this meeting. It must again be reiterated that for success as NCT agents, these structures must not only achieve selectivity for tumor versus normal but must also attain sufficient concentrations in the tumor itself.

At several different laboratories, a second approach in the small molecular weight category has been the design and synthesis of boron-containing nucleosides. The rationale is that the more rapidly-proliferating tumor cells may use the salvage pathway for incorporating such compounds into tumor nucleic acids; or, at the very least, the compound may be converted to its nucleotide. If the nucleotide *per se* were not freely diffusible, then the boron moiety would be trapped intracellularly. Boron-containing groups, either as a single boron moiety or as a carborane function, have been attached both to the pyrimidine base as well as the carbohydrate portion of the molecule. The chemical methodology for such syntheses is now becoming well established and was further described at this meeting. In addition to nucleosides, the synthesis of nucleotides has also been undertaken. The rationale is twofold: (1) to begin to examine whether nucleosides are converted to their corresponding nucleotides under biological conditions and (2) to use them as potential tumor-targeting agents *per se*. Though nucleotides would not be expected to cross cell membrane, the carborane-containing compounds possess considerable lipophilicity, and this may contribute in balancing the hydrophilic properties of the phosphate group thereby permitting cell penetration. *A priori*, the biological results with such nucleotides cannot be predicted. Thus, the nucleotides may serve two functions.

Based upon the selective affinity that porphyrins have for a variety of neoplasms, research has continued on the development of boron-containing porphyrins. The previously synthesized water-soluble BOPP has now been prepared as a manganese chelate. This compound may serve both as a tumor-selective agent as well as a proton MRI contrast enhancement substance. Major distribution studies of BOPP are clearly desirable.

Another very interesting class of compounds is the boron-containing ether lipids. Their development is based upon the fact that both natural and synthetic ether lipids accumulate in significantly elevated amounts in various tumors, based, certainly in part, upon the absence in such cells of the enzyme O-alkylglycerol monooxidase. A carborane-containing glycerophosphocholine has been synthesized, and its biological results are awaited.

Two other carborane structures have also been prepared. These are carboranyl analogues of the hypoxic cell sensitizer, 2-nitroimidazole, and carboranylaziridines. The former may offer the potential for targeting the hypoxic cell fraction in tumors. The basis for the synthesis of the latter structures is their potential, through the alkylation of nucleic acids by the aziridine function, in placing the boron entity on tumor nucleic acids.

Another approach that was described is to use boron labeled receptor-seeking biomolecules as the means for delivering boron. The idea is to take advantage of receptor sites on tumor cells that may be grossly overexpressed or may be deficient/absent from normal cells. This was the reason for the synthesis of boron-containing peripheral type benzodiazepine-binding compounds for malignant brain tumors. Clearly, a question to be answered is are there a sufficient number of such receptors to allow the attainment of adequate boron concentrations?

One of the chemical limitations in the incorporation of the *ortho* carborane moiety ($C_2B_{10}H_{10}$) into tumor targeting boron compounds is its reactivity with basic amines resulting in the removal of one boron atom from the cage generating the anionic *nido* species ($C_2B_9H_{10}^-$). *Meta-* and *para* carboranes are considerably more resistant to alkaline degradation. For this reason, investigators have undertaken to develop the synthetic methodology for preparing suitable amounts of substituted *meta*-carboranes. In addition to these carboranes, the pseudoicosahedral carborane cluster anion [*closo*-1-$CB_{11}H_{12}$]$^-$ and its neutral derivatives have also been considered as potential boron carriers for BNCT. This research is at the developmental chemical stage.

In addition to these carbon-containing boron hydride anions in which the carbon atom provides the opportunity for bridging the boron moiety to tumor seeking structures, there is increasing interest in the more hydrophilic boron-rich polyhedral boranes, such as $B_{10}H_{10}^{2-}$, $B_{12}H_{12}^{2-}$, $B_{20}H_{18}^{2-}$ and $B_{20}H_{18}^{4-}$. While such structures *per se* are not useful, preparing derivatives can lead to boron synthons that may possibly be incorporated into a variety of chemical structures that may be tumor binders. This was the basis for the research which focused on the synthesis and use of the monovalent anion carbonyl nonahydrodecaborate. Another derivative analogue of these polyhedral borane anions is BSH itself, from which alkylated, acylated, and disulfide compounds have been described. The dimer as its potassium salt ($K_4H_{11}B_{12}SSB_{12}H_{11}$) has been periodinated, and this product has been subjected to preliminary evaluation. Besides being a new and potentially useful new agent, this boron compound has potential for the noninvasive clinical imaging. One of the reasons for considering these anions, in contrast with the carboranes, is their appreciable solubility in aqueous systems, obviating the need for a water-soluble organic vehicle for administrating these compounds to biological systems (i.e., cell cultures and tumor-bearing animals). This limitation in the use of carborane-containing and other lipophilic boron compounds has been met by inserting water-solubilizing functionalities into such structures. These include nonionic species such as polyols and readily ionizable entities such as phosphates. Their preparation was described at this meeting.

Another approach, presented at the symposium, is the use of compounds with proven ability to bind to the minor groove of DNA as the basis for carrying boron to this important target site. A bibenzimidazole phenylboronic acid, whose synthesis had been reported at the Fourth International Symposium, was more completely evaluated and demonstrates clear potential relative to other boron compounds.

From the foregoing, it is readily apparent that much effort has been expended in the area of small molecular weight compounds. The boundary line between small molecular weight compounds and the more macromolecular species is arbitrary, and frequently the smaller molecular weight compounds are directly used in the formation of macromolecular and vesicular species.

One of the initial approaches in using boron macromolecules for targeting tumors involved the use of monoclonal antibodies (Mab). Research has continued in attempting to use its tumor selectivity to deliver boron. In order to minimize conformational changes in the antibody while achieving the required boron loading per tumor cell (10^9 boron-10 atoms), boron polymers containing at least 10^3 boron atoms have been attached to the Mab and its fragments. Many laboratories have developed different polymers and various methods for their attachment to Mabs. Though such conjugates have demonstrated significant *in vitro* immunoreactivity, their *in vivo* behavior is invariably characterized by excessive liver uptake relative to tumor. For this reason, greater effort is now being expended at several places to develop bispecific antibodies; one combining site would recognize a tumor-associated antigen, and the second would bind tightly and by noncovalent forces to a boronated polymer.

Related to the use of antibodies and their fragments has been the development of growth and regulatory peptides, whose receptors may be over-expressed on tumor cells. Such entities have been proposed as the means of delivering boron selectively to such receptors. Among these is epidermal growth factor (EGF), which is over-expressed on various human tumors, including glioblastoma multiforme. For this 6045 da, 53-amino acid molecule to be used as a boron carrier, it is obviously important that the attached boron moiety not compromise EGF's binding to this receptor and that the number of boron atoms and the receptor site density on the tumor be such that boron loading will be adequate for BNCT. Studies are underway to use EGF for this purpose.

Two other higher molecular weight species appear to offer significant potential as delivery agents. These are low density lipoproteins (LDLs) and various liposomes. Work has continued on the development of LDLs as boron carriers based upon the higher receptor site density in tumor cells for LDLs compared with normal cells and their high capacity for carrying hydrophobic boron compounds in vesicles. Previous research demonstrated that fatty acid esters containing the carborane moiety could be exchanged for cholesterol esters in reconstituting LDLs. A comparison has now been made of various alkyl and aryl carboranes as LDL reconstituting agents, and this research has identified the desirable properties for such compounds. Animal distribution studies of such encapsulators can be expected.

Other very important encapsulators for boron delivery are synthetic vesicles such as the liposomes. Boron-containing liposomes are endocytosed by tumor cells and deliver their contents intracellularly. Research has continued to identify the appropriate characteristics of these compounds so that boron persistence will occur. Presenters examined a series of compounds and correlated cellular retention with size, charge, and reactivity of the various boron species. Biological studies suggest that compounds that are reactive with and can bind to intracellular organelles or can be metabolized to such reactive entities appear highly promising.

Most of the research discussed at this symposium in the area of compound development has focused on the use of boron. The basis for this is the very extensive covalent chemistry that is available with boron and the ability to synthesize compounds with high boron content. However, in view of the very interesting nuclear properties of gadolinium-157 with a thermal neutron capture cross section approximately sixty times boron-10 and with the ultimate emission of Auger electrons, there is clear interest in the development of Gd-labelled DNA ligands. In contrast with boron (the capture reaction of which is expected to have tumoricidal effects if it occurs on, in, or perhaps even simply very near tumor cells), it has been shown that for Gd NCT to be effective, there must be a very close association between Gd and the DNA helix if Auger electron shower is to be expected to produce the requisite double-strand breaks of DNA.

Analytical Studies

The development of analytical methodologies ranging from purification and quality determination of various chemical compounds to noninvasive quantification and localization of BNCT agents by boron-11 magnetic resonance imaging and spectroscopy (MRI and MRS) continues to be of great importance in BNCT. Since the radiation dose will be directly dependent not only on the total boron in tumor but on its subcellular location, various analytical methods have been proposed, including their use in clinical pharmacokinetic studies.

Since there remains continued interest in BSH and its conjugates and because BSH *per se* undergoes facile oxidation, methods for purification and quality determination of BSH is of continuing importance. High pressure liquid chromatographic (HPLC) procedures have been perfected for quantification of BSH and its two major oxidation products $B_{24}H_{22}S_2^{4-}$ and $B_{24}H_{22}S_2O^{4-}$. Also reported at this meeting, was the analysis of a dextran conjugate of BSH by the use of Inductively Coupled Plasma Mass Spectrometry (ICP-MS).

In addition to chemical separations and analyses, there has been continuing interest in the measurement of boron compounds in biological specimens and in determining their metabolic fate. Prompt gamma and atomic emission spectroscopy (AES) continue to be the major modalities in the measurement gross tissue boron content. A method for the rapid analysis of total boron in blood has been achieved using AES in which samples can be prepared and analyzed within 5-15 minutes. The procedures include Inductively Coupled Plasma (ICP) AES and Direct Current Plasma (DCP) AES. Important questions are what is the metabolic fate of a particular boron compound and what is the basis for its transport and accumulation by malignant cells? In this respect, separation and analysis of various boron species from biological material is a great importance in understanding the mechanism for localization. There has been speculation that the basis by which BSH is transported and incorporated into tumor may be related to its ability to become bound to plasma proteins and that this conjugate is the responsible entity for tumor localization. An HPLC method is reported for the separation of protein complexes from the BSH anion. While the dimer of BSH is strongly associated with proteins such as albumin, there is little significant binding of proteins with BSH *per se*. These results were further supported by NMR studies involving various types of albumin. Additional research is planned to examine whether binding to proteins occurs under *in vivo* conditions.

Another technique for measuring polyhedral boranes in plasma is the use of Fourier Transform Infrared (FTIR) instrumentation. The characteristic B-H absorption of 2600 cm^{-1} has been used for boron quantification. The procedure that was developed could be performed within 30 minutes, at a sensitivity level of 5 ppm. Improving the technique for detecting amounts below 5 μg boron in 5 ml of plasma appears possible.

In addition to measuring various polyhedral boranes, there is significant interest in determining whether BPA is metabolized in biological tissue or remains unchanged. For this purpose, an HPLC system has been developed. The method has sufficient sensitivity so that BPA and other boron species, which may be derived from it, can be measured in clinical samples. This is important in understanding the transitory nature of BPA in tumor.

The above analytical studies have focused on gross boron concentrations as well as products derived from different boron compounds. However, frequently there is a lack of information as to where in tissue the boron compounds and their metabolites are concentrating as well as a paucity of knowledge as to their subcellular distribution. The discussion below will mention various methodologies presented at the meeting that addressed these important issues.

Several laboratories and their researchers are using the Solid State Nuclear Track Detector, CR-39, in order to determine the spatial distribution of boron-10 in tissue. The techniques that are being perfected will provide significant information correlating tissue type with the concentration of various boron compounds and will assist in determining the compound's potential utility in BNCT.

In addition to autoradiographic studies, information regarding the subcellular location of boron is of importance in calculating the radiobiological doses that can be expected. Also such information is useful in assisting the synthetic chemist in determining whether the compounds that have been designed are achieving their biological objectives. The techniques discussed and presented at the meeting for use in localization and quantification of boron in tissues at the subcellular level included laser resonance ionization spectroscopy (RIS), secondary ion mass spectroscopy (SIMS), FTIR, and electron energy loss spectroscopy (EELS). Use of ion microscopy (IM), an imaging technique based upon SIMS, is further along in its application to BNCT; and the research described demonstrated that quantification of boron in biological tissues is not only feasible but useful in evaluating different boron compounds.

Finally, at the time of clinical trials, it would be very desirable to be able to determine the biodistribution and pharmacokinetics of boron compounds noninvasively by magnetic resonance techniques. This would contribute to establishing the optimal timing for initiating BNCT, as well as the duration of the radiation procedure, for a particular patient. Of the two boron isotopes, ^{10}B and ^{11}B, it is clearly apparent that imaging techniques are much more readily adaptable to boron-11. Studies with tissues and in whole animals have shown that B-11 MRS is sensitive even at low boron concentrations and that B-11 MRI is capable of revealing the spatial distribution of boron compounds under *in vivo* conditions.

CONCLUSION

Increasing interest among synthetic boron chemists has led to significant progress in compound development. One of the key deficiencies from a research perspective is the lack of involvement of biochemists and molecular pharmacologists. It is necessary to probe the molecular basis by which compounds or their metabolic products are accreted and retained by tumor cells. At the present time, our knowledge base in these areas is rudimentary at best. A second important need is for a selected number of compounds, which have already demonstrated potential as possible useful NCT agents, to be synthesized in larger amounts for extensive evaluation by many researchers and especially for their toxicological profile to be assessed in animals. Currently, there is no source for compounds that may require a more detailed biological evaluation. Finally, routine screening procedures need to be established so that various compounds and vesicles can be compared with one another. Different laboratories use different assays and different tumor types. Ultimately, the boron vehicles must be compared in a standard biological system so that the relative advantage of each compound can be determined.

With respect to the development of analytical procedures, much remains to be accomplished, especially at the subcellular level and in noninvasive imaging. Procedures are now well established for the gross measurement of boron in tissue samples. In order to assist radiation biologists and clinicians in dose calculations and treatment planning, it will be essential that the spatial distribution of boron in tissue and in subcellular organelles become an established and routine analytical procedure. The extent to which ^{11}B-NMR can be developed and used noninvasively to assist the clinician in dose determinations is unclear at this juncture, but certainly it is worthwhile to attempt to develop this methodology in order to meet that objective.

SUMMING UP: BIOLOGICAL STUDIES RELATING TO BNCT

Rolf F. Barth
Department of Pathology
The Ohio State University, Columbus, Ohio 43210

INTRODUCTION

A total of thirty-seven papers were presented relating to preclinical biological studies on BNCT. The broad subject categories included pharmacokinetic and therapy experiments carried out in small animals; pharmacokinetic and treatment planning experiments carried out in large animals; *in vitro* studies; and finally, a session on the relative biological effectiveness of the $^{10}B(n,\alpha)^7Li$ reaction. Each of these subject areas will be summarized with some selectivity, and the following review is not a comprehensive one. Plenary lectures were presented by J. Coderre[1] and P. Gavin[2] on small and large animal studies relevant to BNCT and since these were in part integrative reviews of work carried out by other investigators, they will not be summarized. A plenary presentation was given by J. Chapman[3] on the relevance of research on hypoxic cell sensitizers and photodynamic therapy (PDT) to BNCT. Since both PDT and BNCT are binary systems, they share the common requirement of delivery of a sensitizer and radiation to an appropriate target. At this point in time, however, the photosensitizers that have been used for PDT appear to be targeting vascular endothelial cells rather than tumor cells[3]. Nevertheless, the design of tumor localizing compounds would be important for both PDT and BNCT, and there clearly is a common ground here. On the otherhand, the photodosimetric principles associated with PDT are not relevant to BNCT. One of the major lessons that PDT basic and clinical research provides is that the tumor vascular endothelium may be a potential target for BNCT, and this is an area worthy of future investigation.

Small Animal Studies: Pharmacokinetics

Schuepbach et al.[4] have used the rat RG2 glioma to establish an animal model that simulated human treatment of glioblastoma, i.e. debulking surgery, followed one week later by administration of sodium borocaptate (BSH). Unfortunately, boron values, which were determined at two different laboratories, showed such a high degree of variability (l0x from one laboratory to the other) that no useable ^{10}B uptake data was obtained. Nevertheless, the groundwork has been laid for a more definitive study, and this should help to better define certain parameters relating to the combination of surgery

Advances in Neutron Capture Therapy, Edited by
A.H Soloway *et al.*, Plenum Press, New York, 1993

and BNCT, the most important of which is the interval between the time of surgery and the initiation of BNCT. Kabalka et al.[5] have described methodology for the detection of boron by means of Magnetic Resonance Spectroscopy (MRS) and Magnetic Resonance Imaging (MRI), and these techniques could detect boron at concentrations as low as 30ppm. Most importantly, they potentially can be used to follow the *in vivo* distribution of boron compounds. Joel et al.[6] have investigated the possibility of reducing the hepatic toxicity of the oxidized disulfide dimer of BSH, $Na_4B_{24}H_{22}S_2$ by the prior administration of glutathione-monoester (GSH-ME). This compound significantly reduced elevations of two hepatic marker enzymes, ALT and AST following administration of dimer and concomitantly there was a 2.4x increase in tumor boron concentration (30.7 v.s 12.6 ppm). Mallesch et al.[7] studied the distribution of the boronated porphyrin BOPP[8] in rats carrying the BC-1 mammary carcinoma. A maximum tumor: blood ratio of 2:1 was observed at 48 hours, but as reported elsewhere, hepatic uptake of this compound can be even higher[8]. Nakagawa et al.[9] have studied blood brain barrier permeability in rats that had been given D_2O. Animals whose mean blood D_2O concentrations exceeded 36.5% showed a significant leakage of ^{125}I labeled albumin into the brain indicating that caution is required when using this technique to increase the depth of penetration of thermal neutrons. It was concluded that as far as blood brain barrier permeability is concerned, levels of deuteration below 30% probably are safe. La Hann et al.[10] have reported on the cardiovascular toxicity associated with i.v. administration of BPA in rats. Infusions of up to 3000 mg/kg were well tolerated, but doses >4,000 mg/kg produced lethal declines in cardiac output, stroke volume and blood pressure.

Small Animal Studies: Therapy

Bailey et al.[11] have described a nude rat model for the treatment of intracerebral melanoma by means of BNCT. A human melanoma cell line was implanted into the brains of nude rats and 4-5 weeks later BPA was administered i.p., followed by irradiation at the Brookhaven Medical Research Reactor (BMRR). BNCT treated animals showed significant prolongations in median survival times with 30-40% long term survivors compared to irradiated controls. Solares et al.[12] have studied the effects of fractionation on the GL261 glioma that was implanted intracerebrally into mice, followed by administration of BPA. All single fraction groups survived significantly longer than controls, but this was not consistently related to dose. No significant improvement in survival time was seen in animals treated with two fractions, and three fractions had a detrimental effect on survival. Matalka et al.[13] have carried out experiments on the effects of prior administration of BPA to nude rats carrying intracerebral implants of the human MRA27 melanoma followed by californium-252 brachytherapy. Significantly longer survival times were observed in animals that received BPA plus brachytherapy compared to ^{252}Cf brachytherapy alone. Matalka et al.[14] also described their studies in rats using BPA as a capture agent for the treatment of intracerebral implants of the F98 glioma by BNCT. In contract to results obtained with the MRA 27 melanoma[11], no significant prolongations in survival times were observed in BNCT treated animals, and all succumbed to their tumors. Mallesch et al.[15] have implanted Harding-Passey melanoma cells into the pancreas of mice and 1-2 weeks later administered BPA-fructose i.p.. At varying time intervals thereafter, boron content in the melanoma and pancreas were determined by inductively coupled plasma-atomic omission spectroscopy. Pancreatic values exceeded those of the melanoma. Laster et al.[16] have described their experience using the boronated porphyrin BOPP[8] as a capture agent for BNCT of mammary cancer in mice. Since a tumor control dose could not be delivered (it was too high, TCD$\approx$54Gy), the endpoint for these experiments was tumor growth delay. Although a growth delay response could be demonstrated, blood levels of BOPP remained high and alternative regimen such as hypofractionation may be more advantageous with BOPP. Morris and Hopewell[17] have studied skin and spinal cord tolerance in rats following administration of BSH at a dose of 100 mg/kg. Radiations were carried out either with the DIDO reactor at Harwell, U.K. or the BMRR at the Brookhaven National Laboratory, N.Y. with total absorbed physical doses of 0.6, 8.7 and 1.4 Gy/MW-min respectively. Moist desquamation of the skin was observed, and in contrast to X-irradiation, the response was biphasic. However, no evidence of neurological dysfunction was seen over a period of 84 weeks with an estimated dose to the cord of 18-31 Gy using the DIDO reactor. Accordingly, studies have been initiated at the BMRR using considerably higher radiation doses.

Large Animal Studies: Normal Tissue Tolerance

Watkins, Moss et al[18] have developed a computational model using MCNP and a processing code TREAT to evaluate the various dose components of BNCT irradiations. The model is still in an early stage of development, but already has been used at Petten for normal tissue tolerance studies. Huiskamp, Gavin et al.[19] and Philipp, Gavin et al.[20] have carried out dose effect comparisons between the Petten High Flux Reactor (HFR) and the BMRR in dogs. Thirty dogs were irradiated at the BMRR, 15 without compound, and 15 who had been given BSH i.v. and had blood boron concentrations of 30-62 ppm. Fifteen dogs have been irradiated at Petten, 12 with compound (blood boron concentrations 25-62 ppm) and 3 without. Of the 15 dogs irradiated at the HFR, two irradiated in 1991 developed lethal brain necrosis. MRI showed massive changes, mainly in the white matter. Seifert et al[21] described the brain histopathology of these two dogs, which showed extensive necrosis, peri-ventricular hemorrhage surrounded by edematous tissue, glial proliferation, swelling of endothelial cells, and thickening of capillary basement membranes. Of 13 dogs irradiated in 1992, two that received a physical dose of 19 Gy with 50 ppm ^{10}B-BSH, showed brain lesions comparable to those irradiated at the BMRR with 27 Gy plus ^{10}B. It was concluded that the use of a dose rate factor for the γ component and derived fast neutron RBE and compound factors for skin and brain, and a dose description beyond the thermal peak allow a good equivalent dose description relative to the observed effects[19].

Large Animal Studies: Pharmacokinetics

Bradshaw et al.[22] have used projection reconstruction MRI techniques to obtain qualitative comparisons of BSH in tissues. With a scan time of 8 min, 28 sec the image data provided a fair snapshot of kinetic changes during elimination phases in the dog head. La Hann et al.[23] have studied the bioavailability of i.v. formulations of BPA in dog and rat. Greatest solubility was achieved by complexing BPA with fructose at a 1:1 molar ratio. Less than 10mg of BPA were soluble in distilled water, pH 7.4. Briefly increasing the pH to ~ 10 for a 200 mg/ml BPA suspension, and then readjusting the pH to 7.4, solubilized 173 mg of BPA per ml. Neither BPA-fructose or BPA-cyclodextrin showed signs of precipitation over 7 days. Using the BPA-fructose formulation doses $\approx$ 3,000 mg/kg could be administered to rats without inducing obvious toxicities. It was postulated that i.v. infusion of high doses of BPA-fructose could increase melanoma uptake of ^{10}B.

Large Animal Studies: Treatment Planning

Gavin, Huiskamp et al.[24,25] have developed a computer model for the compound factor in normal brain tissue. A Monte Carlo computer model was used to determine the radiation doses to nuclei of vascular endothelial cells as a function of distribution pattern and boron levels for BSH and BPA. The compound factor of 0.33-.40 for BSH predicted that ~ 83% of the peak physical of the $^{10}B(n,\alpha)^{7}Li$ capture dose (~ 5-6 fission events) would not contribute to the CNS endothelial cell dose. The uniform distribution of BPA resulted in a higher compound factor (1.96) and higher calculated endothelial cell doses. The brain necrosis seen with BSH most likely is due to the vascular endothelial dose, and this has been supported by histologic analysis of affected canine brains. Nigg, Moran and Wheeler[26] have determined the fast neutron dose distributions in the canine CNS. Studies were carried out in canines (Labrador Retrievers) with or without reflection of scalp and underlying muscles. Calculations were performed using a recently developed BNCT analytic dosimetry software package. Scalp reflection increased the fast neutron component to the brain by ~ 2x, assuming a perpendicular beam. However, this was decreased with an oblique beam. This study illustrated the importance of 3-dimensional planning calculations for all *in vivo* irradiations. Atkinson and Wheeler[27] have summarized the results of a series of irradiations carried out on canines with spontaneously occurring brain tumors. Parameters that were summarized included average ^{10}B concentration, RBE dose and log cell kill. No survival data, however, were presented.

In Vitro Studies

Liu, Barth et al.[28] have presented data on the cellular uptake, distribution and toxicity of a group of carboranyl uridines (CBU). One compound, designated CBU-2, has been intensively studied. Although large amounts (46-75 μg/g) were taken up by cells *in vitro*, there did not appear to be any selectivity for neoplastic cells, nor was uptake dependent upon either nucleoside or electron transport systems. Boron could not be demonstrated in cellular RNA or DNA fractions, and this suggested that insertion of a carboranyl group at the 2' position of ribose in deoxyuridine changed its biochemical properties. Elstad, Meinkoth et al.[29] have studied the *in vitro* and *in vivo* uptake of BPA in mice carrying the BL6 subline of the Bl6 melanoma. *In vitro* concentrations of BPA ranged from 25-100 mg/l, and B uptake ranged from 0.5-0.9μg/gm cells. This was enhanced ~ 10x if cells were propagated in medium that had an 88% reduction in tyrosine and phenylalanine. For *in vivo* studies, 750 mg BPA/kg b.w. were injected i.p. as a slurry. Tumor levels were 2-2.5μgB/g tissue. Liposomal delivery, the details of which were not provided, also was studied. Albertson, Millan et al.[30] have studied the *in vitro* uptake of corticotrophin releasing hormone (CRH) that had been conjugated to a carborane moiety (0-CRH). Uptake by ATT-20 cells was measured indirectly by quantifying the c-AMP response following exposure to O-CRH. Cell irradiations were carried out at the Oregon State University Research Reactor with an unspecified radiation dose. Reduction in surviving AT-T20 cells was seen, but there was no indication of the reduction in S.F. compared to controls. Nelson and Braby[31] have described their plans to study the effects of irradiation of single capillary endothelial cells with individual high LET particles. They described the requirements for single cell irradiation, construction of the apparatus, and testing of experimental procedures. A model has been developed that simulates dose deposited to cells by either high or low LET radiation. To carry out these studies a microbeam irradiation apparatus has been constructed that permits cellular irradiation with a single particle with beams of ~ 500 particles per sec. No data, however, on cell irradiations were presented, although such studies are in progress. Ichihashi, Utsumi et al.[32] have studied NCT using BPA and two cell lines, P-39 sensitive to X-irradiation and G-361 that was resistant. Cells were irradiated with either 50kV X-rays (1100 cGy/mm) or a n_{th} flux of 7.9 x 10^5 n/cm^2 at the Kyoto University Research Reactor in the absence or presence of BPA. Both P-39 and G-361 were sensitive to BNCT suggesting there was no repair of either sublethal damage (SLD) or potentially lethal damage (PLD). Corder and Martin[33] have reported on bibenzimadazoles as radioprotectors against BNCT induced damage. Two compounds, Hoechst 33258 and Hoechst 33342 protect cells against low LET ionizing radiation by virtue of their DNA binding properties. The present studies were carried out to define their radioprotective effects against high LET radiation using purified DNA PBR322 plasmids. There was a clear indication that both agents provided some protection against BNCT induced double strand breaks (dsb). The ratio of dsb per single strand break (ssb) was 1:100 for low LET radiation compared to 12:100 for dsb.

RBE of the $^{10}B(n,\alpha)^7$ Li Reaction

The presentations on biological studies concluded with a series of four papers on the relative biological effectiveness of the $^{10}B(n,\alpha)^7$ Li capture reaction. There has been considerable controversy about this subject, and although some progress has been made, there still are no definitive conclusions. Wambersie, Menzel and Gahbauer[34] have raised the question as to whether RBE is a useful concept in fast neutron therapy and BNCT. In BNCT, the highly inhomogeneous spatial distribution of the boron within the tumor and the short range of the alpha particles raise difficult problems in dosimetry and radiobiology. The concept of absorbed dose for BNCT, therefore, has shortcomings, but, nevertheless, the available radiobiological data on alpha particles, and their oxygen enhancement ratio (essentially unity) and repair capacity (zero), are useful for selecting optimum conditions for BNCT. Laster, Liu et al.[35] have compared dose-RBE and biological dosimetry approaches for treatment planning in BNCT. Preliminary experiments have been carried out irradiating cells (type unspecified) in the presence or absence of ^{10}B-boric acid and quantifying the doses attributable to each component of the BMRR neutron beam. When cells were irradiated in the presence of 30μg^{10}B/gm boric acid, an RBE of 3.6 was calculated for the $^{10}B(n,\alpha)^7Li$ reaction. The use of Monte Carlo calculations and assumed RBE's didn't provide an accurate assessment of effective dose in these experiments, which were designed to simulate patient treatment planning for BNCT. Further studies will be carried out using this system, and it is anticipated that the difficulties that were encountered can be resolved. Petterson, Lindstrom et al.[36] have studied the *in vitro* findings and radiation effects of carboranylalanine. Murine B16 and human IGR-1 melanoma cell lines and human glioma U-343 were studied using CBA. B16 cells in the presence or absence of CBA were irradiated at the R2-0 reactor at Studsvik, Sweden. In addition, B16 and IGR-I cells were irradiated with a ^{60}Co gamma source (0-10 Gy, 1Gy/min) or with 200 MeV helium ions. Binding of CBA to the three cell lines was demonstrated, but the mechanism was unclear. Cells irradiated in the presence of L-CBA or L-BPA showed significant reductions in S.F., but D-CBA was without effect. Finally, Verrijk, Huiskamp and Begg[37] have developed a P-C based computer program for the estimation of radiation dose *in vitro* and *in vivo* BNCT experiments. The program has been written to determine energy deposition in cell nuclei resulting from charged particles emitted after B or N neutron capture in nuclear, cytoplasmic and extracellular compartments. By varying cell geometry and arrangement parameters dynamic modeling was achieved. It was concluded that a three dimensional photon equivalent dose profile could be calculated by combining simulation results with 3-D neutron energy and flux distributions in a tumor or normal tissue volume.

Conclusions

The biological presentations at this meeting further establish the potential efficacy of BNCT as a therapeutic modality. The work carried out with rodent brain tumor models, and the normal tissue tolerance studies in canines should lay the groundwork for more extensive therapy experiments in canines, which are an important prelude to clinical studies in Europe and the United States. *In vitro* and *in vivo* studies are essential for evaluating new boron compounds, and with the increased activity in compound development, there is a need to develop a standardized protocol for their biological evaluation. This logically should begin with *in vitro* studies on cellular uptake, persistence and localization, progress to *in vivo* pharmacokinetic studies in small animals, followed by *in vivo* therapy experiments using rodent models. If the data obtained indicate selective uptake by neoplastic cells with favorable tumor: blood and tumor: normal tissue ratios, and some demonstration of efficacy using rodent tumor models, then studies on distribution in tumor bearing canines are indicated. If favorable pharmacokinetics and tumor selectivity can be demonstrated, then animal therapy experiments would be indicated. Although this protocol has not been formally enunciated, it is evident that this has been put into practice by a number of investigators working either together or independently of one another. Non-human biological experiments should provide a firm basis for clinical studies of new compounds, as well as to further define the usefulness and limitations of the two compounds that currently are in clinical use, BSH and BPA.

REFERENCES

1. Coderre, J. (1992) Small animal studies in BNCT research: usefulness and limitations. Abstracts of the Fifth International Symposium on Neutron Capture Therapy. September 13-17, 1992. Columbus, Ohio. p99.

2. Gavin, P., DeHaan, C., Moore, M., Weidner, J., Swartz, D., Kraft, S., Atkinson, C., Amaro, C., Bauer, W., and Siefert, A. Large animal studies on the use of BNCT for the treatment of brain tumors. in Advances in Neutron Capture Therapy. eds. Soloway, A.H., Barth, R.F., and Carpenter, D. Plenum Press (in press).

3. Chapman, J. Lessons learned from hypoxic sensitizer and photodynamic therapy research of potential use to BNCT. *ibid.*

4. Schuepbach D., Gavin, P., Salford, L., Ceberg, C., and Fankhauser, H. RG2 rat glioma model: influence of the blood-brain barrier on biodistribution of boronsulfhydryl. *ibid.*

5. Kabalka, G., Cheng, G., Anderson, C., Bendel, P., Micca, P., and Slatkin, D. *In vivo* pharmacokinetics of a boron neutron capture agent in a tumor-bearing rat via boron-11 MRS and MRI. *ibid.*

6. Joel, D., Slatkin, D., and Coderre, J. Uptake of ^{10}B in gliosarcomas following the injection of glutathione monoethyl ester and sulfhydryl borane. *ibid.*

7. Mallesch, J., Moore, D., Kahl, S., Thorne, R., and Allen, B. Biodistribution of a boronated porphyrin in BC-1 mammary carcinoma. *ibid.*

8. Kahl, S.B., Joel, D., Nawrocky, M.M., Micca, P.L., Tran, K.P., Finkel, G.C. et al. (1990). Uptake of a nido-carboranylporphine by human glioma xenografts in athymic nude mice and by syngeneic ovarian carcinomas in immunocompetent mice. Proc. Natl Acad. Sci. U.S.A. 87:7265-7269.

9. Nakagawa, Y., Kageji, T., Hatanaka, H., Moritani, M., Kitamura, K., Matsumoto, K., and Kobayashi, M. Blood-brain barrier permeability in partially deutered rat brain in advances in neutron capture therapy. in Advances in Neutron Capture Therapy. eds. Soloway, A.H., Barth, R.F., and Carpenter, D. Plenum Press (in press).

10. LaHann, T., Sills, C., Hematillake, G., Dymock, T., and Daniell, G. Cardiovascular toxicities associated with intravenous administration of p-boronophenylalanine formulations. *ibid.*

11. Bailey, M., Matalka, K., Barth, R., Coderre, J., Soloway, A., Goodman, J., Staubus, A., and Rofstad, E. Treatment of inctracerebral malignant melanoma using a rat model and l-boronophenylalanine as the capture agent. *ibid.*

12. Solares, G., Zamenhof, R., Wazer, D., DiPetrillo, T., Saris, S., Cano, G., Madoc-Jones, H., and Harling, O. Effects of fractionation of boron neutron capture therapy in a mouse glioma model. *ibid.*

13. Matalka, K., Barth, R., Bailey, M., Maruyama, Y., Feola, J., and Wierzbicki, J. Californium-252 brachytherapy of intracerebral melanoma with and without administration of boronophenylalanine utilizing a nude rat model. *ibid.*

14. Matalka, K., Barth, R., Staubus, A., Moeschberger, M., and Coderre, J. Boron neutron capture therapy (BNCT) of F98 glioma bearing rats using boronophenylalanine (BPA) as a capture agent. *ibid.*

15. Mallesch, J., Chiaraviglio, D., Allen, B., and Moore, D. An intra-pancreatic and hepatic nude mouse cancer xenograft model for boron neutron capture therapy. *ibid.*

16. Laster, B., Kahl, S., Warkentien, L., and Bond, V. *In vivo* studies in NCT with boronated porphyrin and tumor growth delay as an end point. *ibid.*

17. Morris, G. and Hopewell, J. Boron neutron capture therapy: skin and spinal cord tolerance. *ibid.*

18. Watkins, P., Moss, R., Siefert, A., Huiskamp, R., Gavin, P., and Konijnenberg, M. Evaluation of dose components for the healthy tissue tolerance studies on dogs at the HFR Petten. *ibid.*

19. Huiskamp, R., Gavin, P., Wheeler, F., Siefert, A., and Philipp, K. Dose effect comparisons between HFR and BMRR irradiated dogs with respect to healthy tissue tolerance. *ibid.*

20. Philipp, K., Gavin, P., Casado, J., Siefert, A., Moss, R., and Huiskamp, R. Developments in the healthy tissue tolerance studies as a precondition for clinical application of boron neutron capture therapy. *ibid.*

21. Siefert, A., Casado, J., Philipp, K., Huiskamp, R., Gavin, P., Dühmke, E., and Moss, R. Brain effects observed in the canine healthy tissue tolerance studies for BNCT with borocaptate sodium at the epithermal neutron beam of the HFR, Petten. *ibid.*

22. Bradshaw, K., Schweizer, M., Glover, G., and Hadley, J. Pharmacokinetics of borocaptate sodium in canine head determined by ^{11}B magnetic resonance imaging and spectroscipy. *ibid.*

23. LaHann, T., Lu, D., Sills, C., Daniell, G., Kraft, S., Gavin, P., and Bauer, W. Bioavailability of intravenous formulations of p-boronophenylalanine in dog and rat. *ibid.*

24. Gavin, P., Huiskamp, R., Wheeler, F., and Griebenow, M. Computer modeling the boron compound factor in normal brain tissue. *ibid.*

25. Gavin, P., Huiskamp, R., Wheeler, F., and Griebenow, M. Computer modeling the boron compound factor in brain tumors. *ibid.*

26. Nigg, D., Moran, J., and Wheeler, F. Determination of fast-neutron dose distributions in the canine central nervous system. *ibid.*

27. Atkinson, C. and Wheeler, F. Dose optimization for boron neutron capture therapy of spontaneous canine brain tumors. *ibid.*

28. Liu, L, Barth, R., Tjarks, W., Soloway, A., and Ives, D. Cellular uptake, subcellular distribution, and *in vitro* toxicity of carboranyl uridines. *ibid.*

29. Elstad, C., Meinkoth, K., Mathison, B., Meadows, G., Kinder, D., Hawthorne, M., Shelly, K., Feakes, D., and Bauer, W. *In vivo* and *in vitro* uptake of boronated compounds by Bl6-BL6 murine melanoma. *ibid.*

30. Albertson, B., Millan, M., Binney, S., Willeke, G., Martsolf, S., Johnson, J., and Loriaux, D. The use of boron neutron capture for the treatment of pituitary tumors. *ibid.*

31. Nelson, J. and Braby L. Irradiation of single cells with individual high-LET particles. *ibid.*

32. Ichihashi, M. Utsumi, H., Tsukamoto, H., Mishima, Y., and Kobayashi, T. Thermal neutron capture therapy using $^{10}B_1$-BPA: inducement of marked lethal effect on melanoma cells irrespective of x-ray sensitivity. *ibid.*

33. Corder, A. and Martin R. Bibenzimidazoles as radioprotectors against BNC-induced DNA damage. *ibid.*

34. Wambersie, A., Menzel, H., and Gahbauer, R. RBE: a useful concept in neutron therapy and BNCT? *ibid.*

35. Laster, B., Liu, H., Gordon, C., and Bond, V. A comparison of the dose: RBE and the biological dosimetry approaches for treatment planning in BNCT. *ibid.*

36. Pettersson, O, Lindström, Olsson, P., Carlsson, J., Sjöberg, and Larsson, B. Characteristics of melanoma cells regarding binding of carboranylalanine and effects of boron neutron capture. *ibid.*

37. Verrijk, R., Huiskamp, R., and Begg, A. A PC-based computer program for the estimation of the radiation dose in *vitro* and *in vivo* boron neutron capture irradiation experiments.

SUMMING UP: CLINICAL PAPERS

Heinz Fankhauser[1] and Patrick R. Gavin[2]

[1]Department of Neurosurgery, CHUV, 1011 Lausanne, Switzerland;
[2]Department of Veterinary Clinical Medicine and Surgery, Washington State University, Pullman, WA 99164-6610

INTRODUCTION

Boron Neutron Capture Therapy (BNCT) can only be conceived in the framework of a large multidisciplinary team. Clinical papers presented at the Fifth International Symposium on Neutron Capture Therapy for Cancer can roughly be divided into 3 groups, according to the team to which the authors belong, each of which has its particular concerns, reflecting the level of advancement of the project. A number of papers stemmed from authors which were already applying BNCT to patients with intracranial tumors and melanomas,[1-8] and these mostly reflected attempts to refine and improve the current treatment methods. An equal number stemmed from authors who intended to start clinical trials shortly,[9-18] dealing with beam and treatment room design, quality control and clinical biodistribution of boron compounds, and protocol design. A small number of clinical papers dealt with those lesions, which may become candidates for BNCT, as well as with new boron compounds, preparing the future beyond intracranial tumors and melanomas, and beyond BSH and BPA.[3,19,20]

APPLIED BNCT

Eight clinical papers stemmed from authors who were already performing BNCT in patients. These mainly dealt with treatment optimization.[1,3,4,5,7,21] Certification problems with health and nuclear safety authorities were reported,[6] but only one paper presented complete interim results of BNCT in a small series of patients [8] and one paper analyzed long term survivors.[2] The absence of reports on systematic follow-up results may have caused some disappointment to the audience, since such results are critical to justify ongoing efforts in those centers not yet applying BNCT.

Oda et al.[8] have given a detailed description of their first 7 patients with glioblastoma treated with intraoperative thermal BNCT since 1990 in the Kyoto University Reactor. They injected 100 mg of BSH per kg. Just prior to irradiation they measured tumor boron concentrations below 10 ppm and tumor-to-blood boron ratios below 1, which were unexpectedly low. The outcome in this series, after less than 2 years, was poor, since 5 out of the 7 patients had died, 3 from their tumor, and 2 from other causes. There was no evidence of radionecrosis in any of these cases, although no systematic investigation into this question had been performed. In the discussion the authors pointed out that most of these patients had

Advances in Neutron Capture Therapy, Edited by
A.H Soloway *et al.*, Plenum Press, New York, 1993

previous external radiation and chemotherapy. Chemotherapy was especially considered responsible for an altered and unfavorable BSH accumulation pattern. Therefore it is suggested that not only external radiation, but chemotherapy, too, should be proscribed before BNCT.

Hatanaka et al.[2] have chosen to give an account of those 9 patients, out of some 120, who are alive or lived for more than 10 years after BNCT. These include 1 glioblastoma, 1 astrocytoma grade III, 3 low grade astrocytomas, 1 oligodendroglioma, 2 meningiomas, and 1 chondrosarcoma. The longest survival was over 20 years. Three other patients were alive for more than 15 years. Two patients have died after more than 10 years. The cause was radiation necrosis, without evidence of recurrent tumor; both patients had had previous conventional radiotherapy. The lessons from these cases were that the life quality after BNCT is good or excellent, unless previous radiation therapy has been given. In the absence of such pretreatment, radiologically huge cerebral defects after treatment were well tolerated, without the global reduction of cerebral performance, as is observed with whole brain radiation. A concern to the audience was pituitary function. Transient diabetes insipidus has been observed, but there was no persistent pituitary deficit. This problem will nevertheless have to be carefully reappraised when epithermal beams will be used, as is the case for ocular function after BNCT, since Hatanaka has observed retinal changes which might be related to the known accumulation of BSH in the choroid. The two patients who died more than 10 years after BNCT from radionecrosis, remind us that we will have to evaluate normal tissue tolerance not only with respect to early radiation effects, but that at some point we have to look for early indicators of late damage.

Fukuda et al.,[1] reporting on BNCT with BPA of skin melanomas, described their method to anticipate the maximum tolerable dose for the peritumoral normal skin, which now determines the radiation parameters for the treatment. In order to improve predictability of boron concentrations and its time course in normal skin, recent patients no longer received subcutaneous injections of BPA, but the drug was given by an intravenous infusion of the compound coupled to fructose. This has allowed, in the last two cases, to reduce the skin reaction to slight and reversible moist desquamation, while optimizing the chances for tumor control.

Partial substitution of body water by heavy water allows enhanced depth penetration of thermal neutrons in tissue and minimizes the gamma ray component from unavoidable hydrogen capture. This method has already been used for intracranial tumors in at least 15 patients in Japan. The physical advantage of using heavy water is rather straight forward, since the advantage depth for thermal neutrons increases roughly in the same proportion as the heavy water fraction in the patient. The validity and clinical significance of this concept has been demonstrated in a transplantable intracerebral rat glioma model, presented by Sakaki et al.[21]

Honda et al.[4] reported that this tool had recently been explored also for the treatment of melanomas. In hamster experiments, substitution rates of 11 to 16% of body water were achieved. With heavy water, a significant advantage in the radiation response to BNCT in a hamster melanoma model could be demonstrated. This principle had recently been applied to a patient with a metastatic melanoma of inguinal lymph nodes.

The question arises whether this substitution has other effects, desired or undesired. Honda et al.[4] have investigated the question whether heavy water may alter the biodistribution pattern of BPA, and found no influence.

Substitution with heavy water in humans reaches a limit at some 30%, which, as discussed by Hatanaka, should not and cannot be crossed for toxicological and practical reasons. Heavy water is expensive, at least in Japan. The question then arose, whether the optimum use of heavy water is cost effective. Does this represent a very expensive way to slightly improve the performance of a suboptimal thermal neutron beam, or, on the contrary, is this a very cheap and elegant way to offer a true alternative to an expensive epithermal beam? Nakagawa et al. [7] certainly supported this last point of view. These authors had administered heavy water to 2 patients with deep seated brain tumors undergoing BNCT. They had further enhanced the depth coverage with thermal neutrons by using a bilateral radiation port. Nevertheless, a convincing comparison between the two methods, thermal BNCT with heavy water and epithermal BNCT, is certainly not possible before the first patients with deep seated tumors have been radiated with an epithermal beam.

PREPARATION FOR BNCT

A number of papers were related to preparations for impending clinical trials with BNCT, including physical,[10,14,16] medical,[9,11,12,15,17,22,23,24,25] and site specific certification problems.[14,18]

Biodistribution studies in humans are certainly important here. Two papers, one presented by Coderre [9] and the other by Mallesch et al.[15] have focussed on 16, and 6 patients respectively, with melanomas or gliomas, absorbing oral doses of BPA up to 189 mg/kg or injected intravenously with 1 gram of BPA-fructose complex. The results have provided valuable information on the pharmacokinetics of BPA in humans and confirm the absence of toxicity at used doses, but as far as tumor concentrations were concerned, the authors were somewhat disappointed. There was a wide heterogeneity of boron concentrations, which did not allow prediction of the approximate time course of boron in tumors. Nevertheless, Coderre [9] had obtained some good isolated boron values in glioblastomas, with a maximum of 11.2 ppm and a tumor-to-blood (t/b) ratio of 7:1 in one patient. For cutaneous melanomas the best tumor value was 11.3 ppm with a t/b ratio of 6:1. Mallesch [15] reported average t/b ratios for metastatic melanoma of 3.8, but, surprisingly, she could not detected boron in the tumor any of the 3 glioma patients.

In 2 papers, Stragliotto et al.[17] and Haritz et al.[13] reported a total of 94 patients with intracranial tumors studied after intravenous administration of BSH, up to 100 mg BSH/kg, including a few cases injected twice 24 hours apart. Intervals between injection and surgical tissue harvesting ranged from 3 to 72 hours. The results indicate a widely heterogeneous uptake, reflected in widely variable tumor boron concentrations and t/b ratios. For malignant gliomas, metastases and meningiomas, one has to wait some 18 hours to have a reasonable, but unpredictable chance to obtain a t/b ratio above 1. In the best cases and in the specimens with maximum concentrations this did nevertheless not rise above 2.5 to 4. When using high doses of BSH one can expect tumor concentrations between 10 and 30 ppm at 18 to 24 hours post administration. There seemed to be no advantage to wait longer than 24 hours, since the t/b ratio did not clearly improve beyond this point, but the absolute concentrations certainly went down, although boron could be detected in the tumor even 72 hours after injection. The highest injected doses caused arm pain, and occasional feeling of chest oppression and general malaise and calls for caution. As has been pointed out during a poster discussion by LaHann [23] concerning toxicity of BPA in small animals, a survey of currently available information shows that the toxicity of BSH and its safety at currently used maximal doses in humans has not yet been fully determined.

In further conclusion, the invaluable strong point of BSH has been reconfirmed by Haritz [13] and Stragliotto,[17] and this is it's virtual exclusion from normal brain, but modifications of dose and dose administration schedule do not seem to turn BSH into a optimal boron carrier from the point of view of absolute tumor concentrations and tumor-to-blood ratio, unless the interesting results obtained by Joel et al. [22] in an animal tumor model, are confirmed in patients. These authors have observed strongly enhanced and persistent BSH accumulation in a rat gliosarcoma after preceding administration of gluthathion-monoesther. This drug was investigated in an attempt to reduce liver toxicity of the BSH dimer BSSH. At this stage the authors were nevertheless unable to report what precisely happens to the critical blood boron level, and if the ratios were changed.

Reliable quality control of BSH and other boron carriers is mandatory, but, as for systematic toxicity studies, it has lagged behind its clinical introduction due to unavailable and difficult application of appropriate technology. Gianotto et al.[12] presented a multi-method approach using high performance liquid chromatography, inductively coupled plasma - atomic emission spectroscopy, nuclear magnetic resonance spectroscopy, and Fourier transformed infrared spectroscopic analysis. Despite this unquestionable progress in BSH purity determination, all newly detected contaminants in commercially available BSH samples had still not been identified and tolerable contamination levels have not yet been defined.

Papers concerning non-invasive indicators for BSH accumulation in intracranial tumors were presented by Kraft,[24] looking at dogs, and by Stragliotto,[25] looking at humans. A degree of correlation between contrast enhancement on CT or MRI and measured tumor boron levels was found, but this was not entirely consistent and hardly quantitative. Both, contrast enhancement and BSH accumulation are strongly time dependent, and both time factors should be controlled before any quantitative information can be expected. The only conclusion therefore was that non-enhancing tumors accumulate little or no boron, if measured as macrodistribution. During the discussion it was pointed out, that this does not automatically mean that there is no microscopic boron in a radiobiologically critical location.

A series of papers have been devoted to the first patient trial which is planned by the European Concerted Action in Petten, including patients with malignant gliomas, using epithermal neutrons, BSH, bilateral radiation and 4 fractions. The history of beam tailoring by Moss et al.,[16] the advancement of the construction of the treatment room by Constantine et al.,[10] and an outline of the initial treatment protocol by Fankhauser et al.[11] were presented and gave rise to interesting and animated discussions. When treating patients, precise decision have to be made. The most critical and uneasy ones include the radiation parameters applied to the first cases. The question then arose, which source of information is most trusted in order to make these decisions. The chosen dose should be safe, but nevertheless efficient to arrive at the goal which has been set for the trial. Are sophisticated computer simulations of delivered doses acceptable despite the fact that human biodistribution studies of boron compounds have not yielded well defined numbers which could be fed into the program? Are radiobiological data obtained from cell culture experiments or the results of the numerous studies with experimental animal tumors more reliable to determine the initial radiation parameters in patients? Should many more experiments be designed and performed before starting clinical trials with epithermal neutrons? The prevailing opinion in the Concerted Action was, that the value of further experiments is limited, since at this stage it is not known what experimental result exactly is needed to guaranty safety and efficacy of BNCT in humans. A typical example, which was extensively discussed during the symposium, is the question whether one has to prove that every single tumor cells takes up 10^9 boron atoms, before proceeding with patient irradiation. For heterogeneous, infiltrating human brain tumors this is virtually impossible and would delay clinical trials forever. Encouraged by the results of Hatanaka and others, the view presented by members of the European Concerted Action was that lack of experimental data should no longer delay the initiation of human radiations, since this is the only way to obtain a definite evaluation of BNCT. The initial clinical trial is to be designed strictly as a normal tissue tolerance study for BSH and epithermal neutrons, considering aspects of tumor control as secondary at this stage. All efforts have to be directed towards collection the desired information with as few patients as possible. A further view was, that the most trustable information concerning the starting parameters for the human study was likely to be obtained from the dose-response curves for asymptomatic MRI changes in the dog brain. Elaboration of these curves is part of the ongoing normal tissue tolerance study with healthy, BSH loaded dogs, in the definite treatment beam in HF-reactor in Petten. A tentative human study protocol, derived from the dog experiments, was presented by Fankhauser et al.,[11] including identical beam design, identical blood boron levels, identical MRI endpoint, but bilateral radiation portal, 4 fractions, and at least a 20% reduction of the total initial radiation time. This should allow elaboration of the normal tissue dose-response curves for humans, and this, in turn, would allow the start of well designed human studies to investigate tumor control with BNCT. During the discussion it has been stressed that in view of the extreme complexity of BNCT radiobiology, and the invariably disastrous prognosis of malignant gliomas this appeared to be the safest and most efficient way to proceed, despite poorly defined effects of BNCT in humans, including the unknown effect of fractionation.

FUTURE BORON CARRIERS AND FUTURE DOMAINS OF APPLICATION OF BNCT

Kallio et. al.[19] suggested that BNCT should not be undertaken until a reasonably specific carrier for boron is available. They have presented the rational for the use of low density lipoproteins (LDL) as boron vehicle: LDL is the main physiological carrier for cholesterol which is taken up by growing cells through receptor mediated endocytosis; an inconsistent increase of the number of LDL receptors has been found on brain tumors and glioma cell lines; it has been possible in the past to substitute cholesterol in the core of the LDL by boronated fatty acids. The authors have performed an in vivo uptake study for unboronated, but radiolabeled LDL in tumor patients. Four patients with recurrent gliomas were studied by gamma camera after intravenous administration of autologous LDL. In 2 cases ^{111}In was used as a label. After 2 and 24 hours only minimal increase radioactivity was observed in the tumor area as compared to the controlateral hemisphere, and this was considered non-specific.

Obviously, there was in vivo release of the ^{111}In label, since strong activity was observed over the kidneys and the bladder. The 2 patients injected with ^{99m}TC-labeled LDL showed an activity ratio in the tumor compared to the normal hemisphere of 1.4 and 1.5 after 2 hours and of 2.7 and 2.5 after 20 hours; in one of these patients an additional control study for non-specific uptake was performed using ^{99m}Tc-labeled human serum albumin, nonspecific ratios obtained were 1.4 at 2 hours and 1.7 at 20 hours, respectively. Although 2 out of 4 patients showed slight but significant and possibly specific accumulation of autologous LDL in the recurrent tumor area, this type of studies needs confirmation with boron loaded LDL and boron determination in tissues, including blood. Only then will a comparison with existing boron carriers be possible. LDL may become part of a multi-component boron delivery system.

Two papers discussed future candidates for BNCT other than intracranial tumors and skin melanomas.

Hatanaka et al.[3] presented the first case subjected to BNCT for a large cerebral arteriovenous malformation, using BSH. The rational is that the theoretically well established vascular sparing effect of BSH, providing protection of endothelial cells in normal brain from high LET particles, does no longer play in the tightly packed vascular conglomerates of the nidus of the malformation. To date, the case showed diminished contrast enhancement on CT as possible indicator of a response.

Sauerwein et al.[20] presented arguments to consider BNCT for some forms of ocular melanoma for which none of the currently used therapies, including coagulation, brachytherapy by plaque application and charged particle beam radiation, is appropriate. These include massive extraocular, intraorbital, and conjunctival extensions. Since enucleation of the eye in these cases does not modify the already extremely low potential for metastatization, salvage of the eye and visual function should by all means be attempted.

ACKNOWLEDGEMENT

This work has been supported by the Fonds National de la Recherche Scientifique, the Ligue Suisse contre le Cancer, the Ligue Vaudoise contre le Cancer, the Zyma Foundation for the Advancement of Medical and Biological Sciences, and the Commission of the European Communities.

REFERENCES

1. H. Fukuda, Y. Abe, J. Takahashi, J. Hiratsuka, T. Kobayashi, K. Kanda, C. Honda, M. Shiono, M. Ichihashi, and Y. Mishima, Determination of appropriate neutron fluence and timing of irradiation for the treatment of malignant melanoma by using ^{10}B-BPA, Fifth International Symposium on Neutron Capture Therapy for Cancer, Columbus, Ohio, September 13-17, 1992 (abstract).

2. H. Hatanaka, J. Goodman, and M. Miner, Analysis of clinical results of long-surviving brain tumor patients undergoing BNCT with Shionogi's mercaptoundecahydrododecaborate, Fifth International Symposium on Neutron Capture Therapy for Cancer, Columbus, Ohio, September 13-17, 1992 (abstract).

3. H. Hatanaka, Y. Nakagawa, and S. Sasaki, Endovascular Boron-Neutron Capture Therapy for giant arteriovenous malformation, *in*: "Advances in Neutron Capture Therapy. Proceedings of the 5th International Symposium on Neutron Capture Therapy," A.H. Soloway, R.F. Barth, and D.E. Carpenter, eds., Plenum Press, New York, N.Y. (1993) (this volume).

4. C. Honda, M. Shiono, N. Wadabayashi, M. Ichihashi, Y. Mishima, K. Yoshino, T. Kobayashi, K. Kanda, and H. Karashima, Deuteriation trials for Neutron Capture Therapy of deep-seated melanoma lesions, *ibidem.*

5. K. Kanda, T. Kobayashi, K. Akuta, and K. Ono, Experience of BNCT using the KUR, *ibidem.*

6. K. Kanda, Procedures of medical application of research reactors, *ibidem*.

7. Y. Nakagawa, T. Kageji, H. Hatanaka, K. Mukai, K. Kitamura, F. Tsuji, K. Kanda, T. Kobayashi, Thermal Neutron Capture Therapy for brain stem and skull base tumors: heavy water and multiportal neutron delivery, *ibidem*.

8. Y. Oda, M. Takagaki, Z.H. Zhang, H. Kikuchi, T. Kobayashi, and K. Kanda, An experience of clinical BNCT on malignant brain tumors, *ibidem*.

9. J.A. Coderre, Boronophenylalanine: A biodistribution study in humans, Fifth International Symposium on Neutron Capture Therapy for Cancer, Columbus, Ohio, September 13-17, 1992 (abstract).

10. G. Constantine, L. Dewit, R.L. Moss, B.J. Mijnheer, K. Ravensberg, and F. Stecher-Rasmussen, Designing a treatment room for BNCT clinical trials at the HFR, Petten, *in*: "Advances in Neutron Capture Therapy. Proceedings of the 5th International Symposium on Neutron Capture Therapy," A.H. Soloway, R.F. Barth, and D.E. Carpenter, eds., Plenum Press, New York, N.Y. (1993) (this volume).

11. H. Fankhauser, P.R. Gavin, A. Siefert, and G. Stragliotto, A treatment protocol for BNCT of malignant gliomas with borocaptate sodium (BSH) and epithermal neutrons, *ibidem*.

12. A.K. Gianotto, and W.F. Bauer, Purity and quality determinations of borocaptate sodium, *ibidem*.

13. D. Haritz, D. Gabel, H. Klein, and R. Huiskamp, Results of continued clinical investigations of BSH in patients with malignant glioma, *ibidem*.

14. O.K. Harling, R.D. Rogus, J.-M. Chabeuf, S.C. Yam, F. Lambert, J.C. Yanch, R.G. Zamenhof, G. Solares, D.E. Wazer, and H. Madoc-Jones, Preparations for clinical trials at the MIT Reactor and the New England Medical Center, *ibidem*.

15. J.L. Mallesch, D.E. Moore, B.J. Allen, W.H. McCarthy, R. Jones, and W.A. Stening, The biodistribution of p-boronophenylalanine-fructose in human patients with glioma and metastatic melanoma, *ibidem*.

16. R.L. Moss, On the progress towards clinical trials at Petten (post-Sydney), *ibidem*.

17. G. Stragliotto, D. Schüpbach, P.R. Gavin, and H. Fankhauser, Update on biodistribution of borocaptate sodium (BSH) in patients with intracranial tumors, *ibidem*.

18. R.G. Zamenhof, D.E. Wazer, J. Bernard, O.K. Harling, T.A. DiPetrillo, H. Madoc-Jones, J.A. Coderre, A.G. Meek, Federal and institutional approvals necessary prior to a U.S. clinical trial of Neutron Capture Therapy, *ibidem*.

19. M. Kallio, J. Leppälä, T. Nikula, P. Nikkinen, H. Gylling, M. Färkkilä, J. Hiltunen, J. Jääskläinen, and K. Liewendahl, ^{111}In- and ^{99m}Tc-labeled low density lipoprotein uptake by high-grade gliomas, *ibidem*.

20. W. Sauerwein, H. Szypniewski, N. Bornfeld, and P. Chauvel, NCT - a therapeutic alternative in the treatment of ocular melanomas? Fifth International Symposium on Neutron Capture Therapy for Cancer, Columbus, Ohio, September 13-17, 1992 (abstract).

21. S. Sakaki, T. Asakura, M. Kusumoto, H. Hatanaka, Y. Nakagawa, and K. Komura, A basic research on Boron Neutron Capture Therapy - application of heavy water and oral boronphenylalanine. Fifth International Symposium on Neutron Capture Therapy for Cancer, Columbus, Ohio, September 13-17, 1992 (abstract).

22. D.D. Joel, D.N. Slatkin, and J.A. Coderre, Uptake of ^{10}B in gliosarcomas following the injection of glutathione monoethyl ester and sulfhydryl borane monomer (BSH), *in*: "Advances in Neutron Capture Therapy. Proceedings of the 5th International Symposium on Neutron Capture Therapy," A.H. Soloway, R.F. Barth, and D.E. Carpenter, eds., Plenum Press, New York, N.Y. (1993) (this volume).

23. T.R. LaHann, C. Sills, G. Hematillake, T. Dymock, and G. Daniell, Cardiovascular and pulmonary toxicities associated with intravenous admininstration of p-boronophenylalanine, *ibidem*.

24. S. Kraft, P.R. Gavin, C.E. DeHaan, and T.L. Richards, Contrast enhancement on computed tomography: a potential indicator of boron concentration in canine intra-axial brain tumors. Fifth International Symposium on Neutron Capture Therapy for Cancer, Columbus, Ohio, September 13-17, 1992 (abstract).

25. G. Stragliotto, H. Fankhauser, P.R. Gavin, and R.A. Meuli, Correlation of BSH uptake with CT-scan contrast enhancement, *in*: "Advances in Neutron Capture Therapy. Proceedings of the 5th International Symposium on Neutron Capture Therapy," A.H. Soloway, R.F. Barth, and D.E. Carpenter, eds., Plenum Press, New York, N.Y. (1993) (this volume).

SYMPOSIUM REGISTRANTS

Abbe, J. Ch.
CNRS
Chimie Nucleaire BP 20
Strasbourg
France
TEL (33) 88-28-64-04
FAX (33) 88-28-64-31

Adams, Diane
Department of Pathology
The Ohio State University
Hamilton Hall, 1645 Neil Ave.
Columbus, OH 43210
TEL 614-292-2436

Aizawa, Otohiko, Professor
Atomic Energy Research Laboratory
Musashi Institute of Technology
Ozenji 971, Asao-ku, Kawasaki-shi
215 Japan
TEL 044-966-6131
FAX 044-955-6071

Aizawa, Toshiko, Mrs.
Atomic Energy Research Laboratory
Musashi Institute of Technology
Ozenji 971, Asao-ku, Kawasaki-shi
215 Japan
TEL 044-966-6131
FAX 044-955-6071

Akine, Y., M.D.
Outpatient Clinic
National Cancer Center Hospital
5-1-1 Tsukiji, Chuou-Ku
Tokyo 104, Japan
TEL 03-3542-2511
FAX 03-3545-3567

Alam, Fazlul
US Borax Research Group
412 Crescent Way
Anaheim, CA 92801
TEL 714-991-8036
FAX 714-776-6014

Albertson, Barry D., Dr.
Division of Endocrinology (L-607)
Oregon Health Sciences University
3181 S.W. Sam Jackson Park Road
Portland, OR 97201
TEL 503-494-6996
FAX 503-494-6990

Alexander, Charles W.
Oak Ridge National Laboratory
Box 2008, MS-6384
Oak Ridge, TN 37831-6384
TEL 615-574-7071
FAX 615-574-6008

Allen, Barry, Dr.
Biomedicine and Health Program,
Ansto
PMB 1, Menai NSW 2234
Australia
TEL 612-717-3426
FAX 612-543-5097

Aldemir, Tunc
The Ohio State University
206 W. 18th Ave.
Columbus, OH 43210
TEL 614-292-4627
FAX 614-292-3163

Alpen, Edward, L., Dr.
Neutron Technology Corporation
1182 Miller Avenue
Berkeley, CA 94708
TEL 510-845-0321
FAX 510-845-7824

Amaral, Michael, Dr.
St. Cloud Neurological Surgery, P.A.
48 North 29th Avenue
St. Cloud, MN 56303
TEL 612-259-8812
FAX 612-259-9538

Anissuzamman, Abul, Dr.
College of Pharmacy
The Ohio State University
500 W. 12th Ave.
Columbus, OH 43202

Atkinson, Carol A., Ms.
INEL BNCT Program
EG&G Idaho, Inc.
P.O. Box 1625
Idaho Falls, ID 83415-1575
TEL 208-526-7723
FAX 208-526-0528

Auterinen, Iiro
Technical Research Centre of Finland
Reactor Laboratory
P.O. Box 200
SF-02151 ESPOO Finland
TEL 358-0-4566353
FAX 358-0-4566390

Bailey, Michael Q.
The Ohio State University
1149 Veterinary Hospital
1935 Coffey Rd.
Columbus, Ohio 43210
TEL 614-292-1040

Balo, Paul
Oak Ridge National Lab
PO Box 2008, Bldg. 7920 MS 6384
Oak Ridge, TN 37831
TEL 615-574-7071
FAX 615-576-6312

Barth, Rolf F., Dr.
The Ohio State University
Department of Pathology
1645 Neil Ave., 165 Hamilton Hall
Columbus, Ohio 43210
TEL 614-292-2177
FAX 614-292-7072

Bauer, William F., Dr.
INEL BNCT Program
EG&G Idaho, Inc.
P.O. Box 1625
Idaho Falls, ID 83415-2208
TEL 208-526-1180
FAX 208-526-8541

Bennett, Brian
Cornell University
Baker Laboratory of Chemistry
Ithaca, NY 14853
TEL 607-255-3884
FAX 607-255-4137

Bergamo, Frank
FRB Corporation
6120 Moorfield Ave.
Colorado Springs, CO 80919
TEL 719-531-6559
FAX 719-531-0192

Bigelow, John E., Dr.
Oak Ridge National Lab
PO Box 2008
Oak Ridge, TN 37831-6385
TEL 615-574-6926
FAX 615-574-6008

Binney, Stephen E., Dr.
Oregon State University
Dept. of Nuclear Engineering
Corvallis, OR 97331-5902
TEL 503-737-7068
FAX 503-737-0480

Blue, James W., Ph.D.
Cleveland Clinic (Retired)
30648 Wolf Road
Bay Village, Ohio 44140
TEL 216-871-1541

Blue, Thomas, Ph.D.
2091 Robinson Laboratory
206 West 18th Avenue
Columbus, OH 43210
TEL 614-292-0629
FAX 614-292-3163

Bradshaw, Kenneth M., Mr.
Neutron Technology
Boise, ID 83701

Broadhurst, John H.
University of Minnesota
116 Church St. SE
Minneapolis, MN 55455
TEL 612-624-2502
FAX 612-624-4578

Broomhead, John, Dr.
Chemistry Dept.
Australian National University
PO Box 4
Canberra, A.C.T.
Australia 2601
TEL 062-49-2934
FAX 062-49-0760

Brownell, Gordon L.
77 Massachusetts Ave.
Mass. Institute of Technology
Cambridge, MA 02139
TEL 617-253-3805
FAX 617-726-5123

Brugger, Robert M., Ph.D.
Medical Department - Building 490
Brookhaven National Laboratory
Upton, New York 11973
TEL 516-282-3625
FAX 516-282-5311

Burian, Jiri, Dr.
Nuclear Research Institute, Dept. 204
25068 REZ near Prague
Czechoslovakia
TEL 0042-2-6857831 ext. 2465
FAX 0042-2-844060

Capala, Jacek
Department of Pathology
168 Hamilton Hall, 1645 Neil Avenue
The Ohio State University
Columbus, OH 43210-1238
TEL 614-292-2436
FAX 614-292-7072

Carlsson, Jörgen
Dept. Radiation Sciences
Uppsala University Box 535
S-75121 Uppsala, Sweden
TEL 46 18 183841
FAX 46-18-183833

Carolan, Martin
Biomedicine and Health Program,
Ansto
PMB 1, Menai NSW 2234
Australia
TEL 6102-717-3351
FAX 6102-717-9261

Cassady, John M.
Dean, College of Pharmacy
College of Pharmacy
Parks Hall, 500 W. 12th Avenue
The Ohio State University
Columbus, Ohio 43210
614-292-2266

Chauvel, Pierre, M.D.
Centre Antoine-Lacassagne
227 Avenue de la Lanterne
06200 Nice, France
TEL (33) 9318-8022
FAX (33) 9318-0433

Chia, Victor K. F., Dr.
Charles Evans and Associates
301 Chesapeake Drive
Redwood City, CA 94063
TEL 415-369-4567
FAX 415-369-7921

Christensen, Richard N.
Department of Nuclear Engineering
2067 Robinson Lab
206 West 18th Avenue
Columbus, OH 43210

Coderre, Jeffrey A.
Medical Department
Brookhaven National Laboratory
Upton, NY 11973
TEL 516-282-2298
FAX 516-282-5311

Combs, Matthew J.
University of Virginia
MR-4 Building, Room 1138-70
Charlottesville, VA 22908
TEL 804-924-2843
FAX 804-924-0492

Constantine, G.
Joint Research Centre of the CEC
POSTB 2
1755 ZG Petten
The Netherlands
FAX (31) 2246-1449

Corder, Anne
Peter MacCullum Cancer Institute
481 Little Lonsdale St.
Victoria 3000
Melbourne, Australia 3054
TEL (03) 641 5034

Crawford, J.F.
Paul Scherrer Institute
CH 5232 Villigen, Switzerland
TEL 41-5699-2080
FAX 41-5699-2309

Dobelbower, Micheal
The Ohio State University
206 W. 18th Ave.
Columbus, OH 43210
TEL 614-292-1074
FAX 614-292-3163

Dorn, Ronald V., III, Dr.
Medical Director
INEL BNCT Program
EG & G Idaho Inc.
Idaho Falls, ID 83415-3519
TEL 208-526-9153
FAX 208-526-0528

Edwards, Benjamin, Ph.D.
VAMC Martinez
150 Muir Rd.
Martinez, CA 94553
TEL 510-372-2000

Eggers, Philip E.
Eggers & Associates
5366 Reserve Dr.
Dublin, OH 43017
TEL 614-791-8118

Ellis, Frank, Dr.
2 Bladen Close, Woodstock Road
Oxford, OX2 8AD
United Kingdom

Elstad, Catherine A., Dr.
Department of Pharmacy
Wagner Hall
Washington State University
Pullman, WA 99164-6510
TEL 509-335-8030
FAX 509-335-0162

Estes, Guy P.
MS B226
Los Alamos National Laboratory
Los Alamos, NM 87545
TEL 505-667-4041
FAX 505-665-5538

Evans, Jeff
The Ohio State University
2040 Robinson Lab
206 W. 18th Ave.
Columbus, OH 43210
TEL 614-292-1074
FAX 614-292-3163

Fankhauser, Heinz, M.D.
Department of Neurosurgery
CHUV
1011 Lausanne, Switzerland
TEL 21-314-48-74
FAX 21-314-48-73

Färkkilä, Markus, Docent
University Hospital of Helsinki
Department of Neurology
Haartmaninkatu 4
Helsinki, Finland 00290
TEL 358-0-4711
FAX 358-0-4714009

Farnworth, Chris
Centronic Ltd.
King Henry Drive, New Addington
Croydon, Surrey CR9 OBG
England
TEL 011-44-689-842121
FAX 011-44-689-845117

Feakes, Debra A.
Department of Chemistry and Biochemistry
University of California Los Angeles
Los Angeles, CA 90024
TEL 310-825-7378
FAX 310-206-4038

Fukuda, Hiroshi, M.D.
Dept. of Radiology & Nuclear Med.
Research Institute for TB and Cancer
Tohoku University
4-1 Seiryomachi Aoba-ku, Sendai
980 Japan
TEL 81-22-274-1111 (x3416)
FAX 81-22-275-7324

Gabel, Barbara
Universitat Bremen
Fachbereich Chemie
Postfach 330 440
D-2800 Bremen 33
Germany

Gabel, Detlef, Prof. Dr.
Universität Bremen
Fachbereich Chemie
Postfach 330 440
D-2800 Bremen 33
Germany
TEL (421) 218-2200
FAX (421) 218-2871

Gahbauer, Reinhard A., M.D.
Division of Radiation Oncology
Arthur G. James Cancer Hospital
The Ohio State University
Columbus, OH 43210
FAX 614-292-4044

Gajewski, Janusz
Department of Radiation Oncology
University Hospital at Stony Brook
Stony Brook, NY 11974-7028
TEL 516-444-2208
FAX 516-689-8801

Gao, Yigong, Dr.
Research Biochemicals Inc.
1 Strathmore Rd.
Natick, MA 01760
TEL 508-651-8151
FAX 508-655-1359

Gavin, Patrick R., Dr.
Dept. of Veterinary Clinical Med. & Surgery, McCoy Hall
Washington State University
Pullman, WA 99164-6610
TEL 509-335-0738
FAX 509-335-0880

Gay, Isabelle
Cornell University
Baker Laboratory
Ithaca, NY 14853-1301
TEL 607-255-3884
FAX 607-255-4137

Gianotto, Anita K., Ms.
INEL BNCT Program
EG&G Idaho, Inc.
P.O. Box 1625
Idaho Falls, ID 83415-2208

Goldstein, Gerald
Department of Energy
10038 Maple Leaf Dr.
Gathersburg, MD 20879

Goodman, Joseph H., M.D.
Division of Neurosurgery, 316 Means Hall
1654 Upham Drive
The Ohio State University
Columbus, OH 43210
614-293-5440

Grecula, John
Ohio State University
James Cancer Hospital
300 West 10th Avenue
Columbus, Ohio 43210
TEL 614-293-8415
FAX 614-293-4044

Greskovich, John, M.D.
The Ohio State Univesity
Robinson Laboratory
206 W. 18th Ave.
Columbus, OH 43210
614-488-1623
614-292-3163

Griebenow, Beth
INEL
436 N. 55 W.
Idaho Falls, ID 83402

Griebenow, Merle
INEL BNCT Program
EG&G Idaho, Inc.
P.O. Box 125
Idaho Falls, ID 83415-3519
TEL 208-526-1296
FAX 208-526-0528

Grusell, Erik, Dr.
Department of Radiation Services
Uppsala University
P. O. Box 535
S-75121 Uppsala, Sweden
TEL 46-182500
FAX 46-183833

Gupta, Nilendu
2148 Robinson Laboratory
206 West 18th Avenue
Columbus, OH 43210
TEL 614-292-9115
FAX 614-292-3163

Hamm, Robert W. Dr.
AccSys Technology, Inc.
1177A Quarry Lane
Pleasanton, CA 94566
TEL 510-462-6949
FAX 510-462-6993

Hariharan, Joya
The Ohio State University
College of Pharmacy
500 W. 12th
Columbus, OH 43210
TEL 614-292-1228

Haritz, Dietrich
Department of Neurosurgery
ZKH St. Jürgenstr.
2800 Bremen Germany
TEL 049-421-4975482
FAX 049-421-4973323

Piscol-Haritz, Caroline
Department of Neurosurgery
ZKH St. Jürgenstr.
2800 Bremen Germany

Harker, Yale W., Dr
INEL BNCT Program
EG&G Idaho, Inc.
P.O. Box 1625
Idaho Falls, ID 83415-7112
TEL 208-526-4251
FAX 208-526-4715

Harling, Otto K. Prof
Massachusetts Institute of Technology
138 Albany St., NW12-204
Cambridge, MA 02139
TEL 617-253-4201
FAX 617-253-7300

Harrington, Baiba
ANSTO
PMB 1, Menai 2234 NSW
Australia
TEL 61-2-7173473
FAX 61-2-7179275

Harsh, Griffith H., IV, M.D.
Neurosurgery
Harvard Medical School, Fruit Street
Massachusetts General Hospital
Boston, MA 02114
TEL 617-789-2494
FAX 6-7-719-3154

Hart, Michelle L.
Nuclear Engineering Dept.
The Ohio State University
206 W. 18th Ave. Room 2040
Columbus, OH 43210
TEL 614-292-1074
FAX 614-292-3163

Hatanaka, Hiroshi, M.D., D.Sc.
Department of Neurosurgery
Teikyo University
2-11-1 Kaga, Itabashi-ku
Tokyo 173, Japan
TEL 03-3964-1211 (ext. 1612)
FAX 03-5375-1716

Hawthorne, M. Frederick, Professor
Department of Chemistry and
Biochemistry
405 Hilgard Ave
University of California/Los Angeles
Los Angeles, CA 90024-1569
TEL 310-825-7378
FAX 310-206-4038

Hayakawa, Yoshinori, Ph.D.
Institute for Basic Medical Research
University of Tsukuba
1-1-1 Tennoodai, Tsukuba
Ibaraki-ken 305 Japan
TEL 0298-64-2571
FAX 0298-64-2575

Hazle, John D., Ph.D.
University of Texas
M.D. Anderson Cancer Center
1515 Holcombe Blvd., Box 547
Houston, TX 77030
TEL 713-792-3226
FAX 713-794-1364

Heimberger, Lisa A.
Westinghouse Henford Company
Richland WA 99352

Henry, Louis G.
SUNY/B BMRC
Rotary Rd.
Buffalo, NY 14214
TEL 716-829-2826
FAX 716-829-2030

Herminghuyser, Kevin
The Ohio State University
2040 Robinson Lab
206 W. 18th Ave.
Columbus, OH 43210
TEL 614-292-1074
FAX 614-292-3163

Hiltunen, Jukka
Map Medical Technologies
Otakaari 3
SF-02150 Espoo Finland

Honda, Chihiro, M.D.,Ph.D.
Department of Dermatology
Kobe University School of Medicine
7-5-1 Kusunoki-cho, Chuo-ku
Kobe 650, Japan
TEL 078-341-7451
FAX 073-382-2497

Hotz, Norman J.
INEL BNCT Program
EG&G Idaho, Inc.
P.O. Box 1625
Idaho Falls, ID 83415-2208

Huiskamp, René, Ph.D.
Radiobiology & Radioecology
Netherlands Energy Research
Foundation
P.O. Box 1
1755 ZG Petten, The Netherlands
TEL 31-2246-4069
FAX 31-2246-3491

Ichihashi, Masamitsu
Department of Dermatology
Kobe University School of Medicine
7-5-1 Kusunoki-cho, Chuo-ku
Kobe 650, Japan
TEL 078-341-7451
FAX 078-382-2497

Igi, Kozo A., Ph.D.
Callery Chemical Company
Box 429
Pittsburgh, PA 15230
TEL 412-967-4184
FAX 412-967-4188

Jiang, Yirun
Nuclear Engineering Program
The Ohio State University
206 W. 18th Ave.
Columbus, OH 43210
TEL 614-292-1074
FAX 614-292-3163

Joel, Darryl D.
Medical Department, Bldg. 490
Brookhaven National Laboratory
Upton, NY 11973
TEL 516-282-3603
FAX 516-282-5311

Kabalka, George W., Dr.
Chemistry Dept., 612 Buehler Hall
University of Tennessee
Knoxville, TN 37996-1600
TEL 615-974-3260
FAX 615-974-2997

Kahl, Stephen B., Dr.
School of Pharmacy, S-926
University of California/San Francisco
513 Parnassus Avenue
San Francisco, CA 94143-0446
TEL 415-476-4684
FAX 415-476-0688

Kakihana, Hidetake, Professor
Dept. Of Chemistry
Faculty of Science
Shinshu University
Matsumoto 390 Japan

Kalef-Ezra, John Ph.D.
University of Ioannina
Medical Physics Lab
Ionnina 45110 Greece
TEL (30) 651-33655
FAX (30) 651-74112

Kallio, Merja, M.D.
Department of Neurology
University of Helsinki
Haartmaninkatu 4
SF-00290 Helsinki
Finland
TEL 358-0-4712261
FAX 358-0-4714009

Kanda, Keiji, Assoc. Prof.
Research Reactor Institute
Kyoto University
Kumatori-cho, Sennan-gun
Osaka 590-04, Japan
TEL 81-724-52-0901
FAX 81-724-53-0360

Karam, R.A., Director
Georgia Institute of Technology
6133 Kines Mountain Ct.
Atlanta, GA 30332-0425
TEL 404-844-3620
FAX 404-853-9325

Kobayashi, Tooru
Research Reactor Institute
Kyoto University
Kumatori-cho, Sennan-gun
Osaka 590-04, Japan
TEL 81-724-52-0901
FAX 81-724-53-5810 (or 1043)

Konijnenberg, M.W., Ph.D.
Netherlands Cancer Institute
Antoni van Leeuwenhoek Huis
Department of Radiotherapy
Plesmanlaan 121
1066 CX Amsterdam The
Netherlands
TEL 31-20-512-2116
FAX 31-20-669-1101

Kraft, Susan L., Dr.
College of Veterinary Medicine
Veterinary Medical Center, Trotter
Hall
Kansas State University
Manhattan, KS 66506
TEL 913-532-4171
FAX 913-532-5884

Krause, Chuck
Boron Biologicals, Inc.
533 Pylon Drive, P.O. Box 33489
Raleigh, NC 27636
TEL 919-832-2044
FAX 919-832-5980

Krüger, Uwe
Schering Aktiengesellschaft
Diagnostic Imaging Division
Kontrast Media Research
D-1000 Berlin 65
Germany
TEL 49-30-468-8948
FAX 49-30-468-8719

Kunze, Jay F.
333 Engineering Building West
University of Missouri
Columbia, MO 65211
TEL 314-882-3550
FAX 314-882-0397

LaHann, Thomas R., Dr.
College of Pharmacy
Idaho State University
Pocatello, ID 83209-8334
TEL 208-236-4456
FAX 208-236-4482

Lake, James A., Dr.
Idaho National Engineering Lab
EG & G Idaho
P. O. Box 1625 MS 1576
Idaho Falls, Idaho 83415
TEL 208-525-5518
FAX 208-525-5562

Laramore, G.E., Ph.D., M.D.
University of Washington
Department of Radiation Oncology, RC-08
Seattle, WA 98195
TEL 206-547-4120
FAX 206-548-6218

Larsson, Börje, Ph.D.
Institute of Medical Radiobiology (IMR)
August Forel-Strasse 7
CH 8029 Zurich
Switzerland
TEL 41-1-385-6511
FAX 41-1-385-6204

Laster, Brenda H., Dr.
Medical Department, Bldg. 490
Brookhaven National Laboratory
Upton, NY 11973
TEL 516-282-4638
FAX 516-282-5311

Laxson, Robert R.
Oak Ridge National Lab
PO Box 2008
Oak Ridge, TN 37831-6384
TEL 615-574-7071

Lemmen, Peter, Dr.
Organisch Chemisches Institut
Technische Universität München
Lichtenbergstrasse 4
D-8046 Garching
Germany
TEL 49-89-3209-3319
FAX 49-89-3209-2727

Lennox, Arlene, J.
Fermilab Neutron Therapy Facility
P.O.Box 400 MS 301
Batavia, IL 60501
TEL 708-840-4850
FAX 708-840-4552

Liu, Hungyuan B.
Brookhaven National Laboratory
Medical Department, Building 490
Upton, NY 11973
TEL 516-282-4041
FAX 516-282-5311

Liu, Liang
Pathology Department
The Ohio State University
1645 Neil Avenue
Columbus, Ohio 43210

Livesey, John H., Dr.
Dept. of Radiation Oncology RC-09
University of Washington
Seattle, Washington 98195
TEL 206-548-4090
FAX 206-548-4829

Loughlin, Therese, Dr.
Wessex Neurological Centre
Southampton General Hospital
Tremona Road S09 4XY
Southampton, UK
TEL 0703-796860
FAX 0703-704236

Loveless, F. C.
Centronic Limited
Centronic House, King Henry's Drive
New Addington, Croydon Surrey
England CR9 OBG
TEL 0689-842121
FAX 0689-849711

Lunato, Anthony
College of Pharmacy
217 Parks Hall
The Ohio State University
Columbus, OH 43210-1291
TEL 614-292-1228

Mahoney, Francis J., Ph.D.
National Cancer Institute
9000 Rockville Pike
Bethesda, MD 20892

Mallesch, Julia, Ms.
Biomedicine and Health Program, Ansto
PMB 1, Menai NSW 2234
Australia
TEL 612-717-3351
FAX 612-717-9262

Malmquist, Jonas
Uppsala University
Institute of Chemistry
Box 531
S-751 21 Uppsala Sweden
TEL 46-18-18-3798
FAX 46-18-50-8542

Marashi, M.K.
Neutron Physics Group
Nuclear Research Center
Atomic Energy Organization of Iran
P.O. Box 11365-8486
Tehran, Iran
021-6139499

Martin, Roger F., Dr.
Molecular Sciences Group
Peter MacCallum Cancer Institute
481 Little Lonsdale Street
Melbourne 3000
Australia
TEL 03-641-5555
FAX 03-670-3357

Maruyama, Yosh, M.D.
Univ. of Kentucky Medical Center
800 Rose Street
Lexington, KY 40536-0084
TEL 606-233-6486
FAX 606-257-4931

Masiakowski, Jerzy T., Ph.D.
Institute of Physics
Adam Mickiewicz University
60-780 Poznan
Poland
FAX 48-61-460-69

Matalka, Khalid
Shileds Warren Radiation Lab.
Harvard Medical School
50 Binney St.
Boston, MA 02115

Mathison, Bruce A.
Washington State University
224 Wegner Hall
Pullman, Washington 99164-6510
509-335-1965

Matsumoto, Tetsuo, Dr.
Atomic Energy Research Laboratory
Musashi Institute of Technology
Ozenji 971, Asao-ku
Kawasaki-shi
215 Japan
TEL 044-966-6131
FAX 044-955-6071

McGregor, John, M.D.
Ohio State University
1654 Uphan Drive - 317 Means Hall
Columbus, Ohio 43212
TEL 614-293-5440

Meriaty, Harry
Biomedicine and Health Program, Ansto
PMB 1, Menai NSW 2234
Australia
TEL 612-717-9257
FAX 612-453-5097

Micca, Peggy
Medical Department
Brookhaven National Laboratory
Upton, NY 11973
TEL 516-282-3566
FAX 516-282-5311

Mishima, Yutaka, M.D., Ph.D.
Mishima Int. for Dermatological Research
17-8-801, 3-Chome, Motomachi-dori
Chuo-ku
Kobe 650 Japan
TEL 81-78-334-1663
FAX 81-78-334-1660

Miura, Michiko, Ph.D.
Medical Department, Bldg. 490
Brookhaven National Laboratory
Upton, NY 11973
TEL 516-282-3618
FAX 516-282-5311

Moore, Douglas, Assist. Prof.
Department of Pharmacy
University of Sidney
NSW 2006
Australia
TEL 612-692-2334
FAX 612-552-3760

Moore, Michael P., Dr.
Dept. of Veterinary Clinical Med. &
Surgery
McCoy Hall
Washington State University
Pullman, WA 99164-6610

Morris, G.M., Dr.
Research Institute
Churchill Hospital
Old Road, Headington
Oxford, Oxon OX3 7LJ
United Kingdom
TEL 0865-225850
FAX 0865-225847

Morris, John H., Dr.
Dept. of Pure and Applied Chemistry
Strathclyde University
295 Cathedral Street
Glasgow, Scotland G1 1XL
United Kingdom
TEL 041-552-4400
FAX 041-552-5664

Morrison, George H.
Department of Chemistry
Baker Laboratory
Cornell University
Ithaca, NY 14853-1301
TEL 607-255-3614

Moss, R.L., Dr.
HFR Division
Joint Research Centre of the CEC
1755 ZG Petten
The Netherlands
TEL 02246-5126
FAX 02246-1449

Mulder, Robert U., Director
University of Virginia Reactor Facility
University of Virginia, Circleville
Department of Nuclear Engineering
Charlottesville, VA 22903
TEL 804-296-6588
FAX 804-982-5473

Mumford-Zisk, Jonathon
Cornell Univerity
Dept. of Chemistry
Ithaca, NY 14853-1301
TEL 607-255-3885
FAX 607-255-4137

Nadler, Jon
Department of Energy
785 Doe Place M/S 1220
Idaho Falls, ID 83401-1562
TEL 208-526-0796
FAX 208-526-5964

Nakagawa, Yoshinobu, M.D.,
Dr.med.Sci.
National Kagawa Children's Hospital
Zentsuji-cho, Zentsuji City
Kagawa 765, Japan
TEL 08-0886-62-0885

Nakama, Hideyuki
Tokyo Rosai Hospital
3-30-1-302 Omori-Minami Ota-Ku
Tokyo, Japan
TEL 03-3742-7301

Nakamura, Hiroyuki
Tohoku University
Department of Chemistry
Faculty of Science
Sendai 980, Japan
TEL 022-222-1800
FAX 022-263-9207

Nelson, John M.
Pacific Northwest Laboratories
P. O. Box 999
Richland, Washington 99301
TEL 509-376-5692
FAX 509-376-9944

Nemoto, Hisao, Dr.
Tohoku University
Dept. of Chemistry, Faculty of
Science
Sendai 980, Japan
TEL 022-222-1800 (x3393)
FAX 022-263-9207

Ng, Chor-Yi
The Ohio State University
206 W. 18th Ave.
Columbus, OH 43210
TEL 614-292-1074
FAX 614-292-3163

Niemkiewicz, John, M.S.
Dept. of Radiation Oncology
Riverside Methodist Hospitals
3535 Olentangy River Road
Columbus, OH 43214
TEL 614-566-5714

Nigg, David, W., Dr.
INEL BNCT Program
EG&G Idaho, Inc.
P.O. Box 1625
Idaho Falls, ID 83415-1575
TEL 208-525-5514
FAX 208-526-5526

Nowak, Tom
Department of Surgery
CCFMG Valley Med Center
445 S. Cedar
Fresno, CA 93702
TEL 209-453-4440

Oda, Yoshifumi. M.D., D.Sc.
Department of Neurosurgery
Kyoto University Hospital
Shogoingawara, Sakyo-ku
Kyoto 606, Japan
TEL 81-75-751-3450
FAX 81-75-751-9501

Olsson, Pär
Department of Radiation Sciences
Uppsala University, Box 535
Uppsala S-75121
Sweden
TEL 46-18-183839
FAX 46-18-183833

Pazirandeh, A. Dr.
Physics Dept. Tehran Univ.
P.O. Box 19395-1943
Tehran, Iran
FAX 0098-21-8006095 or
0098-21-8009439

Perks, Christopher
AEA Environment and Energy
B364 Harwell Laboratory
Oxfordshire OX11 ORA
United Kingdom
TEL 0235 434784
FAX 0235 434431

Pettersson, Orn-Anong
Dept. Radiation Sciences
Uppsala University, Box 535
S-75121 Uppsala, Sweden
TEL 46 18 183841
FAX 46-18-183833

Philipp, Katharina
Netherlands Energy Research
Foundation
P.O. Box 1
1755 ZG Petten
The Netherlands
TEL 31-2246-4280
FAX 31-2246-3491

Phillips, Mark, Ph.D.
University of Washington
1959 N. E. Pacific Street, RC-08
Seattle, Washington 98195
TEL 206-548-6219

Pignol, Jean Philippe, Dr.
Group Strasbourgeois de
Neutrontherapie
23 Rue Du Loess
Strasbourg, France 67037
TEL (33) 88-88-64-00
FAX (33) 88-28-09-90

Prashar, Jog
Department of Pharmacy
University of Sydney
Sydney NSW 2006
Australia
TEL 612-692-2333
FAX 612-552-3760

Qu, Tanxia (Bruce)
The Ohio State University
206 W. 18th Avenue
Columbus, OH 43210
TEL 614-292-9115
FAX 614-292-3163

Quastel, Michael R., M.D., Ph.D.
Institute of Nuclear Medicine
Soroka Medical Center
POB 151
Beer Sheva, Israel 84101
TEL 972-57-40485
FAX 972-57-74696

Radel, Peggy A., Dr.
Department of Pharmaceutical
Chemistry
University of California/San Francisco
513 Parnassus Avenue - Room S-926
San Francisco, CA 94143-0446
TEL 415-476-6781
FAX 415-476-0688

Ramsey, E.B., Ph.D.
University Hospital
Department of Radiation Oncology
State Univ. of NY at Stony Brook
Stony Brook, NY 11794-7028

Randolf, Peter D., Dr.
INEL BNCT Program
EG&G Idaho, Inc.
P.O. Box 1625
Idaho Falls, ID 83415-2208

Reinstein, Lawrence E., Ph.D.
Department of Radiation Oncology
University Hospital at Stony Brook
Stony Brook, NY 11974-7028
TEL 516-444-2208
FAX 516-444-7629

Renton, Diana, Marketing Manager
Theragenics Corporation
5325 Oakbrook Parkway
Norcross, GA 30093
TEL 404-381-8338
FAX 404-381-8447

Reynolds, Albert B.
University of Virginia Reactor Facility
University of Virginia
Charlottesville, VA 22901
TEL 804-982-5446

Rief, Herbert, Dr.
Joint Research Centre of the CEC
Ispra Site, Building 36/015
21020 Ispra (VA), Italy
TEL 39-332-789746
FAX 39-332-785584

Rogus, Ronald D.
Massachusetts Institute of Technology
138 Albany Street, NW13-252
Cambridge, MA 02139
TEL 617-253-5720
FAX 617-253-7300

Rong, F.G.
College of Pharmacy
The Ohio State University
500 W. 12th Ave.
Columbus, OH 43210
TEL 614-292-1228

Rotaru, Joan
The Ohio State University
1645 Neil Ave.
165 Hamilton Hall
Columbus, OH 43210
TEL 614 292-2436

Sakaki, Seigo, M.D.
Department of Neurosurgery
Kagoshima University
1208-1 Ushuku-cho
Kagashima, 890 Japan

Sakurai, Yoshinori
Research Reactor Institute
Kyoto University
Kumatori-cho, Sennan-gun
Osaka 590-04, Japan
TEL 81-724-52-0901
FAX 81-724-53-1043

Sauerwein, Wolfgang
Strahlenklinik des
Universitätsklinikum der GHS Essen
Hufelandstraße 55
W -4300 Essen 1
Germany
TEL 49-201-723-2051
FAX 49-201-723-5908

Saxena, Renu
Radiation Oncology Department
Wayne State University and Harper Hospital
3990 John R
Detroit, MI 48201

Scasnar, Vladimir
University of Sherbrooke
Faculty of Med., Dept. of Nuclear Medicine
Sherbrooke QUE J1H5N4, Canada
TEL 819-563-5555 ext:4679
FAX 819-564-5442

Schinazi, Raymond F., Ph.D.
Veterans Affairs Medical Center
Medical Research 151
1670 Clairmont Road
Decatur, GA 30033
TEL 404-728-7711
FAX 404-728-7726

Schüpbach, Daniel, M.D.
Department of Neurosurgery
CHUV
1011 Lausanne, Switzerland
TEL 21-314-48-74
FAX 21-314-48-73

Schweizer, Martin, Ph.D.
Department of Medicinal Chemistry
University of Utah
308 Skaggs Hall
Salt Lake City, UT 84112
TEL 801-581-7599
FAX 801-581-7087

Seifert, Axel, Dr.
Joint Research Centre of The CEC
P.O. Box 2
1755 ZG Petten
The Netherlands
TEL 31-2246-5126
FAX 31-2246-5132

Shani, Gad
Department of Nuclear Engineering
Ben Gurion University
P.O. Box 653
Beer Sheva, Israel 84105
FAX 972-572-77843

Shelly, Ken
Dept. Chemistry and Biochemistry
University of California Los Angeles
Los Angeles, CA 90024
TEL 310-825-7378
FAX 310-206-4038

Singaram, Bakthan, Ph.D.
Neutron Technology
1182 Miller Ave
Berkeley, CA 94708
TEL 510-849-3954

Sjöberg, Stefan, Dr.
Institute of Chemistry, Box 531
Uppsala University
S-75121 Uppsala, Sweden
TEL 46-18-32-4302
FAX 46-18-50-8542

Slatkin, Daniel
Medical Department
Brookhaven National Laboratory
Upton, NY 11973
TEL 516-282-3361
FAX 516-282-5311

Smith, Michael
University of Sydney
Department of Pharmacy
Sydney NSW 2006
Australia
TEL 612-692-2333
FAX 612-552-3760

Solares, Guido R.
Radiation Oncology Department
New England Medical Center, Box 246
750 Washington Street
Boston, MA 02111
TEL 617-350-8006
FAX 617-350-8008

Soloway, Albert H.
College of Pharmacy
The Ohio State University
500 W. 12th
Columbus, OH 43210
TEL 614-292-2509
FAX 614-292-7766

Sood, Anup, Dr.
Boron Biologicals, Inc.
533 Pylon Dr.
P.O. Box 33489
Raleigh, NC 27636-3489
TEL 919-832-2044
FAX 919-832-5980

Southworth, Geoffrey
Boron Biologicals, Inc.
533 Pylon Dr.
P.O. Box 33489
Raleigh, NC 27636-3489
TEL 919-832-2044
FAX 919-5980

Spielvogel, Bernard, Dr.
Boron Biologicals, Inc.
533 Pylon Drive, P.O. Box 33489
Raleigh, NC 27636-3489
TEL 919-832-2044
FAX 919-832-5980

Staubus, Alfred E., Pharm.D., Ph.D.
College of Pharmacy
The Ohio State University
500 West 12th Avenue
Columbus, OH 43210
TEL 614-292-0891
FAX 614-292-7766

Stecher-Rasmussen, Finn, Dr.
Netherlands Energy Research
Foundation
P.O. Box 1
Petten 1755 ZG
The Netherlands
TEL 31-2246-4537
FAX 31-2246-3490

Stohler, Eric
Medical Isotopes Inc.
120 Pleasant St.
Concord, HN 03301
TEL 603-226-1859

Storr, Greg J., Dr.
INEL BNCT Program
EG&G Idaho, Inc.
P.O. Box 1625
Idaho Falls, ID 83415-1575
TEL 208-526-5517
FAX 208-526-5562

Stragliotto, Giuseppe, M.D.
Department of Neurosurgery
CHUV
1011 Lausanne, Switzerland
TEL 21-314-48-74
FAX 21-314-48-73

Straubinger, Robert M.
Department of Pharmaceutics
SUNY Buffalo
539 Cooke Hall
Amherst, NY 14260-1200
TEL 716-645-2844
FAX 716-645-3693

Strouf, Oldřich, Dr.
Institute of Inorganic Chemistry
Czechoslovak Academy of Sciences
Majakovského 24
16000 Prague 6, Czechoslovakia
TEL 6857831/3141
FAX 422-328496

Sweet, William, M.D.
Massachusetts General Hospital
15 Parkman Street
Boston, MA 02114
TEL 617-726-3547
FAX 617-726-7546

Szypniewski, Heike
Strahlenklinik des
Universitätsklinikum der GHS Essen
Hufelandstraße 55
W-4300 Essen 1, Germany
TEL 49 2097232059
FAX 49 2097235908

Takagaki, Masao, M.D.
Department of Neurosurgery
School of Medicine, Kyoto University
54 Kawara-machi, Shogoin, Sakyo-ku
Kyoto 606, Japan
TEL 075-751-3655
FAX 075-771-6415

Takahashi, Hiroaki
Shinshu University
Matsumoto 390 Japan
TEL 81-263-35-4600 (x4156)
FAX 81-263-33-5323

Tambunchong, Chinda, Dr.
Srinakharinwirot University
Bankok, Thailand
FAX 66-2-2584006 or
66-2-2600128

Tang, Peng-Peng Zhu, Dr.
Department of Medicinal Chemistry
308 Skaggs Hall, University of Utah
Salt Lake City, UT 84112
TEL 801-581-4945
FAX 801-581-7078

Thonnard, Norbert, Dr.
Atom Sciences, Inc.
114 Ridgeway Center
Oak Ridge, TN 37830
TEL 615-483-1113
FAX 615-483-3316

Tjarks, Werner
College of Pharmacy
The Ohio State University
500 West 12th Avenue
Columbus, OH 43210
TEL 614-292-1228
FAX 614-292-2435

Tokita, N. M.D.
c/o Dr. Y. Akine
National Cancer Center
5-1-1 Tsukiji, Chuou-Ku
Tokyo 104 Japan
FAX 03-3545-3567

Tokuuye, K., M.D.
Department of Radiation Therapy
National Cancer Center Hospital
5-1-1 Tsukiji, Chuou-Ku
Tokyo 104, Japan
TEL 81-33-542-2511
FAX 81-33-545-3567

Twilegar, Ron
Neutron Technology Corp.
877 West Main St., Suite 303
Boise, ID 83702
TEL 208-336-8466
FAX 208-336-8669

Ujeno, Y., Prof.
Iwakura-Nagatani-cho 622-37
Sakyo-ku, Kyoto-shi
Kyoto 606 Japan

van den Wollenberg, D.J.M.
Hoofdstraat 105A
2202 EW Noordwijk
The Netherlands

Venhuizen, Jim
INEL BNCT Program
EG&G Idaho, Inc., P.O.Box 1625
Idaho Falls, ID 83415-3519
TEL 208-526-9421
FAX 208-526-0528

Verrijk, R., Dr.
Netherlands Cancer Institute
Plesmanlaan 121
Amsterdam 1066 CX
The Netherlands
TEL 3120-5122047
FAX 3120-6172625

Vilaithong, Thiraphat, Assoc. Prof., Dr.
Dept. of Physics
Chiang Mai University, P.O. Box 111
Chiang Mai 50002, Thailand
TEL 66-53-221699 ext 3379
FAX 66-53-217143

Wallace, Steven
University Wollongong, ANSTO
PMB 1 Menai NSW 2234
Australia
TEL 612-717-3426
FAX 612-543-5097

Wambersie, André, Professor
Unité de Radiothérapie
Cliniques Universitaires St. Luc
Avenue Hippocrate, 10
1200 Brussels
Belgium
TEL 32-2-764-4726
FAX 32-2-764-3703

Wang, C-K Chris, Asst. Prof.
School of Mechanical Engineering
Georgia Institute of Technology
Atlanta, GA 30332-0225
TEL 404-894-3727
FAX 404-894-3733

Weidner, Jay P., Dr.
Dept. of Veterinary Clinical Med. & Surgery
McCoy Hall
Washington State University
Pullman, WA 99164-6610

Weinreich, Regin, Dr. rer. nat. (Ph.D.)
Institute of Medical Radiobiology
August Forel Strasse 7
Zurich
CH-8029 Switzerland
TEL 41-56-99-2940
FAX 41-1-385-6204

Wessol, Daniel E.
INEL BNCT Program
EG&G Idaho, Inc.
P.O. Box 1625
Idaho Falls, ID 83415-2208

Wheeler, Floyd J.
INEL BNCT Program
EG&G Idaho, Inc.
P.O. Box 1625
Idaho Falls, ID 83415-1575
TEL 208-525-5701
FAX 208-525-5562

White, William L., Dr.
Research Biochemicals Inc.
1 Strathmore Rd.
Natick, MA 01760
TEL 508-651-8151
FAX 508-655-1359

Whittemore, William L.
General Atomics
P.O. Box 85608
San Diego, CA 92186
TEL 619-455-3276
FAX 619-455-4169

Wielopolski, Lucian, Dr.
Radiation Therapy Facility
Brookhaven National Laboratory
Medical Building 490
Upton, NY 11973
TEL 516-282-3656
FAX 516-282-7244

Wiersema, Dick
INEL BNCT Research Program
Box 1625
Idaho Falls, ID 83402-3519
TEL 208-526-9153
FAX 208-526-0528

Wierzbicki, Jacek, Ph.D.
Department of Radiation Medicine
University of Kentucky Medical Center
Lexington, KY 40503
TEL 606-233-4813
FAX 606-257-4931

Wilbur, D. Scott, Ph.D.
Dept. of Radiation Oncology, XD-48
2121 N. 35th St.
University of Washington
Seattle, Washington 98103
TEL 206-685-3085
FAX 206-685-3279

Woollard, Jeff
The Ohio State University
206 W. 18th Ave.
Columbus, OH 43210
TEL 614-292-1074
FAX 614-292-3163

Wu, Terry H.S., Dr.
333 Engineering Bldg. West
University of Missouri
Columbia, MO 65211
TEL 314-882-3551
FAX 314-882-0397

Wyzlic, Iwona M., Dr.
College of Pharmacy
The Ohio State University
500 West 12th Avenue
Columbus, OH 43210
TEL 614-292-1228
FAX 614-292-2435

Yamamoto, Yoshinori, Professor
Department of Chemistry
Faculty of Science
Tohoku University
Sendai 980, Japan
TEL 022-222-1800 (x3391)
FAX 022-263-9207

Yanagie, Hironobu, M.D., D.Sci.
Department of Surgery
Institute of Medical Science
University of Tokyo
4-6-1 Shiroganedai Minato-ku
Tokyo 108, Japan
TEL 03-3443-8111
FAX 03-3443-6031

Yanch, Jacquelyn C.
Room E25-330C 45 Carleton Street
Massachusetts Institute of Technology
Cambridge, MA 02139
TEL 617-258-6999
FAX 617-253-3019

Yoshino, Kazuo, Dr.
Dept. of Chemistry, Faculty of Science
Shinshu University
Matsumoto 390, Japan
TEL 81-263-35-4600 (x4156)
FAX 81-263-33-5323

Youtie, Bill
Mitsubishi
2100 Riveredge Pkwy, Suite 475
Atlanta, GA 30328
TEL 404-859-0069
FAX 404-956-7301

Zamenhof, R.G.
New England Medical Center
Box 246
750 Washington Street
Boston, MA 02111
TEL 617-350-8006
FAX 617-350-8008

AUTHOR INDEX

INDEX[1]

[1] This index consists of key words provided by the authors and listed according to the first page of the article in which they appear.